Hydrogen Energy

Edited by
Detlef Stolten

Related Titles

Garcia-Martinez, Javier (ed.)

Nanotechnology for the Energy Challenge

2010
ISBN: 978-3-527-32401-9

Hirscher, M. (ed.)

Handbook of Hydrogen Storage

New Materials for Future Energy Storage

2010
ISBN: 978-3-527-32273-2

Barbaro, P., Bianchini, C. (eds.)

Catalysis for Sustainable Energy Production

2009
ISBN: 978-3-527-32095-0

Mitsos, A., Barton, P. I. (eds.)

Microfabricated Power Generation Devices

Design and Technology

2009
ISBN: 978-3-527-32081-3

Züttel, A., Borgschulte, A., Schlapbach, L. (eds.)

Hydrogen as a Future Energy Carrier

2008
ISBN: 978-3-527-30817-0

Wengenmayr, R., Bührke, T. (eds.)

Renewable Energy

Sustainable Energy Concepts for the Future

2008
ISBN: 978-3-527-40804-7

Sundmacher, K., Kienle, A., Pesch, H. J., Berndt, J. F., Huppmann, G. (eds.)

Molten Carbonate Fuel Cells

Modeling, Analysis, Simulation, and Control

2007
ISBN: 978-3-527-31474-4

Barclay, F. J.

Fuel Cells, Engines and Hydrogen

An Exergy Approach

2006
ISBN: 978-0-470-01904-7

Olah, G. A., Goeppert, A., Prakash, G. K. S.

Beyond Oil and Gas: The Methanol Economy

2006
ISBN: 978-3-527-31275-7

Kockmann, N. (ed.)

Micro Process Engineering

Fundamentals, Devices, Fabrication, and Applications

2006
ISBN: 978-3-527-31246-7

Hydrogen Energy

Edited by
Detlef Stolten

WILEY-VCH Verlag GmbH & Co. KGaA

The Editor

Prof. Detlef Stolten
Forschungszentrum Jülich GmbH
IEF-3: Fuel Cells
Leo-Brandt-Straße
52425 Jülich
Germany

We would like to thank the following companies for the image material used in the cover illustration: Daimler AG, Dynetek Europe GmbH, Forschungszentrum Jülich GmbH, Linde AG, MTU Onsite Energy GmbH, and TOSHIBA Europe GmbH.

All books published by Wiley-VCH are carefully produced. Nevertheless, authors, editors, and publisher do not warrant the information contained in these books, including this book, to be free of errors. Readers are advised to keep in mind that statements, data, illustrations, procedural details or other items may inadvertently be inaccurate.

Library of Congress Card No.: applied for

British Library Cataloguing-in-Publication Data
A catalogue record for this book is available from the British Library.

Bibliographic information published by the Deutsche Nationalbibliothek
Die Deutsche Nationalbibliothek lists this publication in the Deutsche Nationalbibliografie; detailed bibliographic data are available in the Internet at http://dnb.d-nb.de.

© 2010 WILEY-VCH Verlag GmbH & Co. KGaA, Weinheim

All rights reserved (including those of translation into other languages). No part of this book may be reproduced in any form – by photoprinting, microfilm, or any other means – nor transmitted or translated into a machine language without written permission from the publishers. Registered names, trademarks, etc. used in this book, even when not specifically marked as such, are not to be considered unprotected by law.

Composition K+V Fotosatz GmbH, Beerfelden

Printing and Binding Strauss GmbH, Mörlenbach

Cover Design Schulz Grafik-Design, Fußgönheim

Printed in the Federal Republic of Germany
Printed on acid-free paper

ISBN 978-3-527-32711-9

Contents

Foreword *XIX*

Preface *XXI*

List of Contributors *XXV*

Fuel Cell Basics

1 **Electrocatalysis and Catalyst Degradation Challenges in Proton Exchange Membrane Fuel Cells** *3*
Hubert A. Gasteiger, Daniel R. Baker, Robert N. Carter, Wenbin Gu, Yuxin Liu, Frederick T. Wagner, and Paul T. Yu

Abstract *3*
1.1 Introduction *3*
1.2 Voltage Losses in State-of-the-Art Automotive PEM Fuel Cells *4*
1.3 Catalyst Development Needs and Approaches *6*
1.4 Catalyst Degradation via Platinum Dissolution *10*
1.5 Carbon-Support Corrosion *11*
1.6 Conclusion *14*
References *14*

2 **High-Temperature PEM Fuel Cells: Electrolytes, Cells, and Stacks** *17*
Christoph Wannek

Abstract *17*
2.1 Introduction *17*
2.2 Approaches to Increase the Operating Temperature of Sulfonated Membranes *19*
2.3 HT-PEFCs with Phosphoric Acid-Based Polybenzimidazole-Type Membranes *23*
2.4 Alternative Liquid Electrolytes *33*
2.5 Acidic Salts and Oxides *35*
2.6 Conclusion *36*
References *37*

3 Current Status of and Recent Developments in Direct Liquid Fuel Cells 41
Jürgen Mergel, Andreas Glüsen, and Christoph Wannek

Abstract 41
3.1 Introduction 41
3.2 Direct Methanol Fuel Cells 44
3.3 Direct Ethanol Fuel Cells 55
3.4 Conclusion 57
References 57

4 High-Temperature Fuel Cell Technology 61
Gael P. G. Corre and John T. S. Irvine

Abstract 61
4.1 Introduction 61
4.2 Solid Oxide Fuel Cell 65
4.3 Molten Carbonate Fuel Cell 78
4.4 Thermodynamics of Fuel Cells 78
4.5 Fuel Cell Efficiency 80
References 85

5 Advanced Modeling in Fuel Cell Systems: a Review of Modeling Approaches 89
Matthew M. Mench

Abstract 89
5.1 Introduction 89
5.2 State-of-the-Art Computational Models for Low-Temperature Polymer Electrolyte Fuel Cell Systems 98
5.3 Case Study of Water Management in PEFCs 102
5.4 Future Research Needs 112
Acknowledgments 113
References 113

Fuel Infrastructures

6 Hydrogen Distribution Infrastructure for an Energy System: Present Status and Perspectives of Technologies 121
Françoise Barbier

Abstract 121
6.1 Introduction 121
6.2 Hydrogen Transport by Gaseous Pipelines 123
6.3 Hydrogen Transport by Road 129
6.4 Alternative Hydrogen Delivery Systems 133

6.5	Stationary Bulk Storage of Hydrogen	*134*
6.6	Supporting Technologies	*136*
6.7	Hydrogen Fueling Stations	*141*
6.8	Conclusion	*145*
	References	*146*

7 Fuel Provision for Early Market Applications *149*
Manfred Fischedick and Andreas Pastowski

Abstract *149*
7.1 Introduction: Hydrogen Supply Today and Tomorrow *149*
7.2 Balancing New Applications and Hydrogen Supply *151*
7.3 Criteria for Fuel Supply – Short- and Long-Term Requirements *154*
7.4 Hydrogen Production and Distribution *156*
7.5 Conclusion *164*
References *165*

Hydrogen Production Technologies

8 Non-Thermal Production of Pure Hydrogen from Biomass: HYVOLUTION *169*
Pieternel A. M. Claassen, Truus de Vrije, Emmanuel G. Koukios, Ed W. J. van Niel, Ebru Özgür, İnci Eroğlu, Isabella Nowik, Michael Modigell, Walter Wukovits, Anton Friedl, Dominik Ochs, and Werner Ahrer

Abstract *169*
8.1 Introduction *169*
8.2 State of the Art *171*
8.3 Methodology *171*
8.4 The Project's Current Relation to the State of the Art *174*
8.5 Conclusion *185*
Acknowledgments *185*
References *185*

9 Thermochemical Cycles *189*
Christian Sattler

Abstract *189*
9.1 Introduction *189*
9.2 Historical Development *190*
9.3 State of Work *191*
9.4 Conclusion and Outlook *202*
Abbreviations *203*
References *203*

10	**Hydrogen Production:**
	Fundamentals and Case Study Summaries *207*
	Kevin W. Harrison, Robert Remick, Gregory D. Martin, and Aaron Hoskin

	Abstract *207*
10.1	Heating Value, Heat of Reaction, and Free Energy *207*
10.2	Heat of Formation and Free Energy of Formation *209*
10.3	Calculating Fuel Cell System Efficiency *210*
10.4	Water Electrolysis *213*
10.5	Case Studies of Wind/Hydrogen Projects *217*
10.6	Conclusion *225*
	Acronyms and Abbreviations *225*
	Acknowledgment *226*
	References *226*

11	**High-Temperature Water Electrolysis Using Planar Solid Oxide Fuel Cell Technology: a Review** *227*
	Mohsine Zahid, Josef Schefold, and Annabelle Brisse

	Abstract *227*
11.1	Introduction to High-Temperature Electrolysis *228*
11.2	History of High Temperature Steam Electrolysis *230*
11.3	Solid Oxide Electrolyzer Cells *233*
11.4	Solid Oxide Electrolyzer Stacks *239*
11.5	Conclusion *240*
	References *241*

12	**Alkaline Electrolysis – Introduction and Overview** *243*
	Detlef Stolten and Dennis Krieg

	Abstract *243*
12.1	Introduction *243*
12.2	Definition *244*
12.3	The Principle *244*
12.4	History *246*
12.5	Basics of Electrolysis *249*
12.6	Technical Alkaline Concepts *254*
12.7	Status of Technology *265*
12.8	Conclusion *266*
	Acknowledgments *267*
	References *267*

13 Polymer Electrolyte Membrane (PEM) Water Electrolysis *271*
Tom Smolinka, Sebastian Rau, and Christopher Hebling

Abstract *271*
13.1 Introduction *271*
13.2 Fundamentals of PEM Electrolysis *272*
13.3 Membrane Electrode Assembly *278*
13.4 Current Collectors, Bipolar Plates, and Stack Design *280*
13.5 System Design *285*
13.6 Conclusion *286*
13.7 Symbols and Abbreviations *287*
References *288*

14 Reforming and Gasification – Fossil Energy Carriers *291*
Jens Rostrup-Nielsen

Abstract *291*
14.1 Introduction. The Need for H_2 *291*
14.2 Basic Technologies *292*
14.3 Process Schemes *296*
14.4 Hydrogen from Coal *301*
14.5 Conclusion *303*
References *303*

15 Reforming and Gasification – Biomass *307*
Achim Schaadt, Siegfried W. Rapp, and Christopher Hebling

Abstract *307*
15.1 Introduction *307*
15.2 Gasification of Biomass *308*
References *318*

16 State of the Art of Ceramic Membranes for Hydrogen Separation *321*
Wilhelm-A. Meulenberg, Mariya E. Ivanova, Tim van Gestel, Martin Bram, Hans-Peter Buchkremer, Detlev Stöver, and José M. Serra

Abstract *321*
16.1 Introduction *321*
16.2 Microporous Membranes for H_2 Separation *322*
16.3 Dense Ceramic Membranes for H_2 Separation *333*
16.4 Conclusion and Outlook *344*
Acknowledgments *346*
References *346*

17 Hydrogen System Assessment: Recent Trends and Insights *351*
Joan M. Ogden *351*

Abstract *351*
17.1 Introduction *352*
17.2 Survey of Hydrogen System Assessment Models: Recent Trends and Insights *354*
17.3 Towards a Comprehensive Framework for Hydrogen Systems Analysis *367*
References *368*

Storages

18 Physical Hydrogen Storage Technologies – a Current Overview *377*
Bert Hobein and Roland Krüger

Abstract *377*
18.1 Introduction *377*
18.2 General Overview *377*
18.3 Fuel System Design and Specifications *382*
18.4 Conclusion *393*
References *393*

19 Metal Hydrides *395*
Etsuo Akiba

Abstract *395*
19.1 Introduction *395*
19.2 Part I: Fundamentals of Metal Hydrides for Hydrogen Storage *396*
19.3 Part II: Applications of Metal Hydrides *404*
19.4 Conclusion *411*
References *412*

20 Complex Hydrides *415*
Andreas Borgschulte, Robin Gremaud, Oliver Friedrichs, Philippe Mauron, Arndt Remhof, and Andreas Züttel

Abstract *415*
20.1 Introduction *415*
20.2 The Structure of Complex Hydrides *419*
20.3 Thermodynamics of Complex Hydrides *420*
20.4 Organic Hydrides for Hydrogen Storage *424*
20.5 Hydrogen Storage Systems Using Complex and Organic Hydrides *425*
References *427*

21 Adsorption Technologies *431*
Barbara Schmitz and Michael Hirscher

Abstract *431*
21.1 Adsorption *431*
21.2 History of Adsorption *432*
21.3 Hydrogen Adsorption *432*
21.4 Materials *433*
21.5 Hydrogen Storage *436*
21.6 Total Storage Capacity *439*
21.7 Conclusion *441*
References *441*

Policy Perspectives, Initiatives and Cooperations

22 National Strategies and Programs *449*
Jörg Schindler

Abstract *449*
22.1 The Imminent Transition to a Postfossil Energy World *449*
22.2 The Role of Secondary Energy Carriers *454*
22.3 Hydrogen in Transport *455*
22.4 National Strategies and Programs *456*
22.5 Conclusion *462*
Acknowledgment *462*
References *463*

23 Renewable Hydrogen Production *465*
Alan C. Lloyd, Ed Pike, and Anil Baral

Abstract *465*
23.1 Introduction *465*
23.2 Rationale for Renewable Hydrogen *465*
23.3 Renewable Hydrogen Pathways *472*
23.4 Renewable Hydrogen Policy Drivers *480*
23.5 Conclusion *484*
Acknowledgment *485*
References *486*

24 Environmental Impact of Hydrogen Technologies *489*
Ibrahim Dincer and T. Nejat Veziroglu

Abstract *489*
24.1 Introduction *489*
24.2 Sustainable Development *490*
24.3 Sustainable Development and Thermodynamic Principles *493*

24.4	Hydrogen Versus Fossil Fuels	497
24.5	Future Energy Systems	505
24.6	Case Study I	507
24.7	Case Study II	515
24.8	Conclusion	524
	Acknowledgments	524
	Nomenclature	524
	References	526

Strategic Analyses

25 Research and Development Targets and Priorities 533
Clemens Alexander Trudewind and Hermann-Josef Wagner

	Abstract	533
25.1	Introduction	533
25.2	Procedure	534
25.3	Scenarios	534
25.4	Investigation of Technologies	536
25.5	Conclusion	546
	Acknowledgments	547
	References	548

26 Life Cycle Analysis and Economic Impact 551
Ulrich Wagner, Michael Beer, Jochen Habermann, and Philipp Pfeifroth

	Abstract	551
26.1	Introduction	551
26.2	Definitions and Methodology	552
26.3	Extraction, Conversion, and Distribution of Fuels	553
26.4	Results of Process Chain Analyses	555
26.5	Conclusion	563
	References	564

27 Strategic and Socioeconomic Studies in Hydrogen Energy 567
David Hart

	Abstract	567
27.1	Introduction	567
27.2	Defining Socioeconomics	568
27.3	Examples	569
27.4	Economic Analysis	569
27.5	Visions and Futures	570
27.6	Social Behavior	571
27.7	Drivers and Barriers	572

27.8	Finance	*572*
27.9	Business	*573*
27.10	Conclusion	*574*
	Further Reading	*574*

28   Market Introduction for Hydrogen and Fuel Cell Technologies   *577*
Marianne Haug and Hanns-Joachim Neef

	Abstract	*577*
28.1	Introduction	*577*
28.2	Market Introduction of Radical Innovations: What Do We Know from the Literature?	*579*
28.3	The Fuel Cell and Hydrogen Road Maps: from Visions to Public/Private Coalitions	*581*
28.4	International Cooperation: Value Added During Market Introduction?	*583*
28.5	Market Introduction: The Status Quo	*584*
28.6	Conclusion: Co-evolution of Technology and Policy	*593*
	References	*594*

29   Hydrogen and Fuel Cells around the Corner – the Role of Regions and Municipalities Towards Commercialization   *597*
Andreas Ziolek, Marieke Reijalt, and Thomas Kattenstein

	Abstract	*597*
29.1	Introduction	*597*
29.2	The Role of Regional and Local Activities	*599*
29.3	HyRaMP – Organizing Local and Regional Drivers in Europe	*604*
29.4	Conclusion	*605*
	References	*606*

30   Zero Regio: Recent Experience with Hydrogen Vehicles and Refueling Infrastructure   *609*
Heinrich Lienkamp and Ashok Rastogi

	Abstract	*609*
30.1	Introduction	*610*
30.2	Hydrogen Production and Quality	*611*
30.3	Refueling Infrastructure	*614*
30.4	FCV Fleets and Demonstration	*620*
30.5	Socioeconomic Investigations	*623*
30.6	Dissemination	*623*
30.7	Conclusion	*624*
	Acknowledgments	*625*
	References	*626*

Safety Issues

31 Safety Analysis of Hydrogen Vehicles and Infrastructure *629*
Thomas Jordan and Wolfgang Breitung

Abstract *629*
31.1 Motivation of Hydrogen-Specific Safety Investigations *630*
31.2 Phenomena *631*
31.3 Safety Analysis Procedures *635*
31.4 Scenarios *637*
31.5 Outlook *643*
References *644*
Further Reading *647*

32 Advancing Commercialization of Hydrogen and Fuel Cell Technologies Through International Cooperation of Regulations, Codes, and Standards (RCS) *649*
Randy Dey

Abstract *649*
32.1 Introduction *649*
32.2 Hydrogen – a Part of the New Energy Mix *650*
32.3 Regulations, Codes, and Standards (RCS) – a Necessary Step to Commercialization *650*
32.4 International RCS Bodies – Responsible for the Standardization of Hydrogen and Fuel Cell Technologies *651*
32.5 International Cooperation in RCS *652*
32.6 International Cooperation Between RCS and Pre-Normative Research (PNR) *653*
32.7 Hydrogen Refueling Stations (HRS) *653*
32.8 Conclusion *655*
Definitions *655*
References *656*

Existing and Emerging Markets

33 Aerospace Applications of Hydrogen and Fuel Cells *661*
Christian Roessler, Joachim Schoemann, and Horst Baier

Abstract *661*
33.1 Introduction and Overview of Hydrogen and Fuel Cell Use *661*
33.2 Possible Fuel Cell Types for Aviation *663*
33.3 Application in Unmanned Aerial Vehicles (UAVs) *664*
33.4 Applications in General Aviation *670*
33.5 Application to Commercial Transport Aircraft *674*

33.6	Conclusion 677
	References 678

34 Auxiliary Power Units for Light-Duty Vehicles, Trucks, Ships, and Airplanes *681*
Ralf Peters

	Abstract 681
34.1	Operating Conditions for Auxiliary Power Units 681
34.2	System Design 691
34.3	Present Status of Fuel Cell-Based APU Systems 703
34.4	System Evaluation 708
34.5	Conclusion 709
	Acknowledgments 709
	References 710

35 Portable Applications and Light Traction *715*
Jürgen Garche

	Abstract 715
35.1	Introduction 715
35.2	Demand on Fuel Cells for Portable Applications 716
35.3	Fuel Cell Technology 717
35.4	Fuel 720
35.5	Applications 721
	References 732

Stationary Applications

36 High-Temperature Fuel Cells in Decentralized Power Generation *735*
Robert Steinberger-Wilckens and Niels Christiansen

	Abstract 735
36.1	Introduction 735
36.2	Distributed Generation as a Tool to Improve the Efficiency of Electricity Provision 736
36.3	Fuel Cells in Distributed Generation 739
36.4	Designing for High Efficiency 741
36.5	Developments in the United States 744
36.6	Asian and Pacific Developments 746
36.7	European Developments 748
36.8	Economic Prospects in DG Fuel Cell Development 750
36.9	Outlook 751
	References 751

37 Fuel Cells for Buildings   755
John F. Elter

Abstract   755
37.1 Introduction   755
37.2 Voice of the Customer37.2 Voice of the Customer   758
37.3 Fuel Cell Basics and Types37.3 Fuel Cell Basics and Types   761
37.4 Recent Advances   768
37.5 Fuel Cell Systems   772
37.6 System Control   784
37.7 Conclusion   785
References   786

Transportation Applications

38 Fuel Cell Power Trains   793
Peter Froeschle and Jörg Wind

Abstract   793
38.1 Introduction   793
38.2 Layout and Functionality of the Fuel Cell Hybrid Power Train   795
38.3 Technological Leaders of Fuel Cell Drive Train Development   799
38.4 Next Milestones on the Way to Commercialization   808
38.5 Future Outlook   809
References   809

39 Hydrogen Internal Combustion Engines   811
H. Eichlseder, P. Grabner, and R. Heindl

Abstract   811
39.1 A History   811
39.2 State of the Art   814
39.3 New Concepts   818
39.4 Future Perspectives   825
39.5 Conclusion   829
References   829

40 Systems Analysis and Well-to-Wheel Studies   831
Thomas Grube, Bernd Höhlein, Christoph Stiller, and Werner Weindorf

Abstract   831
40.1 Introduction   831
40.2 Platinum Group Metal Requirements for Fuel Cell Systems   832
40.3 Dynamic Powertrain Simulation   836
40.4 Well-to-Wheel Studies   841
Abbreviations   849

Symbols *850*
References *850*

41 Electrification in Transportation Systems *853*
Arndt Freialdenhoven and Henning Wallentowitz

Abstract *853*
41.1 Driving Forces for Electric Mobility *853*
41.2 Design of Battery Electric Vehicles (BEVs) *856*
41.3 Requirements on Players *869*
41.4 Conclusion *872*
References *873*

Index *875*

Foreword

Novel energy technologies are required to avoid climate change and to reduce the dependency on fossil fuels as the predominant primary energy source. Moreover, the energy sector is a major economic factor, and advanced technologies such as fuel cells can create new business opportunities.

Over many years, the government of North Rhine-Westphalia has been substantially supporting hydrogen and fuel cell technologies through the establishment of the Fuel Cell and Hydrogen Network NRW and by funding numerous research and technology projects. By doing so, we have opened up opportunities to achieve the ambitious energy goals the North Rhine-Westphalia government has set itself. In addition, we keep promoting basic research and technological development, especially in the energy sector, as these are essential for our future.

I am therefore happy that the State of North Rhine-Westphalia and the city of Essen, Germany's energy capital, are the hosts of the 18th World Hydrogen Energy Conference 2010.

Zur Vermeidung des Klimawandels und zur Reduzierung unserer Abhängigkeit von fossilen Energieträgen sind neue Energietechnologien erforderlich. Der Energiesektor stellt zudem einen wesentlichen ökonomischen Faktor dar. Fortschrittstechnologien wie die Brennstoffzelle bieten neue Marktchancen.

Die Landesregierung von Nordrhein Westfalen hat die Wasserstoff- und Brennstoffzellentechnik durch die Etablierung des Netzwerks Brennstoffzelle und Wasserstoff NRW sowie durch die Unterstützung von Forschungs- und Technologieprojekten über viele Jahre hinweg gefördert. Damit haben wir Möglichkeiten geschaffen, unsere anspruchsvollen Ziele hinsichtlich Klimaschutz und Energieversorgung zu erreichen. Darüber hinaus unterstützen wir die Grundlagenforschung und technische Entwicklung speziell im Energiesektor, da dies essentiell für unsere Zukunft ist.

Im freue mich daher besonders, dass das Land Nordrhein-Westfalen mit Essen, der Energiehauptstadt Deutschlands, der Gastgeber der 18. World Hydrogen Energy Conference 2010 ist.

April 2010   Dr. Jürgen Rüttgers
                   Ministerpräsident des Landes Nordrhein-Westfalen

Hydrogen Energy. Edited by Detlef Stolten
Copyright © 2010 WILEY-VCH Verlag GmbH & Co. KGaA, Weinheim
ISBN: 978-3-527-32711-9

Preface

Over the past few years, hydrogen has gained importance as an energy carrier for future transport applications as fuel cells have reached a state in which they have been proven to work technically by several major car manufacturers, although major cost reductions still need to be made. Furthermore, renewable power, which inherently fluctuates, has emerged over the past decade, requiring novel means of compensation for these fluctuations. Water electrolysis and solar hydrogen generation as energy storage technologies in the longer term are suitable means of amending these fluctuations in concert with smart grids and smart end use of power. The hydrogen produced may be used preferably for transportation or for reconversion to electric power. As for stationary fuel cells, natural gas will be the predominant energy supply. In this respect, technologies for fuel cell-based household energy systems are at an advanced stage of field testing, particularly in Japan.

In the past, hydrogen as an energy carrier was focused upon from the 1970s until the mid-1990s, when the two oil price spikes had their effect. Back then, remote and hence untappable hydropower was investigated for conversion into hydrogen via electrolysis and transportation over long distances, for example from Canada to Germany via cryo-ships. However, improved power lines, the introduction of DC–DC power lines and increased demand in North America rendered these technologies obsolete. Irrespective thereof, the concept of electrolyzing nuclear power and particularly using high-temperature off-heat from nuclear reactors that were under development at that time was still pursued. In Germany, this was terminated when the political demise of nuclear power finally turned into mainstream policy in the early 1990s.

Meanwhile, additional primary energy sources were developed to a notable level in the grids, such as wind power in Germany, contributing about 6% to power production on average. Other technologies are in the testing phase, such as concentrated solar heat for power production, which may be used for hydrogen generation in the future. These fluctuating power sources require advanced control and storage systems in order to make a substantial contribution to the overall power supply. Hydrogen storage may complement these technologies, constituting a very good means of energy storage in general and for long-term storage in particular. Today's emerging fuel cell technology for transportation

Hydrogen Energy. Edited by Detlef Stolten
Copyright © 2010 WILEY-VCH Verlag GmbH & Co. KGaA, Weinheim
ISBN: 978-3-527-32711-9

will use hydrogen as a fuel. Electrification of transportation will integrate power production and energy use in transportation in the forthcoming decades, be it via hydrogen or electricity itself.

In some special markets, fuel cells are already commercially established, such as in submarines, as a portable power supply, particularly for military and recreational vehicles. For these technologies, supply of the necessary energy carrier, be it hydrogen or methanol, is a major issue.

Whereas fuel cells are a prerequisite for wide spread hydrogen use, the reverse does not hold true. There are major application areas for fuel cells, such as stationary fuel cells, portable fuel cells and auxiliary power units for heavy-duty diesel and jet fuel-powered transportation, that will rely on the incumbent energy supply for good reasons. They exploit either the high energy efficiency that fuel cells provide for larger systems or the higher storage density that these fuels provide and hence the longer potential operating times for portable systems. For propulsion of vehicles, however, hydrogen turns out to be inherently the ideal fuel, emitting no soot, and fitting into electrification schemes for automobiles with regenerative braking, providing long cruising range capability with quick and easy refueling. Moreover, it is a mass market application with the potential for setting up a new infrastructure. This is not to say that hydrogen should or will be the only fuel for automotive propulsion in the future, but there is a good case for it to hold a major share.

Electricity can be made from a variety of primary energies and has wide use in very different applications. Hydrogen adds the capability of storing that energy, hence these two energy supply routes will have complementary roles in the future. They both act as energy hubs separating the generation from the energy use, and hence leaving great potential for further inventions and optimization on either side. The discussion about where the hydrogen will come from is particularly important for novel technologies by means of which hydrogen can be produced, such as bio-approaches. It is often mixed up, however, with the general question of how the energy demand that we are expecting in the next few decades will be covered. This is not a problem that can be solved solely with hydrogen. However, hydrogen can help solve it via high energy pathway efficiency, and it can definitely help reduce CO_2 emissions and local pollution. Being a secondary energy carrier, the environmental record of hydrogen depends very much on the production method; ranging from electrolysis via bio-processes to clean-coal derived hydrogen.

This broad scope of fuel cells and hydrogen is covered by the World Hydrogen Energy Conference, which was held in May 2010 for the 18th time, this year in Essen, Germany. When organizing this conference, I became aware of the necessity for not just having very specific conference contributions on the topics mentioned above and general information for laymen alongside it – additionally, it was clear that overviews of scientific information were needed on specific topics for scientists and engineers who have already been working in this area for some time and want to broaden their scope, or for workers new to the subject. Hence all sessions at the conference were planned to start with an overview on the session topic from basic principles to a brief status review. Full texts of these contributions constitute this book, which I sincerely hope will make it easier for professionals and students in this area to acquire a quick and thorough overview on a particular topics.

Financial and organizational support of this book from the State of North Rhine-Westphalia through the 'EnergieAgentur.NRW' is gratefully acknowledged, and also financial support from the Helmholtz Centers Forschungszentrum Jülich and the Deutsches Zentrum für Luft- und Raumfahrt.

The great efforts that the authors have made in submitting their texts promptly within the very short timeline given is acknowledged, and many thanks are due to Thomas Grube, who helped me considerably in handling the many issues associated with editing this book. Not least the great efforts of the Wiley team should be mentioned – they made it possible for this book to appear in about 12 months from the first idea to the printed version.

I hope that this book will help to extend the knowledge of those already working in this field, will serve as a quick but reasonably comprehensive reference for people needing an introduction and may lay the basis for new avenues of study on the topic of hydrogen and fuel cells.

Detlef Stolten
Editor and Chairman of the 18th WHEC 2010

List of Contributors

Werner Ahrer
PROFACTOR GmbH
Im Stadtgut A2
4407 Steyr-Gleink
Austria

Etsuo Akiba
National Institute of Advanced Industrial Science and Technology (AIST)
Tsukuba Central 5
1-1-1, Higashi
Tsukuba
Ibaraki 305-8565
Japan

Horst Baier
Lehrstuhl für Luftfahrttechnik
TU München
Boltzmannstrasse 15
85747 Garching
Germany

Daniel R. Baker
General Motors Company
Electrochemical Energy Research Laboratory
10 Carriage Street
Honeoye Falls, NY 14472
USA

Anil Baral
International Council on Clean Transportation
1225 Eye Street NW
Suite 900
Washington, DC 20005
USA

Françoise Barbier
Air Liquide
Centre de Recherche Claude Delorme
1 chemin de la Porte des Loges –
Les Loges en Josas
BP 126
78354 Jouy-en-Josas
France

Michael Beer
Forschungsstelle
für Energiewirtschaft eV, München
Bavarian Hydrogen Initiative
Am Blütenanger 71
80995 Munich
Germany

Andreas Borgschulte
EMPA Materials Sciences and Technology
Department of Energy, Environment, and Mobility
Laboratory 138 "Hydrogen & Energy"
Überlandstrasse 129
8600 Dübendorf
Switzerland

Hydrogen Energy. Edited by Detlef Stolten
Copyright © 2010 WILEY-VCH Verlag GmbH & Co. KGaA, Weinheim
ISBN: 978-3-527-32711-9

Martin Bram
Forschungszentrum Jülich GmbH
Institute of Energy Research 1
52425 Jülich
Germany

Wolfgang Breitung
Institut für Kern- und Energietechnik
Karlsruhe Institut für Technologie
76021 Karlsruhe
Germany

Annabelle Brisse
European Institute for Energy Research (EIFER)
Emmy-Noether-Strasse 11
76131 Karlsruhe
Germany

Hans-Peter Buchkremer
Forschungszentrum Jülich GmbH
Institute of Energy Research 1
52425 Jülich
Germany

Robert N. Carter
General Motors Company
Electrochemical Energy Research Laboratory
10 Carriage Street
Honeoye Falls, NY 14472
USA

Niels Christiansen
Topsøe Fuel Cell
Nymøllevej 66
2800 Lyngby
Denmark

Pieternel A. M. Claassen
Wageningen UR
(University and Research Centre)
Agrotechnology and Food Innovations
6708 WG Wageningen
The Netherlands

Gael P. G. Corre
University of St. Andrews
School of Chemistry
North Haugh
St. Andrews
Fife KY16 9ST
UK

Truus de Vrije
Wageningen UR
(University and Research Centre)
Agrotechnology and Food Innovations
6708 WG Wageningen
The Netherlands

Randy Dey
The CCS Global Group
1182 Old Post Dr.
Oakville, Ontario
L6 M1A6
Canada

Ibrahim Dincer
University of Ontario Institute of Technology (UOIT)
Faculty of Engineering and Applied Science
2000 Simcoe Street North
Oshawa
Ontario L1H 7K4
Canada

Helmut Eichlseder
Graz University of Technology
Institute for Internal Combustion Engines and Thermodynamics
Inffeldgasse 21a
8010 Graz
Austria

John F. Elter
University at Albany, State University
of New York
College of Nanoscale Science
and Engineering
Center for Sustainable Ecosystem
Nanotechnologies
1400 Washington Avenue
Albany, NY 12222
USA

İnci Eroğlu
Middle East Technical University
Department of Chemical Engineering
06531 Ankara
Turkey

Manfred Fischedick
Wuppertal Institute for Climate,
Environment and Energy
Research Group 1: Future Energy
and Mobility Structures
Doeppersberg 19
42103 Wuppertal
Germany

Arndt Freialdenhoven
ika, RWTH Aachen
Steinbachstraße 7
52074 Aachen
Germany

Anton Friedl
Vienna University of Technology
Institute of Chemical Engineering
Getreidemarkt 9
1060 Vienna
Austria

Oliver Friedrichs
EMPA Materials Sciences
and Technology
Department of Energy, Environment,
and Mobility
Laboratory 138 "Hydrogen & Energy"
Überlandstrasse 129
8600 Dübendorf
Switzerland

Peter Froeschle
Strategic Energy Projects
and Market Development
Fuel Cell and Battery Electric Vehicles
Daimler AG
Neue Straße 95
73230 Kirchheim/Nabern
Germany

Jürgen Garche
ZSW Ulm
Electrochemical Energy Technology
Division
Helmholtzstrasse 8
89081 Ulm
Germany

Hubert A. Gasteiger
Technical University Munich
Chair of Technical Electrochemistry
Department of Chemistry
Lichtenbergstraße 4
85748 Garching
Germany

Tim van Gestel
Forschungszentrum Jülich GmbH
Institute of Energy Research 1
52425 Jülich
Germany

Andreas Glüsen
Forschungszentrum Jülich GmbH
Institute of Energy Research –
Fuel Cells (IEF-3)
52425 Jülich
Germany

Peter Grabner
Graz University of Technology
Institute for Internal Combustion
Engines and Thermodynamics
Inffeldgasse 21a
8010 Graz
Austria

Robin Gremaud
EMPA Materials Sciences
and Technology
Department of Energy, Environment,
and Mobility
Laboratory 138 "Hydrogen & Energy"
Überlandstrasse 129
8600 Dübendorf
Switzerland

Thomas Grube
Forschungszentrum Jülich GmbH
Institute of Energy Research –
Fuel Cells (IEF-3)
52425 Jülich, Germany

Wenbin Gu
General Motors Company
Electrochemical Energy Research
Laboratory
10 Carriage Street
Honeoye Falls, NY 14472
USA

Jochen Habermann
Forschungsstelle
für Energiewirtschaft eV, München
Bavarian Hydrogen Initiative
Am Blütenanger 71
80995 Munich
Germany

David Hart
E4tech
Avenue Juste-Olivier 2
1006 Lausanne
Switzerland

and

Imperial College London
South Kensington
London SW7 2AZ
UK

Kevin W. Harrison
National Renewable Energy
Laboratory
1617 Cole Boulevard
Golden, CO 80401
USA

Marianne Haug
Universität Hohenheim
70599 Stuttgart
Germany

and

Oxford Institute of Energy Studies
57 Woodstock Road
Oxford OX2 6FA
UK

Christopher Hebling
Fraunhofer Institute for Solar Energy
Systems ISE
Heidenhofstrasse 2
79110 Freiburg
Germany

René Heindl
Graz University of Technology
Institute for Internal Combustion
Engines and Thermodynamics
Inffeldgasse 21a
8010 Graz
Austria

Michael Hirscher
Heisenbergstr. 3
70569 Stuttgart
Germany

Bert Hobein
Ford Forschungszentrum Aachen
Süsterfeldstrasse 200
52072 Aachen
Germany

Bernd Höhlein
Hydrogen and Fuel Cell Network
North Rhine-Westphalia
Korbweg 4
52441 Limick
Germany

Aaron Hoskin
Natural Resources Canada
580 Booth
Ottawa
Ontario K1A 0E4
Canada

John T. S. Irvine
University of St. Andrews
School of Chemistry
North Haugh
St. Andrews
Fife KY16 9ST
UK

Mariya E. Ivanova
Forschungszentrum Jülich GmbH
Institute of Energy Research 1
52425 Jülich
Germany

Thomas Jordan
Institut für Kern- und Energietechnik
Karlsruhe Institut für Technologie
76021 Karlsruhe
Germany

Thomas Kattenstein
HyRaMP
Palais des Academies
Rue Ducale 1
1000 Brussels
Belgium

Emmanuel G. Koukios
National Technical University
of Athens
School of Chemical Engineering
Bioresource Technology Unit
Athens
Greece

Dennis Krieg
Forschungszentrum Jülich GmbH
Institute for Energy Research –
Fuel Cells (IEF-3)
52425 Jülich
Germany

Roland Krüger
Ford Forschungszentrum Aachen
Süsterfeldstrasse 200
52072 Aachen
Germany

Heinrich Lienkamp
Infraserv GmbH & Co. Höchst KG
Industriepark Höchst
65926 Frankfurt am Main
Germany

Yuxin Liu
General Motors Company
Electrochemical Energy Research
Laboratory
10 Carriage Street
Honeoye Falls, NY 14472
USA

List of Contributors

Alan C. Lloyd
International Council on Clean Transportation
18124 Wedge Parkway
Suite 535
Reno, NV 89511
USA

Gregory D. Martin
National Renewable Energy Laboratory
1617 Cole Boulevard
Golden, CO 80401
USA

Matthew M. Mench
The Pennsylvania State University
Department of Mechanical and Nuclear Engineering
University Park, PA 16802
USA

Philippe Mauron
EMPA Materials Sciences and Technology
Department of Energy, Environment, and Mobility
Laboratory 138 "Hydrogen & Energy"
Überlandstrasse 129
8600 Dübendorf
Switzerland

Jürgen Mergel
Forschungszentrum Jülich GmbH
Institute of Energy Research – Fuel Cells (IEF-3)
52425 Jülich
Germany

Wilhelm A. Meulenberg
Forschungszentrum Jülich GmbH
Institute of Energy Research 1
52425 Jülich
Germany

Michael Modigell
RWTH Aachen University
Mechanical Process Engineering
Aachener Verfahrenstechnik
52074 Aachen
Germany

Hanns-Joachim Neef
Fuel Cell and Hydrogen Network
North-Rhine Westphalia
c/o EnergieAgentur NRW
Haroldstrasse 4
40213 Düsseldorf
Germany

Isabella Nowik
RWTH Aachen University
Mechanical Process Engineering
Aachener Verfahrenstechnik
52074 Aachen
Germany

Dominik Ochs
PROFACTOR GmbH
Steyr-Gleink
Austria

Joan M. Ogden
University of California, Davis
Institute of Transportation Studies
1 Shields Avenue
Davis, CA 95616
USA

Ebru Özgür
Middle East Technical University
Department of Chemical Engineering
06531 Ankara
Turkey

Andreas Pastowski
Wuppertal Institute for Climate,
Environment and Energy
Research Group 1: Future Energy
and Mobility Structures
Doeppersberg 19
42103 Wuppertal
Germany

Ralf Peters
Forschungszentrum Jülich GmbH
Institute of Energy Research –
Fuel Cells (IEF-3)
52425 Jülich
Germany

Philipp Pfeifroth
Forschungsstelle
für Energiewirtschaft eV, München
Bavarian Hydrogen Initiative
Am Blütenanger 71
80995 Munich
Germany

Ed Pike
International Council
on Clean Transportation
One Post Street
Suite 2700
San Francisco, CA 94104
USA

Siegfried W. Rapp
Vogesenstrasse 4a
79415 Bad Bellingen
Germany

Ashok Rastogi
Infraserv GmbH & Co. Höchst KG
Industriepark Höchst
65926 Frankfurt am Main
Germany

Sebastian Rau
Fraunhofer Institute for Solar Energy
Systems ISE
Heidenhofstrasse 2
79110 Freiburg
Germany

Marieke Reijalt
HyRaMP
Palais des Academies
Rue Ducale 1
1000 Brussels
Belgium

Arndt Remhof
EMPA Materials Sciences
and Technology
Department of Energy, Environment,
and Mobility
Laboratory 138 "Hydrogen & Energy"
Überlandstrasse 129
8600 Dübendorf
Switzerland

Robert Remick
National Renewable
Energy Laboratory
1617 Cole Boulevard
Golden, CO 80401
USA

Christian Roessler
Lehrstuhl für Luftfahrttechnik
TU München
Boltzmannstrasse 15
85747 Garching
Germany

Jens Rostrup-Nielsen
Haldor Topsøe A/S
Nymoellevej 55
2800 Lyngby
Denmark

Christian Sattler
Deutsches Zentrum für Luft-
und Raumfahrt eV (DLR)
51147 Köln
Germany

Achim Schaadt
Fraunhofer Institute for Solar Energy
Systems ISE
Heidenhofstrasse 2
79110 Freiburg
Germany

Barbara Schmitz
Heisenbergstr. 3
70569 Stuttgart
Germany

Jörg Schindler
Schopenhauerstr. 13
85579 Neubiberg
Germany

Josef Schefold
European Institute for Energy
Research (EIFER)
Emmy-Noether-Strasse 11
76131 Karlsruhe
Germany

Joachim Schoemann
Lehrstuhl für Luftfahrttechnik
TU München
Boltzmannstrasse 15
85747 Garching
Germany

José M. Serra
Instituto de Tecnología Química
(UPV-CSIC)
Av. Los Naranjos s/n
46022 Valencia
Spain

Tom Smolinka
Fraunhofer Institute for Solar Energy
Systems ISE
Heidenhofstrasse 2
79110 Freiburg
Germany

Robert Steinberger-Wilckens
Forschungszentrum Jülich
Institute of Energy Research
52425 Jülich
Germany

Christoph Stiller
Ludwig-Bölkow-Systemtechnik GmbH
Daimlerstrasse 15
85521 Ottobrunn
Germany

Detlef Stolten
Forschungszentrum Jülich GmbH
Institute for Energy Research –
Fuel Cells (IEF-3)
52425 Jülich
Germany

Detlev Stöver
Forschungszentrum Jülich GmbH
Institute of Energy Research 1
52425 Jülich
Germany

Clemens Alexander Trudewind
Ruhr University Bochum
Lehrstuhl für Energiesysteme
und Energiewirtschaft
Universitätsstrasse 150
44801 Bochum
Germany

Ed W.J. van Niel
Lund University
Applied Microbiology
22100 Lund
Sweden

T. Nejat Veziroglu
International Association
for Hydrogen Energy (IAHE)
5794 SW 40 Street #303
Miami, FL 33155
USA

Frederick T. Wagner
General Motors Company
Electrochemical Energy Research
Laboratory
10 Carriage Street
Honeoye Falls, NY 14472
USA

Hermann-Josef Wagner
Ruhr University Bochum
Lehrstuhl für Energiesysteme
und Energiewirtschaft
Universitätsstrasse 150
44801 Bochum
Germany

Ulrich Wagner
TU München
Lehrstuhl für Energiewirtschaft
und Anwendungstechnik
Theresienstrasse 90
80290 Munich

and

Forschungsstelle für Energiewirtschaft
eV, München
Bavarian Hydrogen Initiative
Am Blütenanger 71
80995 Munich
Germany

Henning Wallentowitz
ika, RWTH Aachen
Steinbachstraße 7
52074 Aachen
Germany

Christoph Wannek
Forschungszentrum Jülich GmbH
Institute of Energy Research –
Fuel Cells (IEF-3)
52425 Jülich
Germany

Werner Weindorf
Ludwig-Bölkow-Systemtechnik GmbH
Daimlerstrasse 15
85521 Ottobrunn
Germany

Jörg Wind
Strategic Energy Projects
and Market Development
Fuel Cell and Battery Electric Vehicles
Daimler AG
Neue Straße 95
73230 Kirchheim/Nabern
Germany

Walter Wukovits
Vienna University of Technology
Institute of Chemical Engineering
Vienna
Austria

Paul T. Yu
General Motors Company
Electrochemical Energy Research
Laboratory
10 Carriage Street
Honeoye Falls, NY 14472
USA

Mohsine Zahid
European Institute for Energy
Research (EIFER)
Emmy-Noether-Strasse 11
76131 Karlsruhe
Germany

Andreas Ziolek
HyRaMP
Palais des Academies
Rue Ducale 1
1000 Brussels
Belgium

Andreas Züttel
EMPA Materials Sciences
and Technology
Department of Energy, Environment,
and Mobility
Laboratory 138 "Hydrogen & Energy"
Überlandstrasse 129
8600 Dübendorf
Switzerland

Fuel Cell Basics

Hydrogen Energy. Edited by Detlef Stolten
Copyright © 2010 WILEY-VCH Verlag GmbH & Co. KGaA, Weinheim
ISBN: 978-3-527-32711-9

1
Electrocatalysis and Catalyst Degradation Challenges in Proton Exchange Membrane Fuel Cells

Hubert A. Gasteiger, Daniel R. Baker, Robert N. Carter, Wenbin Gu, Yuxin Liu, Frederick T. Wagner, and Paul T. Yu

Abstract

After a brief review of the kinetics of the cathodic oxygen reduction and the anodic hydrogen oxidation reaction, a fundamental membrane electrode assembly performance model is outlined, which demonstrates that a 4–10-fold reduced amount of platinum is required for commercially viable large-scale vehicle applications. The various catalyst technology roadmaps to achieve this goal are discussed. With the increasing number of prototype proton exchange membrane fuel cell (PEMFC)-powered vehicles, catalyst durability has also become a strong focus of academic and industrial R&D. Therefore, the key issues of platinum sintering/dissolution under dynamic vehicle operation and of carbon-support corrosion during PEMFC startup/shutdown are reviewed.

Keywords: electrocatalysis, platinum catalysts, catalyst degradation, carbon-support corrosion, proton exchange membrane fuel cells, fuel cell-powered vehicles, hydrogen oxidation

1.1
Introduction

Over the past few years, significant R&D efforts have been aimed at meeting the challenging performance and cost targets for the use of proton exchange (PEM) fuel cells in vehicles. Catalyst development and optimization of membrane electrode assemblies (MEAs) increased PEM fuel cell power densities to levels which satisfy vehicle packaging needs (~ 1 W cm^{-2}) at platinum catalyst loadings of ~ 0.5 mg$_{Pt}$ cm$_{MEA}^{-2}$, corresponding to Pt specific power densities of ~ 0.5 g$_{Pt}$ kW^{-1}, which equates to ~ 50 g of platinum in a 100 kW automotive fuel cell stack [1]. This, together with the development of high-pressure hydrogen tanks (70 MPa), permitted the construction of small fuel cell vehicle test fleets for testing under real driving conditions, leading to the demonstration by Toyota and General Motors that a vehicle driving range of 300 miles can indeed be met. However, as shown in an ongoing study by the United States Depart-

Hydrogen Energy. Edited by Detlef Stolten
Copyright © 2010 WILEY-VCH Verlag GmbH & Co. KGaA, Weinheim
ISBN: 978-3-527-32711-9

ment of Energy (DoE), conducted by the National Renewable Energy Laboratory (NREL), the durability of a significant number of fuel cell vehicles from various companies averages only 1200 h (with a maximum of 1900 h), limited mostly by catalyst degradation [2]. Therefore, irrespective of the questions regarding the viability of the future generation of renewable hydrogen and the infrastructure required for its distribution, large-scale commercialization of PEM fuel cell vehicles still requires significant advances in catalyst development in order both to reduce the amount of platinum metals required to run a fuel cell stack and to enhance catalyst durability. Approaches to how one might reach the catalyst activity target of $<0.2\ g_{Pt}\ kW^{-1}$ [1] and the underlying degradation phenomena which currently limit PEM fuel cell life in vehicles to significantly less than the 6000 h life target are discussed in this chapter.

1.2
Voltage Losses in State-of-the-Art Automotive PEM Fuel Cells

In order to determine whether a reduction in the platinum specific power density below the current value of $\sim 0.5\ g_{Pt}\ kW^{-1}$ can be achieved by further optimization of electrode, diffusion medium (DM) and MEA structures, the value of the various voltage loss terms in state-of-the-art MEAs must be quantified. Therefore, over many years, work at General Motors has been focused on developing methods by which the voltage losses due to catalyst kinetics and various transport processes in an MEA could be predicted on the basis of fundamental physical-chemical parameters, in order to provide a model of the cell voltage, E_{cell}, as a function of current density, i:

$$E_{cell} = E_{rev.} - i \times (R_{electronic} + R_{membrane}) - \eta_{HOR} - |\eta_{ORR}|$$
$$- i \times R_{H^+, eff.} - \eta_{tx, gas (dry)} - \eta_{tx, gas (wet)} \tag{1.1}$$

where $E_{rev.}$ is the reversible thermodynamic potential depending on temperature and gas partial pressure and $R_{electronic}$ are the electronic resistances in an MEA (mostly the contact resistance between the DM and the bipolar plate flow-field [3]). Whereas these terms are independent of current density, all other terms vary with current density: $R_{membrane}$ is the relative humidity (RH)-dependent proton conduction resistance of the membrane [4], η_{HOR} is the overpotential loss for the hydrogen oxidation reaction (HOR) [5], η_{ORR} is the overpotential loss for the oxygen reduction reaction (OOR) [3], $R_{H^+, eff.}$ is the effective RH-dependent proton conduction resistance in the anode and the cathode electrode [6, 7], $\eta_{tx, gas (dry)}$ is the gas diffusion overpotential mostly controlled by the diffusion of oxygen in air through the diffusion medium and the cathode electrode in the absence of liquid water (in the case of PEM fuel cell operation with pure hydrogen, voltage losses from hydrogen diffusion are essentially negligible) [8], and $\eta_{tx, gas (wet)}$ are additional gas diffusion overpotential losses caused by the presence of liquid water in the diffusion media and the electrodes.

Using independently measured physical-chemical parameters for the various terms in Equation 1.1, a quasi-two-dimensional down-the-channel model was developed (normal to the MEA surface and along the flow-field channel) to describe the various voltage loss terms as a function of current density for typical automotive operating conditions. This is shown in Figure 1.1, illustrating the well-known fact that the slow ORR kinetics on a state-of-the-art carbon-supported platinum catalyst (Pt/C) are responsible for about two-thirds of the overall voltage losses at high current density, evidenced by comparing the ORR kinetics-limited performance (top line in Figure 1.1) with the actual performance of a full active area PEM fuel cell short-stack (lowest line in Figure 1.1, with error bars indicating the standard deviation in cell voltage for all MEAs in the stack). The voltage losses due to the hydrogen oxidation reaction at the anode are negligible under the conditions shown in Figure 1.1 ($\ll 5$ mV at 1.5 A cm^{-2}) [5, 8]. The ohmic losses at the highest current density of 1.5 A cm^{-2} are ~ 90 mV and are mostly due to electronic contact resistances (~ 60 mV), with only ~ 30 mV losses caused by the proton conduction resistance of the membrane; clearly, the development of low-resistance bonding between the bipolar plate flow-field and the diffusion medium could result in up to 5% efficiency gains at 1.5 A cm^{-2}. The remaining voltage loss terms from proton conduction resistances in the electrodes and from oxygen diffusion resistances through a nominally dry diffusion medium and electrode are small, adding up to only ~ 40 mV at 1.5 A cm^{-2}. It should be mentioned that the last term in Equation 1.1 cannot be determined from measurable materials properties and thus represents the difference between the predicted losses (short dashed line in Figure 1.1) and the measured cell voltage (lowest line in Figure 1.1). The reason why this difference is very small (~ 20 mV at 1.5 A cm^{-2}) is that the maximum local RH in the MEA barely reaches 100% at the highest current density [8], so that blocking of electrode and DM pores by liquid water is not a strong factor under automotive operating conditions characterized by low RH conditions for reasons of systems simplification [9]. It should be noted that the model predictions shown in Figure 1.1 are consistent with the measured high-frequency resistance and were also validated by comparing the predicted current distribution along the flow-field channels with current distribution measurements in full active area hardware [8].

Given the fact that the voltage losses due to electronic, proton, and gas transport resistances are rather small, it is clear that significant increases in the platinum specific power density can only be achieved by reducing the catalyst loadings on both the anode and the cathode electrode. This, however, would lead to lower cell voltages and therefore to lower fuel cell efficiencies, which cannot be tolerated as it would sacrifice the promised high energy efficiency of fuel cell vehicles and would also lead to additional engineering challenges with regards to heat rejection via the vehicle radiator [9].

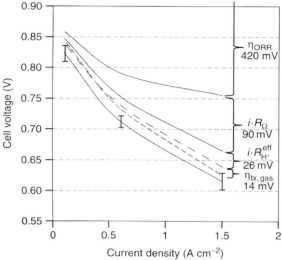

Fig. 1.1 Voltage loss terms in state-of-the-art H$_2$-air PEM fuel cell full active area short-stack (~20 cells) operated under representative automotive conditions. MEAs: 0.2/0.3 mg$_{Pt}$ cm^{-2} (anode/cathode) coated on an 18 μm thick perfluorosulfonic acid membrane and sandwiched between 200 μm thick diffusion media (SGL 25BC) on both the anode and cathode sides. Operating conditions vary with current density: (i) at low current density, P_{inlet}=110 kPa$_a$, $T_{coolant, out}$=72 °C, RH$_{inlet}$=30/45% (anode/cathode), and H$_2$-air feeds at s=2/5.5 (anode/cathode); (ii) at high current density, P_{inlet}=155 ± 20 kPa$_a$, $T_{coolant, out}$=80 °C, RH$_{inlet}$ ~ 30/60% (anode/cathode), and H$_2$-air feeds at s=2/1.8 (anode/cathode). For details, see [8]. Reproduced from W. Gu et al. [8], with permission from John Wiley & Sons, Ltd.

1.3
Catalyst Development Needs and Approaches

Based on the above-outlined considerations, either non-platinum catalysts or more active platinum-based catalysts must be developed, so that the required amount of platinum can be reduced without loss of power density and efficiency. Although significant advances have been made in the area of non-Pt cathode catalysts [10], Pt-based catalysts are currently still the most promising option to provide the high power densities required for vehicle applications. The activity gains that would have to be realized by advanced platinum-based catalysts in order to lower the Pt specific power densities are shown in Table 1.1, considering that the platinum loadings on the anode electrode can be lowered to 0.05 mg$_{Pt}$ cm^{-2} without losses in fuel cell performance in the case of fuel cell operation with pure hydrogen [5].

Table 1.1 shows that the total amount of Pt in the MEA can be reduced significantly without performance loss by lowering the anode Pt loading and by implementation of carbon-supported Pt-alloy catalysts (e.g., Pt$_3$Co/C [1, 11]) with

Table 1.1 Effect of lowering platinum loadings on the anode (L_{anode}) and the cathode ($L_{cathode}$) on the platinum specific power density, P_{Pt}, by assuming that the current high current density performance of the MEA can be maintained at the level shown in Figure 1.1 (i.e., 0.62 V at 1.5 A cm^{-2}, corresponding to 0.93 W cm^{-2}): this requires advanced cathode catalysts for the ORR which have improved mass activity over currently used state-of-the-art Pt/C.

L_{anode} (mg$_{Pt}$ cm^{-2})	$L_{cathode}$ (mg$_{Pt}$ cm^{-2})	P_{Pt} (g$_{Pt}$ kW^{-1})	Cathode catalyst technology
0.20	0.30	0.54	Conventional Pt/C, see Figure 1.1
0.05	0.20	0.27	"2×" conventional Pt alloy/C
0.05	0.10	0.16	"4×" Pt-based cathode catalyst
0.05	0.05	0.10	"10×" Pt-based cathode catalyst

demonstrated close to twofold higher *Pt mass activity* (in units of A mg$_{Pt}^{-1}$ at a given potential; see below) compared with Pt/C. Implementation of this near-term "2×" cathode catalyst technology is expected to lead to Pt specific power densities of $P_{Pt} \sim 0.27$ g$_{Pt}$ kW^{-1}, closely approaching the initially announced target of <0.2 g$_{Pt}$ kW^{-1} [1]. Conceptually, the latter can be reached by the recently developed *dealloyed Pt alloy* catalysts (third row in Table 1.1), which have shown fourfold higher mass activities compared with Pt/C [12, 13]. Although this concept must still be proven for its long-term stability in operating PEM fuel cells, it is a promising path towards reaching the initial <0.2 g$_{Pt}$ kW^{-1} goal. However, considering the significant constraint on world-wide platinum resources, large-scale PEM fuel cell vehicle commercialization really necessitates a reduction in the platinum content per vehicle to the order to 10 g$_{Pt}$, i.e., $P_{Pt} \sim 0.1$ g$_{Pt}$ kW^{-1} for a 100 kW PEM fuel cell vehicle. As shown in Table 1.1 (bottom row), this would require novel Pt-based cathode catalysts with 10-fold higher mass activity compared with conventional Pt/C.

In principle, there are two possible pathways towards platinum-based cathode catalysts with "10×" higher Pt mass activity, $i_{m(0.9\,V)}$ (in A mg$_{Pt}^{-1}$), which is commonly defined at a potential of 0.9 V versus the reversible hydrogen electrode (RHE) potential at an oxygen partial pressure of 100 kPa$_a$ [3]. This is best illustrated by considering that the Pt mass activity for the ORR depends on both the *specific activity* of a Pt-based catalyst, $i_{s(0.9\,V)}$ (in μA cm$_{Pt}^{-2}$), and its specific surface area, A_{Pt} (in m$_{Pt}^2$ g$_{Pt}^{-1}$):

$$i_{m(0.9\,V)}[A\ mg_{Pt}^{-1}] = i_{s(0.9\,V)}[\mu A\ cm_{Pt}^{-2}] \times A_{Pt}[m_{Pt}^2\ g_{Pt}^{-1}] \times 10^{-5} \qquad (1.2)$$

where the specific activity represents the intrinsic activity of a platinum catalyst surface (directly proportional to the so-called turnover frequency in heterogeneous catalysis), while the specific surface area is a measure of the exposed platinum surface area per unit platinum mass (also referred to as platinum dispersion). Hence there are two different pathways to increase platinum mass activity, via [14]: (i) core-shell concepts where a platinum monolayer is supported on a non-platinum nanoparticle leading to very high platinum specific surface areas and, consequently, to high mass activity [15] and also ultra-thin platinum (alloy) coatings supported on nanostructured supports [16]; and/or (ii) increased specific activity through alloying of platinum with transition metals [12, 17, 18], or the extraordinarily high specific activity observed for the (111) surface planes of Pt alloys [17, 19].

Figure 1.2a shows the specific activity gains observed for polycrystalline bulk platinum alloys, with Pt_3Co, Pt_3Ni, and Pt_3Fe yielding two- to threefold higher specific activity over pure platinum, consistent with what was observed for high surface area carbon-supported Pt alloys [1, 11, 20]. Specific activities depend not only on the actual transition metal, but also on the surface structure of the Pt alloy [17]: (i) *skeleton* structures are produced by removal of the acid-soluble transition metal from the Pt alloy surface during acid leaching or contact with the acidic electrolyte (aqueous or ionomeric); (ii) *Pt skin* structures are produced by platinum surface segregation during high-temperature annealing in vacuum or inert gas. It was shown recently that the same structures and a similar effect on specific activity are observed for high surface area carbon-supported Pt alloy catalysts [21].

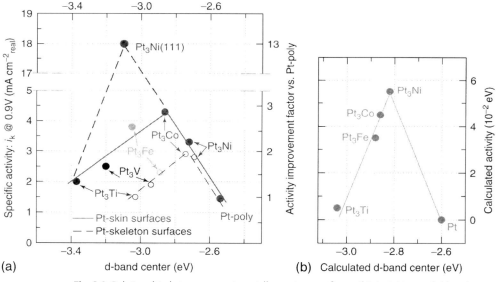

Fig. 1.2 Relationship between experimentally measured specific activities for the ORR on Pt_3M surfaces in 0.1 M $HClO_4$ at 0.9 V versus RHE and 60 °C versus the d-band center position. (a) Experimental specific activities, $i_{s(0.9\,V)}$, for Pt alloy skin and skeleton type surfaces. (b) Activities and d-band center positions calculated by DFT for (111) oriented skin surfaces. Reproduced from V. R. Stamenkovic and N. M. Markovic [17], with permission from John Wiley & Sons, Ltd.

In addition, Figure 1.2a also shows the extraordinarily high specific activity observed for the (111) surface planes of Pt$_3$Ni(111) single crystals [17, 19]. As shown in Figure 1.2b, the activity enhancement by alloying of Pt with transition metals can be understood on the basis of *ab initio* density functional theory (DFT) molecular models, which relate the transition metal-induced shift of the d-band center energy with the binding strength of adsorbed oxygen intermediates, thereby affecting the oxygen reduction activity [18, 22]. While the alloying-induced variation in specific activity for the ORR is well described by DFT (see Figure 1.2b), the large differences observed between various low-index surface planes of Pt skin-terminated Pt$_3$Ni single crystals has been surprising, showing roughly four- and eightfold higher specific activity of Pt$_3$Ni(111) compared with Pt$_3$Ni(110) and Pt$_3$Ni(100), respectively [19]. Even though these activities were measured in aqueous perchloric acid electrolyte using the rotating disk electrode (RDE) method, it was shown in the past that the activity of carbon-supported Pt and Pt alloy catalysts obtained by RDE measurements is in good quantitative agreement with their ORR activity in MEAs tested in fuel cells under comparable conditions [1, 23].

Considering the excellent correspondence between ORR activities measured in MEAs and by RDE in conjunction with the extremely high specific activity obtained on Pt$_3$Ni(111) surfaces, it is tempting to raise the question of whether one could envision its incorporation into MEAs. One possibility would be the shape-controlled synthesis of *large-nano* octahedral Pt$_3$Ni particles which would be terminated by (111) surface planes [14], similarly to what had been demonstrated for Pt$_3$Fe alloys [24]. In this case, however, the size of the Pt$_3$Ni octahedra would have to be large in order to provide large enough (111) surface planes which then might exhibit the same high specific activity as that found on bulk Pt$_3$Ni(111) single crystals. Based on Equation 1.1, however, the low platinum specific surface area, A_{Pt}, of large-nano octahedra would compromise the achievable mass activity, which of course is the ultimate figure of merit for catalyst activity in fuel cells. The latter may be estimated using Equation 1.1 in combination with the measured specific activity of $i_{s(0.9\,V)} = 18\,000\,\mu A\,cm_{Pt}^{-2}$ for Pt$_3$Ni(111) (see Figure 1.2a) and the approximate platinum specific surface area of $A_{Pt} \sim 250\,m_{Pt}^2\,g_{Pt}^{-1} \times (d_{Pt_3Ni}\,[nm])^{-1}$ based on simple geometric arguments. The results are shown in Figure 1.3.

Clearly, Figure 1.3 suggests that the mass activity even of 30 nm large Pt$_3$Ni octahedra would be 10 times larger than that of a state-of-the-art Pt/C catalyst [$i_{m(0.9\,V)} = 0.16\,A\,mg_{Pt}^{-1}$] [14], corresponding to the long sought for "10×" ORR catalyst mentioned in Table 1.1. The future will show whether this concept can indeed be realized, in which case one of the major hurdles for fuel cell commercialization could be resolved.

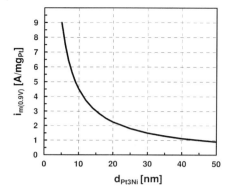

Fig. 1.3 Projected Pt mass activity versus particle diameter for Pt_3Ni octahedra exposing (111) facets, assuming no effect of (111) terrace width on activity.

1.4
Catalyst Degradation via Platinum Dissolution

Catalyst degradation in operating PEM fuel cells occurs mostly under transient conditions, leading both to dissolution of the active catalyst components (i.e., Pt and Pt alloys) [25–27]) and also to carbon-support corrosion (i.e., decomposition into CO_2) [28, 29]. The former is caused by the finite solubility of platinum at the cathode potentials in acidic electrolytes [25], particularly under voltage-cycling conditions (produced by power or load cycling under vehicle operation) [30, 31], since the formation of more dissolution-resistant platinum oxide is a kinetically slow process [32]. Platinum dissolution leads to a loss of active surface area during extended voltage cycling by two different mechanisms [25]: (i) diffusion of dissolved platinum species towards the membrane and platinum precipitation in the membrane phase due to reaction with hydrogen permeating through the membrane from the anode side, leading to a loss of electrically connected platinum surface area; and (ii) Ostwald ripening of platinum inside the cathode electrode, particularly near the cathode/DM interface, leading to a loss of platinum surface area due to nanoparticle growth. The same phenomenon is observed with Pt alloys, with the difference that the dissolved transition metals will remain in the ionomer phase as their reduction potential is below that of hydrogen [26]; if significant amounts of transition metal are dissolved, fuel cell performance is compromised, particularly at high current densities [33]. Since platinum nanoparticle solubility decreases with increasing particle size, following the Gibbs–Thomson relationship, larger particles are more stable towards voltage cycling [34, 35], which gives some hope that large-nano Pt alloys might display improved durability in addition to improved ORR activity [14].

Unfortunately, platinum surface area loss leads to a decrease in platinum mass activity, particularly during voltage cycling. This is shown in Figure 1.4, where 30 000 accelerated voltage cycles lead to a two- to threefold loss in mass activity in the case of both pure platinum and Pt alloy catalysts. Nevertheless, the mass activity advantage of Pt alloys of over pure Pt is maintained during

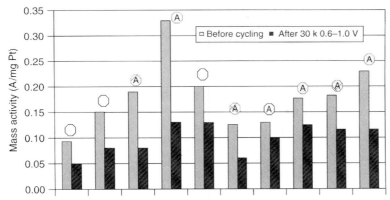

Fig. 1.4 Mass activity, $i_{m(0.9\,V)}$, of Pt/C (symbols without letter) and Pt alloy/C (symbols with letter A) before and after 30 000 voltage-cycles between 0.6 and 1.0 V RHE at 20 mV s^{-1} in H$_2$–N$_2$ (counterelectrode/cathode electrode) at 80 °C, 100% RH, and ambient pressure. Open symbols represent conventional carbon supports (e.g., Vulcan XC72, Ketjen black), shaded symbols with a crossing line represent fully graphitized carbon supports, and shaded symbols without a crossing line are acetylene blacks. Reproduced from F.T. Wagner et al. [36], with permission from John Wiley & Sons, Ltd.

these experiments; similar behavior was observed for dealloyed Pt-Cu/C catalysts [13]. Owing to the strongly degrading effect of voltage cycling, automotive fuel cell systems are generally hybridized with batteries to reduce the number of voltage cycles during fuel cell operation.

1.5
Carbon-Support Corrosion

Under steady-state fuel cell operation, carbon-support corrosion is a minor contributor to voltage degradation. However, carbon-support corrosion is significantly accelerated during start/stop cycles, due to the simultaneous but spatially separated presence of hydrogen and oxygen (air) in the anode flow field, forming a so-called hydrogen-air front: (i) when stopping a fuel cell system, the hydrogen supply to the anode is turned off, and air will leak slowly into the anode compartment; (ii) after a long shut-down, the anode compartment will be filled with air via leaks to the environment, which will be replaced by hydrogen when starting the fuel cell. This effect was first discussed by Reiser et al. [29], and is depicted in Figure 1.5.

Owing to the high electronic conductivity of the bipolar plates, hydrogen electrooxidation in the anode compartment (lower left corner of Figure 1.5) leads to the reduction of spatially separated oxygen in the anode compartment (upper left corner of Figure 1.5). Since the effective proton conduction resistance along the proton-conducting membrane is very high over millimeter length distances, the protons for the oxygen reduction reaction in the anode compartment are

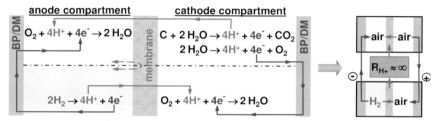

Fig. 1.5 Schematics of the processes leading to carbon-support corrosion during either fuel cell start/stop transients or during local hydrogen starvation. The simultaneous presence of spatially separated regions of hydrogen and oxygen (from air) on the anode side effectively leads to the phenomenon that an H_2-air fuel cell drives a current through an air–air electrolytic cell (see the right-hand side).

supplied by the opposing cathode compartment, leading to an increase in the local cathode potential until oxygen evolution or carbon corrosion occurs. This can be understood more easily by the sketch on the right-hand side of Figure 1.5, conceptually splitting the stack into an H_2-air fuel cell (lower part) which is electrically short-circuited with an electrolytically driven air-air cell. Using the known kinetics for the various reactions (HOR, ORR, oxygen evolution, and carbon corrosion), the start/stop-induced carbon-support corrosion rates can be modeled very accurately [28, 29, 37–39].

An analysis of the detailed start/stop mechanism shows that its effect can be mitigated by both materials and systems solutions. Of course, the most straightforward mitigation strategy is to minimize the residence time of the H_2-air front in the anode compartment, which can be done most effectively for the startup procedure [28, 38]; other system controls-related mitigation strategies were discussed in detail by Perry *et al.* [40]. On the materials side, start/stop degradation can be reduced by replacing conventional carbon supports (e.g., Vulcan XC72 or Ketjen black) with either acetylene blacks or fully graphitized carbon supports [28, 38]. Also, since the rate of oxygen reduction in the anode compartment determines the rate of carbon corrosion on the opposing cathode electrode, lowering of the platinum loading on the anode electrode leads to lower start/stop degradation, one of the rare instances where lowering the platinum loading actually improves durability without affecting performance due to the very high HOR activity of Pt/C [5].

The phenomenon of localized hydrogen starvation is closely related to that of start/stop degradation. In the former, oxygen permeation through the membrane from the cathode side in regions of the anode flow field with poor or interrupted hydrogen supply (e.g., by water droplets blocking anode flow field channels) again leads to oxygen reduction on the anode and to an increase in the local cathode potential to where carbon corrosion can occur [28, 39, 41]. Compared with start/stop degradation, the degradation rates are slower in this case, since carbon corrosion rates are limited by the permeation rate of oxygen through the membrane [28]. The materials-related mitigation strategies, naturally, are the same as in the case of start/stop degradation [28].

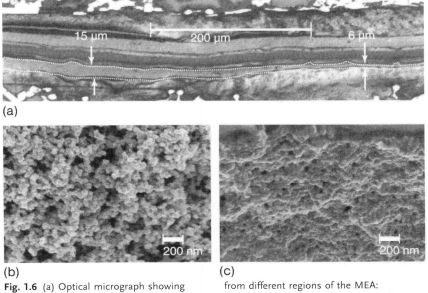

(a)

(b) (c)

Fig. 1.6 (a) Optical micrograph showing the cross-section of an aged MEA (mounted in epoxy) that spans regions that were starved (right-hand side) and not starved (left-hand side) of hydrogen. The cathode electrode is outlined with dotted white lines for clarity. Scanning electron micrograph (SEM) images of freeze-fractured sections of the MEA showing the cathode electrode from different regions of the MEA: (b) cathode electrode after aging in the non-starved region; (c) cathode electrode after aging in the H$_2$-starved region. Note that SEM analysis was done on separate samples that were not mounted in epoxy. Reproduced from R.N. Carter et al. [37], with permission from John Wiley & Sons, Ltd.

Voltage losses from start/stop degradation or local hydrogen starvation are most apparent at high current densities [38], pointing to increased mass transport resistances. This can be understood by the observation that the associated carbon-support corrosion leads to so-called *cathode thinning*, caused by a loss of cathode electrode void volume, which in turn increases oxygen diffusion resistances. This is illustrated in Figure 1.6, showing the thinning of the cathode electrode over hydrogen-starved regions of the anode flow field (a), leading to a loss of cathode electrode void volume in the thinned regions (c) compared with the non-corroded regions (b).

Considering the rapid degradation which can be produced by H$_2$-air fronts occurring during fuel cell startup and shutdown processes, leading to unacceptably large performance losses after only 100 slow H$_2$-air front events [38], it becomes apparent that the fuel cell control system must be extremely rugged to permit long-term durability. Although this can be approached at the systems design level, it would of course be desirable in the long term to develop and implement more durable cathode catalyst support materials based on graphitized carbon supports or more resistant non-carbon supports such as the nanostructured supports developed by 3M [16]. Alternatively, anode catalysts which have no activity for the oxygen reduction reaction would provide a materials-based solution to improve start/stop robustness.

1.6
Conclusion

The current major challenges for developing commercially viable PEM fuel cell vehicles are the need to lower the platinum loading of the MEA without loss of performance and to improve the stability of the catalyst with respect to platinum dissolution and carbon-support corrosion. Based on catalyst development advances in the past few years, several promising concepts exist, which might eventually permit commercially viable platinum loadings [dealloyed Pt alloys, thin Pt (alloy) films on nanostructured supports, large-nano shape-controlled Pt alloys]. Start/stop degradation can to a large extent be mitigated by system controls, but advanced support and catalyst materials to improve system robustness are desired.

References

1 Gasteiger, H.A., Kocha, S.S., Sompalli, B., and Wagner, F.T. (2005) *Appl. Catal. B*, **56**, 9.

2 Wipke, K., Sprik, S., Kurtz, J., and Garbak, J. (2009) Field experience with fuel cell vehicles. In *Handbook of Fuel Cells – Fundamentals, Technology and Applications* (eds W. Vielstich, H.A. Gasteiger, and H. Yokokawa), John Wiley & Sons, Ltd, Chichester, vol. 6, p. 893.

3 Neyerlin, K.C., Gu, W.B., Jorne, J., and Gasteiger, H.A. (2006) *J. Electrochem. Soc.*, **153**, A1955.

4 Mittelsteadt, C.K. and Liu, H. (2009) Conductivity, permeability, and ohmic shorting of ionomeric membranes. In *Handbook of Fuel Cells – Fundamentals, Technology and Applications* (eds W. Vielstich, H.A. Gasteiger, and H. Yokokawa), John Wiley & Sons, Ltd, Chichester, vol. 5, p. 345.

5 Neyerlin, K.C., Gu, W.B., Jorne, J., and Gasteiger, H.A. (2007) *J. Electrochem. Soc.*, **154**, B631.

6 Neyerlin, K.C., Gu, W., Jorne, J., Clark, A., and Gasteiger, H.A. (2007) *J. Electrochem. Soc.*, **154**, B279.

7 Liu, Y.X., Murphy, M.W., Baker, D.R., Gu, W.B., Ji, C.C., Jorne, J., and Gasteiger, H.A. (2009) *J. Electrochem. Soc.*, **156**, B970.

8 Gu, W., Baker, D.R., Liu, Y., and Gasteiger, H.A. (2009) Proton exchange membrane fuel cell (PEMFC) down-the-channel performance model. In *Handbook of Fuel Cells – Fundamentals, Technology and Applications* (eds W. Vielstich, H.A. Gasteiger, and H. Yokokawa), John Wiley & Sons, Ltd, Chichester, vol. 6, p. 631.

9 Masten, D.A. and Bosco, A.D. (2003) System design for vehicle applications. In *Handbook of Fuel Cells – Fundamentals, Technology and Applications* (eds W. Vielstich, A. Lamm, and H.A. Gasteiger), John Wiley & Sons, Ltd, Chichester, vol. 3, p. 714.

10 Lefevre, M., Proietti, E., Jaouen, F., and Dodelet, J.-P. (2009) *Science*, **324**, 71.

11 Yu, P., Pemberton, M., and Plasse, P. (2005) *J. Power Sources*, **144**, 11.

12 Mani, P., Srivastava, R., and Strasser, P. (2008) *J. Phys. Chem. C*, **112**, 2770.

13 Neyerlin, K.C., Srivastava, R., Yu, C.F., and Strasser, P. (2009) *J. Power Sources*, **186**, 261.
14 Gasteiger, H.A. and Markovic, N.M. (2009) *Science*, **324**, 48.
15 Adzic, R.R., Zhang, J., Sasaki, K., Vukmirovic, M.B., Shao, M., Wang, J.X., Nilekar, A.U., Mavrikakis, M., Valerio, J.A., and Uribe, F. (2007) *Top. Catal.*, **46**, 249.
16 Debe, M.K., Schmoeckel, A.K., Vernstrom, G.D., and Atanasoski, R. (2006) *J. Power Sources*, **161**, 1002.
17 Stamenkovic, V.R. and Markovic, N.M. (2009) Oxygen reduction on platinum bimetallic alloy catalysts. In *Handbook of Fuel Cells – Fundamentals, Technology and Applications* (eds W. Vielstich, H.A. Gasteiger, and H. Yokokawa), John Wiley & Sons, Ltd, Chichester, vol. 5, p. 18.
18 Greeley, J., Stephens, I.E.L., Bondarenko, A.S., Johansson, T.P., Hansen, H.A., Jaramillo, T.F., Rossmeisl, J., Chorkendorff, I., and Norskov, J.K. (2009) *Nat. Chem.*, **1**, 552.
19 Stamenkovic, V.R., Fowler, B., Mun, B.S., Wang, G.F., Ross, P.N., Lucas, C.A., and Markovic, N.M. (2007) *Science*, **315**, 493.
20 Thompsett, D. (2003) Platinum alloys as oxygen reduction catalysts. In *Handbook of Fuel Cells – Fundamentals, Technology and Applications* (eds W. Vielstich, A. Lamm, and H.A. Gasteiger), John Wiley & Sons, Ltd, Chichester, vol. 3, p. 467.
21 Chen, S., Sheng, W., Yabuuchi, N., Ferreira, P.J., Allard, L.F., and Shao-Horn, Y. (2009) *J. Phys. Chem. C*, **113**, 1109.
22 Stamenkovic, V., Mun, B.S., Mayrhofer, K.J.J., Ross, P.N., Markovic, N.M., Rossmeisl, J., Greeley, J., and Norskov, J.K. (2006) *Angew. Chem. Int. Ed.*, **45**, 2897.
23 Gasteiger, H.A. and Garche, J. (2008) Fuel cells. In *Handbook of Heterogeneous Catalysis*, 2nd edn (eds G. Ertl, H. Knözinger, F. Schüth, and J. Weitkamp), Wiley-VCH Verlag GmbH, Weinheim, p. 3081.
24 Qiu, J.M. and Wang, J.P. (2007) *Adv. Mater.*, **19**, 1703.
25 Ferreira, P.J., la O', G.J., Shao-Horn, Y., Morgan, D., Makharia, R., Kocha, S., and Gasteiger, H.A. (2005) *J. Electrochem. Soc.*, **152**, A2256.
26 Chen, S., Gasteiger, H.A., Hayakawa, K., Tada, T., and Shao-Horn, Y. (2010) *J. Electrochem. Soc.*, **157**, A82.
27 Ota, K. and Koizumi, Y. (2009) Platinum dissolution models and voltage cycling effects: platinum dissolution in polymer electrolyte (PEFC) and low-temperature fuel cells. In *Handbook of Fuel Cells – Fundamentals, Technology and Applications* (eds W. Vielstich, H.A. Gasteiger, and H. Yokokawa), John Wiley & Sons, Ltd, Chichester, vol. 5, p. 241.
28 Yu, P.T., Gu, W., Zhang, J., Makharia, R., Wagner, F.T., and Gasteiger, H.A. (2009) Carbon-support requirements for highly durable fuel cell operation. In *Polymer Electrolyte Fuel Cell Durability* (eds M.I.F.N. Büchi and T.J. Schmidt), Springer, New York, p. 29.
29 Reiser, C.A., Bregoli, L., Patterson, T.W., Yi, J.S., Yang, J.D.L., Perry, M.L., and Jarvi, T.D. (2005) *Electrochem. Solid State Lett.*, **8**, A273.
30 Kinoshita, K., Lundquist, J., and Stonehart, P. (1973) *J. Electroanal. Chem.*, **48**, 157.
31 Mathias, M.F., Makharia, R., Gasteiger, H.A., Conley, J.J., Fuller, T.J., Gittleman, C.J., Kocha, S.S., Miller, D.P., Mittelsteadt, C.K., Xie, T., Yan, S.G., and Yu, P.T. (2005) *Interface*, **14**, 24.
32 Darling, R.M. and Meyers, J.P. (2003) *J. Electrochem. Soc.*, **150**, A1523.
33 Greszler, T.A., Moylan, T.E., and Gasteiger, H.A. (2009) Modeling the impact of cation contamination in a polymer electrolyte membrane fuel cell. In *Handbook of Fuel Cells – Fundamentals, Technology and Applications* (eds W. Vielstich, H.A. Gasteiger, and H. Yokokawa), John Wiley & Sons, Ltd, Chichester, vol. 6, p. 728.
34 Makharia, R., Kocha, S., Yu, P., Sweikart, M.A., Gu, W., Wagner, F., and Gasteiger, H.A. (2006) *ECS Trans.*, **1** (8), 3.
35 Shao-Horn, Y., Sheng, W.C., Chen, S., Ferreira, P.J., Holby, E.F., and Morgan, D. (2007) *Top. Catal.*, **46**, 285.
36 Wagner, F.T., Yan, S.G., and Yu, P.T. (2009) Catalyst and catalyst-support durability. In *Handbook of Fuel Cells – Fundamentals, Technology and Applications*

(eds W. Vielstich, H. A. Gasteiger, and H. Yokokawa), John Wiley & Sons, Ltd, Chichester, vol. 5, p. 250.

37 Carter, R. N., Gu, W., Brady, B., Yu, P. T., Subramanian, K., and Gasteiger, H. A. (2009) Electrode degradation mechanisms studied by current distribution measurements. In *Handbook of Fuel Cells – Fundamentals, Technology and Applications* (eds W. Vielstich, H. A. Gasteiger, and H. Yokokawa), John Wiley & Sons, Ltd, Chichester, vol. 6, p. 829.

38 Yu, P. T., Gu, W., Makharia, R., Wagner, F. T., and Gasteiger, H. A. (2006) *ECS Trans.*, **3** (1), 797.

39 Gu, W., Carter, R. N., Yu, P. T., and Gasteiger, H. A. (2007) *ECS Trans.*, **11** (1), 963.

40 Perry, M. L., Patterson, T. W., and Reiser, C. (2006) *ECS Trans.*, **3** (1), 783.

41 Patterson, T. W. and Darling, R. M. (2006) *Electrochem. Solid State Lett.*, **9**, A183.

2
High-Temperature PEM Fuel Cells: Electrolytes, Cells, and Stacks

Christoph Wannek

Abstract

Although today, most of the technical requirements which are considered to be relevant before classical (Nafion®-based) low temperature polymer electrolyte membrane (PEM) fuel cells can compete successfully with existing technologies on the market place are already met, at least on the laboratory scale, there are important research efforts to design cells with an operating temperature well above today's 80 °C. Basically, such high-temperature PEM fuel cells could benefit from accelerated electrode kinetics, a simplified system design and a higher tolerance towards impurities in the hydrogen fuel. This paper reviews the state-of-the-art in basic science and industrial development of membranes, electrodes and stacks and points out the main challenges for future development. It first describes various approaches to increase the operating temperature of classical membrane materials up to 120 °C, before it focuses on the development of phosphoric acid-doped polybenzimidazoles-type membrane cells that operate at 160–180 °C. As the technology readiness levels of alternative electrolytes such as phosphonic acids, ionic liquids, acidic salts and mixed oxides are still very low, they will be described only very briefly.

Keywords: high-temperature fuel cells, polymer electrolyte membrane fuel cells, electrolytes, stacks, membrane electrode assembly, membrane materials

2.1
Introduction

Polymer electrolyte membrane fuel cells (PEMFCs) [or polymer electrolyte fuel cells (PEFCs)] fueled with hydrogen have received much attention as highly efficient power sources which could replace today's internal combustion engines and batteries, especially in passenger cars and portable applications. Due to its low operating temperature (room temperature up to 90 °C), the solid electrolyte (which allows dynamic operation), and the fact that ambient air can be used as

Hydrogen Energy. Edited by Detlef Stolten
Copyright © 2010 WILEY-VCH Verlag GmbH & Co. KGaA, Weinheim
ISBN: 978-3-527-32711-9

oxidant, the PEFC is better suited than any other fuel cell type for these applications.

Today, most of the technical requirements which are considered to be relevant before fuel cells can compete successfully with existing technologies in the market place have already been met on the laboratory scale, durability and cost being the greatest issues for further development (Table 2.1).

Even though the commercialization of a significant number of fuel cell passenger cars will only take place from 2015 onwards, many important automobile manufacturers have reconfirmed only recently that they will carry on their fuel cell development and prepare for market introduction [2]. Other uses of PEFCs, such as back-up power stations, combined heat and power units, speciality vehicles, and portable applications, are even considered as nearer term applications. Although the hydrogen fuel for cars will be stored on-board, for some of the latter applications it is more reasonable to produce it on-site by reforming an energy carrier such as natural gas or methanol into a hydrogen-rich gas.

The membrane electrode assembly (MEA), as the heart of the fuel cell, comprises a proton-conducting membrane coated on both sides with platinum-based catalytic layers, which are needed to promote the hydrogen oxidation at the anode and the oxygen reduction at the cathode. Perfluorinated sulfonic acid (PFSA) polymers are the best type of membrane material known to day, DuPont's Nafion® being the most popular representative (Figure 2.1).

Table 2.1 Comparison of targets and 2009 status for the development of PEFCs for transportation according to the United States Department of Energy Hydrogen Program and Vehicle Technologies Program.

Property	Target (2015)	Status	
		Car	Laboratory cell
Efficiency at 25% rated power (%)	60	59	>60
Durability (h)	5000	1977	>7300 with cycling
Platinum metal loading (g kW^{-1})	0.2	–	<0.2
Cost (projection for 500 000 units per year) (US$ kW^{-1})	30	73	Not applicable

Source: data taken from [1].

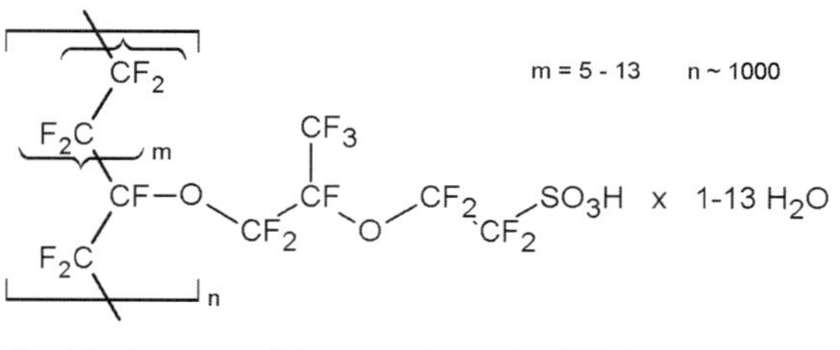

Fig. 2.1 Structure of the repeating unit of Nafion.

Because the most relevant PEFC research today is carried out by industrial research teams, scientifically published information on the state-of-the-art MEA materials and also on fuel cell performance and durability is relatively scarce (e.g., [3, 4]). As PEFC development is already in a pre-commercial stage, much engineering is already being done regarding the stacking of a multitude of MEAs and the design of a complete fuel cell system.

The most important drawback of PFSAs is that they have to be kept fully hydrated to maintain a high proton conductivity and thus good fuel cell performance. This can be achieved in low-pressure systems at temperatures up to around 80 °C, but at high load operation of the fuel cell (= low electrical efficiency) the necessary heat removal from the fuel cell stack in a car requires very large radiators. It is assumed that an operating temperature of at least 95 °C is needed for efficient heat rejection at peak power [5]. Another important driver for a higher operating temperature is the increasing tolerance of the anode catalyst towards "poisoning" of the hydrogen fuel by carbon monoxide (CO) impurities. At an operating temperature of 80 °C, CO has to be removed to levels well below 100 ppm.

As a consequence, during the last decade much effort has been spent on developing high-temperature (HT) PEFCs. As a first step, research has been carried out to identify polymer membranes with high conductivity at relatively low humidity levels and good stability under the demanding operating conditions of a fuel cell. In a second step, the MEA, stack, and system design have to be adapted to the properties of these novel electrolytes.

In the following, a brief overview of the current status of HT-PEFC development is given. Different approaches to increasing the operating window of PFSA materials are detailed in Section 2.2, in Section 2.3 the development of HT-PEFCs with phosphoric acid-loaded benzimidazole-type membranes is discussed and Section 2.4 presents some information and remarks on alternative electrolytes such as ionic liquids and solid acids.

We concentrate on phosphoric acid membrane cells as these are closest to commercialization. In this context, many material issues are discussed that are not specifically related to this special sub-type but to all HT-PEFCs, such as catalyst materials, catalyst supports, and bipolar plates.

2.2
Approaches to Increase the Operating Temperature of Sulfonated Membranes

The operation of MEAs with a normal Nafion membrane is possible at temperatures up to around 120–130 °C, where the polymer approaches its glass transition temperature (T_g) and thus loses mechanical stability. Additionally, the operation of a system with fully humidified gases at 130 °C implies that a water vapor pressure of 2.7 bar has to be established. To achieve a reasonable concentration/partial pressure of the reactants, the system pressure has to amount to at least 3 bar, which has a negative effect on the total system efficiency. Consequently, for high-temperature operation, the properties of classical low-temperature membranes have to be tuned to increase their mechanical stability and their ability to retain water at humidity levels below 100%.

In recent years an impressive amount of membrane research has been carried out to address these issues and a short overview has been given [6]. In the following, only a selection of the numerous approaches can be briefly discussed.

The careful architecture of membranes containing rigid-rod co-monomers with angled or bulky side-groups leads to the creation of interconnected pores filled with sulfonic acid groups over the whole membrane thickness (Figure 2.2). In highly sulfonated polymers with a rigid-rod backbone the sulfonic acid groups cannot pack well, leaving voids that absorb and hold water which yields high conductivity even at relative low humidity (Figure 2.3). Another approach to create robust polymer backbones on the one hand and highly sulfonated pore chains on the other consists in the preparation of hydrophilic-hydrophobic multiblock copolymers [10].

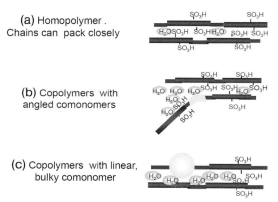

Fig. 2.2 Basic concepts of rigid-rod nematic liquid crystalline polymers with frozen-in free volume. Reproduced from [7].

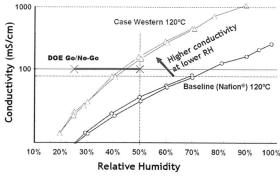

Fig. 2.3 Proton conductivity of rigid-rod membranes in comparison with Nafion and the United States Department of Energy's technical targets. Reproduced from [8].

Alternatively, the PFSA membrane properties can be enhanced by incorporating highly hygroscopic nanoparticles (e.g., TiO_2, SiO_2) or intrinsic proton conductors such as layered zirconium phosphates or heteropoly acids.

As an alternative to perfluorinated membranes, a large variety of sulfonated fluorine-free polymers have been evaluated for both low- and for high-temperature PEFC applications (for a review, see [11]). Among these materials are highly sulfonated polyarylenes [12] and sulfonated polybenzimidazoles [13].

The proton conductivity of sulfonated membranes depends on the density of sulfonic acid groups within the membrane. Increasing this concentration [= lowering the so-called "equivalent weight" (EW)] leads to higher conductivity but also to stronger swelling or even dissolution of the membrane. To circumvent these mechanical problems, low-EW membranes can be reinforced by two- or three-dimensionally stable non-ionic materials (Figure 2.4 [9]).

Among the industrial membrane and MEA manufacturers that are working on HT-PEFCs, 3M and Solvay Solexis [14] are pursuing the concept of PFSA materials with shorter side-chains compared with Nafion (cf., Figure 2.1). These polymers have a higher modulus and tear resistance due to having a higher degree of crystallinity than Nafion at the same equivalent weight. As a consequence, the equivalent weight can be decreased and hence the conductivity increased compared with Nafion. A short stack with MEAs based on short side-chain membranes was successfully operated up to 130 °C at various levels of humidity [15].

Some of the best fuel cell performance data obtained at 120 °C and relative humidity levels significantly below 100% have been published by Asahi Glass [16]. With a 40 µm thick "polymer composite membrane", a power density of 0.5 W cm^{-2} (Figure 2.5) and a durability of 4000 h (Figure 2.6) have been obtained.

Further, it has been reported that membranes with a thickness of 15 µm can survive 27,000 wet-dry cycles (at 80 °C) and that a fuel cell performance of 1.5 W cm^{-2} at 120 °C and a dewpoint of 50 °C has been achieved with an MEA based on a membrane only 5 µm thick [17].

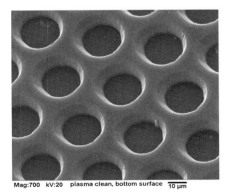

Fig. 2.4 SEM image of the high-strength polymer support used for the fabrication of two-dimensionally stable composite membranes. Reproduced from [9].

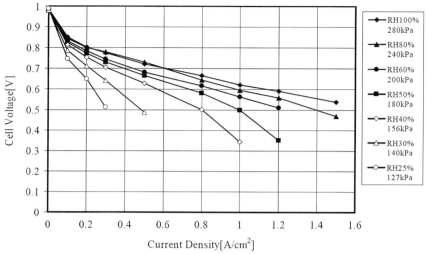

Fig. 2.5 Current–voltage curves (H_2/air operation) of an NPC MEA-I at 120 °C and different degrees of humidification. Reproduced from [16] by permission of ECS – The Electrochemical Society.

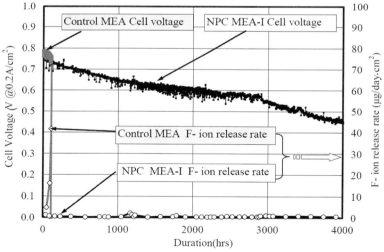

Fig. 2.6 Constant-current durability test at 120 °C and 50% relative humidity (H_2/air at 200 kPa). Reproduced from [16] by permission of ECS – The Electrochemical Society.

2.3
HT-PEFCs with Phosphoric Acid-Based Polybenzimidazole-Type Membranes

As Nafion and other sulfonated membranes lose much of their proton conductivity at temperatures above 100 °C and ambient pressure due to the evaporation of water, one approach to circumvent this problem has been to replace water by less volatile liquids such as phosphoric acid.

In 1994, the first studies on the conductivity of Nafion equilibrated with concentrated phosphoric acid were published [18]. This membrane had a phosphoric acid content equivalent to ~5 mol of phosphoric acid per mole of sulfonic acid groups of the polymer. Nearly independent of temperature between 120 and 175 °C, this membrane showed a proton conductivity of about 20 mS cm^{-1} under a dry atmosphere and about 50 mS cm^{-1} at 20% relative humidity. Due to the insufficient mechanical stability of the polymer, no systematic fuel cell tests with phosphoric acid-doped Nafion have been conducted at temperatures of 150 °C and above.

The use of polybenzimidazole (PBI) impregnated with phosphoric acid as fuel cell membrane was first suggested by Savinell and co-workers [19, 20].

The following discussion first focuses on the membrane systems that have been studied during the last 15 years and the development of electrodes and MEAs (which were developed with some delay compared with the membrane research) adapted to the specific features of the electrolyte. The following two sections give an overview of the major industrial developers of PBI-based HT-PEFCs and the state-of-the-art performance which can be achieved. Finally, the most important scientific and technical challenges that remain to be solved before this type of fuel cell can compete successfully in the market place are discussed. Only some issues concerning stack materials and system design are discussed in this context as experience in stack and system development is still fairly limited.

2.3.1
Membrane Materials, Proton Conductivity, and MEA Design

2.3.1.1 Membrane Synthesis

Polybenzimidazole (PBI), or poly[2,2'-(m-phenylene)-5,5'-dibenzimidazole] (Figure 2.7a), is a heat-resistant, high-performance polymer (melting point >600 °C) which is an isolator in its pure form but which is able to take up large quantities of acids or bases such as phosphoric acid. The proton conductivity of the fully "doped" material (see below) is almost as high as that of perfluorinated membranes and far less dependent on relative humidity, thus allowing its use in PEFCs without humidification of the reactants.

On the other hand, in contrast to classical phosphoric acid fuel cells, the electrolyte system in an H_3PO_4-PBI membrane is essentially solid and is therefore easier to handle, is more tolerant to pressure variations, and raises fewer difficulties concerning the acid management.

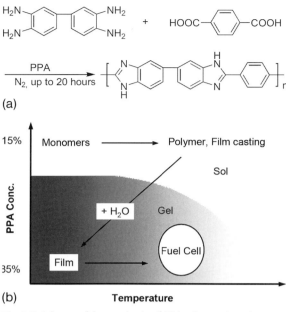

Fig. 2.7 Scheme of the synthesis of PBI polymers in polyphosphoric acid (a) and the state diagram of the sol-gel process (b). Reprinted with permission from [21]. Copyright 2005 American Chemical Society.

PBI is the most popular representative of a whole class of related polymers which can be used in this type of fuel cell; for reviews of HT-PEFC membranes, see [22–24]. PBI can be synthesized by polycondensation of 3,3′-diaminobenzidine and diphenyl isophthalate (Figure 2.7a). In the conventional method to produce a fuel cell membrane, a PBI film is cast from an organic solution of the polymer, dried, purified, and imbibed with phosphoric acid. In contrast, the membrane can also be manufactured by polymerizing the monomers directly in polyphosphoric acid (PPA). The PPA is then hydrolyzed to phosphoric acid, which causes a sol-gel transition of the polymer-electrolyte system and forms a membrane (Figure 2.7b) [21]. By using the latter method, mechanically stable membranes with extremely high acid contents (>85%) can be manufactured.

2.3.1.2 Proton Conductivity

PBI takes up different quantities of phosphoric acid when immersed in baths with different concentrations of aqueous H_3PO_4. The first two equivalents of acid "neutralize" the two basic nitrogen atoms per repeat unit. All subsequent quantities that are taken up contribute to the proton conductivity of the membrane (="free acid"; Figure 2.8a). As Figure 2.8b clearly shows, the conductivity of highly doped PBI is nearly as high as that of Nafion. It increases significantly

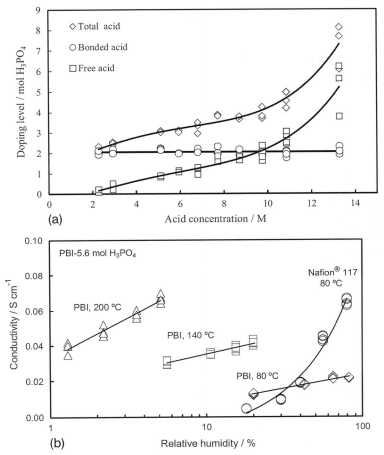

Fig. 2.8 Doping level of PBI in molecules of H_3PO_4 per repeat unit as a function of the acid concentration (a) and conductivity of acid-doped PBI and Nafion as a function of humidity at different temperatures (b). Reproduced from [25] with permission. Copyright Wiley-VCH Verlag GmbH & Co. KGaA.

with temperature (in contrast to the behavior known for Nafion) and is less dependent on the relative humidity of the environment, although there is still a marked influence.

The mechanism of the proton transport within the material has been identified to depend significantly on the doping level, humidity, and temperature [26]. It has been suggested that it is amorphous H_3PO_4 that acts as the species mainly responsible for the proton transport. At high acid (and water) loadings, the proton transfer mainly occurs by a Grotthus mechanism, that is, "proton hopping" through the hydrogen bond network of acid and water molecules through the formation/cleavage of covalent bonds (Figure 2.9).

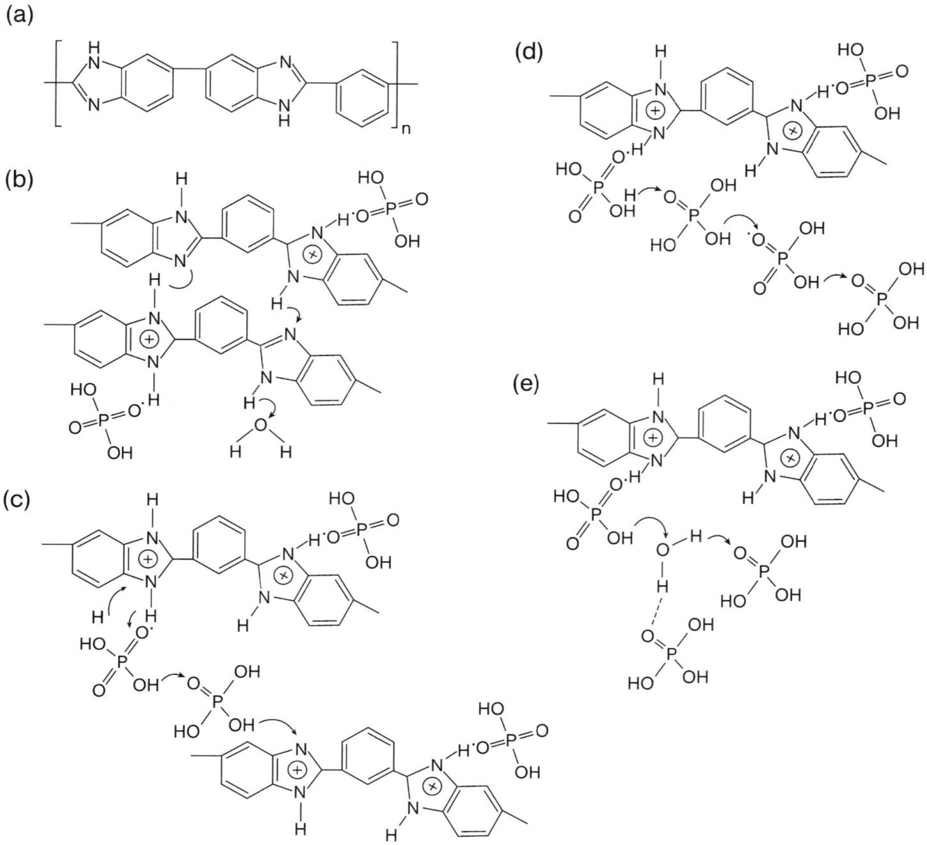

Fig. 2.9 Chemical structures of (a) PBI, (b) H_3PO_4-protonated PBI, (c) proton transfer path acid – benzimidazole unit – acid, (d) proton transfer along acid – acid, and (e) proton transfer along acid – water. Reproduced from [26] by permission of ECS – The Electrochemical Society.

At temperatures above 130–140 °C, the conductivity of phosphoric acid under anhydrous conditions decreases due to the formation of pyrophosphoric acid ($H_4P_2O_7$) by condensation of two molecules of H_3PO_4 and the elimination of water (dehydration) [28]. As a consequence, at 160 °C the cell resistance of PBI-type HT-PEFCs at open-circuit conditions is significantly higher than under electrical load (Figure 2.10 [27]), when the water produced by the fuel cell reaction re-hydrates the membrane and shifts the equilibrium between phosphoric acid and pyrophosphoric acid more towards the more conductive H_3PO_4.

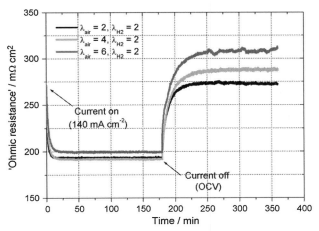

Fig. 2.10 Time dependence of the cell resistance after changing the current density from $j=0$ to 140 mA cm^{-2} and vice versa at 160 °C at different air flow rates. Reprinted from [27] with permission from Elsevier.

2.3.1.3 Optimization of the Electrodes and the MEA Assembly

Although PBI-based HT-PEFCs are operated around 100 K above the temperature level of a classical PEFC, platinum (mostly supported on carbon) is still needed for effective catalysis of the anodic and cathodic reactions. However, due to the high CO tolerance of the cell (cf., Section 2.3.3), no co-catalyst (such as ruthenium in low-temperature PEFCs) is needed on the anode. In terms of durability, pure platinum is still the cathode catalyst of choice, while platinum alloy catalysts can be used to maximize the cell performance (if a larger degradation rate is acceptable). The precious metal loadings that are generally reported for PBI-phosphoric acid HT-PEFCs amount to about 1–2 mg cm^{-2} per cell and are therefore much higher than for the "classical PEFC" (see Section 2.3.4.1). Although PBI has a pronounced film-forming tendency, it has been incorporated successfully into the catalyst layers to fix the phosphoric acid [29]. Additionally or instead, other polymers such as polytetrafluoroethylene (PTFE) (weight fraction between 20 and 60%) can be used as binder and hydrophobic agent [30].

As an alternative to the use of acid-doped membranes (which are difficult to handle), the phosphoric acid can also be introduced into the MEA by H_3PO_4-loaded gas diffusion layers (GDLs) [31] or impregnated gas diffusion electrodes (GDEs) [32].

Because doped PBI has comparably little mechanical stability, the sealing strategy and MEA compression are critical issues. Integrated gasket concepts are being considered by various developers (e.g., [33]), the gasket materials generally being varieties of fluorocarbons, silicones, or nitriles. The compression of the MEA during its assembly and in the cell has to be controlled carefully in order to obtain good contact of the components without squeezing some of the acid out of the MEA. The automated production of the PBI membrane, gas diffusion electrodes, and even the MEA assembly has been started and already advanced remarkably by some developers. Details are given in Section 2.3.4.3.

2.3.2
Industrial and Commercial Products

Intensive development of the HT-PEFC technology with PBI-type membranes has been carried out continuously in both academic and industrial research. As a consequence, several companies offer HT-PEFC products with different levels of integration between single membranes and complete systems. Most of the products discussed in the following are still in a pre-commercial stage and are not available for private customers.

Today, FuMA-Tech appears to be the only relevant supplier of free PBI-type membranes, their main product being AB-PBI (2,5-benzimidazole) [24, 32].

In terms of development time, staff, and sales, BASF Fuel Cell is the biggest player in the area of HT-PEFCs today. As the successor of Hoechst, Celanese, and Pemeas, they have been developing and optimizing Celtec®-MEAs (Figure 2.11a) for many years. Currently three different MEA types are offered: P-1000 (standard product, 20 000 h durability proven), P-2000 (especially designed for temperature cycling), and P-3000 (very thin MEA for micro fuel cells) [34]. The PBI membrane material used for these MEAs, – developed at Rensselaer Polytechnic Institute [21], – is not available independently. Advent Technologies recently started to offer MEAs that are not based on PBI but a different membrane polymer loaded with phosphoric acid [35]. Samsung Advanced Institute of Technology has published excellent performance data obtained with their in-house MEAs [36], but it is not clear whether this development is intended to be progressed into a product.

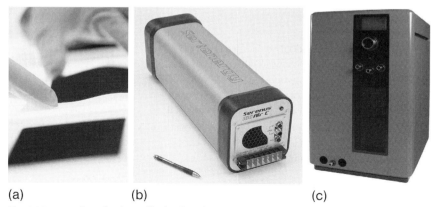

Fig. 2.11 Examples of industrially developed HT-PEFCs with PBI–phosphoric acid electrolyte. (a) Membrane electrode assemblies (BASF Fuel Cell [37]); (b) air-cooled 900 W stack (Serenergy [38]); (c) HT-PEFC system operated with natural gas (Plug Power [39]).

Sartorius had been carrying out HT-PEFC MEA and stack development for a few years [31], but stopped this activity just recently. At the moment, the Danish company Serenergy seems to be the only company publicly offering HT-PEFC stacks. Their air-cooled Serenus modules (nominal power $\sim$900 W and 3 kW; Figure 2.11 b) contain one or three stacks each with 65–90 cells (developed in cooperation with the Technical University of Denmark) [38].

Plug Power Inc. is working on complete combined heat and power (CHP) units using HT-PEFC technology (Figure 2.11 c [40]). Volkswagen AG is evaluating pressurized HT-PEFC stacks for automotive purposes [41]. Small portable systems with a micro reformer are another possible application [42].

2.3.3
Performance and Durability – State of the Art

Even though the absolute performance values of the above-mentioned products may diverge significantly, information from the various developers is used in the following to show some general characteristics of these high-temperature fuel cells.

2.3.3.1 Operating Temperature

In principle, there is no limitation to the operating window in terms of temperature. The upper temperature limit for successful long-term use is estimated to be 200–220 °C, but today's membrane materials degrade fairly quickly at operating temperatures higher than 180 °C. Even at temperatures as low as 40 °C HT-PEFC cells can be operated safely [41] provided that the system design prevents the occurrence of liquid water within the cells which might cause irreversible leaching of phosphoric acid and thus loss of proton conductivity (especially during start-up and shut-down).

2.3.3.2 Performance and CO Tolerance

MEAs manufactured by commercial developers can deliver 0.3 W cm^{-2} at 0.6 V and 160 °C using hydrogen and air in low stoichiometric excesses at ambient pressure (Figure 2.12). At this typical working temperature, a PBI-based cell shows high CO tolerance, which allows the cell to be operated under steam reformate (hydrogen gas prepared by reforming of natural gas, methanol, etc., 70% H_2, 1% CO) without further CO removal steps as it maintains at least 70% of the power density obtained with hydrogen at 0.6 V (Figure 2.12). The temperature dependence of the CO tolerance is much stronger than the effect on the cell performance with hydrogen. With 1% CO in the fuel gas, an HT-PEFC can hardly be operated below 120 °C, whereas at temperatures of 175 °C and above even more than 10% CO in a "reformate"/oxygen cell can be used without affecting the cell performance too strongly [25]. Similar trends are observed with simulated autothermal diesel reformate ($\sim$35% H_2, 1% CO).

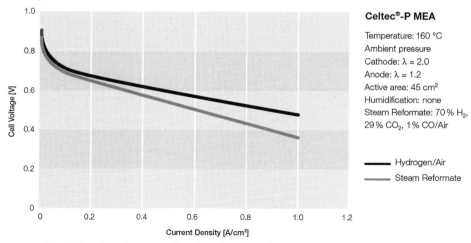

Fig. 2.12 Cell performance of PBI-based MEAs with hydrogen (upper curve) and steam reformate (70% H_2, 1% CO) (lower curve) as fuel ($T=160\,°C$, oxidant air, ambient pressure). Reproduced from [34].

2.3.3.3 Durability

Under steady-state conditions ($j=0.2\,\text{A cm}^{-2}$, H_2-air, ambient pressure, no humidification), a degradation rate of less than $6\,\mu\text{V h}^{-1}$ at $160\,°C$ has been obtained with MEAs of about $50\,\text{cm}^2$ in size over a period of $20\,000$ h [34]. Both load cycling and especially temperature cycling have a detrimental influence on the durability of the cells. At temperatures of $150–160\,°C$, both excellent performance and lifetime can be achieved. Increasing the cell temperature leads to a much faster degradation (e.g., Figure 2.13). From the literature (e.g., [44]), it appears that acid loss is not an important cause of degradation under appropriate operating conditions. Instead, under steady-state conditions, degradation of the cathode catalyst layer (loss of catalytically active Pt area, carbon support corrosion, increasing mass transport resistance) is the main reason for decreasing performance. Other issues such as membrane failure are more important under load and temperature cycling.

2.3.4
Main Challenges for Future Development

2.3.4.1 Performance

Compared with the polarization curves that are obtained with MEAs based on classical PEFC membranes, the PBI-H_3PO_4 system is characterized by a significantly lower open-circuit voltage and a strong decrease in the cell voltage at low current densities, followed by a broad range in which the slope of the polarization curve is nearly constant (Figure 2.12). Because of these features, the techni-

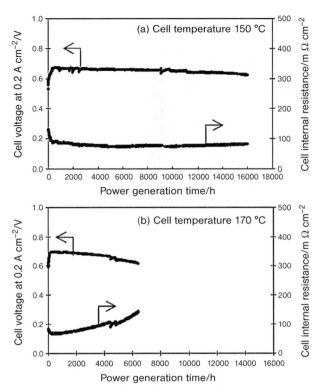

Fig. 2.13 Time course of the cell voltages and cell resistances of PBI-based MEAs operated at a current density of 200 mA cm^{-2} at cell temperatures of 150 and 170 °C. Reprinted from [43] with permission from Elsevier.

cal working range of this type of fuel cell lies around 600 mV and therefore at a significantly lower cell voltage than that of low-temperature PEFCs with an operating voltage of 700–750 mV.

This behavior is caused by both the adsorbing nature of the phosphate anions on the catalyst (reducing its activity) and the comparably low solubility of oxygen in and its slow diffusion through the phosphoric acid which is located not only in the membrane but also in the electrodes (see below). Although this "cathode problem" has been known from phosphoric acid fuel cell (PAFC) technology for decades now, no clear separation of these different effects on the kinetics of the oxygen reduction reaction has been possible yet due to the absence of measurements under real fuel cell operating conditions. The development of new analytical tools, such as a high-temperature rotating-disk electrode, will be necessary in this context.

By the addition of perfluorinated acids (such as trifluoromethanesulfonic acid, F_3CSO_3H) to the cathodic catalyst layer, the oxygen solubility can be increased and the phosphate anion adsorption can be reduced [45–47].

In today's MEAs, there is a fairly rapid exchange of phosphoric acid between the membrane and the catalyst layers, resulting in a dynamic equilibrium. As a consequence, the total amount of H_3PO_4 has to be chosen carefully to be on the one hand sufficiently high to guarantee a high conductivity of the membrane, but on the other low enough to avoid too much acid content in the cathode [48–50]. In future, efficient ways to fix the acid inside the membrane have to be developed.

All novel oxygen reduction catalysts (platinum alloys, core-shell catalysts, etc.) that are currently being developed for low-temperature PEFCs will be applicable in all types of HT-PEFCs.

Regarding stack design, it is of primary importance to establish a homogeneous temperature distribution within the stack, as (i) higher temperatures lead to better performance, (ii) locally too high temperatures may cause rapid degradation, and (iii) the number of heating/cooling cells has to be minimized to reduce stack volume and weight.

2.3.4.2 Durability

Degradation of the cathode has been identified as being the most important cause of performance degradation under normal operating conditions. This is partly due to the loss of catalytically active platinum surface area caused by Pt dissolution and Ostwald ripening [51, 52]. Another important degradation issue is carbon corrosion, especially at high cell voltages. Due to the higher operating temperatures, this issue is more important in PBI-based HT-PEFCs than in cells which are operated between 80 and 120 °C. Alternative catalyst supports which are more stable towards oxidation, such as graphitized carbon, carbon nanotubes, and semiconducting oxides [53], are currently being evaluated to minimize the effect of carbon corrosion.

The massive uptake of H_3PO_4 by PBI logically reduces its mechanical strength, especially at high temperatures. As a consequence, to assure long-term stability under dynamic conditions, the stress-relaxation behavior of membranes has to be investigated systematically. Increasing the average molecular weight of the polymer is advantageous in this context. Blends between acid-doped polybenzimidazoles and a second polymer can help to assure better membrane properties at higher temperatures than pure acid-doped PBI without resulting in significant loss of proton conductivity at either low or high temperatures [54].

The cold-start behavior of PBI-based cells is a critical issue as liquid water can cause leaching of the electrolyte from membranes. To circumvent this problem, either the fuel cell has to be heated to > 100 °C before any current is drawn, or the reactants have to be supplied to the cells in very large excesses to assure that all product water is carried away in gaseous form even at low temperatures where the saturation vapor pressure is low. The cold-start problem will certainly have to be addressed in the future, either by tuning the electrolyte itself or by blending it in an appropriate way with a second phase which fixes the acid and assures reasonable conductivity at low temperature. In this context, it is noteworthy that an autothermal cold start of an HT-PEFC stack from −20 °C has already been demonstrated with pure hydrogen as fuel [31]. Due to the very low CO tolerance below 100 °C, this would not have been possible with reformate gas.

2.3.4.3 Cost Issues

As phosphoric acid in the catalyst layers partly impedes the electrode reactions, the noble metal loading in state-of-the-art MEAs is 1–2 mg cm^{-2} per cell and is thus much higher than in low-temperature PEFCs, although the higher operating temperature generally favors the electrode kinetics. Reducing the catalyst loading and/or the development of an effective catalyst recycling are important factors for reducing the cost of HT-PEFCs. The development of highly active platinum-free oxygen reduction catalysts (e.g., prepared from iron or cobalt compounds and nitrogen-containing organic precursors such as polyaniline, polypyrrole, or cyanamide) might represent an interesting long-term alternative to noble metal-based catalysts.

The bipolar plates in today's HT-PEFCs consist of carbon–polymer composites. Although they perform well, they are thick, heavy, and almost impossible to recycle. To decrease the volume, weight, and cost of the stacks, currently much research is being done to develop metallic bipolar plates that are stable under the demanding operating conditions of HT-PEFCs. Very thin silicon plates represent an interesting third alternative, as the material shows excellent chemical stability against phosphoric acid. Further studies will have to show whether the electrical conductivity of this type of bipolar plates can be sufficiently increased to limit ohmic losses.

Finally, efficient manufacturing techniques for membranes, catalyst layers, MEAs, and stack assemblies which are suitable for mass production have to be developed. BASF Fuel Cell probably has the best facilities and the best know-how among all manufacturers of HT-PEFCs [55, 56], but it can be speculated that this experience is still less than that of companies which have been developing and producing low-temperature PEFCs for many years.

2.4 Alternative Liquid Electrolytes

2.4.1 Phosphonic Acids

Phosphonic acids [R–PO(OH)$_2$; derived from phosphoric acid (H$_3$PO$_4$) by replacing a hydroxy group by an organic chain R] have been intensively studied for use as intrinsic conductive electrolytes [57]. These materials are characterized by relatively strong self-dissociation and high proton mobility. One of the most promising candidates identified so far [P-C7 = $H_3C–(CH_2)_6–PO(OH)_2$, melting point ~105 °C] has an intrinsic proton conductivity of >10 mS cm^{-1} at 200 °C under dry conditions. At lower temperatures and with some degree of external humidification it also takes up water and allows reasonable conductivities. Nevertheless, for use in operational fuel cells, the conductivity of the electrolyte and therefore the density of charge carriers would have to be increased even further. Unfortunately, the individual functional groups tend to condense when the density of phosphonate groups becomes too high, so it seems unlikely that this electrolyte system may be sufficiently optimized [57].

2.4.2
N-Heterocycles as Intrinsic Proton Conductors

Organic compounds with nitrogen atoms in an aromatic ring structure [e.g., imidazole ($C_3H_4N_2$) and triazole ($C_2H_3N_3$)] behave similarly to water by showing amphoteric character, self-dissociation to a certain level, and formation of hydrogen bonds. As a consequence, they can be used as intrinsic proton conductors (i.e., functioning without any additional liquid phase).

Pure monomeric imidazole is a liquid above 90 °C and has a proton conductivity of about 5 mS cm^{-1} at 130 °C [58], and the conductivity of 1H-1,2,3-triazole is 0.13 mS cm^{-1} at room temperature [59]. Significantly higher conductivities of up to 10 mS cm^{-1} at 110 °C under dry conditions can be obtained when these N-heterocyclic compounds are imbibed into sulfonated proton exchange membranes such as Nafion or sPEEK [58, 59]. To obtain a fully polymeric electrolyte, imidazole (or chemically related) groups can be tethered to a polymeric backbone such as polystyrene [60]. Although this concept is very elegant, the proton conductivities of the resulting membranes are not higher than 1 mS cm^{-1} at best.

2.4.3
Ionic Liquids

After the imidazole-type materials in an acidic matrix had been investigated, it was logical to take a step forward and to use ionic liquids (room temperature molten salts) with imidazolium-type cations as electrolyte in a polymer matrix. One of the first examples of this approach consisted in gelification of a neutral polymer matrix [poly(vinylidene fluoride) (PVDF)] in the presence of a small amount of acid [trifluoromethanesulfonic acid (TFA)] in a highly stable ionic liquid solvent [1,2-dimethyl-3-n-propylimidazolium bis(trifluoromethylsulfonyl)-imide (DMPI-Im)] [61]. The resulting electrolytes have a very low vapor-pressure and may offer a Grotthus type of proton conduction mechanism in the absence of water. Furthermore, one can also imagine the design of basic electrolytes based on ionic liquids which would open up new possibilities in electrode design as cheaper catalysts than platinum could be used [62]. Nevertheless, research on ionic liquid-based PEFCs is still in its infancy.

2.5
Acidic Salts and Oxides

Although the "membrane" materials discussed in the following are essentially non-polymer but inorganic, they are included here as their range of application in fuel cells is more or less close to PEFC systems.

2.5.1
Solid Acids

Mixed salts of inorganic acids such as $CsHSO_4$, $Tl_3H(SO_4)_2$, $Rb_3H(SeO_4)_2$, and CsH_2PO_4 are isolators at room temperature but undergo a so-called "superprotonic phase transition" at elevated temperature during which their proton conductivity increases by several orders of magnitude (Figure 2.14a), reaching values of more than 10^{-2} S cm^{-1} by a quasi-liquid mechanism of proton diffusion which results from the thermal rotational motion of the anions (Figure 2.14b).

The best fuel cell results with so-called solid acid fuel cells (SAFCs) have been obtained with a CsH_2PO_4 electrolyte (peak power densities of 415 mW cm^{-2} with H_2–O_2 at 240 °C) [63]. In addition to durability and cost issues, the current SAFC research focuses on the development of thinner membranes (<25 μm) with mechanical integrity and the optimization of the cathode catalyst layer. Although some system requirements (e.g., heating up the stack above the superprotonic phase transition and preventing liquid water) are easy to handle, care has to be taken to avoid completely dry conditions, which cause dehydration of the electrolyte and formation of $CsPO_3$. Superprotonic Inc. is working on the commercialization of SAFCs.

2.5.2
Mixed Oxides

Several diphosphates (MP_2O_7 with, e.g., M=Ce, Zr, or Sn) doped with small amounts of a third cation (Mg, In, or Al) have shown promising results in very small laboratory cells, but face serious scale-up problems such as leak tightness in larger cells [64]. These materials do not contain structural protons but exhibit high proton conductivities of $>10^{-2}$ S cm^{-1} under unhumidified conditions. It has been speculated that protons dissolve in the structures as defects in the presence of hydrogen-containing gases. The most encouraging fuel cell results so far have been obtained with a 350 μm thick $In_{0.1}Sn_{0.9}P_2O_7$ electrolyte, yielding a power density of 264 mW cm^{-2} at 250 °C with dry hydrogen and air [65].

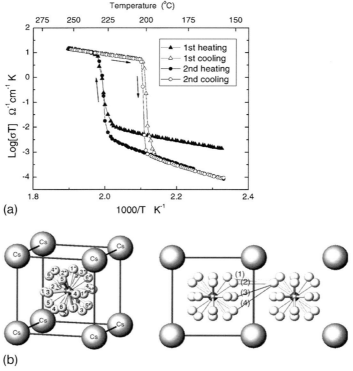

Fig. 2.14 Conductivity of CsH_2PO_4 in humidified air (a) and its high-temperature structure (b). The anion is orientationally disordered. As a consequence, short oxygen–oxygen distances between neighboring cells occur (as indicated), consistent with the length of strong hydrogen bonds. Reprinted from [63] with permission of The Royal Society of Chemistry.

2.6
Conclusion

Impressive progress has been made in the field of high-temperature PEM fuel cells in recent years. Nevertheless, further membrane, MEA, stack, and system development is essential to allow commercial success of HT-PEFCs.

Different approaches have shown that an increase in the operating temperature of sulfonic acid/water-based systems up to 90–100 °C or even 120 °C may be possible and attractive, especially in fuel cells running on pure hydrogen for transportation applications.

The development of cells with proton conduction relying on phosphoric acid in polybenzimidazole-type membranes and operating at 160–180 °C has already reached a pre-commercial stage and will be used preferably in applications where the hydrogen is produced on-site and/or where the heat produced by the cell can be used as a by-product.

All alternative concepts of high-temperature PEFCs known today (e.g., phosphonic acids, ionic liquids, solid acids, or mixed oxides as electrolytes) will probably not be optimized sufficiently to become technically attractive within the next few years. An electrolyte showing attractive conductivity and stability over the whole temperature range between −40 and +200 °C is not (yet) in sight and remains to be identified.

References

1 Satyapal, S. (2009) Hydrogen program overview. Paper presented at the DOE Hydrogen Program Annual Merit Review, Arlington, VA, 18 May 2009; available at http://www.hydrogen.energy.gov/pdfs/review09/program_overview_2009_amr.pdf (accessed 8 January 2010).

2 Daimler AG, Ford Motor Company, General Motors Corporation/Opel, Honda Motor Company, Hyundai Motor Company, Kia Motors Corporation, Renault SA, Nissan Motor Company, and Toyota Motor Corporation (2009) Letter of Understanding (LoU) signed on 9 September 2009 and press release.

3 Ettingshausen, F., Kleemann, J., Michel, M., Quintus, M., Fuess, H., and Roth, C. (2009) Spatially resolved degradation effects in membrane-electrode-assemblies of vehicle aged polymer electrolyte membrane fuel cell stacks. *J. Power Sources*, **194** (2), 899–907.

4 Chen, S., Gasteiger, H. A., Hayakawa, K., Tada, T., and Shao-Horn, Y. (2010) Platinum-alloy cathode catalyst degradation in proton exchange membrane fuel cells: nanometer-scale compositional and morphological changes. *J. Electrochem. Soc.*, **157** (1), A82–A97.

5 Gittleman, C. (2009) Paper presented at the High Temperature Membrane Working Group Meeting, Arlington, VA, 19 May 2009; available at http://www1.eere.energy.gov/hydrogenandfuelcells/pdfs/htmwg_may09_automotive_perspective.pdf (accessed 8 January 2010).

6 Epping-Martin, K. and Kopasz. J. P. (2009) The US DOEs high temperature membrane effort. *Fuel Cells*, **9** (4) 356–362.

7 Litt, M. (2009) Rigid rod polyelectrolytes: effect on physical properties. Frozen-in free volume: high conductivity at low RH. Paper presented at the DOE Hydrogen Program Annual Merit Review, Arlington, VA 19 May 2009; available at http://www.hydrogen.energy.gov/pdfs/review09/fc_08_litt.pdf (accessed 8 January 2010).

8 Papageorgopoulos, D. (2009) Fuel cell technologies. Paper presented at the DOE Hydrogen Program Annual Merit Review, Arlington, VA, 19 May 2009; available at http://www.hydrogen.energy.gov/pdfs/review09/fc_0_papageorgopoulos.pdf (accessed 8 January 2010).

9 Mittelsteadt, C. K., Braff, W., VanBlarcom, S., and Liu, H. (2009) Dimensionally stable membranes. Paper presented at the DOE Hydrogen Program Annual Merit Review, Arlington, VA, 19 May 2009; available at http://www.hydrogen.energy.gov/pdfs/review09/fc_02_mittelsteadt.pdf (accessed 8 January 2010).

10 Lee, H.-S., Roy, A., Lane, O., Dunn, S., and McGrath, J. E. (2008) Hydrophilic–hydrophobic multiblock copolymers based on poly(arylene ether sulfone) via low-temperature coupling reactions for proton exchange membrane fuel cells. *Polymer*, **49** (3), 715–723.

11 Roziere, J. and Jones, D. J. (2003) Non-fluorinated materials for proton exchange membrane fuel cell. *Annu. Rev. Mater. Res.*, **33**, 503–555.

12 Schuster, M., Kreuer, K.-D., Andersen, H. T., and Maier, J. (2007) Sulfonated poly(phenylene sulfone) polymers as hydrolytically and thermooxidatively

stable proton conducting ionomers. *Macromolecules*, **40** (3), 598–607.

13 Peron, J., Ruiz, E., Jones, D.J., and Roziere, J. (2008) Solution sulfonation of a novel polybenzimidazole. A proton electrolyte for fuel cell application. *J. Membr. Sci.*, **314** (1–2), 247–256.

14 Arcella, V., Ghielmi, A., and Tommasi, G. (2003) High performance perfluoropolymer films and membranes. *Ann. N. Y. Acad. Sci.*, **984**, 226–244.

15 Arico, A.S., Di Blasi, A., Brunaccini, G., Sergi, F., Antonucci, V., Asher, P., Buche, S., Fongalland, D., Hards, G.A., Sharman, J.D.B., Bayer, A., Heinz, G., Zuber, R., Gebert, M., Corasaniti, M., Ghielmi, A., and Jones, D.J. (2009) High temperature operation of a solid polymer electrolyte fuel cell stack based on a new ionomer membrane. *ECS Trans.*, **25** (1), 1999–2007.

16 Endoh, E., Kawazoe, H., and Nakagawa, H. (2006) Development of highly durable MEA for PEMFC under high temperature operations (2). *ECS Trans.*, **1** (8), 221–227.

17 Kinoshita, S. (2008) Performances of highly durable PFSA polymer based MEAs at high temperature low RH and dry–wet conditions. Paper presented at Progress MEA08 – First CARISMA International Conference, La Grande Motte, France, 22 September 2008.

18 Savinell, R., Yeager, E., Tryk, D., Landau, U., Wainright, J., Weng, D., Lux, K., Litt, M., and Rogers, C. (1994) A polymer electrolyte for operation at temperatures up to 200 °C. *J. Electrochem. Soc.*, **141** (4), L46–L48.

19 Savinell, R.F. and Litt, M.H. (1994) Proton conducting polymers used as membranes. US Patent 5 525 436, filed 1 November 1994, issued 11 June 1996.

20 Wainright, J.S., Wang, J.-T., Wenig, D., Savinell, R.F., and Litt, M. (1995) Acid-doped polybenzimidazoles: a new polymer electrolyte. *J. Electrochem. Soc.*, **142** (7), L121–L123.

21 Xiao, L., Zhang, H., Scanlon, E., Ramanathan, L.S., Chou, E.-W., Rogers, D., Apple, T., and Benicewicz, B.C. (2005) High-temperature polybenzimidazole fuel cell membranes via a sol-gel process. *Chem. Mater.*, **17** (21) 5328–5333.

22 Mader, J., Xiao, L., Schmidt, T.J., and Benicewicz, B.C. (2008) Polybenzimidazole/acid complexes as high-temperature membranes. *Adv. Polym. Sci.*, **216**, 63–124.

23 Li, Q., Jensen, J.O., Savinell, R.F., and Bjerrum, N.J. (2009) High temperature proton exchange membranes based on polybenzimidazoles for fuel cells. *Prog. Polym. Sci.*, **34** (5), 449–477.

24 Asensio, J.A. and Gomez-Romero, P. (2005) Recent developments on proton conducting poly(2, 5-benzimidazole) (ABPBI) membranes for high temperature polymer electrolyte membrane fuel cells. *Fuel Cells*, **5** (3), 336–343.

25 Li, Q., He, R., Jensen, J.O., and Bjerrum, N.J. (2004) PBI-based polymer membranes for high temperature fuel cells – preparation, characterization and fuel cell demonstration. *Fuel Cells*, **4** (3), 147–159.

26 Ma, Y.-L., Wainright, J.S., Litt, M.H., and Savinell, R.F. (2004) Conductivity of PBI membranes for high-temperature polymer electrolyte fuel cells. *J. Electrochem. Soc.*, **151** (1), A8–A16.

27 Wippermann, K., Wannek, C., Oetjen, H.-F., Mergel, J., and Lehnert, W. (2010) Cell resistances of poly(2,5-benzimidazole)-based high temperature polymer membrane fuel cell membrane electrode assemblies: time dependence and influence of the operating parameters. *J. Power Sources*, **195** (9), 2806–2809.

28 Lobato, J., Canizares, P., Rodrigo, M.A., and Linares, J.J. (2007) PBI-based polymer electrolyte membrane fuel cells. Temperature effects on cell performance and catalyst stability. *Electrochim. Acta*, **52** (12), 3910–3920.

29 Belack, J., Kundler, I., Schmidt, T., Uensal, O., Kiefer, J., Padberg, C., and Weber, M. (2005) Proton-conducting polymer membrane coated with a catalyst layer, said polymer membrane comprising phosphonic acid polymers, membrane/electrode unit and the use thereof in fuel cells. International Patent WO 2005/023914, filed 4 September 2003, issued 17 March 2005.

30 Pawlik, J., Uensal, O., Schmidt, T., Padberg, C., and Hoppes, G. (2006) Membrane-electrode unit and fuel ele-

ments with increased service life. International Patent WO 2006/015806, filed 5 August 2004, issued 16 March 2006.
31 Foli, K., Gronwald, O., Haufe, S., Kiel, S., Mähr, U., Melzner, D., Reiche, A., Walter, F., and Weisshaar, S. (2006) Sartorius HT-PEMFC membrane electrode assembly. Paper presented at 2006 Fuel Cell Seminar, Honululu, HI, 13–17 November 2006.
32 Wannek, C., Lehnert, W., and Mergel, J. (2009) Membrane electrode assemblies for high-temperature polymer electrolyte fuel cells based on poly(2,5-benzimidazole) membranes with phosphoric acid impregnation via the catalyst layers. *J. Power Sources*, **192** (2), 258–266.
33 Schmidt, T., Padberg, C., Hoppes, G., Ott, D., Rat, F., and Jantos, M. (2006) Membrane electrode units and fuel cells with an increased service life. International Patent WO 2006/008158, filed 21 July 2004, issued 26 January 2006.
34 Product information of BASF Fuel Cell GmbH, http://www.basf-fuelcell.com/en/projects/celtec-mea.html (accessed 6 January 2010).
35 Product information of Advent Technologies SA, http://www.adventech.gr/products.html (accessed 6 January 2010).
36 Kwon, K., Kim, T.Y., Yoo, D.Y., Hong, S.-G., and Park, J.O. (2009) Maximization of high-temperature proton exchange membrane fuel cell performance with the optimum distribution of phosphoric acid. *J. Power Sources*, **188** (2), 463–467.
37 BASF Fuel Cell (2010) Press photograph; available at http://www.basf.com/group/corporate/de/news-and-media-relations/press-photos/dvd/energy-management (accessed 8 January 2010).
38 Product information of Serenergy A/S, http://www.serenergy.dk/Serenus_166_390_Air_C.htm (accessed 6 January 2010).
39 Plug Power Inc. (2010) Brochure; available at http://www.plugpower.com/userfiles/GenSys%20HT%20MK.pdf (accessed 8 January 2010).
40 Vogel, J. (2008) International stationary fuel cell demonstration. Paper presented at the DOE Hydrogen Program Annual Merit Review, Arlington, VA, 12 June 2008; available at http://www.hydrogen.energy.gov/pdfs/review08/fc_40_vogel.pdf (accessed 6 January 2010).
41 Hübner G. (2007) High-temperature PEM Fuel Cells. Paper presented at the "FUNCHY: Functional Materials for Mobile Hydrogen Storage" workshop, Karlsruhe, 21 November 2007; available at http://www.gkss.de/imperia/md/content/gkss/institut_fuer_werkstoffforschung/wtn/h2-speicher/funchy/funchy–2007/9_vw_huebner_funchy–2007.pdf (accessed 6 January 2010).
42 Technology description of UltraCell corporation, http://www.ultracellpower.com/sp.php?rmfe (accessed 8 January 2010).
43 Oono Y., Fukuda, T., Sounai, A., and Hori, M. (2010) Influence of operating temperature on cell performance and endurance of high temperature proton exchange membrane fuel cells. *J. Power Sources*, **195** (4), 1007–1014.
44 Yu, S., Xiao, L., and Benicewicz, B.C. (2008) Durability studies of PBI-based high temperature PEMFCs. *Fuel Cells*, **8** (3–4) 165–174.
45 Gang, X., Hjuler, H.A., Olsen, C., Berg, R.W., and Bjerrum, N.J. (1993) Electrolyte additives for phosphoric acid fuel cells. *J. Electrochem. Soc.*, **140** (4), 896–902.
46 Mamlouk, M. and Scott, K. (2010) The effect of electrode parameters on performance of a phosphoric acid-doped PBI membrane fuel cell. *Int. J. Hydrogen Energy*, **35** (2), 784–793.
47 Hong, S.-G., Kwon, K., Lee, M.-J., and Yoo, D.Y. (2009) Performance enhancement of phosphoric acid-based proton exchange membrane fuel cells by using ammonium trifluoromethanesulfonate. *Electrochem. Commun.*, **11** (6), 1124–1126.
48 Wannek, C., Konradi, I., Mergel, J., and Lehnert, W. (2009) Redistribution of phosphoric acid in membrane electrode assemblies for high-temperature polymer electrolyte fuel cells. *Int. J. Hydrogen Energy*, **34** (23), 9479–9485.
49 Kwon, K., Park, J.O., Yoo, D.Y., and Yi, J.S. (2009) Phosphoric acid distribution in the membrane electrode assembly of high temperature proton exchange

membrane fuel cells. *Electrochim. Acta*, **54** (26), 6570–6575.

50. Lobato, J., Canizares, P., Rodrigo, M. A., Linares, J. J., and Pinar, F. J. (2010) Study of the influence of the amount of PBI-H_3PO_4 in the catalytic layer of a high temperature PEMFC. *Int. J. Hydrogen Energy*, **35** (3), 1347–1355.

51. Dam, V. A. T., Jayasayee, K., and deBruijn, F. A. (2009) Determination of the potentiostatic stability of PEMFC electro catalysts at elevated temperatures. *Fuel Cells*, **9** (4), 453–462.

52. Wang, X., Kumar, R., and Myers, D. J. (2006) Effect of voltage on platinum dissolution. *Electrochem. Solid-State Lett.*, **9** (5), A225–A227.

53. Liu, G., Zhang, H., Zhai, Y., Zhang, Y., Xu, D., and Shao, Z.-G. (2007) Pt_4ZrO_2/C cathode catalyst for improved durability in high temperature PEMFC based on H_3PO_4 doped PBI. *Electrochem. Commun.*, **9** (1), 135–141.

54. (a) Kerres, J., Schönberger, F., Chromik, A., Häring, T., Li, Q., Jensen, J. O., Pan, C., Noye, P., and Bjerrum, N. J. (2008) Partially fluorinated arylene polyethers and their ternary blend membranes with PBI and H_3PO_4. Part I. Synthesis and characterisation of polymers and binary blend membranes. *Fuel Cells*, **8** (3–4) 175–187; (b) Li, Q., Jensen, J. O., Pan, C., Bandur, V., Nilsson, M. S., Schönberger, F., Chromik, A., Hein, M., Häring, T., Kerres, J., and Bjerrum, N. J. (2008) Partially fluorinated arylene polyethers and their ternary blend membranes with PBI and H_3PO_4. Part II. Characterisation and fuel cell tests of the ternary membranes. *Fuel Cells*, **8** (3–4) 188–199.

55. (a) Harris, T. A. L., Walczyk, D. F., and Weber, M. M. (2010) Manufacturing of high-temperature polymer electrolyte membranes – Part I: system design and modeling. *J. Fuel Cell Sci. Technol.*, **7** (1), 011007-1–011007-9; (b) Harris, T. A. L., Walczyk, D. F., and Weber, M. M. (2010) Manufacturing of high-temperature polymer electrolyte membranes – Part II: implementation and system model validation. *J. Fuel Cell Sci. Technol.*, **7** (1), 011008-1–011008-8.

56. Puffer, R. H. and Rock, S. J. (2009) Recent advances in high temperature proton exchange membrane fuel cell manufacturing. *J. Fuel Cell Sci. Technol.*, **6** (4), 041013-1–041013-7.

57. Schuster, M., Rager, T., Noda, A., Kreuer, K. D., and Maier, J. (2005) About the choice of the protogenic group in PEM separator materials for intermediate temperature, low humidity operation: a critical comparison of sulfonic acid, phosphonic acid and imidazole functionalized model compounds. *Fuel Cells*, **5** (3), 355–365.

58. Kreuer, K. D. (2001) On the development of proton conducting polymer membranes for hydrogen and methanol fuel cells. *J. Membr. Sci.*, **185** (1), 29–39.

59. Zhou, Z., Li, S., Zhang, Y., Liu, M., and Li, W. (2005) Promotion of proton conduction in polymer electrolyte membranes by 1H-1,2,3-triazole. *J. Am. Chem. Soc.*, **127** (31), 10824–10825.

60. Herz, H. G., Kreuer, K. D., Maier, J., Scharfenberger, G., Schuster, M. F. H., and Meyer, W. H. (2003) New fully polymeric proton solvents with high proton mobility. *Electrochim. Acta*, **48** (14–16), 2165–2171.

61. Navarra, M. A., Panero, S., and Scrosati, B. (2005) Novel, ionic-liquid-based, gel-type proton membranes. *Electrochem. Solid-State Lett.*, **8** (6), A324–A327.

62. Armand, M., Endres, F., MacFarlane, D. R., Ohno, H., and Scrosati, B. (2009) Ionic-liquid materials for the electrochemical challenges of the future. *Nature Mater.*, **8** (8), 621–629.

63. Haile, S. M., Chisholm, C. R. I., Sasaki, K., Boysen, D. A., and Uda, T. (2007) Solid acid proton conductors: from laboratory curiosities to fuel cell electrolytes. *Faraday Discuss.*, **134**, 17–39.

64. Genzaki, K., Heo, P., Sano, M., and Hibino, T. (2009) Proton conductivity and solid acidity of Mg-, In-, and Al-doped SnP_2O_7. *J. Electrochem. Soc.*, **156** (7), B806–B810.

65. Nagao, M., Takeuchi, A., Heo, P., Hibino, T., Sano, M., and Tomita, A. (2006) A proton-conducting In^{3+}-doped SnP_2O_7 electrolyte for intermediate-temperature fuel cells. *Electrochem. Solid-State Lett.*, **9** (3), A105–A109.

3
Current Status of and Recent Developments in Direct Liquid Fuel Cells

Jürgen Mergel, Andreas Glüsen, and Christoph Wannek

Abstract

Direct liquid fuel cells, such as the direct methanol fuel cell (DMFC) or the direct ethanol fuel cell (DEFC), convert liquid fuel directly into electric current. In comparison to fuel cell systems that operate with pure hydrogen or hydrogen-rich gases from reforming processes, the fuel in the DMFC is supplied directly via liquid methanol. Apart from the very high energy density of methanol, the DMFC is characterized by easy handling and trouble-free refueling. As the reforming step is by-passed in direct fuel cells, compensation in the form of higher overvoltages (i.e. electrochemical losses) is acceptable. Despite the resulting moderate power densities, direct fuel cells are more attractive for a variety of applications in the low to medium power range than PEM fuel cells powered by hydrogen. Examples of their use include replacement of batteries in portable applications and for light traction, as there is no need for the relatively expensive and time-consuming charging of batteries or for a spare battery for multiple-shift operation. Furthermore, the high energy density of the liquid energy carrier permits much longer operating times than batteries or fuel cell systems based on hydrogen.

This paper outlines the level of development of different direct liquid fuel cells based on current research findings and trends.

Keywords: direct liquid fuel cell, direct methanol fuel cell, direct ethanol fuel cell

3.1
Introduction

Direct liquid fuel cells, such as the direct methanol fuel cell (DMFC) and the direct ethanol fuel cell (DEFC), convert the liquid fuel directly into electric current. In comparison with fuel cell systems that operate with pure hydrogen or hydrogen-rich gases from reforming processes, the fuel in the DMFC, for example, is supplied directly via liquid methanol. In comparison with gaseous fuels, liquids such as ethanol and methanol have the advantage of having a higher energy density (Figure 3.1). This has a particularly strong effect in a fuel cell system if long operating times are to be achieved with one tank of fuel (Figure 3.2). In addition to the very high energy density of the liquid energy carriers, the direct liquid fuel cells are also characterized by easy handling and unproblematic refueling. However, since there is no reforming step with direct liquid fuel cells, users have to accept higher electrochemical losses in comparison with hydrogen-operated polymer electrolyte fuel cells (PEFCs). In spite of the resulting moderate power densities, direct fuel cells are attractive for various applications in the small to medium power range up to 5 kW in comparison with PEFCs operated with hydrogen. For example, they are used to replace batteries or accumulators in portable applications and for light traction since there is no need for the comparatively laborious and time-consuming recharging of the batteries or a second battery for multiple-shift operation. Moreover, their range is greater than that of conventional lead acid batteries by a factor of 5 (Figure 3.2) and they can reach the operating range of an internal combustion engine (ICE).

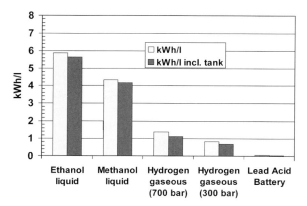

Fig. 3.1 Energy content of various energy carriers with and without consideration of the tank required.

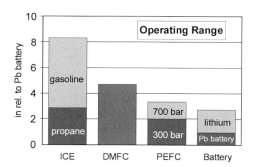

Fig. 3.2 Operating range of forklift trucks with various energy carriers in comparison with a lead acid battery.

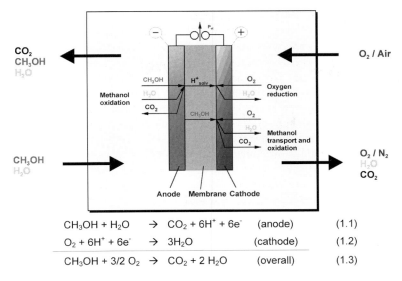

Fig. 3.3 Reaction scheme of a direct methanol fuel cell with a proton-conducting electrolyte.

$$CH_3OH + H_2O \rightarrow CO_2 + 6H^+ + 6e^- \quad \text{(anode)} \qquad (1.1)$$

$$O_2 + 6H^+ + 6e^- \rightarrow 3H_2O \quad \text{(cathode)} \qquad (1.2)$$

$$CH_3OH + 3/2\, O_2 \rightarrow CO_2 + 2H_2O \quad \text{(overall)} \qquad (1.3)$$

The core component of the DMFC, and also the DEFC, is the membrane electrode assembly (MEA). In the case of DMFCs, Figure 3.3 shows the details of the catalytically active partial reactions if a proton-conducting electrolyte is used for the membrane. Typical power densities at a cell voltage of 0.4 V at moderate temperatures of about 50 °C are currently in the region of 50 mW cm^{-2} [1, 2]. If the operating temperature of the MEA in the stack can be raised to 70–80 °C under autothermic operating conditions, then 80–100 mW cm^{-2} can be achieved.

Alternatively, recent studies have once again examined alkaline DMFC and DEFC systems making use of anion-exchange membranes. If an alkaline electrolyte is used, then the following partial reactions proceed in the DMFC:

$$CH_3OH + 6\,OH^- \rightarrow CO_2 + 5\,H_2O + 6e^- \quad \text{(anode)} \qquad (3.4)$$

$$3/2\,O_2 + 3\,H_2O + 6e^- \rightarrow 6\,OH^- \quad \text{(cathode)} \qquad (3.5)$$

$$CH_3OH + 3/2\,O_2 \rightarrow CO_2 + 2\,H_2O \quad \text{(overall)} \qquad (3.6)$$

An advantage of such systems in comparison with classical DMFCs in an acidic medium would be a more favorable overall catalysis of the methanol oxidation and thus lower voltage losses [3]. However, ongoing investigations indicate low power densities of <10 mW cm^{-2} [4] in comparison with DMFCs with proton-conducting membranes (Nafion) under similar operating conditions (air operation, ambient pressure, ~60 °C) (see Section 3.2.4).

In contrast to DEFCs, DMFC technology is very mature due to the progress made worldwide in recent years and is already commercially available for some applications. For example, Smart Fuel Cell AG has to date sold more than 14 000 DMFC systems in the power range up to 250 W [5]. These devices are used as battery chargers for various applications, especially in the leisure sector. Applications for DMFCs as battery replacements or range extenders in the portable sector of < 50 W include mobile phones, personal digital assistants (PDAs), and laptops, and also back-up power packs for military applications. For light traction, DMFC systems with an electric power of up to 2 kW have been developed, usually as DMFC hybrid systems. However, before it is possible to commercialize DMFCs extensively, apart from cost reduction, two other major challenges remain to be met: increasing the low durability caused by relatively rapid aging of the MEA components and improving the as yet inadequate performance of the MEAs, which is partly responsible for the low overall efficiency of DMFC energy systems. This includes, for example, reducing the methanol permeation and water permeability of existing electrolyte membranes.

Ethanol is also a promising energy carrier for direct fuel cells due to its high energy density (Figure 3.1) and its simple handling since, in contrast to methanol, ethanol is not toxic. In recent years, work on DEFCs has been intensified worldwide, but in comparison with DMFCs their power data are similar only in alkaline cells, although it has still not been possible to achieve 100% conversion (Section 3.3).

3.2
Direct Methanol Fuel Cells

In recent years, the development of DMFCs has been mainly concentrated on the field of "portables" as a possible replacement for batteries, since due to increasing energy requirements, especially for modern mobile phones (more functions, bigger displays, etc.), operating times with Li batteries are limited. In contrast to, for example, Li batteries, due to the higher energy density of DMFC systems, more energy can be made available from the same volume, thus achieving longer operating times [6].

3.2.1
Designs

There are two different concepts for designing DMFC systems: "active" and "passive" systems. With active systems (Figure 3.4a), for example, pumps, fans and heat exchangers are used in order to provide the DMFC stack with a controlled supply of reactants and to remove the waste heat and product water. In this way, active systems can run under optimized operating conditions with respect to temperature, methanol concentration, and flow rate so that the mass transfer and thus the electric power of the cells can be improved compared with

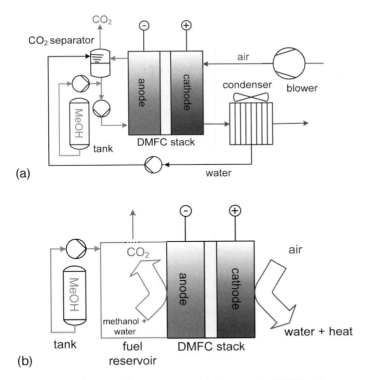

Fig. 3.4 Flow diagram of (a) an active and (b) a passive DMFC system.

a passive system. This gain in output for the DMFC stack is achieved at the expense of greater system complexity and a reduction of overall system efficiency due to the energy required for blowers and pumps [7]. For this reason, active systems are more suitable for DMFC systems in the higher power range.

In passive systems (Figure 3.4b), the MEAs or cells are usually only supplied with reactants by means of diffusion, convection, evaporation, and capillary forces. There is therefore no need for electric consumers such as blowers or pumps, which reduce the overall efficiency of the DMFC system. The advantage of passive systems is the lower system complexity. Passive systems normally operate with lower power densities than active systems and are therefore more suitable for small portable systems. Methanol is usually fed to the anode in a gaseous form, that is, the methanol is evaporated in the anode compartment via an appropriate device [8].

With respect to both systems, however, water management must be designed in such a way that only concentrated methanol is used as energy carrier in the fuel reservoir. That is to say, assuming that the electrochemical reaction, methanol permeation, and water evaporation are the dominant processes in MEA operation, it can be concluded that for the water balance only the reaction water (Equation 3.3) (in addition to the water that flows into the stack with the moist ambient air) may leave the DMFC with the waste air, whereas the water that enters the cathode from the anode (Figure 3.3) has to be fed back into the anode loop in order to operate the system autonomously with respect to the water balance. In an active system, other water fractions can be recovered from the cathode off-gas by a condenser (Figure 3.4a). For the passive system, this means that some of the water produced at the cathode (Equation 3.2) has to be transported back to the anode through the membrane in order to make this water available there for the anodic reaction (Equation 3.1). This can be achieved by an innovative MEA design [9, 10].

3.2.2
Technical Products on the Market

As already mentioned in the Introduction, at the moment the only DMFC products commercially available are manufactured by Smart Fuel Cell AG (SFC). In contrast to most of the other fuel cell companies, who are still at the development phase or who operate subsidized demonstration facilities, SFC has been marketing fully commercial DMFCs to industry and end consumers for more than 4 years. These fuel cells were specifically developed for the reliable provision of electric power for mobile and portable applications in the power range up to 250 W. The main applications are in the leisure sector for supplying power to electric consumers on mobile homes, caravans, and sailing boats. A lead acid battery is usually charged by a DMFC with a continuous output of 25–90 W. The EFOY family (Figure 3.5) comprises five models with a charging capacity of 600–2200 W h per day. A frequently cited disadvantage of methanol is its toxicity. What is forgotten here, however, is that methanol is considerably less toxic than the established energy carrier gasoline, which due to its benzene content is, moreover, also carcinogenic and teratogenic. Methanol, in contrast,

Fig. 3.5 The EFOY fuel cell family from SFC.

although acutely toxic, is neither teratogenic nor carcinogenic and, moreover, is much more readily biodegradable than gasoline or diesel. It is therefore not harmful to handle methanol if hermetically sealed containers are used, as in the case of EFOY cartridges. EFOY fuel cartridges, for example, have the seal of approval from the German Technical Control Board (TÜV-GS Siegel) and are licensed for transport purposes on board land vehicles, ships, and aircraft.

The first commercial products in the portable sector up to 50 W were launched on the Japanese market by Toshiba in October 2009. This is an external charger that can charge the batteries of cell phones or other mobile digital devices via a USB flash drive. The palm-sized Dynario^{TM} (Figure 3.6) is only available on the Japanese market in a limited edition of 3000, weighs 280 g, and can recharge a mobile phone for a maximum of two times with one cartridge of methanol (14 ml) [11]. The development of the Dynario illustrates the fact that Toshiba and some other companies have in the past few years abandoned their ambitious goal of developing an integrated DMFC system for laptops and are instead working on external chargers [12]. In the same way, Samsung SDI in Korea, which in the past developed DMFC systems for laptops, presented a DMFC module for military applications in June 2009 that can supply 1.8 kW h of electric energy with one cartridge. According to Samsung SDI, the device should be available for military applications in 2010 [13].

In the light traction sector, the only marketable products are also from SFC, based on the EFOY Pro Series. These devices are so-called "range extenders" or internal battery chargers, which charge the batteries of electric scooters or small electric cars with a maximum electric power of 90 W and thus extend the range of these vehicles [5]. In contrast, Forschungszentrum Jülich is developing DMFC systems in the power range 2–5 kW for the light traction sector which will be used not as range extenders but rather as replacements for batteries or accumulators. A market and feasibility study demonstrated that for forklift trucks these hybridized DMFC systems can reach a cost level that is comparable to that of conventional battery systems in the field of material handling [14]. Not only do such systems ensure longer operating times, they also dispense with the need for the relatively time-consuming recharging of the battery and therefore spare batteries are no longer required for multiple-shift operation. As

Fig. 3.6 Dynario from Toshiba.

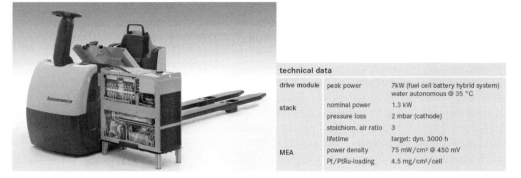

Fig. 3.7 Forklift truck (class 3) with Forschungszentrum Jülich's DMFC system V 3.1.

part of a project initiated in 2007, optimized DMFC systems with long-term stability are being developed in cooperation with industrial partners. These systems are to be used for pilot applications in electrically powered forklifts. Figure 3.7 shows the first DMFC system constructed as part of a feasibility study.

A detailed overview of the worldwide status of DMFC technology for portable applications and light traction not yet on the market can be found in the literature [15].

3.2.3
Main Problem Areas

In the commercialization of direct methanol fuel cells, apart from the continuing need for cost reduction, there are currently two major challenges: increasing the as yet low long-term stability due to the relatively rapid aging of the DMFC components and improving the still inadequate performance of the MEAs, partly responsible for the low overall efficiency of DMFC energy systems of ≤0.25.

3.2.3.1 Performance

Several factors are responsible for the lower overall efficiency of DMFCs in comparison with PEFCs, and these factors also exert a mutual influence on each other. The overall efficiency of a fuel cell, η_{tot}, is the product of the voltage efficiency, η_U, the faradic efficiency, η_F (=fuel utilization level), and the system efficiency, η_S, which describes the fraction of useful energy of the electricity generated minus the power consumed by the system components. Due to the methanol permeation in the DMFC (Figure 3.3), the faradic efficiency η_F is always <1 and under typical application conditions it is currently between 0.5 and 0.7. The voltage efficiency η_U is the ratio of the cell voltage and the thermoneutral voltage. A high cell voltage is therefore an essential requirement for high efficiency of a direct methanol fuel cell. The heating value of methanol corresponds to a cell voltage $E_0 = 1.090$ V [16]. In fact, however, DMFCs are operated at cell voltages of 0.3–0.6 V, which correspond to a voltage efficiency η_U of 0.27–0.54. The greatest losses result from the slow electrochemical oxidation of methanol at the anode and from mixed-potential formation at the cathode [17] (Figure 3.3).

The water permeation (see below) indirectly influences the overall efficiency since the evaporation enthalpy of the water arising at the cathode withdraws heat from the system so that certain cell temperatures (typically 50–70 °C) cannot be exceeded. Since the electrode reactions progress more effectively in DMFCs at higher temperatures, if too much water arises at the cathode this has a negative effect on cell voltage [18]. Water permeation is also of significance in maintaining the simplest and most compact water-autonomous system possible (i.e., small condenser).

Methanol and Water Permeation

Membranes Perfluorinated sulfonic acid membranes, such as produced for example by DuPont under the trade name Nafion®, are most frequently used for DMFCs. However, these membranes take up a relatively large amount of water and methanol so that there may be considerable leakage of water and methanol to the cathode. This leakage takes place, on the one hand, in the form of diffusion and, on the other, as electroosmotic flow since every proton that migrates through the membrane is hydrated by a certain number of solvent molecules [19], which therefore migrate through the membrane with the proton. Depending on the temperature and chemical structure of the membrane, the number of entrained solvent molecules – the so-called electroosmotic drag coefficient – varies within a range of about 1–5 and therefore provides an opportunity for considerably influencing the permeation of water and methanol.

Intensive efforts are therefore being focused on developing membranes to reduce the permeation of water and methanol. On the one hand, the structure of the perfluorinated sulfonic acid membranes is being adapted by, for example, varying the side-chains by means of which the sulfonic acid groups are attached to the main chain [20], or by replacing the sulfonic acid group by a different acid group [21].

In a second approach, composite membranes of perfluorinated polysulfonic acids and inorganic additives are being produced in order to reduce the water transport [22]. In particular oxides, such as SiO_2, ZrO_2, and TiO_2, and also phosphonates and heteropoly acids of molybdenum and tungsten are used as inorganic additives.

However, the most intensive efforts are being focused on the development of completely new membrane polymers. Sulfonic acid groups are also usually used here as protogenic groups, and occasionally phosphonic acid groups or carboxylic acid groups. The polymer main chain must be resistant to the oxidizing potentials that occur in the direct methanol fuel cell. The location of oxidative attack was identified as the aliphatic C–H bonds [23] so that in addition to the perfluorinated polymers mentioned above, investigations are also being made, in particular, of polymers with aromatic main chains. In the literature, descriptions can be found of polyetherketones, polyetheretherketones, polysulfones, polyethersulfones, and other polyaromatics [24]. The acid groups are usually directly bound to the phenyl groups. These polymers are often characterized by lower methanol and water permeation, but they generally also have lower proton conductivity, poorer mechanical properties, and lower chemical stability than the perfluorinated polysulfonic acids, so that in practical applications it is difficult to exploit the benefits of their lower water and methanol permeation in direct methanol fuel cells. However, there are promising approaches [24].

MEA Design A suitable MEA design makes it possible to reduce the transport of methanol and water. A compact anode decreases the diffusive transport of methanol and water from the flow distributors to the membrane surface and thus the volume of the solvent that may be transported through the membrane [25]. In particular, the permeation of methanol can be greatly reduced by demand-oriented delivery of methanol. If, due to diffusion, only slightly more methanol enters the catalyst layer from the flow channel than is consumed there by the electrochemical reaction, then only little methanol can be transported through the membrane. However, there is the risk of depletion of methanol, which can lead to irreversible damage to the anode (see Section 3.2.3.2). This is why precise dosing of the methanol supply is absolutely essential. In the same way, a compact cathode can decrease the permeation of water and methanol [26], since the removal of water arising at the cathodic surface of the membrane is inhibited and thus the activity gradient between the anode and cathode – as the driving force of diffusion – is reduced. There is, however, a danger that the water arising at the cathode will fill the pores of the cathode and thus inhibit the access of atmospheric oxygen to the cathodic catalyst layer. It therefore appears more promising at present to control diffusion via the anode.

Cell Voltage

Methanol-Tolerant Catalysts As a rule, platinum is currently used as the catalyst in the cathode. However, not only oxygen is reduced at its surface but also permeated methanol is oxidized (Figure 3.3). Methanol oxidizes only slowly at the cathode so that a considerable proportion of the catalyst is occupied by interim stages of methanol oxidation and is therefore not available for oxygen reduction. Furthermore, the oxidation of methanol takes place at a lower potential than the reduction of oxygen so that a mixed potential is formed. The cell voltage is thus reduced which, in turn, leads to lower efficiency. Attempts are being made to avoid this problem by using catalysts that only catalyze the oxygen reduction and do not absorb the methanol. In the literature, Pt alloys, transition metal complexes with macrocycles and transition metal chalogenides – especially transition metal selenides – have been described for this purpose [27]. However, the methanol tolerance of the Pt alloys is little better than that of pure platinum and the catalytic activity of the transition metal complexes and transition metal chalogenides is generally significantly poorer than that of platinum with respect to oxygen reduction (cf., Section 3.2.3.3), so that further efforts are required to increase the cathode potential of DMFCs by this method.

Anode Catalysts Pure platinum is not very suitable as an anode catalyst since the oxidation of methanol first only leads to an intermediate carbonyl species, which is firmly adsorbed on the Pt and is only further oxidized to CO_2 at very high potentials. To circumvent this problem, PtRu alloys are generally used. In this case, methanol oxidation takes place on the Pt and further oxidation of the intermediate carbonyl is promoted by the OH groups adsorbed on the ruthenium, which reduces the anodic overvoltage by about 150 mV in comparison with pure platinum. Different Pt alloys are being investigated in current research. Better results than with PtRu were obtained, however, if at all, with ternary PtRuM alloys [28]. In this respect, Ir and Sn are especially promising as the third alloying component. The structures of these trimetallic catalysts are currently under discussion, and the stability of these and other novel materials (such as so-called "core–shell" or "dealloyed" catalysts) is doubtful in DMFC operation. In most of these cases, the differences from PtRu in a real operating DMFC are still marginal and depend on the composition of the alloy and the operating conditions so that only slight benefits have been gained to date.

Catalyst Supports The performance of DMFC catalysts can be further improved by the use of new substrate materials. In comparison with carbon black, carbon nanotubes (CNTs) are characterized by an extremely high specific surface (accessible for Pt deposition and also for the half-cell reactions of DMFCs) and also by electrical conductivity higher by roughly two orders of magnitude [29]. Accordingly, Pt/CNT and PtRu/CNT catalysts display higher electrochemical activities [30, 31] and MEAs with these catalysts have greater power densities [32, 33] than reference specimens with Vulcan XC-72 as the substrate. Due to the greater oxidation resistance of the CNTs [34], it is to be expected, moreover, that these MEAs will also display increased long-term stability. Since CNTs are now produced in quantities of several tonnes [35] and – depending on the quality – are already available at a price of just a few hundred US dollars per kilogram [36], their application in commercial DMFCs seems to be realistic in the medium term.

Electrically conducting or semiconducting oxides and carbides (such as Ti_xO_y, SnO_x, WO_x, and WC/WC_2) can also be used as a substrate instead of carbon black or as surface promoters, which improve the properties of XC-72 to such an extent that $Pt/MO_x/XC\text{-}72$ catalysts display higher electrochemical activities than Pt/XC-72 [37].

3.2.3.2 Long-Term Stability

In comparison with hydrogen-operated PEFCs, there is little in the literature on the aging of direct methanol fuel cells (DMFCs). Essentially, the degradation ef-

Table 3.1 Recoverable and unrecoverable degradation effects in DMFC MEAs.

Permanent degradation/unrecoverable performance loss	Temporary degradation/recoverable performance loss
Loss of electrochemically active surface area of electrodes [39–42]	Cathode Pt catalyst surface oxidation [38]
Ruthenium crossover (migration) from the anode to the cathode [39, 41, 43]	Temporary membrane dehydration
	Incipient cathode flooding
Irreversible loss of cathode hydrophobicity [44]	
Electrode delamination [45, 46]	

fects described for DMFC membrane electrode assemblies can be divided into recoverable and unrecoverable processes (Table 3.1) [38, 39]. The recoverable losses disappear after stopping and restarting the experiment, while the unrecoverable losses cause a steady decrease of the starting value during the aging experiment.

The reasons for the loss of electrochemically active surface in DMFCs are similar to those described for PEFCs. First, the particle size of catalysts increases during fuel cell operation. Second, an insufficient distribution of methanol solution within the anode flow fields will lead to a local depletion of fuel, causing an increase in anode overpotential and subsequent corrosion of the catalyst, which includes platinum, ruthenium [43], and the carbon support. Another irreversible process, ruthenium crossover (migration) from the anode through the polymer membrane to the cathode [39, 41, 43], also leads to performance degradation. Piela *et al.* [41] attributed this to an intrinsic instability of the commercial Pt–Ru–carbon black catalyst. The ruthenium migration does occurs not only due to Ru corrosion in the case of methanol depletion (see above), but also under normal operating conditions, in the course of which Ru migrates even without any current flow. The presence of Ru in the catalyst leads to slower oxygen reduction and accelerated methanol oxidation, thus reducing the cathode potential, cell voltage, and also performance.

It is known from the literature that ruthenium corrosion is one of the main causes of aging in DMFC MEAs. Furthermore, it is known that chloride ions act as complexing agents and promote the corrosion of ruthenium and platinum. Intensified degradation of DMFC MEAs in the presence of 4 ppm of chloride ions in the anode liquid was thus observed under real operating conditions in a real operating system, which was attributable to increased corrosion of ruthenium and platinum [47, 48]. In order to achieve long-term stability in cell operation, it is therefore necessary to prevent the electrochemically active areas from being contaminated by foreign ions. Such ions could be transported by an insufficiently filtered air supply, for example, or via other media flowing into the stack. Yasuda *et al.* [49], for instance, were able to demonstrate that, apart from other organic compounds, also metallic ions were transported via the methanol fuel, which thus led to a decrease in performance. The degradation was proportional to the input of metallic ions into the electrolyte membrane.

Since, in particular, the corrosion of noble metal catalysts and the input of foreign ions as a degradation effect contribute to performance losses in DMFCs, it is of particular importance to reduce the degradation effects described above by a suitable choice of operating conditions and by optimized materials. Only in this way is it possible to achieve the required operating time of 3000–5000 h [50] under real operating conditions. As yet, there have been very few examples of successful DMFC tests for several thousand operating hours under real operating conditions.

The best DMFC stacks and systems currently achieve lifetimes of at least 3000 h. SFC thus guarantees an operating life of 3000 h within 36 months for their commercial DMFC systems, although these systems have a maximum power of only 65 W. However, these 3000 h can also include the replacement of a stack [5].

3.2.3.3 Cost

Due to the expensive materials used and the small numbers produced, the cost of fuel cell systems is still very high in comparison with established technologies so that they are only used in a few niche applications with special additional benefits. This is particularly the case with DMFCs, since due to the increased use of noble metals (4 mg cm^{-2} per cell) in comparison to PEFCs (approx. 0.3 mg cm^{-2} per cell), the cost of the overall system is determined by the considerably higher stack cost. DMFCs can therefore only be used economically under certain boundary conditions. For example, as part of a market and feasibility study [51], it was shown that, in spite of the high system costs, hybridized fuel cell systems based on direct methanol fuel cells can potentially achieve a cost level which is comparable to that of conventional battery systems for certain applications with forklift trucks. The reason for this is the greater range of the DMFC systems and the fact that spare batteries are not needed for multi-shift operation, thus increasing ease of servicing. The production costs of the 1 kW DMFC hybrid system (Figure 3.7) are decisively influenced by two components: the MEA and the hybrid batteries [52]. The stack thus accounts for about 60% of the overall cost. In a comparable PEFC system, the stack cost would account for only 10% of the total. In development work on DMFCs, the major priority is therefore given to reducing the MEA cost since this amounts to about 85% of the cost of the stack at a specific power density of about 60 mW cm^{-2}. Possible approaches would be to reduce the amount of material used by employing more efficient MEAs with the same noble metal loading (Section 3.2.3.1), the use of platinum-free catalysts, or recycling the platinum from used stacks so that the major proportion of the investments could be recovered.

Although non-noble metal catalysts can be used in alkaline systems (Section 3.2.4), there is, even in the longer term, no alternative in sight for platinum(–ruthenium) catalysts with high noble metal loadings for methanol oxidation with H$^+$-conducting electrolytes in order to achieve good performance data. In contrast, above all on the basis of PEFC development, there are promising developments with highly active, in part methanol-tolerant, cathode catalysts (see also Section 3.2.3.1). Transition metals such as iron or cobalt and heteroatom-containing organic precursors such as polyaniline (PANI), polypyrrole (PPy), or cyanamide (CM) can, after carbonization, be used to produce active cathode catalysts, in which the electrochemically active center consists of a transition metal ion interacting with one or more pyrrolic or pyridinic nitrogen atoms in the interstices of the graphitic sheets [53, 54]. A PEFC-MEA has already been constructed with this type of cathode catalyst and it displayed better performance than a commercial MEA with a platinum catalyst [55].

Particularly from the cost aspect, it is necessary to recycle efficiently the (noble metal) catalysts used in MEAs at the end of their useful life. To this end, in the classical approach, after dismantling the stack, shredding the MEAs, and removing large pieces of the membrane in order to remove all organic components, the MEA material is incinerated and then the metals are recycled separately by means of various extraction, solution, and precipitation steps. Since incineration of the fluorinated membrane/ionomer material currently employed (e.g., Nafion) releases the aggressive and toxic hydrofluoric acid (HF), this process requires the use of special linings for the incinerators and also laborious scrubbing of the off-gas in order to separate the HF. The plant costs for catalyst recycling could possibly be considerably reduced by already binding the hydrofluoric acid in the solid by the addition of inorganic metallic oxides (such as CaO and thus the formation of CaF_2, which is poorly soluble and thermally stable) [56]. As an alternative to this pyrometallurgical route, leaching the catalyst out of the shredded MEA without a preliminary combustion step is being evaluated (hydrometallurgical procedure) [57].

Although catalyst recycling with a recovery of more than 95% of the noble metals is possible, due to the fluorine-containing polymers still used at present it is time consuming (because there are several stages) and cost intensive. As part of a feasibility study [51], it was shown that the stack cost and hence the cost of the overall system could be cut by 32% if the platinum was recycled after the stack had operated for 5 000 h and re-used as catalyst material.

3.2.4
Alkaline DMFCs

Due to the lack of a suitable anion-exchange membrane (AEM), for a long time the development of alkaline polymer membrane fuel cells lagged behind that of acid DMFCs although they appear attractive for a number of reasons: (i) methanol oxidation in an alkaline medium proceeds faster than in an acidic medium by an order of magnitude and thus, in principle, permits the application of non-platinum catalysts (anode Ni, Co, Au, Sn, Zn, or Pd; cathode Ag, Fe/Co/Ni), (ii) ion transport from the cathode to the anode counteracts fuel crossover, and (iii) the problem with the formation of crystalline carbonates familiar from alkaline fuel cells with liquid electrolytes does not occur if the addition of bases in the anode can be dispensed with.

Accordingly, development of this approach has accelerated significantly since Tokuyama and Solvay have now developed materials with long-term stability under operating conditions at 60–80 °C on the basis of polymers with quaternary ammonium groups and mobile hydroxide ions, and also since base-doped polybenzimidazole (PBI) membranes have come into use. At present, however, the performance data for alkaline DMFCs are considerably poorer than those for acidic DMFCs. Thus, for example, in spite of the use of Pt catalysts at 60 °C with a fluorinated AEM in air operation and without any alkaline solution in the anode, values of less than 10 mW cm^{-2} are achieved as peak performance [4]; an MEA test with a PBI membrane, 2 M KOH in the anode, oxygen operation at 90 °C, and slight overpressure also only reached $P_{max} = 30$ mW cm^{-2} [58].

The major obstacles to the development of alkaline DMFCs are, in particular, the comparatively low membrane conductivity of the AEM (as a consequence of the smaller degree of dissociation in comparison with Nafion and lower ion mobility of the hydroxide ions in comparison with protons), the fact that the development of ionomer solutions for electrode optimization [59] is still in its infancy, and experimental experience that the addition of a base to the anode solution (and thus the above-mentioned "carbonate problem") is necessary to achieve high power densities.

At present, it cannot yet be foreseen whether alkaline DMFCs will in the medium term equal or even improve on the performance of their acidic counterparts. In contrast, it is already possible to achieve high power densities with alkaline DEFCs (see the next section).

3.3
Direct Ethanol Fuel Cells

The complete oxidation of ethanol proceeds as a 12-electron process via several intermediates. Scheme 3.1 shows a simplified scheme. However, even at temperatures above 100 °C, the CO_2 yield is still only 20–40% [60]. Particularly active anode catalysts for an acidic medium have been identified in the Pt–Sn–Rh system. In this context, tin considerably increases the activity of ethanol oxidation whereas the rhodium component promotes splitting of the carbon–carbon bond [61]. The same catalysts as used in the acidic DMFC are employed here at the cathode. Although by using this combination of catalysts in oxygen operation at a slight overpressure a cell performance of 100 mW cm^{-2} can be obtained in a DEFC, the CO_2 yield, however, amounts to only 2% and instead largely acetic acid is formed as the dead-end product of ethanol oxidation [62].

Even higher power densities can be reached in alkaline DEFCs – also using platinum-free catalysts – but the CO_2 yield is almost zero in this case also (the acetate anion being the main product). In this way, at a cell temperature of

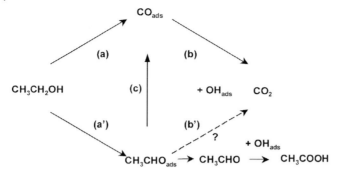

Scheme 3.1 Schematic mechanism of ethanol oxidation. Reproduced from Leger et al. [63] with kind permission of Elsevier.

80 °C in oxygen operation using an MEA with a Pd–(Ni–Zn)/C anode, AEM A006 from Tokuyama and an Fe–Co–HYPERMEC cathode catalyst from Acta, power densities of up to 170 mW cm^{-2} have been achieved [64]. Here also, 2 mol l^{-1} KOH was added to the anode solution (10 wt% ethanol in water) to maximize performance (see above for the "carbonate problem" arising from the addition of alkaline solution) [65].

In spite of the advantages of ethanol as a fuel (30% higher energy density than methanol, can be produced from biomass, nontoxic), the incomplete oxidation of ethanol to CO_2 (and thus a lower faradic efficiency) is regarded as a disqualifying criterion for the use of DEFCs in the kilowatts class. However, the very good performance data (especially in alkaline solutions) make DEFCs very promising for applications in the low power range. This also becomes apparent if present DEFC products and demonstrator systems are considered: (i) ISE Freiburg is developing a passive DEFC system with an acid electrolyte (present MEA performance: 12 mW cm^{-2} at 50 °C, 7 mW cm^{-2} under system-oriented operating conditions) [66], (ii) Horizon is commercially marketing a "Bio-Energy Discovery Kit" for teaching purposes [67], and (iii) NDCPower presented a 100 W stack at the 2009 Hannover Fair [68].

Alternative fuels as replacements for methanol and ethanol are regularly discussed in the literature. As in the case of ethanol, the lack of total oxidation prevents the use of all energy carriers with a C–C bond in fuel cells of the kilowatts class.

Among further liquid fuels without a C–C bond, formic acid (HCOOH) fuel cells – in spite of a high thermodynamic potential ($E_0 = 1.40$ V) and remarkable MEA performance or system efficiency [69] – do not seem to be of interest for large-scale applications due to the low volumetric power density of formic acid (only 40% of that of methanol). The status of the development of direct dimethyl ether (H_3C–O–CH_3) fuel cells has recently been reviewed by Serov und Kwak [70]. With hydrazine (H_2N–NH_2), impressive cell power can be achieved with a very high energy density of the fuel (12% higher than methanol) and emission-free operation [71]. However, its application seems doubtful in view of the toxicity of hydrazine, the energy required for its production, and also the fact that present fuel cell catalysts also catalyze its decomposition in the anode compartment.

3.4
Conclusion

Direct liquid fuel cells are attractive for different applications, above all as replacements for batteries or accumulators. They can be used in various power classes up to about 5 kW. Important and impressive progress has been made with this technology due to intensive development work worldwide. The first commercial products for applications in the leisure sector have been on the market for several years. However, the properties of the materials available on the market for use in DMFC systems are not yet optimal. Further research and development work is therefore urgently required in order to realize direct liquid fuel cells and systems with high efficiencies and power densities and also with the necessary long-term stability. Only in this way can the great potential of this technology be fully exploited.

References

1 Reshetenko, T.V., Kim, H.-T., Lee, H., Jang, M., and Kweon, H.-J. (2006) *Journal of Power Sources*, **160** (2), 925–932.
2 Kim, H.-T., You, D.J., Yoon, H.-K., Joo, S.H., Pak, C., Chang, H., and Song, I.-S. (2008) *Journal of Power Sources*, **180** (2), 724–732.
3 Gasteiger, H.A. and Garche, J. (2008) In *Handbook of Heterogenous Catalysis*, Vol. 6 (eds G. Ertl, H. Knönzinger, F. Schütt, and J. Weitkamp), Wiley-VCH Verlag GmbH, Weinheim, pp. 3081–3121.
4 Scott, K., Yu, E., Vlachogiannopoulos, G., Shivara, M., and Duteanu, N. (2008) *Journal of Power Sources*, **175** (1), 452–457.
5 SFC Smart Fuel Cell AG (2009) http://www.sfc.com (accessed 6 October 2009).
6 Rashidi, R., Dincer, I., and Berg, P. (2009) *Journal of Power Sources*, **187** (2), 509–516.
7 Qian, W., Wilkinson, D.P., Shen, J., Wang, H., and Zhang, J. (2006) *Journal of Power Sources*, **154** (1), 202–213.
8 Eccarius, S., Krause, F. Beard, K., and Agert, C. (2008) *Journal of Power Sources*, **182** (2), 565–579.
9 Ren, X.M., Gottesfeld, S.B., and Hisch, R.S. (2004) Fluid management component for use in a fuel cell, US Patent Application 2004/0062980A1.
10 Ren, X.M., Kovacs, F.W., Shufon, K.J., and Gottesfeld, S. (2004) Passive water management techniques in direct methanol fuel cells, US Patent Application 2004/0209154A1.
11 Toshiba (2009) *News Release*, 22 October 2009, Toshiba Corporation, Tokyo, http://www.toshiba.co.jp/about/press/2009_10/pr2201.htm (accessed 22 October 2009).
12 Butler, J. (2009) Portable fuel cell survey 2009, *Fuel Cell Today*, April 2009, http://www.fuelcelltoday.com/media/pdf/surveys/2009-portable-free.pdf (accessed 23 October 2009).
13 *Samsung SDI develops military portable DMFC*, Fuel Cells Bulletin, 2009, (6), 6–7.

14 Mergel, J., Müller, M., Janssen, H., and Stolten, D. (2008) In *Tagungsband 4. Deutscher Wasserstoff Congress 2008*, Schriften des Forschungszentrums Jülich, Reihe Energie und Umwelt, Vol. 12, pp. 81–93.

15 Aricò, A. S., Baglio, V., and Antonucci, V. (2009) In *Electrocatalysis of Direct Methanol Fuel Cells* (eds H. Liu and J. Zhang), Wiley-VCH Verlag GmbH, Weinheim, pp. 1–78.

16 Bandi, A. and Specht, M. Calculated from lower heating value of methanol = 5.47 kW h/kg. In *Landolt-Börnstein Group VIII, Vol. 3C, Renewable Energy*, pp. 478ff.

17 Kulikovsky, A. A. (2005) *Journal of the Electrochemical Society*, **152** (6), A1121–A1127.

18 Nölke, N. (2007) *Entwicklung eines Direkt-Methanol-Brennstoffzellensystems der Leistungsklasse kleiner 5 kW*, Schriften des Forschungszentrums Jülich, Reihe Energietechnik, Vol. 64, p. 59.

19 Ren, X. and Gottesfeld, S. (2001) *Journal of the Electrochemical Society*, **148** (1), A87–A93.

20 Beattie, P. D., Orfino, F. P., Basura, V. I., Zychowska, K., Ding, J., Chuy, C., Schmeisser, J., and Holdcroft, S. (2001) *Journal of Electroanalytical Chemistry*, **503** (1–2), 45–56.

21 Yamabe, M., Akiyama, K., Akasuta, Y., and Kato, M. (2000) *European Polymer Journal*, **36** (5), 1034–1041.

22 Staiti, P., Aricò, A. S., Baglio, V., Lufrano, F., Passalacqua, E., and Antonucci, V. (2001) *Solid State Ionics* **145** (1–4), 101–107.

23 Yu, J., Yi, B., Xing, D., Liu, F., Shao, Z., Fu, Y., and Zhang, H. (2003) *Physical Chemistry Chemical Physics*, **5** (3), 611–615.

24 Kim, D. S., Guiver, M. D., and Kim, Y. S. (2009) Proton exchange membranes for direct methanol fuel cells. In *Electrocatalysis of Direct Methanol Fuel Cells* (eds H. Liu and J. Zhang), Wiley-VCH Verlag GmbH, Weinheim, pp. 379–416.

25 Liu, F., Lu, G., and Wang, C. Y. (2006) *Journal of the Electrochemical Society*, **153** (3), A543-A553.

26 Xu, C. Zhao, T. S. and He, Y. L. (2007) *Journal of Power Sources*, **171** (2), 268–274.

27 Lamy, C., Coutanceau, C., and Alonso-Vante, N. (2009) Methanol-tolerant cathode catalysts for DMFC. In *Electrocatalysis of Direct Methanol Fuel Cells* (eds H. Liu and J. Zhang), Wiley-VCH Verlag GmbH, Weinheim, pp. 257–314.

28 Cooper, J. S. and McGinn, P. J. (2006) *Journal of Power Sources*, **163** (1), 330–338.

29 Wang, C. H., Chen, L.-C., and Chen, K.-H. (2009) Carbon nanotube-supported catalysts for the direct methanol fuel cell. In *Electrocatalysis of Direct Methanol Fuel Cells* (eds H. Liu and J. Zhang), Wiley-VCH Verlag GmbH, Weinheim, pp. 315–354.

30 Prabhuram, J., Zhao, T. S., Tang, Z. K., Chen, R., and Liang, Z. X. (2006) *Journal of Physical Chemistry B*, **110** (11), 5245–5252.

31 Liu, H., Song, C., Zhang, L., Zhang, J., Wang, H., and Wilkinson, D. P. (2006) *Journal of Power Sources*, **155** (2), 95–110.

32 Steigerwalt, E. S., Deluga, G. A., and Lukehart, C. M. (2002) *Journal of Physical Chemistry B*, **106** (4), 760–766.

33 Gan, L., Lu, R., Du, H., Li, B., and Kang, F. (2009) *Electrochemistry Communications*, **11** (2), 355–358.

34 Li, L., and Xing, Y. (2006) *Journal of the Electrochemical Society*, **153** (10), A1823–A1828.

35 Bayer MaterialScience (2008) *Baytubes* brochure, www.baytubes.com/downloads/bms_cnt_baytubes_en.pdf (accessed 6 November 2009).

36 Carbon Solutions, list of available products, www.carbonsolution.com/products/products.html (accessed 6 November 2009).

37 Shao, Y., Liu, J., Wang, Y., and Lin, Y. (2009) *Journal of Materials Chemistry*, **19** (1), 46–59.

38 Zelenay, P. (2006) *ECS Transactions*, **1** (8), 483–495.

39 Cha, H.-C., Chen, C.-Y., and Shiu, J.-Y. (2009) *Journal of Power Sources* **192** (2), 451–456.

40 Knights, S.D., Colbow, K.M., St-Pierre, J., and Wilkinson, D.P. (2004) *Journal of Power Sources*, **127**, (1–2), 127–134.
41 Piela, P., Eickes, C., Brosha, E., Garzon, F., and Zelenay, P. (2004) *Journal of the Electrochemical Society*, **151** (12), A2053-A2059.
42 Cheng, C.X., Peng, C., You, M., Liu, L., Zhang, Y., and Fan Q. (2006) *Electrochimica Acta*, **51** (22), 4620–4625.
43 Lai, C.-M., Lin, J.-C., Hsueh, K.-L., Hwang, C.-P., Tsay, K.-C., Tsai, L.-D., and Peng Y.M. (2008) *Journal of the Electrochemical Society*, **155** (8), B843–B851.
44 Sarma, L.S., Chena, C.-H., Wang, G.-R., Hsueh, K.L., Huang, C.-P., Sheu, H.-S., Liu, D.-G., Lee, J.-F., and Hwang, B.-J. (2007) *Journal of Power Sources*, **167** (2), 358–365.
45 Kim, Y.S. and Pivovar, B. (2006) *ECS Transactions*, **1** (8), 457.
46 Jiang, R., Rong, C., and Chu, D. (2007) *Journal of the Electrochemical Society*, **154** (1), B13–B19.
47 Forschungszentrum Jülich (2008) *IEF-3 Report 2007*, Forschungszentrum Jülich, Energy Technology, Vol. 63, p. 47.
48 Steinberger-Wilkens, R., Mergel, J., Glüsen, A., Wippermann, K., Vinke, I., Batfalsky, P., and Smith, M.J. (2008) Performance degradation and failure mechanisms of fuel cell materials. In *Materials for Fuel Cells* (ed. M. Gasik), Woodhead Publishing Great Abington, Chapter 11, pp. 425–465.
49 Yasuda, K., Nakano, Y., and Goto, Y. (2007) *ECS Transactions*, **5** (1), 291–296.
50 US Department of Energy (2007), *Fuel Cell Technologies Program, Multi-Year Research, Development and Demonstration Plan*, pp. 3.4–19, http://www1.eere.energy.gov/hydrogenandfuelcells/mypp/pdfs/fuel_cells.pdf (accessed 13 November 2009).
51 Management Engineers (2007) *Feasibility Study Direct Methanol Fuel Cells*, Report, Forschungszentrum Jülich, IEF-3.
52 Werhahn, J. (2008) *Kosten von Brennstoffzellensystemen auf Massenbasis in Abhängigkeit von der Absatzmenge*, Schriften des Forschungszentrums Jülich, Reihe Energie und Umwelt, Vol. 35.
53 Zelenay, P. (2009) Paper presented at the DOE Hydrogen Program 2009 Annual Merit Review, Arlington, VA, 18–22 May 2009, www.hydrogen.energy.gov/pdfs/review09/fc_21_zelenay.pdf (accessed 6 November 2009).
54 Birry, L. and Dodelet, J.-P. (2009) 215th ECS Meeting, San Francisco, CA, 24–29 May 2009, Abstract 493, http://ecsmeet7.peerx-press.org/ms_files/ecsmeet7/2008/11/24/00000245/00/245_0_art_0_k55f0q.pdf (accessed 13 November 2009).
55 Lefevre, M., Proietti, E., Jaouen, F., and Dodelet, J.-P. (2009) *Science* **324** (5923), 71–74.
56 Zuber, R., Privette, R., Seitz, K. Fehl, K., and Hagelüken, C. (2004) Paper presented at the Fuel Cell Seminar 2004, San Antonio, TX, 1–5 November 2004, www.preciousmetals.umicore.com/publications/presentations/fuel_cells/show_FCSeminar_2004.pdf (accessed 6 November 2009).
57 Shore, L. (2009) Paper presented at the DOE Hydrogen Program 2009 Annual Merit Review, Arlington, VA, 18–22 May 2009, www.hydrogen.energy.gov/pdfs/review09/fc_33_shore.pdf (accessed 6 November 2009).
58 Hou, H., Sun, G., He, R., Sun, B., Jin, W., Liu, H., and Xin, Q. (2008) *International Journal of Hydrogen Energy*, **33** (23), 7172–7176.
59 Varcoe, J.R., Slade, R.C.T., and Yee, E.L.H. (2006) *Chemical Communications*, (13), 1428–1429.
60 Wang, J., Wasmus, S., and Savinell, R.F. (1995) *Journal of the Electrochemical Society*, **142** (12), 4218–4224.
61 Antolini, E. (2007) *Journal of Power Sources*, **170** (1), 1–12.
62 Wang, Q., Sun, G.Q., Cao, L., Jiang, L.H., Wang, G.X., Wang, S.L., Yang, S.H., and Xin, Q. (2008) *Journal of Power Sources*, **177** (1), 142–147.
63 Leger, J.-M., Rousseau, S., Coutanceau, C., Hahn, F., and Lamy, C. (2005) *Electrochimica Acta*, **50** (25–26), 5118–5125.
64 Acta, SpA, list of products (2009) www.acta-nanotech.com (accessed 6 November 2009).

65 Bianchini, C., Bambagioni, V., Filippi, J., Marchionni, A., Vizza, F., Bert, P., and Tampucci, A. (2009) *Electrochemistry Communications*, **11** (5), 1077–1080.

66 Fraunhofer ISE (2008) *Annual Report 2008*, www.ise.fraunhofer.de/publications/information-material/annual-reports/fraunhofer-ise-annual-report-2008 (accessed 6 November 2009).

67 Horizon Fuel Cell Technologies, www.horizonfuelcell.com/store/bioenergy.htm (accessed 6 November 2009).

68 NDCPower, www.ndcpower.com/?english/products/eos (accessed 6 November 2009).

69 Choi, J.-H., Jeong, K.-J., Dong, Y., Han, J., Lim, T.-H., Lee, J.-S., and Sung, Y.-E. (2006) *Journal of Power Sources*, **163** (1), 71–75.

70 Serov, A. and Kwak, C. (2009) *Applied Catalysis B*, **91** (1–2), 1–10.

71 Asazawa, K., Yamada, K., Tanaka, H., Oka, A., Taniguchi, M., and Kobayashi, T. (2007) *Angewandte Chemie International Edition*, **46** (42), 8024–8027.

4
High-Temperature Fuel Cell Technology

Gael P. G. Corre and John T. S. Irvine

Abstract

This chapter reviews high temperature fuel cell technology. The two main concepts are the Solid Oxide Fuel Cell (SOFC) and the molten carbonate Fuel Cell (MCFC). SOFCs feature a solid oxide electrolyte, and operate in the range 600–1000 °C, while MCFCs feature a molten carbonate electrolyte, and operate in the range 500–700 °C. Due to their high temperatures, such fuel cells offer different characteristics, applications, requirements and advantages to that of low temperature fuel cells. Key advantages over low temperature fuel cells are the fuel flexibility, offering the possibility to operate with practical fuels, and an improved overall efficiency.

After fuel cells principles and characteristics have been introduced, the SOFC technology is reviewed. The history, applications and materials used in SOFCs are presented. An overview of MCFCs is given as well. Finally, a study of fuel cell thermodynamics and efficiency is provided.

Keywords: high-temperature fuel cells, solid oxide fuel cell, molten carbonate fuel cell, hydrogen fuel, thermodynamics

4.1
Introduction

4.1.1
Principle

A fuel cell is an energy conversion device that produces electricity, and heat, by electrochemical combination of a fuel with an oxidant, which can be viewed as a battery with an external fuel supply. A fuel cell consists of four components: two electrodes, the *anode* and the *cathode*, separated by an *electrolyte*, and connected by an external circuit or *interconnect*, as shown in Figure 4.1. Fuel is fed to the anode, where it is oxidized, releasing electrons to the external circuit. Oxidant is fed to the cathode, where it is reduced using the electrons delivered by

Hydrogen Energy. Edited by Detlef Stolten
Copyright © 2010 WILEY-VCH Verlag GmbH & Co. KGaA, Weinheim
ISBN: 978-3-527-32711-9

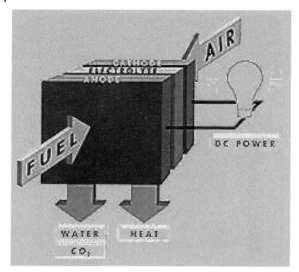

Fig. 4.1 Principle scheme of an individual fuel cell [1].

the external circuit. The electrons flow through the interconnect, from the anode to the cathode, producing direct current electricity.

In theory, any gases capable of electrochemical oxidation and reduction can be used as fuel and oxidant in a fuel cell. Oxygen is the most common oxidant for fuel cells since it is readily and economically available from air. Hydrogen, which offers high electrochemical reactivity, is the most common fuel. However, fuel cells can be developed to work with alternative fuels to hydrogen.

The key feature of a fuel cell is its high energy conversion efficiency. Because fuel cells convert the chemical energy of the fuel directly to electrical energy without the intermediate of thermal energy (unlike indirect conversion in conventional systems), their conversion efficiency is not subject to the Carnot limitation. Further energy gains can be achieved when the heat produced is used in combined heat and power, or gas turbine applications. The improved efficiency as compared with conventional energy conversion devices is the main reason why fuel cells are attracting considerable attention. In addition to their high energy conversion efficiency, fuel cells offer several additional advantages over conventional methods of power generation. They offer a much lower production of pollutants. A fuel cell fuelled with H_2 and air produces only water. Other significant advantages offered are modular construction and size flexibility, which make them well suited for decentralized applications, high efficiency at part load, fuel flexibility, and remote/unattended operation. Moreover, their vibration-free operation eliminates noise usually associated with conventional power generation systems. More details concerning general features of fuel cells can be found in the literature [1, 2].

4.1.2
History

The fuel cell concept, ascribed to Sir Humphrey Davy, dates from the beginning of the 19th century. The first hydrogen–oxygen cell was successfully operated by Sir William Grove in 1839 [3] and is generally referred to as the first fuel cell. While investigating the electrolysis of water, Grove observed that when the current was switched off, a small current flowed through the circuit in the opposite direction, as a result of a reaction between the electrolysis products, hydrogen and oxygen, catalyzed by the platinum electrodes. Grove recognized the possibility of combining several of these in series to form a gaseous voltaic battery [4], and also made the crucially important observation that there must be a "notable surface of action" between the gas, the electrolyte, and the electrode phases in a cell. Maximizing the area of contact between these three phases remains at the forefront of fuel cell research and development. Some 50 years after Grove's "gas battery", Mond and Langer introduced the term fuel cell [5] to describe their device, which had a porous platinum black electrode structure and used a diaphragm made of a porous non-conducting substance to hold the electrolyte.

Despite the fact that the fuel cell was discovered over 160 years ago, and the high efficiencies and environmental advantages offered, it is only now that fuel cells are approaching commercial reality.

4.1.3
Different Types of Fuel Cells

There is a wide range of fuel cells in different stages of development. Although all types of fuel cells have the same basic operating principle, they have different characteristics that stem from the nature of electrolyte involved. The five main types are summarized in Table 4.1. The nature of the electrolyte dictates the nature of ions transferred and the direction of this transport, which in turn determines on which side of the electrolyte water is produced. Moreover, each electrolyte requires to be operated in a specific temperature range, which is a major difference in characteristics between different types of fuel cells. The

Table 4.1 Different types of fuel cells and characteristics [1].

Type	Temperature (°C)	Fuel	Electrolyte	Charge carrier
Polymer electrolyte membrane (PEM)	70–110	H_2, methanol	Sulfonated polymers	$(H_2O)_n H^+$
Alkali fuel cell (AFC)	100–250	H_2	Aqueous KOH	OH^-
Phosphoric acid fuel cell (PAFC)	150–250	H_2	H_3PO_4	H^+
Molten carbonate fuel cell (MCFC)	500–700	H_2, hydrocarbons, CO	$(Li, K)_2CO_3$	CO_3^{2-}
Solid oxide fuel cell (SOFC)	600–1000	H_2, CO, hydrocarbons, alcohols	$(Zr, Y)O_{2-\delta}$	O^{2-}

molten carbonate fuel cells (MCFC) and solid oxide fuel cell (SOFC) have elevated operating temperatures, compared with much lower operating temperatures for alkaline (AFC), polymer electrolyte membrane (PEMFC), and phosphoric acid (PAFC) fuel cells. The operating temperature dictates in turn the physicochemical and thermomechanical properties of materials to be used as cell component and also the type of fuel on which the cell can be operated. Moreover, this difference in operating temperatures has a number of implications for the applications for which particular fuel cell types are most suited.

Large differences exist in application, design, size, cost, and operating range for the different type of fuel cells. Of the available fuel cell technologies, the PEFC and SOFC are thought to have the most potential to achieve cost and efficiency targets for widespread use in power generation, and have been the most investigated types.

In general, high-temperature fuel cells exhibit higher efficiency and are less sensitive to fuel composition. PEMFC systems require a pure H_2 fuel stream because the precious metal anode catalysts are poisoned by even low levels of CO or other compounds such as those containing sulfur. Current PEFC systems operate below 100 °C, but there is a great deal of ongoing research to find polymer electrolytes that can operate at higher temperatures, since increasing the operating temperature relaxes the fuel purity requirements relative to catalyst poisoning. In contrast, due to their high operating temperature, CO is a fuel rather than a poison for SOFCs. Hence high-temperature fuel cells can be operated on fuels other than H_2.

4.1.4
Applications

The potential applications of fuel cells in society are ever increasing, driven by the various benefits that the implementation of fuel cells would bring over current technologies, such as environmental and efficiency improvements. Applications being considered range all the way from very small scale, requiring only a few watts, to larger scale distributed power generation of hundreds of megawatts.

The small-scale power supply is a well-suited market for fuel cells. Indeed, fuel cells offer significantly higher power densities than batteries, in addition to being smaller and lighter and having much longer lifetimes. Hence an increasing number of applications are emerging where only a few watts are required, such as palm and laptop computers, mobile phones, and other portable electronic devices.

Their potential high reliability and low maintenance coupled with their quiet operation and modular nature make fuel cells well suited to localized "off-grid" power generation, for either high-quality uninterrupted power supplies or remote applications. High-temperature fuel cells (MCFCs and SOFCs) are suitable for continuous power production, where the cell temperature can be maintained. If the heat released is used to drive a gas turbine to produce extra energy, the system efficiency can be increased to levels as high as 80%, significantly higher than any conventional electricity generation process. Moreover, the heat produced makes SOFCs particularly suited to combined heat and power (CHP) applications ranging from less than 1 kW to several megawatts, which covers individual households, larger residential units and businesses, and industrial premises, providing all the power and hot water from a single system.

The combination of their high efficiency and significantly reduced emissions of pollutants means that fuel cell-powered vehicles are a very attractive proposition, especially in heavily populated urban areas. The efficiency is to be compared with about 20% for a combustion engine. Low-temperature fuel cells, in particular the PEMFC, are the most suited to transport applications, because of the need for only a short warm-up. The concept of a fuel cell-powered vehicle running on hydrogen, the so-called "zero emission vehicle", is a very attractive one and is currently an area of intense activity for almost all the major motor manufacturers. As an example, fuel cell-powered buses running on compressed hydrogen are successfully operated in several cities around the world.

4.2
Solid Oxide Fuel Cell

4.2.1
General Considerations

A SOFC is defined by its solid ceramic electrolyte, which is a nonporous metal oxide. Such electrolytes are oxygen ion (O^{2-}) conductors, impervious to gas flow, and have negligible electronic conductivity. Solid oxide electrolytes require a high operating temperature to display suitable conductivities, typically in the range 700–1000 °C, which has a number of consequences for SOFC operation [6]. SOFCs involve multiple complex physicochemical processes. The principle of an SOFC, involving hydrogen as a fuel, is illustrated in Figure 4.2.

Oxygen is electrochemically reduced at the cathode–electrolyte–gas interface. Electrons are delivered to the cathode through the interconnect, where they react with oxygen molecules in the gas phase to deliver oxygen ions to the electrolyte via a charge-transfer reaction:

$$\frac{1}{2}O_2(g) + 2e^-(c) \rightleftharpoons O^{2-}(e) \tag{4.1}$$

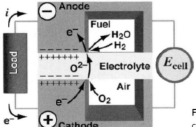

Fig. 4.2 Principle of a solid oxide fuel cell involving H_2 as a fuel [7].

where the three phases are denoted (g) for the gas, (c) for the cathode and (e) for the electrolyte. Oxygen ions migrate through the electrolyte via a vacancy hoping mechanism towards the anode–electrolyte–fuel interface, where they participate in the fuel oxidation, written as follows for hydrogen:

$$H_2(g) + O^{2-}(e) \rightleftharpoons H_2O(g) + 2e^-(a) \tag{4.2}$$

where (a) denotes the anode. The hydrogen in the gas phase reacts with the oxygen ions provided by the electrolyte to deliver electrons to the anode. Provided that a load is connected between the anode and the cathode, the electrons from the anode will flow through the load back to the cathode, and electric current will flow through the circuit.

The overall electrochemical cell reaction in a SOFC, based on oxygen and hydrogen, is written as follows:

$$O_2 + H_2 \rightleftharpoons H_2O \tag{4.3}$$

Another accurate way to represent the overall reaction occurring in the cell, regardless of the fuel involved, is to describe the oxygen transfer:

$$O_2(c) \rightleftharpoons O_2(a) \tag{4.4}$$

An SOFC can therefore be considered as an oxygen pump. The amount of oxygen transported to the anode will depend on the type of fuel used and the reactions occurring at the anode.

4.2.2
History of the SOFC

The SOFC was first conceived following the discovery of solid oxide electrolytes in 1899 by Nernst [8]. Nernst discovered that the very high electrical resistance of pure solid oxides could be greatly reduced by the addition of certain other oxides. The most promising of these mixtures consisted mainly of zirconia (ZrO_2) with small amounts of added yttria (Y_2O_3). This is still the most widely used electrolyte material in the SOFC.

The first working SOFC was demonstrated by Baur and Preis in 1937, using stabilized zirconia as electrolyte and coke and magnetite respectively as a fuel and oxidant [9]. The current produced by their cell was too low for any practical purposes, but the possibility of operated SOFCs had been demonstrated. Unfortunately, the high operating temperature and the reducing nature of the fuel led to serious materials problems and, despite very significant efforts by Baur and other researchers, the search for suitable materials was unsuccessful.

This effectively hindered the development of the SOFCs until the 1960s, when a first period of intensive activity in SOFC development began. Intensive research programs that were driven by new energy needs mainly for military, space, and transport applications addressed mainly the electrolyte conductivity improvement and the first steps in SOFC technology. A second period of intense activity began in the mid 1980s and still continues today. These research efforts have led SOFC commercialization close to reality. Different companies have developed different concepts, and several demonstration units have been operated for significant periods of time.

4.2.3
Characteristics

The advantages of the high operating temperature include the possibility of running directly on practical hydrocarbon fuels without the need for a complex and expensive external fuel reformer and purification systems. Internal reforming can be performed at high temperatures and SOFCs are not poisoned by CO, which can be oxidized at the anode and act as a fuel. When practical fuels are used, the environmental impact is better than for combustion technologies, in the sense that less CO_2 and NO_x are produced per unit of power generated. Looking at the overall system efficiency, the high-quality exhaust heat released during operation can be used as a valuable energy source, either to drive a gas turbine when pressurized or for CHP applications.

Originally, SOFCs were developed for operation primarily in the temperature range 900–1000 °C, which is beneficial for the fuel reforming, electrochemistry kinetics, and the added value of the exhaust heat. However, some important drawbacks stem from such elevated temperatures. The materials that can be used are limited with respect to their chemical stability in oxidizing and/or reducing environments and their chemical and thermomechanical compatibility with adjacent components. Hence considerable efforts are being made to lower the operating temperature by 200 °C or more, which would allow the use of a broader set of materials, with less demands on seals and balance-of-plant components, simplifying thermal management, aiding in faster start-up and cool-down, and resulting in less degradation of cell and stack components [2]. Because of these advantages, activity in the development of SOFCs capable of operating in the temperature range 600–800 °C has increased dramatically in the last few years. However, at lower temperatures, electrolyte conductivity and electrode kinetics decrease significantly.

In terms of applications, the length of time that is generally required to heat up and cool down the system restricts the use of SOFCs in applications that require rapid temperature fluctuations. This is a consequence of the need to use a relatively weak, brittle component as the substrate material and because of problems associated with thermal expansion mismatches. This restriction applies particularly for transport applications, where rapid transport start-up is essential.

Reducing the cost of SOFCs is a crucial issue for their commercialization. Currently, the high cost-to-performance ratio limits the introduction SOFC in the energy market. In this respect, lower operating temperatures also make pos-

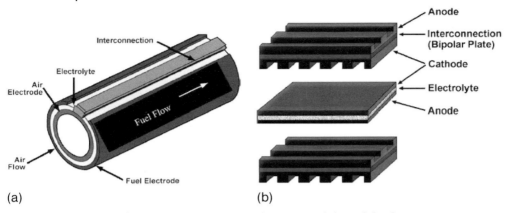

Fig. 4.3 The two most common SOFC designs: (a) tubular and (b) planar [2].

sible the use of inexpensive metallic interconnections in place of lanthanum chromite-based ceramic interconnections.

4.2.4
Design

The solid-state character of all SOFC components means that, in principle, there is no restriction on the cell configuration. Instead, it is possible to shape the cell according to criteria such as overcoming design or applications issues. As for other fuel cell concepts, it is necessary to stack SOFCs to increase the voltage and the power produced. A stack can in principle comprise any number of cells depending on the desired power, and a fuel cell plant can be designed in modules of stacks in series and parallel connections. To construct an electric generator, individual cells are connected in both electrical parallel and series to form a semi-rigid bundle that becomes the basic building block of a generator.

The two most common designs of SOFCs, the tubular and the planar, are represented in Figure 4.3. In the tubular cells, the cell components are deposited in the form of thin layers on a cathode tube. In the planar design, the cell components are configured as thin, flat plates. The interconnection, which is ribbed on both sides, forms gas flow channels and serves as a current conductor.

Alternative designs have been proposed, such as the planar segmented design developed at Rolls Royce [10] and the SOFC roll developed at the University of St. Andrews [11]. The latter is an innovative design that takes advantage of both planar and tubular designs.

4.2.5
Materials for SOFCs

4.2.5.1 **Electrolytes**

SOFC electrolyte materials were recently reviewed by Goodenough [12] and Skinner and Kilner [13], and are detailed in reviews addressing SOFCs [6, 14]. Much of the research carried out on SOFCs in the 1960s focused on optimizing the ionic conductivity of the solid electrolyte. Research on electrolyte materials is still active today, in an effort to lower operating temperatures.

The most commonly used electrolyte in SOFCs is yttria-stabilized zirconia (YSZ). This material possesses an adequate level of oxygen ion conductivity and exhibits desirable stability in both oxidizing and reducing atmospheres. Stabilized zirconia is usually unreactive towards other components used in the SOFC and has negligible electronic conductivity. It is also abundant, relatively inexpensive, and mechanically strong while being easy to fabricate. The properties of stabilized zirconia have been extensively studied and several reviews dedicated to this material have been published [15–17]. The most commonly used stabilizing oxides or dopants are CaO, MgO, Y_2O_3, Sc_2O_3 and certain rare earth oxides. These oxides exhibit relatively high solubility in ZrO_2, which is stable over wide ranges of composition and temperature.

ZrO_2, in its pure from, does not serve as a good electrolyte because its ionic conductivity is too low. The addition of certain aliovalent oxides stabilizes the cubic fluorite structure of ZrO_2 from room temperature to its melting point and, at the same time, creates a high concentration of oxygen vacancies by charge compensation according to the following equation (written for Y_2O_3 stabilization using Kroger–Vink notation):

$$Y_2O_3 \rightarrow 2\,Y'_{Zr} + V_O^{\cdot\cdot} + 3\,O_O^x \tag{4.5}$$

The high oxygen vacancy concentration gives rise to high oxygen ion mobility and leads to an extended oxygen partial pressure range of ionic conduction, making stabilized zirconia suitable for use as an electrolyte in SOFCs (the oxygen partial pressure range covers the conditions 1–10^{-18} atm) to which a SOFC electrolyte is exposed in the fuel cell during operation. Oxygen ion conduction takes place in stabilized ZrO_2 by movement of oxygen ions via vacancy sites. It is generally found that that the ionic conductivity is a maximum near the minimum level of dopant oxide required to stabilize the cubic phase fully. At higher dopant levels, the ionic conductivity decreases. Typically, the level of Y_2O_3 present in YSZ is around 8 mol%.

The operating temperature is principally governed by the nature of the electrolyte, that is, its ionic conductivity, and the thickness of the electrolyte layer. Indeed, the oxygen ionic conductivity for electrolyte materials is usually expressed as

$$\sigma = \sigma_0 T^{-1} \exp(-E_{el}/RT) \tag{4.6}$$

For YSZ, $\sigma_0 \approx 3.6 \times 10^5$ S K cm^{-1} and $E_{el} \approx 8 \times 10^4$ J mol^{-1} [18].

Therefore, increasing the operating temperature or making thinner electrolyte layers can reduce the resistivity to oxygen ion movement. Conventional zirconia-based SOFCs generally require an operating temperature above 850 °C. This high operating temperature places severe demands on the material used as interconnects and for manifolding and sealing, and necessitates the use of expensive ceramic materials and specialist metal alloys. There is therefore considerable interest in lowering the operating temperature of SOFCs to below 750 °C to permit the use of cheaper materials, such as stainless steel, and reduce fabrication costs, whilst maintaining high power outputs. Reducing the electrolyte thickness will obviously allow a reduction of the operating temperature, but this approach is of course limited. The alternative route consists in developing new electrolyte materials showing higher conductivity than doped zirconia.

The search for, and study of, alternative solid electrolyte materials has been an active research area for many years. Figure 4.4 presents conductivity plots of different electrolyte materials that have been developed [14]. Among those various materials, two promising alternative electrolytes to YSZ are gadolinia-doped ceria [19, 20] and lanthanum gallate-based perovskites [21]. Both of these electrolytes offer the possibility of lower temperature operation for SOFCs between 500 and 700 °C. Scandia-doped zirconia has also attracted particular attention, since it has similar properties to YSZ but exhibits higher ionic conductivities, although it is also more expensive.

Gadolinia-doped ceria (CGO) offers an ionic conductivity substantially higher than that of YSZ. However, at elevated temperatures in a reducing atmosphere,

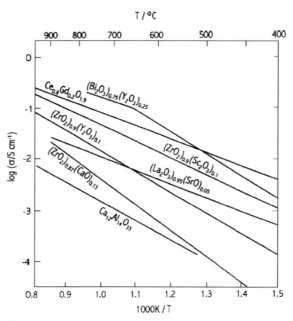

Fig. 4.4 Specific conductivities of selected solid oxide electrolytes [14].

such as that present at the anode, ceria undergoes partial reduction to Ce^{3+}, which introduces electronic conductivity that significantly lowers the efficiency of the SOFC, and also an undesirable structural change. Considerable effort has been devoted to minimizing the electronic conductivity of doped ceria under reducing conditions. One solution is to use an additional ultra-thin interfacial electrolyte layer which prevents electronic transport, and can suppress the reduction of ceria under reducing conditions.

Increased conductivity can be obtained from lanthanum gallate ($LaGaO_3$) by replacing both the trivalent lanthanum and gallium with divalent cations, generally strontium and magnesium [19]. A favored composition in terms of ionic conductivity is $La_{0.9}Sr_{0.1}Ga_{0.8}Mg_{0.2}O_{2.85}$. The ionic conductivity of LSGM, although significantly higher than that of YSZ, is slightly less than that of CGO at 500 °C. However, the potential range of operating temperatures of LSGM is greater than that of CGO because it does not suffer from the problems exhibited by CGO at higher temperatures associated with electronic conduction. Hence there is interest in the possibility of using LSGM at temperatures around 600–700 °C, which are currently too low to obtain adequate power densities with zirconia-based SOFCs. However, there are problems associated with the stability of certain compositions of LSGM. It has also proved difficult to prepare pure single-phase electrolytes of LSGM, additional phases raising doubts over the long-term durability of SOFCs with LSGM electrolytes, and much further research into this material is still required.

4.2.5.2 Electrodes

General Requirements
Electrodes for SOFC must fulfill some important requirements to ensure high and durable power output. Electrodes serve to provide sites for electrochemical oxidation of the fuel (anode) and reduction of the oxidant (cathode). Since electrochemical reactions in a SOFC involve gaseous reactants, electrons and oxygen ions (Equations 4.1 and 4.2), reactions will occur only on sites possessing conductivities for those three phases. Those active sites, located at electrode–electrolyte–gas three-phase interface as depicted in Figure 4.5, are commonly referred to as the triple phase boundary (TPB). Practical systems must be designed with an extended active surface area. To extend the TPB area, electrodes are fabricated as mixed ionic and electronic conductor (MIEC) porous ceramics or ceramic–metallic composites. The electrode microstructure provides interpenetrating, continuous three-dimensional electron, ion, and gas transport networks. The pore, metal, and ceramic phase typically occupy one-third of the volume. An ideal microstructure would offer the highest TPB length for electrochemical reactions and optimized contact between the electrolyte and the anode, and be dimensionally stable during operation.

Chemical and thermal compatibility between electrode materials and the adjacent components is essential. No solid-state reaction should occur between the

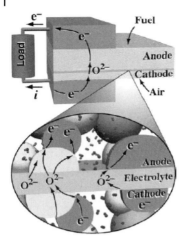

Fig. 4.5 Illustration of fuel cell chemical and transport processes in the three-phase regions [7].

electrode materials and the interconnect or electrolyte, while the thermal expansion coefficient of these materials should be close enough to allow stable long-term operation. Electrodes must show high electrocatalytic activity towards oxidation/reduction of fuel/oxidant gases. High wettability with respect to the electrolyte surface is highly advantageous for active electrodes. In addition, ease of fabrication and low cost are of tremendous importance for a wide range of commercial applications. The ability to maintain the microstructure over time is equally important. The high operating temperature can cause the microstructure to evolve with time with potential loss of connectivity in the conductive phases.

Anode Materials

The standard anode material used in SOFC is the Ni/YSZ cermet. Such cermets have been extensively studied and their performance optimized. Nickel–zirconia cermets were introduced by Spacil as a response to the failure of all-metal anodes [22]. Ni constitutes the best transition metal option, but all-nickel anodes suffer from two major drawbacks. First, Ni has a significant thermal expansion mismatch to stabilized zirconia, which can result in large stresses at the anode–electrolyte boundary, causing cracking or delamination during fabrication and operation [23]. Second, at high temperatures, the metal aggregates by grain growth, finally obstructing the porosity of the anode and eliminating the three-phase boundaries required for cell operation. To address these problems, Spacil associated nickel with the stabilized zirconia ceramic material of the electrolyte. YSZ acts as an inhibitor for the coarsening of Ni powders during both consolidation and operation, and therefore retains the dispersion of metal particles and the porosity of anode during long-term operation. Furthermore, the introduction of YSZ in the anode provides an anode thermal expansion coefficient acceptably close to those of other cell components. YSZ also offers significant ionic contribution to the overall conductivity, thus effectively broadening the TPB length. Nickel serves as an excellent reforming catalyst and electrocatalyst for electrochemical oxidation of hydrogen. It also provides predominant electronic conductivity for the anode. These cermets are chemically stable in reducing atmospheres at high temperatures. More importantly, the intrinsic charge transfer resistance that is associated with the electrocatalytic activity at Ni/YSZ boundary is low.

Despite being used in most SOFC applications Ni/YSZ anodes suffer from a few significant limitations. A severe limitation is their inability to operate on hydrocarbons, with the possible exception of methane diluted in large amount of steam. The possibility of using practical fuels in SOFCs, without the need for a pre-reforming step, could greatly accelerate their commercialization. Further, the ability to operate reduction and oxidation cycling of the anode would be a definite advantage for the development of practical systems. Therefore, most research aimed at overcoming the limitations of nickel-based anodes has focused on the development of alternative anode materials that are catalytically active for the oxidation of methane and higher hydrocarbons, and inactive for cracking reactions that lead to carbon deposition. Other desirable properties include tolerance to sulfur; this tolerance is required for the use of practical fuels such as natural gas and biogas that contain sulfur impurities. The modification of Ni/YSZ cermets, either by doping or addition of an extra phase, has been attempted but most of the research dedicated to alternative anodes has been focused on the development of nickel-free anodes.

The inactivity towards cracking reactions is difficult to achieve and this requirement rules out most transition metals, with the possible exception of Cu, Ag and Au [24]. Although these metals make good current collectors, they are not highly active oxidation catalysts. As many metal oxides are not active for hydrocarbon cracking and are less likely to suffer from sulfur poisoning than Ni/YSZ anodes, most of the work in this area has been focused on the development of electronic or mixed ionic–electronic conducting (MIEC) single oxides. The mixed conductivity is an attractive feature that can potentially extend the TPB area to the whole anode surface area, as compared with Ni/YSZ cermets where the TPB area is limited. This, however, requires the oxide to have sufficient conductivities (both ionic and electronic) and to display a high enough electrocatalytic activity. Finding a single material that gathers all required properties has proven to be a difficult task. Therefore, alternative designs of porous composites are often made.

Gorte and co-workers [25] have successfully synthesized and demonstrated Cu-based SOFC cermet anodes for direct electrochemical utilization of a large variety of hydrocarbon fuels with little carbon deposition. Cu has been proposed to replace Ni as an electronic conductor in the anode due to its poor catalytic activity for C–C and C–H bond activation, which therefore inhibits carbon formation but means that copper does not serve as a catalyst. An important limitation is the migration of Cu ions into the YSZ following high-temperature calcinations. To overcome those limitations, a novel fabrication method has been developed in which the porous YSZ part of the cermet is prepared first, and the Cu is added in a separate step that does not require high-temperature processing [25]. Direct utilization of various hydrocarbons has been reported on Cu-ceria-YSZ composite anodes. Stable power generation of $\sim 0.1\,\mathrm{W\,cm^{-2}}$, for at least 12 h, from the direct utilization, without reforming, of toluene, n-decane and synthetic diesel fuel at 700 °C has been demonstrated [26].

Marina et al. [27] studied GDC anodes ($Ce_{0.6}Gd_{0.4}O_{1.8}$) with steam-diluted CH_4 as a fuel. No carbon deposition was observed after 1000 h of operation at 1000 °C, with H_2O/C ratios as low as 0.3. Further, these anodes sustained several rapid thermal cycles and a full redox cycle without degradation However, studies have indicated that GDC is not such a good electrocatalyst for direct electrochemical CH_4 oxidation and also shows low activity for CH_4 reforming [28], and it was suggested to add a catalyst (Ni, Ru, Rh) to break down the C–H bond more easily. Adding a small amount of catalyst has been found to be very efficient in improving the performance of GDC anodes by Primdhal and Mogensen [29]. Joerger and Gauckler [30] studied Ni/GDC cermets for H_2 and CH_4 oxidation reaction. A large improvement in the electrocatalytic activity was found compared with that of Ni/YSZ cermet anodes, for both H_2 and CH_4. Activation energies for the H_2 and CH_4 oxidation reactions were lower on Ni/GDC anodes than those on Ni/YSZ anodes, exhibiting the catalytic activity and electrocatalytic activities of ceria for the reforming and oxidation of hydrocarbons and hydrogen.

Drawbacks of GDC anodes seem to arise from the chemical compatibility between GDC and YSZ, which react and diffuse into each other during the sintering process at 1200 °C, forming a Gd-rich phase with ionic conductivity two orders of magnitude lower than that of YSZ at 800 °C [31]. Further studies seem to show that the solid-state reaction can be suppressed. One way suggested is to use an interlayer of $Ce_{0.43}Zr_{0.43}Gd_{0.10}Y_{0.04}O_2$. Another way is to rely on the ability of Ni to inhibit the reactivity between YSZ and ceria, and an Ni–GDC composition 40:60 vol.% has been recommended to inhibit solid-state reactions [32].

Perovskites have also been widely investigated as potential SOFC anode materials, and among these materials, chromites and titanates appear promising [33, 34]. Interesting results have been obtained with lanthanum strontium titanates [35] and especially cerium-doped lanthanum strontium titanate [36]; however, it is now thought that the cerium-doped anodes are in fact two phase, consisting of a ceria–perovskite assemblage [36]. It was also reported that Y-doped $SrTiO_3$ exhibits high electrical conduction under SOFC anodic conditions [37]. For example, the optimized composition of $Sr_{0.86}Y_{0.08}TiO_{3-\delta}$ exhibits a conductivity of 82 S cm^{-1} at a pO_2 of 10^{-19} atm at 800 °C. However, the sample was pre-reduced in pure argon or 7% H_2–Ar at 1400 °C before conductivity measurements. It is assumed that the conductivity of the materials would be significantly lower if the sample was only reduced below 1000 °C; in this case less Ti^{4+} would be reduced to Ti^{3+}. The high-temperature pre-reduction process for such titanates makes it difficult to co-fire the anode and cathode.

Lanthanum strontium titanates are usually treated in the literature as simple cubic perovskites, although the presence of extra oxygen beyond the ABO_3 stoichiometry plays a critical role in both the structure and the electrochemical properties [38, 39]. The presence of such disordered defects appears to affect strongly the redox characteristics of the oxide, as indicated by marked effects on conductivity induced by mild reduction. Unfortunately, although the materials in this oxygen-excess lanthanum strontium titanate series are much easier to reduce, and hence exhibit much higher electronic conductivity than their oxygen stoichiometric analogs, they do not exhibit very good electrochemical performance [40]. This is attributed to the inflexibility of the coordination demands of titanium, which strongly prefers octahedral coordination in the perovskite environment. In order to make the B-site coordination more flexible and hence to improve electrocatalytic performance, Mn and Ga were introduced to replace Ti in $La_4Sr_8Ti_{12}O_{38-z}$-based fuel electrodes. Mn supports p-type conduction in oxidizing conditions, and has been shown previously to promote electroreduction under SOFC conditions [41]. Furthermore, Mn is known to accept lower coordination numbers in perovskites [42], especially for Mn^{3+}, and hence it may facilitate oxide ion migration. Similarly, Ga is well known to adopt coordination lower than octahedral in perovskite-related oxides. By optimization of the electrode microstructure, polarization resistances in wet H_2 were 0.12 $\Omega\,cm^2$ in wet H_2, 1.5 $\Omega\,cm^2$ in wet 5% H_2 and a remarkably low value, 0.36 $\Omega\,cm^2$, in wet CH_4, at 950 °C. These polarization resistances were attained after about 24 h in fuel conditions; initial polarization resistances were 2–3 times higher. This long period to achieve equilibration is fairly typical for donor-doped strontium titanates that are not cation vacancy compensated and we attribute this to reorganization of a complex defect structure.

$LaCrO_3$-based materials have been investigated as interconnect materials for SOFCs; however, they are also potential anode materials for SOFCs due to their relatively good stability in both reducing and oxidizing atmospheres at high temperatures [43]. The reported polarization resistance using these materials is too high for efficient SOFC operation, although significant improvements have been achieved using low-level doping of the B-site. When the B-sites are doped by other multi-valent transition elements that tolerate reduced oxygen coordination, such as Mn, Fe, Co, Ni, and Cu, oxygen vacancies may be generated at the B-site dopants in a reducing atmosphere at high temperature. Of the various dopants, nickel seems to be the most successful, and the lowest polarization resistances have been reported for 10% Ni-substituted lanthanum chromite [44]; however, other workers have found nickel exsolution from 10% Ni-doped lanthanum chromites in fuel conditions [45]. A composite anode of 5% Ni with a 50:50 mixture of $La_{0.8}Sr_{0.2}Cr_{0.8}Mn_{0.2}O_3$ and $Ce_{0.9}Gd_{0.1}O_{1.95}$ was successfully used for SOFCs with different fuels [46].

As perovskites with one cation occupying the B-site have not generally yielded good enough properties for efficient anode operation in SOFCs, an extensive series of studies looking at the possibility of enhancing performance by using two different B-site ions, both with concentration in excess of the percolation limit (i.e., >30%), were performed. The objective was to obtain complementary functionality from appropriate cation combinations, hopefully without seriously degrading the good properties induced by the individual ions. Not surprisingly, many of the tested combinations did compromise properties, but in some important instances good complementary functionality was achieved. Improved performance has been obtained with complex perovskites based upon Cr and Mn at the B-sites forming compositions $(La, Sr)Cr_{1-x}M_xO_{3-\delta}$ [47]. The electrode polarization resistance approaches $0.2\,\Omega\,cm^2$ at $900\,°C$ in 97% H_2–3% H_2O. Very good performance is achieved for methane oxidation without using excess steam. The anode is stable in both fuel and air conditions and shows stable electrode performance in methane. Thus both redox stability and operation in low steam hydrocarbons have been demonstrated, overcoming two of the major limitations of the current generation of nickel zirconia cermet SOFC anodes. Catalytic studies of LSCM demonstrate that it is primarily a direct oxidation catalyst for methane oxidation as opposed to a reforming catalyst [48], with the redox chemistry involving the Mn–O–Mn bonds [49]. Although oxygen ion mobility is low in the oxidized state, the diffusion coefficient for oxide ions in reduced LSCM is comparable to that for yttria-stabilized zirconia [50]. Recently, very high performance has been achieved using LSCM based anodes [51, 52]. Through the addition of a small amount of catalyst, and using the wet impregnation fabrication method, power densities of 1.1 and 0.7 W/cm^2 were achieved at 800C in humidified (3% H_2O) hydrogen and methane respectively. This high performance has been shown to stem from the high TPB area created by the interactions between LSCM and YSZ during the electrode fabrication and activation processes [53]. Another important double perovskite is $Sr_2MgMoO_{6-\delta}$, which has recently been shown to offer good performance, with power densities of $0.84\,W\,cm^{-2}$ in H_2 and $0.44\,W\,cm^{-2}$ in CH_4 at $800\,°C$, and good sulfur tolerance. [54]. The molybdenum-containing double perovskite was initially prepared at $1200\,°C$ in flowing 5% H_2 and then deposited on top of a lanthanum ceria buffer layer before testing [55].

Cathode Materials
The reduction of oxygen occurs at the SOFC cathode. Due to the high temperature of YSZ-based SOFCs, materials that can be used as the cathode are limited to noble metals and electron-conducting oxides. Nobel metals such as platinum or palladium are unsuitable for practical applications due to prohibitive costs. Several doped oxides and mixed oxides have been investigated [6, 14]. There seem to be two predominant types of behavior: triple phase boundary (TPB) dominated and mixed ion–electron conductors (MIEC). In TPB-type cathodes, the main role of the oxide material is as an electron conductor and the oxygen reduction reaction takes place at the cathode–electrolyte–gas phase boundary. For this type of electrode, it is very important to optimize the microstructure to maximize the TPB length per unit area of the cathode. This type of cathode is also very prone to contamination from impurity species from all sources [56]. In MIECs, the oxygen reduction is extended over part of the surface of the cathode material and hence superior performance can be obtained especially at lower temperatures. Although there is significant extension of the electrochemical re-

action from the interface in MIEC materials, microstructure is still very important and models are being developed to understand SOFC cathode behavior better, addressing both microstructural features and oxygen exchange and diffusion kinetics [57].

Perovskite oxides have been employed as fuel cell cathodes for many years. The best known of these are materials in the family $(La_{1-x}Sr_x)_{1-y}MnO_3$, commonly abbreviated to LSM. This material was used as an early cathode on the first zirconia-based high-temperature SOFCs and proved to be very effective. The LSM family of materials behave as TPB cathodes because of their low diffusivity for oxygen [58]. MIECs with perovskite structure have greater promise for intermediate temperature (IT) SOFCs [59–61]. Most of the best perovskite cathodes, particularly for IT-SOFC operation, have Co on the B-site and a high degree of oxygen non stoichiometry, for example, $Sm_{0.5}Sr_{0.5}CoO_{3-\delta}$. Unfortunately, these cobalt perovskites have two drawbacks: they tend to be react with YSZ, the most commonly used electrolyte material, and they suffer from very high expansion coefficients as result of a chemical expansion component to the thermal expansively [62]. $(La, Sr)(Co, Fe)O_{3-\delta}$ (LSCF) has been investigated because of the rapid surface exchange kinetics, the mixed conductivity and high oxygen vacancy concentration [63]. $Ba_{0.5}Sr_{0.5}Co_{0.8}Fe_{0.2}O_{3-\delta}$ has also been successfully investigated as a cathode material for IT-SOFC application by Shao et al. [64]. Recently, layered perovskite oxides showing the chemical formula $LnBaMn_2O_{5+\delta}$ were investigated in the field of colossal magnetoresistance (CMR). These layered perovskites with the chemical formula $LnBaCo_2O_{5+\delta}$, Ln = Pr, Nd, Sm, Eu, Gd, Tb, Dy, Ho, and Y, can be used effectively for SOFC cathode materials [64, 65].

4.2.5.3 Interconnects

The interconnect in an SOFC stack is a very important component, which has two functions: first to provide the electrical contact between adjacent cells, and second to distribute the fuel to the anode and the air to the cathode [14]. This requires that the interconnect has a high electronic conductivity in both oxidizing and reducing atmospheres at high temperatures, must not react with any of the anode, cathode or electrolyte materials at the high operating temperatures, and must be impermeable. These requirements severely restrict the choice of materials for the interconnect, especially at the higher operating temperatures of most zirconia-based SOFCs.

Most zirconia-based SOFCs use lanthanum chromite ($LaCrO_3$) as the interconnect. $LaCrO_3$ has a perovskite structure and is a p-type conductor, and satisfies all the above-mentioned criteria. The electrical conductivity of $LaCrO_3$ can be enhanced by substituting the La^{3+} with a divalent cation, such as strontium, calcium, or magnesium. The main drawback associated with lanthanum chromite interconnects is their manufacturing costs due their difficult sinterability.

For SOFCs operating in the intermediate temperature range, 500–750 °C, it becomes feasible to use certain ferritic stainless-steel composites which fulfill

4.3
Molten Carbonate Fuel Cell

The second, and further commercially developed, main type of high-temperature fuel cell system is the molten carbonate fuel cell (MCFC). The MCFC uses a eutectic mixture of alkali metal carbonates, Li_2CO_3 and K_2CO_3, immobilized in an $LiAlO_2$ ceramic matrix as the electrolyte. At high operating temperature, for example 650–700 °C, the alkali metal carbonates form a highly conductive molten salt, with carbonate ions, CO_3^{2-}, providing ionic conduction. Carbon dioxide, CO_2, and oxygen, O_2, must be supplied to the cathode to be converted to carbonate ions, which provide the means of ion transfer between the cathode and the anode. The anode and the cathode reactions are as follows:

$$\text{Anode:} \quad H_2 + CO_3^{2-} \rightarrow CO_2 + H_2O + 2e^-$$

$$\text{Cathode:} \quad \frac{1}{2}O_2 + CO_2 + 2e^- \rightarrow CO_3^{2-}$$

In the MCFC, CO_2 is produced at the anode and consumed at the cathode. Therefore, MCFC systems generally feed the CO_2 from the anode to the cathode. The electrodes are nickel based, the anode usually consisting of a nickel–chromium alloy, and the cathode is made of a lithiated nickel oxide. At both electrodes, the nickel phase provides catalytic activity and conductivity. At the anode, the chromium additions maintain high porosity and protect against corrosion [67].

Because they operate at high temperatures, MCFCs have flexibility in the chosen fuel, there is no need to use precious metals as catalyst, and they have high-quality waste heat for co-generation applications. There are also some disadvantages: a CO_2 recycling system must be implemented, the molten electrolyte is corrosive which gives degradation issues, and the materials are relatively expensive. The main application area for this technology is distributed power generation, often utilizing biogas.

4.4
Thermodynamics of Fuel Cells

A comprehensive study of fuel cell thermodynamics has been performed by Kee et al. [7]. The thermodynamic limit on fuel cell performance can be understood by considering the energy and entropy accounting associated with a generic steady flow process.

4.4 Thermodynamics of Fuel Cells

The rate at which work is done by a system, $\dot{W}$, can be obtained by combining the first and second laws of thermodynamics:

$$\dot{W} = -\dot{m}(\Delta h - T_0 \Delta s) - T_0 \dot{P}_s \tag{4.7}$$

where $\Delta h = h_{out} - h_{in}$ and $\Delta s = s_{out} - s_{in}$ are respectively the net enthalpy and entropy differences associated with the flow streams entering and leaving the system and $\dot{m}$ is the mass flow rate through the system. Entropy is produced within the system due to internal irreversible processes at a rate $\dot{P}_s$. Since $\dot{P}_s$ is required to be positive by the second law of thermodynamics, the greatest power is produced by a reversible process and equals

$$\dot{W}_{rev} = -\dot{m}(\Delta h - T_0 \Delta s) \tag{4.8}$$

Although this general expression applies to any steady-flow process, a relevant case for fuel cells is one in which the temperature remains fixed at T_0 and the pressure is constant, but the composition changes due to internal chemical reactions. In this case, the greatest work production rate achievable is

$$\dot{W}_{rev} = -\sum_k \Delta(\dot{N}_k \mu_k) \tag{4.9}$$

where μ_k and N_k are the species chemical potentials and molar flow rate, respectively, and the sum runs over all species. If depletion effects are small enough, so that $\mu_{k,in} = \mu_{k,out}$ and if $\dot{N}_k = \nu_k \dot{N}$, where ν_k is the stoichiometric coefficient of species k and $\dot{N}$ is a rate of progress variable for a global oxidation reaction, Equation 4.9 reduces to

$$\dot{W}_{rev} = -\dot{N} \sum_k \nu_k \mu_k \tag{4.10}$$

If the fluid stream is an ideal gas mixture, the reversible work production rate can be written as

$$\dot{W}_{rev} = -\dot{N}\left(-\Delta G^0 - RT \ln \prod_k P_k^{\nu_k}\right) \tag{4.11}$$

where ΔG^0 represents the free-energy change between reactants and products in the global reaction and P_k is the partial pressure. This expression also holds in the case where one or more reactants are supplied in separate streams. All that is required is to evaluate the partial pressures of each species in the stream in which it is present.

Since the cell potential can be expressed as $E_{cell} = \dot{W}/I$, Equation 4.11 can be used to determine the reversible cell potential. The electric current generated in a fuel cell as a direct consequence of the reactions that result in oxidation of the fuel, is given by

$$I = nF\dot{N} \tag{4.12}$$

where F is Faraday's constant and n the number of exchanged electrons. Hence the potential developed by a reversible cell is

$$E_{rev} = -\frac{\Delta G^0}{nF} - \frac{RT}{nF} \ln \prod_k P_k^{v_k} \qquad (4.13)$$

When applying Equation 13 to the reaction accounting for the oxygen transfer (Equation 4), the reversible potential simplifies to

$$E_{rev} = -\frac{RT}{4F} \ln \frac{P_{O_2(a)}}{P_{O_2(c)}} \qquad (4.14)$$

This reversible potential is known as the Nernst potential. This ideal potential depends on the electrochemical reactions that occur with different fuels and oxygen. The Nernst equation provides a relationship between the ideal standard potential E^0 for the cell reaction and the ideal equilibrium potential (E) at other temperatures and partial pressures of reactants and products.

4.5
Fuel Cell Efficiency

Different efficiencies must be combined to produce the overall efficiency of a fuel cell. The overall efficiency is defined by the product of the electrochemical efficiency, ε_E, and the heating efficiency, ε_H. The electrochemical efficiency is, in turn, the product of the thermodynamic efficiency, ε_T, the voltage efficiency, ε_V, and the current or Faradic efficiency, ε_J [1, 7, 67]:

$$\varepsilon_{FC} = \varepsilon_E \varepsilon_H = \varepsilon_T \varepsilon_V \varepsilon_J \varepsilon_H \qquad (4.15)$$

4.5.1
Heating Efficiency

The heating efficiency applies to cases where the fuel contains more species than the electrochemically active ones, such as gases, impurities, and other combustibles. The heating value efficiency, ε_H, is defined as

$$\varepsilon_H = \frac{\Delta H^0}{\Delta H_{com}} \qquad (4.16)$$

where ΔH^0 represents the amount of enthalpy in the electrochemically active species and ΔH_{com} represents the amount of enthalpy included in all combustible species in the fuel gases fed to the fuel cell. A pure fuel will obviously give a heating efficiency of 100%.

4.5.2
Thermodynamic Efficiency

The thermodynamic efficiency of a process measures how efficiently chemical energy extracted from the fuel stream is converted to useful power, rather than heat:

$$\varepsilon_T = \frac{\dot{W}}{\dot{W} + \dot{Q}} \quad \text{or} \quad \varepsilon_T = \frac{\dot{W}}{\dot{m}|\Delta h|} \tag{4.17}$$

where $\dot{Q}$ is the heat rate production of the cell. Using the maximum work production rate achievable, which is the one of a reversible process (Equation 4.8), one can write the maximum theoretical efficiency of a system as

$$\varepsilon_T = \frac{\Delta(h - T_0 s)}{\Delta h} \tag{4.18}$$

The thermodynamic efficiency is an extremely important feature when analyzing a fuel cell. This efficiency justifies the need for fuel cell development. Indeed, since the chemical energy is transferred directly to electricity, the free enthalpy change of the cell reaction may be totally converted to electrical energy. In a conventional heat engine, where only the temperature is changes, ε_{rev} is limited to the familiar Carnot efficiency:

$$\varepsilon_{T,\,Carnot} = 1 - \frac{T_0}{T} \tag{4.19}$$

For a constant-temperature fuel cell, the intrinsic maximum thermodynamic efficiency is given by

$$\varepsilon_{rev} = \frac{\Delta G}{\Delta H} = 1 - \frac{T \Delta S}{\Delta H} = \frac{\Delta G^0}{\Delta H^0} + \frac{RT}{\Delta H^0} \ln \prod_k p_k^{v_k} \tag{4.20}$$

4.5.3
Current Efficiency

The current efficiency can be commonly expressed as the fuel utilization efficiency. The efficiency of a SOFC drops if all the reactants are not converted to reaction products. For 100% conversion of a fuel, the amount of current density, i_F, produced is given by Faraday's law:

$$i_F = zF \left(\frac{df}{dt} \right) \tag{4.21}$$

where (df/dt) is the molar flow rate of the fuel. For the amount of fuel actually consumed, the current density produced is given by

$$i = zF\left(\frac{df}{dt}\right)_{consumed} \tag{4.22}$$

The current efficiency, ε_J, is the ratio of the actual current produced to the current available for complete electrochemical conversion of the fuel:

$$e_J = \frac{i}{i_F} \tag{4.23}$$

This efficiency can also be expressed in terms of fuel consumption:

$$e_J = \frac{h_{in} - h_{out}}{h_{in} - h_{ox}} \tag{4.24}$$

where h_{ox} corresponds to the enthalpy change when all the useful fuel has been consumed.

4.5.4
Voltage Efficiency

When an electric current is drawn from a fuel cell, part of the chemical potential available must be used to overcome the irreversible internal losses. Hence the actual cell potential is decreased from its equilibrium potential, meaning that in an operating SOFC, the cell voltage is always less than the reversible voltage. The voltage efficiency, δ_V, is defined as the ratio of the operating cell voltage under load, E, to the reversible cell voltage, E_R, and is given as

$$\varepsilon_V = \frac{E}{E_R} \tag{4.25}$$

A voltmeter connecting the anode and cathode can measure the cell electrical potential E, which depends on the current flow. Indeed, many irreversibilities in a fuel cell scale with current density. If no current flows, the cell voltage is the open-circuit potential or open-circuit voltage (OCV). In most cases, the OCV will equal the potential developed by a reversible cell. The difference between the operating cell voltage and the expected reversible voltage is termed polarization or overpotential and is represented by η. This cell overpotential comprises the total ohmic losses for the cell and the polarization losses associated with the electrodes. The useful voltage under load conditions can therefore be expressed as

$$V = E^0 - IR - \eta_{anode} - \eta_{cathode} \tag{4.26}$$

where I is the current passing through the cell. The electrical resistance R encompasses the ohmic resistance of all components. The polarization resistances, η_{anode} and $\eta_{cathode}$, for the anode and cathode account for non-ohmic losses in each electrode. The polarization loss of each electrode is composed of (i) activation overpotential due to energy barriers to charge transfer reactions, (ii) concentration overpotential associated with diffusion of gas-phase species resistance through the electrodes, and (iii) contact resistance which is caused by poor adherence between the electrode and the electrolyte [7, 67]. Although polarizations cannot be eliminated, material choice and electrode designs can contribute to their minimization. Figure 4.6 shows a typical voltage–current polarization curve for an SOFC. The voltage loss increases with increase in the current density. Activation overpotentials contribute the most at low current, whereas at high currents concentration polarizations become important. Ohmic losses dominate the losses in the intermediate currents zone.

The concave portion at low current density corresponds to activation-related losses. When the current density increases, the losses are dominated by the ohmic polarization. When the current density approaches its highest values, losses are dominated by concentration polarizations that cause the steep drop in the cell voltage. Each of the different types of losses is described hereafter.-2

4.5.4.1 Internal Resistance

The size of the voltage drop due to ohmic losses is simply proportional to the current:

$$V = IR \tag{4.27}$$

The internal resistance R encompasses the contributions from the electrodes, electrolyte, interconnect, and bipolar plates:

$$R = R_{electronic} + R_{ionic} + R_{contact} \tag{4.28}$$

In most fuel cells, the electrolyte contribution to this resistance is the most important, due to the ionic nature of its conductivity. The interconnect and bipolar plates contributions can also be important. To minimize the ohmic losses, the increasingly preferred practice is to fabricate dense, gas-tight electrolyte membranes as thin as possible.

4.5.4.2 Charge-Transfer or Activation Polarization

The activation polarization is related to the charge-transfer processes occurring during the electrochemical reactions on electrode surfaces. The losses are caused by the slowness of the reactions taking place on the surface of the electrodes. Electrochemical reactions involve an energy barrier that must be overcome by the reacting species. A proportion of the voltage generated is hence lost in driving the electron transfer. This energy barrier, called the activation energy, results in activation or charge-transfer polarization, η_A. Activation polarization is related to current density, i, by the Butler–Volmer equation:

$$i = i_0 \exp\left(\frac{(1-\beta)\eta_A F}{RT}\right) - i_0 \exp\left(-\frac{\beta \eta_A F}{RT}\right) \tag{4.29}$$

where β is the symmetry coefficient and i_0 the exchange current density. The symmetry coefficient is considered as a fraction of the change in polarization

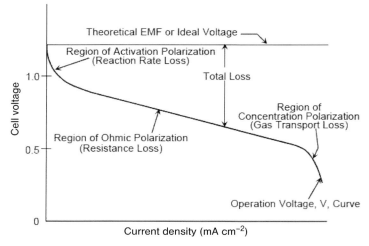

Fig. 4.6 Ideal and actual fuel cell voltage–current characteristics [1].

which leads to a change in the reaction rate constant. The exchange current density is related to the balanced forward and reverse electrode reaction rates at equilibrium. A high exchange current density means a high electrochemical reaction rate and, in that case, a good fuel cell performance is expected.

The exchange current density can be determined experimentally by extrapolating plots of $\log i$ versus η to $\eta = 0$. For large values of η (either negative or positive). one of the bracketed terms in equation 4.29 becomes negligible. After rearranging, one obtains

$$\eta = a \pm b \log i \qquad (4.30)$$

which is usually referred to as the Tafel equation. Parameters a and b are constants which are related to the applied electrochemical material, type of electrode reaction and temperature.

The constant a [in the form $v = a \ln(i/i_0)$] is higher for an electrochemical reaction which is slow. The constant i_0 is higher if the reaction is faster. The current density i_0 can be considered as the current density at which the overvoltage begins to move from zero. The smaller is i_0, the greater is the voltage drop.

The exchange current density is a crucial factor in reducing the activation overvoltage. The cell performance can be improved through an increase of the exchange current density. This can be done in the following ways [1]:
- raising the cell temperature
- using more effective catalysts
- increasing the roughness of the electrodes
- increasing reactant concentration, for example, pure O_2 instead of air
- increasing the pressure.

In low- and medium-temperature fuel cells, activation overvoltage is the most important irreversibility and cause of voltage drop, and occurs mainly at the cathode. Activation overvoltage can be important at the anode when fuels other than hydrogen are involved.

4.5.4.3 Diffusion or Concentration Polarization

Concentration polarization, η_D, is related to the transport of gaseous species through the porous electrodes and, therefore, its magnitude is dictated by the microstructure of the electrode, specifically the volume percent porosity, the pore size, and the tortuosity factor. It becomes significant when the electrode reaction is hindered by mass transport effects, that is, when the supply of reactant and/or the removal of reaction products by diffusion to or from the electrode are slower than those corresponding to the charging/discharging current i. When the electrode process is governed completely by diffusion, the limiting current, i_L, is reached. In such a case, the demand for reactants exceeds the capacity of the porous anode to supply them by gas diffusion mechanisms. High tortuosity (bulk diffusion resistance) is often assumed to explain this behavior.

The voltage drop due to the mass transport limitations can be expressed as

$$\Delta V = \frac{RT}{2F} \ln\left(1 - \frac{i}{i_L}\right) \tag{4.31}$$

where i_L is postulated to be the limiting current density at which the fuel is used up at a rate equal to its maximum supply speed. The current density cannot rise above this value because the fuel gas cannot be supplied.

References

1. US Department of Energy (2004) *Fuel Cell Handbook*, 7th edn, US Department of Energy, Washington, DC.
2. Singhal, S. C. (2002) *Solid State Ionics*, 152–153, 405.
3. Grove, W. R. (1839) *Philos. Mag.*, 14, 127.
4. Grove, W. R. (1843) *Philos Mag.*, 21, 417.
5. Mond, L. and Langer, C. (1889) *Proc. R. Soc. London*, 46, 296.
6. Minh, N. Q. (1993) *J. Am. Ceram. Soc.*, 76, 563.
7. Kee, R. J., Zhu, H., and Goodwin, D. G. (2005) *Proc. Combust. Inst.*, 30, 2379.
8. Nernst, W. (1899) *Z. Electrochem.*, 6, 41.
9. Baur, E. and Preis, H. Z. (1937) Über Brennstoffketten mit Festleitern. *Z. Electrochem.*, 43, 727.
10. Gardner, F. J., Day, M. J., Brandon, N. P., Pashley, M. N., and Cassidy, M. (2000) *J. Power Sources*, 86, 122.
11. Jones, F. G. E., Connor, P. A., Feighery, A. J., Nairn, J., Rennie, J., and Irvine, J. T. S. (2007) *J. Fuel Cell Sci. Technol.*, 4, 1.
12. Goodenough, J. B. (2003) *Annu. Rev. Mater. Res.*, 33, 91.
13. Skinner, S. J. and Kilner, J. A. (2003) *Mater. Today*, 6, 30.
14. Ormerod, M. (2003) *Chem. Soc. Rev.*, 32, 17.
15. Stevens, R. (1986) *Zirconia and Zirconia Ceramics*, 2nd edn, Magnesium Elektron, Twickenham.
16. Etsell, H. and Flengas, S. N. (1970) *Chem. Rev.*, 70, p. 317.

17 Subbarao, E.C. and Maiti, H.S. (1984) *Solid State Ionics*, **11**, 317.
18 Sasaki, K. and Maier, J. (2000) *Solid State Ionics*, **134**, 303.
19 Mogensen, M., Sammes, N.M., and Tompsett, G.A. (2000) *Solid State Ionics*, **129**, 63.
20 Steele, B.C.H. (2000) *Solid State Ionics*, **129**, 95.
21 Steele, B.C.H. and Heinzel, A. (2001) *Nature*, **414**, 345.
22 Spacil, S. (1970) Electrical device including nickel-containing stabilized zirconia electrode. US Patent 3558360.
23 Toebes, M.L., Bitter, J.H., van Dillen, A.J., and de Jong, K.P. (2002) *Catal. Today*, **76**, 33.
24 Atkinson, A., Barnett, S., Gorte, R.J., Irvine, J.T.S., McEvoy, A.J., Mogensen, M., Singhal, S.C., and Vohs, J. (2004) *Nature materials*, **3**, 17–27.
25 Kim, H., Park, S., Vohs, J.M., and Gorte, R.J. (2001) *J. Electrochem. Soc.*, **148**, A693.
26 Vohs, R.J. Gorte, J.M. (2003) *J. Catal.*, **216**, 477.
27 Marina O.A., Bagger C., Primdahl S., and Mogensen M. (1999) *Solid State Ionics*, **123**, 199.
28 Marina O.A. and Mogensen M. (1999) *Appl. Catal. A*, **189**, 117.
29 Primdahl S. and Mogensen M. (2002) *Solid State Ionics*, **152–153**, 597.
30 Joerger M.B. and Gauckler L.J. (2001) In *SOFC-VII* (eds H. Yokokawa and S.C. Singhal), proceedings of the international symposium, vol. 7. Vol. 2001–2016, Electrochemical Society, Pennington, NJ, p. 662.
31 Tsoga A., Naoumidis A., Gupta A., and Stover D. (1999) *Mater. Sci. Forum*, **308–311**, 234.
32 Tsoga A., Gupta A., Naoumidis A., and Nikolopoulos P. (2000) *Acta Mater.*, **48**, 4709.
33 Primdahl, S., Hansen, J.R., Grahl-Madsen, L., and Larsen, P.H. (2001) *J. Electrochem. Soc.*, **148**, A74.
34 Pudmich, G., Boukamp, B.A., Gonzalez-Cuenca, M., Jungen, W., Zipprich, W., and Tietz, F. (2000) *Solid State Ionics*, **135**, 433.
35 Marina, O.A., Canfield, N.L., and Stevenson, J.W. (2002) *Solid State Ionics*, **149**, 21.
36 Marina, O.A. and Pederson, L.R. (2002) In *Proceedings of the 5th European Solid Oxide Fuel Cell Forum* (ed. J. Huijsmans), European SOFC Forum, Lucerne, p. 481.
37 Hui, S.Q. and Petric, A. (2002) *J. Electrochem. Soc.*, **149**, 1.
38 Canales-Vazquez, J., Smith, M.J., Irvine, J.T.S., and Zhou W.-Z., (2005) *Adv. Funct. Mater.*, **15**, 1000.
39 Ruiz-Morales, J.C., Canales-Vazquez, J., Savaniu, C., Marrero-Lopez, D., Zhou W.-Z., and Irvine, J.T.S. (2006) *Nature*, **439**, 568.
40 Canales-Vazquez, J., Tao, S.W., and Irvine, J.T.S. (2003) *Solid State Ionics*, **159**, 159–165.
41 Holtappels, P., Bradley, J., Irvine, J.T.S., Kaiser, A., and Mogensen, M. (2001) Electrochemical characterization of ceramic SOFC anodes. *J. Electrochem. Soc. A*, **148**, 923.
42 Poeppelmeier, K.R., Leonowicz, M.E., and Longo, J.M. (1982) $CaMnO_{2.5}$ and $Ca_2MnO_{3.5}$ – new oxygen-defect perovskite-type oxides. *J. Solid State Chem.*, **44**, 89.
43 Yokokawa, H., Sakai, N., Kawada, T., and Dokiya, M. (1992) *Solid State Ionics*, **52**, 43.
44 Sfeir, J., Herle, J.V., Vasquez, R. (2002) In *Proceedings of the 5th European Solid Oxide Fuel Cell Forum* (ed. J. Huijsmans), European SOFC Forum, Lucerne, p. 570.
45 Sauvet, A.-L. and Irvine, J.T.S. (2002) In *Proceedings of the 5th European Solid Oxide Fuel Cell Forum* (ed. J. Huijsmans), European SOFC Forum, Lucerne, p. 490.
46 Liu, J., Madsen, B.D., Ji, Z.Q., and Barnett, S.A. (2002) *Electrochem. Solid State Lett.*, **5**, A122.
47 Tao, S.W. and Irvine, J.T.S. (2003) *Nat. Mater.*, **2**, 320.
48 Tao, S.-W., Irvine, J.T.S., and Plint, S.M. (2005) *J.Phys.Chem.*, **110**, 21771.
49 Plint, S.M., Connor, P.A., Tao, S.-W., and Irvine, J.T.S. (2006) *Solid State Ionics*, **177**, 2005.
50 Raj, E.S., Kilner, J.A., and Irvine, J.T.S. (2006) *Solid State Ionics*, **177**, 1747.
51 Kim, G., Corre, G., Irvine, J.T.S.,Vohs, J.M., and Gorte, R.J. (2008) Engineering composite oxide SOFC anodes for efficient oxidation of methane, *Electrochemi-*

cal and solid state letters, **11**, Issue 2, B16–B19.
52 Kim, G., Lee, S., Shin, J., Corre, G., Irvine, J.T.S., Vohs, J.M., and Gorte, R.J. (2009) A catalyst effect on engineered ceramic anodes for SOFCs, *Electrochemical and solid state letters,* **12**, Issue 3, B48–B52
53 Corre, G., Kim, G., Cassidy, M., Vohs, J.M., Gorte, R.J., and Irvine, J.T.S. (2009) Activation and Ripening of Impregnated Manganese Containing Perovskite SOFC Electrodes under Redox Cycling, *Chemistry of materials,* **21** (6), 1077–1084
54 Huang, Y.H., Dass, R.I., Xing, Z.L., and Goodenough, J.B. (2006) *Science,* **312**, 254.
55 Huang, Y.H., Dass, R.I., Denyszyn, J.C., and Goodenough, J.B. (2006) *J. Electrochem. Soc.,* **153**, A1266.
56 Mogensen, M., Jensen, K.V., Jorgensen, M.J., and Primdahl, S. (2002) *Solid State Ionics,* **150**, 123.
57 Adler, S.B. (2004) *Chem. Rev.,* **104**, 4791.
58 Carter, S., Selcuk, A., Chater, R.J., Kajda, J., Kilner, J.A., and Steele, B.C.H. (1992) *Solid State Ionics,* **53–56**, 597.
59 Steele, B.C.H. (1996) *Solid State Mater. Sci.,* **1**, 684.
60 Bouwmeester, H.J.M. (2003) *Catal. Today,* **82**, 141.
61 Liu, M.L. (1997) *J. Electrochem. Soc.,* **144**, 1813.
62 Kilner, J.A. and Irvine, J.T.S. (2008) New oxide cathodes and anodes. In *Handbook of Fuel Cells. Advances in Electrocatalysis, Materials, Diagnostics and Durability* (eds W. Vielstich, H.A. Gasteiger, and H. Yokokawa), Wiley–Blackwell, Chichester, Chapter 32.
63 Adler, S.B., Lane, J.A., and Steele, B.C.H. (1996) *J. Electrochem. Soc.,* **143**, 3554.
64 Shao, Z.P., Yang, W.S., Cong, Y., Dong, H., Tong, J., and Xiong G.X. (2000) *J. Membr. Sci.,* **172**, 177.
65 Troyanchuk, I.O., Kasper, N.V., and Khalyavin, D.D. (1998) *Phys. Rev B,* **58**, 2418.
66 Kim, J.-H., and Manthiram, A. (2008) *J. Electrochem. Soc.,* **155** (4), B385.
67 Larminie, J. and Dicks, A. (2005) *Fuel Cell Systems Explained,* 2nd edn, John Wiley & Sons, Ltd, Chichester.
68 de Boer, B. PhD thesis (1998) University of Twente.

5
Advanced Modeling in Fuel Cell Systems: a Review of Modeling Approaches

Matthew M. Mench

Abstract

A review of fuel cell modeling approaches across various length scales is given, with special focus on low-temperature hydrogen-fed polymer electrolyte fuel cell (PEFC) systems. The overall scope of fuel cell modeling is incredibly broad, and ranges from fundamental atomistic modeling of catalytic processes and transport to nearly completely empirically based online system control models. In the first part an overview of the various scales of modeling is given, with some examples from the literature, along with a summary of the common limitations and requirements for successful predictive capabilities. In the second part, a critical analysis of the present state of polymer electrolyte fuel cell performance and water management models is given, and the inability of the existing framework of models to accurately predict the liquid water distribution and transport in the PEFC is demonstrated and discussed. Based on the discussion of the various modeling approaches and the limitations in the current framework, the needs of future research to provide precise fundamental and engineering models are discussed.

Keywords: fuel cells, modeling, low-temperature fuel cells, hydrogen polymer electrolyte fuel cell, computer modeling, water management

5.1
Introduction

The field of fuel cell modeling is too broad to describe completely in a single chapter. There are literally hundreds if not thousands of computational models of various aspects of fuel cells, ranging from purely fluidics models of flow into an internal manifold, to steady-state and transient performance models of individual cells, to models of entire stacks and systems. Indeed, simply to cover the available literature of one type of fuel cell in detail would be an enormous task. The purpose of this review is therefore not to detail the particular models at every system and level, but rather broadly to frame and categorize the entire field of fuel cell modeling, and identify the outcomes, limitations, and future needs of the field. In the second part, the present state of multi-phase flow modeling in polymer electrolyte fuel cell systems is explored in detail, to understand the limitations and future research requirements to achieve more accurate models. It must be understood that although examples of published models are presented, the lists are not meant to be exhaustive, and therefore some excellent contributions whose basic approaches are already covered may not be included. Where appropriate, significant review articles are cited, to provide the reader with a starting point to gather additional details and references.

5.1.1
Classification of Models

There are many ways in which models can be classified, based on computational domain, type of fuel cell, system, stack or cell, modeling approach, purpose, physicochemical phenomena investigated, and others. It is useful first to separate the myriad modeling approaches into categories, based on the length scale of the computational domain. Fuel cell systems range from very minute power microbial fuel cells to megawatt-sized molten carbonate systems. In general, however, there is a basic continuum of modeling efforts that can be shown which can be applied to all systems, which ranges from the most fundamental atomistic model to empirical control-based system models. Each length scale and level of approximation involved are particularly suited to investigate specific phenomena. On going from the microscopic molecular-based modeling towards a full system model, the research vector goes from the very fundamental academic research to fully empirical modeling, as described in Table 5.1. Although it should be noted that rather than a true continuum, the computational domains and modeling approaches used typically vary by orders of magnitude, so that really there are discrete model varieties which focus on specific length scales. There is also an ongoing effort to develop methods to include phenomena which span the various length scales. This is developed in greater degree in other fields and, to date, computational modeling efforts in fuel cells are not normally considered to be state-of-the-art in terms of computational complexity; rather, they are fuel cell applications of modeling techniques which have previously been applied to other systems. Within each of the various levels, there are further separations and approaches, mostly designed to exploit some computational advantage of the approach to explore a particular phenomena.

Table 5.1 shows a summary of the various levels of modeling approached, and also the purposes and outcomes of the approaches. Additionally, some examples taken from literature of these models applied to various fuel cell systems are given, although the list is not intended to be exhaustive. Where available, review articles are given in Table 5.1 as a starting point for the reader interested in more detail. All model types have been used to explore steady-state and transient effects, although by nature dynamic effects are more computationally complex, and therefore of more recent interest. The most models have been

Table 5.1 Summary of various modeling scales and purposes.

Modeling level	Purpose and outcome	Some examples
Molecular level approaches	Attempts to model transport of charge, mass, or heat at a molecular level, so that basic limitations and material transport limitations can be assessed	Used to understand catalysis and electrode microstructure (e.g., [5–8]) Used to understand transport in membranes: (e.g., [9–12])
Macroscopic mechanistic approach for single or partial fuel cell	Attempts to understand transport issues or predict behavior with a macroscopic resolution but highly resolved control volume that does not encompass an entire fuel cell. Models can be one- to three-dimensional, single-phase, multi-phase, steady-state or transient, and involve many different manifestations for different purposes	Used to predict water distribution and performance in PEFCs (e.g., [13–28] and more) Used to examine design choices (e.g., [29–39] and more)
Full stack modeling	Attempts to analyze full stack performance, flow distribution, start-up/shut-down, other protocol or material choices. Models can be one- to three-dimensional, and are mostly steady-state	Examples (e.g., [40–49] and more)
System models	Attempts to integrate submodels of other components into an overall system model to understand efficiencies, losses, and relative trade-offs between design choices. Can be for online control or not	Examples (e.g., [50–60] and more)
Some review articles	Coverage varies	Review articles: [61–65]

developed in the literature at the single cell level. These are generally mechanistic, or semiempirical, meaning that the models are designed to solve for anticipated *I–V* performance, or examine a particular phenomenon via discretization and solution of analytical expressions which describe the fundamental physicochemical processes involved. Other examples of this include specific types of degradation modeling and freeze–thaw modeling. It should be noted that computational details such as grid generation and algorithm type are not a focus of this review, although there are articles available which review the computational details of fuel cell modeling available (e.g. [1]). At the stack and system level, models become more empirical in nature due to increasing computational complexity, and at the system level, control-level models are almost completely empirical in nature, and utilize some method of pattern recognition such as neural networks or other means. There is also an emerging class of model-based sensing and diagnostics (either *in situ* or *ex situ* for a cell, stack, or system). These models generally are semiempirical in nature, The area of model-based sensing is fascinating and extremely broad, and involves using some fundamental or empirical model-based approach to learn additional diagnostic information from measured data. An example includes use of symbolic dynamics to reconstruct the level of ambient CO in the fuel flow based on the transient response to a voltage perturbation to deduce carbon monoxide content in fuel streams [2], the use of particle swarm optimization to deduce certain polymer electrolyte fuel cell (PEFC) model parameters [3], and detection of hydrogen leaks in a fuel cell stack [4].

5.1.2
Limitations of Modeling

Although the ultimate computational tool would be a completely 3D, transient, multi-physics model with all length scales from the nano to macro level accurately quantified, no real computational model can predict the complete physics of a fuel cell system. The proper application and use of modeling are based on a full understanding of the inherent limitations of the basic analysis and the desired output. That is, the model output is only useful within the imposed constraints and assumptions of the model itself. It is critical not to overstate the conclusions of models and ignore the basic limitations implicit in the formulation.

The main limitations which currently prevent researchers from obtaining a complete description of the systems modeled include:
- inclusion of the complete controlling physicochemical phenomena
- knowledge of transport parameters
- computational power
- proper validation.

It is interesting to note that even if an infinite level of computational power were available, there would still be limitations that prevent the models from achieving desired levels of accuracy.

5.1.2.1 Inclusion of the Complete Controlling Physicochemical Phenomena
In order to produce results, each model must have certain limiting assumptions and neglect various phenomena which may be unimportant in the particular domain of interest. This practice is necessary, and extremely useful; however, in many cases as the field progresses, more is revealed that renders the initial state or implicit assumptions incorrect. As an example, early PEFC models commonly used an isothermal assumption that has been shown to be inaccurate, except at very low current densities [66]. It is vital that the community understands the implications of the explicit of implicit assumptions made to interpret results correctly. Additionally, it is vital that the modeler has a firm knowledge of the available experimental evidence. If these assumptions and ignored phe-

nomena are indeed important, then the extrapolation of the model is inaccurate. An example of the failure of models to predict accurately important phenomena based on implicit assumptions is given in Section 5.3. In single cell PEFC multi-phase models, the interfaces between the electrode and joining microporous layer have been ignored, and implicitly assumed to be regions of perfect contact, which is now shown to lead to significant error in the prediction of the internal water distribution.

Assumptions which ignore various potential phenomena are a natural and useful part of the advancement of science and can be valid depending on the phenomenon of interest or operating parameters. However, they should be continually examined based on new experimental evidence, so that they can continue to be improved.

5.1.2.2 Knowledge of the Transport Parameters

Although a model can be fundamentally sound in terms of its analytical description, a practical limitation of many models is the lack or uncertainty in the use of the relationships which describe the transport of heat, mass, and charge. For nonempirical models, there is a need for a complete and accurate description of transport parameters in order to accurately predict the performance and internal distributions of species, heat, and mass. Complicating the matter is the fact that the transport is a function of material properties, which vary with manufacturer and design. Table 5.2 lists some of the basic charge, mass, and heat transport parameters which must be described in order to implement a fuel cell performance model. In principle, fuel cell degradation could be modeled with knowledge of how the heat, mass, and charge transport parameters slowly change over time, but these relationships are difficult to determine and typically involve additional phenomena and disparate time scales that are not included in a typical performance model framework. An example of this is the degradation in electrolyte conductivity in PEFCs as a result of ionic impurities. Relationships between impurities and the increase in ionic resistivity of the electrolyte can be described based on experiment and, in theory, these relationships can be integrated into the model framework. As another example, if a relationship between the loss of hydrophobicity of the diffusion media with time in the PEFC system could be determined, it could be incorporated into the multi-phase flow parameters in the performance model. However, the time scales of impurity-based degradation are orders of magnitude longer than those of other changes in the fuel cell, so that it would be computationally challenging to integrate anything besides an empirical relationship into a performance-based model.

The uncertainty in many of these transport values fundamentally limits the ability of modelers to predict the true behavior in any real quantitative fashion, and forces fuel cell developers to rely on less insightful (but more rapid) empirical models, and experimental testing, which is much more costly and less generically applicable. An excellent example of the uncertainty in transport parameters is the mass diffusivity of water in Nafion^{TM} used in the PEFC. Several re-

Table 5.2 List of transport parameters needed for single-cell PEFC performance models.

Charge transport parameter needed	Typical functionality
Ionic conductivity of the electrolyte	Temperature, water content, ambient species
Electrolyte electrical conductivity (transference number, SOFCs)	Material, environment
Kinetic relationship for reaction	Temperature, reactant concentration, morphology of electrode, catalyst type, age, impurities
Ionic conductivity in the electrode	Electrolyte content, morphology, porosity, connectivity of electrolyte, temperature, water content, ambient species
Electrical conductivity in the electrode	Material content, morphology, porosity, connectivity, temperature, water content, ambient species, orientation (if anisotropic), age, impurities
All electrical contact resistances in cell interfaces	Compression, age

Mass transport parameters needed	Typical functionality
Gas-phase diffusivities • Adjustment for porous media	Pressure and temperature, porous media porosity and tortuosity
Liquid-phase transport (low-temperature fuel cells): • Saturation versus capillary pressure relationship for porous media • Liquid- and gas-phase permeability as a function of saturation	Porous media material properties (tortuosity, hydrophobic content, pore size distribution, material fiber orientation)
Mass transport relationships in electrolyte besides charge transport (e.g., diffusion or permeation of water, reactant crossover, impurities, etc.)	Material, age, water content, temperature

Heat transport parameters needed	Typical functionality
Specific heat and thermal diffusivity of all materials (transient simulation only)	Temperature, orientation (for anisotropic materials)
Thermal conductivity of all materials	Temperature, orientation
Latent heat (multi-phase simulations only)	Temperature
Thermal contact resistance of all materials	Compression, age
Convective heat transport in flow channels	Operating parameters and species

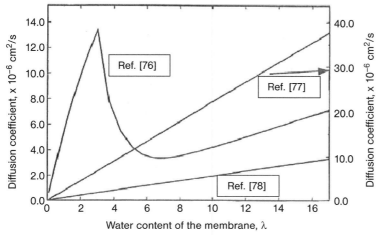

Fig. 5.1 Comparison of water diffusivities in Nafion electrolyte material found in the literature. Adapted from [75].

searchers have independently measured this with different approaches, and found entirely different values, as illustrated in Figure 5.1. Since the value of diffusivity is critical in determining the overall water balance in the fuel cell, the fact that there is so much uncertainty in the data renders the quantitative conclusions based on use of any of the known relationships in doubt. Electro-osmotic drag is another polymer transport parameter that has been extensively researched, with very different relationships emerging [67–71]. Thermo-osmotic and drag-assisted dissolved reactant diffusion are also phenomena that can impact performance and degradation [72, 73], but fall within the uncertainty associated with electroosmotic drag and diffusion parameters, so that they are commonly ignored. An excellent discussion of various transport parameters which have been measured and deduced for PEFC systems was given by Weber and Newman [74].

5.1.2.3
Computational Power

Although continual computing power and parallelized code architecture are commonplace, basic fundamental limitations exist in the computing power involved as soon as the grid structure becomes too fine, or the nonlinearities in the model become too great. In comparison with the state-of-the-art combustion or turbulence models, however, standard fuel cell, stack, and system model algorithms and solution approaches are fairly basic. Consider a typical chemical combustion simulation of a flame spreading in a cylinder of an internal combustion engine. Modern computational simulations include multi-dimensional turbulence, wall effects, soot formation, and hundreds of reversible chemical reactions which render the reaction matrix extremely stiff. In fuel cells, computa-

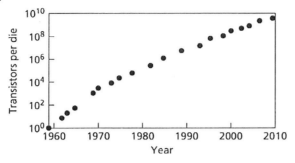

Fig. 5.2 Plot of Moore's law since 1960. From [79].

tional limitations are normally only a result of the disparity in length or time scales, multi-phase flow, and the resolution of the computational domain. Generally, steady-state fuel cell model computational limitations are a result of the number of grid points, whereas transient simulations are limited by the time scale and degree of grid resolution, and not fundamental limitations of the model itself. In theory, this limitation will be overcome in the future by use of highly parallelized and efficient computing and continually increasing computing power. Moore's law is still holding at an exponential pace, with transistors becoming a factor of two smaller, faster, and less expensive every 18–22 months since the 1960s, as shown in Figure 5.2 [79].

5.1.2.4
Proper Validation

The need for proper and complete validation is related to the use of limiting assumptions. One cannot use any model to predict phenomena which fall outside a validated range of measured parameters. Almost all fuel cell models in the literature have been validated by comparison with measured performance data, such as a polarization curve. Additionally, other forms of models such as degradation or mechanical stress models are normally validated *in situ* or *ex situ* by comparison with some experimental evidence. However, even from a performance model perspective, a mere comparison with a polarization curve is not sufficient to say a model is validated. Indeed, considering that most performance models have at least three adjustable parameters to describe kinetic, ohmic, and transport losses, it is not surprising that good agreement between a model and a simple polarization curve can be achieved. Instead, proper model validation includes comparison of measured and predicted data over the complete computational scope of the output expected. In other words, if the model is simply to predict the polarization curve and nothing more, then validation over a range of operating conditions based simply on the polarization curve can be sufficient, but extrapolation beyond simple I–V data is doubtful. Consider Figure 5.3 [80], where the results show that the basic model, in the range of op-

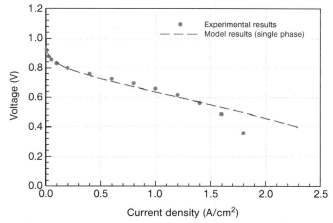

Fig. 5.3 Model from the literature shown with actual polarization data. From a pure performance perspective under the operating conditions shown, this model is accurate. Adapted from [80].

erating parameters shown, can adequately predict the polarization behavior of the system. Although this sort of preliminary validation is important, it is not sufficient evidence to conclude that the predicted internal distributions of heat, water, and charge are accurate, since the model could be over-predicting one polarization and under-predicting another at the same time. Additionally, any experimentalist knows that the next time the cell is assembled with a different electrode assembly, the performance could be significantly different although the model predictions would be the same. Many performance models in the literature attempt to make predications or draw conclusions based on unvalidated data. This can provide useful insight, but should be viewed only as an extrapolation. In this case, the uncertainty in these conclusions should be indicated, and complete validation cannot be assumed until sufficient supporting data are available. A case study of this can be found in Section 5.3, where water management and a comparison of the models' predictive capabilities in PEFCs are discussed.

What normally prohibits the proper validation is a lack of existing detailed and reliable experimental diagnostics. There is always a tremendous need to develop new and unintrusive real-time diagnostics which can accurately provide reliable validation data. Techniques for measuring the distribution of heat, mass, and charge are always needed. At the microscopic level, this is particularly challenging, however, as tools simply to not exist in many cases to measure what is being modeled.

5.2
State-of-the-Art Computational Models for Low-Temperature Polymer Electrolyte Fuel Cell Systems

Low-temperature fuel cells are probably the most developed and modeled fuel cell system, due to its utilization for a wide variety of applications and the number of researchers in the field. In this section, a brief summary of the level of sophistication of the computational models at each of the levels along the scale shown in Table 5.1 is discussed. Full system online control models are necessarily developed for all commercial and prototype systems. For most fuel cell systems, the literature is abundant with basic full stack, performance, and specialty models of varying complexity.

5.2.1
Microscopic Models

Molecular level modeling in fuel cells generally consists of examining the transport in a specific material in order to examine the transport or material limitations involved. In the PEFC, the two main limiting regions of transport are the catalysis reaction and ionic transport in the membrane electrolyte. Figure 5.4 shows a representative microscopic level modeling result of oxygen concentration around a catalyst particle coated on a semipermeable ionomer, supported by a carbon particle from [81]. Although not truly at the molecular level, this model represents the common scope of microscopic models. Many microscopic models are based on molecular dynamics simulations of molecular interactions, as listed in Table 5.1.

5.2.2
Single-Cell Models

Single-cell modeling for polymer electrolyte fuel cells can roughly be split into two categories of single-phase and multi-phase flow. Additionally, the mechanistic models can be split into multi-domain (governing equations are prescribed for each material region and appropriate adjoining boundary conditions are considered) and single-domain (a single governing equation is prescribed for the entire computational domain and unique source terms are prescribed within each material layer) approaches. Early performance models were based on single-phase flow, and are also appropriate for high-temperature membrane systems [such as polybenzimidazole (PBI)-based membranes], as discussed below. In general, due to the lack of liquid-phase treatment, single-phase models are useful only in a limited range to predict operational performance, but are convenient to predict general trends. They can also be useful for modeling and analysis of high-temperature PEFC systems, such as PBI-membrane-based PEFCs [82–85], although a two-phase model of the PBI-based PEFC, where aqueous phase electrochemical reactions are taken into account, has also been developed

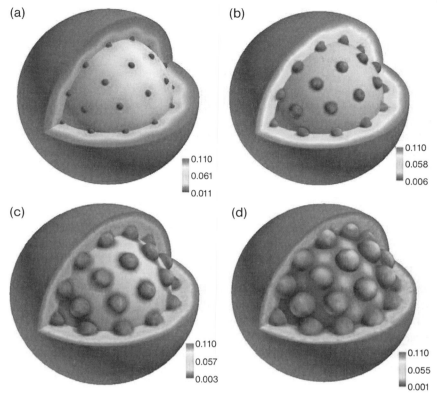

Fig. 5.4 Example of results for the microscopic model used to examine charge transport and mass distribution at a catalyst particle. The system is modeled as a carbon support with multiple embedded active catalyst particles, covered by a thin ionomer layer. In this figure, the effects of particle size on the distribution of oxygen concentration around catalyst particles from (a) 1.25 to (d) 7.5 nm are examined. Image from [81].

[86]. Empirical or semiempirical stack level and system performance models can also be fitted into this category, and generally fall into the purview of industrial models. As shown in Figure 5.5, PEFCs have a variety of multi-phase water transport phenomena. Multi-phase modeling, which includes the characterization of some or all of the liquid water transport, is generally appropriate for low-temperature PEFCs under all operating regimes, including start-up and shut-down behavior. The additional complexity of adding multi-phase flow permits some investigation of the flooding losses and liquid water storage in the PEFC, which is critical in various degradation mechanisms.

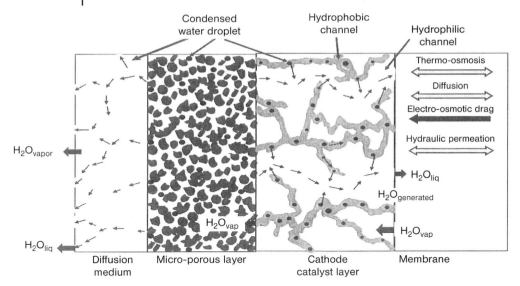

Fig. 5.5 Schematic of various water transport modes in the cathode-side PEFCs. Image from [66].

5.2.3
Single-Cell Models

Although the computational complexity of adding multiple plates into a stack generally prohibits the use of a fully fundamental mechanistic model of an entire stack, a semiempirical and fundamental focused study is essential to stack development. For example, purely fluidic models of flow distribution into the individual stack plates from the main manifold are an essential part of design optimization, although generally not of the level of complexity required for academic publication. Additionally, study of expected temperature transients, coolant flow, and plate-to-plate variations can be made using simplifying assumptions and some form of lumped analysis. Various stack-level models are given in Table 5.1, and there are dozens more available in the literature. Figure 5.6 [87] shows the results of a stack level thermal transient simulation for two different designs of stack. The model correctly predicted observed trends concerning liquid water transport and redistribution upon shut-down to an ambient state.

System-level models, as discussed in Table 5.1, normally deal with online system control. Due to the rapid nature of the sensing and response needed, the models are normally not fundamentally based, and rely on some form of pattern recognition or artificial intelligence to diagnose rapidly any departure from design points and apply corrective action. Figure 5.7 illustrates a typical control system flow diagram for an automotive application. Immediate response to control flow, humidification, temperature control, and power systems is needed to ensure acceptable operation and system longevity. Some system-level modeling

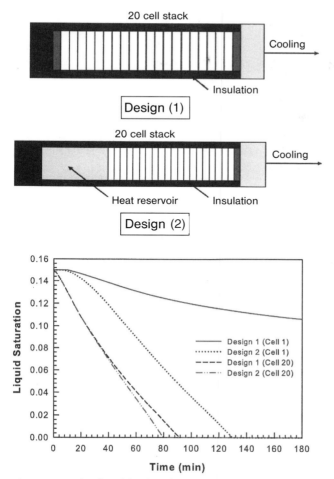

Fig. 5.6 Example of stack-level model output. In this multi-phase model, water transport via phase change was shown to result in significant motion of water in a fuel cell stack during shut-down to an ambient state. Design (1), which induced a steeper temperature gradient in the stack, was shown to increase total water motion and decrease final liquid saturation under ambient conditions. Images from [87].

is designed on a more fundamental basis, and instead had the goal of overall system optimization or analysis. Thus, the need for immediate response is removed, and a coupled and more fundamental examination of the system integration and operational trade-offs can be made. Many examples of these types of studies can be found in the literature.

Although significant progress has been made in understanding and quantifying many physicochemical phenomena using PEFC single-cell level models, there are still many unresolved issues that need further investigation. More comprehensive modeling efforts, from molecular to stack level, are needed to

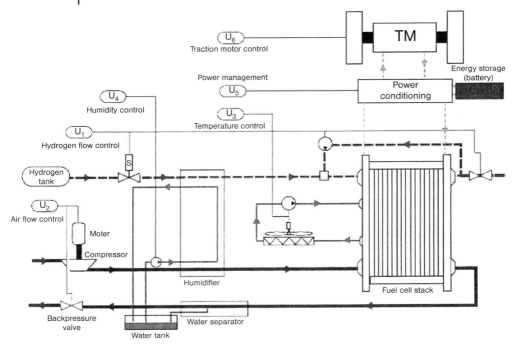

Fig. 5.7 Example of a system level control process flow diagram. All subsystems of humidification, cooling, reactant flow, and recirculation must work in conjunction to avoid unwanted off-design transients which are responsible for rapid stack degradation. Courtesy of Manish Sinha of General Motors.

link the structural and transport characteristics of the fuel cell materials with the durability and performance of the overall system. Such fundamental modeling effort is essential for developing intelligent mitigation strategies and for engineering new materials for prolonged operations.

5.3
Case Study of Water Management in PEFCs

In this section, a detailed analysis and comparison of the model predictions and experimental evidence concerning the multi-phase flow liquid distribution in PEFCs is given. The discrepancies in the results illustrate that detailed experimental validation is necessary, and continual re-evaluation of common modeling assumptions, both implicit and explicit, is needed to assure progress towards accurate solutions.

Improper water management (e.g., liquid water accumulation at various locations in the fuel cell, as illustrated in Figure 5.8) is responsible for real-time performance loss (a phenomenon generically known as flooding), performance in-

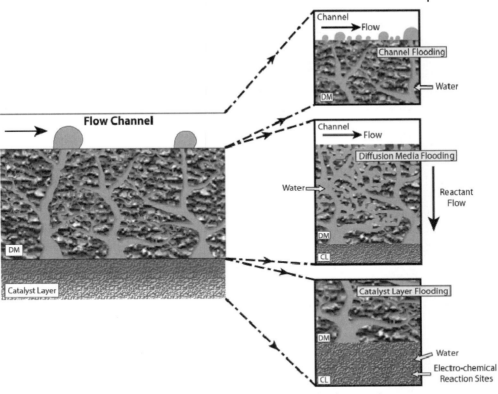

Fig. 5.8 Illustration of possible regions experiencing flooding in a PEFC. Liquid water accumulation in the channels (not shown), diffusion media, catalyst layer, or along interfaces can result in performance decay and degradation. Images from [66].

stability, and loss of efficiency via increased flow channel pressure drop [66]. In addition to this performance loss, however, there is a longer term impact of excess liquid water overhead stored in the fuel cell. Excess liquid water storage in the fuel cell media, including in the catalyst layer, diffusion media [or gas diffusion layer (GDL)], channels, and along the interfaces between the catalyst layer (CL) and microporous layers (MPLs), has been found to be responsible for a majority of fuel cell degradation mechanisms including catalyst, carbon support, ionomer loss, impurity poisoning of the ionomer, freeze–thaw damage, and loss of hydrophobicity [88]. Given that it has been definitively shown by many researchers that liquid water is the key controlling factor in most important forms of fuel cell degradation, much effort has been expended in developing computational models and experimental diagnostics to describe the liquid water distribution in the fuel cell media. Due to experimental limitations, however, the computational models have historically developed at a faster pace than the rate at which detailed experimental validation data could be provided. Recently, the use of high-resolution neutron [89, 90] and X-ray imaging [91] has revealed the true liquid water distribution in the diffusion media and flow channels and along the interfaces of the materials during regular operation in an unaltered fuel cell with true thermal boundary conditions. The results have led to the realization that several of the previously used assumptions and theories about liquid water distribution and behavior in porous media are inappropriate under many normal operating conditions, and new models are needed which include more accurate physical representation of the materials and physics involved.

5.3.1
Experimental Setup

For many of the results described, neutron radiography (NR) has been utilized as a tool for liquid water visualization. Neutron imaging is a nondestructive testing technique that has been shown to be an excellent and precise diagnostic tool for providing quantified images of liquid water distribution in an operating and unaltered fuel cell [e.g., 89, 90, 92–97]. Neutron imaging facilities exist at several locations in the world, including the National Institute of Standards and Technology (NIST) and Pennsylvania State University in the United States, the Paul Scherrer Institute in Switzerland, and others. X-ray imaging is also an emerging technique which can be used to visualize liquid water in PEFCs, although the sensitivity is reduced. The specific details of the various experimental facilities involved in the studies described are given in detail in the cited references.

5.3.2
Evidence of Experimental and Computational Discrepancy

5.3.2.1 Implicit Computational Assumption 1: a Perfect CL/MPL Interface
For computational simplicity, and due to the lack of an available method to model the interfacial region accurately, the interface between the microporous layer and the catalyst layer is ignored, and this is therefore implicitly assumed to be an infinitely thin region with perfect contact. In this way, the microporous layer is perfectly bound by the catalyst layer with no interfacial roughness or gaps, and transport properties change instantaneously between regions. A constant capillary pressure across this boundary is also commonly assumed, and also isotropic in-plane properties of the MPL and CL media. When the actual interfacial morphology of the mating surfaces is scanned using optical profilometry [98], two key microfluidic phenomena become apparent:
1. *There can be a significant interfacial accumulation between the MPL and CL*
   Surface roughness is of the same order of magnitude as the catalyst layer thickness, and much larger than the normal pore size of either the catalyst layer or micro-porous layer (see Figures 5.9 and 5.10) [98]. Therefore, the interaction of these layers, even under compression, will result in void regions capable of storing relatively large amounts of water not accounted for in the computational models [99]. Even when plastic deformation is assumed between the mating

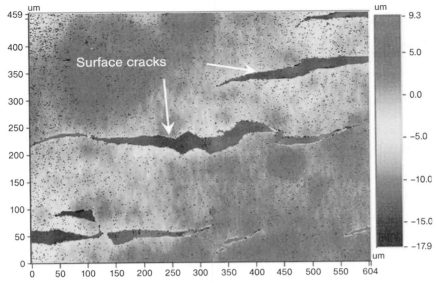

Fig. 5.9 Surface scan of virgin commercially available CL surface using optical profilometry. Images from [99].

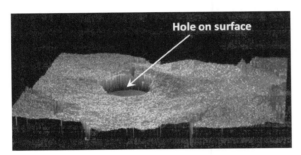

Fig. 5.10 Surface scan of virgin commercially available MPL surface using optical profilometry. Images from [99].

surfaces, the interfacial roughness still exists. Interfacial water has also been shown to be responsible for causing delamination under freeze–thaw conditions by Kim and Mench [100, 101], as shown in Figure 5.11.

2. *Large-scale cracks on the MPL and CL can completely alter the water storage and flow behavior*

As also seen in Figures 5.9 and 5.10, cracks of the order of 20 μm wide and extending through the entire MPL and CL can be found covering up to ∼8% of the surface area of some commercially available materials. These cracks, not accounted for in models, can significantly alter the flow behavior along this interface by providing conduits for capillary flow from the in-

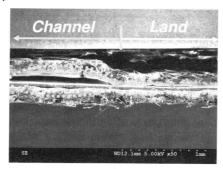

Fig. 5.11 Scanning electron micrograph (SEM) image of membrane electrode assembly delaminated through repeated freeze–thaw cycles, as described in [100, 101].

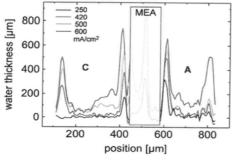

Fig. 5.12 Results of X-ray imaging of operating fuel cell showing liquid accumulation at interfaces in PEFC, and overall liquid distribution very different from that predicted by computational models. Image from [91].

terfacial accumulations, and even act as storage pooling locations for liquid water. At the least, they are direct evidence of highly non-isotropic transport behavior.

Recent experimental evidence from X-ray imaging [91] has definitively shown significant water accumulation at the cathode MPL/CL interface, as shown in Figure 5.12 [91]. A computational model which predicts the interfacial morphology compression at the MPL/CL interface using advanced tools of tribology and directly measured surface morphology showed that between 5 and 20% of the total water content stored in a PEFC can be accumulated at the MPL/CL interface for completely elastic compression [99]. Therefore, a more accurate representation of this region must be included in future multi-phase modeling efforts. Due to the small scale features of the real interfaces involved (on the order of microns), however, multi-dimensional meshes capable of simulating the true interfacial morphology must be extremely fine, and greatly complicate

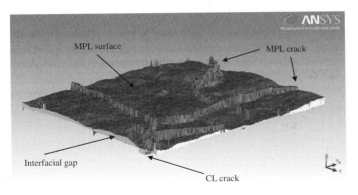
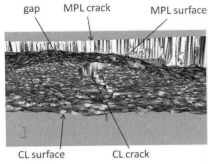

Fig. 5.13 Top view and side view of computational mesh required to capture micron-level morphological features between a MPL/CL surface taken from experimental data at Pennsylvania State University. Domain size is 604×459 μm with an interfacial gap distance of 0–42 μm. The maximum element size is 5 m, with a shell mesh (surface) size of 446 184 elements. From work by Yi Zhang, Pennsylvania State University Fuel Cell Dynamics and Diagnostics Laboratory.

the ability of the computer code to converge. An example of a computer mesh of the interfacial region between the catalyst layer and microporous layer in a typical PEFC, based on actual measured surface morphology. is shown in Figure 5.13. To capture the measured morphological features, the computational domain, which measures only about 500 μm^2 in area by 50 μm thick has nearly half a million mesh elements.

5.3.2.2 Computational Assumption 2: Capillary Flow Dominance

Many computational models assume a basic Leverett function borrowed from the field of soil science to describe the critical relationship between capillary pressure and saturation [102]. This relationship is critical, as it controls the flow of liquid water through the porous media via capillary action. Although this was a reasonable assumption at the time, as no alternative description was available, recent evidence and direct measurements show this approximation to yield significant errors [102]. Several groups have now presented new relationships derived from directly measured experimental data [103–105], yet the use of the flawed relationship persists.

Additionally, many modelers continue to ignore transport from phase change-induced flow by assuming isothermal flow or neglecting phase change. Models that do integrate the impact of phase change do not yet include evaporation or condensation kinetics, and instead predict motion based solely on the thermodynamic saturation pressure relationship with temperature and diffusion. This may or may not be accurate, as it has not yet been validated by experimental data. Some recent models do include this important phenomena, but many still do not. New experimental work has determined that significant PCI flow can result in the fuel cell from temperature differences of even small fractions of a degree [72, 73]. This is especially relevant at shut-down [106–108], when water redistribution can result in amplification of freeze damage if the temperature gradients created are unfavorable.

Additionally, a fundamental misunderstanding of the physical nature of capillary flow and a lack of proper inclusion of the two regimes of condensate accumulation in porous media in many models have led to an improper treatment of the capillary flow, and concomitant poor prediction of the water distribution profile. During operation, water vapor will condense in the diffusion media as it cools toward the channel. In porous media, the flow will condense in a dropwise manner, in a pendular regime without capillary flow until the saturation value reaches a critical threshold when drops coalesce and a continuous conduit is formed [109]. In most PEFC materials, this threshold appears to be around 15–20% [89]. Above this saturation level, the flow enters a connected, funicular regime where the capillary connection between the droplets is established, and a rapidly flowing capillary channel develops. In this regime, the capillary flow is dominant in the porous media. However, the source of liquid is still from a condensation of the product vapor. As a result, models which neglect phase-change effects predict a monotonically decreasing profile of water saturation through the GDL from the CL side to the channel side, as shown in Figure 5.14. This is not observed experimentally, however, as shown in Figure 5.15. Instead, there is a peak profile within the diffusion media, also confirmed by results of Weber and Hickner [90], that is a result of the condensation source term and flow away from the peak by capillary action. There is a periodic build-up of water until a funicular regime is reached, then drainage flow cycle that is observed [109]. A maximum saturation in the GDL is reached for a variety of operating conditions, indicating that the GDL-based capillary flow is very rapid, and dominates removal once the irreducible saturation and funicular regime is reached. Therefore, the peak value of the saturation observed is at the transition saturation for the particular GDL material. This physical model is consistent with all data and observation to date.

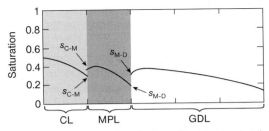

Fig. 5.14 Computational prediction of liquid water distribution in the catalyst layer, microporous layer, and diffusion media when only capillary flow with perfect interfaces is assumed. Image from [110].

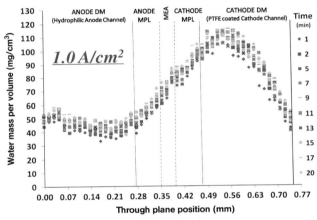

Fig. 5.15 Direct experimental measurement of water distribution in an operating PEFC, showing GDL peak water content in the cathode diffusion media (DM) resulting from condensation source. Image from [111].

5.3.2.3 Computational Assumption 3: a "Saturation Jump" Exists at the MPL/CL Interface

As a result of the perfect MPL|CL interface simplification used in most modeling efforts, the boundary condition applied in multi-phase fuel cell models as at the abrupt transition from the CL to the MPL is most often that of a continuity of capillary pressure. Capillary pressure is inversely proportional to the pore radius, and the MPL and GDL have very different pore radii (GDL is much larger). Hence, with this simplification, a "saturation jump" at the MPL|GDL interface is predicted in all such models and is a commonly held belief. The theory holds that there should be a higher saturation value in the GDL side due to the larger pore sizes. However, experimental evidence shows conclusively that, at least in terms of average saturation values, this does not happen, as shown in Figure 50.16 from data taken with neutron imaging under normal operating

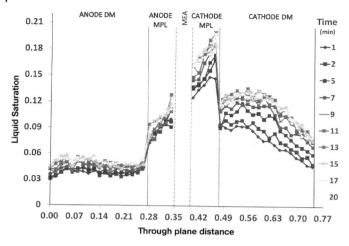

Fig. 5.16 Through-plane liquid saturation profile inside a GDL at 0.2 A cm^{-2} with an assumed MPL porosity of 0.5. Notice the reverse saturation jump at the MPL/DM interface. Image from [111].

conditions [111]. In fact, there is a *reverse* saturation jump along this interface, so that the effective saturation in the MPL is *higher* than in the GDL at this boundary location.

The reasons for this behavior appear to be related to storage of liquid in the cracks in the MPL and CL and along the interface, but this is not yet confirmed. That is, the non-cracked MPL portion may still have a low saturation, as predicted from capillary theory, but the presence of the liquid in the cracks dwarfs this water content by comparison in the global average saturation calculation. Additionally, the presence of the cracks erodes the normally envisioned functioning of the MPL as a liquid water barrier. The void space in the MPL from the cracks is sufficient to produce the increased average saturation values, which are calculated by assuming the void space in the MPL without cracks. More data are needed with increased spatial resolution to elucidate the reasons for this behavior, but it is clear from the repeated data from different groups that the predicted saturation jump does not occur on a global average basis, and additional consideration of the physics is needed to model the fuel cell liquid distribution properly.

5.3.2.4 Computational Assumption 4: the Bruggeman Relationship is Appropriate

In a majority of fuel cell models, a Bruggeman relationship to account for the flow restriction of reactant diffusion through partially saturated pores is used, as shown below [66]:

$$D_{\text{eff}} = D^0 \phi (1-s)^{1.5} \tag{5.1}$$

where D_{eff} is the effective diffusivity, ϕ is the dry porosity, and s is the liquid saturation. This relationship is typically used to modify diffusivity through all porous media in macroscopic fuel cell models. Here, the saturation is assumed merely to reduce the effective porosity. However, this relationship implicitly assumes that the pore itself is segregated into liquid-filled and gas-filled sections, with a straight pathway for gas flow remaining available until 100% saturation, as illustrated in Figure 5.17. This is clearly not the case if some small portion of liquid is aligned to block a small pore opening, or remains as a film across an access location. As a result, models using this approach without some modification must either restrict the initial porosity to unrealistically low values, or vastly over-predict the water saturation levels needed to flood the cell and cause performance decay. If, instead, a thin-film model were used requiring oxygen penetration though a continuous liquid layer on the catalyst, the water content is vastly under-predicted compared with measured results [111]. Clearly, a mixed approach accounting for some film blockage and some equivalent pore-filling effect is appropriate. It should be noted that some of the more advanced computational multi-fluidic models now do include some impact of film formation to account for the vast discrepancy.

The fact that the saturation levels required to cause the voltage level decay seen experimentally is predicted to be significantly higher than what is observed experimentally (5–15 mg cm^{-2} total liquid is commonly observed) is proof alone that some fundamental limitation exists in the model that must be resolved.

Recently, advanced diagnostics have been developed which can finally achieve the necessary resolution in time and space to validate common assumptions and beliefs in multi-phase single-cell performance models in PEFCs. These accumulated tools have revealed that many of the basic assumptions and common beliefs regarding water distribution and flux in PEFCs are false, and must be

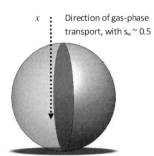
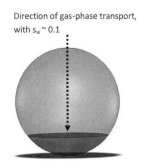

Fig. 5.17 Illustration of failure of Bruggeman relationship for multi-phase flow.
(a) The partially filled pore illustrates the implicit assumption of the Bruggeman relationship, that the multi-phase flow is segregated parallel to the direction of transport. That is, even 99% saturation allows a fraction of gas-phase connected transport through only the gas-phase. (b) An illustration of how significantly less water saturation in a pore can result in complete shut-down of gas-phase transport. Image from [111].

updated to resolve the discrepancies observed. In particular, it has been determined that an unbonded MPL|CL interface can be critical in terms of water storage and transport via interfacial roughness and relatively large-scale cracking. As a result of this and phase change-induced flow, the so-called saturation jump at the MPL|GDC interface is not observed. In the GDL, a peak of water content is observed, with a maximum saturation equal to the transition saturation from the dropwise pendular to the connected funicular regime. From this peak, capillary action is believed to flow liquid periodically out into the channel and back towards the interface of the MPL facing the GDL.

5.4
Future Research Needs

In order for fuel cells to become a commercial reality for portable, stationary, and transportation applications, they must become at least as durable and efficient as their conventional competitors with a similar or reduced cost. This can only be achieved through accurate and robust models which are are needed across a continuum of length scales to help understand physical phenomena, engineer materials and design, and achieve optimal control.

Microscopic-level modeling holds perhaps the greatest promise for understanding fundamental charge transport processes and engineering the inexpensive and high-performance electrolyte materials that are needed to achieve long-term cost and performance objectives. At this stage of development, significant (e.g., order of magnitude) increases in catalyst reactivity with reduced precious metal loading and ionomer conductivity can probably only be achieved through engineering at the molecular level. Advanced models which examine fundamental transport in the electrolyte and reaction at the catalyst have shown great promise for understanding these limitations, but are still limited in utility due to the small computational domain and inability to manufacture and study experimentally nano-structured materials, although great strides are being made in these areas.

The most active level of modeling in terms of academic study has been the single-cell performance-based model. At this stage of development, basic performance models are replete, and more complex models with detailed validation are needed to help understand and engineer materials with high performance and to promote longevity. Although these studies have contributed a great deal of understanding to the community, there is still much work to be done, particularly for multi-phase flow models where recent experimental evidence has shown the many classical modeling predictions to be inaccurate.

Stack-based models suffer from an inability to include a full range of physics due to the computational complexity, and are therefore useful for examining specific phenomena such as shut-down and fluidics in a manifold structure, or as semiempirically based models used for narrow purposes.

Much of the durability concerns in PEFC systems have been linked to off-design transients which occur due to imperfect load following behavior from the control systems. Therefore, proper control models are absolutely critical to enhance efficiency and durability. Therefore, online control-level models must continue to be developed as systems mature. Additionally, long-term scale degradation sensors must be developed as predictive and mitigative sensing technology to assist online control systems.

Finally, the classical problem of integrating the models across length scales has not been resolved, and the incorporation of microscopic approaches into macroscopic models and beyond is both computationally and analytically challenging. No single model can satisfy every need, and in the end, many different models across length scales and purpose are needed to push technology forward. The field of fuel cell modeling will continue to look to other fields with greater computational sophistication to adopt new techniques to study various nonlinear phenomena.

Durability and lifetime modeling presents special challenges in that the uncertainty in the parameters and the phenomena themselves which are responsible for the degradation is high, in many cases a result of local transients, and occurs over a significantly longer time scale than normal parametric change-related performance variations. Based on the more developed but similar scientific field of batteries, lifetime performance degradation modeling will remain a major challenge for some time to come, although the fundamental mechanisms for degradation are becoming more understood and predictive based on fundamental modeling.

Acknowledgments

This work is based on the cumulative understanding gained from research programs sponsored by various government and industrial sources. M.M. Mench wishes to thank his excellent team of dedicated students and engineers for their help. He also acknowledges the National Science Foundation CAREER Award No. 0644811 for partial support of this work, and the various industrial sponsors of his laboratory.

References

1 Siegel, C. (2008) Review of computational heat and mass transfer modeling in polymer-electrolyte-membrane (PEM) fuel cells. *Energy*, **33** (9), 1331–1352.
2 Bhambare, K.S., Gupta, S., Mench, M.M., and Ray, A. (2008) A carbon monoxide sensor in polymer electrolyte fuel Gupta cells based on symbolic dynamic filtering. *Sens. Actuators B: Chem.*, **134** (2), 803–815.
3 Ye, M., Wang, X., and Xu, Y. (2009) Parameter identification for proton exchange membrane fuel cell model using particle swarm optimization. *Int. J. Hydrogen Energy*, **34** (2), 981–989.
4 Ingimundarson, A., Stefanopoulou, A.G., and McKay, D.A. (2008) Model based detection of hydrogen leaks in a fuel cell stack. *IEEE Trans. Control Syst. Technol.*, **16** (5), 1004–1012.

5 Cheng, C. H., Malek, K., Sui, P. C., and Djilali, N. (2010) Effect of Pt nano-particle size on the microstructure of PEM fuel cell catalyst layers: insights from molecular dynamics simulations. *Electrochim. Acta*, **5**, 1588–1597.

6 Goddard, W., Merinov, B., van Duin, A., Jacob, T., Blanco, M., Molinero, V., Jang, S. S., and Jang, Y. H. (2006) Multi-paradigm multi-scale simulations for fuel cell catalysts and membranes. *Mol. Simul.*, **32**, 251–268.

7 Kanako, H., Takashi, T., and Katsuhide, O. (2005) Molecular dynamics simulation for the behavior of H_2 on the electrode catalyst of fuel cell. In *Proceeedings of the International Symposium for Advanced in Fluids*, Vol. 5, pp. 57–58.

8 Malek, K., Eikerling, M., Wang, Q., Navessin, T., and Liu, Z. (2007) Self-organization in catalyst layers of polymer electrolyte fuel cells. *J. Phys. Chem. C*, **111** (36), 13627–13634.

9 Brandell, D., Karo, J., Liivat, A., and Thomas, J. O. (2007) Molecular dynamics studies of the Nafion®, Dow® and Aciplex® fuel-cell polymer membrane systems. *J. Mol. Model.*, **13** (10), 1039–1046.

10 Urata, S., Irisawa, J., Takada, A., Shinoda, W., Tsuzuki, S., and Mikami, M. (2005) Molecular dynamics simulation of swollen membrane of perfluorinated ionomer. *J. Phys. Chem. B*, **109** (9), 4269–4278.

11 Oh, K. S., Kim, D. H., and Park, S. (2008) Behaviour of water molecules in Nafion 117 for polymer electrolyte membrane fuel cell by molecular dynamics simulation. *Mol. Simul.*, **34** (10), 1237–1244.

12 Brunello, G., Lee, S., Jang, S., and Qi, Y. (2009) A molecular dynamics simulation study of hydrated sulfonated poly(ether ether ketone) for application to polymer electrolyte membrane fuel cells: effect of water content. *J. Renew. Sustain. Energy*, **1**, 033101–033115.

13 Niu, X.-D., Munekata, T., Hyodo, S.-A., and Suga, K. (2007) An investigation of water-gas transport processes in the gas-diffusion-layer of a PEM fuel cell by a multi-phase multiple-relaxation-time lattice Boltzmann model. *J. Power Sources*, **172** (2), 542–552.

14 You, L-X. and Liu, H.-T. (2006) A two-phase flow and transport model for PEM fuel cells. *J. Power Sources*, **155** (2), 219–230.

15 Park, J. and Li, X. (2008) Multi-phase micro-scale flow simulation in the electrodes of a PEM fuel cell by lattice Boltzmann method, *J. Power Sources*, **178** (1), 248–257.

16 Wang, Y. (2008) Modeling of two-phase transport in the diffusion media of polymer electrolyte fuel cells. *J. Power Sources*, **185** (1), 261–271.

17 Sun, P.-T., Xue, G.-G., Wang, C.-Y., and Xu, J.-C. (2009) A domain decomposition method for two-phase transport model in the cathode of a polymer electrolyte fuel cell. *J. Comput. Phys.*, **228** (16) 6016–6036.

18 Ge, J.-B. and Liu, H.-T. (2007) A three-dimensional two-phase flow model for a liquid-fed direct methanol fuel cell. *J. Power Sources*, **163** (2) 907–915.

19 Natarajan, D. and Nguyen, T. V. (2003) Three-dimensional effects of liquid water flooding in the cathode of a PEM fuel cell. *J. Power Sources*, **115** (1), 66–80.

20 Weber, A. Z. and Newman, J. (2006) Coupled thermal and water management in polymer electrolyte fuel cells, *J. Electrochem. Soc.*, **153**, A2205.

21 Shimpalee S. and Van Zee J. W. (2007) Numerical studies on rib and channel dimension of flow-field on PEMFC performance. *Int. J. Hydrogen Energy*, **32** (7) 842–856.

22 Shimpalee S., Beuscher U., and Van Zee J. W. (2007) Analysis of GDL flooding effects on PEMFC performance. *Electrochim. Acta*, **52** (24), 6748–6754.

23 Meng, H. (2007) Numerical investigation of transient responses of a PEM fuel cell using a two-phase non-isothermal mixed-domain model. *J. Power Sources*, **171** (2), 738–746.

24 Meng, H. (2009) Multi-dimensional liquid water transport in the cathode of a PEM fuel cell with consideration of the micro-porous layer (MPL). *Int. J. Hydrogen Energy*, **34** (13), 5488–5497.

25 Jung, H.-M., Lee, K.-S., and Um, S. (2008) Macroscopic analysis of characteristic water transport phenomena in polymer electrolyte fuel cells. *Int. J. Hydrogen Energy*, **33** (8), 2073–2086.

26 Um, S. and Wang, C.-Y. (2006) Computational study of water transport in proton exchange membrane fuel cells. *J. Power Sources*, **156** (2) 211–223.

27 Pasaogullari, U. and Wang, C.-Y. (2004) Two-phase transport and the role of microporous layer in polymer electrolyte fuel cells. *Electrochim. Acta*, **49** (25), 4359–4369.

28 Natarajan, D. and Nguyen, T.V. (2003) Three-dimensional effects of liquid water flooding in the cathode of a PEM fuel cell. *J. Power Sources*, **115** (1), 66–80.

29 Jeon, D.H., Greenway, S., Shimpalee, S., and Van Zee, J.W. (2008) The effect of serpentine flow-field designs on PEM fuel cell performance. *Int. J. Hydrogen Energy*, **33** (3), 1052–1066.

30 Yan, Y.-M., Liu, H.-C., Soong, C.-Y., Chen, F., and Cheng C.H. (2006) Numerical study on cell performance and local transport phenomena of PEM fuel cells with novel flow field designs. *J. Power Sources*, **161** (2), 907–919.

31 Wang, X.-D., Zhang, X.-X., Yan, W.-M., Lee, D.-J., and Su, A. (2009) Determination of the optimal active area for proton exchange membrane fuel cells with parallel, interdigitated or serpentine designs. *Int. J. Hydrogen Energy*, **34** (9) 3823–3832.

32 Wang, X.-D., Duan, Y.-Y., Yan, W.-M., Lee, D.J., Su, A., and Chi, P.H. (2009) Channel aspect ratio effect for serpentine proton exchange membrane fuel cell: role of sub-rib convection. *J. Power Sources*, **193** (2) 684–690.

33 Grujicic, M. and Chittajallu, K.M. (2004) Design and optimization of polymer electrolyte membrane (PEM) fuel cells. *Appl. Surf. Sci.*, **227** (1–4), 56–72.

34 Li, P.-W., Schaefer, L., Wang Q.-M., Zhang, T., and Chyu, M.K. (2003) Multi-gas transportation and electrochemical performance of a polymer electrolyte fuel cell with complex flow channels. *J. Power Sources*, **115** (1), 90–100.

35 Nguyen, P.T., Berning, T., and Djilali, N. (2004) Computational model of a PEM fuel cell with serpentine gas flow channels. *J. Power Sources*, **130** (1–2), 149–157.

36 Maharudrayya, S., Jayanti, S., and Deshpande, A.P. (2005) Flow distribution and pressure drop in parallel-channel configurations of planar fuel cells. *J. Power Sources*, **144** (1), 94–106.

37 Yuan, W., Tang, Y., Pan, M.-Q., Li, Z.-T., and Tang, B. (2010) Model prediction of effects of operating parameters on proton exchange membrane fuel cell performance. *Renew. Energy*, **35** (3), 656–666.

38 Zhukovsky, K. and Pozio, A. (2004) Maximum current limitations of the PEM fuel cell with serpentine gas supply channels. *J. Power Sources*, **130** (1–2), 95–105.

39 Lai, Y.-H., Rapaport, P.A., Ji, C.-X., and Kumar, V. (2008) Channel intrusion of gas diffusion media and the effect on fuel cell performance. *J. Power Sources*, **184** (1), 120–128.

40 Promislow, K. and Wetton, B. (2005) A simple, mathematical model of thermal coupling in fuel cell stacks. *J. Power Sources*, **150** (4), 129–135.

41 Chang, P.A.C., St-Pierre, J., Stumper, J., and Wetton, B. (2006) Flow distribution in proton exchange membrane fuel cell stacks. *J. Power Sources*, **162** (1), 340–355.

42 Karimi G. and Li, X.-G. (2006) Analysis and modeling of PEM fuel cell stack performance: Effect of *in situ* reverse water gas shift reaction and oxygen bleeding. *J. Power Sources*, **159** (2), 943–950.

43 Karimi, G., Jafarpour, F., and Li, X. (2009) Characterization of flooding and two-phase flow in polymer electrolyte membrane fuel cell stacks. *J. Power Sources*, **187** (1), 156–164.

44 Meiler, M., Andre, D., Schmid, O., and Hofer, E.P. (2009) Nonlinear empirical model of gas humidity-related voltage dynamics of a polymer-electrolyte-membrane fuel cell stack. *J. Power Sources*, **190** (1), 56–63.

45 Khandelwal, M., Lee, S., and Mench, M.M. (2007) One-dimensional thermal model of cold-start in a polymer electrolyte fuel cell stack. *J. Power Sources*, **172** (2), 816–830.

46 Sundaresan, M. and Moore, R.M. (2005) Polymer electrolyte fuel cell stack thermal model to evaluate sub-freezing startup. *J. Power Sources*, **145** (2), 534–545.

47 Meiler, M., Schmid, O., Schudy, M., and Hofer, E.P. (2008) Dynamic fuel cell

stack model for real-time simulation based on system identification. *J. Power Sources*, **176** (2), 523–528.

48 Zhang, Y., Mawardi, A., and Pitchumani, R. (2007) Numerical studies on an air-breathing proton exchange membrane (PEM) fuel cell stack. *J. Power Sources*, **173** (1), 264–276.

49 Philipps, S.P. and Ziegler, C. (2008) Computationally efficient modeling of the dynamic behavior of a portable PEM fuel cell stack. *J. Power Sources*, **180** (1), 309–321.

50 Meiler, M., Schmid, O., Schudy, M., and Hofer, E.P. (2008) Dynamic fuel cell stack model for real-time simulation based on system identification. *J. Power Sources*, **176** (2), 523–528.

51 Pathapati, P.R., Xue, X., and Tang, J. (2005) A new dynamic model for predicting transient phenomena in a PEM fuel cell system. *Renew. Energy*, **30** (1), 1–22.

52 Wishart, J., Dong, Z., and Secanell, M. (2006) Optimization of a PEM fuel cell system based on empirical data and a generalized electrochemical semiempirical model. *J. Power Sources*, **161** (2), 1041–1055.

53 Brown, T.M., Brouwer, J., Samuelsen, G.S., Holcomb, F.H., and King, J. (2008) Dynamic first principles model of a complete reversible fuel cell system. *J. Power Sources*, **182** (1), 240–253.

54 Hatti, M. and Tioursi, M. (2009) Dynamic neural network controller model of PEM fuel cell system. *Int. J. Hydrogen Energy*, **34** (11), 5015–5021.

55 Rouss, V. and Charon, W. (2008) Multi-input and multi-output neural model of the mechanical nonlinear behaviour of a PEM fuel cell system. *J. Power Sources*, **175** (1), 1–17.

56 Zhang, Y.-J., Ouyang, M.-G., Lu, Q.-C., Luo, J.-X., and Li, X.-H. (2004) A model predicting performance of proton exchange membrane fuel cell stack thermal systems. *Appl. Thermal Eng.*, **24** (4), 501–513.

57 Meiler, M., Andre, D., Schmid, O., and Hofer, E.P. (2009) Nonlinear empirical model of gas humidity-related voltage dynamics of a polymer-electrolyte-membrane fuel cell stack. *J. Power Sources*, **190** (1), 56–63.

58 Xue, X., Tang, J., Smirnova, A., England, R., and Sammes, N. (2004) System level lumped-parameter dynamic modeling of PEM fuel cell. *J. Power Sources*, **133** (2), 188–204.

59 Muller, E.A. and Stefanopoulou, A.G. (2006) Analysis, modeling, and validation for the thermal dynamics of a polymer electrolyte membrane fuel cell sytems. *ASME J. Fuel Cell Sci. Technol.*, **3** (2), 99–110.

60 Vasu, G. and Tangirala, A.K. (2008) Control-orientated thermal model for proton-exchange membrane fuel cell systems. *J. Power Sources*, **183** (1), 98–108.

61 Haraldsson, K. and Wipke, K. (2004) Evaluating PEM fuel cell system models. *J. Power Sources*, **126** (1–2), 88–97.

62 Cheddie, D. and Munroe, N. (2005) Review and comparison of approaches to proton exchange membrane fuel cell modeling. *J. Power Sources*, **147** (1–2), 72–84.

63 Smitha, B., Sridhar, S., and Khan, A.A. (2005) Solid polymer electrolyte membranes for fuel cell applications – a review. *J. Membr. Sci.*, **259** (1–2), 10–26.

64 Djilali, N. (2007) Computational modelling of polymer electrolyte membrane (PEM) fuel cells: challenges and opportunities. *Energy*, **32** (4), 269–280.

65 Wang, C.Y. (2004) Fundamental models for fuel cell engineering, *Chem. Rev.*, **104** (10), 4727–4766.

66 Mench, M.M. (2008) *Fuel Cell Engines*, John Wiley & Sons, Inc., Hoboken, New Jersey.

67 Zawodzinski, T.A., Davey, J., Valerio, J., and Gottesfeld, S. (1995) The water content dependence of electroosmotic drag in proton-conducting polymer electrolytes. *Electrochim Acta*, **40** (3), 297–302.

68 Ise, M., Kreuer, K.D., and Maier, J. (1999) Electroosmotic drag in polymer electrolyte membranes: an electrophoretic NMR study. *Solid State Ionics*, **125**, 213–223.

69 Ren, X.M. and Gottesfeld, S. (2001) Electro-osmotic drag of water in poly(perfluorosulfonic acid) membranes. *J. Electrochem. Soc.*, **148** (1), A87–A93.

70 Park, Y.H. and Caton, J.A. (2008) An experimental investigation of electro-

osmotic drag coefficients in a polymer electrolyte membrane fuel cell, *Int. J. Hydrogen Energy*, **33** (24), 7513–7520.

71 Luo, Z.-P., Chang, Z.-Y., Zhang, Y.-X., Liu, Z., and Li, J. (2009) Electro-osmotic drag coefficient and proton conductivity in Nafion® membrane for PEMFC. *Int. J. Hydrogen Energy,* DOI: 10.1016/j.ijhydene.2009.09.013

72 Kim, S. and Mench, M.M. (2009) Investigation of temperature-driven water transport in polymer electrolyte fuel cell: thermo-osmosis in membranes. *J. Membr. Sci.*, **328**, 113–120.

73 Kim, S. and Mench, M.M. (2009) Investigation of temperature-driven water transport in polymer electrolyte fuel cell: phase-change induced flow. *J. Electrochem. Soc.*, **156**, B353–B362.

74 Weber, A.Z. and Newman, J. (2003) Transport in polymer-electrolyte membrane – I. Physical model. *J. Electrochem. Soc.*, **150** (7), A1008–A1015.

75 Motupally, S., Becker, A.J., and Weidner, J.W. (2000) Diffusion of water in Nafion 115 membranes. *J. Electrochem. Soc.*, **147** (9), 3171–3177.

76 Zawodzinski, T.A., Neeman, M., Sillerud, L.O., and Gottesfeld, S. (1991) *J. Phys. Chem.*, **95**, 6040.

77 Fuller, T.F. (1992) PhD thesis, University of California, Berkeley, CA.

78 Nguyen, T.V. and White, R.E. (1993) *J. Electrochem. Soc.*, **140**, 2178.

79 Bishop, B. (2005) Nanotechnology and the end of Moore's law? *Bell Lab. Tech. J.*, **10** (3), 23–28.

80 Manahan, M.P., Kim, S., Kumbur, E.C., and Mench, M.M. (2009) Effects of surface irregularities and interfacial cracks on polymer electrolyte fuel cell performance. *ECS Trans.*, **25**, 1745–1754.

81 Yan, Q.-G. and Wu, J.-X. (2008) Modeling of single catalyst particle in cathode of PEM fuel cells. *Energy Convers. Manage.*, **49**, 2425–2433.

82 Cheddie, D. and Munroe, N. (2006) Parametric model of an intermediate temperature PEMFC. *J. Power Sources*, **156**, 414.

83 Cheddie, D. and Munroe, N. (2006) Mathematical model of a PEMFC using a PBI membrane. *Energy Convers. Manage.*, **47**, 1490.

84 Cheddie, D. and Munroe, N. (2006) Analytic correlations for intermediate temperature PEM fuel cells. *J. Power Sources*, **160**, 299.

85 Cheddie, D. and Munroe, N. (2006) Three-dimensional modeling of high temperature PEMFCs. *J. Power Sources*, **160**, 215.

86 Cheddie, D. and Munroe, N. (2007) A two-phase model of an intermediate temperature PEM fuel cell. *Int. J. Hydrogen Energy*, **32**, 832.

87 Khandelwal, M. (2008) PhD thesis, Pennsylvania State University.

88 Borup, R., Meyers, J., Pivovar, B., Kim, Y.S., Mukundan, R., Garland, N., Myers, D., Wilson, M., Garzon, F., Wood, D., Zelenay, P., Moore, K., Stroh, K., Zawodzinski, T., Boncella, J., McGrath, J.E., Inaba, M., Miyatake, K., Hori, M., Ota, K., Oguni, Z., Miyata, S., Nishikata, A., Siroma, Z., Uchimoto, Z., Yasuda, K., Kinijima, K., and Iwashita, N. (2007) Scientific Aspects of Polymer Electrolyte Fuel Cell Durability and Degradation, *Chem. Rev.*, **107** (7), 3904.

89 Turhan, A., Kim, S., Hatzell, M., and Mench, M.M. (2009) Impact of channel wall hydrophobicity on through-plane water distribution and flooding behavior in a polymer electrolyte fuel cell. *Electrochim. Acta*, **55** (08), 2734–2745.

90 Weber, A.Z. and Hickner, M. (2008) Modeling and high-resolution-imaging studies water-content profiles in a polymer-electrolyte-fuel-cell membrane-electrode assembly, *Electrochim. Acta*, **26**, 7668–7674.

91 Hartnig, C., Manke, J., Kuhn, R., Kardjilov, N., Banhart, J., and Lehnert, W. (2008) Cross-sectional insight in the water evolution and transport in polymer electrolyte fuel cells, *Appl. Phys. Lett.*, **92**, 134106.

92 Owejan, J.P., Trabold, T.A., Jacobson, D.L., Arif, M., and Kandlikar, S.G. (2007) Effects of flow field and diffusion layer properties on water accumulation in a PEM fuel cell, *Int. J. Hydrogen Energy*, **32** (17), 4489–4502.

93 Bolliat, P., Kramer, D., Seyfang, B.C., Frei, G., Lehmann, E., Scherer, G.G., Wokaun, A., Ichikawa, Y., Tasaki, Y., and Shinohara, K. (2008) In situ observation

of the water distribution across a PEFC using high resolution neutron radiography, *Electrochem Commun.*, **10** (4), 546–550.

94 Park, J., Li, X., Tran, D., Abdel-Baset, T., Hussey, D. S., Jacobson, D. L., and Arif, M. (2008) Neutron imaging investigation of liquid water distribution in and the performance of a PEM fuel cell, *Int. J. Hydrogen Energy*, **33** (13), 3373–3384.

95 Pekula N., Heller, K., Chuang, P. A., Turkan, A., Mench, M. M., Brenizer, J. S., and Ünlü, K. (2005) Study of water distribution and transport in a polymer electrolyte fuel cell using neutron imaging, *Nucl. Instrum. Methods Phys. Res. Sect. A*, **542** (1–3), 134–141.

96 Turhan, A., Heller, K., Brenizer, J. S., and Mench, M. M. (2006) Quantification of liquid water accumulation and distribution in a polymer electrolyte fuel cell using neutron imaging. *J. Power Sources*, **160**, 1195–1203.

97 Hussey, D., Jacobson, D. L., Arif, M., Owejan, J. P., Gagliardo, J. J., and Trabold, T. A. (2007) Neutron images of the through-plane water distribution of an operating PEM fuel cell, *J. Power Sources*, **172** (1), 225–228.

98 Hizir, F. E., Ural, S. O., Kumbur, E. C., and Mench M. M. (2009) Characterization of interfacial morphology in polymer electrolyte fuel cells: micro-porous layer and catalyst layer surfaces. *J. Power Sources*, **195**, 3463–3471.

99 Swamy, T., Kumbur, E. C., and Mench, M. M. (2010) Characterization of interfacial structure in PEFCs: water storage and contact resistance model, *J. Electrochem. Soc.*, **157**, B77–B85.

100 Kim, S., Ahn, B. K., and Mench, M. M.(2008) Physical degradation of membrane electrode assemblies undergoing freeze/thaw cycling: diffusion media effects, *J. Power Sources*, **179**, 140–146.

101 Kim, S. and Mench, M. M. (2007) Physical degradation of membrane electrode assemblies undergoing freeze-thaw cycling: microstructure effects. *J. Power Sources*, **174**, 206–220.

102 Kumbur, E. C., Sharp, K. V., and Mench, M. M. (2007) On the effectiveness of the Leverett approach to describe water transport in fuel cell diffusion media. *J. Power Sources*, **172**, 816–830.

103 Kumbur, E. C., Sharp, K. V., and Mench, M. M. (2007) A validated Leverett approach to multi-phase flow in polymer electrolyte fuel cell diffusion media. Part 1. Hydrophobicity effect. *J. Electrochem. Soc.*, **154**, B1295–B1304.

104 Kumbur, E. C., Sharp, K. V., and Mench, M. M. (2007) A validated Leverett approach to multi-phase flow in polymer electrolyte fuel cell diffusion media. Part 2. Compression effect and capillary transport. *J. Electrochem. Soc.*, **154**, B1305–B1314.

105 Kumbur, E. C., Sharp, K. V., and Mench, M. M. (2007) A validated Leverett approach to multi-phase flow in polymer electrolyte fuel cell diffusion media. Part 3. Temperature effect and unified approach. *J. Electrochem. Soc.*, **154**, B1315–B1324.

106 He, S., and Mench, M. M. (2006) One-dimensional transient model for frost heave in polymer electrolyte fuel cells. *J. Electrochem. Soc.*, **153**, A1724–A1731.

107 He, S., Kim, S. H., and Mench, M. M. (2007) 1-D transient model for frost heave in polymer electrolyte fuel cells: II. Parametric study. *J. Electrochem. Soc.*, **154**, B1024–B1033.

108 He, S., Lee, J. H., and Mench, M. M. (2007) 1-D transient model for frost heave in polymer electrolyte fuel cells: III. Heat transfer, microporous layer, and cycling effects. *J. Electrochem. Soc.*, **154**, B1227–B1236.

109 Litster, S. and Djilali, N. (2005) *Transport Phenomena in Fuel Cells*, WIT Press, Southampton.

110 Nam, J. H., Lee, K.-J., Hwang, G.-S., Kim, C.-J., and Kaviany M. (2009) Microporous layer for water morphology control in PEMFC, *Int. J. Heat Mass Transfer*, **52** (11–12), 2779–2791.

111 Turhan, A. (2009) Ph. D. thesis, Pennsylvania State University.

Fuel Infrastructures

6
Hydrogen Distribution Infrastructure for an Energy System: Present Status and Perspectives of Technologies

Françoise Barbier

Abstract

Hydrogen as a clean energy carrier (its use produces no CO_2 emissions) offers effective solutions for a secure energy supply and a clean environment. To make hydrogen readily available, a hydrogen infrastructure allowing for its convenient distribution to end users is essential. The choice of relevant modes of hydrogen transport and distribution from the production plant is therefore critical. Safe, cost-effective, and energy-efficient solutions are needed. Factors such as the degree of hydrogen penetration, the environmental performance of the different pathways, and geographic parameters will influence the infrastructure deployment. Advances in delivery technologies will allow the existence of a transition infrastructure to support the commercial introduction of fuel cell vehicles and hydrogen stations. In this chapter, an overview of the distribution technologies already mastered by gas companies such as pipelines, gaseous trailers, and cryogenic trucks is given, with insights into costs, advantages or disadvantages, and advanced solutions. Further, supporting technologies such as compressors, pumps, and hydrogen quality requirements for fuel cell applications are presented. The large-scale storage to regulate hydrogen consumption and production is discussed. Finally, recent progress in refueling at filling stations is presented.

Keywords: hydrogen distribution, hydrogen infrastructure, transport, storage

6.1
Introduction

Today, energy and transport systems are mainly based on fossil energy carriers, giving rise to the two major problems of energy supply and environmental impact. As a consequence, the world has to reduce its rate of fossil fuel consumption, build more efficient and less polluting energy systems, and find renewable substitutes. Among the various options, hydrogen as a clean energy carrier offers effective solutions for a secure energy supply and a clean environment. It

Hydrogen Energy. Edited by Detlef Stolten
Copyright © 2010 WILEY-VCH Verlag GmbH & Co. KGaA, Weinheim
ISBN: 978-3-527-32711-9

has the potential to reduce the environmental impact of automobiles in well to wheel comparisons. Significant numbers of hydrogen-fueled fuel cell vehicles are expected to be deployed within the next several years [1]. However, the successful development of hydrogen as a transportation fuel will depend on the presence of a convenient and safe hydrogen delivery infrastructure to make it available to consumers.

Delivery methods for hydrogen are determined by the production volume and the delivery distance. Because hydrogen can be produced from a variety of domestic resources, production can take place in large, centralized plants or in a distributed manner – directly at refueling stations and stationary power sites. Due to the higher capital investment required for centralized production, distributed production is expected to play a particularly important role while hydrogen is gaining public acceptance. Hydrogen delivery systems must include not only transport from central production operations, but also the storage, compression, and dispensing operations, which are essential no matter where production takes place.

Hydrogen is a gas with a low volumetric energy density (the lower heating value is 10.8 MJ N^{-1} m^{-3} at standard temperature and pressure, compared with 35.2 MJ N^{-1} m^{-3} for natural gas) [2, 3]. Improving hydrogen's volumetric energy density is accomplished with either high pressures (compressed gas) or extremely low temperatures (cryogenic liquid). Today, gaseous or liquid trucks and gaseous pipelines are mature technologies for delivering industrial hydrogen to the point of end use [4].

The current industrial hydrogen system has been used safely for many years in chemical and metallurgical applications, the food industry, and space programs. It therefore provides a technical starting point for building a future hydrogen energy infrastructure. However, the energy-intensive process of liquefaction, the large capital investment for pipelines, and the low hydrogen-carrying capacity of current gaseous trucks result in energy inefficiencies and high delivery costs. Therefore, understanding the factors that contribute to delivery costs and improving the technologies are essential when planning the deployment of a hydrogen energy infrastructure. Alternative pathways such as hydrogen carriers in a form other than free H_2 molecules, such as liquid hydrocarbons, metals hydrides, and adsorbents, could be useful but extensive analysis is needed to evaluate the promising materials. Rail and ships are also potential transport modes, but are not typically used today. The use of the natural gas pipeline network to deliver pure hydrogen or mixed gas is another option to be evaluated.

Compression is also an integral aspect of hydrogen delivery. Gas compressors and liquid pumps are used today but newer approaches are being studied.

Hydrogen production may be variable and is unlikely to match the instantaneous demand for hydrogen. Because the user demand will vary with time, hydrogen storage on a daily, weekly, and seasonal basis will likely be required. The storage of large quantities of hydrogen is therefore a key step in the delivery process to regulate consumption and production. Underground facilities in various types (geological storage, buried tanks) are being considered.

Finally, hydrogen stations to refuel vehicles are part of the delivery infrastructure. There is a need to develop refueling stations with safe filling procedures, reliable and compatible with public expectations, and delivering a quality of hydrogen appropriate for fuel cell use.

This chapter provides an overview of the multiple hydrogen delivery pathways from the location where it is produced to the point of use. The state-of-the art of the technologies in the delivery system is presented. The recent developments made in research and the perspectives are then discussed. The building up of a hydrogen delivery infrastructure and its evolution also require consideration of various scenarios in order to estimate its design and cost, and to understand which factors (hydrogen demand, transport distances, geographic conditions, market fraction, etc.) are most important. It is outside the scope of this chapter to present the distribution models currently being developed to choose the most appropriate delivery mode.

6.2
Hydrogen Transport by Gaseous Pipelines

6.2.1
State of the Art on Hydrogen Pipelines

6.2.1.1 Overview of Hydrogen Networks

A number of commercial hydrogen pipelines are used today to distribute large quantities (tens of thousands of cubic meters per hour) of gaseous hydrogen to the industrial market. Their lengths range from less than 1 km to several hundred kilometers. The major actors are the industrial gas companies, namely Air Liquide, Air Products, Linde, and Praxair. In response to an increased demand for hydrogen by refining customers, existing networks are expanding and new portions are being built (in March 2009, Air Products, as an example, announced a 60 km extension to the US Gulf Coast hydrogen pipeline network in Louisiana). The hydrogen network is estimated at around 1600 km in Europe and 1100 km in North America [5, 6]. Most of the pipelines are located where large quantities of hydrogen are consumed in the refining and chemical sectors. These include systems in northern Europe (covering The Netherlands, northern France and Belgium), Germany (Ruhr and Leipzig areas), the United Kingdom (Teesside), and North America (Gulf of Mexico, Texas–Louisiana, California, Alberta). Smaller systems also exist in South Africa, Brazil, Thailand, Korea, Singapore, and Indonesia. Overall, these pipeline lengths are tiny when compared with the worldwide natural gas transport pipeline system, which would exceed 2 000 000 km.

Figure 6.1 displays parts of the worldwide H_2 pipeline network. For example, the 240 km long pipeline in the Ruhr area of Germany (Figure 6.1a) acquired by Air Liquide in 1998 has been in operation since 1938.

Within the "Zero Regio" European project for hydrogen energy applications, Linde has installed a 900 bar hydrogen pipeline (of 1 in diameter) over a dis-

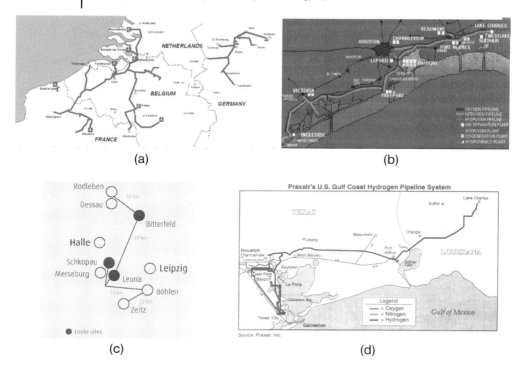
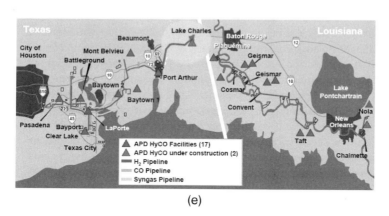

Fig. 6.1 Main hydrogen pipelines in the world. (**a**) Air Liquide hydrogen pipelines in Benelux, France and Germany (Ruhr area). (**b**) Air Liquide hydrogen pipelines in the Gulf Coast (USA). (**c**) Linde hydrogen pipelines in Germany. (**d**) Praxair hydrogen pipelines in the Gulf Coast (USA). (**e**) Air Product hydrogen pipelines in the Gulf Coast (USA). From [4].

tance of 1.7 km in the Frankfurt–Hoechst industrial park to supply fuel cell passenger vehicles [7].

6.2.1.2 Pipeline Characteristics

Pipelines require adequate design, installation, and maintenance procedures. The operating pressure of hydrogen pipelines is generally lower than 100 bar (most commonly between 40 and 70 bar) and the diameter of the pipelines (D) usually ranges from 10 to 300 mm. Current pipelines are made of steel. A technical concern is hydrogen embrittlement of metallic pipelines and welds, characterized by a loss of ductility and rupture when subjected to stress. The steels used for H_2 pipelines are therefore low-carbon, low-alloy, and low-strength steels to reduce the risk of embrittlement (e.g., API X42 steel with C < 0.2 and Mn < 1.3 wt%) [8]. These steels combine economic affordability with an adequate range of physical properties such as strength, toughness, ductility, and weldability. For safety reasons, most pipelines are buried so steels are protected by coatings or cathodic protection to prevent corrosion issues.

Pipeline construction involves extensive welding for joining, with a minimum of inspections before operation for safety considerations. The exploitation of a pipeline network also requires compressor stations as hydrogen is generally available at low pressure. Hydrogen compressors feeding the pipeline system are usually found at locations where hydrogen is produced. The compressors are expensive and require high maintenance, so they are actually not installed if another alternative is possible. For instance, when hydrogen is produced using natural gas (steam methane reforming), the natural gas feedstock can be compressed and the production plant operated at a higher pressure. Friction losses in pipelines with hydrogen are much lower than those with natural gas as the viscosity of hydrogen is lower (the energy loss during transportation of hydrogen is about 4% of the energy content). As a consequence, there should be no need for compressor stations along the pipelines, which should significantly reduce the compressor capital cost.

6.2.1.3 Pipeline Costs

The transport of gaseous hydrogen via existing pipelines is currently the lowest-cost option for delivering large volumes of gas. It is capital intensive and the delivery cost varies with distance at different pipeline capacities (Figure 6.2). The high initial capital costs taking into account new pipeline construction and compressor capital costs constitute a major barrier to expanding the hydrogen pipeline delivery infrastructure for the emerging energy market – particularly while current demand for hydrogen is low.

The total cost of a pipeline for high-pressure transport (given by unit of length) depends on [10]:
- materials, accounting for 15–35% of the total cost depending on the diameter of the pipe;

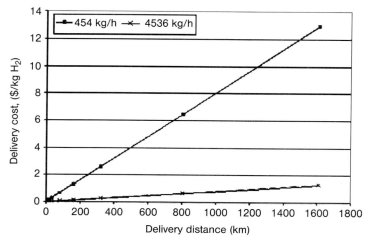

Fig. 6.2 Cost of pipeline delivery with distance at different pipeline capacities. From [9].

- installation (including labor), accounting for 40–50% of the total cost;
- rights of way, corresponding to the land costs for installing pipelines, usually a small fraction of the total cost;
- miscellaneous costs, which include surveys, engineering, supervision, interest, administration and overheads, contingencies, and allowance for funds used during construction (20–30% of the total cost).

Different studies estimate the pipeline capital cost as roughly between US$0.3 and 1.5 million per kilometer, depending on its location and its size (diameter, length). The pipeline cost increases with the diameter of the pipe as the materials cost increases with diameter. Some analytical formulas for capital cost estimations based on correlations with natural gas pipeline costs and as a function of pipe diameter and in some cases pipeline length have been reported [11–12]. Based on these studies, the hydrogen pipeline capital costs were assumed to be 10% higher than those for natural gas (for a given diameter). Figure 6.3 displays the estimated cost of installed pipelines per kilometer as a function of the pipe diameter.

Pipelines thus represent a high initial investment and it increases strongly with distance and flow rates. Cost also increases with the pipeline diameter but as the pipeline's capacity increase is proportional to diameter $D^{2.5}$, the cost of hydrogen transported decreases rapidly with increasing pipe diameter. Together with economy of scale, this makes pipelines a good choice for long-distances, large diameter transmission, and large amounts of hydrogen.

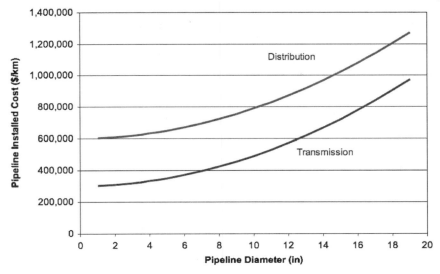

Fig. 6.3 Pipeline model installed cost ($ per kilometer) dependence on pipeline diameter. The total cost installed includes materials costs, installation costs, rights of way, and miscellaneous costs. The gap between transmission and distribution accounts for a mean difference in terms of rights of way and installation costs, but can vary greatly with location. From [11].

6.2.2
Perspectives of Evolution for H_2 Pipelines

A hydrogen pipeline carries about 30% less energy than a natural gas pipeline due to the lower heating value of hydrogen. The distribution of larger energy quantities in hydrogen pipelines requires an increase in flow pressure (>100 bar). This increase in pressure may have implications for the materials that could be used in the pipeline construction.

Furthermore, the operating conditions of a hydrogen pipeline for energy applications would be different from those of an industrial pipeline which today operates at nearly constant pressures, without significant pressure cycles or swings. Hydrogen energy pipelines would have to bear variations of pressure. This may be a concern due to the susceptibility of steels to hydrogen embrittlement, which affects their mechanical properties and decreases their resistance to fatigue crack [13].

To address these challenges, there is renewed interest in research on new pipeline materials compatible with hydrogen and their use at higher operating pressures, and to reduce capital costs.

New steels are being explored to develop a better understanding of hydrogen embrittlement and to identify steel compositions and processes suitable for the construction of a new pipeline infrastructure or potential use of the existing steel pipeline infrastructure [14, 15].

Research is also concentrated on alternatives to metallic pipelines to achieve cost and performance targets for hydrogen transmission and distribution. Polymeric and fiber-reinforced polymer (FRP) pipelines, which have the advantages of being light compared with steels, easier to handle, join, and weld, insensitive to corrosion, and insensitive to hydrogen embrittlement, are under investigation.

Polymeric pipes currently used in the natural gas distribution network are made of polyethylene and have a pressure rating limited to 10 bar. Polymers such as polyamide (and more particularly polyamide-12) are of more interest as the permeability of hydrogen is significantly reduced and its thermo-mechanical properties allow pipes to sustain a 20 bar operating pressure and an 80 °C operating temperature [16]. Therefore, plastic pipes can be an alternative to steel thanks to savings in installation and maintenance costs. However, material supply can represent a high proportion of the total cost.

Pipes in composite materials (FRP) are composed of a thermoplastic liner (mainly polyethylene) wrapped with high-strength fibers (most commonly aramid fibers), then coated with a thermoplastic layer. This last layer provides protection against environmental attacks and helps to retain the wrapping mainly responsible for the mechanical properties. Compared with simple plastic pipes, wrapping with aramid fibers allows pressures up to 100 bar. These reinforced plastic pipes are already used for natural gas and crude oil distribution in the Middle East and their development for H_2 delivery is currently part of the DOE Hydrogen Program (Figure 6.4) [17, 18]. According to the literature [19], FRP pipes could be a cost-effective option to metallic pipes when long lengths can be installed (200–300 m). However, the manufacturing process does not allow the production of plastic pipes with diameters as large as those of steel pipes (100 and 150 mm are the most common diameters). Further developments are still needed to evaluate the feasibility of large-scale manufacturing operations, assess joining technology, and developing codes of practice and standards for hydrogen-service FRP pipelines.

Fig. 6.4 Composite pipeline (FRP) instrumented for testing. From [17].

6.2.3
Use of Existing Natural Gas Pipelines

Another alternative for the expansion of the hydrogen pipeline network would be to use the existing natural gas network (at least for a transition period). There have been many studies (e.g., Nordic H_2 Program, Naturalhy European Project) on the adaptation of the natural gas network for the transportation of hydrogen, pure or mixed with natural gas. The main issue is the compatibility of the pipeline material (but also seals, valves, etc.) with hydrogen as it causes embrittlement in steel pipelines and permeability in polymeric pipelines. Considering the current most common grades of natural gas pipelines steels (API X52 and X60), injection of mixtures with up to 20–25% of hydrogen seems to be practicable but higher proportions could raise serious problems. It is possible to introduce a 3% maximum content of hydrogen in the existing natural gas network with limited modifications, but higher contents require more substantial adaptations [10]. In addition, to achieve this distribution mode and end-use of pure hydrogen, low-cost and efficient means of separating hydrogen from the gas mixture must be available. The search for separator technologies which offer low cost, high recovery, and high purity is one of the key issues.

6.3 Hydrogen Transport by Road

6.3.1
Gaseous Trucks

6.3.1.1 Status
Truck fleets are currently used by industrial gas companies to transport seamless steel vessels of compressed gaseous hydrogen for short distances (200–300 km) and small users (1–50 $m^3\ h^{-1}$) from centralized production. Single cylinder bottles, multi-cylinder bundles, or long cylindrical tubes are installed on trailers (Figure 6.5). Storage pressures range from 200 to 300 bar and a trailer can carry 2000–6200 $N\ m^{-3}$ of H_2 for trucks subject to a weight limitation of 40 t. The amount of hydrogen carried out is thus relatively small (from 180 to 540 kg, depending on the number of tubes or bundles), which represents $\sim$1–2% of the total mass of the truck. Current trailers utilize Type I storage cylinders (all-metal). To increase performance, bundles of lightweight composite hoop-wrapped cylinders or tubes (Type II) can be used.

The main cost factors in compressed gas truck delivery are capital costs, operation, and maintenance, including drivers' labor and fuel costs. The amount of time the trailer is stored at the customer site is also a factor affecting delivery cost. The capital investment is low for small quantities of hydrogen but it does not benefit from economy of scale with increasing demand and the costs increase linearly with delivery distance. This mode of delivery is relatively easy but it has to be adapted to hydrogen quantities and distances in order to be cost competitive.

Fig. 6.5 Two types of compressed gas hydrogen trailers operated by Air Liquide in Europe: tube trailer carrying 2000–3000 Nm^3 of H$_2$ (depending on the number of tubes) and Type II composite cylinder trailer carrying 6200 N m^{-3} of H$_2$ (540 kg). From [4].

6.3.1.2 Perspectives

The supply by gaseous truck (tube trailer, cylinders) is one of the most mature modes, preferred for short distances and small amounts of hydrogen. Limitations are the low weight storage capacity for high customer consumption (requiring frequent delivery) and the low pressure of hydrogen delivered, which requires additional compression at the fueling station site. Therefore, alternative technologies with higher pressure, higher hydrogen-carrying capacity, and lower cost systems are being investigated, as described below.

Lincoln Composites has developed higher volume tubes of composite structure (plastic liner completely wrapped with epoxy-impregnated carbon fiber) for hydrogen gaseous tube trailer delivery [20]. The TITAN™ tank (1.08 m in diameter, 11.5 m in length, 8400 l in water volume, and 2087 kg in weight) operating at 250 bar can deliver 2–3 times the amount of hydrogen as a steel tank of similar mass. Figure 6.6 shows the storage unit holding four tanks capable of storing 600 kg of hydrogen at 250 bar. Higher pressure tanks up to 350 bar are planned for 2010.

Hybrid technologies are being explored at the Lawrence Livermore National Laboratory (LLNL), such as cryo-compression combining pressure and low temperature to increase the amount of hydrogen that can be stored per unit volume

Fig. 6.6 Container with four composite tanks developed by Lincoln Composites. Source: Lincoln Composites [20].

(a) (b)

Fig. 6.7 700 bar cylinder prototype developed and tested within the STORHY European Project: (a) Type III technology; (b) Type IV technology. From [23].

and avoid the energy penalties associated with hydrogen liquefaction. Compressed hydrogen gas at cryogenic temperatures is much denser than in regular compressed tanks at ambient temperatures. These new vessels would have the potential to store hydrogen at temperatures as low as 80 K under pressures of 200–400 bar. This approach requires the development of insulated pressure composite tanks [21]. Alternatively, one could consider using cold hydrogen gas tanks that would require less cooling. There may be some optimum combination of pressure and temperature over the range 80–200 K. Recently, LLNL has identified inexpensive glass fiber materials for cold hydrogen gas storage (~ 150 K and up to 500 bar), expecting a 50% trailer cost reduction [22].

Lightweight compressed gas cylinders at 700 bar have also been developed to increase storage capacity. They consist of a metallic (Type III) or polymeric (Type IV) liner in a fiber-reinforced composite structure. An improvement in the gravimetric system storage density (around 5 wt%) is achieved with this high-pressure technology (Figure 6.7) [23]. Developments are on-going to reduce cost.

6.3.2
Cryogenic Liquid Trucks

Hydrogen can be transported by road in liquid form (cooled to 20 K or −253 °C) to distribute larger quantities (hundreds of cubic meters per hour). In terms of weight capacity, super-insulated liquid hydrogen trucks can transport up to 10 times more hydrogen than the tube trailers used for conveying compressed gas. Liquid hydrogen trucks (Figure 6.8) operating at atmospheric pressure have volumetric capacities of about 50 000–60 000 l and can transport up to 4000 kg with a truck mass of ~40 t. It is a preferred distribution mode for medium–large amounts of hydrogen and long distances, which explains why the liquid hydrogen business has been developed most extensively in North America (the hydrogen liquefaction capacity in North America is about 10 times larger than that in Europe). The liquid hydrogen transported in the truck is then vaporized to a high-pressure product for use at the customer site.

A main issue with this pathway is the liquefaction plant, which is capital intensive, hence the liquefaction process is costly. The electricity input for lique-

Fig. 6.8 Road tanker operated by Air Liquide for conveying liquid hydrogen to users. Source: Air Liquide Image Bank.

faction accounts for ~35% of the lower heating value of hydrogen (compared with ~10% for gas compression). Electricity costs account for 50–80% of the liquefaction costs.

Distance is the chief deciding factor between liquid and gaseous hydrogen. The number of liquid trucks will depend on the hydrogen demand and the localization of the liquefaction point. However, the liquid truck capacity being much higher than that of a compressed gas truck, this mode of delivery is less dependent upon the transport distance. The truck capital cost and operating cost (fuel, labor) are much lower. As a consequence, liquid trucking is more economical than gaseous trucking for long distances (from approximately 400 km to thousands of kilometers) and medium amounts of hydrogen.

However, one has to consider the availability of liquid hydrogen. Currently, the industrial hydrogen market is served by three liquefiers in Europe (Germany's second hydrogen liquefaction plant started up in 2007) and 10 in North America. Larger markets would justify the construction of new liquid plants.

Significant cost reductions due to scaling effects of liquefaction equipment are possible. However, this mode of delivery relies on the price of electricity and on the decision to install new liquefaction units. Better technologies could offer opportunities to reduce capital cost, improve the energy efficiency of the liquefaction process and reduce the amount of hydrogen lost due to boil-off during storage and transportation (the evaporation rate, which depends on the size, shape, insulation of the container, and time of storage, is typically of the order of 0.2% per day for a 100 m^3 container). Studies are under way to improve liquefaction technologies and propose novel approaches (for example, improvement of *ortho–para* conversion, development of magnetic refrigeration, etc.) [24–26].

6.4
Alternative Hydrogen Delivery Systems

6.4.1
Rail and Maritime Transport

Cryogenic tanks such as those used for trucking can be adopted for railway transport. At present, however, almost no hydrogen is transported by rail. Reasons include the lack of timely scheduling and transport to avoid excessive hydrogen boil-off and the lack of rail cars capable of handling cryogenic liquid hydrogen. A recent study presented by NREL shows an interest in rail delivery for long distances and large demands [27].

Hydrogen transport by barges or ships faces similar issues in that few vessels are designed to handle the transport of hydrogen via inland waterways. Storage methods and terminal technologies must also be developed to support the transport of hydrogen by rail or water. During the 1990s, several concepts for transatlantic transport of liquid hydrogen were developed in Canada, after the impulse given by the Euro–Quebec–Hydro–Hydrogen Pilot Project (EQHHPP), but no vessels have been constructed.

6.4.2
Chemical Carriers

Hydrogen can also be transported using hydrogen-rich carrier compounds. Chemical carriers such as hydrocarbons, methanol, ammonia, and dimethyl ether are attractive because they are liquid at room temperature and could be delivered via existing and/or low-cost infrastructure, and usually are easier to handle than cryogenic hydrogen. However, they require an extra transformation step to release hydrogen, and hydrogen carriers such as methanol and ammonia may present some additional safety and handling challenges. Hydrogenation and dehydrogenation of hydrocarbons are operations requiring high amounts of energy and high temperatures. Projects are under way in the DOE Hydrogen Program to demonstrate the feasibility of dehydrogenation of some liquid carriers (e.g., *N*-ethylcarbazole [28]).

Solid carriers (hydrides, adsorbents) are also a means of transporting, storing, and delivering hydrogen in a chemical state other than free hydrogen molecules. However, new materials must be developed to provide greater hydrogen capacity and optimized energy. This topic is directly linked to the development of hydrogen storage in materials.

To sum up, a carrier with high energy density and simple transformation (both hydriding and dehydriding) can deliver hydrogen using trucks and has the potential to alter the distribution system radically. However, carriers are not yet well understood, and extensive engineering and economic analysis is still needed to identify promising materials. Carrier pathways also require the return of spent fuel for reprocessing. In addition, there is a lack of knowledge on safety and environmental issues.

6.4.3
Other Concepts

A very different approach is to consider transporting liquid hydrogen in pipes. In the context of research devoted to future energy networks, new concepts for energy transmission based on several simultaneously utilized energy carriers have been proposed in various projects [29]. This is the case with the ICEFUEL (Integrated Cable Energy System for Fuel and Power) project supported by the German Federal Ministry for Education and Research. The aim of this project is the development of a system for the combined transmission of cryogenic hydrogen, electricity, and data. The technical and economic feasibility of this hybrid system is being investigated [30]. Key parts are the design of a flexible and super-insulated tube which can be handled like an underground cable and allows a cost-efficient installation and operation over distances of up to 10 km. The design also allows for the integration of electric power and data. Demonstration has still to be proven.

6.5
Stationary Bulk Storage of Hydrogen

The storage of large quantities of hydrogen for long periods is a key step in the build-up of infrastructure in order to regulate the hydrogen consumption and production and ensure continuity in supply. Various underground hydrogen storage schemes have been investigated. One option is to store gaseous hydrogen in geological formations including depleted gas fields or aquifers, caverns, and so on. Another is underground storage in buried tanks, either in compressed gas form or in liquid form. Geological storage is generally close to the hydrogen production site, whereas buried tanks are close to the point of use, such as refueling stations.

6.5.1
Geological Storage

Many geological sites have the potential to store gases, including salt caverns, mined caverns, natural caves, and aquifer structures. The underground storage of hydrogen draws on experience from natural gas, routinely used to provide seasonal and surge capacity. However, hydrogen is rather more difficult to contain than natural gas due to the smaller size of its molecule and its higher diffusion coefficient. Candidate sites for geological storage must have promising permeability characteristics. The chemistry between hydrogen and minerals must be known for integrity of the storage unit and identification of contaminants that may be introduced. Geological storage may also suffer from hydrogen leakage. All of these points and the potential cost of geological storage have to be addressed.

Underground storage of hydrogen is not a new concept. A major study of underground storage of gaseous hydrogen was conducted in 1979 by the Gas Technology Institute in the USA [31]. Over recent decades, there have been several examples of underground storage of pure hydrogen or synthetic gas H_2–CO mixtures [32].

Salt caverns, which offer the advantage of being almost impermeable to gases, are currently the only underground facility used to store hydrogen. The city of Kiel, Germany, has been storing town gas (60–65% hydrogen) in a gas cavern since 1971. In the UK, at Teesside, ICI has stored 10^6 N m^{-3} of nearly pure hydrogen (95% H_2 and 3–4% CO_2) in three salt caverns at a depth of about 400 m for a number of years. In North America, mainly in Texas, there are two hydrogen-filled caverns. ConocoPhillips has had a Syngas (95% hydrogen) storage cavern at Clemmons salt dome near Sweeny, TX, since the 1980s. This cavern is connected directly to the refinery at Old Ocean, TX. Praxair also operates an industrial hydrogen storage cavern facility located in Texas and integrated in the Gulf Coast hydrogen pipeline system [33].

6.5.2
Buried Tanks

Underground storage may also consist in burying compressed gas or liquid hydrogen tanks that are usually placed at ground level, in order to save ground space. Of course, this option should only apply for smaller hydrogen quantities (typically 10^3–10^4 N m^{-2}) and at a much smaller depth (typically a few meters) than direct gas storage in caverns.

6.5.2.1 Compressed Gas Tanks
At present the largest manufactured compressed hydrogen tanks in the world (about 15 000 m^3) can be pressurized only up to 12–16 bar [4].

Although a compressed gas storage installation is usually located outdoors, at or above ground level, it can be buried to save ground space and also to provide improved protection from external influences such as radiation from adjacent fires or damage caused by explosions. However, this alternative is rarely used because it makes inspection of the vessels and interconnection of pipes less easy, and it requires preventive measure to prevent corrosion. Nevertheless, to overcome the difficulty of inspection and pipe interconnection it has been proposed to place the tanks in a basin and submerge them afterwards with a liquid such as water, so that tank protection from heating and explosion is compatible an easy inspection simply by lowering the water level in the basin. Alternatively, the tank can be placed in a protection system consisting of a sack that contains the tank and a gas-permeable material arranged in such a manner as to constitute a layer around the tank.

6.5.2.2 Liquid Tanks

The technology of liquid hydrogen storage is extensively used for space applications. Liquid hydrogen tanks for long-term storage have perlite vacuum or multilayer insulation. Common stationary tanks have capacities ranging from 1500 l ($\sim$100 kg H_2) up to 75 000 l ($\sim$5 t H_2) with radii of 1.4–3.8 m and heights of 3–14 m. The largest tank belongs to NASA and is located at Cape Canaveral. This tank at ground level has an outer spherical diameter of 20 m, a storage volume of about 3800 N m^{-3} (about 270 t liquid H_2) and its evaporation rate is less than 0.03% per day, allowing a storage period of several years.

Currently there are a few examples of liquid hydrogen tank being either placed in a room underground or actually buried underground. The first example is the liquid H_2 tank of the bus refueling station built by Air Liquide, installed by BOC, and operated by BP in London within the HyFleet-CUTE European project. A second example is the concept proposed by Linde for the hydrogen filling station opened in 2007 in Munich; another one is planned to start in 2010 in Berlin. Thanks to the underground storage tank, the hydrogen filling station looks no different than a conventional filling station. The tanking operation is also as quick and easy as in a conventional filling station.

6.6
Supporting Technologies

6.6.1
Gaseous Hydrogen Compressors

Compressors used to increase the pressure of gaseous hydrogen are indispensable components of the supply chain. Although the basic principles of compression are generic to most gases, there are differences in materials and design when dealing with hydrogen. Figure 6.9 shows the operating characteristics of various compressors, all of them requiring mechanical driving power.

There are three basic types of mechanical compressors:
- The reciprocating (piston) compressor uses pistons with a back-and-forth motion to compress the gas, and contains inlet and outlet check valves (Figure 6.10). It is commonly used for compression of H_2 gas for most flow rates. However, there are some issues, including poor efficiency and reliability, contamination from lubricants, and high capital costs due to expensive materials to prevent embrittlement and the risk of failures during use. The large number of moving parts also tends to increase maintenance issues and costs.
- The diaphragm compressor is a mature technology for very low flow rates. H_2 is isolated from the mechanical parts of the compressor by a set of metallic diaphragms. It is a good choice for compressing without incurring contamination of gas or leakage but limited to small-volume applications.

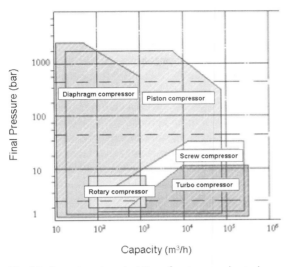

Fig. 6.9 Operating characteristics of various mechanical compressors [34].

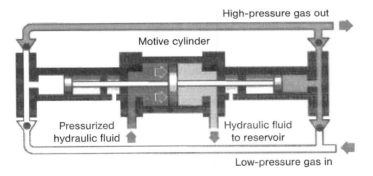

Fig. 6.10 Piston compressor. Source: Hydropac, Inc. [35].

- The centrifugal compressor is a rotative type compressor that uses velocity/momentum changes to generate pressure. It is seldom used for hydrogen applications due to the molecule's low molecular weight, which causes seal design problems and limits the outlet pressure. It also has limitations on pressures that can be developed and flows that can be handled. However, it is used in cryogenic H_2 applications where flow is relatively high and the pressure desired is relatively low. Reliability is greater than for a piston compressor.

Heat is produced during hydrogen compression, reducing the performance of the compressor. The main challenges nowadays are the improvement of energy efficiency and reliability and reduction of the cost. New approaches are being considered to develop alternative compressor designs. Some examples of progress are presented below.

NREC in collaboration with Praxair is developing an advanced hydrogen centrifugal compressor using "off-the-shelf" parts [36].

Non-mechanical compressors are also being investigated. These less mature technologies have several advantages, including lower capital and maintenance costs, the absence of moving parts and contamination:

- Among them, one has a solid-state electrochemical compressor, the principle of which is a membrane electrode assembly. This compressor, used when a small quantity of hydrogen has to be compressed, shows higher efficiency (Figure 6.11).
- Metal hydride-based hydrogen compressors are also under investigation [38]. They are thermally powered systems that use the properties of reversible metal hydride alloys to compress hydrogen. They have potential for special niche applications.
- In order to increase efficiency, Linde has introduced ionic compression systems for hydrogen [39]. Unlike conventional mechanical systems, the ionic compressor uses an ionic liquid in direct contact with hydrogen instead of a piston in the pressurizing process. The ionic liquid is an organic salt that does not mix with the gas and with no vapor pressure.

None of the commercially available compressor technologies presented here above meets the requirements necessary to develop a "hydrogen-based economy"

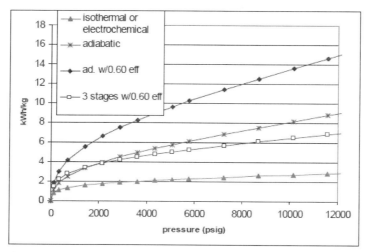

Fig. 6.11 Comparison of efficiencies of different H_2 compression technologies. The efficiency of an electrochemical compressor is higher than that of mechanical compressors. From [37].

in terms of costs, scalability, and efficiency. Compression technology is adapted to the needs of current on-going projects which are at the demonstration scale and require low flow rates. However, improvements and potential breakthroughs in this domain are necessary to meet the challenges of the widespread use of hydrogen as an energy carrier.

6.6.2
Liquid Hydrogen Pumps

Pumping liquid hydrogen is a challenge for the following reasons:
- Pumps must operate under extremely cold temperatures to maintain the hydrogen in a liquid state at all times. Any vaporization will cause damaging cavitation in the pump and thus reduction of performance and premature wear. Installation of the system must be done with care to avoid any trap of gaseous hydrogen bubbles in the liquid lines which could create cavitation.
- The NPSH parameter (Net Positive Suction Head) of the pump must be low enough by design to avoid cavitation, as it is impossible to increase the available hydrogen NPSH at the pump inlet by physically elevating the source tank, due to the low density of the liquid.
- The small size of the molecule makes the design of the pump dynamic seals a challenge.
- Losses in efficiency due to heat entries or friction result in evaporation of hydrogen which is vented at low pressure, therefore spoilt.

However, recent advances in cryopumping technologies allow proposals for high-output hydrogen refueling station schemes (more than 1000 kg per day at the pump) based on liquid hydrogen delivery and hydrogen compression by cryopumps to customers.

At hydrogen fueling stations, cold filling is required for fast 700 bar filling (see Section 6.7). There are two approaches to meet that requirement: either indirect filling with high-pressure buffers and a heat exchanger for cold filling, or direct filling with a liquid hydrogen pump and partial bypass of the atmospheric vaporizer for cold filling. Air Liquide has developed a high–pressure hydrogen cryogenic pump [40, 41]. The first application of the concept will be for the commercial refueling station BC TRANSIT in Whistler, Canada, for the 2010 Winter Olympics.

6.6.3
Hydrogen Quality Management

It is well known that proton exchange membrane fuel cells (PEMFCs) can easily be poisoned by trace impurities that may be present in the hydrogen (e.g., CO, NH_3, and sulfur [42]). As a result, hydrogen producers are increasingly being asked to guarantee the levels of impurities in delivered hydrogen that have not been requested by the more traditional hydrogen users, such as refiners.

Several organizations are addressing this fuel quality issue, including the International Organization for Standardization (ISO), the Society of Automotive Engineers (SAE), the California Fuel Cell Partnership (CaFCP), and the New Energy and Industrial Technology Development Organization (NEDO)/Japan Automobile Research Institute (JARI). Interim guidelines and a draft version of an ISO Specification on hydrogen quality have been issued (e.g., SAE J2719, ISO/CD 14687-2 prepared by ISO/TC 197/WG 12). The existing recommendations for H_2 specifications indicate very low impurity levels (e.g., 4 ppb sulfur).

Although it is possible for the industry to produce extremely pure hydrogen with existing purification processes, such hydrogen would incur penalties in terms of energy efficiency and cost as the allowable limits of the contaminants are lowered. For example, in the frame of the DYNAMIS European project, Air Liquide has assessed the capabilities of hydrogen production from decarbonized fossil fuels and hydrogen purification using pressure swing adsorption (PSA) to meet the purity requirements dictated by use in fuel cells [43]. The concentrations of impurities in the product gas are not independent and depend on various factors such as the feed gas composition, the amount and composition of adsorbent, and the operating pressure. For the case of hydrogen produced by steam methane reforming (SMR), the simulation results show that reducing the impurity concentration causes an increase in the adsorbent volume and consequently a reduction in the H_2 yield (Figure 6.12). The level of impurity specification for H_2 thus has an impact on the production process and production

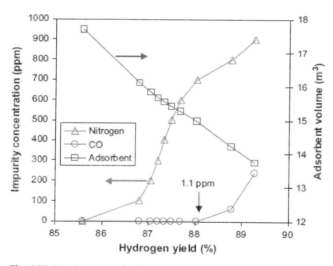

Fig. 6.12 Simulation results showing the relationship between impurity concentration, PSA adsorbent volume and hydrogen yield for a steam methane reformer. For H_2 yields less than 88%, the CO concentration is at sub-ppm levels [43].

costs. Using the H2A hydrogen delivery model, the sensitivity of the cost of hydrogen to the level of impurities has been studied by Argonne National Laboratory [44].

The commercial H_2 supplier has to maintain the high purity required during distribution (or to re-purify as a result of potential contamination). Another aspect is the ability to verify/certify that the regulatory requirements have been met at these very low contaminant levels. This requires having standardized sampling and analytical methods for H_2 impurity analysis. The gas composition analysis, especially for some of the impurities that have very low allowable concentrations, may add significantly to the cost of the delivered H_2.

Specifications for H_2 should be oriented to the most critical impurities and based on a "cost–benefit" analysis to establish the optimum trade-off between hydrogen fuel quality and cost in order to contribute favorably to fuel cell development. Studies are under way to prioritize recommendations on fuel quality to the standardization process.

6.7
Hydrogen Fueling Stations

Hydrogen refueling stations are part of the delivery infrastructure. With the development of hydrogen vehicles, new refueling stations devoted to high public use and adapted to this new market have been developed. Today, a total of about 180 stations have opened worldwide [45]. Their deployment is focused on three geographic regions: North America (around half of the stations), Western Europe, and South-East Asia. The major actors are Air Products, Linde and Air Liquide.

In the USA, most developments are in California, where 250 demonstration vehicles – passenger and transit buses – have been placed on roads with about 26 hydrogen stations. This should expand according to the 2009 California Fuel Cell Partnership Action Plan [1]. In Japan, the development of stations follows the JHFC plan (Japan Hydrogen Fuel Cell). In Europe, the first European deployment of high magnitude has occurred in the early 2000s as part of the CUTE project (Clean Urban Transport for Europe), with a captive fleet of 47 buses Citaro FC 350 bar spread over ten cities. Stations in the CUTE demonstration program highlighted the problems of rapid filling, heating of tanks and hydrogen embrittlement of high-pressure hoses. It also showed the need for harmonization of safety requirements, and the recommended process for installation of hydrogen refueling stations. In this frame, the HyApproval project sponsored by the European Commission within the Sixth Framework Programme was launched, contributing to the establishment of a guide issued in 2008 to facilitate the process of issuing permits for stations [46]. In 2009, the "H2 Mobility" initiative launched in Germany with major companies and public policy makers is aimed at building up a hydrogen infrastructure with significant expansion of the hydrogen fueling stations network by the end of 2011.

A fueling station includes a hydrogen source (gaseous or liquid), a compressor, a storage unit and a dispenser. Most of the stations deliver gaseous hydrogen compressed at 350 or 700 bar in order to fill vehicle tanks quickly (refueling time less than 4 min). Composite tanks consisting of a metallic or polymeric liner (namely Type III and Type IV) in a fiber-reinforced composite structure have been developed to store hydrogen at high pressure and to increase the autonomy of vehicles [23]. However, fast filling at high pressure results in a very high temperature increase in the composite vessel (because of the near-adiabatic compression of the gas), which can damage it. As a consequence, there are a number refueling issues when combining high pressure (e.g., 700 bar), fast filling, composite cylinder use and safety.

6.7.1
Fueling Protocols

Experimental and modeling studies have been carried out to understand the evolution of parameters during filling and develop safe and reliable procedures to achieve rapid high-pressure filling of composite storage systems.

Figure 6.13 displays the temperature evolution during filling [47]. The gas temperature follows roughly a two-step evolution: a rapid increase during the first seconds of the filling (due to the high compression ratio and low heat transfer) and a lower increase during the rest of the filling.

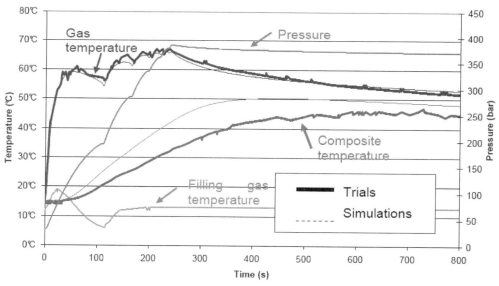

Fig. 6.13 Evolution of gas temperature, pressure, composite temperature (outer layer) and filling gas temperature as a function of time; recorded at an Air Liquide filling station. Comparison between experiment and simulation [47].

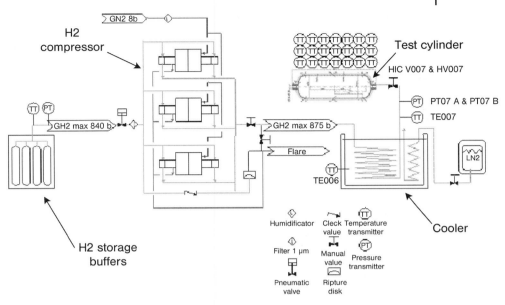

Fig. 6.14 Fast filling test bench at Air Liquide [48].

In the frame of the STORHY European project [48], a test program has been set up to assess fast filling possibilities for different 700 bar storage systems (Type III and Type IV). Figure 6.14 shows the experimental bench to perform filling tests at Air Liquide. The results show that 700 bar fast filling is feasible if a slight cooling of the gas is applied (about 0 °C for Type III cylinders and about −20°C for Type IV cylinders). Preliminary calculations show that the cooling energy in this application would be about 0.05 kW h kg^{-1} H$_2$ for Type III cylinders and 0.12 kW h kg^{-1} H$_2$ for Type IV cylinders, which is low compared with the H$_2$ energy content (33 kW h kg^{-1}). Applying slight cooling would therefore not be a limitation for the 700 bar fast filling process. A technology with a heat exchanger has been developed to cool the gas.

6.7.2
Dispenser

The dispenser is the interface that will allow the user to fill up their vehicle. The dispenser includes a fill line for connecting the vehicle, a communication protocol with the vehicle, and an operator interface to select the parameters of filling. To secure the filling of hydrogen vehicles, an industry consortium has developed a standardized communication protocol between the hydrogen station and the vehicle: the California Fuel Cell Protocol (CaFCP). The communication protocol allows the sharing of information on pressure, temperature, and volume of the hydrogen tank of the vehicle with the filling station. In addition, the

CaFCP allows for the grounding of the vehicle and can be ordered to fill from the vehicle. The filling is monitored continuously and ensures secure and optimized filling, avoiding the problem of over-filling and anticipating the effects of thermal expansion in the hydrogen tank of the vehicle. The main evolution of the protocol is the development of an infrared communication version in contrast to the existing wired version. Air Liquide has actively participated in the development of this protocol and its deployment through projects with General Motors.

6.7.3
Operator Interface

The operator interface is the relationship between the hydrogen station and the user/operator. Initially, interfaces with multiple adjustment possibilities and flexibility of use (e.g., touch screen) were proposed. Today, the trend is to have simpler interfaces to reduce the risk of errors. The functions available are thus limited: choice of the final pressure, choice of filling rate (slow or fast), and choice of communication mode (with or without CaFCP).

6.7.4
Hydrogen Station Development

The trend of the past 2 years has been the integration of hydrogen stations in standardized containers of 10, 20, or 40 ft. This design allows mobile equipment that is easy to handle. Air Liquide today offers part of its hydrogen stations integrated in a container. The example in Figure 6.15 shows the integration performed on the station supplied to Renault.

Fig. 6.15 Air Liquide hydrogen station integrated into a container.

Fig. 6.16 Air Liquide hydrogen station (semi-automatic version).

The feedback capitalized on past projects has enabled hydrogen stations to be developed that are adapted to different markets. Air Liquide is now proposing two families of hydrogen stations: semi-automatic and automatic.

The semi-automatic stations (Figure 6.16) are fully pneumatic equipment without electric utilities. They are intended for specific and mobile uses. Their main assets are an attractive selling price and high mobility of use. These stations are packaged in stainless-steel panels of small sizes and can be handled by using a truck. They are available in versions of 350 and 700 bar, with or without communication with the vehicle, and exclusively offered with pneumatic compressors.

The automatic stations are facilities with electric utilities. They are designed for stationary applications requiring a high daily compression capacity. These stations are typically integrated into containers of 10, 20, or 40 ft depending on the size of the compressors and buffers. The automatic stations are available in versions of 350 and 700 bar, with or without communication with the vehicle, and provided with pneumatic or membrane compressors.

6.8
Conclusion

The introduction of hydrogen as a fuel requires the development of an infrastructure for distribution and refueling. In this chapter, the various components of the hydrogen delivery infrastructure incorporating multiple pathways capable of handing hydrogen in various forms have been presented.

The infrastructure relies on a combination of hydrogen delivery options, whose share of application will evolve with time, depending on the development of the hydrogen market. The current industrial hydrogen delivery modes provide a technical starting point for building the infrastructure. However, new solutions for higher pressure hydrogen are needed to adapt industrial technologies to an energy system that is more efficient and less costly for customers. There will be a timeline for transition. Early markets such as back-up power and fuel cell-powered forklifts will play an important role in developing and testing improved delivery solutions and will pave the way for the commercialization of hydrogen-powered vehicles.

Hydrogen has been used widely in large-scale industrial applications for many decades. Experience shows that it can be handled safely provided that users stick to the appropriate standards, regulations and best practices. It is the result of a long learning process within these technologies. Future infrastructure systems for hydrogen applications, such as new storage media and refueling stations, need to have at least the same high safety standards as the established technologies for acceptance by consumers. There are efforts under way worldwide to develop appropriate safety procedures and codes and standards for hydrogen use in energy applications. Active work is being carried out by the technical committee responsible for the development of international standards for hydrogen energy technologies (ISO/TC 197). This is a decisive step to achieve standardized and harmonized regulations for hydrogen infrastructure deployment and societal acceptance.

References

1 Fuel Cell Partnership (2009) *Hydrogen Fuel Cell Vehicle and Station Deployment Plan: a Strategy for Meeting the Challenge Ahead. CaFCP Action Plan, February 2009*, http://www.fuelcellpartnership.org (accessed 2 February 2010).

2 Air Liquide (1976) *Encyclopédie des Gaz*, Elsevier, Amsterdam.

3 Züttel, A., Borgschulte, A., Schlapbach, L., Chorkendorff, Ib., and Suda, S. (2008) Properties of hydrogen. In *Hydrogen as a Future Energy Carrier* (eds A. Züttel, A. Borgschulte, and L. Schlapbach), Wiley-VCH Verlag GmbH, Weinheim, pp. 71–147.

4 Weber, M. and Perrin, J. (2008) Hydrogen transport and distribution. In *Hydrogen Technology – Mobile and Portable Applications* (ed. A. Léon), Springer, Berlin, pp. 129–149.

5 Perrin, J. (2007) *Industrial Distribution Infrastructure*. Roads2HyCom WP2, July 2007, http://www.roads2hy.com.

6 Association Française de l'Hydrogène. *Les Réseaux de Pipelines d'Hydrogène dans le Monde. Mémento de l'Hydrogène*. Association Française de l'Hydrogène, Paris, http://www.afh2.org (accessed 2 February 2010).

7 Boening, A., Lienkamp, H., and Rastogi, A. (2008) *Demonstration of H_2 Infrastructure and Fuel-Cell Passenger Cars: Project Zero Regio*, http://www.zeroregio.com (accessed 2 February 2010).

8 San Marchi, C. and Somerday, B.P. (2008) *Effects of High-Pressure Gaseous Hydrogen on Structural Metals*, SAE 2007 Transactions, Paper 2007-01-0433, SAE International, Warrendale, PA.

9 Balat, M. (2008) Potential importance of hydrogen as a future solution to environmental and transportation problems. *Int. J. Hydrogen Energy*, **33**, 4013–4029.

10 Cuni, A., Weber, M., and Guerrini, O. (2008) *Distribution Issues*. Roads2HyCom WP2, June 2008, http://www.roads2hy.com.

11 Yang, C. and Ogden, J. (2007) Determining the lowest-cost hydrogen delivery mode. *Int. J. Hydrogen Energy*, **32**, 268–286.

12 Tzimas, E., Catello, P., and Peteves, S. (2007) The evolution of size and cost of a hydrogen delivery infrastructure in Europe in the medium and long term. *Int. J. Hydrogen Energy*, **32**, 1369–1380.

13 Murakami Y. (2006) The effect of hydrogen on fatigue properties of metals used for fuel cell system. *Int. J. Fracture*, **138**, 167–195.

14 Stalheim D. (2008) Materials solutions for hydrogen delivery in pipeline. *2008 DOE Hydrogen Program Annual Merit Review Proceedings*, http://www.hydrogen.energy.gov.

15 Sofronis, P., Robertson, I.M., and Johnson, D.D. (2009) A combined materials science/mechanics approach to the study of hydrogen embrittlement of pipeline steels. *2009 DOE Hydrogen Program Annual Merit Review Proceedings*, http://www.hydrogen.energy.gov.

16 Lohmar, J. (2006) Polyamide-12 for high pressure gas installations. Presented at the 2006 World Gas Conference.

17 Smith, B., Frame, B., Anovitz, L., and Armstrong, T. (2008) Composite technology for hydrogen pipelines. *2008 DOE Hydrogen Program Annual Merit Review Proceedings*, http://www.hydrogen.energy.gov.

18 Adams, T. and Rawls, G. (2009) Fiber reinforced composite pipelines. *2009 DOE Hydrogen Program Annual Merit Review Proceedings*, http://www.hydrogen.energy.gov.

19 Smith, B., Frame, B., Eberle, C., Anovitz, L., and Armstrong, T. (2007) Fiber-reinforced polymer pipelines for hydrogen delivery. *2007 DOE Hydrogen Program Annual Merit Review Proceedings*, http://www.hydrogen.energy.gov.

20 Baldwin D. (2009) Development of high pressure hydrogen storage tank for storage and gaseous truck delivery. *2009 DOE Hydrogen Program Annual Merit Review Proceedings*, http://www.hydrogen.energy.gov.

21 Aceves, S., Berry, G., Martinez-Frias, J., and Espinosa-Loza, F. (2006) Vehicular storage of hydrogen in insulated pressure vessels. *Int. J. Hydrogen Energy*, **31**, 2274–2283.

22 Weisberg, A., Aceves, S., Myers, B., and Ross, T. (2009) Inexpensive delivery of cold hydrogen in high performance glass fiber composite pressure vessels. *2009 DOE Hydrogen Program Annual Merit Review Proceedings*, http://www.hydrogen.energy.gov.

23 Colom, S., Weber, M., and Barbier, F. (2008) STORHY: a European development of composite vessels for 70 MPa hydrogen storage. *Proceedings of the 17th World Hydrogen Energy Conference (WHEC2008)*, 15–19 June 2008, Brisbane.

24 Rahman, M., Hong Ho, S., and Rosario, L. (2005) Review and some research results on hydrogen liquefaction and storage. *Proceedings of the International Conference on Mechanical Engineering 2005 (ICME2005)*, 28–30 December 2005, Dhaka, Bangladesh.

25 Berstad, D., Stang, J., and Neksa, P. (2009) Comparison criteria for large-scale hydrogen liquefaction processes. *Int. J. Hydrogen Energy*, **34**, 1560–1568.

26 Jankowiak, J. and Schwartz, J. (2009) Advanced hydrogen liquefaction process. *2009 DOE Hydrogen Program Annual Merit Review Proceedings*, http://www.hydrogen.energy.gov.

27 Sozinova, O. (2009) H2A Delivery Components Mode. *2009 DOE Hydrogen Program Annual Merit Review Proceedings*, http://www.hydrogen.energy.gov.

28 Cooper, A., Scott, A., Fowler, D., Cunningham, J., Ford, M., Wilhelm, F., Monk, V., Cheng, H., and Pez, G. (2008) Hydrogen storage by reversible hydrogenation of liquid-phase hydrogen carriers. *FY 2008 DOE Program Annual Progress Report*, p. 603, http://www.hydrogen.energy.gov.

29 Favre-Perrod P. (2008) *Hybrid Energy Transmission for Multi-Energy Networks*. Dissertation, ETH Zurich, October 2008.

30 Elliger, T. (2008) *Wasserstoff on ice*, Brennstoffzellen Magazin 1. 2008, http://www.icefuel.eu (accessed 2 February 2010).

31 Foh, S., Novil, M., Rockar, E., and Randolph, P. (1979) *Underground Hydrogen Storage*. Final Report, BNL 51275, Institute of Gas Technology, Chicago, December 1979.

32 Panfilov, M., Gravier, G., and Fillacier, S. (2006) Underground storage of H_2 and

H_2–CO_2–CH_4 mixtures. *Proceedings of the 10th European Conference on the Mathematics of Oil Recovery (ECMOR)*, Amsterdam, September 2006.

33 Praxair Inc., http://www.praxair.com (accessed 2 February 2010).

34 Freedom Car and Fuel Partnership (2007) *Hydrogen Delivery Technology Roadmap*, DOE Hydrogen Program, February 2007, http://www1.eere.energy.gov.

35 http://www.hydropac.com (accessed 2 February 2010).

36 Di Bella, F. and Osborne, C. (2009) Development of a centrifugal hydrogen pipeline gas compressor. *2009 DOE Hydrogen Program Annual Merit Review Proceedings*, http://www.hydrogen.energy.gov (accessed 2 February 2010).

37 Lipp, L. (2008) Development of highly efficient solid state electrochemical hydrogen compressor. *2008 DOE Hydrogen Program Annual Merit Review Proceedings*, http://www.hydrogen.energy.gov.

38 Laurencelle, F., Dehouche, Z., Morin, F., and Goyette, J. (2009) Experimental study on a metal hydride based hydrogen compressor. *Int. J. Hydrogen Energy*, **475**, 810–816.

39 The Linde Group, http://www.linde.de (accessed 2 February 2010).

40 Allidières, L. (2005) Cryogenic fluid pumping system. International Patent WO2005/085637A1 (EP1723336B1).

41 Allidières, L., Bourgeois, P., and Drouvot, P. (2006) 700 bar liquid hydrogen pump. *Proceedings of the 16th World Hydrogen Energy Conference (WHEC2006)*, 15–19 June 2006, Lyon.

42 Cheng, X., Shi, Z., Glass, N., Zhang, L., Zhang, J., Song, D., et al. (2007) A review of hydrogen fuel cell contamination: impacts, mechanisms and mitigation. *J. Power Sources*, **165**, 739–756.

43 Besancon, B., Hasanov, V., Imbault-Lastapis, R., Benesch, R., Barrio, M., and Mølnvik, M. (2009) Hydrogen quality from decarbonized fossil fuels to fuel cells. *Int. J. Hydrogen Energy*, **34**, 2350–2360.

44 Papadiasa, D., Ahmed, S., Kumara, R., and Joseckb, F. (2009) Hydrogen quality for fuel cell vehicles – a modeling study of the sensitivity of impurity content in hydrogen to the process variables in the SMR–PSA pathway. *Int. J. Hydrogen Energy*, **34**, 6021–6035.

45 TÜV-Süd, http://www.netinform.net/H2/H2Stations (accessed 2 February 2010).

46 European Commission. *HyApproval – Handbook for Hydrogen Refuelling Station Approval*, European Commission FP6, http://www.hyapproval.org (accessed 2 February 2010).

47 Pregassame, S., Barral, K., Allidières, L., Charbonneau, T., and Lacombe, Y. (2004) Operation feedback of hydrogen filling station. *Hydrogen and Fuel Cells, 2004 Conference and Trade Show*.

48 Colom, S., Allidières, L., Vinard, T., Bur, S., and Trompezinski, S. (2008) Hydrogen refuelling stations: benchmark of storage systems to define fueling protocols, *Proceedings of the 17th World Hydrogen Energy Conference (WHEC2008)*, 15–19 June 2008, Brisbane.

7
Fuel Provision for Early Market Applications

Manfred Fischedick and Andreas Pastowski

Abstract

Early market applications of fuel cells need to be supplied with hydrogen, comprising production and distribution. Production of hydrogen can be based on a multitude of processes and energy carriers and can be assessed using criteria such as systemic requirements, economic efficiency, environmental sustainability, and user acceptability. While the long-term sustainable vision is based on renewable energy, cost considerations may be prevalent for early applications. Global production of industrial hydrogen is substantial, almost entirely based on fossil fuels and generally unavailable for other use. However, excess production capacity and by-product hydrogen and also the industrial hydrogen infrastructure might be used for the supply of limited volumes of hydrogen for early applications of fuel cells. Depending on the industrial structure, potential regional supply of industrial hydrogen varies. However, with growing demand and in order to put the long-term vision into practice, new supply from renewable sources will need to be set up.

Keywords: fuel cells, hydrogen production, industrial hydrogen

7.1
Introduction: Hydrogen Supply Today and Tomorrow

Currently, hydrogen is used for various mainly industrial applications. Mostly, hydrogen serves as an input for chemical processes. So far, except for space missions, hydrogen has hardly been used as an energy carrier. Nevertheless, its potential to be produced from a multitude of primary energy sources, to serve as a means for storing energy, and to be used cleanly in various applications has raised interest in hydrogen as a potential major energy vector in a future sustainable energy system [1]. Depending on application, any such utilization needs a quantitatively and qualitatively sufficient hydrogen supply, which involves production and distribution. Owing to the physical properties of hydrogen, distribution is possible in gaseous form via pipelines or by rail in tank wa-

Hydrogen Energy. Edited by Detlef Stolten
Copyright © 2010 WILEY-VCH Verlag GmbH & Co. KGaA, Weinheim
ISBN: 978-3-527-32711-9

gons and using specialized ships or trucks. For other than transport via pipeline, cryogenic transport is an important alternative because the volume-related energy content of gaseous hydrogen is very low. This would result in enormous and costly vehicle mileage per unit of energy transported as compared with today's liquid fuels. In the case of pipelines, dedicated hydrogen networks are an option. Moreover, it is possible to transport hydrogen by pipeline mixed with natural gas, applying membranes to separate the hydrogen close to its destination [2].

New fields for energetic utilization of hydrogen include stationary heat and power and fuel supply for vehicles or for portable devices. It is obvious that the applications mentioned require varying volumes of hydrogen and can be served by differing production and distribution facilities, infrastructures, and logistics. While most of the current industrial use of hydrogen is based on on-site production and some volumes are distributed via pipelines, at least hydrogen supply for vehicles and portable devices will require more or less concentrated production but a spatially dense distribution system. This means that existing industrial hydrogen production and distribution scarcely fit the needs of spatially distributed fuel requirements for vehicles and portable devices with high degrees of market penetration. Notwithstanding the aforementioned, for early applications cost considerations may be prevalent and hydrogen supply may rely on existing supply which may serve as a nucleus for setting up a new infrastructure for the future.

One of the main fields of potential future application of hydrogen and fuel cells might be transportation. The transport sector plays a crucial role in both limiting the emissions of greenhouse gases and breaking the dependence of mobility on crude oil. Globally, transportation accounts for roughly one-quarter of energy-related CO_2 emissions [3]. Moreover, global transport-related use of fossil fuels is on a steep growth path with 25% more energy used in 2000 than in 1990. Further strong growth of nearly 90% is projected for the period 2000 to 2030. Thus transportation has been key for the overall increase in global oil use during the last three decades and will be for the next 20 years in a business-as-usual case.

The dependence of transport on oil is around 97% for most countries and will remain largely unchanged by recent efforts to introduce biofuels through 2030 [4, 5]. Biofuels can only supply relatively small quantities limited by arable land, food production, and the protection of diminishing habitats such as rainforests [6]. Opposed to this, hydrogen could be produced based on a variety of renewable sources of energy in enormous volumes without comparable limitations of available land area. However, a shift from the traditional oil-based transport energy supply to renewable energy is challenging and will require significant efforts over a longer period of time. One of those efforts is setting up a new infrastructure for hydrogen supply.

The existing infrastructure for the production and distribution of fossil fuels used for transportation is substantial. This is usually not considered a burden because it is in place and has been built up over decades. Costs of operation and amortization of this infrastructure for road fuels make up a limited share of the pump price and, therefore, are of no concern. Opposed to this, the perception of investment is completely different for infrastructure which currently does not exist. In fact, high initial financing for production and distribution facilities of hydrogen need to be expected. Depending on the geographic coverage, time horizon, and the assumed share in overall energy use, the financing required for distribution-related infrastructure alone can be enormous. For instance, it has been estimated that for a very high degree of market penetration and 25 member states of the European Union, the cumulative capital requirements for a hydrogen distribution infrastructure might amount to 700–2200 billion euros until 2050 [7]. Investment volumes of such a size usually provide the impression that it might be very difficult to put a sufficient infrastructure in place. However, due to insufficient knowledge of further development of the various energy-related technologies, it is unclear whether a very high market penetration of hydrogen is likely and which cost reductions can be achieved with regard to production and distribution during such a long period of time.

Besides the sheer volume of long-term cumulative capital expenditure, at an early stage of switching to new fuels the initial capacity utilization of supply infrastructure may be very poor, which is a strong disincentive for any such investment. Further complications may arise from an unclear perspective of the future number of fuel cell-equipped vehicles in operation and thus derived demand for hydrogen. In addition, the current debate and implementation efforts dedicated to electric cars with batteries are another source of uncertainty as it remains to be seen which kind of electric vehicle will dominate in the future or whether the two will coexist.

Thus investment in the provision of new road fuels such as hydrogen looks very challenging in the long run and is intrinsically difficult at an early stage of switching from one fuel to another [8]. At the same time, the mass market ramp-up of vehicles with fuel cells depends on a sufficient refueling infrastructure which is crucial for the trust of customers in the usability of hydrogen-fueled vehicles. This situation is often referred to as a typical chicken and egg dilemma. Therefore, making hydrogen supply match evolving demand is far from trivial. The following discussion is important for early applications of fuel cells and takes a look into the future.

7.2
Balancing New Applications and Hydrogen Supply

Figure 7.1 provides an overview of the possible interplay between some new applications of fuel cells, different phases of market introduction, and the kind of infrastructure required [9]. Keeping in mind that the numbers in Figure 7.1 are only for illustration, the various applications start at different points in time and the infrastructure required for the production and distribution of hydrogen for most early applications can still be provisional.

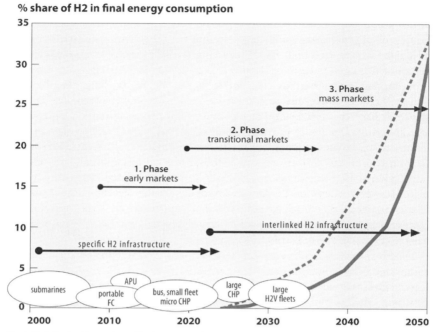

Fig. 7.1 Principal interplay of hydrogen applications, phases of market introduction, and infrastructure. Adapted from [9].

Generally, three typical phases can be distinguished. The first phase is characterized by basic research in fuel cells specialized for particular applications. In this phase, early products and markets emerge and prototypes are introduced ideally starting with applications where at this stage of development fuel cells are already relatively competitive in terms of cost. Typical early applications are portable devices and small-scale auxiliary power units (APUs) for telecommunication equipment or micro combined heat and power (CHP). Initially, the sizes and numbers of fuel cells used are relatively small. However, small-sized equipment may be appropriate for developing advanced manufacturing techniques. As fuel consumption of the fuel cells used is spatially scattered and relatively small, there is no need for a dense hydrogen infrastructure. Hydrogen demand for those applications remains limited and the hydrogen infrastructure used is specific.

In the second phase, the early markets grow in size and first pilot applications of fuel cell-equipped vehicles or larger stationary fuel cells are introduced. However, the range of the vehicles is limited and they are typically operated in fleets of dedicated users. Larger stationary fuel cells tend to be located where excess industrial hydrogen exists. This phase is particularly important for the ramp-up of fuel cell applications, that is, the scaling up of fuel cells for mobile use and for the further reduction of fuel cell production costs. During most of phase 2, the first filling stations are set up but there is still no need for a spatially dense hydrogen supply infrastructure. However, transitional markets signal that phase 2 is partly overlapping with phase 3. The main difference between phase 2 and phase 3 is that during the latter fuel cell-equipped cars are started to be sold to the general public. Therefore, this phase is particularly important for hydrogen supply, which needs to be increased in terms of volume. Further, this is the time when an initially sufficient infrastructure for distribution needs to be in place. In fact, the setting up of infrastructure for the production and distribution of hydrogen will need to take place somewhat earlier during phase 2.

Early applications can be distinguished in two ways. First, some applications are of general importance in bringing about greater production numbers of fuel cells in order to enhance progress on the learning curve of development and manufacturing. Those applications are often smaller in size but with otherwise similar characteristics, such as fuels cells for CHP or for road vehicles. Second, early applications also include prototypes of fuel cells for CHP and for initial use of road vehicles in fleet operation.

Basic research and development usually take place at universities and private companies where some prototypes are tested. The first step towards the market introduction of fuel cell vehicles will involve a limited number of prototypes that are tested under real-world conditions. What follows, which can be perceived of as a first step of market introduction, is the roll out of larger numbers of vehicles to dedicated fleet operators. At a very early stage, nonexistent infrastructure for refueling and maintenance of equipment makes it necessary to introduce early fuel cell vehicle fleets with dedicated fleet operators and limited operating ranges and to set up refueling and maintenance at the places where the vehicle fleets are located.

Taking account of the current technical status of fuel cell vehicles, early fleet applications are designed primarily for achieving practical experience beyond the usual testing procedures undertaken by fuel cell and vehicle manufacturers. Typically, the cost of equipment at this stage is on a level that precludes even the technical pioneers from private use.

During this initial phase of market penetration, there is, therefore, no need to provide full geographic coverage of refueling and maintenance. As a consequence, both can be kept limited in terms of spatial density and cost. In most respects this is similar for stationary use of fuel cells which is somewhat less complicated given its spatially fixed operation. In the case of larger fuel cells for CHP, location at places where excess industrial hydrogen is available can be an option. Another option could be to combine such fuel cells with new technologies for producing hydrogen sustainably at the same site. The advantage of such an approach is the complete implementation of sustainable production and use of hydrogen for demonstration.

As soon as marketing and sales of vehicles begins to address the general public, vehicle range and the density of refueling stations will need to be on a level that makes it rather unlikely for users to end up with no more hydrogen in the tank and no filling station within reach. Therefore, provision of hydrogen will need to be in place to an extent of spatial dispersion that early mass market customers of hydrogen-fueled vehicles will perceive as being sufficient. Hence a workable combination of the operating range of early fuel cell vehicles and the spatial density of filling stations is a prime determinant of when fuel cell vehicles can be sold to the general public.

It is for this reason that major actors from related German industries signed up to a hydrogen infrastructure build-up plan with the participation of the German Minister of Transport in September 2009. The plan includes the objective of the car industry to commercialize electric vehicles with fuel cells amounting to several hundred thousand units at the global scale starting from 2015. Before this, a significant expansion of the number of hydrogen filling stations in Germany is intended by the end of 2011 [10].

7.3
Criteria for Fuel Supply – Short- and Long-Term Requirements

It appears unlikely that future hydrogen supply may be based solely on one energy carrier as it has been mostly the case with current road fuels and industrial hydrogen (see Section 7.4.1). In principle, hydrogen can be supplied based on a variety of production technologies (see Section 7.4.2). It is the wide range of technical options for hydrogen provision and the need to match demand with supply in addition to meeting sustainability criteria which make the issues involved challenging. Some of the potential primary energy carriers for hydrogen production are in tune with the long-term vision of renewable energy for road vehicles. However, those are partly at an early stage of technical development and/or costly.

Those currently used in industry for hydrogen production are based almost entirely on fossil fuels which do not meet sustainability criteria but are technically well established, relatively cost-effective and already have substantial related infrastructure in place. However, with further rises in crude oil prices the production costs of fossil-based hydrogen will see corresponding increases whereas hydrogen produced from renewable sources will be less affected and will become more competitive in terms of relative cost.

Generally, hydrogen supply for fuel cells may be assessed from different angles based on various criteria:
- systemic requirements such as the mentioned match of supply and demand or resulting from features of the related application
- economic efficiency as a prerequisite of keeping costs limited
- environmental sustainability of supply as a fundamental requirement for meeting long-term objectives
- user acceptability as a prerequisite of the roll-out to mass markets.

These criteria may be weighed differently according to the application considered and the stage of market introduction.

From a systemic point of view, supply solutions may differ depending on the stationary and/or mobile type of fuel cell application. Larger stationary applications can be relatively unsophisticated regarding the supply logistics required. For keeping logistics costs low, stationary fuel cells for CHP may simply be installed where the hydrogen is or can be produced. Currently, some early applications of stationary fuel cells have already been located at plants for chlorine production (see Section 7.4.3). Thus the continuous supply of hydrogen from electrolyzers can be piped over very short distances to the stationary fuel cells without the need to set up a supply infrastructure covering longer distances from production to utilization and for compressed or cryogenic transport of hydrogen. However, existing industrial hydrogen production can be only a limited source for greater numbers of larger stationary fuel cells. For other small-scale applications of stationary fuel cells such as auxiliary power units, it may not be useful to locate hydrogen production at the place of use owing to the limited hydrogen volumes required.

Economic efficiency of hydrogen supply is a prerequisite of user acceptability. Efficiency considerations play an important role both for initial market introduction of early applications and in particular the potential roll-out to mass markets. For early applications, it might make sense to source the limited volumes of hydrogen cheaply from existing sources instead of setting up a costly new supply infrastructure which remains heavily underutilized. However, sticking to fossil-based industrial hydrogen is not up to the sustainable vision and might reduce acceptability with early users.

The main objective of introducing hydrogen as an energy carrier is to improve the environmental balance of energy use and the security of energy supply of the relevant applications. Hence early applications also need to demonstrate their contribution to environmental sustainability. This means that there is a strong contradiction between economic efficiency and environmental sustainability considerations. Both aspects are important and might impact on user acceptability. Further to cost and environmental sustainability considerations, users' acceptability may be influenced by technical features of early applications that are important for usability and secure use and have an important influence on users' trust in the relevant products.

It is obvious that in the case of hydrogen supply for early applications, a decision needs to be taken as to whether priority is to be given to cost-effectiveness or environmental sustainability considerations. Both have their particular merits but it may not always be feasible to combine them in early applications, given the premature stage of development of many of the sustainable hydrogen production options. However, beyond early applications, in the medium to long term supply will need to be in tune with sustainability criteria.

7.4
Hydrogen Production and Distribution

7.4.1
Industrial Hydrogen Production

In 2006, around 630.8 billion normal cubic meters of hydrogen were used to a large extent in various branches of the petrochemical industry, equaling 52.6 million metric tons globally. A significant fraction of that hydrogen has been consumed by refineries. Further to this, large volumes of hydrogen end up in the production of ammonia and methanol [11]. While in some regions there are pipeline systems in operation for distribution (see, e.g., Section 7.4.2), most industrial hydrogen is produced at the site of its utilization.

Nearly 96% of all industrial hydrogen is directly processed from fossil fuels, with natural gas being by far the most intensely used energy carrier with an estimated share of 49%, followed by liquid hydrocarbons (29%), coal (18%) and about 4% from chlorine electrolysis and other by-product sources of hydrogen [11]. Depending on the production of the electricity used for chlorine electrolysis, a related fraction of that hydrogen is also based on fossil energy carriers.

Figure 7.2 shows the hydrogen production sites in Europe that have been identified in the Roads2HyCom Project. As the map reveals, industrial hydrogen production sites can be found all over Europe with somewhat lower concentrations in Northern Europe, Ireland, and France. As was to be expected, production sites are spatially concentrated in those areas of Europe where industrial activity is particularly strong.

Three categories are depicted: merchant, captive, and by-product hydrogen. Merchant hydrogen, which is produced on-site or at neighboring plants which require hydrogen in larger volumes as an input or to be traded as smaller specified volumes. The most important hydrogen merchants are Air Products, Air Liquide, Praxair, and Linde Gas. Captive hydrogen is produced by the owners of the plants which use the hydrogen. However, outsourced on-site production of hydrogen by merchants can often be considered to be as captive as the category captive itself. The main difference is who produces the hydrogen that is dedicated as an input to specified production sites. The third category is by-product hydrogen that differs from the other categories in that it is produced in one process and not needed for further steps of the same overall production process. It is obvious that by-product hydrogen seems to be more likely to be made available for energetic application in fuel cells because hydrogen from the other categories is bound to particular uses via ownership of production plants or long-term contracts with merchants. However, most by-product hydrogen is currently used also. Mostly it is used for process heat or to generate electricity and could be substituted for natural gas provided that it can easily be made available. In any case, for early applications of fuel cells, it can be assumed that the volume of hydrogen demand is so low compared with industrial production that nearly all industrial hydrogen production sites can contribute to supply. At such low

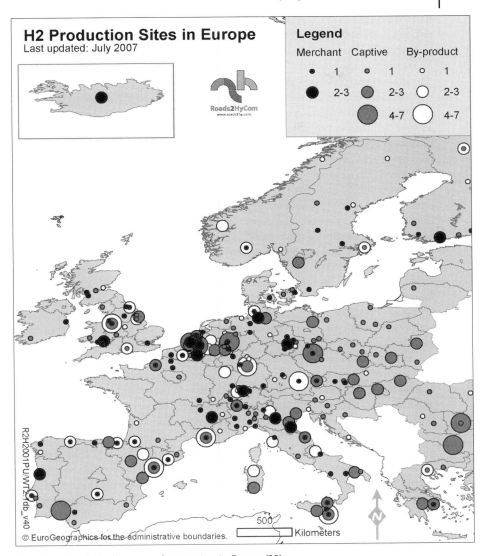

Fig. 7.2 Industrial hydrogen production sites in Europe [12].

volumes of demand, it is primarily a matter of logistics cost which supply structure is the most cost-effective. However, this is subject to change with increasing demand for hydrogen for use in fuel cells.

Processes in refineries involve hydrogen production as a by-product in addition to utilization as a chemical input for other processes. Up to now, most refineries have been net hydrogen producers. However, refineries are expected to turn into net hydrogen consumers or have already undergone that transformation [13]. One reason for this is decreasing qualities of crude oil, which reduce the volumes of hydrogen that arise from refining. Further, air quality regulations that limit the sulfur content in road fuels give rise to increasing desulfurization at refineries, which is a significant sink for hydrogen [14]. Ammonia and methanol production have always been hydrogen sinks and the hydrogen produced for these processes is entirely captive.

Chlorine electrolysis produces hydrogen as a by-product at a high level of purity. Existing processes for chlorine electrolysis generate hydrogen in fixed volumes according to the laws of stoichiometry. However, a technology for electrolysis currently under development using oxygen-depolarized cathodes will allow for chlorine electrolysis without producing hydrogen and result in a reduction in electricity consumption by around 30%. Depending on technical availability, retrofitting costs, and the market value of hydrogen, this new process can be phased in [11]. In the meantime, those volumes of hydrogen from chlorine production which are currently burnt for process heat and in power plants or vented off might be used for early applications of fuel cells. In the future, industrial hydrogen will become less available than it was in the past.

7.4.2
Principle Options for Hydrogen Production

Hydrogen production processes can be distinguished in terms of the feedstock, the kind of energy used, and whether or not direct emissions are caused. Table 7.1 provides an overview of the most important hydrogen production processes.

Steam reforming of natural gas is the dominant process for current production of industrial hydrogen. The main reasons for this are technical availability for large-scale application and cost-effectiveness. However, even with natural gas as feedstock, substantial emissions remain. Moreover, the price of gas is more closely related to the price of crude oil and cost-effectiveness is subject to rises in the crude oil price compared with options that are based on non-fossil feedstocks.

Thermochemical splitting of water might become a very important source for sustainable hydrogen production. Reasons for this are existing large areas of deserts which cannot be used for food or other biomass production but where solar radiation is very intense. Newly planned projects such as Desertec [18] are aimed at electricity production based on concentrated solar radiation, which could also be used for thermochemical splitting of water. Thermochemical splitting of water could be more efficient than producing electricity in the first instance and then using that electricity for water electrolysis, because the former involves one fewer conversion process. However, optimal location in the Earth's solar belt may require long-distance transport of hydrogen by pipeline or ship. Thermochemical splitting of water could also be used with nuclear power, which is more flexible with regard to location but increases the volumes of unwanted radioactive waste.

Gasification of coal results in high greenhouse gas emissions unless carbon capture and storage are applied, which are still at a very early stage and reduce

Table 7.1 Properties of hydrogen production processes.

Primary method	Process	Feedstock	Energy	Direct emissions
Thermal	Steam reforming	Natural gas	High-temperature steam	Some emissions
	Thermo-chemical Water splitting	Water	High-temperature heat from concentrated solar radiation	No emissions
	Gasification	Coal Biomass	Steam and oxygen at high temperature and pressure	Emissions depending on feedstock
	Pyrolysis	Biomass	Moderately high-temperature steam	Some emissions
Electrochemical	Electrolysis	Water	Electricity from wind, solar, hydro and nuclear	No emissions
	Electrolysis	Water	Electricity from coal or natural gas	Some emissions
	Photo-electro-chemical	Water	Direct sunlight	No emissions
Biological	Photo-biological	Water and algae strains	Direct sunlight	No emissions
	Anaerobic digestion	Biomass	High-temperature heat	Some emissions
	Fermentative Micro-organisms	Biomass	High-temperature heat	Some emissions

Source: adapted from [15–17].

conversion efficiency. Gasification of biomass is an option to which the same limitations apply as for most energetic use of newly produced biomass (limitation of arable land area, interference with food production).

Electrolysis is an available technology which may serve to convert electricity into hydrogen as a means of energy storage or for other purposes with fuel cells. However, direct forms of hydrogen production with less conversion processes may be more efficient.

In principle, most of the processes for hydrogen production mentioned in Table 7.1 could serve to provide hydrogen to early applications of fuel cells, given their technical availability. Demand for hydrogen for early applications may be too small for some processes to be cost-effective, but existing plants might step in to cover the relatively small demand.

7.4.3
Industrial Hydrogen – the Example of North Rhine-Westphalia

The aforementioned has provided the insight that industrial hydrogen is much more common than has probably been thought in the past. However, most industrial hydrogen is only produced to be used as an input in the chemical industry, hence it is unavailable for other uses.

Volumes of hydrogen available for other uses can in principle be identified as excess production capacity and by-product hydrogen that is currently not used or used as an energy carrier that could be substituted for natural gas or other fuels. The former potential is difficult because excess capacity of hydrogen production may be important for phases of economic prosperity and for growth of the industry which has set up the production site. The willingness to provide hydrogen from such sources might therefore be restricted to phases of economic slowdown and, therefore, cannot serve as a basis for continuous supply. In contrast, by-product hydrogen offers the greatest potential to make contributions to the supply for new energetic applications. Nevertheless, for very early and small hydrogen volume applications, all industrial sources of hydrogen may be considered. For this purpose, it is useful to take stock of all existing sources of industrial hydrogen in a region considered for the location of early applications. It is clear that the extent of available by-product hydrogen depends on the industrial structure and is therefore subject to substantial variety. With the following example we will take a look at North Rhine-Westphalia, where an ample supply of by-product hydrogen exists. This is, however, specific for this particular region and findings might differ substantially for other regions.

In addition to what has been examined in the Roads2HyCom [12] and GermanHy [19] projects, a study that took stock of industrial hydrogen in North Rhine-Westphalia has provided evidence on the existing potential for making use of it as an energy carrier for new applications. Based on estimates on industrial hydrogen production, a survey was performed concentrating on particular production sites of interest.

With regard to the industrial processes considered, a selection was made regarding the likelihood of making significant contributions to the supply of hydrogen for new energetic applications. Ammonia and methanol production sites were neglected because related large-scale hydrogen production is entirely captive, leaving hardly any room for other uses. Limited consideration was applied to coke oven gas because of the low level of hydrogen purity. Moreover, two out of three existing coke oven sites in North-Rhine Westphalia are integrated in steel works where the coke oven gas is used for internal processes or sold to a utility based on a long-term agreement that has recently been renewed. In the case of the only coke oven site in North Rhine-Westphalia where the coke oven gas is not completely captive or otherwise unavailable, a fraction is further processed and fed into the gas grid.

In order to estimate the volume of industrial hydrogen available for fuel cell application, four categories of use were defined based on the three applied in the

Roads2HyCom project. Captive hydrogen is produced exclusively for industrial processes and, therefore, unavailable for other kinds of utilization. Merchant hydrogen is usually produced using steam methane reformers or comes from processes where hydrogen is a by-product. Owing to the value added by merchants, this hydrogen will most likely be too costly for other use. As the fourth category there is still a certain share of by-product hydrogen that cannot reasonably be used and that is still being vented off into the atmosphere. The hydrogen potentially available for early application of fuel cells can be defined as those volumes which are currently used as an energy carrier or vented off into the atmosphere.

Figure 7.3 shows a map of the industrial production sites in North Rhine-Westphalia which have been identified as potential sources of hydrogen for new applications of fuel cells. Geographically, the Cologne and Düsseldorf areas clearly have an important share in the overall potential. Further plants can be found along the river Rhine and in the northern part of the Ruhr area. Ibbenbüren is located in the northern part of North Rhine-Westphalia close to Lower Saxony. Overall, the sites identified cover the Rhine–Ruhr area fairly well and might serve as a nucleus of an emerging hydrogen fueling infrastructure.

Figure 7.4 provides an overview of the quantitative findings. Accounting for 84%, refineries and the other plants make up the major share in total production of hydrogen of plants considered in the survey for North Rhine-Westphalia. However, those plants can only contribute 16% to the volume potentially available for new applications. The reason for this is that for both categories only a

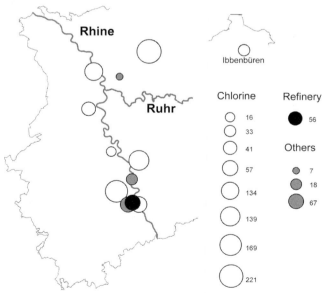

Fig. 7.3 Selected industrial hydrogen production sites in North Rhine–Westphalia and volumes of hydrogen available for early applications of fuel cells [H2NRW].

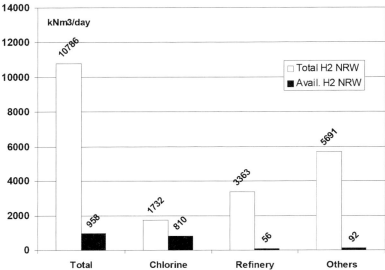

Fig. 7.4 Selected industrial hydrogen production in North Rhine-Westphalia and potential for new energetic use of industrial hydrogen [1].

minor excess capacity of about 2% is not used for chemical processes. The main potential can be found in plants for chlorine production, where about 50% could be made available.

As compared with a German study from 1995 [20], there are some reservations regarding the volume of hydrogen that could be made available for early applications of fuel cells. In the meantime, many companies have substantially reduced the volumes of hydrogen vented off into the atmosphere. This has mainly resulted from higher energy prices, and an important share of the hydrogen formerly vented off is nowadays used for generating process heat or in power plants. Further to this, there are already cases where the hydrogen from chlorine electrolysis is used in stationary fuel cells to generate electricity. Hence the volumes vented off have shrunk and the occurrence of the remaining hydrogen vented off can hardly be planned. It mostly would need to be stored, making its utilization more costly. Therefore, this category of hydrogen is available to a far lesser extent than it was in former times.

Under conservative model assumptions, a projection of the identified potential hydrogen supply from industrial production to new applications of fuel cells provided the insight that industrial hydrogen might be sufficient in some regions of North Rhine-Westphalia for at least 10 years to come. However, this number is primarily for illustration and subject to the uncertainties surrounding trajectories of industrial hydrogen production and the build-up of a fleet of fuel cell vehicles which actually cannot be predicted. Nevertheless, the contribution from industrial by-product hydrogen could be substantial for a considerable time.

In addition to figures on the volume of hydrogen produced, other relevant information on purity, pressure, and the existing hydrogen-related infrastructure was obtained through the survey. The industrial hydrogen produced cannot match the need for purity for use in fuel cells without further processing. Hydrogen from chlorine electrolysis has the highest level of purity. Generally, pressures of transported and stored industrial hydrogen are much lower compared with what is needed for non-stationary applications of fuel cells. Scarce exceptions are small and highly specified volumes of merchant hydrogen. For industrial use, there is simply no need to handle hydrogen at higher pressures. Therefore, in addition to purification, processing of industrial hydrogen for fuel cell applications needs to include compressing. Depending on logistic considerations, this could finally be done at the filling station, where pressurized distribution to the filling station makes no sense.

In addition to production facilities, the most important infrastructure for industrial hydrogen that could play a key role in the installation of a fueling infrastructure for early fuel cell vehicles in North Rhine-Westphalia is the existing

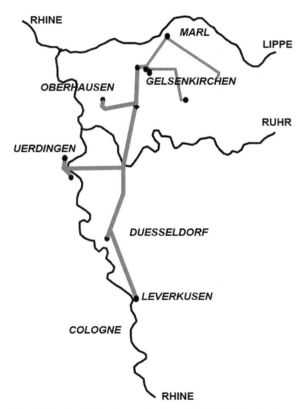

Fig. 7.5 Industrial hydrogen pipelines in North Rhine-Westphalia. Adapted from Air Liquide (personal communication).

network of pipelines. Since 1998, Air Liquide has operated the hydrogen pipeline network in North Rhine-Westphalia, which in total stretches over 240 km (Figure 7.5). The pipeline network connects 14 production sites, of which four feed in hydrogen. The total capacity of the pipelines is estimated to amount to 250×10^6 N m^3 of hydrogen per year.

Summing up the results of the survey, there are substantial volumes of industrial hydrogen produced in North Rhine-Westphalia of which a significant proportion could be made available for fuel cell vehicles. Most of the available industrial hydrogen in North Rhine-Westphalia is a by-product of chlorine production with its typical high purity from the process. Refineries and other production plants only offer very limited volumes that could be used for fuel cells because that hydrogen is almost entirely captive. In North Rhine-Westphalia, a substantial pipeline network for industrial hydrogen exists that could serve as a backbone for hydrogen supply for early applications of fuel cells and beyond.

7.5
Conclusion

Long-term availability of energy and technical measures to supply energy sustainably are the challenges where hydrogen use might form part of the answer. However, this hydrogen will need to be produced and distributed in ways that are different from today's industrial hydrogen use. In order to explore the real potential of fuel cells, it is necessary to set up early applications and to bridge the gap regarding the required supply infrastructure. Any hydrogen supply infrastructure needs to fit the needs of the respective state of market introduction of fuel cell applications. This is, of course, substantially different for early applications than at a stage of complete market penetration. For many early applications, the hydrogen supply infrastructure can be provisional. However, for the roll-out of fuel cell vehicles to the general public, depending on the vehicle range, a sufficiently dense refueling infrastructure is a prerequisite.

For very early applications in fleet operation, the supply infrastructure can be minimal and it might be cost-effective to rely on the existing industrial hydrogen supply instead of setting up something new. Notwithstanding this, combining new applications of fuel cells with new processes for producing hydrogen sustainably has its particular merits in allowing one to demonstrate the complete sustainable technological vision of hydrogen production and use. In order to take smart decisions, defining the preferences and assessing the particular circumstances are key for success.

Almost all current global industrial hydrogen is based on fossil fuels and thus unsustainable. Whereas hydrogen supply for ammonia and methanol production is completely captive, other production may include some excess capacity that could be used. Some hydrogen is produced as an unwanted by-product that is often used as an energy carrier and could be replaced by natural gas. Whereas in earlier times substantial volumes of hydrogen were vented off into the atmosphere, energetic use of such hydrogen has now become common. Future industrial hydrogen supply will most likely be influenced by refineries becoming net hydrogen consumers and the potential retrofitting of chlorine electrolysis for less electricity consumption and with hydrogen no longer being produced as a by-product.

A survey undertaken for North Rhine-Westphalia has revealed that the volume of industrial hydrogen that could be made available for fuel cell vehicles is substantial. It can be estimated at 350×10^6 N m^3 of hydrogen per year. With 85 of the industrial hydrogen identified as potentially available for new users originating from chlorine production, this particular industry makes up the major share. This reflects the industrial structure of this state and cannot be assumed for other regions. As an added advantage, current processes utilized for chlorine production come with hydrogen as a relatively pure by-product by industrial standards. Even hydrogen from that source will require further purification for use in fuel cells and will need to be pressurized because hydrogen transport and storage at high pressures typical for storage in vehicles is non-existent in industry. Moreover, an industrial pipeline network for hydrogen exists in North Rhine-Westphalia which might serve as an early backbone of hydrogen delivery for early applications of fuel cells.

However, the main task for bridging the gap into a sustainable energy future goes beyond this and concentrates on hydrogen supply from renewable sources.

References

1 Pastowski, A. and Grube, Th. (2010) *Scope and Perspectives of Industrial Hydrogen Production and Infrastructure for Fuel Cell Vehicles in North Rhine-Westphalia*, Energy Policy, article in press, http://dx.doi.org/10.1016/j.enpd2009.11.058.
2 http://www.naturalhy.net (accessed 3 December 2009).
3 International Energy Agency (2009) *Transport, Energy and CO_2: Moving Towards Sustainability*, IEA, Paris.
4 Fulton, L. (2004) *Reducing Oil Consumption in Transport: Combining Three Approaches*, IEA/EET Working Paper EET/2004/01, IEA, Paris.
5 Bringezu, S., Schütz, H., O'Brien, M., Kauppi, L., Howarth, R.W., and McNeely, J. (2009) *Towards Sustainable Production and Use of Resources: Assessing Biofuels*, UNEP, Nairobi.
6 Gärtner, S., Pastowski, A., Reinhardt, G., and Rettenmaier, N. (2007) *Rain Forest for Biodiesel? Ecological Effects of Using Palm Oil as a Source of Energy*, WWF Germany, Frankfurt am Main.
7 Tzimas, E., Castello, P., and Peteves, S. (2007) The evolution of size and cost of a hydrogen delivery infrastructure in europe in the medium and long term. *Int. J. Hydrogen Energy*, **32**, 1369–1380.
8 Agnolucci, P. (2007) Review: hydrogen infrastructure for the transport sector, the high-tech strategy for Germany, *Int. J. Hydrogen Energy* **32**, 526–544.
9 Arnold, K., vor der Brüggen, T., Fischedick, M., Merten, F., Nitsch, J., Viehbahn, P., Pietzner, K., and Ramesohl, S. (2005) *Strategie zur Förderung einer Wasserstoffenergiewirtschaft in NRW*, Wuppertal Institute, Wuppertal.
10 Linde AG (2009) *Joint Press Release of Linde, Daimler, EnBW, NOW, OMV, Shell, Total and Vattenfall: Initiative "H_2 Mobility" – Major companies sign up to hydrogen infrastructure built-up plan in Germany*,

http://www.linde.com. (accessed 15 March 2010).

11 Bala, S., Masahiro, Y., and Schlag, S. (2007) *CEH Marketing Research Report: Hydrogen*, SRI Consulting, Menlo Park, CA.

12 Maisonnier, G., Perrin, J., Steinberger-Wilckens, R., and Trümper, S.C. (2007) *European Hydrogen Infrastructure Atlas and Industrial Excess Hydrogen Analysis. Part II: Industrial Surplus Hydrogen and Markets and Production*, Oldenburg, http://www.roads2hy.com (accessed 7 January 2010).

13 CONCAWE Refinery Technology Support Group (2007) *Oil Refining in the EU in 2015*, CONCAWE Refinery Technology Support Group, Brussels.

14 Garvey, M.D. (2008) Hydrogen market growth – no end in sight. *CryoGas International*, 22–24.

15 National Hydrogen Association (2004) *Hydrogen Production Overview. Factsheet 1.008*, National Hydrogen Association, Washington, DC, http://www.hydrogenassociation.org/ (accessed 7 January 2010).

16 Winter, C.J. and Nitsch, J. (eds) (1998) *Hydrogen as an Energy Carrier*, Springer, Berlin.

17 Ogden, J.M. and Williams, R. H. (1989) *Solar Hydrogen – Moving Beyond Fossil Fuels*, World Resources Institute, Washington, DC.

18 Desertec Foundation, http://www.desertec.org (accessed 7 January 2010).

19 GermanHy, http://www.germanhy.de (accessed 7 January 2010).

20 Zittel, W. and Bünger, U. (1995) *Untersuchung über eine Einstiegsstrategie für den Aufbau einer Wasserstoffinfrastruktur für mobile Anwendungen in Deutschland*, LBST, Ottobrunn.

Hydrogen Production Technologies

Hydrogen Energy. Edited by Detlef Stolten
Copyright © 2010 WILEY-VCH Verlag GmbH & Co. KGaA, Weinheim
ISBN: 978-3-527-32711-9

8
Non-Thermal Production of Pure Hydrogen from Biomass: HYVOLUTION

Pieternel A. M. Claassen, Truus de Vrije, Emmanuel G. Koukios, Ed W. J. van Niel, Ebru Özgür, İnci Eroğlu, Isabella Nowik, Michael Modigell, Walter Wukovits, Anton Friedl, Dominik Ochs, and Werner Ahrer

Abstract

HYVOLUTION is the acronym for the Integrated Project (IP) "Non-Thermal Production of Pure Hydrogen from Biomass," which was defined in the 6th EU Framework Programme on Research, Technological Development and Demonstration, Priority 6.1 Sustainable Energy Systems. This IP started on 1 January 2006 and will end on 31 December 2010. Its aim, "Development of a blueprint for an industrial bioprocess for decentral hydrogen production from locally produced biomass," adds to the number and diversity of hydrogen production routes giving greater security of supply at the local and regional level. Moreover, this IP contributes a complementary strategy to fulfill the increased demand for renewable hydrogen expected in the transition to the Hydrogen Economy. The novel approach adopted in the project is based on a combined bioprocess employing thermophilic and phototrophic bacteria, to provide the highest hydrogen production efficiency in small-scale, cost-effective industries. In HYVOLUTION, 10 EU countries, Turkey, Russia, and South Africa are represented to assemble the critical mass needed to make a breakthrough in cost-effectiveness.

Keywords: hydrogen production, non-thermal processing, biomass, HYVOLUTION, biohydrogen, biosynthesis

8.1
Introduction

Hydrogen will be an important energy carrier in the future, according to several reports in prominent journals such as *Science* [1] and *Scientific American* [2]. However, to make the future Hydrogen Economy fully sustainable, renewable resources instead of fossil fuels have to be employed for hydrogen production. The concept of HYVOLUTION is based on the exploitation of bacteria, which freely and efficiently produce pure hydrogen as a by-product during growth on biomass. This approach, which started in the FP (Framework Programme) 5 project BIOHYDROGEN, allows a great reduction in CO_2 emission and provides independence of fossil imports. Both topics are dominant in all global agreements on climate protection and urgent in mitigating the greenhouse effect. The technologies developed as a result of the research in this Integrated Project (IP) will be commercialized post-2020. This will be in time to facilitate the transition to mass hydrogen markets, even more so since the European Commission has set an objective of 20% substitution by bio-fuels in the road transport sector in 2020. Hydrogen as an alternative motor fuel is seen in FP 6 as a market coming to maturity in 2015–2020.

The main scientific objective of this project is the development of a two-stage bioprocess for the cost-effective production of pure hydrogen from multiple biomass feedstocks. The bioprocess starts with a thermophilic fermentation of feedstock to hydrogen, CO_2, and intermediates. In a consecutive photoheterotrophic fermentation, all intermediates will be converted to more hydrogen and CO_2, to achieve an overall efficiency for the bioprocess of 75% (nearly 9 mol of hydrogen per mole of hexose).

Several sub-objectives contributing to the main scientific objective are:
- pretreatment technologies for optimal biodegradation of energy crops and bioresidues
- maximum efficiency in conversion of fermentable biomass to hydrogen and CO_2
- assessment of dedicated installations for optimal gas cleaning and gas quality protocols
- minimal energy demand and maximal product output through innovative system integration
- identification of market opportunities for a broad feedstock range.

The main technological objective is the construction of prototype modules of the plant which, when assembled, form the basis of a blueprint for the whole chain for converting biomass to pure hydrogen.

The sub-objectives to be achieved are prototypes of:
- equipment for mobilization of fermentable feedstock
- reactors for thermophilic hydrogen production
- reactors for photoheterotrophic hydrogen production
- devices for monitoring and control of the hydrogen production processes
- equipment for optimal gas cleaning procedures.

In addition to scientific and technological objectives, socio-economic activities are included to increase public awareness and societal acceptance, and to promote the identification of future opportunities, key stakeholders, and legal consequences of this specific bioprocess for decentral hydrogen production.

8.2
State of the Art

Distinct advanced strategies for the production of hydrogen from biomass are currently being studied:
- thermal processes such as gasification or supercritical water gasification
- non-thermal (biological or fermentative) processes, which are the issue in HYVOLUTION.

Non-thermal processes have a specific advantage for the efficient conversion of biomass with high moisture content to pure hydrogen. Second, fermentative processes do not require large installations for economy of scale. In this way, small-scale installations can be constructed for on-site cost-effective conversion of the locally produced biomass preventing energy loss through transport. Most research on the biological or fermentative production of hydrogen has been performed with hydrogen-producing bacteria, which have optimum growth and hydrogen production at ambient temperatures. The main drawback of these bacteria is that, in addition to hydrogen, they produce other reduced intermediates which compete with hydrogen production. As a result, the efficiency of this type of biomass conversion to hydrogen is low.

In a Dutch project, a conceptual design has been made for the biological production of hydrogen from potato steam peels [3]. The conceptual design was based on hydrogen production rates, regarded as feasible after further R&D. The outcome was a cost level of about €4 kg^{-1} hydrogen (comparable to €30 GJ^{-1}), which is about 3–4 times higher than the present price of hydrogen, derived from fossil fuels produced in large-scale installations.

The consortium for HYVOLUTION was formed to exploit the acquired knowledge and to make a breakthrough with a new taskforce aimed at the development of an industry producing hydrogen at a cost competitive with other biofuels. The price target will be achieved by efforts to reduce costs in the biomass pretreatment, optimize the efficiency and productivity in the fermentations, enabling low-cost thermo- and photobioreactors, develop dedicated, low-cost gas upgrading procedures, and introduce optimum system integration for making economic balances with respect to energy and heat utilization.

8.3
Methodology

In HYVOLUTION, the approach is based on the combination of a thermophilic fermentation (also called dark fermentation) with a photoheterotrophic fermentation. The novel issue is the application of thermophilic bacteria to start the bioprocess. This offers two important benefits in non-thermal hydrogen production. First, thermophilic fermentation at $\geq 70\,°C$ is superior in terms of hydrogen yield compared with fermentations at ambient temperature [4]. In thermophilic fermentations, glucose is converted to, on average, ≥3 mol of hydrogen and ≤2 mol of acetic acid as the main by-product. In other, mesophilic fermentations at ambient temperature, the average yield is only 1–2 mol of hydrogen, at the most, per mole of glucose. This is due to the production of more reduced by-products such as butyrate, propionate, ethanol, or butanol under mesophilic growth conditions. The second advantage lies in the production of acetic acid as the by-product of the first fermentation. Acetic acid is a prime substrate for photoheterotrophic bacteria. Energy from light enables photoheterotrophic bacteria to overcome the thermodynamic barrier in the conversion of acetic acid to hydrogen [5]. Through the combination of thermophilic fermentation with photoheterotrophic bacteria, complete conversion of the substrate to hydrogen and CO_2 can be established, resulting in a 75% conversion efficiency or 9 mol of hydrogen per mole of glucose, which is the main scientific objective of this IP. The various activities are assembled in workpackages (WPs) and integrated to a coherent project as shown in Figure 8.1, with both fermentations forming the core of the bioprocess.

WP 1, Biomass, addresses the efficient conversion of a range of agricultural produce and bioresidues to feedstocks specifically suitable for hydrogen fermentation by:

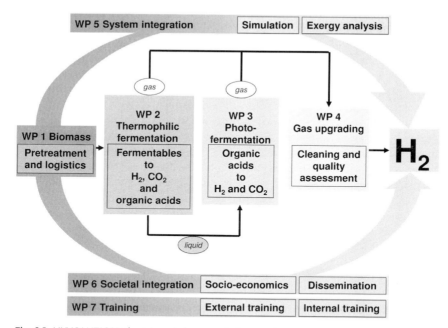

Fig. 8.1 HYVOLUTION: the integrated approach for non-thermal hydrogen production, which covers the whole chain from biomass to hydrogen, including societal integration for implementation in society.

- optimal selection of the most promising biomass sources
- high efficiency in the conversion of biomass to fermentable feedstocks.

The simultaneous utilization of hexose sugars and pentose sugars in addition to oligomeric carbohydrates by hydrogen-producing, thermophilic bacteria has been well documented [6–9]. These observations confer new opportunities for the development of agro-industrial chains which will utilize primary and secondary bioresidues besides energy crops. The generally large contribution of the biomass cost price to the final production cost of hydrogen requires the development of tailor-made pretreatment procedures. Furthermore, HYVOLUTION is specifically aimed at small-scale hydrogen production units with new logistic opportunities which have not been studied before to create new prospects for European rural areas.

WP 2, Thermophilic Fermentation, is aimed at maximum efficiency in the conversion of fermentable biomass to hydrogen, with as little by-product formation as possible. Therefore, the overall objective of WP 2 is to construct a stable thermophilic fermentation process for hydrogen production from fermentable biomass feedstocks investigated in WP 1.

The specific objective of WP 3, Photofermentation, is the utilization of the effluent of dark fermentation for highly efficient hydrogen production. The approach in WP 3 is to investigate the optimization of photofermentative hydrogen production from organic acids with high yields.

Through fundamental research on the physiology, biochemistry, and genomics of pure cultures of thermophilic and phototrophic bacteria, an insight in metabolic pathways is obtained [10, 11]. This is needed to model fermentations for optimum productivity and adjustment of the two consecutive fermentations. This insight will be the basis for identifying and/or developing improved strains and creating mixed cultures which are generally known for robustness, an important asset for industrial performance. The development of dedicated bioreactors also resides in WP 2 and WP 3 with the construction of prototype bioreactors being part of the technological objectives. Since the thermophilic bacteria are inhibited by hydrogen, starting at a partial concentration of 20% hydrogen [12], the challenge is to design a special thermobioreactor allowing easy gas removal [13]. For the photobioreactor, the emphasis is on the configuration of the bioreactor allowing maximum light capture on minimal surface area.

The goal of WP 4, Gas Upgrading, is the purification and assessment of the gas that is produced in the bioreactors. The aim is to produce hydrogen of constant quality. Therefore, the objectives are the development of an appropriate gas upgrading system and coupling to the bioreactors of WPs 2 and 3. The most important boundary condition for this system is the minimum energy demand needed to keep the overall process efficiency high. The gas upgrading will be specifically designed to remove the fairly high concentration of CO_2 in the raw gas and to handle a relatively small and fluctuating quantity of hydrogen. Further, the removal of other potential contaminants will be addressed, even though these are considered to contribute little to the final composition. It is the area in between production and application that will be addressed in WP 4 to deliver technically and economically feasible gas cleaning devices [14], with handling and safety procedures suitable to a small-scale hydrogen production plant.

In WP 5, System Integration, the focus is on ensuring maximum product output at minimum energy demand and minimum cost for production of hydrogen from biomass. The main objective of WP 5 is therefore the development of an integrated system representing the optimum combination of units and process routes. This will be achieved by innovatively integrating the different units developed and investigated in WPs 1–4 to a process route using process simulation, exergy analysis, cost evaluation, and detailed process engineering, including safety aspects and process control strategy.

The specific objectives of WP 6, Societal Integration, are the definition of the economic and social impact of hydrogen production from biomass and the promotion of use of hydrogen from biomass by enhancing awareness at the level of biomass providers and end-users. Two sub-objectives are:
- development of a methodology to analyze socio-economic and environmental impacts of biohydrogen production
- development of a strategy for dissemination and training activities at biomass providers, end-users, stake-holders and policy makers.

System integration and societal integration form a basis to secure the scientific and technical activities in WPs 1–4. These issues are fundamental to the development of this new bioprocess for small-scale hydrogen production and to make it viable in terms of process economics and socio-economics, including environmental impact. Both disciplines are prominently addressed in HYVOLUTION to identify the necessary adjustments right from the start. This is done to avoid routes which will have no economic future or do not adhere to sustainability, and to make optimum use of the integrated approach.

The activities in WP 7, Training, are directed at promoting the use of hydrogen from biomass and supporting the growth and implementation of the technology developed within the HYVOLUTION project. This will be achieved by the development of materials for training of industry, small and medium-sized enterprises, public organizations and policy makers to awaken public interest and to elucidate the advantage of the production of hydrogen from biomass.

8.4 The Project's Current Relation to the State of the Art

8.4.1
WP 1, Biomass

The work in WP 1, Biomass, started with the collection of data on the suitability and availability of various types of biomass spread over the 27 EU countries, which has been followed by combined cost and technical suitability mapping. The current biomass-based hydrogen production potential in the EU countries is estimated at 31 Mt H_2 annually using 10% of the crops and 100% of the agro-industrial residues under consideration [15]. The superposition of regional income expectation and technical development scenarios, involving key parameters determining the costs for biomass production, transport, and refinery, has been applied to identify the crucial factors in the biomass to hydrogen logistic chains. More specifically, this type of modeling has been applied to logistic chains for production (per crop and region), transport (spatial dispersion), and refining parameters (co-product exploitation). For an outlook to the future, 12 sustainability parameters of HYVOLUTION have been identified, which will be the basis for a sustainability index map of the long-term potential of biomass to biohydrogen chains in the EU. In close collaboration with WP 6, Societal Integration, extensive data acquisition, including data regarding the relevant socio-economic parameters, has been performed to conduct case studies for the implementation of a biohydrogen plant in both a rural (Thessaly in Greece) and an industrialized environment (South Holland in The Netherlands).

From the broad spectrum of biomass studied, four types have been selected for further experimental studies: sugar beet thick juice or molasses, potato steam peels, wheat bran, and barley straw, representing primary and secondary agro-industrial bioresidues (Figure 8.2). The experimental work has been focused on pretreatment and hydrolysis for the mobilization of sugars from biomass with the aim of reducing costs. For the conversion of the selected feedstocks to fermentable feedstock, the dosages of chemicals and enzymes have been reduced, and also the duration of the hydrolysis. A start has been made for the assessment of co-products from feedstocks such as sugar beet crop residues, sugar beet pulp, and the solid residues remaining after hydrolysis of potato steam peels, exploring several potential applications, for example as an additional energy source or, alternatively, as a nutritious animal feed.

Fig. 8.2 Biomass for hydrogen production.
(a) Potato steam peels; (b) barley straw.

8.4.2
WP2, Thermophilic Fermentation

In WP 2, Thermophilic Fermentation, the focus has been on *Caldicellulosiruptor saccharolyticus* as a representative of extreme thermophilic bacteria (Figure 8.3) [16]. The annotation of the genome of *C. saccharolyticus* has been completed [17]. Modern techniques such as transcriptomics, proteomics, bioinformatics, and genome-wide modeling have yielded further insight into the metabolism of this bacterium, which appears special in giving high substrate to hydrogen conversion efficiencies when using a great variety of carbohydrates [11, 16]. The observed presence of unusual enzymes in glycolysis suggests a prime role of ATP, indirectly governing the metabolic switch to lactate. Together with modeling studies, these findings indicate that simple removal of lactate dehydrogenase activity in *C. saccharolyticus* may not produce the desired mutants. Despite numerous different approaches, transformation of *C. saccharolyticus* has still not succeeded although the genetic tools have allowed the analysis of co-cultures of different *Caldicellulosiruptor* species. The observed stable coexistence with superior hydrogen production, yield, and productivity is an important asset for future industrial application. Designed co-cultures of *Caldicellulosiruptor* species seem to create a synergy achieving hydrogen yields close to the theoretical maximum of 4 mol H_2 mol^{-1} hexose, equivalent to almost 100% conversion efficiency in thermophilic fermentation [18]. The hydrogen production by pure cultures has been studied in defined media, in media containing molasses, or hydrolyzates pre-

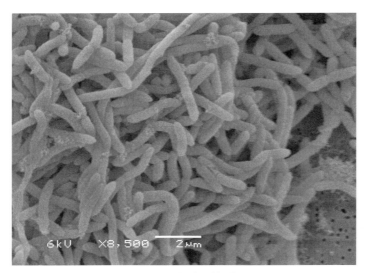

Fig. 8.3 Scanning electron micrograph of *Caldicellulosiruptor saccharolyticus*. Bar, 2 μm. Source: A. Pereira and M. Verhaart, Wageningen University, The Netherlands.

pared in WP 1 from *Miscanthus*, potato steam peels, barley grain and straw, or corn grain and stalk [9, 19, 20].

With increasing sugar concentrations to ~ 30 g glucose l^{-1}, a decrease in the substrate conversion efficiency has been observed, especially during growth on hydrolyzates. This may be due to the limited osmotolerance observed in some thermophilic hydrogen producers [12]. The penalty on maximum volumetric H_2 productivity was much less severe, remaining at ~ 15 mmol H_2 l^{-1} h^{-1}. Using continuous cultivation on molasses, it has been shown that addition of yeast extract was not needed. Apparently, this natural substrate contains sufficient nutrients for maintaining growth and efficient hydrogen production by *C. saccharolyticus*.

A novel fluidized and trickle bed bioreactor has been tested with *C. owensensis*, showing fair results. Hydrogen production has been established, albeit at low efficiency and productivity, without the need for hydrogen removal by using a stripping gas. These results have been supported with a comparable bioreactor configuration inoculated with an anaerobic thermophilic consortium, growing at 65 °C. Here, extremely high hydrogen productivities of >200 mmol H_2 l^{-1} h^{-1}, have been obtained. The challenge is to achieve similar productivities with the high yields obtainable with extreme thermophilic bacteria growing at temperatures ≥ 70 °C. This is important for reducing the burden on the photofermentative step where the reactor costs are decisive. An upscaled modular bioreactor, total volume 30 l, equipped with tailor-made control devices has been constructed to mimic the system used at 65 °C. This modular prototype has also served as the model for the 600 l bioreactor which is currently being built. In addition to these fluidized bed bioreactors, the application of a newly developed membrane bioreactor has permitted promising hydrogen productivities at 70 °C, in cultures grown on cellobiose as substrate. This was the first step towards the development of a membrane bioreactor with *in situ* removal of hydrogen and CO_2 without stripping annihilating the sometimes detrimental effects [21].

8.4.3
WP 3, Photofermentation

As in WP 2, the work in WP 3, Photofermentation, has been focused on physiological and genetic parameters and tools, new bacteria, use of real substrates such as the effluents coming from the thermophilic fermentation, and construction and use of bioreactors. In WP 3, the model organisms are Purple Non Sulfur (PNS) bacteria, coming from the Rhodobacteriaceae, and one Purple Sulfur bacterium, *Thiocapsa roseopersicina*, well-known for its inexhaustible supply of hydrogenase genes and enzymes but lack of growth on acetate [10, 22]. In PNS bacteria, hydrogen productivity and yield are affected by organic acid concentration [23], temperature [24], and varying nitrogen sources under continuous light or light-dark cycles [25].

The links between acetate metabolism and electron flow in various PNS bacteria and in *T. roseopersicina* have been mapped to near completeness. Further analysis of the intricate hydrogenase system in *T. roseopersicina* has pinpointed the candidate genes for successful improvement of Rhodobacteriaceae without adding antibiotic resistance genes. A hup^- mutant of *Rhodobacter sphaeroides* with improved hydrogen production during growth on malate has been obtained [26, 27], but active expression of hydrogenase has remained unsuccessful. Through natural selection, a heat-resistant *Rhodobacter capsulatus* has been obtained which showed hydrogen production at high temperatures. Some progress has been made in adjustment of the media used for photofermentation with the emphasis on ammonia as the main inhibitor of nitrogenase-mediated hydrogen production [28]. Using a natural zeolite such as clinoptilolite, a decrease in ammonium concentration of >60% has been achieved, giving more degrees of freedom in feedstock use in WP 1 and 2. Other pretreatments such as centrifugation, filtration, dilution, sterilization, and supplementations with buffer, molybdenum, and iron have also improved hydrogen yields [29].

Photofermentation using effluents from thermophilic fermentations on potato steam peels and molasses has been performed very successfully, showing productivities and yields which were comparable to those obtained with defined media. Among various PNS bacteria, *R. capsulatus* (DSM1710) gave the highest yield (24%) and productivity (0.57 mmol $H_2\,l^{-1}\,h^{-1}$) after adjustment by adding buffer, iron, and molybdenum to the effluent from a fermentation on potato steam peels [30]. Successful sequential operation of thermophilic and photofermentation has also been shown with sugar beet molasses as the initial feedstock. The *R. capsulatus hup*$^-$ strain gave the highest yield (58%) and productivity (1.37 mmol $H_2\,l^{-1}\,h^{-1}$) of several PNS bacteria tested [31]. Therefore, the overall yield in the integrated bioprocess using sucrose in molasses achieved 6.1 mol mol^{-1} hexose (51%) [20]. Long-term continuous hydrogen production has been achieved by three strains of *R. capsulatus* (wild-type, Hup$^-$, and heat adapted) in indoor and outdoor fed-batch panel reactors (4–8 l) using defined medium or effluents from thermophilic fermentations. Stable biomass concentration and hydrogen production in long-term continuous operations depend greatly on the carbon to nitrogen ratio. Stability can be further improved by keeping the acetate concentration and carbon to nitrogen ratio in the photobioreactor at certain values by changing the feed composition. Progress has also been achieved with the prototype photobioreactors in terms of simple design, low material and production costs, and high utilization of sunlight within the optimum wavelength range [32]. The performances of panel and tubular photobioreactors have been compared under outdoor conditions (Figure 8.4) and led to the conclusion that panel photobioreactors are suitable for low light intensity areas and tubular reactors for high light intensity. Hydrogen production has been established in 65 and 80 l panel photobioreactors with reasonable productivities over more than 50 days, under outdoor conditions.

Pilot-scale panel (25×4 l) and tubular photobioreactors (65 l) have been constructed and operated side-by-side, outdoors, in Aachen during summer. Comparable productivities per unit illuminated surface area were achieved for both reactors in fed-batch mode by *R. capsulatus* (DSM155) on an acetate-, lactate-, and glutamate-containing defined medium [34]. The illuminated surface area

Fig. 8.4 Comparison of hydrogen production parameters in an 80 l tubular and a 40 l panel photobioreactor placed in a greenhouse in Ankara, Turkey, during winter 2008. The hydrogen productivity per unit illuminated area during growth of *Rhodobacter capsulatus* (DSM1710) on a defined medium with acetate and glutamate was 0.3 mmol H_2 m^{-2} h^{-1} [33].

Table 8.1 Hydrogen productivities in outdoor fed-batch cultures of *Rhodobacter capsulatus* (DSM155 and DSM1710) growing on a defined medium containing acetate, lactate, and glutamate in pilot scale bioreactors.

Rhodobacter capsulatus strain	Bioreactor type	Location	Volumetric productivity (mol H_2 m^{-3} h^{-1})	Productivity per illuminated surface area (mmol H_2 m_{IS}^2 d^{-1})	Productivity per ground area (mmol H_2 m_G^2 d^{-1})
DSM155	Panel	Aachen	0.36	56	450
DSM155	Tubular	Aachen	0.15	49	54
DSM1710	Tubular	Ankara	0.31	112	78

per unit ground space area is about 8.5 times higher in the panel reactor than in the tubular system (Table 8.1).

8.4.4
WP 4, Gas Upgrading

In WP 4, Gas Upgrading, downstream processing of the product gas consisting of the gas upgrading and subsequent storage is addressed. Theoretically, the concentration of hydrogen in the raw gas produced in the thermophilic fermentation amounts to 66%. The concentration of CO_2 is ∼33% and additional impurities (H_2S, NH_3) are estimated to be in the 100 ppm range when using higher concentrations of hydrolyzates. The concentration of hydrogen in the raw gas produced in the photofermentation is 90–95%, due to solution of CO_2. The drawback here is the day-night cycle with high and virtually no hydrogen production. Different conventional gas purification systems have been assessed in detail in terms of energy, operational, and investment costs. Furthermore, a novel gas purification system based on membrane contactor technology which is expected to consume less energy than the standard gas purification system is under development.

From the conventional gas upgrading options, vacuum swing adsorption (VSA) has been selected based on the assumption that the gas mixture produced in the bioreactors will contain not more than 33% of CO_2. The technological applicability of the VSA in the biohydrogen process has been demonstrated and evaluated by connecting a test device to the photobioreactor. Two zeolite-based adsorbing substances (13X and 5A), which are commercially available and widely used in industrial applications for hydrogen purification, have been selected. The trade-off study on the conventional gas upgrading systems has shown energy losses of around 20% when applying a hydrogen-rich stream contaminated by 30% of CO_2 to the VSA. When comparing gas upgrading of mixtures containing 30–40% CO_2, the results strongly favor the application of membrane separation, especially in terms of energy demand.

The membrane contactor which is under development within HYVOLUTION employs a dense membrane resulting in hermetically separated phases [35]. Thus, evaporation of the carrier solution into the gaseous feed phase is avoided, resulting in a clean product and no loss of carrier solution. The selectivity of the membrane has proven to be irrelevant since the separation of the gaseous feed takes place mainly by selective chemical absorption of one component in the liquid phase. Polyvinyltrimethylsilane (PVTMS) has been chosen as the membrane material due to its high permeability [14]. Monoethanolamine, K_2CO_3, and K_2CO_3 promoted by piperazine have been evaluated as absorbent solutions in terms of desorption energy demand and reactivity. K_2CO_3 promoted by piperazine has been selected for the HYVOLUTION process since it has a low energy demand and a higher theoretical capacity than monoethanolamine [36] and a considerably higher reactivity than K_2CO_3 [37].

The cassette-type membrane contactor module itself and its manufacture have been optimized continually. The hydrodynamics of the liquid phase, which are important for the convective transport of absorbent to the membrane and therefore the performance of the module, have been improved by, for example, the application of new spacer types. The current estimation of the size required for the membrane when applied in a 60 kg H_2 h^{-1} production plant is 10 000 m^2, equal to 12.5 m^3.

In order to regenerate the absorbent, a desorption step has been installed to remove CO_2 from the absorbent. The loaded absorbent was transported to the desorber, which also consisted of a membrane contactor, and the regenerated absorbent was pumped back to the absorber [14]. The application of sweep gas together with an increase in temperature has been identified as the most efficient approach for desorption.

A storage system has been designed to make the hydrogen availability independent of the production process and to compensate for the fluctuating flow rates. Within the classical storage systems for molecular hydrogen, compressed gas at a pressure of 200 bar involves an energy penalty of $\sim 11\%$ and liquid storage of $\sim 50\%$. Low-pressure storage at 10–15 bar, commonly applied with helium, decreases the overall thermal energy output by less than 5% and thus presents a low-energy consuming alternative for hydrogen storage. The storage of hydrogen in, for example, metal hydride or chemical hydride systems has also been investigated and might be an option for the future.

In addition to the gas upgrading systems, a gas sensor system has been developed to control the quality of the product gas. A prototype of the gas sensor system for continuous or semi-continuous analysis (H_2, CO_2, H_2S, and O_2) has been developed, manufactured and tested. The system is based on cheap commercial sensors and is temperature controlled. An electrochemical hydrogen sensor with a measurement range of up to 5% together with an innovative dilution method has been selected for measurement of concentrations up to 100% of hydrogen. CO_2 is measured by means of a commercial near-infrared (NIR) absorbance sensor. Detection of H_2S is effected using an electrochemical sensor together with a dilution method or with UV absorbance and O_2 by a galvanic fuel cell sensor.

8.4.5
WP5, System Integration

The fine tuning of the calculation models with defined parameters to fulfill the demands for WP 5, System Integration, for simulation, basic engineering, costing and, exergy analysis, has been finalized. The current simulation studies and sensitivity analyses have been based on experimental data from WPs 1–4 and include the first process and heat integration steps. The commercial software package ASPENplus® has been selected to predict the behavior of HYVOLUTION using basic mass balance, energy balance, phase and chemical equilibrium, and reaction kinetics. A software package and simulation models using experimental data have been described [38, 39]. The process has been scaled to produce 60 kg h^{-1} of pure hydrogen (equivalent to 2 MW thermal power).

Case studies and sensitivity analyses have been carried out to improve process performance in terms of utility and energy demand. The need for increasing substrate concentrations in the thermophilic fermentation and also the photofermentation has become a prime issue. However, a reduction of heat and water demand by making an increase in substrate concentration turned out to be problematic due to exceeding a critical limit of osmolality in the thermophilic fermentation (THF) [21]. Investigation of different recirculation options has indicated the following (Figure 8.5):

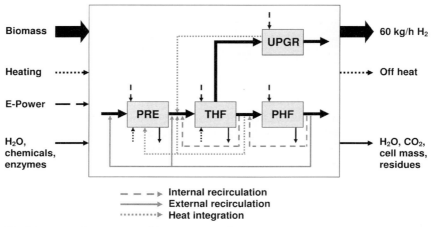

Fig. 8.5 Options for process and heat integration by recirculating process water in HYVOLUTION. PRE, pretreatment; THF, thermophilic fermentation; PHF, photofermentation; UPGR, gas upgrading.

- Internal recirculation of effluents in the thermophilic reactor (THF) gives a considerable reduction in heat demand but also a prohibitive increase in acetate concentration.
- Internal recirculation in the photobioreactor (PHF) gives a strong decrease in water demand.
- External recirculation of PHF effluent to the THF achieves a partial reduction in heat and water demand and only slightly increases the concentration and osmolality in the thermophilic reactor.
- External recirculation in the THF combined with internal recirculation in the PHF gives a strong reduction in water and chemicals demand without a significant change in osmolality.
- Heat exchanger between the inlet and outlet of the THF seems the only solution to reduce strongly the heat demand in the THF. Heat integration in the THF and also in the pretreatment step (PTR) reduces the heat duty in the case of feedstock potato steam peel by 85% and 30%, respectively [40].

The exergy of compounds and streams is calculated as the sum of three components, chemical and physical exergy and the exergy change of mixing, using an MS Excel spreadsheet. The calculation of the necessary thermodynamic properties has been based on integrated polynomial functions for the values of specific heat, entropy, and enthalpy, using the same correlations as in the process simulation tool [41, 42]. Exergy losses of the investigated process options are between 7 and 9%. The efficiency based on the chemical exergy of the biomass feed and pure hydrogen produced is 36–45% depending on the feedstock used and the configuration of the overall process. These results correspond to data for anaerobic digestion of biomass to hydrogen and also to biogas production of 36 and 46%, respectively [43].

The exergetic efficiency achieved depended strongly on the evaluation of the various products. A significant increase in exergetic efficiency of the overall process is achieved when defining residual biomass (produced cell mass and nonfermentable residues from pretreatment) as a usable product as the exergetic efficiency almost doubles. The possible contribution of heat integration to the increase in the exergetic efficiency of the process is negligible due to the strong impact of chemical exergy compared with physical exergy in this low-temperature process. Nevertheless, heat integration plays an important role from the point of view of energy demand (see above) and economic evaluation.

Process engineering has been addressed by the design of a hydrogen production plant connected with a sugar factory, including a process flow sheet of the hydrogen plant and interfaces of the sugar factory, selection and sizing of equipment, and a plot plan of the equipment. Detailed heat integration followed, with comparison of the impact of a stand-alone biohydrogen plant using steam stripping for hydrogen removal. The process engineering comprised risk and safety analysis of the HYVOLUTION based on a HAZOP study and a dedicated process control system including full instrumentation.

A first cost evaluation has addressed different scenarios concerning substrate concentration, yield, and productivities based on a model developed with the Aspen Icarus Process Evaluator® (IPE) in parallel with the simulation model. The cost calculations showed that the cost of the feedstock is of major importance. The objective in HYVOLUTION is about € 6 GJ^{-1}, contributed by biomass production and pretreatment, for the total cost for hydrogen from an industrial HYVOLUTION plant. The current estimations range from € 4 to € 12.7 GJ^{-1} for wheat bran and sugar beet in various regions, respectively. Preliminary cost evaluation of the overall process also showed the decisive effect of the cost of enzymes for pretreatment, yeast extract, and the material for the photobioreactor.

The purpose of this early cost evaluation was to identify the technical and economic bottlenecks of the present process and to give an understanding of the economic aspects of HYVOLUTION. The data will be updated at the end of HYVOLUTION.

8.4.6
WP 6, Societal Integration

The socio-economic studies in WP 6, Societal Integration, have evaluated biomass production- and feedstock-related cost data, using EU database data with superposition of four regional zones, namely high income-high technological development, low income-high technological development, high income-low technological development, and low income-low technological development. The cost data for the overall process have been used in WP 6 to permit filling of the HYVOLUTION Social Economic Model (HYSEM) assuming a reduction to 10%

of the current cost of enzymes, yeast extract, and plastic material for the photobioreactor. HYSEM has been filled with data provided by WP 1 and 5 to assess the socio-economic impact of a hydrogen plant in two case studies in a rural (Thessaly in Greece) and an industrialized environment (South Holland in The Netherlands). This model considered the socio-economic impacts of (i) construction of the hydrogen plant, (ii) operation of the plant, and (iii) production of feedstock. The first results showed significant indirect and induced effects. A tentative sensitivity analysis has been made to identify the effect of key variables, such as chemical costs and hydrogen market price, on the profitability. The current analysis was based on several anticipated gains and a more conclusive socio-economic analysis will be delivered at the end of the project.

To ensure sustainability and optimum performance in the environmental profile of HYVOLUTION, all technological research was accompanied by Life Cycle Assessment (LCA). As the major methodology, the Eco Indicator 99 procedure was applied following the ISO Standard 14040. The production of biogas (methane containing) followed by a steam reforming process was chosen as most suitable reference system.

During the course of the project, the major factors responsible for an environmental impact were identified and suggestions for lowering these impacts were made. Important contributors to a weak environmental profile are the base and phosphate buffer solutions for pH control in both reactors. Following an LCA-based engineering approach, these factors were continuously monitored throughout the project and steadily improved. The first LCA results showed a tremendously high impact factor of 54 points. As a result of several improvements achieved within the environmental performance, this impact factor has currently decreased to 1.5 points, with the major increment arising from the use of fossil fuels to maintain the process. The final goal will be to make the technology as environmentally friendly as the reference system. This technology exhibits the lowest impact (0.2 points) of all investigated technologies and is therefore regarded as the ultimate benchmark for HYVOLUTION in terms of environmental performance. It should be pointed out that the HYVOLUTION technology in terms of making it technically feasible is very young and has to be further developed.

8.4.7
WP 7, Training

The target groups for WP 7, Training, have been identified as users of renewable energy, engineering companies, local and rural communities, and policy makers and local authorities, including the companies that have registered in the Industrial Interest Group of HYVOLUTION. Since sufficient information is available to clarify the advantages of hydrogen and also the advantages and drawbacks of the hydrogen economy, the main focus of training has been on production of biohydrogen and HYVOLUTION, including technological, ecological, and economical issues. Draft papers on general training topics, dedicated for training of the general public and making an introduction to more specific training topics, have been developed. The main issues are (i) conventional and novel hydrogen production processes and (ii) why and how to produce biohydrogen. These materials will be become available for electronic presentations after evaluation.

8.5
Conclusion

The progress in HYVOLUTION has clearly shown that the concept of non-thermal production of pure hydrogen from biomass is realistic. The potential of agro-industrial resources, bioresidues, and dedicated crops to contribute to hydrogen production in the EU is considerable. The yield obtained with thermophilic fermentation has met the predicted target and the yield of the photofermentation has been steadily increasing. Prototypes of new bioreactor configurations have been shown to permit enormous gains in productivity and decreases in process costs. However, this promise still needs to be confirmed during upscaling. The same is true for the novel gas upgrading technology.

Currently, the long term scenario (2030) predicts a cost of 6 kg^{-1} H_2, provided that an overall yield of 85% and productivity of 53 and 3.3 mmol H_2 l^{-1} h^{-1} for the thermophilic and photofermentation, respectively, are achieved. Since the present situation with molasses as feedstock is an overall yield of 51% with productivities of 16 and 0.86 mmol H_2 l^{-1} h^{-1}, respectively, HYVOLUTION seems well on its way.

Acknowledgments

All partners in HYVOLUTION are acknowledged for their contribution to the work and results presented here. The research in HYVOLUTION is financially supported by the Commission of the European Communities, 6th Framework Programme, Priority 6, Sustainable Energy Systems (019825).

References

1 Toward a hydrogen economy (2004) *Science*, **305** (5686), 957–976.
2 Wald, M. L. (2004) Questions about a hydrogen economy. *Sci. Am.*, **42**, 41–47.
3 Claassen, P. A. M. and de Vrije, T. (2004) *Biological Hydrogen Production – Final Report*. Holtzbrinely Nature Publishing Group, New York City, USA, ISBN 90-6754-753-0.
4 de Vrije, T. and Claassen, P. A. M. (2003) Dark hydrogen fermentations. In *Biomethane and Bio-hydrogen. Status and Perspectives of Biological Methane and Hydrogen Production* (eds J. H. Reith, R. H. Wijffels, and H. Barten), Smiet Offset, The Hague, pp. 103–123.
5 Akkerman, I., Janssen, M., Rocha, J. M. S., Reith, J. H., and Wijffels, R. H. (2003) Photobiological hydrogen

production: photochemical efficiency and bioreactor design. In *Bio-methane and Biohydrogen. Status and Perspectives of Biological Methane and Hydrogen Production* (eds J. H. Reith, R. H. Wijffels, and H. Barten), Smiet Offset, The Hague, pp. 124–145.

6 de Vrije, T., de Haas, G. G., Tan, G. B., Keijsers, E. R. P., and Claassen, P. A. M. (2002) Pretreatment of *Miscanthus* for hydrogen production by *Thermotoga elfii*. *Int. J. Hydrogen Energy*, **27**, 1381–1390.

7 Kádár, Zs., de Vrije, T., van Noorden, G. E., Budde, M. A. W., Szengyel Zs., Réczey, K., and Claassen P. A. M. (2004) Yields from glucose, xylose, and paper sludge hydrolysate during hydrogen production by the extreme thermophile *Caldicellulosiruptor saccharolyticus*. *Appl. Biochem. Biotechnol.*, **114**, 497–508.

8 Kengen, S. W. M., Goorissen, H. P., Verhaart, M., van Niel, E. W. J., Claassen, P. A. M., and Stams, A. J. M. (2009) Biological hydrogen production by anaerobic microorganisms. In *Biofuels* (eds W. Soetaert and E. J. Vandamme), John Wiley & Sons, Ltd, Chichester, pp. 197–221.

9 de Vrije, T., Bakker, R. R., Budde, M. A. W., Lai, M., Mars, A. E., and Claassen, P. A. M. (2009) Efficient hydrogen production from the lignocellulosic energy crop *Miscanthus* by the extreme thermophilic bacteria *Caldicellulosiruptor saccharolyticus* and *Thermotoga neapolitana*. *Biotechnol. Biofuels*, **2**(12), 1–15.

10 Kovács, K. L., Marótia, G., and Rákhely, G. (2006) A novel approach for biohydrogen production. *Int. J. Hydrogen Energy*, **31**, 1460–1468.

11 de Vrije, T., Mars, A. E., Budde, M. A. W., Lai, M. H., Dijkema, C., de Waard, P., and Claassen, P. A. M. (2007) Glycolytic pathway and hydrogen yield studies of the extreme thermophile *Caldicellulosiruptor saccharolyticus*. *Appl. Microbiol. Biotechnol.*, **74**, 1358–1367.

12 van Niel, E. W. J., Claassen, P. A. M., and Stams, A. J. M. (2003) Substrate and product inhibition of hydrogen production by the extreme thermophile *Caldicellulosiruptor saccharolyticus*. *Biotechnol. Bioeng.*, **81**, 255–262.

13 van Groenestijn, J. W., Hazewinkel, J. H. O., Nienoord, M., and Bussmann, P. J. T. (2002) Energy aspects of biological hydrogen production in high rate bioreactors operated in the thermophilic temperature range. *Int. J. Hydrogen Energy*, **27**, 1141–1147.

14 Modigell, M., Schumacher, M., Teplyakov, V. V., and Zenkevich, V. B. (2008) A membrane contactor for efficient CO_2 removal in biohydrogen production. *Desalination*, **224**, 186–190.

15 Karaoglanoglou, L. S., Diamantopoulou, L. K., and Koukios, E. G. (2008) At the crossroads of feasibility and sustainability: building biomass-to-biohydrogen supply chains. In *From Research to Industry and Markets* (ed. A. Grassi), Proceedings of 16th European Biomass Conference and Exhibition, 2–6 June 2008, Valencia, Spain, ETA-Florence, Florence and WIP-Munich, Munich, pp. 435–438.

16 van Niel, E. W. J., Budde, M. A. W., de Haas, G. G., van der Wal, F. J., Claassen, P. A. M., and Stams, A. J. M. (2002) Distinctive properties of high hydrogen producing extreme thermophiles, *Caldicellulosiruptor saccharolyticus* and *Thermotoga elfii*. *Int. J. Hydrogen Energy*, **27**, 1391–1398.

17 van de Werken, H. J. G., Verhaart, M. R. A., Van Fossen, A. L., Willquist, K. U., Lewis, D. L., Nichols, J. D., Goorissen, H. P., Mongodin, E. F., Nelson, K. E., van Niel, E. W. J., Stams, A. J. M., Ward, D. E., de Vos, W. M., van der Oost, J., Kelly, K. M., and Kengen, S. M. W. (2008) Hydrogenomics of the extremely thermophilic bacterium *Caldicellulosiruptor saccharolyticus*. *Appl. Environ. Microbiol.*, **74**, 6720–6729.

18 Zeidan, A. A. and van Niel, E. W. J. (2009) Developing a thermophilic hydrogen-producing co-culture for efficient utilization of mixed sugars. *Int. J. Hydrogen Energy*, **34**, 4524–4528.

19 Panagiotopoulos, I. A., Bakker, R. R., Budde, M. A. W., de Vrije, T., Claassen, P. A. M., and Koukios, E. G. (2009) Fermentative hydrogen production from

pretreated biomass: a comparative study. *Biores. Technol.*, **100**, 6331–6338.

20 Özgür, E., Mars, A. E., Peksel, B., Louwerse, A., Yücel, M., Gündüz, U., Claassen, P. A. M., and Eroğlu, İ. (2009) Biohydrogen production from beet molasses by sequential dark and photofermentation. *Int. J. Hydrogen Energy*, **35**, 511–517.

21 Willquist, K., Claassen, P. A. M., and van Niel, E. W. J. (2009) Evaluation of the influence of CO_2 as stripping gas on the performance of the hydrogen producer *Caldicellulosiruptor saccharolyticus*. *Int. J. Hydrogen Energy*, **34**, 4718–4726.

22 Rákhely, G., Laurinavichene, T. V., Tsygankov, A. A., and Kovács, K. L. (2007) The role of Hox hydrogenase in the H_2 metabolism of *Thiocapsa roseopersicina*. *Biochim. Biophys. Acta*, **1767**, 671–676.

23 Uyar, B., Eroğlu, İ., Yücel, M., and Gündüz, U. (2008) Photofermentative hydrogen production from volatile fatty acids present in dark fermentation effluents. *Int. J. Hydrogen Energy*, **34**, 4517–4523.

24 Özgür, E., Uyar, B., Öztürk, Y., Yücel, M., Gündüz, U., and Eroğlu, İ. (2009) Biohydrogen production by *Rhodobacter capsulatus* on acetate at fluctuating temperatures. *Resour. Conserv. Recycl.*, **54**, 310–314.

25 Uyar, B., Eroğlu, İ., Yücel, M., Gündüz, U., and Türker, L. (2007) Effect of light intensity, wavelength and illumination protocol on hydrogen production in photobioreactors. *Int. J. Hydrogen Energy*, **32**, 4670–4677.

26 Kars, G., Gündüz, U., Rakhely, G., Yücel, M., Eroğlu, İ., and Kovacs, K. L. (2008) Improved hydrogen production by hydrogenase deficient mutant strain of Rhodobacter sphaeroides O. U.001. *Int. J. Hydrogen Energy*, **33**, 3056–3060.

27 Kars, G., Gündüz, U., Yücel, M., Rakhely, G., Kovacs, K. L., and Eroğlu, İ. (2009) Evaluation of hydrogen production by *Rhodobacter sphaeroides* O. U.001 and its *hupSL* deficient mutant using acetate and malate as carbon sources. *Int. J. Hydrogen Energy*, **34**, 2184–2190.

28 Akköse, S., Gündüz, U., Yücel, M., and Eroğlu, İ. (2009) Effects of ammonium ion, acetate and aerobic conditions on hydrogen production and expression levels of nitrogenase genes in *Rhodobacter sphaeroides* O. U.001, *Int. J. Hydrogen Energy*, **34**, 8818–8827.

29 Uyar, B., Schumacher, M., Gebicki, J., and Modigell, M. (2009) Photoproduction of hydrogen by *Rhodobacter capsulatus* from thermophilic fermentation effluent. *Bioprocess Biosyst. Eng.*, **32**, 603–606.

30 Afşar, N., Özgür, E., Gürgan, M., de Vrije, T., Yücel, M., Gündüz, U., and Eroğlu, İ. (2009) Hydrogen production by *R. capsulatus* on dark fermentor effluent of potato steam peel hydrolysate, *Chem. Eng. Trans.*, **18**, 385–390.

31 Öztürk, Y., Yücel, M., Daldal, F., Mandacı, S., Gündüz, U., Türker, L., and Eroğlu İ. (2006) Hydrogen production by using *Rhodobacter capsulatus* mutants with genetically modified electron transfer chains. *Int. J. Hydrogen Energy*, **31**, 1545–1552.

32 Modigell, M., Schumacher, M., and Claassen, P. A. M. (2007) Hyvolution – Entwicklung eines zweistufigen Bioprozesses zur Produktion von Wasserstoff aus Biomasse, *Chem. Ing. Tech.*, **79**, 637–641.

33 Boran, E., Özgür, E., Gebicki, J., van der Burg, J., Yücel, M., Gündüz, U., Modigell, M., and Eroğlu, İ. (2009) Investigation of influencing factors for biological hydrogen production by *Rhodobacter capsulatus* in tubular photobioreactors, *Chem. Eng. Trans.*, **18**, 357–362.

34 Gebicki, J., Modigell, M., Schumacher, M., van der Burg, J., and Roebroeck, E. (2009) Development of photobioreactors for anoxygenic production of hydrogen by purple bacteria, *Chem. Eng. Trans.*, **18**, 363–366.

35 Beggel, F., Modigell, M., Shalygin, M., Teplyakov, V., and Zenkevitch, V. (2009) Novel membrane contactor for gas upgrading in biohydrogen production. *Chem. Eng. Trans.*, **18**, 397–402.

36 Cullinane, J. T. and Rochelle, G. T. (2004) Carbon dioxide absorption with aqueous potassium carbonate promoted by

piperazine. *Chem. Eng. Sci.*, **59**, 3619–3630.

37 Cullinane, J.T. and Rochelle, G.T. (2005) Thermodynamics of aqueous potassium carbonate, piperazine, and carbon dioxide. *Fluid Phase Equilibria*, **227**, 197–213.

38 Wukovits, W., Friedl, A., Schumacher, M., Modigell, M., Urbaniec, K., Ljunggren, M., Zacchi, G., and Claassen, P.A.M. (2007) Identification of a suitable process route for the biological production of hydrogen. In *From Research to Market Deployment* (ed. A. Grassi), Proceedings of the 15th European Biomass Conference and Exhibition, 7–11 May 2007, Berlin, Germany, ETA-Florence, Florence and WIP-Munich, Munich, pp. 1919–1923, DVD, ISBN 3-936338-21-3.

39 Wukovits, W., Friedl, A., Markowski, M., Urbaniec, K., Ljunggren, M., Schumacher, M., Zacchi, G., and Modigell, M. (2007) Identification of a suitable process scheme for the non-thermal production of biohydrogen, *Chem. Eng. Trans.*, **12**, 315–320.

40 Foglia, D., Wukovits, W., Friedl, A., Ljunggren, M., Zacchi, G., Urbaniec, K., Markowski, M., and Modigell, M. (2009) Identification of a suitable process route for the biological production of hydrogen. In *From Research to Industry and Markets* (ed. A. Grassi), Proceedings of the 17th European Biomass Conference and Exhibition, 29 June–3 July 2009, Hamburg, Germany, DVD, ISBN 978-88-89407-57-3.

41 Modarresi, A. (2007) *Exergy Analysis of Non-Thermal Biological Hydrogen Production from Biomass*. MSc dissertation, Vienna University of Technology.

42 Modarresi, A., Wukovits, W., and Friedl, A. (2008) Exergy analysis of biological hydrogen production. *Comput.-Aided Chem. Eng.* **25**, 1137–1142.

43 Ptasinski, K.J., Prins, M.J., and van der Heijden, S.P. (2006) Thermodynamic investigation of selected production processes of hydrogen from biomass. In *Proceedings of the AIChE 2006 Annual Meeting, San Francisco*, pp. 363c/1–363c/7.

9
Thermochemical Cycles

Christian Sattler

Abstract

Thermochemical cycles promise to be one of the most efficient possibilities for large-scale hydrogen production. Therefore, nuclear and solar heated thermochemical cycles have been under development since the oil crises of the 1970s. During low oil prices in the 1980s and 1990s the interest on these technologies was low but since the ratification of the Kyoto Protocol thermochemical cycles are back in the focus. This chapter gives an overview on the different classes of thermochemical cycles and their state of development. It also describes where the limits of these technologies are and gives a view on possibilities for a future industrial application. Where possible it also discusses the economics of the processes.

Keywords: thermochemical cycles, hydrogen production, fuel cells

9.1
Introduction

Producing hydrogen for mass markets [1–3] means implementing processes that are as efficient as possible to convert available energy sources such as renewables or nuclear into the chemical energy vector. By using thermochemical cycles, very high conversion efficiencies can be achieved, making it possibly the most effective alternative for large-scale hydrogen production. Therefore, nuclear and solar heated thermochemical cycles have been under development since the oil crises of the 1970s. Since the signing of the Kyoto Protocol, a new boom for thermochemical cycles took place. The thermal dissociation of water (equation 9.1) is, theoretically, the simplest reaction to split water [4–6]. However, because of its thermodynamics, the process is the most challenging with respect to practical realization. Although water thermolysis is conceptually simple, it has at least two points that preclude its implication for large-scale hydrogen production: first, its need for an extremely high-temperature heat source, above 2500 K, to achieve a reasonable degree of water dissociation; and second, the re-

Hydrogen Energy. Edited by Detlef Stolten
Copyright © 2010 WILEY-VCH Verlag GmbH & Co. KGaA, Weinheim
ISBN: 978-3-527-32711-9

quirement for an effective technique to separate hydrogen and oxygen at high temperatures to avoid an explosive gas mixture.

$$H_2O \rightarrow H_2 + 1/2\, O_2 \tag{9.1}$$

To avoid the problems of thermal water dissociation, different consecutive sets of chemical reactions, so-called thermochemical cycles, were proposed to divide the single water thermolysis reaction into two or more steps. Thermochemical cycles avoid the H_2–O_2 separation problem and thereby reduce the temperature level. The cycles were originally developed to use high-temperature heat from nuclear power plants for hydrogen production. Since very high temperature nuclear reactors are presently not the favoured thechnology to be developed for the next generation of nuclear power-plants, concentrating solar technologies are in the focus now.

9.2
Historical Development

Research on thermochemical cycles started in the 1960s and a large number of possible theoretical cycles were proposed [2, 7, 8]. In the 1970s and early 1980s, numerous studies and comparisons were carried out to identify the most promising cycles based on their thermodynamics, theoretical efficiencies, and projected cost (e.g., [9–12]). Most of the work done during those years was promoted by the nuclear energy sector, with the goal of diversifying the use of high-temperature thermal energy supplied by nuclear reactors. However, even then the thermal limits of very high-temperature nuclear reactors and their availability were taken into account so that the high-temperature reactions of the cycles were often developed for the use of concentrated solar radiation. The main results were reported from the programs carried out by the Joint Research Centre of the European Union in Ispra, Italy [13], by General Atomics [14] and Westinghouse [15] in the United States, and by the Japanese Atomic Energy Research Institute. In the late 1980s, because of the availability of cheap fossil fuels and the doubts about nuclear energy raised by the accidents at Chernobyl and Three Mile Island, interest in thermochemical cycles decreased. Since then, for about 10 years only a few new results were reported, mainly on the UT-3 cycle developed by and named after the University of Tokyo [16], and on the sulfur-iodine cycle originally proposed and named after the company General Atomics. A revival in the R&D of thermochemical cycles has taken place in the late 1990s and 2000s. The driving force is the production of hydrogen as a greenhouse gas free energy carrier to fulfill the requirements of the Kyoto Protocol and hopefully its successors. As in the 1970s, prospective studies and comparative analyses of cycles have been performed but, in addition, hardware has been developed and tested. In contrast to the work conducted 30 years ago, concentrated solar radiation is now a more important heat source than nuclear heat. The first studies on thermochemical cycles coupled to nuclear fusion have been carried out to evaluate whether this might be a long-term alternative.

9.3
State of Work

The term "thermochemical cycle" is well accepted in the literature, but it does not describe the overall process precisely. Many of the cycles employ an electrochemical step, especially for hydrogen production. As a subclass, these variants are named "hybrid thermochemical cycles." The most prominent is the hybrid sulfur cycle or Westinghouse cycle [15]. The interest in thermochemical cycles is due to their possibly very high efficiencies. Several of them offer theoretical thermodynamic efficiencies of 50%, or more, but practical values are less due to thermal losses and some irreversible formation of by-products. Since no industrial plant has yet been constructed, the efficiency values are based on small-scale experiments, models, and flow sheets. Nevertheless, the processes are able to convert solar energy more efficiently than converting it into electricity and split water by low-temperature electrolysis competing only with high-temperature electrolysis [17].

The main groups of thermochemical cycles are classified in Table 9.1. The most promising are listed and are divided into four groups, namely those based on sulfur, volatile oxides, non-volatile oxides, and low-temperature steps. Nuclear heat is, and most probably always will be, limited to temperatures below 900 °C, which might be possible with the so-called "very high-temperature reactor (VHTR)" that is one of the six options to be developed in the GEN IV program [18]. For this reason, concentrated solar energy is the only possible heat source to run the thermo-

Table 9.1 Thermochemical cycles currently under consideration.

	Steps	Maximum temperature (°C)	LHV efficiency (%)
Sulfur cycles			
Hybrid sulfur	2	900	43
(Westinghouse, ISPRA Mark 11)		(1150 without catalyst)	
Sulfur–iodine	3	900	38
(General Atomics, ISPRA Mark 16)		(1150 without catalyst)	
Volatile metal oxide cycles			
Zinc/zinc oxide	2	1800	45
Hybrid cadmium		1600	42
Non-volatile metal oxide cycles			
Iron oxide	2	2200	42
Cerium oxide	2	2000	68
Ferrites	2	1100–1800	43
Low-temperature cycles			
Hybrid copperchloride	4	530	39

dynamically most efficient cycles. Today, temperatures of more than 1000 °C could be demonstrated on solar towers on the scale of several hundred kW$_{th}$. Due to the limited operating temperatures of nuclear reactors, recent developments are focusing on low-temperature cycles that can be coupled more easily to available nuclear heat sources. Although good theoretical efficiencies have been reported, the thermodynamics are less favorable. The theoretical efficiencies of all the cycles given in Table 9.1 lie in the range 35–68% and the required temperature range of the high-energy step is between 300 and 2200 °C.

9.3.1
Sulfur Cycles

Sulfur cycles were developed primarily to couple them with high-temperature heat from nuclear fission power plants. The main work has been carried out by General Atomics and Westinghouse in the United States, the European Joint Research Centre in Ispra, Italy, and the Japan Atomic Energy Research Institute (JAERI, now Japan Atomic Energy Agency, JAEA). Most of the processes are based on the thermal splitting of sulfuric acid into sulfur dioxide and oxygen. Even using efficient catalysts, this process requires temperatures above 850 °C (equation 9.2 b). This is close to the maximum output temperature level of very high-temperature nuclear reactors. The small temperature gap and the corrosive atmosphere are the major problems for realizing the cycles. Additionally, the heat transfer from the primary helium loop of the VHTR over a heat exchanger to a secondary helium loop over another heat exchanger to the sulfuric acid loop causes major safety issues. Especially the helium to sulfuric acid heat exchanger is a crucial component that has not yet been demonstrated. Since neither a VHTR is in operation to deliver heat for a chemical process nor the safety of the plant can be guaranteed, today concentrated solar energy is the only available heat source to be coupled to the cycles on a large scale with the constraint that also no solar plant coupled to a chemical process is working yet either.

9.3.1.1 Hybrid Sulfur Cycle (Westinghouse, Ispra Mark 11)
The hybrid sulfur cycle is a hybrid electrochemical–thermochemical cycle [19]. It consists of two steps: the splitting of sulfuric acid, which is divided into two sub-steps, the vaporization (equation 9.2 a) and the dissociation of sulfuric acid (equation 9.2 b) and decomposition of the resulting SO_3 (equation 9.3):

$$H_2SO_4 \rightarrow H_2O + SO_3 \qquad >450\,°C \qquad (9.2\,a)$$

$$SO_3 \rightarrow SO_2 + 1/2\,O_2 \qquad >850\,°C \text{ with catalyst}$$
$$(1150\,°C \text{ without catalyst}) \qquad (9.2\,b)$$

$$2\,H_2O + SO_2 \rightarrow H_2SO_4 + H_2 \qquad \text{electrolysis, } 80\,°C \qquad (9.3)$$

Electrical power is required for the electrolysis, but the electrochemical oxidation of SO_2 is far more efficient than the electrolytic splitting of water. The overall efficiency of the process is calculated to be about 40%. Carbon-supported platinum electrodes are used for the SO_2 oxidation. Cells made from ceramics such as silicon carbide, silicon nitrite, and cermets possess excellent resistance to corrosion by sulfuric acid at ambient temperature and at low acid concentration. Catalysts mainly based on iron oxide are available for accelerating the reaction rate of the SO_3 reduction at "low" temperature (850 °C). The kinetics of the reaction are much faster if higher temperatures are available as in solar tower installations. Therefore, the use of catalysts might be reduced or even be unnecessary if the sulfuric acid splitting is coupled to concentrated solar radiation. It has to be evaluated whether the higher temperatures are more efficient than the catalyzed reaction on an annual basis. Reactors used in laboratory tests have been made of glass or fused silica; solar reactors are mostly constructed from ceramics such as silicon carbide, but gold-coated steel has also been used [20]. Costs were estimated by JRC Ispra for a plant with a capacity of 100 000 $Nm^3\,h^{-1}$ of H_2 coupled to an HTR. The total capital cost were estimated to be US$ 420 million and the operating costs amounted to US$ 75 million per year. The estimation gives US$ 7.3 GJ^{-1} as the operating cost component for producing H_2. Capital cost plus operating cost gives a breakdown for the hydrogen production cost of US$ 8.4 GJ^{-1} [13, 21].

9.3.1.2 SO_2 Cycle (ISPRA Mark 13 and Mark 13 V2)

The Mark 13 SO_2 cycle was developed for coupling to a nuclear reactor, whereas the Mark 13 V2 was intended for application with a concentrating solar power plant [13]. A 1.4×10^6 $GJ\,yr^{-1}$ solar power system was simulated for this purpose. The decomposition of hydrobromic acid can be performed by a chemical subcycle or in an electrolytic cell. The energy demand of this reaction is high, since HBr has a high free energy of formation. However, in contrast to the sulfur-iodine cycle, H_2SO_4 and HBr can be obtained at high concentration and in separate phases (equation 9.4).

$$SO_2(g) + Br_2(l) + 2\,H_2O(l) \rightarrow 2\,HBr(g) + H_2SO_4(l) \quad 100\text{–}140\,°C \tag{9.4}$$

$$2\,HBr(l) \rightarrow H_2(g) + Br_2(l) \quad \text{(electrochemical step)} \tag{9.5}$$

$$H_2SO_4(g) \rightarrow H_2O(g) + SO_2(g) + {}^1/_2\,O_2(g) \quad 850\text{–}1120\,°C \tag{9.6}$$

The Mark 13 cycle is an example of the problems of thermochemical cycles. Although theoretically feasible, the removal of SO_2 from the O_2–SO_2 gas stream (equation 9.6) is very difficult in terms of corrosion, efficiency, and cost. This separation is carried out in two steps: The first step consists of cooling the gas mixture to –48 °C, to condense the bulk of SO_2. The gas leaving the cooling trap still contains 5–7 vol.% SO_2. In the second step, the remaining SO_2 is removed from by contacting the gas with a dilute aqueous solution of bromine. SO_2 is oxidized to sulfuric acid, which is dissolved in the liquid phase. The oxygen produced contains only a few ppm of SO_2. The HBr and H_2SO_4 formed are recycled. The calculated process efficiency of the Mark 13 cycle is 37%, which results in an overall solar H_2 production efficiency of 21%. The production cost for the process is calculated as US$ 51.57 GJ^{-1}. This is about six times the calculated cost for hydrogen production by the Westinghouse cycle.

9.3.1.3 Advanced SO_2 Cycle (ISPRA Mark 13A)

The ISPRA Mark 13A cycle was first designed as a desulfurization process for refineries. It is an enhancement of the Mark 13 cycle and is based on two of the latter's three reactions (equations 9.4 and 9.5).

The successful operation of a bench-scale unit was followed by the construction and operation of a pilot plant at the Saras refinery in Sarroch, Sardinia, Italy [22]. This plant was designed for a 32 000 Nm^3 h^{-1} flue gas flow that emanated from an internal power station that operated on a mixture of heavy fuel oil and refinery gas. After some additional modifications, the plant started regular operation in 1990. Since the sulfuric acid was not recycled, the cycle was not closed but the energy stored in the flue gas was used for hydrogen production, as a by-product of the desulfurization. Nevertheless, this is still the largest application ever to produce hydrogen by a technology developed for thermochemical hydrogen production.

9.3.1.4 Sulfur–Iodine or General Atomics Process (ISPRA Mark 16)

The second widely investigated sulfur process is the sulfur–iodine water-splitting cycle, also known as the ISPRA Mark 16 or General Atomics process [23]. Like the hybrid sulfur cycle, it was originally developed to split water using high-temperature heat from nuclear power plants. The three cycle steps are as follows:

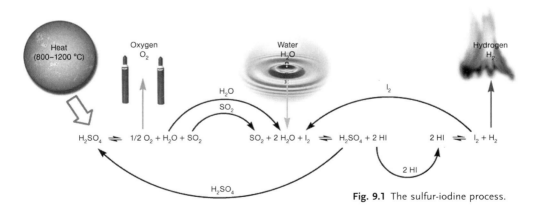

Fig. 9.1 The sulfur-iodine process.

$$2\,H_2O + SO_2 + x\,I_2 \rightarrow H_2SO_4 + 2\,HI_x \quad (27\text{--}127\,°C) \tag{9.7}$$

$$2\,HI_x \rightarrow x\,I_2 + H_2 \quad (127\text{--}727\,°C) \tag{9.8}$$

$$H_2SO_4 \rightarrow H_2O + SO_2 + 1/2\,O_2 \quad (847\text{--}927\,°C) \tag{9.9}$$

All three reactions are performed in separate sections of the plant; the Bunsen reaction (equation 9.7) and the sulfuric acid decomposition (equation 9.9) are run in parallel to avoid SO_2 storage (Figure 9.1).

General Atomics has made great efforts in research and development on the sulfur–iodine process. From the mid-1970s to the mid-1980s, a very active research program on the process was conducted. The feasibility of the individual sections was successfully demonstrated in a glass, quartz, and Teflon laboratory-scale apparatus. The sulfuric acid decomposition step was tested on the solar power tower of the Georgia Institute of Technology. The sulfuric acid splitting step was also demonstrated recently within the European project "Hydrogen by Thermochemical Cycles," HYTHEC [24], and its successor HyCycleS (in the solar furnace of the DLR (German Aerospace Center) in Cologne, Germany) [25]. For industrial application, a 2.4 GW_{th} modular helium reactor (H2-MHR)-driven plant was designed. The reactor should be able to provide temperatures up to 900 °C [26]. The overall efficiency of the cycle was calculated to be 42% at 850 °C. It is assumed that efficiencies of the hydrogen production process of up to 52% might be achievable at a temperature of 950 °C, which seems to be difficult to realize with a nuclear reactor. The decomposition of H_2SO_4 and especially of HI are reported to cause severe corrosion problems. The experimental problems described and the less promising economic calculations led to a reduction in activities on the cycle by General Atomic in the 1980s. In the 2000s, it was in the focus of intensive research again. An electrically heated pilot plant was set up in a joint venture of General Atomics, Sandia National Laboratories, and the French CEA at the General Atomics site in San Diego, CA, USA. The three partners designed one of the steps each. The project is finished but some of the problems described in the 1980s could not yet be solved. Also, a new economic evaluation did not lead to better results than 20 years ago. Therefore, no continuation is currently planned.

9.3.2
Metal/Metal Oxide Thermochemical Cycles

Even more than with the sulfur cycles, thermochemical hydrogen production research is currently focused on materials that can act as effective redox pairs for water splitting such as metal/metal oxides. Since two-step cycles are possible, this group promise to achieve even better efficiencies. The reactants are less corrosive than those of the sulfur cycles. but the cycles require even higher temperatures. This excludes heat sources other than concentrated solar energy at present. The separation of the chemicals, especially the gases involved, is also challenging in many of the cycles. The principle reactions of the two step cycles are equations 9.10 and 9.11, incorporating a metal or metal oxide that is able to abstract the oxygen of the water molecule at temperatures significantly lower than those of the one-step thermal water splitting. A large variety of redox materials have been evaluated, either oxide pairs of multivalent metals (e.g., Fe_3O_4/FeO, Mn_3O_4/MnO) [12, 27] or metal oxide/metal systems (e.g., ZnO/Zn) [28, 29]. In all cycles the thermodynamically limiting step is the reduction of the oxidized metal oxide to generate its reduced form by releasing oxygen (equation 9.11).

$$MO_x + \gamma H_2O \rightarrow MO_{(x+y)} + \gamma H_2 \tag{9.10}$$

$$MO_{(x+y)} \rightarrow MO_x + \gamma/2 O_2 \tag{9.11}$$

The advantage of the cycles is the production of pure hydrogen without separating it from oxygen. Water splitting takes place at temperatures below 800 °C whereas the reduction of the metal oxide, that is, regeneration, takes place at temperatures above 1100 °C. The concept has been proven both experimentally and in laboratory mock-ups as in solar reactors operated in solar furnaces with a scale-up to 10 kW$_{th}$ [30, 31]. The high temperatures of the reduction step impose a barrier to the integration of a number of two-step water splitting processes and the use of a concentrating solar system such as a solar tower. Metal oxide cycles have stood in the shadow of the sulfur base cycles for a long time. With the demonstration of high-power solar collector systems, temperatures above that of nuclear reactors have become possible. The limits for large applications are calculated to be between 1300 and 2000 °C. Nevertheless, in-depth analysis of annual efficiencies have to be carried out to compare very high-temperature thermochemical cycles with their competitors such as efficient electrolysis processes.

9.3.2.1 CeO_2/Ce_2O_3

Cerium is well established as a catalyst in several thermochemical cycles for H_2 production. Experiments on the reaction with $Na_4P_2O_7$ at 950 °C have been carried out and thermal efficiencies of up to 39% were obtained [32–34]. A study of the reaction of CeO_2 and HCl has been reported by Los Alamos National Laboratory [35]. In this two-step thermochemical cycle, H_2 and Cl_2 are the products to be separated. Furthermore, one possible modification of the sulfur-iodine process uses CeO_2 to form insoluble $Ce_2(SO_3)2SO_4 \cdot 4H_2O$, which can be removed and separately dehydrated. Therefore, reactive distillation to remove water is not necessary. Although the cycle consists of eight steps, a maximum overall efficiency of 54% has been calculated [36]. The pure CeO_2/Ce_2O_3 cycle needs very high temperatures. It was recently demonstrated in a solar furnace of CNRS-PROMES in Odeillo, France, at 2000 °C under reduced pressure. A remarkable calculated thermal efficiency of 68% was reported [37].

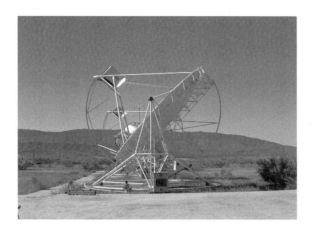

Fig. 9.2 Solar dish, Plataforma Solar de Almería, Spain.

$$2\,CeO_2(s) \rightarrow Ce_2O_3(s) + 1/2\,O_2(g) \qquad 2000\,°C,\ 100\text{--}200\,mbar \qquad (9.12)$$

$$Ce_2O_3(s) + H_2O(g) \rightarrow 2\,CeO_2(s) + H_2(g) \qquad 400\text{--}600\,°C \qquad (9.13)$$

Again, the very high temperature causes concerns if scale-up of the cycle is feasible and efficient. However, since the efficiency is so high even small-scale installations in solar dishes might be an option (Figure 9.2).

9.3.2.2 FeO/Fe$_3$O$_4$

Pure and mixed iron oxides are the systems mostly studied for two-step thermochemical cycles, for reasons of their availability, performance, and low cost. The simplest cycle is based on pure FeO/Fe$_3$O$_4$. Using concentrated solar energy at temperatures above 1600 °C, Fe$_3$O$_4$ can be reduced to FeO. Followed by the exothermic reaction of FeO and H$_2$O at much lower temperatures, H$_2$ and Fe$_3$O$_4$ are generated [38].

$$Fe_3O_4(l) \rightarrow 3\,FeO(l) + 1/2\,O_2 \qquad >1600\,°C \qquad (9.14)$$

$$3\,FeO + H_2O \rightarrow Fe_3O_4 + H_2 \qquad \text{exothermic} \qquad (9.15)$$

The overall efficiency of the ideal Carnot system is reported to be between 50.7% and 62.5%. For an operating plant, losses would lower it to 20.4–25.1%, depending on the concentration of the solar radiation coupled into the reactor. The lower values are for a concentration factor of 5000 and the higher values are for 10 000. Such concentrations are very difficult to achieve on a solar tower. Complicated optics such as secondary concentrators are necessary (Figure 9.3). This will limit the yearly availability of the plant and therefore its annual efficiency. However, with activated iron-based redox systems, the reaction temperature for the two-step water-splitting process can be reduced and less advanced and therefore less expensive solar systems could be used.

tion. Several reactor technologies have been developed for the ferrite cycle, and can be divided into three groups. (i) Static reactors: the redox material is fixed on a structure or is used as the structure itself. The two reactions have to be carried out one after the other. Continuous hydrogen production is possible only if at least two reactors are used in parallel. Examples are the HYDROSOL reactors and the reactors based on ceramic foams developed at Niigata University, Japan [40]. (ii) Rotating reactors: to avoid several reactors, moving reactors use the same redox systems as the static reactors but rotate them through the focus of the concentrating solar system. Therefore, the high temperatures for the thermal reduction are achieved while in the focus whereas the water splitting takes place on the back side of the reactor at reduced temperature. Two examples are the counter-rotating ring receiver reactor recuperator CR5 of the Sandia National Laboratory [41] and the rotary-type reactor of the Tokyo Institute of Technology [42]. (iii) Fluidized bed reactors: recently, work on fluidized bed reactors was started, but it still is difficult to introduce solar radiation into such a reactor efficiently and separate the reactands and products in parallel [43].

9.3.2.4 Manganese Ferrite Plus Activated Sodium Carbonate

To decrease the temperature of the ferrite process further, a three-step water-splitting cycle using the $Na_2CO_3/MnFe_2O_4/Fe_2O_3$ system is under development. Within the context of experiments, with $MnFe_2O_4$ and Na_2CO_3 nano-structured materials H_2 was successfully produced from H_2O at relatively low temperatures ($<800\,°C$).

$$2\,MnFe_2O_4 + 3\,Na_2CO_3 + (1+6\delta)H_2O \rightarrow 6\,Na(Mn_{1/3}Fe_{2/3})O_{2+\delta} \\ + 3\,CO_2 + (1+6\delta)H_2 \qquad 800\,°C \qquad (9.16)$$

$$6\,Na(Mn_{1/3}Fe_{2/3})O_{2+\delta} + 3\,Fe_2O_3 \rightarrow 2\,MnFe_2O_4 + 6\,NaFeO_2 \\ + (1+6\delta)\times 2\,O_2 \qquad 1000\,°C \qquad (9.17)$$

$$2\,NaFeO_2 + CO_2 \rightarrow Na_2CO_3 + Fe_2O_3 \qquad 600\,°C \qquad (9.18)$$

where $\delta = (n-3)/6$ with $3 < n < 4$.

The system consisting of manganese ferrite mixed with "mechano-chemically activated" Na_2CO_3 is reported to produce H_2 with kinetics faster than for other ferrites [44]. At temperatures higher than the melting point of Na_2CO_3, the reactivity among species rises and the quartz walls of the reactor react with the system, forming unwanted side-products that abruptly reduce the H_2 production. This problem is related to those reported in HYDROSOL. Therefore, the reaction has to be kept at a very constant temperature, which is difficult in a solar heated reactor. Also, for a technical apparatus, quartz should be replaced by an inert material or protected by an inert layer.

9.3.2.5 Zn/ZnO

Hydrogen production using Zn can be carried out as a pure two-step water-splitting cycle (Zn/ZnO). The first step in producing hydrogen is the thermal dissociation of ZnO into Zn and O_2 at about 1900 °C. The second step, hydrolysis of Zn, is carried out at 450 °C, generating H_2 and ZnO.

$$Zn + H_2O \rightarrow ZnO + H_2 \qquad 450\,°C \qquad (9.19)$$

$$ZnO \rightarrow Zn + 1/2\, O_2 \qquad 1900\,°C \qquad (9.20)$$

The Zn/ZnO cycle can reach a maximum exergy conversion efficiency of 29% and a solar thermal efficiency of 40%. A study of a 90 MW solar reactor operated at 2000 °C has been carried out [45]. Solar H_2 production costs for a large-scale plant with an annual production rate of 61×10^6 kWh$_{th}$ were estimated to be between US$ 0.13 and 0.15 kWh$_{th}^{-1}$. Due to the extreme conditions regarding temperature and the separation of zinc and oxygen, another process was developed using additional carbon-containing raw materials to promote the reduction of ZnO. This constitutes a so-called "open thermochemical cycle" which is not free of CO_2 emissions. It is therefore an option for a transition period leading from fossil fuel-based hydrogen to renewable hydrogen. In the reduction process, a mixture of ZnO and carbon is converted into Zn and CO at temperatures above 1200 °C. The Zn is condensed from the flow of waste gas. The stored energy and the CO produced can be used fore H_2 production by zinc hydrolysis and the water gas shift reaction. The zinc can also be used in zinc-air fuel cells for direct power generation. In both cases, ZnO is formed, which can be returned to the reactor. After laboratory-scale experiments and the construction and testing of a small-scale solar chemical reactor, a 300 kW pilot plant was successfully tested within the European project SOLZINC. By using biomass (beech charcoal) as the carbon source it was proven to be CO_2 neutral [46–48].

9.3.3
Low-Temperature Thermochemical Cycles – the Chloride and Bromide Family

This family of thermochemical cycles was historically developed to use heat from nuclear power plants for water splitting. The first cycle to be worked on was the iron chloride cycle [49]. After some drawbacks, more complex candidates have been in the focus, such as the UT-3 and the hybrid copper chloride cycle. They were specially developed to lower the temperature of the cycles to adjust them to actually available nuclear reactors providing heat at about 500 °C. The latest proposal to go as low as 300 °C is the uranium–europium cycle.

9.3.3.1 UT3 (Ca/Fe/Br Cycle)
The UT-3 cycle has been investigated extensively for more than 30 years. It was proposed in 1978 by the University of Tokyo (UT), after which it is named [50].

The process consists of four gas–solid reactions, two reactions based on Ca compounds and the other two based on Fe compounds. It is crucial for the cycle to obtain the solid reactants with high reactivity and durability, because it is easy to separate gas from solid products. Bromination of CaO included in the UT-3 cycle was analyzed to clarify the mechanism of the cycle [51].

$$CaBr_2 + H_2O \rightarrow CaO + 2HBr \qquad 680\text{--}730\,°C \qquad (9.21)$$

$$CaO + Br_2 \rightarrow CaBr_2 + 1/2\,O_2 \qquad 480\text{--}530\,°C \qquad (9.22)$$

$$Fe_3O_4 + 8HBr \rightarrow 3FeBr_2 + 4H_2O + Br_2 \qquad 230\text{--}280\,°C \qquad (9.23)$$

$$3FeBr_2 + 4H_2O \rightarrow Fe_3O_4 + 6HBr + H_2 \qquad 680\text{--}730\,°C \qquad (9.24)$$

The overall cycle was demonstrated at UT first on a laboratory scale and later as a pilot plant. Even a solar UT-3 process was proposed [52]. The major areas of research are in the stability of the solids and in the membrane separation processes. Membranes must be developed that are suitable for large-scale use. There is limited potential for future process improvements as the adiabatic implementation is already fairly simple [53, 54]. The great disadvantage of the UT-3 process is that $CaBr_2$ as $FeBr_2$ have low melting points of 740 and 690 °C. The reaction temperature has to be kept below the melting temperature because otherwise unwanted side reactions would occur, but as high as possible to accelerate the reactions. Conventional materials were used for the UT-3 pilot-plant reactors [54, 55]. Low reactivity of Fe_3O_4 to $FeBr_2$ and sublimation of $FeBr_2$ were reported to be the main problems in running the UT-3 cycle. The results of work performed in the 1980s and 1990s indicated that the UT-3 cycle might have the potential to become a competitive process for producing hydrogen, especially if the process could be improved by a membrane separator. This would raise the thermal efficiency of the process and improve the cost. In this case, the thermal efficiency was calculated to be 45%, but recently the overall efficiency of the cycle was reported to be <13%. Therefore, water electrolysis would be more efficient than the UT-3 process.

9.3.3.2 Hybrid Copper Chloride Cycle

Mainly worked on at Argonne National Laboratory (ANL) in the United States, this four-step cycle is one of the most promising currently under development [56]. It consists of three thermal reactions in which H_2, O_2, and HCl are generated, and an electrochemical step in which CuCl is disproportionated to yield copper metal and $CuCl_2$. The oxygen is released from the copper oxychloride between 450 and 530 °C, which is the highest temperature limit for this cycle.

$$2\,Cu(s) + 2\,HCl(g) \rightarrow 2\,CuCl(l) + H_2(g) \qquad 430\text{--}475\,°C \qquad (9.25)$$

$$4\,CuCl(s) \rightarrow 2\,CuCl_2(aq) + 2\,Cu \quad \text{(electrochemical)} \qquad 25\text{--}75\,°C \qquad (9.26)$$

$$2\,CuCl_2(s) + H_2O(g) \rightarrow CuO\cdot CuCl_2(s) + 2\,HCl(g) \qquad 350\text{–}400\,°C \qquad (9.27)$$

$$CuO\cdot CuCl_2(s) \rightarrow 2\,CuCl(l) + {}^1\!/_2\,O_2(g) \qquad 530\,°C \qquad (9.28)$$

The idealized efficiency of the CuCl cycle was calculated to be 41–43% (LHV). The ANL has recently completed an ASPEN simulation to perform sensitivity studies [56]. Important work is being done within the Generation IV International Forum to prepare the next generation of nuclear fission reactors. Especially Canada is in the lead in reporting new developments on different copper-chlorine thermochemical cycles [57–59].

9.3.3.3 Uranium–Europium Cycle

The search for even lower temperature cycles has led to interesting proposals such as this one. The cycle is based on heavy element halide chemistry to reach a maximum temperature of 300 °C.

$$2\,(UO_2Br_2\cdot H_2O) \rightarrow 2\,"UO_3\cdot H_2O(s)" + 4\,HBr(g) + 2\,H_2O(g) \quad 300\,°C \qquad (9.29)$$

$$4\,EuBr_2 + 4\,HBr \rightarrow 4\,EuBr_3 + 2\,H_2(g) \qquad \text{exothermic} \qquad (9.30)$$

$$4\,EuBr_3 \rightarrow 4\,EuBr_2 + 2\,Br_2(g) \qquad 300\,°C \qquad (9.31)$$

$$2"UO_3\cdot H_2O(s)" + 2\,Br_2 + 4\,H_2O \rightarrow 2\,(UO_2Br_2\cdot H_2O) + O_2(g) \quad \text{exothermic} \qquad (9.32)$$

The exact stoichiometry of "$UO_3\cdot H_2O(s)$" has not yet been determined. Work on this cycle is continuing. Major research has to be carried out to demonstrate its feasibility. Nevertheless, low-temperature cycles are relevant because the outcome of the Generation IV International Initiative might be that no new high-temperature nuclear reactors might be developed [60].

9.4
Conclusion and Outlook

The revival of thermochemical cycles is directly related to the discussion on climate change and of the future energy economy. The first peak was caused by the oil crises of the 1970s. Now limited fossil resources, rising prices, and the environmental impact of fossil energy carriers are the main drivers. Although research has been going on for nearly 40 years, no thermochemical cycle is yet in operation on an industrial scale. Neither has it been proven that thermochemical cycles are the way to provide large amounts of hydrogen. However, the efficiency of the cycles makes it very likely that this will eventually be proven, especially since no alternative technologies will come close to thermochemical cycles in relation to their efficiency and to their land use, and therefore also to the production costs. Technologically, the market introduction of solar thermal power plants will make a heat source available to power a wide range of thermochemical cycles. This paves the way for a successful future for these challenging technologies. If VHTR nuclear reactor technologies become available, it would also push thermochemical cycles for hydrogen production.

Abbreviations

ANL	Argonne National Laboratory, USA
CEA	Commissariat á l'Énergie Atomique (French Atomic Energy Commission)
DLR	Deutsches Zentrum für Luft- und Raumfahrt eV (German Aerospace Center)
GEN IV	Worldwide initiative for the development of the next generation (IV) nuclear power plants
H2-MHR	Modular helium gas-cooled reactor
HTR	High-temperature reactor
JAEA	Japan Atomic Energy Agency
JAERI	Japan Atomic Energy Research Institute
JRC	Joint Research Centre of the European Union
LHV	Lower heating value, defined as the amount of heat released by combusting a specified quantity (initially at 25 °C or another reference state) and returning the temperature of the combustion products to 150 °C
UT-3	Thermochemical cycle named after the University of Tokyo
VHTR	Very high-temperature reactor

References

1 De Beni, G. and Marchetti., C. (1970) Hydrogen, key to the energy market. *Eurospectra*, **9** (2), 46–50.
2 Marchetti, C. (1973) Hydrogen and energy. *Chem. Econ. Eng. Rev.*, **5**, 7–25.
3 Winter, C.-J. (2000) *On Energy of Changes The Hydrogen Solution*, Gerling Akademie Verlag, Munich.
4 Russel, J.-L. Jr, McCorkle, K. H., Norman, J.-H., Porter, J.-T., Roemer, T.S, Schuster, J. R., and Sharp, R. S. (1976) Water splitting a progress report. In *First World Hydrogen Energy Conference Proceedings*, 1A, 105–124.
5 Brecher, L. E. (1982) Decomposition of water. US Patent 4 330 523, issued 18 May 1982.
6 Etievant, C. (1991) Solar high-temperature direct water splitting a review of experiments in France. *Solar Energy Mater*, **24**, 413–440.
7 Abraham B. and Schreiner F. (1974) General principles underlying chemical cycles which thermally decompose water into the elements. *Ind. Eng. Chem. Fundam.*, **13** (4), 305–310.
8 Chao, R. (1974) Thermochemical water decomposition processes. *Ind. Eng. Chem. Prod. Res. Dev*, **13**, 94–101.
9 Fletcher, E. A. and Moen, R. L. (1977) Hydrogen and oxygen from water. *Science*, **197**, 10501056.
10 Bilgen E., Ducarroir. M., Foex, M., and Sibieude, F. (1977) Use of solar energy

11 Knoche, K. F. (1997) Entropy production, efficiency, and economics in the thermochemical generation of synthetic fuels. 1. The hybrid sulfuric acid process. *Int. J. Hydrogen Energy*, **2**, 377–385.

12 Nakamura, T. (1997) Hydrogen production from water utilizing solar heat at high temperatures. *Solar Energy*, **19**, 467–475.

13 Beghi, G. E. (1986) A decade of research on thermochemical hydrogen at the joint research centre, ISPRA *Int J Hydrogen Energy*, **11**, 761–771.

14 Besenbruch, G. E. and McCorkle, K. H. (1981) *Thermochemical Water Splitting with Solar Thermal Energy: Final Report*, GA-A16022, General Atomics, San Diego, CA.

15 Brecher, L. E., Spewock, S., and Warde, C. J. (1976) The Westinghouse sulfur cycle for the thermochemical decomposition of water. In *Proceedings of the 1st World Hydrogen Energy Conference*, Miami Beach, FL, 13 March 1976 (ed. T. N. Veziroglu), pp. 9A-19A-16.

16 Sakurai, M., Tsitsumi, A., and Yoshida, K.)1995) Improvement of Ca-pellet reactivity in UT-3 thermochemical hydrogen production cycle. *Int. J. Hydrogen Energy*, **20** (4), 297–301.

17 Kolb, G. J. and Diver, R. B. (2008) *Screening Analysis of Solar Thermochemical Hydrogen Concepts*, SAND2009-1900, Sandia National Laboratories, Albuquerque, NM.

18 Momirlan, M. (2002) Current status of hydrogen energy. *Renew Sustain Energy Rev*, **6** (12), 141–179.

19 Bilgen, E. (1988) Solar hydrogen production by hybrid thermochemical processes. *Solar Energy*, **41** (2), 199–206.

20 Noglik, A., Roeb, M., Rzepczkyk, T., Hinkley, J., Sattler, C., and Pitz-Paal, R. (2009) Solar thermochemical generation of hydrogen: development of a receiver reactor for the decomposition of sulfuric acid. *J Solar Energy Eng*, **131**, 011003-1–011003-7.

21 Farbman, G. H. (1979) Hydrogen production by the Westinghouse sulfur cycle process: program status. *Int. J. Hydrogen Energy*, **4**, 111–122.

22 van Velzen, D. (1991) Desulphurization and denoxing of waste gases producing hydrogen as a by-product. 2nd IEA Technical Workshop on Hydrogen Production, Jülich, Germany, 4–6 September 1991, 1–13.

23 Brown, L. C., Besenbruch, G. E., Lentsch, R. D., Schultz, K. R., Funk, J. F., Pickard, P. S., Marshall, A. C., and Showalter, S. K. (2003) *High Efficiency Generation of Hydrogen Fuels Using Nuclear Power, Final Technical Report for the Period August 1, 1999 Through September 30, 2002*, GA-A24285, General Atomics, San Diego, CA.

24 Roeb, M., Noglik, A., Rietbrock, P. M., Mohr, S., de Oliveira, L., Sattler, C., Cerri, G., de Maria, G., Giovanelli, A. Buenaventura, A., and de Lorenzo, D. (2005) HYTHEC: Development of a dedicated solar receiver-reactor for the decomposition of sulphuric acid. Presented at the European Hydrogen Energy Conference, EHEC 2005, Zaragoza, 2225 November 2005.

25 Roeb, M., Noglik, A., Sattler, C., and Pitz-Paal, R. (2009) Experimental study on sulfur trioxide decomposition in a volumetric solar receiver-reactor. *Int. J. Energy Res.*, **33**, 799–812.

26 Schultz, K. R. (2003) Use of the modular helium reactor. In *World Nuclear Association Annual Symposium*, London, 35 September 2003, World Nuclear Association, London, pp. 111.

27 Kodama, T. and Gokon, N. (2007) Thermochemical cycles for high-temperature solar hydrogen production. *Chem. Rev.*, **107**, 4048–4077.

28 Steinfeld, A. (2002) Solar hydrogen production via a two-step water-splitting thermochemical cycle based on Zn/ZnO redox reactions. *Int J Hydrogen Energy*, **27**, 611–619.

29 Steinfeld, A. (2005) Solar thermochemical production of hydrogen – a review. *Solar Energy*, **78**, 603–615.

30 Roeb, M., Sattler, C., Klüser, R., Monnerie, N., de Oliveira, L., Konstandopoulos, A. G., Agrafiotis, C., Zaspalis, V. T., Nalbandian, L., Steele, A. M., and

Stobbe, P. (2006) Solar hydrogen production by a two-step cycle based on mixed iron oxides. *J. Solar Energy Eng.*, **128**, 125–133.

31. Schunk, L. and Steinfeld, A. (2009) Kinetics of the thermal dissociation of ZnO exposed to concentrated solar irradiation using a solar-driven thermogravimeter in the 1800–2100 K range. *AIChE J.*, **55** (6), 1497–1504.

32. Bowman, M. G. (1981) *The LASL Thermochemical Hydrogen Program: Status on October 31, 1976*. Los Alanos Scientific Laboratory, Los Alanos, NM, USA.

33. Robinson, R. P. (1981) Acidbase models as conceptual aids to the develoment of thermochemical cycles for water splitting. I. Consideration of simple cycles involving oxides. *Inorg Chem*, **20**, 10–11.

34. Bamberger, C. E. and Robinson, R. P. (1980) Thermochemical splitting of water and carbon dioxide with cerium compounds. *Inorg. Chim. Acta*, **42**, 133.

35. Onstott, E. I. (1997) Cerium dioxide as a recycle reagent for thermochemical hydrogen production by splitting hydrochloric acid into the elements. *Int. J. Hydrogen Energy*, **22** (4), 405–408.

36. Onstott, E. I. (1995) Cerium dioxide as a recycle reagent for thermochemical hydrogen production by modification of the sulfur dioxideiodine cycle. *Int. J. Hydrogen Energy*, **20**, 693–695.

37. Abanades S. and Flamant G. (2006) Thermochemical hydrogen production from a two-step solar-driven water-splitting cycle based on cerium oxides. *Solar Energy*, **80**, 16111623.

38. Steinfeld, A., Sanders, S., and Palumbo, R.(1999) Design aspects of solar thermochemical engineering a case study: two-step water-splitting cycle using the Fe_3O_4/FeO redox system. *Solar Energy*, **65**, 43–53.

39. Agrafiotis, C. Roeb, M., Konstandopoulos, A. G., Nalbandian, L., Zaspalis, V. T., Sattler, C., Stobbe, P., and Steele, A. M. (2005) Solar water splitting for hydrogen production with monolithic reactors. *Solar Energy*, **79** (4), 409–421.

40. Kodama, T., Kondoh, Y., Yamamoto, R., Andou, H., and Satou, N. (2005) Thermochemical hydrogen production by a redox system of ZrO_2-supported Co(II)-ferrite. *Solar Energy*, **78** (5), 623–631.

41. Diver, R.B, Siegal, N. P., Moss, T. A., Miller, J. E., Evans, L., Hogan, R. E., Allendorf, M. D., Stuecker, J. N., and James, D. L. (2008) *Innovative Solar Thermochemical Water Splitting*, SAND2008-0878, Sandia National Laboratories, Albuquerque, NM.

42. Ishihara, H., Kaneko, H., Hasegawa, N., and Tamaura, Y. (2008) Two-step water-splitting at 1273–1623 K using yttria-stabilized zirconia-iron oxide solid solution via co-precipitation and solid-state reaction. *Energy*, **33**, 1788–1793.

43. Gokon, N., Kondo, N., Mataga, T., and Kodama, T. (2009) Internally circulating fluidized bed reactor with $NiFe_2O_4$ particles for thermochemical water splitting. In *SolarPACES International Symposium*, Berlin, 15–18 September 2009 (eds T. Mancini and R. Pitz-Paal), Deutsches Zentrum für Luft- und Raumfahrt, Stuttgart, p. 11.

44. Padella, F., Alvani, C., La Barbera, A., Ennas, G., Liberatore, R., and Varsano, F. (2005) Mechanosynthesis and process characterization of nanostructured manganese ferrite. *Mater Chem. Phys.*, **90** (1), 172–177.

45. Fahrni, R. (2002) Hydrogen production, an overview of hydrogen production methods and costs today, *Term. Paper WS01/02*, ETH Zürich, Institue of Energy Technologies, Zürich.

46. Kräuplm S. and Wieckertm C. (2007) Economic evaluation of the solar carbothermic reduction of ZnO by using a single sensitivity analysis and a Monte Carlo risk analysis. *Energy*, **32**, 1134–1147.

47. Epstein, M., Olalde, G., Santén, S., Steinfeld, A., and Wieckert, C. (2008) Towards the industrial solar carbothermal production of zinc. *J. Solar Energy Eng*, **130** (February), 014505-1–014505-4.

48. Wieckert, C., Frommherz, U., Kräupl, S., Guillot, E., Olalde, G., Epstein, M., Santén, S., Osinga, T., and Steinfeld, A. (2007) A 300 kW solar chemical pilot plant for the carbothermic production of zinc. *J Solar Energy Eng*, **129** (May), 190–196.

49 Cremer, H., Hegels, S., Knoche, K. F., Schuster, P., Steinborn, G., Wozny, G., and Wüster, G. (1980) Status report on thermochemical iron/chlorine cycles: a chemical engineering analysis of one process *Int. J. Hydrogen Energy*, **5**, 231–252.

50 Yalcin, S. (1989) A review of nuclear hydrogen production. *Int. J. Hydrogen Energy*, **14** (8), 551–561.

51 Sakurai, M., Miyake, N., Tsutsumi, A., and Yoshida, K. (1996) Analysis of a reaction mechanism in the UT-3 thermochemical hydrogen production cycle. *Int. J. Hydrogen Energy*, **21** (10), 871–875.

52 Sakurai, M., Bilgen, E., Tsutsumi, A., and Yoshida, K. (1996) Solar UT-3 thermochemical cycle for hydrogen production. *Solar Energy*, **57** (1), 51–58.

53 Sakurai, M., Bilgen, E., Tsutsumi, A., and Yoshida, K. (1996) Adiabatic UT-3 thermochemical process for hydrogen production. *Int. J. Hydrogen Energy*, **21** (10), 865–870.

54 Tadokoro, Y., Kajiyama, T., Yamaguchi, T., Sakai, N., Kameyama, H., and Yoshida, K. (1997) Technical evaluation of UT-3 thermochemical hydrogen production process for an industrial scale plant. *Int. J. Hydrogen Energy*, **22** (1), 49–56.

55 Aochi, A., Tadokoro, A. Yoshida, K., Kameyama, H., Nobue, M., and Yamaguchi, T. (1989) Economical and technical evaluation of UT-3 thermochemical hydrogen production process for an industrial scale plant. *Int. J. Hydrogen Energy*, **14** (7), 421–429.

56 Lewis, M. and Masin, J. (2005) An assessment of the efficiency of the hybrid copperchloride thermochemical cycle, 3. AlChE Conference Proceedings 2005, available online at http://aiche.confex.com/aiche/2005/techprogram/P20544.htm.

57 Naterer, G. F., Daggupati, V. N., Marin, G., Gabriel, K. S., and Wang, Z. L. (2008) Thermochemical hydrogen production with a copperchlorine cycle. II. Flashing and drying of aqueous cupric chloride. *Int. J. Hydrogen Energy*, **33**, 5451–5459.

58 Wang, Z. L., Naterer, G. F., Gabriel, K. S., Gravelsins, K. S., and Daggupati, V. N. (2009) Comparison of different copperchlorine thermochemical cycles for hydrogen production. *Int. J. Hydrogen Energy*, **34**, 3267–3276.

59 Naterer, G., Suppiah, S., Lewis, M., Gabriel, K., Dincer, I., Rosen, M. A., Fowler, M., Rizvi, G., Easton, E. B., Ikeda, B. M., Kaye, M. H., Lu, L., Pioro, I., Spekkens, P., Tremaine, P., Mostaghimi, J., Avsec, J., and Jiang, J. (2009) Recent Canadian advances in nuclear-based hydrogen production and the thermochemical CuCl cycle. *Int. J. Hydrogen Energy*, **34**, 2901–2917.

60 Petri, M. C., Yildiz, B., and Klickman, A. E. (2006) US work on technical and economic aspects of electrolytic, thermochemical, and hybrid processes for hydrogen production at temperatures below 550 °C. *Int. J. Nucl. Hydrogen Prod. Appl.*, **1** (1), 7991.

10
Hydrogen Production:
Fundamentals and Case Study Summaries

Kevin W. Harrison, Robert Remick, Gregory D. Martin, and Aaron Hoskin

Abstract

It is important that parties interested in hydrogen technologies standardize methods of evaluating the performance and efficiency of these technologies. A detailed description of the chemical and electrical processes for electrolysis and fuel cells is presented. Hydrogen and hydrocarbon fuels also pose a source of confusion about whether the efficiencies are based on the higher or lower heating value. A discussion of fundamental principals for fuel cell and electrolyzer systems is presented along with recommendations. Renewable hydrogen production is being researched across the globe to enable the environmental benefits of this energy carrier to be realized. The focus of many of these projects is coupling wind energy with hydrogen production (via water electrolysis) in an effort to use all available wind energy and to store that energy to be used during times of high electricity demand. Summaries of projects from Canada, Greece, Spain, United Kingdom and the United States are provided.

Keywords: hydrogen production, fuel cells, electrolysis, heat, wind

10.1
Heating Value, Heat of Reaction, and Free Energy

One of the issues that arises when discussing the calculation of the electrical efficiency of a fuel cell or an electrolysis cell is the confusion among the terms: heat of combustion (often called the heating value), heat of reaction, heat of formation, free energy of reaction, and free energy of formation. Hydrogen and hydrocarbon fuels also pose an additional source of confusion about whether the water vapor produced during combustion is condensed back into liquid water or is lost as a vapor diluted in the combustion products. Condensing the water vapor produces additional heat. Finally, there is the confusion over standard conditions.

Total energy is composed of both electrical and thermal energy known as enthalpy (H). The amount of electrical energy is known as the Gibbs free energy

(G) and corresponds to the maximum amount of useable electrical energy available when hydrogen recombines with oxygen. Irreversible energy (S) is the "cost of doing business" and is dependent on the temperature the reaction takes place. The loss due to entropy is similar to how a bouncing ball loses energy on hitting the floor, as friction from the action of bouncing causes a transfer of thermal energy to atoms in the floor. The energy transferred to those floor atoms dissipates and is not recoverable. Consequently, the change (Δ) in these quantities from a standard set of conditions follows the form

$$\Delta H = \Delta G + T\Delta S \tag{10.1}$$

10.1.1
Heat of Combustion and Heat of Reaction

By definition, the heat of combustion is the amount of heat released when 1 g molecular weight of a substance is burned in oxygen. (Heat of reaction is a more generic term and refers to the heat released in any chemical reaction, combustion or otherwise.) Heat of combustion measurements are usually made in a calorimeter under controlled conditions in which the water vapor, if produced, is condensed and the additional heat of condensation is included. The heat of combustion of methane, for example, is 890.3 kJ mol^{-1} when measured at 25 °C in a calorimeter:

$$CH_4 + 2O_2 \rightarrow CO_2 + 2H_2O_{(liquid)} + 890.3 \text{ kJ mol}^{-1} \text{ heat} \tag{10.2}$$

If the water vapor produced is not condensed, then the heat of combustion of methane is 802.3 kJ mol^{-1}, with the difference being the latent heat of condensation of the water vapor. The higher value, based on forming liquid water, is known in the industry as the higher heating value (HHV) of methane, and the lower value, based on forming water vapor, is the lower heating value (LHV) of methane. Most standard natural gas appliances (e.g., hot water heaters, gas furnaces, and gas ranges) operate with an excess of air and the water vapor does not condense but remains dissolved in the exhaust stream. This is also true of internal combustion engines and gas turbines. In the United States, however, the efficiencies of appliances and heat engines are usually rated based on the HHV, whereas in the European Communities LHV is used. The use of LHV in calculating heat engine efficiencies yields higher efficiency numbers than HHV.

10.1.2
Standard Conditions

The heat of combustion changes with temperature. For example, the heat of combustion of methane (LHV) at 1000 K (727 °C or 1340 °F) is 800.9 kJ mol^{-1}, slightly less than the LHV value at 25 °C. It is useful, therefore, to use a standard set of conditions for calculating efficiency. The usual practice in chemical thermodynamics is to choose 1 atm pressure and 25 °C (298 °K), although other conditions are used as standards by different industries and in other parts of the world.

10.1.3
Free Energy

There are two definitions of free energy, both related to work done by the system. The Helmholtz free energy is the maximum amount of work that can be obtained from a system under perfectly reversible conditions and is used with thermodynamic calculations of heat engines. Gibbs free energy is the net work that can be done by a system. In an electrochemical cell working reversibly at constant temperature and pressure, the net work is equal to the electrical work. Because we want electrical work, the Gibbs free energy is important to fuel cell calculations.

In the combustion of hydrogen in oxygen, heat is released. In the electrochemical reaction between hydrogen and oxygen in a fuel cell, electricity and heat are produced. Although the chemical equation is written in the same way for both reactions, the energy values are not equal. The heat produced by combustion does not equal the electricity produced in the fuel cell.

$$H_2 + \tfrac{1}{2} O_2 \rightarrow H_2O_{(liquid)} \tag{10.3}$$

The HHV for the combustion of hydrogen is 285.8 kJ mol^{-1}, but the Gibbs free energy for the reaction, and therefore the maximum electricity produced by a fuel cell, is only 237.2 kJ mol^{-1}. The difference, 48.6 kJ mol^{-1}, appears as heat produced in the fuel cell. We can show this by the following equation for the fuel cell reaction:

$$H_2 + \tfrac{1}{2} O_2 \rightarrow H_2O_{(liquid)} + 237.2 \text{ kJ mol}^{-1} \text{ electricity} + 48.6 \text{ kJ mol}^{-1} \text{ heat} \tag{10.4}$$

In sum, all fuel cells operating on hydrogen and oxygen produce heat in addition to the electricity. The distribution of energy produced between electricity and heat shown above, however, is for a perfect fuel cell operating in a thermodynamically reversible manner. Actual, practical, fuel cell devices incur losses due to inefficiencies of the electrochemical reactions and to electrical and ionic resistance as the current flows through the fuel cell. These generically are classed as internal resistance losses, and manifest as additional heat produced by the fuel cell at the cost of the electrical generation. Nevertheless, the sum of the electricity and the heat produced by the fuel cell must equal the HHV (or LHV, if the water vapor produced is not condensed).

10.2
Heat of Formation and Free Energy of Formation

The definition of the heat of formation is the heat released or required when products are formed from the reactants in their standard state. The heat of formation of the elements in their standard state is, by definition, zero.

The equation for the formation of methane from its elements is

$$2H_2 + C \rightarrow CH_4 \qquad (10.5)$$

The heat of formation for this reaction at 25 °C is listed in the Joint Army–Navy–Air Force (JANAF) Thermodynamic Tables as -74.9 kJ mol^{-1}. The minus sign indicates that the methane contains less energy than the elements do, and that heat would be released by this reaction were it possible to cause this reaction under standard conditions. Likewise, the free energy of formation of a compound is the work that could be recovered from a reaction in which the compound is formed from its elements under standard conditions.

In the unique case of hydrogen reacting with oxygen to form water, the heat of formation of water has the same values as the heat of combustion of the hydrogen. This is only true because, in the combustion reaction, water is formed from its elements. This is a unique situation that is shared by only a few other chemical reactions. Although the numerical values are the same, the concepts of heat of formation and heat of combustion are distinctly different. In the discussion that follows, we will use the heats of combustion, HHV and LHV, without further reference to heat of formation or the heat or reaction.

10.3
Calculating Fuel Cell System Efficiency

The standard method for calculating the efficiency of a fuel cell power plant or other electrical generation device is to divide the electricity produced by the higher heating value of the fuel used. This is a reasonable method for calculating power plant efficiency, because the power plant operator purchases fuel (natural gas is sold by heating value) and sells electricity.

$$\text{electrical efficiency} = \frac{\text{electricity produced}}{\text{HHV of fuel used}} \qquad (10.6)$$

Therefore, there is a maximum theoretical limit to the electrical efficiency attainable by a fuel cell system represented by the Gibbs free energy divided by the heat of combustion of the fuel. In the case of the hydrogen fuel cell, this value is the Gibbs free energy/HHV (237.2 kJ mol^{-1}/285.8 kJ mol^{-1} = 83%). The use of HHV here is in keeping with the method used in the United States to calculate efficiency for internal combustion (IC) engines/generators and gas turbine/generator systems.

Some United States developers of high-temperature fuel cell, however, prefer the European convention and instead use the LHV of hydrogen for efficiency calculation. The JANAF Tables list the Gibbs free energy for the formation of water vapor from hydrogen and oxygen as 228.6 kJ mol^{-1}, so the maximum theoretical efficiency of a complete fuel cell system based on the LHV of hydrogen is 228.6 kJ mol^{-1}/241.8 kJ mol^{-1} = 94.5%. Using the LHV convention for calculating the efficiency of an electrical generator always yields higher numbers greater than those yielded by calculations using the HHV for the same system. When quoting electrical efficiency of an electric generator, it is important to indicate whether this is based on the HHV or the LHV calculation method.

Practical fuel cells cannot achieve these maximum electrical efficiency numbers because of internal resistance losses. For example, a practical fuel cell operating near its maximum power output may only be able to produce 154 kJ of electricity per mole of hydrogen consumed, with the rest of the heating value appearing as heat produced by the fuel cell. The calculation for such a fuel cell is 154 kJ mol^{-1}/285.8 kJ mol^{-1} = 54% efficient (HHV). The remaining 46% of the energy produced can be recovered from the fuel cell system as co-generated heat.

10.3.1
Voltage Efficiency of Fuel Cells and Stacks

The electrical system efficiency calculations discussed above are applied globally to complete fuel cell systems that include many individual components, such as fuel processors, humidifiers, fuel cell stacks, power conditioners, and controls. Many experimenters and developers, however, also wish to assess the efficiency of the fuel cell stack separate from the efficiency of the system. For this reason, it is convenient to use the concept of voltage efficiency, which is defined as the actual cell (or cell stack) operating voltage divided by the thermodynamic cell voltage:

$$\text{voltage efficiency} = \frac{\text{operating voltage }(V)}{\text{thermodynamic voltage }(E)} \tag{10.7}$$

The thermodynamic cell voltage can be calculated using the Gibbs free energy and the Nernst equation:

$$E = E^0 - \frac{RT}{nF} \ln \frac{[H_2O]}{[H_2][O_2]^{1/2}} \tag{10.8}$$

where
E = the thermodynamic voltage under prevailing conditions
E^0 = the thermodynamic voltage under standard conditions
R = the gas constant (8.314 J K^{-1} mol^{-1})
T = absolute temperature (K) (25 °C = 298 K)

n = the number of electrons transferred ($n = 2$ in this case)
F = Faraday constant (96 485 C mol^{-1})
[ ] = the thermodynamic activity of the reactants and products which, for gases, can be approximated by their partial pressure in atmospheres.

Gibbs free energy can be converted into a thermodynamic voltage using the equation in which free energy is in joules per mole:

$$\text{Gibbs free energy (standard conditions)} = nFE^0 \tag{10.9}$$

Under standard conditions, the Gibbs free energy for the hydrogen–oxygen reaction is 237.2 kJ mol^{-1} for the production of liquid water at 25 °C. Therefore, the thermodynamic voltage for a hydrogen–oxygen fuel cell operating at standard temperature and pressure is 1.229 V:

$$E^0 = \frac{237\,200\text{ J}}{2(96\,485)} = 1.229\text{ V} \tag{10.10}$$

The Nernst equation is useful for calculating the thermodynamic voltage at varying pressures and reactant concentrations, but it should be noted that the Gibbs free energy of the hydrogen–oxygen reaction changes with temperature. Therefore, to calculate the thermodynamic voltage of a cell at other than standard temperature requires looking up the Gibbs free energy for the reaction at that temperature in a thermodynamic table. For example, the thermodynamic voltage for a high-temperature fuel cell operating on hydrogen–oxygen at atmospheric pressure at 1000 K is 0.998 V. Experimenters and developers of higher temperature fuel cells typically use the free energy for the formation of water vapor rather than liquid water in their calculations. Both are listed in most thermodynamic tables.

A good fuel cell with well-sealed components and a properly functioning electrolyte should exhibit a voltage close to the thermodynamic voltage when it is not producing power (no load). This is also known as the open-circuit voltage. Hence a fuel cell operating at 25 °C with 1 atm of hydrogen at the anode and 1 atm of pure oxygen at the cathode should exhibit a voltage of 1.23 V with no load. Comparison of the actual open-circuit voltage with the thermodynamic voltage can be used to determine the integrity of the cell. Pinholes in the electrolyte that allow fuel and oxidant to mix, for example, reduce the open-circuit voltage and indicate a problem.

The approximate efficiency for a fuel cell stack that is producing electrical power can be calculated by dividing the operating voltage by the thermodynamic voltage. Thus, a polymer electrolyte membrane (PEM) fuel cell operating at 0.800 V under standard conditions has a voltage efficiency of 0.800 V/1.229 V = 65%.

10.3.2
Other Efficiency Calculations

Another calculation that is often used to describe fuel cell stack performance is fuel cell stack efficiency, which is calculated as the direct current (d.c.) electrical output of the fuel cell stack divided by the LHV of the fuel consumed in the stack. This calculation is similar to the global system efficiency calculation except that parasitic electrical losses due to auxiliary systems are not included in the calculation and the LHV of the fuel used by the stack does not include parasitic fuel use upstream or downstream of the fuel cell. It is also a more difficult calculation to perform because accurate measurements of the amount of fuel consumed by the fuel cell stack are not easy to obtain.

10.4
Water Electrolysis

Water electrolysis is the reverse of the fuel cell reaction. In fact, many fuel cells based on PEM and solid oxide technology can work both as a fuel cell and as a water electrolysis cell, depending on the direction of the electrical current. The equation for the water electrolysis reaction is simply the reverse of the fuel cell equation:

$$H_2O_{(liquid)} + 237.2 \text{ kJ mol}^{-1} \text{ electricity} + 48.6 \text{ kJ mol}^{-1} \text{ heat} \rightarrow H_2 + {\textstyle\frac{1}{2}}O_2 \tag{10.11}$$

The efficiency calculations, therefore, can also be inverted. The efficiency of an electrolysis system, for example, can be calculated as the heating value of the hydrogen produced divided by the electrical energy input:

$$\text{electrial efficiency}_{(HHV)} = \frac{\text{HHV of } H_2 \text{ produced}}{\text{electricity used}} \tag{10.12}$$

or

$$\text{electrial efficiency}_{(LHV)} = \frac{\text{LHV of } H_2 \text{ produced}}{\text{electricity used}} \tag{10.13}$$

Other efficiency measures are often listed on product brochures of electrolyzer manufacturers, the most frequently used being kilowatt hours (kW h) per normal cubic meter (Nm^3) of dry hydrogen produced (kW h Nm^{-3}). The kilowatt hours per kilogram also appears frequently in the literature. These measures have more meaning to the customers of electrolysis units who, after all, want hydrogen gas, not heat.

10.4.1
Voltage Efficiency of Electrolysis Cells and Stacks

A problem exists, however. Using this approach to calculate the maximum thermodynamic efficiency of an electrolysis cell operating reversibly produces nonsense numbers that exceed 100%:

$$(\text{HHV}) \ 285.8 \ \text{kJ mol}^{-1}/237.2 \ \text{kJ mol}^{-1} = 120.5\% \quad (10.14)$$

$$(\text{LHV}) \ 241.8 \ \text{kJ mol}^{-1}/228.6 \ \text{kJ mol}^{-1} = 105.8\% \quad (10.15)$$

The problem is that it takes both electricity and heat to split water electrochemically and the heat is not being included in the above calculations of the energy input.

Although the thermodynamic voltage for splitting water under standard conditions is the same 1.229 V as for the fuel cell reaction, practical electrolysis cells, like fuel cells, do not operate near this voltage. Whereas the practical fuel cell operates well below 1.23 Vs, in the range 0.750–0.900 V, the practical electrolysis cell operates above this voltage, in the range 1.60–2.00 V. System efficiencies of practical systems calculated using the above approach, although inflated, are always less than 100%, and therefore the problem is not obvious. It is when we attempt to develop a method for calculating individual cell and multiple cell stack efficiencies that we see the problem.

Splitting 1 mol of liquid water to produce 1 mol of hydrogen at 25 °C requires 285.8 kJ of energy: 237.2 kJ as electricity and 48.6 kJ as heat; there is no way around this fact. In PEM and alkaline electrolysis cells, the heat requirement is supplied from the extra heat generated, due to internal resistance as the electric and ionic currents flow through the cell. This heat requirement is directly traceable back to the electricity supplied. In other words, 285.8 kJ of electricity, not 237.2 kJ, is the minimum required to split water in these cells. This translates into a cell voltage of 1.481 V, not the 1.229 V used in calculating the theoretical maximum electrical efficiency of a fuel cell.

The electrochemical potential (standard potential) corresponding to the HHV is 1.48 V per cell, as shown below. This represents the thermoneutral voltage where hydrogen and oxygen would be produced with 100% thermal efficiency (i.e., no waste heat produced from the reaction). This is determined using Faraday's law, and dividing the HHV (285 840 J mol^{-1}) by the Faraday constant ($F = 96\,485$ C mol^{-1}) and the number of electrons needed to create a molecule of H$_2$ ($z = 2$):

$$E_0 = \frac{\Delta_f H^0}{zF} = \frac{285\,840 \ \text{J mol}^{-1}}{2 \times 96\,485 \ \text{C mol}^{-1}} = 1.481 \ \text{V per cell} \quad (10.16)$$

This voltage, 1.481 V, is required for splitting liquid water. It is the voltage at which an electrolysis cell operating at 25 °C can operate without producing excess heat. (Practical cells operate above this voltage and produce excess heat.) It is also the voltage that corresponds to the HHV of hydrogen and so represents a more reasonable value to use when calculating cell and stack voltage efficiency. The equation for calculating the voltage efficiency of a cell or cell stack now becomes

$$\text{voltage efficiency} = \frac{\text{thermal neutral voltage } (E)}{\text{cell operating voltage } (V)} \qquad (10.17)$$

A similar calculation can be performed for water vapor using the LHV. The thermal neutral voltage for splitting water vapor at 25 °C is 1.253 V.

10.4.2
Steam Electrolysis and High-Temperature Cells

The above discussion applies primarily to electrolysis cells operating at temperatures below the boiling point of water; these include PEM and alkaline electrolysis cells. There is a class of high-temperature steam electrolysis cells under development, however, that operate in the 800–1000 °C temperature range, where the thermodynamics are significantly different. As the temperature climbs, the LHV of hydrogen increases and the Gibbs free energy decreases. At 1000 °C, for example, the LHV of hydrogen is 249.2 J mol^{-1} and the Gibbs free energy for the reaction is 179.9 kJ mol^{-1}. The water splitting reaction at 1000 °C can thus be written as

$$H_2O_{(steam)} + 179.9 \text{ kJ mol}^{-1} \text{ electricity} \\ + 69.3 \text{ kJ mol}^{-1} \text{ heat} \rightarrow H_2 + \tfrac{1}{2} O_2 \qquad (10.18)$$

The thermodynamic voltage for this reaction, which corresponds to both the open-circuit voltage for the solid oxide fuel cell and the solid oxide electrolyzer cell, is 0.932 V. The thermal neutral voltage for the electrolysis reaction is 1.291 V. These high-temperature cells are considerably more efficient in that they have lower internal resistance losses and improved reaction kinetics compared with their low-temperature PEM counterparts.

It is well within the realm of possibility that a practical high-temperature electrolysis cell could operate below the thermal neutral voltage. In this case, the heat requirement would have to be made up by an external heat source. A high-temperature fuel cell operating at 1000 °C and 1200 V, for example, would not generate sufficient heat via internal resistance to keep the electrochemical reaction going. As the cell operated, the electrochemical reaction would withdraw heat from the cell components and cool the cell to the point that it ceases to operate. Therefore, sensible heat must be supplied to the cell components from an outside source.

In the high-temperature case, calculating the voltage efficiency of the cell can be straightforward and the thermodynamic voltage can be used. In the case of global system efficiency, however, both the electrical input and the heat input from the external source must be included, otherwise the calculation will produce a nonsensical answer and an efficiency that is greater than 100%.

$$\text{voltage efficiency}_{(\text{cell})} = \frac{\text{thermodynamic voltage } (E)}{\text{operating voltage } (V)} \quad (10.19)$$

$$\text{electrial efficiency}_{(\text{system})} = \frac{\text{HHV of H}_2 \text{ produced}}{\text{electricity used + heat supplied}} \quad (10.20)$$

10.4.3
Recommendations

The following quotation is drawn from a 2002 publication on hydrogen by the Bellona Foundation, Oslo, Norway: "When calculating the efficiency in a fuel cell, the lower heating value is used. In the electrolysis process, the high heating value is used." The full report is available at http://bellona.org/filearchive/fil_Hydrogen_6-2002.pdf, and it basically sums up the European point of view.

In the United States, however, the situation is more equivocal. Developers of natural gas fuel cells do use LHV. Gas turbine manufacturers and manufacturers of IC engine generator sets, however, use HHV to calculate electrical efficiency. These groups complain that comparisons of fuel cells with heat engines are unfair because of the different bases for the efficiency calculations. In the United States, natural gas is sold by the therm (with 1 therm = 100 000 Btu), but it is measured by the cubic foot. The conversion from cubic feet to therms used by a gas supplier in billing customers uses either the measured gas composition and the sum of the HHV of each fractional component, or actual calorimeter measurements of the HHV of the gas being supplied to the customer. In either case, the customer pays based on the HHV of the gas used. It makes sense, therefore, that calculations of system efficiency for all electric generators should use the HHV of the fuel.

Electrolyzer manufacturers appear to have standardized on kW h Nm^{-3} or kW h kg^{-1} as a measure of system efficiency, which sidesteps the LHV versus HHV controversy. As noted above, Europeans prefer HHV for calculating electrolyzer efficiency on the basis of heating value.

As stated, splitting 1 mol of liquid water to produce 1 mol of hydrogen at 25 °C requires 285.8 kJ of energy: 237.2 kJ as electricity and 48.6 kJ as heat. It then follows that the ratio of reversible free energy potential (1.229 V) to the thermoneutral voltage (1.481 V) is 83%. This represents the highest efficiency attainable when using the LHV to determine stack voltage efficiency. Likewise, the same can be said for electrolyzer system efficiency calculations. Therefore, it is worth stating that the highest attainable efficiency is 83% when referencing electrolyzer system and stack efficiencies to the LHV.

The HHV is easily converted into more common forms of the higher heating value:

$$285\,840\text{ J mol}^{-1} \times \frac{1\text{ mol H}_2}{2.0158\text{ g}} \times \frac{1000\text{ g}}{1\text{ kg}} = 141\,799\,781\text{ J kg}^{-1}$$
$$= 141.8\text{ MJ kg}^{-1} \qquad (10.21)$$

$$141.8\text{ MJ kg}^{-1} \times 1\text{ W s J}^{-1} \times \frac{1\text{ W}}{1000\text{ W}} \times \frac{1\text{ h}}{3600\text{ s}} = 39.4\text{ kW h kg}^{-1} \qquad (10.22)$$

The United States Department of Energy (DOE) Fuel Cell Technologies Program Multi-Year Plan [1] includes targets for distributed water electrolysis and for central wind water electrolysis using three measures of efficiency: HHV, LHV, and kW h kg^{-1}. It also sets goals [2]. In all other hydrogen production schemes, however, for example in natural gas reforming and biomass gasification production scenarios, "energy efficiency is defined as the energy in the hydrogen produced (on a LHV basis) divided by the sum of the feedstock energy in (LHV) plus all other energy used in the process" [3].

10.5
Case Studies of Wind/Hydrogen Projects

10.5.1
Canada

10.5.1.1 Ramea Island
Ramea Island is a small island located in the Atlantic Ocean off the southern coast of Newfoundland and Labrador in easternCanada. It was the site of Canada's first wind–diesel demonstration project, in 2004. In an attempt to increase the proportion of the island's electrical generation coming from wind, a project to integrate hydrogen storage with the existing wind–diesel system was initiated. The system consists of three new 100 kW turbines in addition to the six existing 65 kW wind turbines and a 90 m^3 h^{-1} alkaline electrolyzer with 2000 m^3 hydrogen storage (10 bar). The stored hydrogen will be converted back to electricity, as need be, via four 62.5 kW hydrogen IC engine generators. All system components have now been delivered to this remote location, with integration efforts under way.

10.5.1.2 Prince Edward Island
Prince Edward Island is home to Canada's Wind Energy Institute, and boasts one of the strongest wind regimes in Canada, with 44 MW installed and another 30 MW planned. The Wind Energy Institute is located in the North Cape area of the island and is well positioned to host a wind-to-hydrogen demonstration project.

The system's components consist of
- wind turbine: Vergnet 60 kW variable speed, or the local grid
- electrolyzer: unipolar alkaline (66 Nm^3 h^{-1})

- storage: 4000 Nm^3 h^{-1} (at 17 bar), with plans to add 90 kg at 450 bar
- 120 kW retrofitted diesel genset.

This project will provide valuable data, resulting from the direct connection of the variable-speed turbine to the electrolyzer. It is also the only ongoing project that currently employs a unipolar alkaline electrolyzer and should provide information about the durability of this type of electrolyzer with variable input.

In addition to the scientifically interesting data that will be generated, the project has a number of applications for the hydrogen that will be produced. Hydrogen is used with a fuel cell as part of a back-up/auxiliary power unit and is also being used to fuel two 12-seat hydrogen IC engine buses. These buses are an integral part of the public transit system in Charlottetown (the provincial capital).

10.5.2
Greece

10.5.2.1 RES2H2 Project

For the Greek test site of RES2H2, a Casale Chemicals 25 kW electrolysis unit operating at pressures of up to 20 bar is connected to a 500 kW gearless, synchronous, multi-pole Enercon E40 wind turbine. The electrolysis unit has been developed with special cells to be able to withstand rapid changes of input power (15–100% capacity in 1 s). The electrolyzer operates in various modes (percentage of wind turbine production, "peak shaving", etc.), with excess energy from the wind turbine being fed to the grid. The electrolytic hydrogen is purified prior to entering a buffer tank. Part of the hydrogen produced is stored in novel metal hydride tanks having capacities of approximately 40 Nm^{-3} H$_2$. The rest of the hydrogen produced will be compressed to 220 bar and fed to cylinders at a filling station.

Several conclusions were drawn from this undertaking:
- In addition to meeting technical and cost targets and addressing safety issues, the design of a hydrogen energy system must be done in relation to what is market ready – there is no point in optimizing a system specifying units having capacity that is not available or that are still at an early development phase.
- The transportation and installation of hardware are something to be considered for such installations that are in many cases remote and with poor access. In terms of size and weight in combination with the poor quality of the access road, the capacities of the systems involved at the present site were the limit for conventional trucks and lifting equipment.

10.5.3
Spain

10.5.3.1 RES2H2 Project
A second component of the RES2H2 project is located at the Instituto Tecnológico de Canarias, on Gran Canaria Island, Spain. The major components are:
- wind turbine: 500 kW (ENERCON)
- electrolyzer: 5 Nm^3 h^{-1} H$_2$ at 20 bar
- hydrogen compressor: 7.5 kW at 220 bar.

The aim of this site is to optimize the energy produced by a wind turbine by providing electricity to the grid, producing drinking water through a reverse osmosis plant and hydrogen through an electrolyzer (this will be stored in a tank and used in a fuel cell for re-electrification purposes).

10.5.3.2 ITHER Project
The aim of the ITHER Project is the start-up of an installation that allows tests of hydrogen generation by electrolysis, with electricity obtained from renewable sources, with the most diverse available technologies. The project tries to cover all of the hydrogen chain (production, management and efficient use), obtaining the primary energy from renewable sources by means of processes currently available (photovoltaic and wind).

The project consists of three turbines, each with a type of technology and on an average range of powers (80, 225, and 330 kW). With this infrastructure, it is expected to be pioneering not only in Spain, but also at an international level, due to the range of powers that are handled.

For hydrogen storage, the project includes three metal hydride tanks each with a capacity of 7 Nm^3, used to store 100 kW h of energy. In addition to the metal hydride tanks, two other options exist: one is to store hydrogen gas at 45 bar in a tank and the other is to use a trailer of cylinders at 200 bar, with a total capacity of 400 Nm^3. The hydrogen is used in two Ballard 1.2 kW and a Plug Power 5 kW fuel cells, which are integrated as a back-up system in the building.

10.5.3.3 Tahivilla Project
This project involves a wind–hydrogen pilot plant located near Cadiz, which is a part of a research project led by ENDESA Generation with Green Power Technologies, AICIA, and INERCO as partners. After a preliminary study of electricity production by the wind farm where the pilot plant is located (by comparing the production and prediction curves for the last 3 years), simulations were made in order to optimize wind energy generation by means of an integrated system of hydrogen and electric energy generation. This system, the main components of which are an electrolyzer, a fuel cell, and a hydrogen tank, allows

the generation of hydrogen by using part of the energy produced by a variable-speed wind turbine.

The system is located on-site at an 80 MW park located in the south of Spain (1900 equivalent production hours per year, MADE 800 wind turbines). It is composed of an electrolyzer with a maximum electricity consumption of 41 kWe. Once the hydrolysis is complete, the resulting oxygen is vented and the hydrogen is stored at medium pressure (15 bar) in a storage tank. The system also has a compressor, for storage at 200 bar, and a fuel cell capable of generating 12 kWe.

10.5.4
United Kingdom

10.5.4.1 HARI Project

The Hydrogen and Renewables Integration (HARI) project was established in 2001, on the site of an existing renewable energy system at West Beacon Farm, Leicestershire, UK. The two main objectives of this project were to demonstrate and gain experience in the integration of hydrogen energy storage systems with renewable energy systems, and to develop software models which could be used for the design of future systems of this type.

Prior to the installation of the hydrogen energy system, the existing renewable energy systems at the site included two 25 kW wind turbines, 13 kW photovoltaics, and two micro-hydroelectric turbines with a combined output of 3 kW. The addition of a hydrogen energy storage system to the existing renewable energy (RE) supply network was seen as a means of balancing the varying supply with the fluctuating demand, allowing the evaluation of the feasibility of a stand-alone RE system. Three key components added to the existing network were a 36 kW alkaline electrolyzer (with 25 bar output pressure), 2856 Nm^3 of pressurized (137 bar) hydrogen storage, and two fuel cells (2 and 5 kW).

During the operation of this site between 2001 and 2006, a number of lessons were learned, which suggested a number of ways in which the overall efficiency could be optimized when designing similar systems. Importantly, matching the output and input requirements of all components ensures the most efficient energy conversion and hydrogen production. Additionally, the power conversion electronics were found to be the most significant parasitic losses in the system. Over time, the electrolyzer module's efficiency declined. The variable input from the wind turbine caused the electrolyzer to cycle, which led to degradation of the stacks. When it was first installed, the electrolyzer was rated at 36 kW, but over 2 years this had risen to 39 kW for the same hydrogen output.

10.5.4.2 PURE Project

The stand-alone small-sized wind hydrogen energy system PURE Project was a joint project of UNST (community of the Shetland Islands), siGEN (system integrator), and AccaGen SA for the PURE Community of Shetland Islands, and is

supported by the European Union. The project aims to demonstrate how wind power and hydrogen technology can be combined to provide the energy needs for a remote rural industrial estate. PURE was conceived to test and demonstrate safe and effective long-term use and storage of hydrogen produced by renewable energy using wind-powered electrolysis of water, and to regenerate the stored energy into electric energy with a fuel cell. The key components of the system are:
- wind turbines: two 15 kW (Proven Ltd)
- electrolyzer: 15 kW alkaline operating at 55 bar (AccaGen SA)
- hydrogen storage: 44 Nm^3 in H$_2$ cylinders
- PEM fuel cell: 5 kW (Plug Power).

The electrolyzer section consists of an AccaGen electrolyzer unit assembled with advanced cells specifically designed and manufactured by AccaGen SA for wind application, capable of operating up to 55 bar. Apart from high energy efficiency and good dynamic performance in variable operation, a particularly important requirement for a wind-operated water electrolyzer is the possibility of operating it over a wide range with high current yields and sufficient gas purities.

10.5.5
United States

10.5.5.1 Basin Electric, Wind-to-Hydrogen Energy Pilot Project
The goal of this project was to research the application of hydrogen production from wind energy, allowing for continued wind energy development in remote wind-rich areas and mitigating the necessity for electrical transmission expansion [4].

Four modes of operation were considered in the feasibility report to evaluate technical and economic merits. It should be noted that all the modes studied represent hydrogen production efficiencies less than those achievable if the system were operated at full production on "grid" electricity. The modes of operation studied were the following:
- mode 1 – scaled wind
- mode 2 – scaled wind with off-peak
- mode 3 – full wind
- mode 4 – full wind with off-peak.

In summary, the feasibility report, completed on 11 August 2005, found that the proposed hydrogen production system would produce between 8000 and 20 000 kg of hydrogen annually, depending on the mode of operation. This estimate was based on actual wind energy production from one of the North Dakota wind farms of which the Basin Electric Power Cooperative (BEPC) is the electrical off-taker. The cost of the hydrogen produced ranged from $ 20 to $ 10 kg^{-1} (again, depending on the mode of operation).

The hydrogen production system utilizes a bipolar alkaline electrolyzer nominally capable of producing 30 Nm^3 h^{-1} (2.7 kg h^{-1}). The hydrogen is compressed to 6000 psi and delivered to an on-site three-bank cascading storage assembly with 80 kg of storage capacity. Vehicle fueling is made possible through a Hydrogenics-provided gas control panel and dispenser able to fuel vehicles to 5000 psi.

A key component of this project was the development of a dynamic scheduling system to control the wind energy's variable output to the electrolyzer cell stacks. The dynamic scheduling system received an output signal from the wind farm, processed this signal based on the operational mode, and dispatched the appropriate signal to the electrolyzer cell stacks.

Unfortunately, chronic shutdown issues prevented consistent operation and, therefore, did not allow for any accurate economic analysis as originally intended. Much valuable experience was gained in the form of "lessons learned," however, and the project served as an extremely valuable platform for educating the public.

10.5.5.2 National Renewable Energy Laboratory and Xcel Energy, Wind-to-Hydrogen Project

Xcel Energy and the United States Department of Energy's (DOE) National Renewable Energy Laboratory (NREL) have collaborated to design, install, and operate the Wind-to-Hydrogen (Wind2H2) project. As the largest provider of wind-generated electricity in the United States, Xcel Energy is working with NREL to establish and understand state-of-the-art renewable electrolysis equipment and the operation of a renewable hydrogen production facility. Hosted at NREL's National Wind Technology Center (NWTC), the Wind2H2 system was approved for initial operation in March 2007 and is enjoying success as a demonstration project, producing hydrogen directly from renewable energy sources. This unique research-oriented project uses solar and wind energy to produce and store hydrogen. The stored hydrogen can be used both as a transportation fuel and as an energy storage medium, effectively allowing renewable energy to be stored and converted back to electricity at a later time.

The Wind2H2 project is helping researchers understand the hurdles and potential areas for improvement in emerging renewable electrolysis technologies. By allowing engineers to operate and configure an integrated electrolysis facility, this project has permitted the investigation and analysis of hydrogen production, compression, storage, and electricity generation. This project is providing valuable data that are being used to improve the designs of future renewable electrolysis systems. The Wind2H2 project provides important guidance to industry and key stakeholders for development of future renewable electrolysis systems. The Wind2H2 project is the only renewable hydrogen production facility in the world that can operate multiple electrolyzers in any of the following configurations:

1. grid connected
2. directly connected from the output of a photovoltaic array to the electrolyzer stack

3. real-time electrolyzer stack current control based on a power signal from a wind turbine
4. closely coupled photovoltaic (PV) and wind energy sources to the electrolyzer stack with custom designed and built power electronics.

NREL and Xcel Energy have undertaken the Wind2H2 project with several key objectives in mind. First and foremost, the Wind2H2 project is being used to demonstrate operation of a renewable electrolysis system, allowing researchers to evaluate actual system performance and costs and to identify areas for cost and efficiency improvements. Additionally, the project provides operational experience with a renewable electrolysis hydrogen production facility, enabling project engineers to investigate operational challenges and to explore system-level integration issues and opportunities for performance and cost improvements resulting from system-level optimization. The project also seeks to investigate how to maximize the use of renewable energy resources in renewable hydrogen production systems by optimizing energy transfer from PV arrays and wind turbines to the stacks of commercial electrolyzers. Finally, the project is designed to explore operational challenges and opportunities related to energy storage systems and their potential for addressing electric system integration issues inherent with high penetrations of variable renewable energy resources.

To help enable greater penetration of renewable energy sources, hydrogen production from renewable electrolysis must be cost competitive. The DOE has a target of reducing the cost of central production of hydrogen from wind-based water electrolysis to \$ 3.10 kg^{-1} by 2012; by 2017, the DOE seeks to reduce this cost to under \$ 2 kg^{-1} [1].

Electrolyzer manufacturers are improving performance and reducing the capital cost of electrolyzer systems. At the same time, the complete renewable electrolysis system, including the renewable power source, electrolyzer, and interfacing power electronics, must be integrated and optimized to improve system performance and lower costs. The Wind2H2 project presents an excellent research platform to investigate these integration and optimization opportunities.

To achieve the objectives of the Wind2H2 project, NREL engineers have been working to complete a number of project tasks. These tasks include:
- designing, building, and testing dedicated wind- and PV-to-electrolyzer stack power electronics to integrate more closely the renewable energy resources and electrolyzer stacks
- modeling and simulating renewable electrolysis system performance to permit improved hydrogen production system designs
- characterizing renewable energy system impacts on commercial electrolyzer technology and their ability to accommodate the varying energy input from wind and PV sources
- sequencing multiple electrolyzer systems to improve overall system efficiency, responsiveness, and performance with varying renewable energy sources.

The challenge of renewable electrolysis is designing and implementing systems that can cost-effectively produce hydrogen from renewable sources using streamlined, robust, and efficient processes. When the wind turbine or solar array is co-located with the electrolysis system, more direct connection between the source and the electrolyzer stack is possible. This close coupling eliminates the need for long-distance transportation of electricity and reduces the number of electrical conversions, resulting in a more efficient, cost-effective system.

Valuable operational experience is shared through running, testing, daily operations, and troubleshooting the Wind2H2 system. Equipment errors are being logged to help evaluate the reliability of the system. The valuable lessons from this system operational experience will lead to improved design, implementation, and operational plans of renewable electrolysis systems. For example, integrated renewable electrolysis systems require that system components from different manufacturers be configured to function together smoothly. Consequently, programmable logic control systems must be able to communicate with all major elements of the system (e.g., electrolyzers, compressors, power converters, and load transfer switches).

As another primary goal of the project, NREL engineers investigated methods to maximize renewable energy use and to optimize energy transfer within the system. Such system optimization efforts can significantly reduce the cost of renewable hydrogen. In an analysis of the potential improvements to a wind electrolysis system, project engineers estimated that optimized power electronics would result in a cost improvement of 7%, reducing the cost of hydrogen produced from wind to $ 5.83 kg^{-1} from a baseline of $ 6.25 kg^{-1}. For reference, the DOE has set a target for reducing the cost of central production of hydrogen from wind-based electrolysis to $ 3.10 kg^{-1} by 2012. To investigate such optimization opportunities, NREL developed multiple power electronics configurations that convert varying electricity from a solar PV array and wind turbines into the electricity used by the electrolyzer stacks directly. These power converters have the added benefit of executing maximum power point tracking (MPPT) from the wind turbine or PV array to allow higher energy transfer to the electrolyzer stack.

The PV array connected to the electrolyzer stack both with and without an intermediate power converter was tested and analyzed. The PV array was configured to supply different input voltages to a step-down, d.c.-to-d.c. (d.c./d.c.) power converter over many days of testing. It was found that the use of the power converter increases the energy delivered to the electrolyzer stack by 10–20%, depending on the PV array input voltage to the power converter. The efficiency of the power converter decreased as the input voltage from the PV array increased. The greatest input voltage from the PV array to the power converter still provided the maximum energy capture to the electrolyzer stack. In other words, although this configuration has the largest difference between PV voltage and electrolyzer stack voltage (ΔV), it provided the most energy to the stack over a given day.

A 10 kW wind turbine was connected to one of the PEM electrolyzer stacks through an MPPT alternating current to direct current (a.c./d.c.) power converter. This configuration represents a non-grid-tied (i.e., stand-alone) configuration closely coupling a wind turbine to the electrolyzer stack. Testing and analysis shows that the a.c./d.c. converter has the ability to maintain optimal operation of the turbine while delivering power to the electrolyzer stack without a battery link. It eliminates several power electronics conversions inherent in grid-tied electrolyzer configurations as well as a battery and associated maintenance. This a.c./d.c. power converter is undergoing upgrades that are expected to increase further the energy capture from the wind turbine.

One of the major tasks going forward will be to determine how the findings of the Wind2H2 project can improve performance and reduce the cost of renewable electrolysis production systems. The results of system optimization efforts, performance measurements, and evaluation data will be used in economic models to understand better how these system improvements can lower the cost of hydrogen produced via renewable electrolysis.

10.6
Conclusion

It is important that parties interested in hydrogen technologies standardize methods of evaluating the performance and efficiency of these technologies. A detailed description of the chemical and electrical processes for electrolysis and fuel cells is presented in this chapter. Important terminology, units of measure, constants, and chemical reactions are discussed. Recommendations for calculating electrolyzer and fuel cell performance are suggested.

As interest grows in using hydrogen for grid energy storage and as a transportation fuel, pilot projects and research efforts are under way to experiment with hydrogen production and utilization technologies. The focus of many of these projects is coupling wind energy with hydrogen production (via water electrolysis) in an effort to use all available wind energy and to store that energy to be used during times of high electricity demand. The body of knowledge and lessons learned from designing and operating renewable electrolysis plants is growing. As interested parties learn more about the best way to utilize hydrogen as an energy carrier, the cost of hydrogen produced from renewable sources is decreasing.

Acronyms and Abbreviations

a.c.	alternating current
a.c./d.c.	a.c.-to-d.c.
BEPC	Basin Electric Power Cooperative
d.c.	direct current
d.c./d.c.	d.c.-to-d.c.
DOE	US Department of Energy
F	Faraday constant

G	Gibbs free energy
H	enthalpy
HARI	Hydrogen and Renewables Integration
HHV	higher heating value
IC	internal combustion
JANAF	Joint Army–Navy–Air Force
kWe	kilowatt electrical
LHV	lower heating value
MPPT	maximum power point tracking
NREL	National Renewable Energy Laboratory
Nm^3	normal cubic meter
NWTC	National Wind Technology Center
PEM	polymer electrolyte membrane
PV	photovoltaic
RE	renewable energy
S	entropy
Wind2H2	Wind-to-Hydrogen project

Acknowledgment

Employees of the Alliance for Sustainable Energy, LLC, under Contract No. DE-AC36-08GO28308 with the US Department of Energy, have authored this work. The United States Government retains and the publisher, by accepting the article for publication, acknowledges that the United States Government retains a non-exclusive, paid-up, irrevocable, worldwide license to publish or reproduce the published form of this work, or allow others to do so, for United States Government purposes.

References

1 US Department of Energy (2009). *Hydrogen, Fuel Cells and Infrastructure Technologies Program: Multi-Year Research, Development and Demonstration Plan*, Office of Energy Efficiency and Renewable Energy, Washington, DC; www.eere.energy.gov/hydrogenandfuelcells/mypp/ (accessed 27 December 2009).

2 US Department of Energy (2009). *Hydrogen, Fuel Cells and Infrastructure Technologies Program: Multi-Year Research, Development and Demonstration Plan*, Office of Energy Efficiency and Renewable Energy, Washington, DC; www.eere.energy.gov/hydrogenandfuelcells/mypp/pdfs/production.pdf, Table 3.10.4 footnote f, and Table 3.10.5 footnote e (accessed 29 December 2009).

3 US Department of Energy (2009). *Hydrogen, Fuel Cells and Infrastructure Technologies Program: Multi-Year Research, Development and Demonstration Plan*, Office of Energy Efficiency and Renewable Energy, Washington, DC; www.eere.energy.gov/hydrogenandfuelcells/mypp/pdfs/production.pdf, Table 3.10.2 footnote f, and Table 3.1.8, footnote g (accessed 29 December 2009).

4 Rebenitsch et al. (2009) *Wind-to-Hydrogen Energy Pilot Project: Basin Electric Power Cooperative*; www.osti.gov/bridge/servlets/purl/951588-ejIotC/951588.pdf (accessed 29 December 2009).

11
High-Temperature Water Electrolysis Using Planar Solid Oxide Fuel Cell Technology: a Review

Mohsine Zahid, Josef Schefold, and Annabelle Brisse

Abstract

The present global hydrogen demand is met to a large extent by hydrogen from fossil fuels, but hydrogen production via water electrolysis has always been an alternative for production, in niche applications or, on a larger scale, when the electric energy required for the electrolysis reaction was readily available. Conventional alkaline electrolysis or electrolysis with proton-exchange membrane fuel cells are technologically mature processes but limited in energy efficiency to 60–80%. Hydrogen production via water electrolysis could play an increasingly important role in future, for industrial and transportation application and also as a means for storing energy from renewable sources, as a consequence of cost and availability limitations of fossil fuels, if a higher energetic efficiency can be reached. Water electrolysis at high temperature using protonic or ionic conducting electrolytes constitutes an advanced concept aimed at increased electrical-to-chemical energy conversion efficiency. At high temperature (600–900 °C), higher efficiencies are achievable owing to favorable thermodynamic conditions and also because of improved kinetics for the electrode reactions, even without the use of precious metal catalysts. The thermodynamic reason for higher efficiencies is a decrease in the molar Gibbs energy of the reaction with increasing temperature while the molar enthalpy remains essentially unchanged. Reversible operation of solid oxide fuel cells as H_2O electrolyzer cells (SOECs) is well known from pioneering work in the 1980s on tubular cells. In recent years, renewed interest has arisen in SOECs, driven by the availability of (tubular or planar) cells with improved performance at lower temperatures. In this chapter, the current state of research on SOECs is briefly summarized, the main actors in the field are mentioned, and potential hurdles for the future development are identified.

Keywords: water electrolysis, high-temperature electrolysis, hydrogen production, fuel cell, planar solid oxide fuel cell, solid oxide electrolyzer cell

---

Hydrogen Energy. Edited by Detlef Stolten
Copyright © 2010 WILEY-VCH Verlag GmbH & Co. KGaA, Weinheim
ISBN: 978-3-527-32711-9

11.1
Introduction to High-Temperature Electrolysis

Hydrogen is used in many industrial and chemical processes. The first commercial technology of hydrogen production was the alkaline electrolysis of water, developed in the 1920s. At present, around 600 billion N m^3 of hydrogen are produced worldwide per year [1]. About 96% of this production is based on fossil fuels, mainly natural gas and coal. The use of these fuels results in significant greenhouse gas emissions. Moreover, there is a general agreement that global hydrocarbon production will peak in the near future, a factor which should favor the development of hydrogen as an energy carrier.

Hydrogen can be produced via the classical electrolysis of water at low temperature or, alternatively, by using the different fuel cell technologies. These technologies are based on (i) proton-exchange membrane fuel cells (PEMFCs) (referring to the solid polymeric electrolyte membrane), (ii) fuel cells using solid oxide proton conductors, and (iii) fuel cells with a solid oxide ion (O^{2-}) conductor (SOFCs).

In a fuel cell, electrical energy is generated by the exothermic oxidation of hydrogen. In the reverse electrolysis operation of such a cell, steam is reduced in an endothermic reaction using electrical energy, according to the reaction scheme

$$H_2O \xrightarrow{\text{electricity + heat}} H_2 + \frac{1}{2}O_2 \qquad (11.1)$$

The operating temperatures of fuels cells vary widely, from around 80–120 °C for PEMFCs to 700–1000 °C for SOFCs. The free energy required for the reaction (ΔG) decreases with increasing temperature whereas the free enthalpy (ΔH) remains almost constant (Figure 11.1). This thermodynamic relation, in principle unfavorable for the fuel cell mode at high temperatures, explains the particular interest in performing electrolysis at high temperatures. Since the

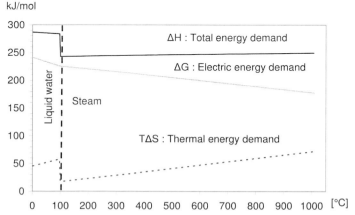

Fig. 11.1 Electrical, heat, and total energy requirements for water electrolysis as a function of temperature.

SOFCs achieve competitive (chemical-to-electrical) energy conversion efficiencies despite the less favorable thermodynamic conditions, one can *a priori* expect that high-temperature electrolysis (HTE) cells achieve much higher (electrical-to-chemical) energy conversion efficiencies (the term energy conversion efficiency for the HTE refers to the electrical-to-chemical energy conversion throughout this chapter). Cell voltages of about 1.1–1.3 V under operation are currently achieved with state-of-the-art cells. Cell voltages are about 1.6 V for advanced PEM electrolyzers (PEM electrolyzers in the kilowatt range are already commercially available, but at a very high cost). Because ionic transfer numbers are close to one for both cell types, the difference in cell voltage translates linearly to the energy consumption for the reaction.

The primary motivation for HTE is the above-mentioned potential of a reduced demand for electrical energy compared with electrolysis at low temperature. This may allow electrical-to-chemical energy conversion efficiencies even exceeding 100%, as already recognized in early work [8]. The free energy of the reaction ΔG decreases from ~ 1.23 eV (237 kJ mol^{-1}) at ambient temperature to ~ 0.95 eV (183 kJ mol^{-1}) at 900 °C, while the free enthalpy term remains essentially unchanged ($\Delta H \approx 1.3$ eV or 249 kJ mol^{-1} at 900 °C; cf. Figure 11.1). Part of the energy required for an ideal (loss-free) HTE can thus be provided by heat. Increasing ohmic and/or reaction losses in a real HTE system increase the demand for electrical energy and decrease the demand for an external heat supply until, finally, the reaction becomes exothermic. Hence three modes of operation are distinguishable in HTE: thermoneutral, endothermic, and exothermic. HTE operates at thermal equilibrium (the thermoneutral mode) when the electrical energy input equals the enthalpy of the reaction. In that case, the entropy necessary for water splitting equals the heat generated by the loss reactions, and the energy conversion efficiency is 100%. In the exothermic mode, on the other hand, the electric energy input exceeds the ΔH term, which corresponds to an efficiency below 100%. Finally, in the endothermic mode, the electric energy input remains below the enthalpy term. Therefore, heat must be supplied to maintain the cell temperature. This mode means that energy conversion efficiencies of the cell or the stacks are above 100%.

An HTE system can be operated with and without an external heat supply. This is different to low-temperature electrolyzers, which run in the exothermic mode, because the energy losses, which arise mainly from the electrochemical reactions, exceed the small difference between ΔH and ΔG at low temperature. The availability of an external heat source influences the design of an HTE system:

Without a heat source, the goal is to approach the thermoneutral mode, that is, to limit the thermal losses to a value required to compensate for the endothermic reaction. This leaves a wide margin for cell overvoltages and, therefore, for an increase in the current density or a lowering of the temperature. Operating temperatures in the range 600–700 °C known from the SOFC development may therefore also be accessible for electrolysis.

With an external heat source of high temperature, on the other hand, the goal is to reduce the overvoltages as far as possible to allow for a significant uptake of heat. This implies, at least with present cell technology, operation under higher temperatures (800 °C or above) and lower electrode overvoltages (i.e., current densities somewhat lower than those achieved in thermoneutral operation).

The operation of SOFCs in electrolysis mode has been demonstrated in several research projects since 2004 [10]. Cells of both commercial and research types and including the common designs were tested (electrolyte-, hydrogen electrode-, and metal substrate-supported). As for fuel cell operation, the hydrogen electrode-supported cells showed the highest performance owing to the low resistance of the thin electrolyte layer. A high current density of -3.6 A cm^{-2} at a cell voltage of 1.48 V and a cell temperature of 950 °C, for example, was reached with such a cell at DTU-Risoe (Denmark) [11].

Cell and stack performance are addressed in this contribution, and also durability, system demonstration, and development issues.

11.2
History of High Temperature Steam Electrolysis

11.2.1
The HOTELLY Project

The HOTELLY project (High Operating Temperature ELectroLYsis) was done by Dornier and Lurgi in the years 1970 to 1987 [2, 3, 5, 6, 7, 12]. The project was based on a tubular technology of SOEC composed of the traditional materials described in paragraph 2. The project demonstrated the reversibility of operation under H_2/H_2O and CO/CO_2 [2] (Figure 11.2).

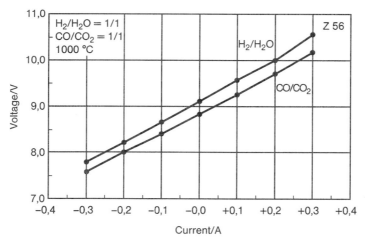

Fig. 11.2 Voltage-current curves at 1000 °C of a 10 tube SOEC stack in the fuel cell (j > 0 A) and the electrolysis (j < 0 A) modes. The figure is taken from the ref [2].

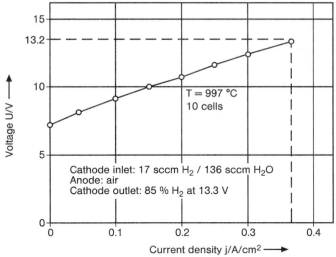

Fig. 11.3 Voltage-current curves of the 10 cell stack tested in the project HOTELLY. The figure is taken from the ref [2].

A production of 1 Nm^3 per hour of hydrogen was obtained with a 10 cell stack operating at 1000 °C. The average cell potential was 1.32 V at a current density of about –0.37 A/cm^2. A hydrogen concentration of 85% at the cathode outlet means that the stack was operated with a rather high steam-conversion rate (Figure 11.3).

The maximum current density [5] achieved at 1.32 V for a short period was –0.6 A/cm^2. Further tests were performed for a 100 cell stack. The longest test duration of 2500 hours [2] was realised for a cell voltage of 1.3 V and current densities between –0.3 and –0.4 A/cm^2. One test was done with a pressurised stack of 2 kW$_{el}$. Due to the high cost of the electrolyser and the lack of a perspective for a short-term industrial application, the work was stopped in the year 1987.

11.2.2
US American Projects

During about the time period of the HOTELLY project, the company Westinghouse also performed high temperature electrolysis using tubular SOEC [8, 9]. The cells were composed of traditional electrolyte and hydrogen electrode materials. The oxygen electrode was a composite of indium oxide and YSZ. Westinghouse decided to stop the development for similar reasons to those stated above.

In 1999, the Lawrence Livermore National Laboratory (LLNL) started the development of electrolysers within the NGASE project (Natural Gas-Assisted Steam Electrolyser) [13, 14]. In that project, the air in the oxygen electrode is replaced by natural gas, which lowers the open-circuit voltage, and thereby the electricity consumption. The goal of the project was to develop a prototype 3–5 kW-equivalent NGASE system for technology validation in the year 2006. Within the proof-of-concept a lowering of the cell voltage by 1 V was observed when methane was fed to the oxygen electrode. The electricity consumption was estimated to reach values one order of magnitude below the one in conventional electrolysers. With novel catalyst materials, a current density up to 1 A/cm^2 at 0.5 V was measured at 700 °C.

The Idaho National Laboratory (INL) started its activities on HTE in 2003 in cooperation with Ceramatec, Argonne National Laboratory and Oak Ridge Laboratory [15, 16, 17, 18, 19]. The objectives are to develop and test combined components including stacks in a so-called Integrated Laboratory Scale Demonstrator. In a first step, a 15 kW demonstrator was constructed and successfully operated. Planned follow-up steps of this project are the upscaling of the demonstrator up to a 200 kW pilot plant and a 1 MW Engineering Demonstration Facility to be coupled with a nuclear power reactor.

11.2.3
The Japanese Program of the Japan Atomic Energy Research Institute (JAERI)

The JAERI project is based on the coupling of a high temperature electrolyser with a nuclear power plant supplying steam at high temperature [20]. The HTTR (High Temperature engineering Test Reactor) was constructed at Oarai, the location of the project. Tubular cells were operated at temperatures ranging from 850 °C to 1000 °C. For the highest temperature, a hydrogen production rate of 152 N cm^3/cm$^2 \cdot$ h was achieved at 0.37 A/cm^2, a result comparable to the one obtained in the HOTELLY project (154 N cm^3/cm$^2 \cdot$ h at that temperature). The use of planar electrolyte supported cells led to comparable results, but for 100 °C lower cell temperature.

11.2.4
European Projects

The successful operation of SOFCs in the electrolysis mode was demonstrated within several research projects since 2004 [10]. Cells of both commercial and research type and including the common designs were tested (electrolyte-, hydrogen electrode-, and metal-substrate supported). As for fuel-cell operation, the hydrogen electrode supported cells showed the highest performance owing to the low resistance of the thin electrolyte layer. A high current density of –3.6 A cm^{-2} at a cell voltage of 1.48 V and 950 °C cell temperature was reached with such a cell by DTU-Risoe (Denmark) [11] (Figure 11.4).

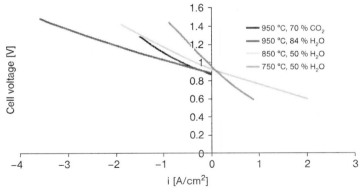

Fig. 11.4 Voltage-current curves for cells from DTU-Risoe in the electrolysis and the fuel cell mode at different temperatures and absolute humidities [11].

11.3 Solid Oxide Electrolyzer Cells

11.3.1 General

Virtually all development work on SOECs done in recent years relates to SOFCs operated in the electrolysis mode, that is, the fuel cells are used without modification [21, 22, 23, 24]. As with SOFC technology in general, much emphasis is put on planar cell technology. Compared with tubular cells, state-of-the-art electrode-supported cells with classical yttria-stabilized zirconia (YSZ) as a thin-film electrolyte provide at least maintained performance at reduced temperature (800 °C or lower) in the SOFC and also in the SOEC mode.

The standard cell design, commonly included in the mentioned research projects, is an Ni/YSZ cermet as hydrogen/steam electrode on an Ni/YSZ supporting substrate and a composite of strontium-doped lanthanum manganite (LSM) or ferrite (LCSF) as oxygen electrode. The performance of cells with LSM electrodes is described in the following.

11.3.2 Results From Cell Testing

Voltage-current density (U–j) data at the research cell in Figure 11.5 show similar performances in the SOFC and SOEC modes, that is, similar curve slopes or area-specific resistances (ASRs). This demonstrates the reversibility of the operation. The voltage offset at zero current stems from a change in the absolute humidity (AH) in the feed gas to the hydrogen/steam electrode. The cell voltage in the SOEC remains below 1.3 V. Hence endothermic operation is feasible even at large current density magnitudes (here $|j|$ up to 1.4 A cm^{-2}). The SOEC cell

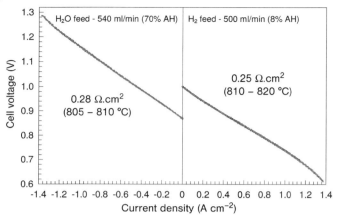

Fig. 11.5 Initial voltage-current behavior of an anode-supported SOFC from research cell reversibly operated in the SOFC and SOEC modes (circular cell with 45 cm² area). Measurements are taken in a non-sealed configuration.

performance fulfills requirement for practical operation, such as high current density, a relatively large cell area (45 cm^{-2}), and moderate cell temperature ($\sim 810\,°C$).

The U–j behavior depends on the humidity of the feed gas to the hydrogen/steam electrode (Figure 11.6). Higher humidity rates would reduce the cell voltage, in agreement with the Nernst equation. Under practical operation, the AH value of the H_2 feed gas should be as high as possible; it is usually limited to about 90% in order to keep a reducing environment, which protects the cells in the absence of a current flow. Again, endothermic operation is feasible over a wide current density range with cell voltages of around 1.0 V. Fast electrode kinetics are ensured by the higher operating temperature.

A leveling-off of the cell voltage occurs when high steam conversion rates are approached (Figure 11.6). The maximum possible steam conversion depends somewhat on the cell provider and it is linked with the porosity of the H_2/steam electrode [10, 26, 27]. Steam conversion rates under constant-current operation are usually set to 40–70%. Higher steam conversion rates would mean a reduced amount of water to be recirculated in an electrolyzer system.

11.3.3
Nonionic Electrolyte Conduction Under Steam Starvation

Although precautions can be taken to prevent operation under steam starvation, one may wonder about the cell behavior beyond the range of voltage rise as a consequence of steam starvation as shown in the Figure 11.6. Higher cell voltages might cause cell destruction via electrolyte reduction and they might lead to increased electronic conduction in the electrolyte [26, 28]. To discuss this

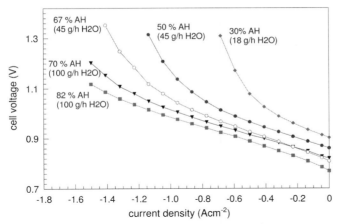

Fig. 11.6 Voltage-current behavior of a commercial anode-supported SOFC from HTceramix in the electrolysis mode as a function of the absolute humidity AH at the hydrogen/steam inlet ($T = 890\,°C$). Measurement conditions as in Figure 11.2 [25].

point for cells with YSZ electrolyte, operation was extended to current densities j corresponding to steam conversion rates above 100% and, alternatively, the steam supply was interrupted under constant-current conditions [26]. In both cases, cell voltages saturated at ∼1.9 V at 810 °C, without indications of electrolyte decomposition during experimental runs lasting several days. The mechanism limiting the cell voltage was therefore attributed to electronic conduction in the YSZ electrolyte. Increased electronic conduction originates from the rearranged defect equilibrium in the electrolyte caused by the very low oxygen partial pressure at the H_2/steam electrode side. These results also indicate that the ionic transfer number in the SOEC mode is lower than that in the SOFC mode. Due to the high activation energy of electronic conduction in YSZ (3.9 eV) [26], the electronic conduction in the YSZ electrolyte may no longer be negligible for temperatures approaching 1000 °C.

An example of operation beyond the range of voltage rise due to steam starvation is shown in the Figure 11.7. The decrease in the slope of the S-shaped U–j curve at the highest current density magnitudes is a consequence of increasing electronic conduction. Such intrinsic limitation of the cell voltage provides cell protection against insufficient steam supply up to the limit given by the drastic increase in Joule heating in the electrolyte.

The operating voltages and, hence, the voltage drop across the electrolyte in the electrolyzer mode are larger in the electrolyzer mode than in the fuel cell mode. This implies that electrolytes with a smaller (voltage) operation window compared with YSZ limited by electronic conduction may not be usable in the electrolysis mode.

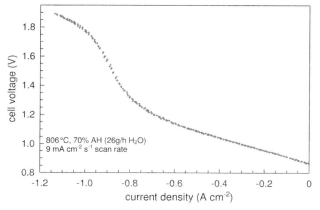

Fig. 11.7 Voltage-current behavior of a commercial anode-supported SOFC from HTceramix in the electrolysis mode (forward and reverse scans virtually coincide). Measurement conditions as in Figure 11.2 (compare [26]).

11.3.4
Cell Degradation

A so far critical issue with respect to the use of SOFC in the SOEC mode is long-term stability. Whereas voltage degradation rates below 1% under constant-current operation are demonstrated for SOFCs (and stacks) running for tens of thousands of hours, degradation in the SOEC mode tends to be higher. Moreover, only a limited number of reported experiments had durations longer than 1000 h. Degradation rates in the 1% range have been reported at small current densities ($|j|<0.3$ A cm^{-2}) [10]. Higher $|j|$ values cause faster degradation. Sudden failure of the SOECs, on the other hand, is rarely observed.

In Figure 11.8, long-term operation data over 400 h at −0.5 A cm^{-2} are shown for a commercial electrode-supported cell (from HTceramix). The degradation rate is about 16% per 1000 h, a high value that is not suitable for practical operation. Significantly lower degradation rates at comparable or higher current densities were found with research-type cells [10].

A number of degradation features are under discussion, most of them known from SOFC operation. The problem consists in the identification of the specific degradation processes responsible for the faster degradation in the SOEC mode. Major treated degradation features are (i) delamination of the oxygen electrode, (ii) oxygen evolution in closed pores in that electrode, (iii) deactivation of the electrodes by changes in the microstructure (e.g., Ni coarsening or oxidation at the H$_2$/steam electrode) or by poisoning, and (iv) electrolyte aging. Poisoning may occur by impurities in the steam supply or by material evaporation, for example from the seals, the cell housing, or the tubing. In view of the reported widely varying degradation rates, one can expect that the influence of a specific degradation feature depends largely on the cell type and the operating conditions.

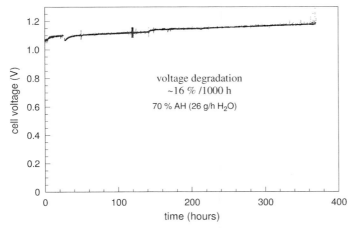

Fig. 11.8 Long-term SOEC operation at $j = -0.5$ A cm^{-2} of a commercial anode-supported SOFC from HTceramix (810 °C, 45 cm^2 cell area) [27]. Measurement conditions as in Figure 11.2.

Delamination of the oxygen electrode has frequently been reported, but it seems to be less critical in cells using composite electrodes of the LSM/YSZ type or the mixed ionic–electronic conductor LSCF. The degradation of the hydrogen electrode composed of a nickel/YSZ cermet has been examined in several studies, for example using impedance spectroscopy (IS) [22, 27]. That electrode seems to be particularly sensitive to silica poisoning from glass sealings [22]. The higher cell voltage in the SOEC mode means a more oxidizing environment for the oxygen electrode, which might reduce the electrode stability. It also means a larger gradient in the oxygen partial pressure across the cell, which could affect the known aging of the YSZ electrolyte. The influence of poisoning of the oxygen electrode with chromium vapor species, well known from SOFC operation, on the other hand, seems somewhat less critical owing to the inverted flow direction of the oxygen.

IS is essentially the only *in situ* tool used so far for the monitoring of cell degradation during aging experiments. IS data can later be complemented by the results from post-mortem analysis. *In situ* tool here means that a small a.c. current signal of varying frequency is superimposed to the steady-state d.c. current. Operation therefore remains in the steady state, unlike during U–j scanning. As an example, the time evolution of the impedance of an HTceramix cell is shown in Figure 11.9. Voltage degradation amounts to about 10% during the time window shown. Impedance at the high-frequency limit remains almost independent of time, that is, the ohmic contributions such as electrolyte and contact resistances hardly vary with time. Also, the impedance term at around 3 Hz, which stems from steam conversion in the hydrogen/steam electrode, shows only a minor time dependence. Cell degradation is mainly reflected by an increase in the impedance contribution at intermediate frequencies, where interfacial

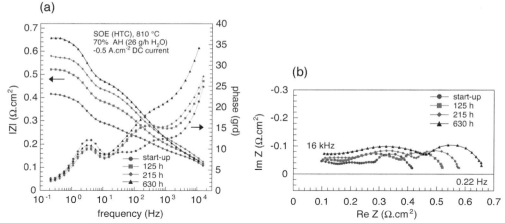

Fig. 11.9 Impedance of a commercial SOFC (from HTceramix) as a function of the operating time in the electrolysis mode under $j=-0.5$ A cm^{-2} in the magnitude/phase (a) and the complex plane (b) presentations [27].

charge-transfer processes commonly appear. That increase is responsible for the increase in the (ohmic) impedance at low frequency, which represents the ASR value at $j=-0.5$ A cm^{-2}. It rises from 0.42 Ω cm^{-2} after start-up to 0.66 Ω cm^{-2} after 630 h of operation.

11.3.5
General Aspects of Cell Testing

The comparison of the behavior of the SOECs requires a certain standardization in the measurement conditions, in order to account for the fact that cell performance at a given temperature depends on the humidity of the feed gas, the stability of the feed gas flow, and the steam conversion rate. Realistic SOEC performance is measured with AH values of the feed gas to the hydrogen/steam electrode exceeding 70%. This allows effective AH values at the electrode surface to be reached under a current flow of roughly 50%, which should approach optimum conditions for the electrochemical reaction. A small percentage of hydrogen in the feed gas is used to protect the hydrogen/steam electrode under zero current flow. Fluctuations in the steam flow in most cases dominate by far the voltage noise, even for a well-adjusted steam supply system. This translates to a lower voltage resolution and implies longer measurement times for the determination of degradation rates as compared with SOFC testing. Larger fluctuations in the steam supply or pressure pulses owing to condensation/re-evaporation must be suppressed, as they may accelerate degradation.

11.3.6
Mid-Term Development Perspectives of SOECs

The large variety of existing SOFCs, and also the ongoing SOFC development, mean that much experimental work remains to be done to gain a broader overview of their behavior in the electrolysis mode.

Nonetheless, the use of SOFCs as electrolyzer cells may, seen on a longer time scale, only represent an initial development step, justified by the cell availability and the promising results encountered with many cell types. One has to consider that the electrolyzer application comes along with changes in the operating conditions, such as an inverted current flow, a high humidity of the feed gas, higher operation voltages, and, in most cases, substantially reduced heat flows. The need for a predominantly ionic electrolyte conduction as an example may impede the use of SOFCs with a narrow voltage window for the electrolyte operation. Therefore, mature SOEC technology may finally work with specifically designed cells which differ from the SOFC in several properties. In particular, the negligibly small heat production in the thermoneutral operation mode should allow larger cell areas.

11.4
Solid Oxide Electrolyzer Stacks

High-temperature electrolysis stacks have been developed based on the SOEC technology discussed above. Among the relatively few experimental results reported in the literature are those of the Idaho National Laboratory (INL) on stacks developed by Ceramatec Inc. [29]. A 10-cell stack was operated in 2004 between 800 and 900 °C under different absolute humidity, gas flow rates, and current densities [24]. Thereafter, a durability test of 1000 h was performed at INL on a 25-cell stack with an active cell area of 64 cm^{-2} [15]. An initial rapid degradation followed by a subsequent slowing of the degradation rate was observed. This two step degradation led to a degradation rate close to 20% per 1000 h. Internal resistances were significantly higher within the stack compared with single cells since the per cell ASR at 800–850 °C ranged between 1.3 Ω cm^{-2} at the beginning of the test to 2.5 Ω cm^{-2} at the end, whereas ASR values below 0.5 Ω cm^{-2} are classically obtained on single cells at similar temperatures. This difference could be due to the joining of cells and interconnects with a degradation of the intermediate conductive layer made of lanthanum strontium chromite [15].

The INL has also developed a test facility for a complete HTE system to address balance-of-plant (BOP) issues such as heat recuperation, hydrogen recycling, and materials of construction [30, 31].

An electrolysis stack was tested in the European Institute for Energy Research for 5000 h at −0.3 A cm^{-2} at about 800 °C (Figure 11.10) [27]. The stack was an R-design unit from HTceramix [32] that integrated five rectangular second-generation InDEC (H.C. Starck) [33] anode-supported SOFC cells (active area per

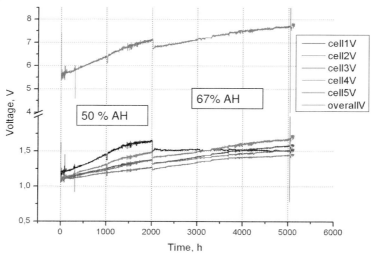

Fig. 11.10 Time evolution of stack and cell voltages under SOEC operation with a current density of $-0.3\ A\ cm^{-2}$.

cell of 50 cm^{-2}) and 1 mm thick coated F17TNb interconnect plates with an MnCo2O4 protective layer. The electrolyte was a 5 µm thick YSZ layer and the oxygen electrode consisted of LSCF. The stack operated during the first 2000 h with 50% absolute humidity in the inlet gas and then with 67% AH. Cell degradation rates varied widely from cell to cell during the first 2000 h with an average stack degradation rate of approximately 14% per 1000 h. That rate thereafter dropped to around 7% per 1000 h, a value closer to those typically found in cell testing at the current density used. In principle, the differences in the cell voltages at 50% AH can arise from an insufficiently homogeneous steam distribution within the stack, and also from non-uniformity of the cell properties. However, the fact that the cell behavior is fairly uniform at 67% AH indicates that the steam distribution is at the origin of these differences.

11.5
Conclusion

In recent years, renewed interest has arisen in the reversible operation of SOFCs as H_2O electrolyzer cells (SOECs). The availability of SOFCs with improved performance at lower temperatures, notably from planar cell technology, means that the favorable operational properties expected from the thermodynamics of the reaction are approached in the experiment. Results for several representative cell types have been given in this chapter. Cell voltages under operation of around 1.0–1.3 V at 800 °C are achieved, which translates to a considerable saving of electrical energy compared with low-temperature electrolysis.

Many solid oxide cells developed specifically for the fuel cell mode seem to operate well in the SOEC mode. Therefore, the development of high-temperature electrolysis may rely for a considerable time on SOFC technology, before a specific SOEC technology is developed.

SOECs can be operated with and without external heat supply; even without such heat supply the energy conversion efficiency is considerably above that of low-temperature electrolyzers. Reaching 90% energetic efficiency would open up new opportunities for hydrogen produced by electricity with low or no CO_2 emissions, in comparison with classical fossil fuel production means. This renders SOEC technology attractive for use in the context of the storage of renewable energy.

The main bottleneck at present for industrial application of SOECs is the cell degradation, which is generally faster than that observed in the fuel cell mode. Therefore, research efforts will be required on this issue, dealing with the identification of degradation features and their suppression via (i) the choice of appropriate cell types or (ii) cell modifications, or (iii) optimization of the operation conditions.

References

1 International Partnership for Hydrogen Economy (IPHE), Implementation Liaison Committee (2005) *Scoping Papers*, www.iphe.net/docs/Scoping_Papers/Combined_Scoping_Papers_pdf (accessed 15 February 2010).
2 Erdle, E., Dönitz, W., Schamm, R., and Koch, A. (1992) *Int. J. Hydrogen Energy*, 17, 817.
3 Dietrich, G. and Schafer, W. (1984) *Int. J. Hydrogen Energy*, 9, 747–752.
4 Dönitz, W. and Erdle, E. (1985) *Int. J. Hydrogen Energy*, 10, 291.
5 Quandt, K.H. and Streicher, R. (1986) *Int. J. Hydrogen Energy*, 11, 309.
6 Dönitz, W. and Erdle, E. (1985) *Int. J. Hydrogen Energy*, 10, 291.
7 Dönitz, W., Schmidberger, R., Steinheil, E., and Streicher, R. (1980) *Int. J. Hydrogen Energy*, 5, 55.
8 Isenberg A.O. (1981) *Solid State Ionics*, 3/4, 431.
9 Maskalick, N.J. (1986) *Int. J. Hydrogen Energy*, 11, 563.
10 European Institute For Energy Research (EIFER) (2010) Hi_2H_2 *Project ("Highly Efficient, High Temperature, Hydrogen Production by Water Electrolysis")*, within the European Framework Program 6, the Deutsches Zentrum für Luft- und Raumfahrt (DLR), the Danish Technical University (DTU) with the Risoe National Laboratory (DTU-Risoe), and the Swiss Federal Laboratories for Materials Testing and Research (EMPA), http://hi2h2.com (accessed 15 February 2010)
11 Mogensen, M., Jensen, S.H., Hauch, A., Chorkendorff, I., and Jacobsen, T. (2006) *Proceedings of 7th Lucerne Fuel Cell Forum* (ed. U. Bossel), 3–7 July 2006, Lucerne, P0301.
12 Dönitz, W. et al. (1988) *Int. J. Hydrogen Energy*, 13, 283.
13 Pham, A.Q., Wallman, H., and Glass, R.S., US patent no 6051125 (April 2000).
14 Martinez-Frias, J., Pham, A.Q., Aceve, S.M. (2003) *Int. J. Hydrogen Energy*, 28, 483.
15 Stoots, C., O'Brien, J.E., Hawkes, G.L., Herring, J.S., and Hartvigsen, J.J. (2006) presented at the *Workshop on High Temperature Electrolysis at the Risoe National Laboratory*, Roskilde, Denmark, 18–19 September 2006.
16 Herring, J.S., O'Brien, J.E., Stoots, C.M., Lessing, P., Hartvigsen, J., and Elangovan, S. (2003) *Second Information*

Exchange Meeting on Nuclear Production of Hydrogen, Argonne National Laboratory, Argonne, Illinois.

17 Herring, J. S., O'Brien, J. E., Stoots, C. M., Hawkes, G., McKellar, M., Sohal, M., Harvego, E., DeWall, K., and Hall, D. (2007) *2007 DOE Hydrogen, Fuel Cells & Infrastructure Technologies Program Review*, Washington D.C.

18 Herring, J. S., O'Brien, J. E., Stoots, C. M., Hawkes, G. L. (2007) *Int. J. Hydrogen Energy*, **32**, 4641.

19 Herring, J. S., O'Brien, J. E.,Stoots, C. M., DeWall, K., McKellar, G. M., Harvego, E., Sohal, M., Hawkes, G. L., Jones R. (2006) Annual Progress Report, http://192.174.58.43/pdfs/progress06/ii_g_1_herring.pdf.

20 Hino, R., Haga, K., Aita, H., and Sekita, K. (2004) *Nuclear Engineering and Design*, **233**, 363.

21 Shin, Y., Park, W., Chang, J., and Park, J. (2007) *Int. J. Hydrogen Energy*, **32**, 1486.

22 Hauch, A., Jensen, S. H., Ramousse, S., and Mogensen, M. (2006) *J. Electrochem. Soc.*, **153**, A1741.

23 Herring, S., O'Brien, J. E., Stoots, C., Hawkes, G., Kellar, M. M., Sohal, M., Harvego, E., DeWall, K., Hall, D., Hartvigsen, J. J., Elangovan, S., Larsen, D., Petri, M., Myers, D., Yildiz, B., Carter, D., Docto, r R., and Bischoff, B. (2006) *2006 DOE Hydrogen, Fuel Cells & Infrastructure Technologies Program Review*, Washington, DC, 17 May 2006.

24 Herring, J. S., O'Brien, J. E., Stoots, C. M., Hawkes, L., Hartvigsen, J. J., and Shahnam, M. (2007) *Int. J. Hydrogen Energy*, **32**, 440.

25 Schefold, J., Garcia, M. J., Brisse, A., Perednis, D., and Zahid, M. (2008) *Proceedings of 8th Lucerne Fuel Cell Forum* (ed. U. Bossel), 30 June–5 July 2008, Lucerne, A1101.

26 Brisse, A., Schefold, J., and Zahid, M. (2008) *Int. J. Hydrogen Energy*, **33**, 5375.

27 Schefold, J., Brisse, A. and Zahid, M. (2009) *J. Electrochem. Soc.*, **156**, 897.

28 Mogensen, M. and Jacobsen, T. (2009) *ECS Trans.*, **25**, 131.

29 Ceramatec Inc. (2010) http://www.ceramatec.com/ (accessed 15 February 2010)

30 Stoots, C. M., O'Brien, J. E., Condie, K., Moore-McAteer, L., Housley, G. K., Hartvigsen, J. J., and Herring, J. S. (2009) *Nucl. Technol.* **166**, 32.

31 Stoots, C. M., Condie, K. G., O'Brien, J. E., Herring, J. S., and Hartvigsen, J. J. (2009) *Proceeding of the International Conference on Nuclear Energy*, Brussels, ICONE17-75417.

32 HTceramix SA, Yverdons, Switzerland (2010) http://www.htceramix.ch/ (accessed 15 February 2010).

33 InDEC (2010) http://www.hcstarck.com/ (accessed 15 February 2010).

12
Alkaline Electrolysis – Introduction and Overview
Detlef Stolten and Dennis Krieg

Abstract

This review describes the principles and history of electrolysis and elucidates the reasons for the development of certain varieties of electrolyzers, namely alkaline electrolyzers, solid polymer electrolyzers, high-temperature electrolyzers, and high-pressure electrolyzers, by outlining the physical and electrochemical basics applying to electrolysis. Materials, design, and operating conditions for alkaline electrolysis are described in further detail, concluding with a brief discussion of the status of electrolysis today.

Keywords: alkaline electrolysis, thermodynamics, kinetics, high-pressure electrolysis, high-temperature electrolysis, electrolyzers

12.1
Introduction

Electrolysis has always been important to produce pure hydrogen for chemical processes. Rising awareness of limited resources, including the report *Limits to Growth* [1] by the Club of Rome in 1972 and the first oil crisis in 1973, made hydrogen emerge as a fuel for the future, potentially replacing oil. In the meantime, cheap oil resulting from the abundance of oil shifted the focus away from changes in the energy system. Today, as societies are focusing increasingly on renewable and clean energy for environmental reasons, electrolysis is becoming more important again for energy use since it paves the way for effective long-term energy storage from renewable sources and for compensation of intermittent energy generation from renewables. In terms of application, hydrogen is widely considered a viable energy carrier for future energy technologies, particularly in the transport sector [2, 3, 4]. Even quantitatively electrolysis still has great potential since worldwide just about 4% of the hydrogen is produced via electrolysis, 18% originates from coal, and 77% from natural gas and oil [5]. Today, water electrolysis is applied chiefly for the production of pure hydrogen to be used in chemical processing. Hence most commercially available electro-

Hydrogen Energy. Edited by Detlef Stolten
Copyright © 2010 WILEY-VCH Verlag GmbH & Co. KGaA, Weinheim
ISBN: 978-3-527-32711-9

lyzers are tailored to these requirements today. This chapter is intended to provide an overview of the history of electrolysis, the electrochemical basics, and the technology of alkaline electrolysis in particular.

12.2
Definition

The word electrolysis was created in 1834 and describes "the producing of chemical changes by passage of an electric current through an electrolyte" [6]. The reactions are the reverse of those occurring in a galvanic element, hence they need the input of energy.

12.3
The Principle

The basic chemical equation of water electrolysis is

$$H_2O_{(l)} \rightarrow H_{2(g)} + 1/2\, O_{2(g)} \tag{12.1}$$

with $\Delta H_R = 286.0$ kJ mol^{-1}, translating into $E = 1.48$ V (cf. Equation 12.4).

The reaction enthalpy and the resulting voltage, called the heating voltage or enthalpy voltage, are given for standard conditions, namely 25 °C and 1 bar pressure, under equilibrium. Since the reaction is endothermic, it requires energy input. Because of the poor conductivity of pure water, dilute alkaline or acidic solutions are used as an electrolyte. Owing to the kinetics of the reaction, overpotentials occur when the reaction sets in. Therefore, for technical electrolysis, a voltage substantially above the heating voltage or enthalpy voltage of 1.48 V at equilibrium is needed. The fact that not all of the enthalpy of the reaction needs to be provided by electricity can be exploited in electrolysis. This is described in detail below.

Today, there are three basic lines of technology. First, there is alkaline electrolysis, which is a mature technology for hydrogen production for use in chemical processes up to megawatt capacities. Applying this process to hydrogen production for energy use requires further cost reduction of the process, increased efficiency, and the capability for dynamic operation. Second, polymer membrane electrolysis utilizing sulfonated, perfluorinated membranes is being intensively investigated. These acidic systems contain acid groups which are attached to the polymer via covalent bonding of a side chain, hence the acid is immobilized. The technology is particularly investigated for smaller electrolyzers up to the 100 kW range. Third, other than those electrolytes which are operated at about 80 °C, there are ceramic electrolytes to be operated at 700 °C and above. The prevalent electrolyte material is yttria-doped zirconia with ceramic electrodes. This technology provides great potential in efficiency and a simple systems design.

Higher electrical efficiency can be achieved owing to a thermodynamic advantage of the reaction at higher temperature and because of better reaction kinetics. Most materials investigated today are being derived from the solid oxide fuel cell (SOFC). Despite its great potential, the technology is still in the laboratory stage owing to the brittleness of its components and the resulting difficulties in making larger cell stacks.

This overview focuses on the principles of electrolysis and alkaline electrolyzers. In alkaline electrolysis, three components provide the basic function for the process in conjunction with the electrolyte: the two electrodes and the separator or diaphragm (Figure 12.1).

The cathode provides the electrons for the decomposition of water into gaseous hydrogen; the hydroxide ions travel through the separator towards the anode, closing the electric circle. At the anode, the hydroxide ions decompose into gaseous oxygen, water, and electrons. The two excess water molecules travel back through the separator to the cathode. By adding up the two electrode reactions, the overall decomposition reaction of water is obtained:

$$4e^- + 4H_2O \rightarrow 2H_2\uparrow + 4OH^- \quad (12.2)$$
$$\underline{4OH^- \rightarrow O_2\uparrow + 2H_2O + 4e^-} \quad (12.3)$$
$$2H_2O \rightarrow 2H_2\uparrow + O_2\uparrow \quad (12.4)$$

The separator has the function of keeping the gases produced apart from one another for the sake of efficiency and safety. In addition, it needs to let the hydroxide ions and water molecules pass through it. In technical systems, the

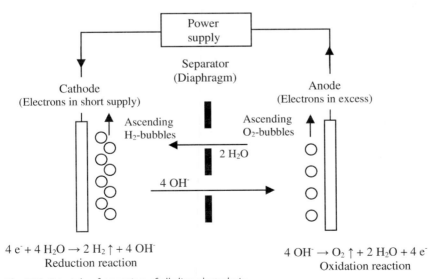

Fig. 12.1 Principle of operation of alkaline electrolysis.

electrolyte circulates through the cells in order to prevent enrichment of KOH at the anode, which would occur otherwise; cf. Figure 12.1 and Equation 12.3.

Diffusion of oxygen into the cathode chamber reduces the efficiency of the electrolysis since it will be catalyzed back to water with the hydrogen present. Extensive mixing, particularly hydrogen diffusion to the anode chamber, must be prevented for safety reasons. For their basic function, however, electrolytic cells would not require the electrodes to be in separate compartments, other than fuel cells.

Depending on the electrolyte, different ions are formed, involving different electrode reactions. The case in point refers to hydroxide ions being utilized in alkaline electrolysis. Acid systems use protons (H^+), and ceramic high temperature electrolysis works with oxygen ions (O^{2-}). Materials investigations of proton-conducting ceramics at medium temperatures around 500 °C are reported.

12.4
History

The phenomenon of electrolysis was first discovered in 1789 by the Dutch merchant Peats van Troostwijk and the Dutch medical doctor Johan Rudolph Diemann using an electrostatic generator [7, 8]. The discovery received scant attention since it was a decade before the emergence of electrochemical science. In March 1800, Alessandro Volta provided with an "electric pile," actually a battery, an inexpensive source of nearly constant voltage with which electrochemistry took off [9]. From 1820 on, Michael Faraday worked scientifically on electrolysis and in 1834 he used the term electrolysis for the first time [10, 11].

In 1900, the first industrial electrolyzer based on a filter-press design was presented by O. Schmidt in Zurich [10]. This was the same year in which Nernst developed the Nernst mass [12] – a solid electrolyte for high-temperature application – consisting of zirconia doped with 15% yttria {15 mass% amounts to 9.1 mol% of Y_2O_3; hence it is fully cubic stabilized zirconia (cf. Phase Diagrams [15], p. 104, Fig. Zr-150) at about the highest point of conductivity of yttria-doped zirconia (cf., Stevens [16]). Give or take some percentage in yttria doping, this is still the predominant electrolyte for use in SOFCs and the workhorse in most solid oxide electrolysis investigations today.

In 1902, there were already 400 electrolysis units in operation [13] and in the 1920s and 1930s electrolysis boomed along with the newly created and leaping demand for ammonia as a fertilizer precursor. Plants exceeding 100 MW of installed power were built, mainly using hydro-power, in Norway and Canada [14].

In 1939, a production rate of 10 000 $Nm^3 h^{-1}$ for a single electrolyzer was achieved for the first time [13]. At that time, many design patents were filed addressing the issues of easy maintenance [15], cost [16] and reduction in size [17]. The issue of bubble formation at the electrodes and their efficient removal was one of the electrochemical focus points of patents at that time. Increasing the operating temperature to 120 °C was proposed much later in the 1950s because of harsher corrosion conditions and the need to develop adequate materials [18].

Conversely, high-pressure operation of electrolyzers was invented and patented for electrolysis at a very early stage by Jacob Emil Noeggenrath. He held a German patent on an "Elektrolyseur, insbesondere Druckelektrolyseur" [electrolyzer, particularly pressure electrolyzer] from 1924 on [19, 20, 21], from which he generated a notable patent family across Switzerland, Austria, Great Britain and the United States [22]. These electrolyzers were already designed for pressures up to 100 atm. Later, Ewald Arno Zdansky turned the high-pressure electrolyzer, strongly supported by patents, into an industrial design at Lonza, Switzerland, in 1948 [13]. It was Lurgi, in Germany, taking up Lonza's technology, however, who first tuned a high-pressure electrolyzer at 30 bar into a product in 1951 [23]. At this pressure level, GHW, in Germany, built pressure electrolyzers until recently [13]. The technology is now offered by StatoilHydro after a reshuffle of their structure and their stake in GHW.

Most developments in those days relied on designs derived from filter presses and were named accordingly. Predominantly, electrolyzers were and still are operated today with an alkaline electrolyte consisting of a caustic potash solution at a level of 20–30 wt% KOH. A notable invention in catalysis was a fine-grained nickel catalyst by Murray Raney. He achieved a high surface area by "fusing together metallic nickel and metallic silicon" and leaching the silicon out of the crushed powder later with caustic soda solution. This was patented in 1925 [24]. A subsequent patent 2 years later described the use of aluminum instead of silicon [25]; this is the base catalyst which is still in use today for alkaline electrolyzers. For economic reasons, it was suggested later to substitute alumina by zinc [26]. Winsel and Justi filed a particular patent application for Raney nickel in 1954 and it was granted in 1957 for use in electrodes of alkaline electrolyzers [27]. The electrode design embraced Raney nickel in a matrix of another metal, which was chosen to improve the electrical conductivity and mechanical stability of the electrode structure. Thereby, lower operating temperatures of about 80 °C were achieved owing to lower overvoltages taking advantage of the Raney nickel catalyst. This development compares with earlier electrodes consisting of nickel metal sheets or later nickel plated metal, so-called nonactivated electrodes without a fine-grained catalyst. In 1967, R. L. Costa and P. G. Grimes introduced the "zero gap" electrode design [28], aiming to reduce the cell resistance by reducing the distance between the two electrodes, which were now fully attached to the separator.

A whole new chapter of electrolysis was opened when the Gemini program (1962–1966) first and then subsequently the Apollo program utilized polymer electrolytes [29, 30]. First sulfonated polystyrene was used as an electrolyte; subsequently, Nafion, a sulfonated perfluorinated material developed by DuPont [31], and its derivatives have prevailed as polymer electrolytes until today.

The attractive features of this solid polymer electrolyte technology include the following. First, only pure water, with no added electrolyte, is pumped to the cell. Second, the electrolyte is only 250 µm thick (or even less). The ohmic loss of the total cell voltage is just about 200 mV at substantial current densities, namely 1000 mA cm^{-2} [32]. Moreover, the membrane is considered an effective barrier against mixing of the product gas [30]. The requirement to use noble metals as catalysts and the compromised longevity are disadvantages compared with alkaline electrolyzers [33]. Finally, in 1987, the first 100 kW PEM electrolyzer was delivered by BBC, who were subsequently succeeded by ABB [10]. Today, critical R&D issues are related to longevity and performance, indirectly involving efficiency. Cell materials such as electrodes and interconnect materials are being investigated [34].

Since electric power is expensive, electrolysis with higher efficiency and electrical efficiency in particular was sought. High-temperature electrolysis mainly based on yttria-stabilized zirconia has been investigated since the 1960s [35]. For the sake of thermodynamic and kinetic advantages, a higher electrical efficiency can be achieved alongside higher current densities [36]. When nuclear power was at its prime in the 1970s, electrolysis was fostered, assuming that electric power would be abundantly available. The heat demand of high-temperature electrolysis fits well in the temperature level of high-temperature nuclear reactors that produce at about 1000 °C. Early projects were pursued by Dornier in the 1970s in Germany and terminated in the mid-1980s when nuclear power in Germany proved not to be socially acceptable. Owing to the brittleness of the ceramic components, larger high-temperature electrolysis units were never built [37]. High-temperature electrolysis, or in more recent terms solid oxide electrolysis (SOEC), is pursued worldwide today [38, 39, 40]. Figure 12.2

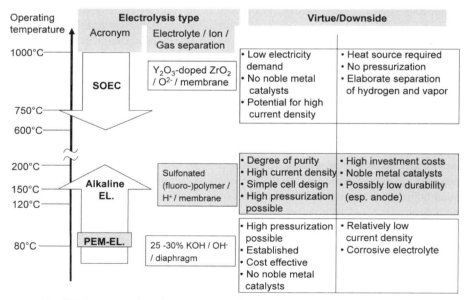

Fig. 12.2 Important electrolysis types at a glance.

summarizes the basic features of the different types of electrolysis by displaying them on a temperature scale and providing information about the electrolyte. Most properties of the specific electrolysis processes originate directly from these two parameters.

12.5
Basics of Electrolysis

12.5.1
Some Thermodynamics

Faraday's law allows us to convert the enthalpy of a reaction into the resulting voltage:

$$E = -\frac{\Delta H}{nF} \tag{12.5}$$

where E is the potential = voltage, ΔH the reaction enthalpy, n the number of electrons involved in the reaction, and F the Faraday constant, $96\,485$ C mol^{-1}.

In Equation 12.1, the enthalpy of liquid water, the higher heating value, is used taking into account the energy needed to split the water and the evaporation enthalpy ΔH_v. The latter can in principle be taken up from the environment as heat. Then, 1.25 V is the resulting enthalpy voltage, referring to the lower heating value. When exploiting this in high-temperature electrolysis with steam, naturally the lower heating value applies and heat can be taken up for from the environment to cover the entropy term, the difference between ΔG^0 and ΔH^0, cf. Equation 12.6. The standard electrolysis voltage for high temperature electrolysis is 1.19 V, assuming that the energy of the entropy term is completely covered by heat.

$$\Delta G^0 = \Delta H^0 - T\Delta S^0 \tag{12.6}$$

where ΔG^0 is the Gibbs free enthalpy, T the absolute temperature, and ΔS^0 the entropy, all at standard conditions.

All of these voltages were as usual taken for sake of comparison at standard conditions and they apply for 25 °C. Regrettably, in most publications the term "standard" is omitted for the electrolysis voltage, giving rise to confusion with the voltage to be calculated using the Nernst equation for real conditions, cf. Equation 12.9. Table 12.1 compares the four possible cases for standard conditions.

The advantage of using heat lies in the lower economic value of heat compared with electricity in most cases. This holds true to a certain extent for overheated steam and even more so for low-temperature heat. Figure 12.3 considers the temperature dependence of the energy and voltage for water electrolysis.

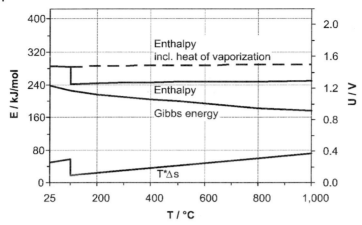

Fig. 12.3 Temperature dependence of the thermodynamic data. Calculated after data from [41].

At 1000 °C, the total energy necessary for electrolysis, ΔH, is slightly higher than at room temperature amounting to 293.3 kJ mol^{-1} or 1.52 V. In contrast to most diagrams in the literature (cf. [43], p. 40), in this diagram ΔH_v is not taken for granted. Hence there is not the typical drop in energy requirement for electrolysis above 100 °C; this is marked by the dashed line. The important thermodynamic driver for high-temperature electrolysis is that ΔG decreases with temperature, providing an additional opportunity for use of thermal energy. At 1000 °C, the necessary electrical energy is just 0.91 V instead of the 1.19 V referred to above (cf. Table 12.1). Hence thermodynamically 40.1% of the energy [(1.52 V–0.91 V)/1.52 V] can be supplied as heat at 1000 °C.

The efficiency of electrolyzer stacks is defined by

$$\eta_{\text{el, LHV}} = \frac{\text{output of hydrogen energy}}{\text{input of electrical energy}} = \frac{\Delta H^0_{298}}{nFU}$$

$$= \frac{\Delta H^0_{298,\,\text{LHV}}(\text{kW h Nm}^{-3})}{\text{input of electrical energy (kW h Nm}^{-3})} \quad (12.7)$$

this equation refers to the lower heating value (LHV) of hydrogen as the chemical energy produced, since in most hydrogen applications later the condensation enthalpy cannot be made use of. The LHV is 3.0 kW h Nm^{-3} and the higher heating value (HHV) is 3.54 kW h Nm^{-3} at standard conditions.

$$\eta = \frac{\text{heating voltage}}{\text{cell voltage}} = \frac{E^0_H}{U} = \frac{1.48\,(\text{V})}{U\,(\text{V})} \quad (12.8)$$

at 298 K and 1 bar.

Table 12.1 Thermodynamic data for the four theoretically possible operating modes. As usual and for comparison, all data are given at standard conditions, namely 25 °C, even the section which technically applies for high-temperature electrolysis, cf. Figure 12.3.

	ΔH^0 (kJ mol^{-1})	E_H^0 (V)	ΔG^0 (kJ mol^{-1})	E^0 (V)
HHV Liquid water	285.6 All electric energy input Base case for electrolysis	1.48	237.1 $T\Delta S$ via heat ΔH_v via electricity (liquid media) Water splitting via electricity Low-temperature electrolyzer takes up its own waste heat and/or heat provided by the environment	1.23
LHV Steam	241.8 $T\Delta S$ via electricity ΔH_v via heat (steam provided) Water splitting via electricity	1.25	228.6 $T\Delta S$ via heat ΔH_v via heat (steam provided) Water splitting via electricity High-temperature electrolysis	1.19

In Equation 12.7, ancillary losses can be considered if the electrical input is defined as the electrical energy input to the system. Equation 12.8 just considers the stack efficiency, neglecting ancillary losses. Here the HHV of the water decomposition reaction was taken, which again is a more stringent view in terms of efficiency as shown in Table 12.1.

As the evaporation energy of the liquid water during electrolysis is considered, then the higher energy consumption of the electrolysis of liquid water is properly addressed.

The Nernst equation describes the electrolysis voltage at equilibrium, that is, at zero current, for conditions deviating from standard conditions for temperature and pressure, 298 K and 1 bar, respectively. Other than in fuel cells, the gases can be assumed to be pure hydrogen and pure oxygen, when humidification is neglected. The Nernst voltage E_N is the voltage at which the system is in equilibrium and from which on electrolysis sets with a further voltage rise. It can ideally be measured as the so-called open-circuit voltage as is done for fuel cells.

$$E_N = E^0 - \frac{RT}{2F} \cdot \ln \prod_i a_i^{v_i} \qquad (12.9)$$

with

$$E^0 = -\frac{\Delta G^0}{2F} \qquad (12.10)$$

Table 12.2 Electrolysis voltage depending on temperature and pressure; for 80 °C $\Delta G_{80°C} = 224.9$ kJ mol^{-1} is assumed by linear interpolation of ΔG values for 300 and 400 K from the *Handbook of Physics and Chemistry* [42].

p (bar)	Electrolysis voltage (V)		
	ΔV (mV)	At 25 °C	At 80 °C
1	0	1.229	1.165
10	44	1.273	1.209
30	65	1.294	1.230
100	88	1.317	1.253

$$E_N = E^0 - \frac{RT}{2F} \ln \frac{(a_{H_2O})^2}{\left(\frac{p_{H_2}}{p_0}\right)\left(\frac{p_{O_2}}{p_0}\right)^{1/2}} \tag{12.11}$$

The activity of liquids is unit and the activity of gases can be simplified as the partial pressure divided by unit pressure, using the common assumption of ideal gases, that is, not too high pressures. The gases are assumed to be pure, rendering the partial pressure identical with the absolute pressure. Table 12.2 provides the Nernst voltages for pressures up to 100 bar at 25 °C as the standard temperature for comparison and at 80 °C as the prevalent operating temperature.

As can be seen from Table 12.2, high-pressure electrolysis requires just minor additional energy thermodynamically. In turn, it provides a kinetic advantage in the cell and no or less compression energy is needed later to compress the hydrogen for storage.

12.5.2
Some Kinetic Aspects

Since thermodynamics is just valid under equilibrium conditions at these voltages, no hydrogen is produced. Once a current is forced through the cell, kinetic losses set in. There are nonlinear losses at each electrode that increase with the current provided plus an ohmic loss due to the internal resistance of the materials, mainly resulting from the ionic transfer in the electrolyte. In Figure 12.4, these losses are superposed.

Overpotentials at the electrodes arise when the electronic transfer occurs between the ionic component and the electrode surface at the catalyst. The slowest transfer process in a reaction sequence is rate determining and can be described by the Butler-Vollmer equation [44, 45]. The overpotential decreases when the exchange current density is higher, that is, the catalyst is better or the temperature is higher, and when the surface area of the catalyst is higher. In addition to these electrochemical effects, there is an impact of the gas bubbles produced at

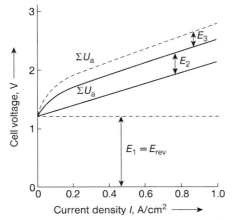

Fig. 12.4 Overpotential of conventional alkaline electrolysis with 30% KOH at 90°C [43]. E_1 is the reversible cell voltage, here given as 1.23 V, and E_2 and E_3 are the anodic and cathodic overpotentials, respectively.

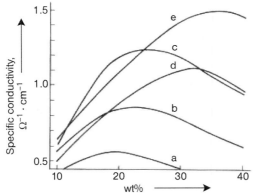

Fig. 12.5 Conductivity of electrolytes. (a) NaOH, 40°C; (b) NaOH, 60°C; (c) NaOH, 80°C; (d) KOH, 60°C; (e) KOH, 80°C [43].

the electrodes. They tend to block the interface and lead to an effect which is similar to the concentration polarization in fuel cells. Hence it is important to make the bubbles detach quickly and easily from the electrode surface via design and choice of materials.

To a great extent, the ohmic overpotential is due to the resistance of the electrolyte. Potassium hydroxide exhibits a much better conductivity than sodium hydroxide at the same concentration and temperature of the aqueous solution. Figure 12.5 shows the conductivity of aqueous solutions of sodium and potassium hydroxide in wt%.

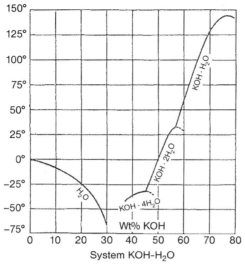

Fig. 12.6 Phase diagram of potassium hydroxide and water [49]. A first phase diagram was published in 1893 [50]; a revised diagram from a KOH supplier based on [49] can be found on the Internet [51].

KOH at a concentration of 30 wt% is the usual electrolyte in alkaline electrolysis today owing to its high conductivity. Increasing the temperature and decreasing the distance between the electrodes further reduce the ohmic resistance. Operating temperatures up to 120 °C and KOH concentrations of up to 40–47 wt% are reported as a concentration maximum [46], since above that concentration KOH will precipitate from the solution at room temperature, when the device is shut down. Additionally, the conductivity decreases with increase in concentration, even at 120 °C. NaOH, which has also been proposed [47], has not been established as an electrolyte (cf. Table 125).

Potassium hydroxide is readily soluble over a wide range of concentrations even at low temperatures, and at 30 wt% even well below −30 °C without a phase change, making start-up and shut-down procedures easy (see Figure 12.6). The boiling point is 120 and 141 °C for 35 and 50 wt% aqueous KOH solution, respectively [48].

12.6
Technical Alkaline Concepts

The alkaline electrolyzer concepts will be discussed in the main categories Design, Materials, and Operating Conditions. Figure 12.7 provides a systematic overview over the topics to be discussed.

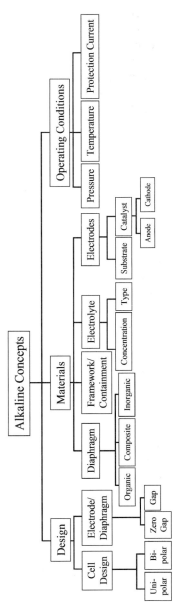

Fig. 12.7 Main variables of electrolyzers. For a further breakdown of operating conditions, see Figure 12.11.

12.6.1
Design

An easy and therefore reliable way to design electrolysis cells is the unipolar design, in which each cell consists of the electrodes, the diaphragm, and containment. The cells are connected with each other outside the containment and the current is collected outside each cell, contributing to conduction losses within the cell. Bell-type cells do not necessarily need a diaphragm for their particular gas collector [43]. The design is simple and provides easy maintenance, even during operation [14].

Bipolar cells prevail today. They are characterized by having a metallic separator between two cells which serves as the cathode in one cell and as the anode in the next cell. Since the design resembles that of a filter press, it is often referred to as the filter press design, particularly in the earlier literature. Advantageously, the bipolar design is more compact and lightweight. Moreover, less wiring is needed as the current passes through the cells and the voltage adds up over the cells. The containment is formed by the cell frames that are stacked together. Bipolar electrolyzers can be operated under pressure [23]. In contrast to unipolar cells. shunt currents occur during operation [52] and contribute to losses. Improved designs minimize these losses. Figure 12.8 shows the principle of unipolar and bipolar electrolysis.

Earlier designs exhibit two separate electrodes and a separator or diaphragm in between. Later, it was suggested for the first time to have one composite component in order to reduce the notable resistance between the two electrodes [28]; subsequently this was called zero gap geometry or zero gap configuration. In addition to the electrolytic resistance between the electrodes (cf. Figure 12.5), gas bubbles forming between the electrodes raise the ohmic resistance further by virtually cutting out part of the sectional area between the electrodes [53]. The zero gap configuration requires electronically highly insulating diaphragms otherwise short-circuiting occurs. To prevent that and to have a wider choice of materials, a design with a micro-gap was suggested [46, 54].

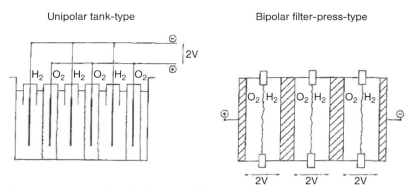

Fig. 12.8 Unipolar and bipolar designs of electrolytic cells [23].

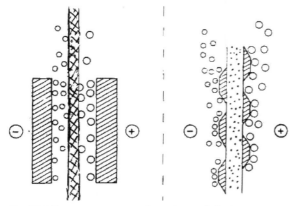

Fig. 12.9 Zero-gap geometry of an advanced diaphragm electrode compound in comparison with a conventional geometry [57].

Moreover, gases increase in amount when bubbling up along the electrochemically active electrode area involving an inhomogeneous current density over the electrolyte if it occurs between the electrodes. The electrolytic resistance between the electrodes increases from bottom to top and the effective local current density also increases, since the electrode is an equipotential surface, involving additional overpotentials [55, 56]. Hence zero gap or micro-gap designs prevail today (Figure 12.9).

Perforated metal sheets are pressed on to the diaphragm when making a zero gap electrode. Strictly, the electrochemically active area is at the fringes of the small holes formed by perforation at the three-phase boundary where ionic conduction, electronic conduction, and gas formation are assured. Hence small holes are required for a highly active electrode. Too small holes, however, will not allow gas bubbles to grow sufficiently to detach from the surface easily by gravity [58]. Trading off these two effects and considering the average diameter of the bubbles of 500 µm for oxygen and 20–30 µm for hydrogen, holes of 1.2 mm at the anode and 0.3 mm at the cathode have been suggested [57, 59].

12.6.2
Materials

Materials in alkaline electrolyzers need to be highly corrosion resistant; 30 wt% KOH brine has evolved as a standard of today. The operating temperature is mostly in the region of 80 °C (cf. Table 12.5). For these conditions, nickel, nickel alloys, asbestos, some oxide ceramics, and the highly aromatic, thermoplastic material polysulfone (Figure 12.10) provide sufficient corrosion resistance, making them the materials of choice for containment materials, electrodes, and diaphragms.

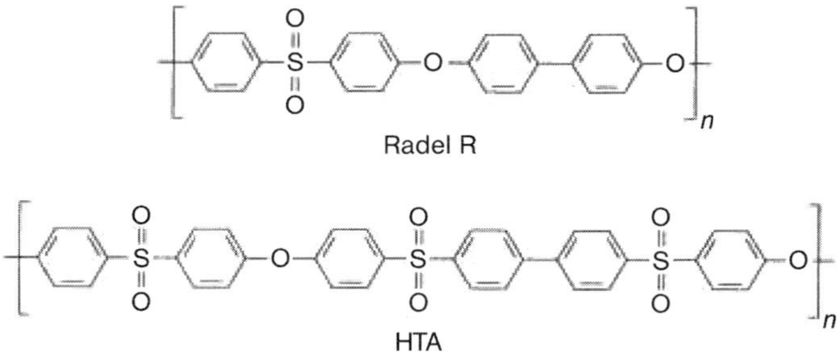

Fig. 12.10 Repeating unit of two polysulfones: Radel R from Union Carbide and Vitrex HTA from ICI [60]. Polysulfones exhibit a wide range of properties which need to be considered; some are not even thermoplastic.

The containment is generally made of nickel-plated steel. The electrodes are made of a substrate made of nickel or nickel-plated steel with a catalyst coating [46, 61]. The requirements for diaphragms are complex in that they need a high permeability for the electrolyte, good gas separating properties, which are crucial for safety reasons, mechanical stability, stability against electrochemical corrosion, plain chemical stability, and not least an acceptable cost [46, 62]. Diaphragms are important for the efficiency of the electrolyzer since the ohmic loss that the diaphragm causes translates linearly into a higher voltage for electrolysis. A material that fulfills the requirements well is asbestos, but it is no longer used since it causes lung cancer if fibers are inhaled. Asbestos is banned in more than 40 countries, including all EU countries[63], and asbestos, except for the chrysotile form, is listed as a chemical substance subject to the PIC Procedure under the Rotterdam Convention involving restrictions in trade and use [64]. Nonetheless, the world production of asbestos is still over 2 million tonnes per annum [63]. In modern electrolyzers asbestos has been replaced, mostly by composite materials such as oxide ceramics with a metal grid reinforcement [59, 65]. Table 12.3 provides an overview of the materials used.

Cathodic catalysts are mainly nickel based. Raney nickel has proved to be a good and long-term stable catalyst at 80 °C. Alternatively, electrochemical deposition of nickel can be used to form the catalyst layer [59, 72]. Nickel whiskers [73] and nickel foams [74] have also been investigated as catalysts owing to their high specific surface. Fine-grained, high-porosity nickel catalysts recrystallize at elevated operating temperatures above 120 °C within a couple of thousand hours and consequently lose their efficiency [65]. Nickel alloyed with Mo and doped with ceramics such as TiO_2 or ZrO_2 exhibits improved longevity [46, 75]. Adding 13 at.% Mo to nickel improves the catalytic activity even further and reduces corrosion during shut-down [13, 59, 76]. Applying these measures, the cathodic overpotential at 1000 mA cm^{-2} and 90 °C was reduced by 1990 to 150–200 mV for activated cathodes with catalyst compared with 350 mV for nonactivated nickel cathodes [46].

Table 12.3 Overview of diaphragm materials.

Material	Type	Temperature (°C)	Thickness (μm)	Specific resistance (Ω cm^2)	Remarks	References
Plain asbestos	Inorganic	<100	2000–5000	0.74	Hazardous	23, 29, 59, 46, 59, 65, 67
Polymer-reinforced asbestos	Composite	<100	200–500	0.15–0.2	Superior chemical resistance and mechanical stability in comparison with plain asbestos	
PTFE-bonded potassium titanate	Composite	120–150	300	0.1–0.15	Excellent stability in hot caustic environment	46, 59, 68, 69
Polymer-bonded zirconia	Composite	<160	200–500	0.25	ZrO_2 on polyphenylsulfone lattice	46, 70
Polysulfone-bonded polyantimonic acid (PAM)/polysulfone impregnated with Sb_2O_5 polyoxides	Composite	120–150	200–500	0.15–0.25	PAM is extremely stable in concentrated alkaline solutions	46, 68, 69
Sintered nickel	Inorganic	<200	200–400	0.05–0.07	Zero gap geometries impossible due to high electrical conductivity Not sufficiently resistant against oxidation and corrosive deterioration during electrolysis	46, 68, 69
Ceramic/oxide-coated nickel material	Composite	<170	25–50	0.07–0.1	Use of nickel titanate first, which was, however, not stable as it was slowly reduced at the cathodic side Chemically very stable, but expensive NiO in Ni matrix	53, 59, 61, 68, 69
Polysulfone	Organic	<125			Excessively low ionic conductivity	65, 69

Indentifying long-term-stable anode catalysts proved to be more difficult than for cathode catalysts. Efforts with anodic catalysis are worthwhile, however, since the anodic overpotential of a plain nickel surface in operation is about 400 mV, representing about 20% of the total cell voltage. A wide variety of materials have been investigated. At the end of the 1970s, $NiCo_2O_4$ and Ba_2MnReO_6 were favored. With PTFE-bonded $NiCo_2O_4$, overpotentials of 150 mV less than that of nickel were achieved, but the material proved not to be long-term stable above 100 °C. In contrast to most other materials, nickel anodes exhibit improved longevity at elevated temperatures around 100 °C. That is explained by a shift of the equilibrium between the more soluble Ni^{4+} and Ni^{3+} towards the less soluble Ni^{3+} [29]. In the 1980s, mixed oxides such as perovskites and spinels were investigated. From choices such as $LaNiO_3$, $La_{1-x}Sr_xCoO_3$, $NiCo_2O_4$, Ni_2CoO_4, and Co_3O_4, the pure cobalt spinel proved to be the most effective. Nickel anodes with catalyst loadings of 2–3 mg cm^{-2} of cobalt spinel were reported to run at an 80 mV lower overvoltage than plain nickel electrodes at 1 A cm^{-2} and 90 °C [59, 77]. In the mid-1980s, perovskites and spinels containing transition metals such as Ni, Co, Mn, and Ag were identified as catalysts owing to their effectiveness, long-term stability, and affordability [78, 79]. Noteworthy examples are Ni_2CoO_4, $La_{0.2}Sr_{0.8}CoO_3$, and Co_3O_4, the last being reported to be the most effective and long-term stable [65]. Earlier suggested lanthanum perovskites such as $LaNiO_3$ and $La_{1-x}Sr_xCoO_3$ lost out over time in favor of cobalt oxides and Raney nickel [80, 81, 82].

Although in the 1980s high current densities of up to 2 A cm^{-2} were demonstrated at low catalyst loadings of 3–6 mg cm^{-2} at the anode [57, 61], for sake of reasonable operating voltage and in turn high efficiency, current densities of 200–300 mA cm^{-2} are preferable [29].

12.6.3
Operating Conditions

The main parameters are operating temperature and operating pressure (Figure 12.11). For the pressure, there are three categories: unpressurized atmospheric electrolyzers, pressure electrolyzers at about 25–30 bar, and high-pressure electrolyzers such as those produced by Hydrogenics and StatoilHydro. The pressure is efficiently generated by the electrolysis process itself through the evolving gases in an isothermal process. The exit valves for the product gases control the pressure. Whereas the fresh water supply pump has to overcome the absolute pressure difference at a low flow rate, the electrolyte circulation pump just has to overcome the pressure drop within the system for pumping the pressurized brine. Second, the compressed gas bubbles represent a lower volume fraction at the electrodes involving less overvoltage. Hence alkaline pressure electrolyzers save substantially on gas compression losses in the system when gaseous storage applies.

There are three temperature regimes in which electrolyzers can be operated. At low temperatures of about 80–90 °C, the electrolyzer needs to be cooled since

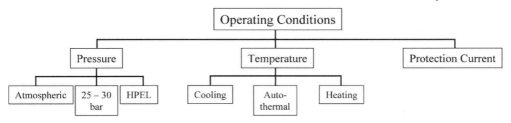

Fig. 12.11 Parameters for operating conditions.

the waste heat produced is greater than can be taken up to balance the entropy term of the reaction (cf. Figure 12.3). With increasing temperature, the entropy term increases, opening a window for further caloric energy input. At a temperature of 150 °C, alkaline electrolyzers are reported to operate autothermally, that is, without cooling or heat input. At even higher temperatures of about 200 °C, the electrolyzer works as a heat sink, requiring external heat, and therefore it operates with less electrical energy input [83].

Disadvantageously, a small current protecting the catalyst [26] is needed for bipolar electrolyzers when the electrolyzer is out of operation [84]. This was proven experimentally in the HYSOLAR Project in Germany. A 10 kW electrolyzer that was coupled with a PV panel needed 1.1 kW h over the average time of idle operation from sunset to dawn [85].

The effect occurs only in bipolar designs which allow for shunt currents, since the anode and cathode are connected via the electrolyte. The effect is negligible in operation, but important when the electrolyzer is shut down. Therefore, a more sophisticated design of the electrolyte and gas ducts and the necessity for engaging a small protection current during operational shut-down times are used in bipolar water electrolyzers [13]. The internal shunt currents discharge the electrodes. When the cathode potential repeatedly falls below a threshold value, the residual aluminum of the cathode made of Raney nickel dissolves, destroying the cathode. Below 0.5 at.% of Al in the Raney nickel, a strong incline in overvoltage is observed [86]. Molybdenum was found to decrease the cathodic overpotential as such and to alleviate leaching. However, only precious metal catalysts were found to be fully resistant. [86]. A protective current keeping the cells at a voltage of 1.3 V was demonstrated to be sufficient [87, 88].

12.6.4
Advanced Concepts

Combining the zero gap geometry, a thin diaphragm in the range 0.2–0.3 Ω cm^2, and improved catalysts resulted in a major reduction in overpotentials (Table 12.4). These advanced electrolyzers achieved voltages of 1.8–1.9 V at 1000 mA cm^{-2} or 1.6–1.7 V at 200 mA cm^{-2}, rendering about 80% and 90% efficiency based on the higher heating value, respectively. For comparison, a con-

Table 12.4 Comparison of overpotentials in conventional and advanced electrolysis dated 1986 [81].

Conventional (200 mA cm^{-2})		Advanced (400 mA cm^{-2})	
Condition	Losses (V)	Condition	Losses (V)
Temperature 80 °C	0.2	Temperature 100–200 °C	0.08
Resistance 1 Ω cm^2		Resistance 0.2 Ω cm^2	
Catalytic nonactivated cathode	0.25	Catalytic activated cathode	0.07
Catalytic nonactivated anode	0.30	Catalytic activated anode	0.20
Total	0.75	Total	0.35

Table 12.5 Design and performance data for alkaline electrolyzers from different companies. This table refers to the technology level of 1985 [23].

Property	The Electrolyser Corp., Ltd	Brown Boveri & Cie	Norsk Hydro AS	De Nora SpA	Lurgi GmbH
Cell type:	Monopolar tank	Bipolar filter press	Bipolar filter press	Bipolar filter press	Bipolar filter press
Operating pressure (psig)	Ambient	Ambient	Ambient	Ambient	450
Operating temperature (°C)	70	80	80	80	90
Electrolyte	28% KOH	25% KOH	25% KOH	29% KOH	25% KOH
Current density (A m^{-2})	1340	2000	1750	1500	2000
Cell voltage (V)	1.90	2.04	1.75 (after 1 year of operation)	1.85 (increases to 1.95 after 2 years)	1.86
Current efficiency (%)	>99.9	>99.9	>98	~98.5	98.75
Oxygen purity (%)	99.7	>99.6	99.3–99.7	99.6	99.3–99.5
Hydrogen purity (%)	99.9	>99.8	98.8–999	99.9	99.8–99.9
Power consumption (d.c.-kW h per Nm^{-3} H$_2$)	4.9	4.9	4.3	4.6	4.5

ventional electrolyzer from BBC operating at 2.0 V achieved an efficiency of 74% [59, 61].

Published data on electrolyzers from the 1980s are summarized in Table 12.5. This table is still of technological relevance although research in electrolysis has been cut back since then owing to cheap oil and the demise of nuclear energy.

Increasing the operating temperature reduces the overpotentials. Although it is the most effective way for efficiency improvement, the stability of materials

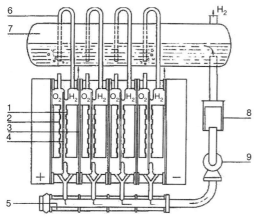

Fig. 12.12 Example of the design of an alkaline electrolyzer for ambient pressure [89]. The overflow tubes are shown for the cathode, not for the anode.

1 Cathode
2 Anode
3 Bipolar plate
4 Diaphragm
5 Electrolyte pipe
6 Overflow tube
7 Gas separator
8 Electrolyte filter
9 Electrolyte pump

limits the operating temperature to the range 120–150 °C [36, 69]. To maintain long-term service, operating temperatures above 150 °C have been abandoned. Two examples are Compagnie Electromécanique in France who wanted to achieve 200 °C and Creusot Loire who reduced the operating temperature to about 150 °C [14].

Figure 12.12 depicts an advanced design for atmospheric operation introduced by BBC in 1979. The cells are connected in series in this bipolar design. The bipolar plates generate an equipotential surface. The gases ascend through the overflow tube (6) into the gas separator (7), which is necessary since some brine is carried up with the gases. The brine flows back through an electrolyte filter (8) and the brine pump (9). Mixing the electrolyte in the gas separator prevents the electrolyte concentration from becoming enriched at the anode (see the reaction in Figure 12.1).

12.6.5
High-Pressure Electrolysis

As shown in Table 12.2, there is just a minor thermodynamic penalty on the electrolysis voltage in high-pressure electrolysis. As mentioned in Section 12.5.2 describing kinetic aspects, bubbles contribute to the overvoltage. The decrease in the size of bubbles under pressure alleviates the losses caused by the bubbles. With a view to a system comprising hydrogen generation and storage, notable energy savings can be achieved by pressure electrolysis as compression occurs in the most efficient isothermal way. Figure 12.13 shows that overall the thermodynamic and kinetic effects level off, so that there is no statistically significant shift in power demand with pressure. However, a major shift occurs with increasing current density, as to be expected (cf. Figure 12.4).

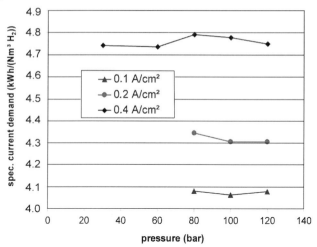

Fig. 12.13 Specific demand of electrical energy of a 5 kW high-pressure electrolyzer from the research center Jülich 90.

Particularly the first stage of compression inside the electrolyzer leads to substantial energy savings compared with the use of compressors. Owing to decreasing bubble size with increase in pressure, the effect that the bubbles increase with increase in operating temperature is compensated for [29]. Hence operating temperatures of 120–150 °C are suggested. Higher pressure prevents boiling [48], and reduces the amount of water vapor carried out with the product gases. Hence it linearly reduces the heat loss of the system [54].

It was engineering reasons that caused the slow market penetration of pressure electrolyzers [23]. Until the 1970s, Lurgi, Germany, commercialized the only pressure electrolyzer at a level of 30 bar [29]. External compression is sometimes perceived to be more easily achievable in view of systems engineering [52], but the trade-off is still contentious [33]. The main arguments put forward against high-pressure electrolysis are the more difficult control of pressure, the necessary safety provisions, and thus the higher investment costs. As discussed above, pressure electrolyzers exhibit increased catalyst corrosion under dynamic operation. The bulky design of the pressure containment poses another disadvantage. According to the boiler equation, the wall thickness d of the electrolyzer containment increases linearly with the operating pressure p and the radius r of the device. Materials with high yield strengths $\sigma_{0.2}$ allow for a leaner design. The safety factor S must also be considered.

$$d = \frac{rp}{\sigma_{0.2}} S \tag{4.12}$$

12.7
Status of Technology

The status of the technology will only be briefly described by some summarizing diagrams (Figure 12.14).

Figure 12.14a shows that electrolyzers with high production capacities attain the best efficiency, that is, the lowest specific power consumption. Achieving a consumption of 4.3 kW h Nm^{-3}, they attain 70% efficiency based on the lower heating value of hydrogen produced. This translates into the less relevant value

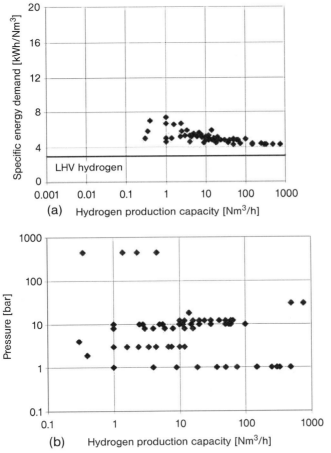

Fig. 12.14 Data collected from commercially available alkaline electrolyzers via a comprehensive Internet survey in 2008. Each electrolyzer type offered is represented by one dot. It is not considered how often or whether at all these electrolyzers were sold and/or installed. (a) Specific power consumption for hydrogen production depicted over the production capacity of the electrolyzers. (b) Operating pressure depicted over the production capacity.

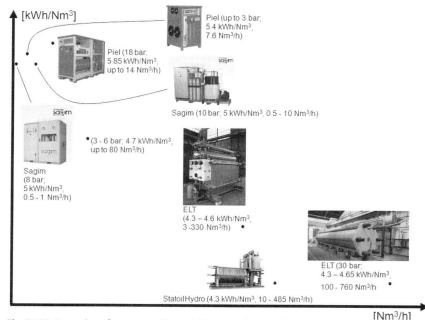

Fig. 12.15 Examples of commercially available electrolyzers. The dots near the icons represent the position of the respective technology in terms of kW h Nm^{-3} of electricity consumed over the capacity of the electrolyzer in Nm^3 h^{-1} (Nm^3 ≙ normal cubic meter).

of 82% based on the higher heating value of 3.54 kW h Nm^{-3}. Figure 12.14b shows the pressure levels of commercial alkaline electrolyzers on a logarithmic scale. Most electrolyzers are pressurized between 2 and 10 bar; just a few operate in the range of 30 bar and above.

Increasing efficiency with increase in size of the electrolyzers can chiefly be attributed to the improved efficiency of the ancillary components with increase in size. Existing commercially available alkaline electrolyzer technology (Figure 12.15) spans three orders of magnitude in capacity.

12.8
Conclusion

This chapter has described the principles and history of electrolysis and elucidated the reasons for the development of certain varieties of electrolyzers, namely alkaline electrolyzers, solid polymer electrolyzers, high-temperature electrolyzers, and high-pressure electrolyzers, by outlining the physical and electrochemical basics applying to electrolysis. Materials, design, and operating conditions for alkaline electrolysis have been described in further detail, concluding with a brief discussion of the status of electrolysis today.

Acknowledgments

The authors would like to thank Jürgen Mergel, Heinz Schmitz, Andrej Kulikovsky, Holger Janssen, and Reinhard Menzer for fruitful discussions and valuable input to this chapter. They also thank Birgit Derksen and Christina Voron for their administrative support.

References

1 Meadow, D.H., Meadows, D.L., Randers, J., and Behrens W.W. (1972) *The Limits to Growth. A Report for the Club of Rome's Project on the Predicament for Mankind*, Universe Books, New York.
2 GermanHy (2009) Studie zur Frage "Woher kommt der Wasserstoff in Deutschland bis 2050?", http://www.germanhy.de/page/fileadmin/germanhy/media/090826_germanHy_Abschlussbe richt.pdf (accessed 28 October 2009).
3 Kroposki, B., Levene, J., Harrison, K., Sen, P.K., and Novachek, F., National Renewable Energy Laboratory (2009) Electrolysis: information and opportunities for electric power utilities, http://www.nrel.gov/docs/fy06osti/40605.pdf (accessed 28 October 2009).
4 Edwards, P.P., Kuznetsov, V.L., and David, W.I.F. (2007) Hydrogen energy. *Philos. Trans. R. Soc. London A*, **365**, 1043–1056.
5 Woehrle, D. (1991) Wasserstoff als Energieträger – Eine Replik. *Nachr. Chem. Tech. Lab.*, **39** (11), 1256–1266.
6 Mish, F.C. (ed.) (2003) *Merriam Webster's Collegiate Dictionary*, 11th edn, Merriam-Webster Incorporated, Springfield, Massachuetts, USA.
7 De Levie, R. (1999) The electrolysis of water. *J. Electroanal. Chem.*, **476**, 92–93.
8 Trasatti, S. (1999) Water electrolysis: who first? *J. Electroanal. Chem.*, **476**, 90–91.
9 Trasatti, S. (1999) 1799–1999: Alessandro Volta's "electric pile": two hundred years, but it doesn't seem like it. *J. Electroanal. Chem.*, **460**, 1–4.
10 Sandstede, G. (1992) *Stand der Technik und Entwicklung der Wasserelektrolyse-Verfahren*, Dechema-Monographien, Band 125, VCH Verlagsgesellschaft mbH, Weinheim.
11 Laidler, K.J. (1997) The chemical history of a current. *Can. J. Chem.*, **75**, 1552–1565.
12 Möbius, H.-H. (1965) Die Nernst-Masse, ihre Geschichte und heutige Bedeutung. *Naturwissenschaften*, **52** (19), 529–536.
13 Kreuter, W. and Hofmann, H. (1998) Electrolysis: the important energy tranformer in a world of sustainable energy. *Int. J. Hydrogen Energy*, **23** (8), 661–666.
14 Le Roy, R.L. (1983) Industrial water electrolysis: present and future. *Int. J. Hydrogen Energy*, **8** (6), 401–417.
15 Ondik, M., and Helen, M. (eds.) (1998) *Phase diagrams for zirconium and zirconia systems*, The American Ceramic Society, Westerville, Ohio.
16 Stevens, R. (1986) Zirconia and Zirconia Ceramics, published by Magnesium Elektron Ltd., Twickenham, UK.
17 Ryman, C. (1947) Pressure electrolyzer. US Patent 2 494 264, filed 26 April 1947, Serial No. 744 168.
18 Zdansky, E.A. (1951) Water electrolyzer. US Patent 2 739 936, filed 23 February 1951.
19 Noeggerath, J.E. (1930) Elektrolyseur, insbesondere Druckelektrolyseur. German Patent 508 480, issued 11 September 1930.
20 Noeggerath, J.E. (1932) Elektrolytischer Druckzersetzer für die Gewinnung von Wasserstoff und Sauerstoff. German Patent 543 757, issued 27 October 1932.
21 Noeggerath, J.E. (1935) Elektrolytischer Druckzersetzer. German Patent 578 705, issued 1 June 1933.

22 Noeggerath, J. E. (1931) Electrolytic apparatus. US Patent 1 799 116, filed 20 December 1927.
23 Vandenborre, H. (1985) *New Developments in Alkaline Water Electrolysis*, Monographien Band 98, VCH Verlagsgesellschaft mbH, Weinheim.
24 Raney, M. (1925) Method of preparing catalytic material. US Patent 1 563 587, filed 25 September 1924.
25 Raney, M. (1927) Method of producing finely-divided nickel. US Patent 1 628 190, filed 14 May 1926.
26 Divisek, J., Malinowski, P., Mergel, J., and Schmitz, H. (1988) Improved components for advanced alkaline water electrolysis. *Int. J. Hydrogen Energy*, **13** (3), 141–150.
27 Justi, E., Scheible, W., and Winsel, A. (1954) Doppelskelett-Katalysator-Elektrode. German Patent, Application No. 1 019 361, filed 23 October 1954.
28 Costa, R. L. and Grimes, P. G. (1967) Electrolysis as a source of hydrogen and oxygen. *Chem. Eng. Prog.*, **63** (4), 56–58.
29 Lu, P. W. T. and Srinivasan, S. (1979) Advances in water electrolysis technology with emphasis on use of the solid polymer electrolyte. *J. Appl. Electrochem.*, **9**, 269–283.
30 Bushnell, S. W. and Purkis, P. M. (1984) Solid polymer electrolyte systems for electrolytic hydrogen production. *Chem. Ind. (London)*, **2**, 61–68.
31 Mauritz, K. A. and Moore, R. B. (2004) State of understanding of Nafion. *Chem. Rev.*, **104**, 4535–4585.
32 Sedlak, J. M., Lawrence, R. J., and Enos, J. F. (1981) Advances in oxygen evolution catalysis in solid polymer electrolyte water electrolysis. *Int. J. Hydrogen Energy*, **6**, 159–165.
33 Dutta, S. (1990) Technology assessment of advanced electrolytic hydrogen production. *Int. J. Hydrogen Energy*, **15** (6), 379–386.
34 Badwal, S., Giddey, S., Ciacchi, F., Clarke, R., and Kao, P. (2007) Research and developments in hydrogen technologies. *Adv. Appl. Ceram.*, **106** (1–2), 40–44.
35 Rohr, F. J. (1979) Hochtemperatur-Elektrolyse. *Chimia*, **33** (9), 343–347.
36 Dönitz, W. (1976) Verbesserungsmöglichkeiten von Elektrolyse-Verfahren im Hinblick auf kostengünstige Wasserstoff-Herstellung. *Chem.-Ing.-Tech.*, **48** (2), 159.
37 Lessing, P. A. (2007) Material for hydrogen generation via water electrolysis. *J. Mater. Sci.*, **42**, 3477–3487.
38 European Union (2008) *EU Project RELHY. Innovative Solid Oxide Electrolyser Stacks for Efficient and Reliable Hydrogen Production*, Start date: 1 January 2008.
39 Yu, B., Zhang, W., Chen, J., Xu, J., and Wang, S. (2008) Advance in highly efficient hydrogen production by high temperature steam electrolysis. *Sci. Chin. Ser. B: Chem.*, **51** (4), 289–304.
40 US Department of Energy (2009) Hydrogen Program, http://www.hydrogen.energy.gov/production.html (accessed 12 November 2009).
41 Barin, I. and Platzki, G. (1995) *Thermochemical Data of Pure Substances*, 3rd edn, Vol 1, VCH Verlagsgesellschaft mbH, Weinheim.
42 Lide, D. R. (ed.) (2009–2010) *CRC Handbook of Chemistry and Physics*, 90th edn, CRC Press, Boca Raton, FL.
43 Häussinger, P., Lohmüller, R., and Watson A. M. (2009) Hydrogen. In *Ullmann's Encyclopedia of Industrial Chemistry*, Wiley-VCH Verlag GmbH, Weinheim, Vol 7, p. 41.
44 O'M Bockris, J., Reddy, A. K. N., and Gamboa-Adeco, M. (2000) *Modern Electrochemistry. Vol. 2A. Fundamentals of Electrodics*, 2nd edn, Kluwer Academic/Plenum, New York, Chapter 7.2 f.
45 Hamann, C. H. and Vielstich, W. (1998) *Elektrochemie*, 3rd edn, Wiley-VCH Verlag GmbH, Weinheim, pp. 76 ff.
46 Wendt, H. and Imarisio, G. (1988) Nine years of research and development on advanced water electrolysis. A review of the research programme of the Commission of the European Communitites. *J. Appl. Electrochem.*, **18**, 1–14.
47 Vandenborre, H., Leysen, R., Nackaerts H., and van Asbroeck, Ph. (1984) A survey of five year intensive R&D work in Belgium on advanced alkaline water electrolysis. *Int. J. Hydrogen Energy*, **9** (4), 277–284.

48 Carbon Group (2007) *Safety Data Sheet – Potassium Hydroxide Solution*. Date of issue: 7 June 2007.
49 Pietsch, E. (ed.) (1938) Kalium. In *Gmelins Handbuch Der Anorganischen Chemie*, 8th edn, Deutsche Chemische Gesellschaft, Verlag Chemie GmbH.
50 Pickering, S.U. (1893) The hydrates of sodium, potassium, and lithium hydroxides. *J. Chem. Soc.*, **63**, 890–909.
51 Olin Corporation (2009) http://koh.olinchloralkali.com/TechnicalInformation/PhaseDiagram.aspx (accessed 12 November 2009).
52 LeRoy, R.L. and Stuart A.K. (1981) Advanced unipolar electrolysis. *Int. J. Hydrogen Energy*, **6** (6), 589–599.
53 Hofmann, H., Plzak, V., Fischer, J., Luft, G., and Wendt, H. (1982) *Elektrodenaktivierung, Diaphragmen geringen elektrischen Widerstandes und eine optimierte Zellkonfiguration für die fortgeschrittene alkalische Wasserelektrolyse*, Dechema-Monographien, Band 92, VCH Verlagsgesellschaft mbH, Weinheim.
54 Hofmann, H., Brand, R., Hildebrandt, J., and Bucher, L. (1993) *Moderne Zellen zur Wasserelektrolyse*, Dechema-Monographien Band 128, VCH Verlagsgesellschaft mbH, Weinheim.
55 Wendt, H., Hofmann, H., and Plzak, V. (1984) Anode and cathode activation, diaphragm construction and electrolyzer configuration in advanced alkaline water electrolysis. *Int. J. Hydrogen Energy*, **9** (4), 297–302.
56 Appleby, A.J., Crepy, G., and Jacquelin, J. (1978) High efficiency water electrolysis in alkaline solution. *Int. J. Hydrogen Energy*, **3**, 21–37.
57 Fischer, J., Hofmann, H., Luft, G., and Wendt, H. (1980) Fundamental investigations and electrochemical engineering aspects concerning an advanced concept for alkaline water electrolysis. *AIChE J.*, **26** (5), 794–802.
58 Kreysa, G. and Kuelps, H.-J. (1981) Experimental study of the gas bubble effects on the IR drop at inclined electrodes. *J. Electrochem. Soc.*, **128** (5), 979–984.
59 Wendt, H. (1984) Neue konstruktive und prozesstechnische Konzepte für die Wasserstoff-Gewinnung durch Elektrolyse. *Chem.-Ing.-Tech.*, **56** (4), 265–272.
60 Parker, D., Bussink, J., van de Grampel, H.T., Wheatley, G.W., Dorf, E.-U., Ostlinning, E., and Reinking, K. (2002) Polymers, high-temperature. In *Ullmann's Encyclopedia of Industrial Chemistry*, Wiley-VCH Verlag GmbH, Weinheim, pp. 14 ff.
61 Kreysa, G. (1983) Aktuelle Entwicklungslinien der elektrochemischen Prozesstechnik. *Chem.-Ing.-Tech.*, **55** (4), 267–275.
62 Ismail, M.I., White, N.P., Dhar, H.P., and Das Gupta, S. (1980) New trends in electrolytic reactor materials: diaphragms. *Polym. Plast. Technol Eng.*, **15** (1), 61–82.
63 WHO (2006) *Elimination of Asbestos-related Diseases*, http://wholib-doc.who.int/ (accessed 11 November 2009).
64 Rotterdam Convention Annex III (2009) www.pic.int see Chemicals, Annex III, Asbestos (accessed 11 November 2009).
65 Wendt, H. (1986) *Wasserstoffgewinnung durch Alkalische Wasserelektrolyse – Technik, Entwicklungslinien, ökonomische Chancen*, Dechema-Monographien, Band 106, VCH Verlagsgesellschaft mbH, Weinheim.
66 Bowen, C.T., Davis, H.J., Henshaw, B.F., Lachance, R., LeRoy, R.L., and Renaud, R. (1984) Developments in advanced alkaline water electrolysis. *Int. J. Hydrogen Energy*, **9** (1/2), 59–66.
67 Beaver, R.N. and Becker, C.W. (1975) Bonded asbestos diaphragms. US Patent 4 093 533, filed 12 December 1975 and issued 6 June 1978.
68 Imarisio, G. (1981) Progress in water electrolysis at the conclusion of the first hydrogen programme of the European Communities. *Int. J. Hydrogen Energy*, **6**, 153–158.
69 Renaud, R. and LeRoy, R.L. (1982) Separator materials for use in alkaline water electrolysers. *Int. J. Hydrogen Energy*, **7** (2), 155–166.
70 Vermeiren, Ph., Adriansens, W., and Leysen, R. (1996) Zirfon®: a new separator for Ni–H_2 batteries and alkaline fuel cells. *Int. J. Hydrogen Energy*, **21** (8), 679–684.

71 Divisek, J., Malinowski, P., Mergel, J., and Schmitz, H. (1985) Improved construction of an electrolytic cell for advanced alkaline water electrolysis. *Int. J. Hydrogen Energy*, **10** (6), 383–388.

72 Divisek, J., Schmitz, H., and Mergel, J. (1980) Neuartige Diaphragmen und Elektrodenkonstruktionen für die Wasser- und Chlorkali-Elektrolyse. *Chem.-Ing.-Tech.*, **52** (5), 465.

73 Vielstich, W. (1980) Moderne Aspekte der elektrochemischen Energiespeicherung, *Ber. Bunsen-Ges. Phys. Chem.*, **84**, 951–963.

74 Abe, I., Fujimaki, T., and Matsubara, M. (1982) Hydrogen production by high-temperature high-pressure water electrolysis, results of test plant operation. *Adv. Hydrogen Energy*, **3**, 167–178.

75 Prigent, M., Nenner, T., Martin, L., and Roux, M. (1983) The development of new electrocatalysts for advanced water electrolysis. In *Hydrogen as an Energy Carrier. Proceedings of the 3rd International Seminar, Lyon, 25–27 May 1983* (eds G. Imarisio and A.S. Strub), EUR 8651, Commission of the European Communities, Brussels, pp. 256–266.

76 Brown, D.E. and Mahmood, N.M. (1980) Metal electrodes for use in electrochemical cells and method of preparation thereof. European Patent EP 0 009 406, filed 2 April 1980 and issued 27 January 1982.

77 Wendt, H., Hofmann, H., Berg, H., Plzak, V., and Fischer J. (1983) Alkaline water electrolysis at enhanced temperatures (120 to 160 °C): basic and material studies, engineering and economics. In *Hydrogen as an Energy Carrier. Proceedings of the 3rd International Seminar, Lyon, 25–27 May 1983* (eds G. Imarisio and A.S. Strub), EUR 8651, Commission of the European Communities, Brussels, pp. 267–285.

78 Burke, L.D., Lyons, M.E., McCarthy, M. (1983) Optimization of thermally prepared ruthenium dioxide-based anodes for use in water electrolysis cells. In *Hydrogen as an Energy Carrier. Proceedings of the 3rd International Seminar, Lyon, 25–27 May 1983* (eds G. Imarisio and A.S. Strub), EUR 8651, Commission of the European Communities, Brussels, pp. 128–138.

79 Wendt, H. (1986) Ziele und Methoden der elektrochemischen Verfahrens- und Reaktionstechnik. *Chem.-Ing.-Tech.*, **58** (8), 644–654.

80 Wendt, H. and Plzak, V. (1983) Electrocatalytic and thermal activation of anodic oxygen- and cathodic hydrogen-evolution in alkaline water electrolysis. *Electrochim. Acta*, **28** (1), 27–34.

81 Divisek, J. (1986) Wasserstoffherstellung durch fortgeschrittene Elektrolysen. *BWK*, **38** (11), 512–515.

82 Divisek, J. and Schmitz, H. (1989) Ni and Mo coatings as hydrogen cathodes. *J. Appl. Electrochem.*, **19**, 519–530.

83 LeRoy, R.L. (1983) Hydrogen production by the electrolysis of water – the kinetic and thermodynamic framework. *J. Electrochem. Soc.*, **130** (11), 2158–2163.

84 Schug, C.A. (1998) Operational characteristics of high-pressure, high-efficiency water-hydrogen-electrolysis. *Int. J. Hydrogen Energy*, **23** (12), 1113–1120.

85 Brinner, A., Bussmann, H., Hug, W., and Seeger, W. (1992) Test results of the HYSOLAR 10 kW PV-electrolysis facility. *Int. J. Hydrogen Energy*, **17** (3), 187–197.

86 Divisek, J., Schmitz, H., and Steffen, B. (1994) Electrocatalyst materials for hydrogen evolution. *Electrochim. Acta*, **39** (11/12), 1725–1731.

87 Mergel, J. and Barthels, H. (1994) Auslegung, Bau und Inbetriebnahme eines 26 kW-Wasserelektrolyseurs fortgeschrittener Technik für den Solarbetrieb. Presented at the 9th Internationales Sonnenforum, 28 June–1 July 1994, Stuttgart.

88 Hug, W., Divisek, J., Mergel, J., Seeger, W., and Steeb, H. (1992) Highly efficient advanced alkaline electrolyzer for solar operation. *Int. J. Hydrogen Energy*, **17** (9), 699–705.

89 Braun, M. (1979) Wasserelektrolyse – Basis einer künftigen Wasserstoffwirtschaft. *Chimia*, **3**, 99–104.

90 Janssen, H., Emonts, B., and Stolten, D. (2008) Alkalische Hochdruck-Elektrolyse – Status am Forschungszentrum Jülich. In *NOW Workshop 2008: Regenerativer Wasserstoff aus der Elektrolyse*, WBZU/Ulm, 07 July 2008 (eds. Garche, J., Bonhoff, K.).

13
Polymer Electrolyte Membrane (PEM) Water Electrolysis

Tom Smolinka, Sebastian Rau, and Christopher Hebling

Abstract

Water electrolysis represents one of the simplest approaches to produce hydrogen and oxygen in a zero-pollution process by using electricity for the electrochemical decomposition of water. In Polymer Electrolyte Membrane (PEM) electrolysis cells, an acidic ionomer is used as the electrolyte. During the past decade, considerable progress has been made to advance this technology. Today, PEM electrolyzers can be regarded as a well-established industrial technology and are close to broader commercialization. An overview of the technical implementation of PEM water electrolysis is given in this chapter. Efficiency values and a selection of materials will be presented for a single cell and at the stack level. Lifetimes and degradation mechanisms will be considered. Finally, production rates and the power consumption for the system will be discussed.

Keywords: Polymer Electrolyte Membrane, water electrolysis, hydrogen production

13.1
Introduction

Water electrolysis is an electrochemical process applying electricity to split water into hydrogen and oxygen. The principle has been known for more than 200 years and has been applied for different military and industrial purposes for more than 100 years. In particular, alkaline water electrolyzers are used for the commercial production of hydrogen based on low-cost hydroelectricity, primarily for the manufacture of ammonia-based fertilizers. Other processes include methanol synthesis, metallurgical processes, fat hardening, and welding. However, hydrogen production by water electrolysis remained only a niche application as hydrogen could be produced at lower cost by steam reforming or as a by-product in refineries.

New interest in water electrolysis was stimulated over the last few decades by the ongoing development of renewable energy sources such as wind and solar

Hydrogen Energy. Edited by Detlef Stolten
Copyright © 2010 WILEY-VCH Verlag GmbH & Co. KGaA, Weinheim
ISBN: 978-3-527-32711-9

power and the present initial commercialization of fuel cells. Water electrolysis can be used to produce hydrogen for temporary energy storage, which helps to overcome the mismatch between the intermittent power supply from renewable energy sources and the power demand in the grid, and as a fuel for automotive or other fuel cell applications. Hence, hydrogen generation by water electrolysis is regarded as an important component of a future energy industry.

Acidic water electrolysis based on a Polymer Electrolyte Membrane (PEM) (also known as SPE® water electrolysis) is a fairly young technology compared with alkaline water electrolysis. The process, initially developed at General Electric in the 1970s, offered several advantages over alkaline electrolyzers. Nevertheless, due to the expensive materials involved, PEM electrolyzers have so far become established only in military and aerospace applications. They serve to generate oxygen for life-support systems on-board nuclear submarines and international space station, and to recharge tanks with high-pressure oxygen on-board commercial aircraft [1]. Another established market for PEM electrolyzers with small hydrogen and oxygen production capacities is laboratory equipment. Apart from these fields of operation/implementation, efforts are being made at present to develop PEM water electrolyzers suitable for fuel cell applications or as components in hydrogen storage systems coupled to renewable energy. The main features of PEM electrolysis leading to significant advantages compared with the alkaline technology are:

- higher power densities and efficiencies (and thus compactness) of the stack
- simpler system design due to the absence of a lye
- excellent partial-load range and rapid response to fluctuating power inputs
- compact stack design allowing high-pressure operation.

For these reasons, PEM electrolyzers offer the possibility of on-site and on-demand hydrogen and oxygen generation in small, highly efficient units that are suitable for autonomous and distributed operation. Nevertheless, challenges remain in terms of lowering the high cost of materials and achieving durability comparable to that of an alkaline electrolyzer.

In this chapter, the main features of PEM water electrolysis are described. Starting with the fundamental principle, the discussion continues with the latest developments in materials for the electrolysis cell and stacks, presents different system designs, and concludes with performance and endurance data.

13.2
Fundamentals of PEM Electrolysis

The general principle of water electrolysis can be expressed by the following equation:

$$H_2O_{(l/g)} + \Delta H_R \rightarrow H_{2(g)} + \tfrac{1}{2}O_{2(g)} \tag{13.1}$$

Table 13.1 Half-cell reactions, typical temperature ranges, and ionic charge carriers for different types of water electrolysis technology.

Technology	Temperature [°C]	Cathode (HER)	Charge carrier	Anode (OER)
Alkaline electrolysis	40–90	$2H_2O + 2e^- \rightarrow H_2 + 2OH^-$	OH^-	$2OH^- \rightarrow \frac{1}{2}O_2 + H_2O + 2e^-$
PEM electrolysis	20–100	$2H^+ + 2e^- \rightarrow H_2$	H^+	$H_2O \rightarrow \frac{1}{2}O_2 + 2H^+ + 2e^-$
HT steam electrolysis	700–1000	$H_2O + 2e^- \rightarrow H_2 + O^{2-}$	O^{2-}	$O^{2-} \rightarrow \frac{1}{2}O_2 + 2e^-$

For this overall reaction, three main technologies can be distinguished according to the electrolyte used in the electrolysis cell [2]: alkaline electrolysis with a liquid electrolyte, PEM electrolysis with an acidic ionomer, and high-temperature (HT) steam electrolysis with a solid oxide as the electrolyte. Table 13.1 provides an overview of the different equations for the hydrogen evolution reaction (HER) and the oxygen evolution reaction (OER), the typical temperature window, and the ions acting as charge carriers through the diaphragm/membrane. The underlying general process is the same in all three technologies: water is fed to an electrochemical cell, in which hydrogen evolves at the negative electrode (cathode) and oxygen at the positive electrode (anode) if a sufficiently high voltage is applied to the cell. Table 13.1 compares the different half-cell reactions, operating temperatures and ionic charge carriers. It is evident that PEM water electrolysis is a low-temperature process comparable to the alkaline process.

The basic design of a PEM water electrolysis cell with its main components is depicted in Figure 13.1. Two half-cells are separated by a solid acidic polymer, the PEM. In most cell designs, the electrodes are deposited directly on the membrane, creating the key component of a PEM electrolysis cell, the membrane electrode assembly (MEA). The MEA is sandwiched between porous current distributors/collectors which enable an electric current to flow from the bipolar plates to the electrodes and, simultaneously, the removal of the generated gas bubbles from the electrodes. The bipolar plates are adjacent to the two half-cells and usually include flow-field structures to enhance the transport of liquid water to the electrodes and oxygen and hydrogen out of the cell.

Equation 13.1 should indicate that the decomposition of water into hydrogen and oxygen is an endothermic reaction, as energy (ΔH_R: enthalpy of reaction, corresponding to the enthalpy of formation of water) has to be applied for this process. Under standard conditions at 25 °C and 1 atm (STP), the standard enthalpy of reaction is $\Delta H_R^\circ = 286$ kJ mol^{-1} for liquid water. According to the expression

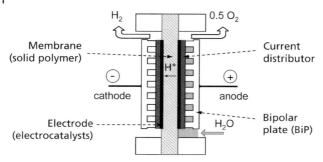

Fig. 13.1 Operating principle of a PEM electrolysis cell.

$$\Delta H_R = \Delta G_R + T\Delta S_R \tag{13.2}$$

the enthalpy of reaction consists of two parts. ΔG_R is the free energy of reaction and corresponds to the minimum share of ΔH_R which has to be applied as work, for example as electricity. The term $T\Delta S_R$ is the product of the thermodynamic temperature T and the entropy of reaction ΔS_R and represents the maximum share of ΔH_R which can be applied as thermal energy to the process.

Under reversible conditions (no losses in the process), the difference between the electrode potentials is the reversible cell voltage V_{rev}. For STP with $\Delta G_R^0 = 237$ kJ mol^{-1}, the standard reversible cell voltage V_{rev}^0 for the decomposition of liquid water can be calculated according to

$$V_{rev}^0 = \frac{\Delta G_R^0}{zF} = \frac{237 \text{ kJ mol}^{-1}}{2 \times 96\,485 \text{ C mol}^{-1}} = 1.23 \text{ V} \tag{13.3}$$

where F is the Faraday constant and z the number of electrons transferred per molecule of hydrogen produced. V_{rev} indicates the minimum cell voltage which is required for the decomposition of liquid water as long as heat corresponding to $T\Delta S_R$ can be integrated into the process. However, PEM water electrolysis is a low-temperature process and thermal energy cannot be added from the surroundings. In this limiting case, the missing energy also has to be supplied as electricity. Therefore, the minimum cell voltage under ideal conditions is called the thermoneutral cell voltage, V_{th}, and for liquid water at STP it is derived as

$$V_{th}^0 = \frac{\Delta H_R^0}{zF} = \frac{286 \text{ kJ mol}^{-1}}{2 \times 96\,485 \text{ C mol}^{-1}} = 1.48 \text{ V} \tag{13.4}$$

From the thermodynamic point of view, high-temperature electrolysis of water is preferred as ΔG_R decreases at higher temperatures and the splitting process then requires less electricity (see Figure 13.2). According to Equation 13.3, values of V_{rev}^0 are also reduced. However, ΔH_R (and thus V_{th}^0) remains nearly constant with higher temperature, resulting in a higher demand for heat $T\Delta S_R$. As PEM water electrolysis is limited roughly to the temperature range from

13.2 Fundamentals of PEM Electrolysis

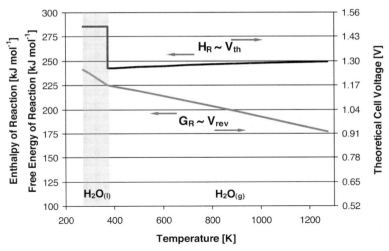

Fig. 13.2 Temperature dependence of the thermodynamic properties ΔH_R and ΔG_R and the corresponding theoretical cell voltages V_{th} and V_{rev}, respectively, for the decomposition of liquid and gaseous water at 101.325 kPa (1 atm).

room temperature to 393 K (see the next section), only slight changes in ΔH_R and ΔG_R, and therefore V_{th} and V_{rev}, respectively, can be observed (see Figure 13.2).

In a working electrolysis cell, irreversibility of various processes is unavoidable, which leads to a cell voltage V_{cell} that is always higher than the reversible cell voltage V_{rev}. Irreversibility is due to different causes; overpotentials at the anode and cathode, and a voltage drop iR_A due to ohmic losses within the cell and its components can be distinguished. With R_A as the cell resistance normalized by the area of the cell (area-specific ohmic resistance), the cell voltage of an operating cell is described by the following equation:

$$V_{cell} = V_{rev} + |\eta_{cath}| + |\eta_{an}| + iR_A \qquad (13.5)$$

The irreversible overpotentials at the electrode–electrolyte interfaces result in a more positive potential of the anode and a more negative potential of the cathode, resulting in a higher cell voltage. Figure 13.3 presents the contributions of different losses as a function of the current density applied to the electrolysis cell. Although catalysts are employed at the electrodes to accelerate the oxidation at the anode and the reduction at the cathode, electrokinetic limitations cause overpotentials at each electrode. Due to the more complex oxidation at the anode (the OER contains an overall transfer of four electrons), the overpotential at this electrode is dominant. By contrast, the kinetics of the HER are fast enough to cause only small overpotentials [3]. Apart from the activation overpotential, additional losses can be caused by a concentration overpotential due to limitations in the mass transport of the reaction species. A so-called "bubble over-

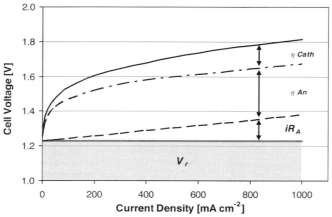

Fig. 13.3 Cell voltage versus current density (*Vi* characteristic = steady-state polarization curve) for a PEM electrolysis cell, with typical contributions due to different losses.

potential", as is known from alkaline electrolyzers, is only of minor importance in a PEM electrolysis cell. At a given operating temperature, the internal ohmic resistance in a PEM electrolysis cell remains nearly stable, independent of the current density. The voltage drop caused by this ohmic resistance increases linearly with the current density.

Note that irreversibility always causes heat dissipation corresponding to the heat flow $\dot{Q}_{diss}$, which is either consumed internally for the endothermic process or released as the heat flow $\dot{Q}_{loss}$ to the surroundings. The simplified energy balance of the electrolysis process can be expressed as

$$P_{el} = \dot{n}_{H_2}\Delta G_R + \dot{Q}_{diss} = \dot{n}_{H_2}\Delta H_R + \dot{Q}_{loss} = \dot{n}_{H_2}\Delta H_R + I(V_{cell} - V_{th}) \quad (13.6)$$

where P_{el} is the electric power input, I the electric current of the cell and $\dot{n}_{H_2}$ the molar hydrogen production rate. Equation 13.6 assumes that the electrolysis process takes place with complete conversion according to Faraday's law. However, the molar flow rate $\dot{n}_{H_2}$ of the hydrogen produced in a real electrolysis cell or stack is lower than the theoretical value, since stray currents in the cell and losses of hydrogen (e.g., diffusion through the membrane, leakage of the cell) occur. The current efficiency or so-called Faraday efficiency ε_I accounts for these internal losses:

$$\varepsilon_I = \frac{\dot{n}_{H_2}}{I(zF)^{-1}} \quad (13.7)$$

Values of ε_I are close to 1.0 in a well-designed PEM electrolysis cell or stack. The voltage efficiency ε_V, which is based on the simplified energy balance of

Equation 13.6, is more important for the evaluation of the water electrolysis process at a cell level. This efficiency can be measured easily according to

$$\varepsilon_V = \frac{V_{th}}{V_{cell}} \tag{13.8}$$

The product of the two efficiencies from Equations 13.7 and 13.8, respectively, results in the cell efficiency ε_{cell} of a real operating electrolysis cell:

$$\varepsilon_{cell} = \varepsilon_I \varepsilon_V \tag{13.9}$$

At the system level, it is useful to take a more general expression for the efficiency derived from the energy balance. The energy efficiency ε_E is the ratio of the energy content of the hydrogen produced in the system per unit time and the electric power needed for its production and thus refers to the efficiency of converting electricity into chemical energy of hydrogen:

$$\varepsilon_E = \frac{\dot{n}_{H_2} \Delta H_R}{P_{el}} \tag{13.10}$$

For the calculation of this energy efficiency, it is important to indicate whether the lower heating value (LHV = 3.00 kWh Nm^{-3} H$_2$) or the higher heating value (HHV = 3.54 kWh Nm^{-3} H$_2$) of ΔH_R is applied. In PEM water electrolysis, liquid water is decomposed into gaseous hydrogen and oxygen. Part of the electricity for this process is spent for the phase change from liquid to gas and consequently the HHV should be used to evaluate the process. However, if hydrogen is converted back to electricity in a fuel cell or to work in a cyclic process, only the chemical energy based on the LHV is used for this conversion. Mixing values for ε_E with different reference points, for example, for calculating the overall efficiency of a hydrogen storage system, leads to erroneous results. To avoid misunderstanding, only the amount of energy employed to produce one standard cubic meter of hydrogen (power consumption expressed in kWh Nm^{-3} H$_2$) is often given as a measure for the efficiency of an electrolysis unit.

Note that the energy efficiency of an electrolyzer refers not only to the power consumed by electrolysis stack but also to the power consumption of all peripheral devices needed to produce hydrogen, such as circulation and feed water pumps, solenoid valves, and power conditioning (rectifier). Therefore, when energy efficiencies of different alkaline and/or PEM electrolyzers are compared, the system boundaries and operating conditions (e.g., hydrogen output pressure) must be clearly defined. Often the energy demand of secondary subsystems such as the cooler is not taken into account. Furthermore, hydrogen losses due to operation of the gas chiller or due to start/stop operation are not included in this efficiency.

13.3
Membrane Electrode Assembly

The general design of a PEM electrolysis cell is similar to that of a PEM fuel cell (see Figure 13.1). The cell consists of an anode side (oxygen production) and a cathode side (hydrogen production), which are separated by a proton exchange membrane. In most cases, perfluorosulfonic acid (PFSA) membranes, such as Nafion® from DuPont or Fumapem® from FuMA-Tech, are used for PEM electrolysis. The thickness of these membranes ranges from 50 µm to approximately 200 µm. The use of reinforcement membranes is state of the art. A well-balanced membrane exhibits high proton conductivity for the transport of the ions and low permeability for gases to prevent mixing of the gases produced. Additionally, the membrane must be an electric insulator to prevent short-circuits within the cell from electrode to electrode and the ionomer must offer high chemical and mechanical stability to withstand the harsh conditions in a PEM electrolysis cell. The motion of protons through the membrane is accompanied by water transport from the anode to the cathode due to the polar character of the molecules and is called electroosmotic drag. The concentration gradient across the membrane leads to diffusion of hydrogen to the oxygen side and vice versa. The smaller molecular size of hydrogen is responsible for considerably more hydrogen permeation than the permeation of oxygen. More hydrogen migrates through the membrane if the partial pressure gradient is increased by a higher operating pressure and temperature [4]. Typical values measured in laboratory cells operating under high pressure up to 7 MPa are in the range between 0.25 and 2.0 vol.% [4–6]. Consequently, cross-permeation has a strong influence on the gas purity and can become a safety issue since the amount of hydrogen can reach values close to 4 vol.%, the lower explosion limit of hydrogen in air at STP. On the hydrogen side, gas purity is less critical. In general, the amount of oxygen in hydrogen is not higher than 2000 ppm for the dry gas and hydrogen purities up to 99.99 vol.% at the cell outlet have been reported [5, 7]. PSFA membranes can be used only at temperatures below 373 K as dehydration of the ionomer takes place at higher temperatures. Loss of mechanical stability and proton conductivity is the result. Inorganic fillers such as SiO_2 and TiO_2 have been found to improve the water retention inside a PSFA membrane, allowing operation at temperatures above 373 K with improved cell performance [8, 9].

Catalysts with a porous structure are deposited on each side of the membrane and serve as electrodes. They are partly filled with an ionomer in order to enlarge the area of the three-phase boundary where half-cell reactions take place. Precious metals are essential for the electrocatalysis because of the acidic nature of the cation-exchange membrane. State-of-the-art MEAs have a catalyst loading of 2–6 mg cm^{-2} on the anode and about 1–2 mg cm^{-2} on the cathode. Platinum is the most efficient catalyst for hydrogen production. On the oxygen side, catalysts such as iridium, ruthenium, and their oxides and mixtures are often used [3, 10–12]. Ruthenium oxide is the most active catalytic material for the anodic

13.3 Membrane Electrode Assembly

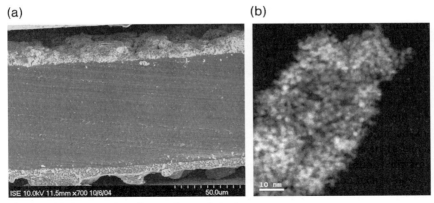

Fig. 13.4 (a) Cross-section of a membrane electrode assembly consisting of a Nafion® N117 membrane and electrodes made of pure platinum (cathode) and pure iridium (anode) (Fraunhofer ISE). (b) High-resolution TEM image of a 1:1 60 wt% Pt–Ru OER catalyst supported on anatase TiO_2. Data courtesy of the University of South Carolina [20].

oxygen reaction but it is highly unstable [13, 14]. Iridium is less active but is still considered to be the most promising material, either pure or mixed with other precious or non-precious materials. Figure 13.4a shows the cross-section of a standard MEA for PEM water electrolysis produced by the hot-pressing method. Pt and Ir with metal loadings of 2 mg cm^{-2} are used for the cathode and anode, respectively. The catalysts are unsupported and the particle size ranges between 3 and 30 µm. Hence the electrochemical activity for HER and OER is fairly moderate compared with supported catalyst systems as presented in Figure 13.4b. The image shows a 1:1 60 wt% Pt–Ru catalyst supported on anatase TiO_2 with an average particle size of 1.2 nm. In general, ambitious research activities have been observed in recent years concerning electrocatalysis for PEM water electrolysis, mainly concerning the following:

- Improvement of the electrocatalytic activity for OER using different binary and ternary supported catalyst systems [14–16].
- Reduction of the metal loading of Pt-based catalysts for HER to values smaller than 1 mg cm^{-2}, mainly by using supported catalysts such as Pt/C or Pd/C [17, 18].
- Substitution of Pt as a catalyst for HER: use of nickel or cobalt glyoximes as low-cost compounds which are stable in strong acidic media. However, the performance data do not compete with platinum [19].

In Figure 13.5, data from MEAs with binary and ternary unsupported catalyst systems at the anode and unsupported Pt at the cathode are shown for a 50 cm^2 single cell. Using mixed-oxide anode catalysts, voltage efficiencies ε_V (if not stated otherwise, all cell efficiency values reported here are based on the HHV) of up to 88% are possible at 1.0 A cm^{-2}, 0.1 MPa and 353 K [16]. Similar

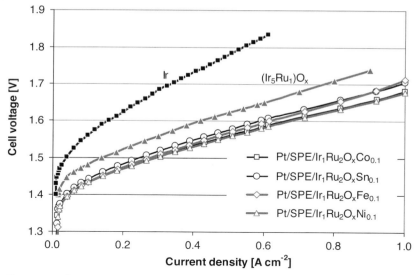

Fig. 13.5 Vi characteristics comparing the performance of different binary and ternary unsupported electrocatalysts for the OER (353 K, 0.1 MPa; cathode: unsupported Pt).

or even better results at 363 K with voltage efficiencies ε_V of 90–96% have been reported by other groups [21–23]. To our knowledge, the lowest cell voltage of $V_{cell} = 1.567$ V measured at 1.0 A cm^{-2} and 353 K (0.1 MPa) was achieved with an $Ir_{0.6}Ru_{0.4}O_2$ anode catalyst and a 20 wt% Pt/C cathode on a Nafion® 115 membrane [14].

13.4
Current Collectors, Bipolar Plates, and Stack Design

Porous current collectors are located between the MEA and the flow fields of the bipolar plates (BiP) to allow a uniform distribution of the electric current between the electrodes and the bipolar plates. A well-designed BiP has to feature high bulk electrical conductivity, low electrical contact resistance, and high gas permeability, and must permit water flow to the electrode. On the hydrogen side, the electrode potentials are close to zero. It is possible to use carbon-based materials such as carbon papers or felts as known from PEM fuel cells [24]. On the oxygen side, graphite and also carbon-based materials are not stable at the working potentials of the anode. Carbon undergoes electrochemical oxidation at voltages higher than 0.9 V, according to [25, 26]

$$C + 2\,H_2O \rightarrow CO_2 + 4\,H^+ + 4e^- \tag{13.11}$$

Instead of carbon, porous structures made from titanium are mainly used as current collectors. Titanium offers acceptable bulk electrical conductivity and it is corrosion resistant on the oxygen side. The main drawbacks of Ti are the formation of oxide layers on the surface at typical potentials of the anode and the high cost of the material. Different types of current collectors made from Ti and carbon are presented in Figure 13.6. The performance of sintered titanium powder is superior to that of other structures, but this form is also the most expensive choice, followed by felt and expanded mesh. However, the porosity should be not too small as water and the gases produced have to flow through the current collector in opposite directions [5].

In single electrolysis cells, the MEA and current collectors are sandwiched between two electrically conductive endplates. To yield higher hydrogen flow rates, several single cells are connected electrically in series and hydraulically in parallel as a stack. Here, a bipolar plate separates two adjacent cells, so that it simul-

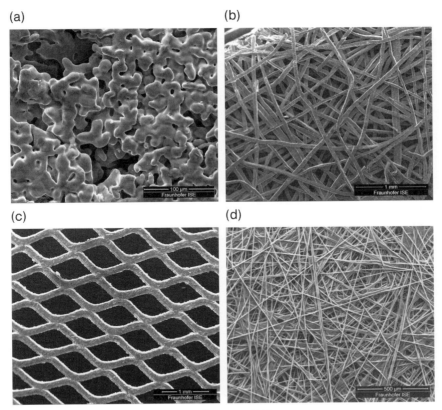

Fig. 13.6 Different materials for current collectors in PEM electrolysis cells. (a) Sintered titanium powder; (b) sintered titanium felt; (c) expanded titanium mesh; (d) carbon-based paper.

taneously acts as the anode of one cell and the cathode of the adjacent cell. The most common stack type is the filter press bipolar pressurized cell stack for PEM electrolysis. The number of cells typically ranges from 10 to about 120. The active area of a PEM electrolysis cell is considerably smaller than that of an alkaline electrolysis cell. All currently known developments aim for an active cell area that is clearly less than 1000 cm^2.

As for current collectors, bipolar plates have to be constructed from corrosion-resistant materials such as expensive titanium or coated stainless steel. Cheaper graphite composite materials as known from PEM fuel cells can be used for the hydrogen side. However, in this case the BiP becomes more complex since at least the surface of the anode has to be corrosion resistant due to application of an electrochemically impenetrable coating. Unfortunately, few data are available concerning the construction and materials employed for BiPs, sealants, and cells, as most companies have developed proprietary stack designs for their products.

Nearly all BiPs have internal structures such as channels or serpentines to form a flow field which facilitates the water supply and gas evacuation in each cell. An example of a conventional BiP made of titanium is shown in Figure 13.7a. The flow-field structure is milled on both sides. Further typical features of a BiP are gasket grooves/applied sealants, and manifolds with a connection to the flow field, which serve as the inlet and outlet for the water and gases, respectively.

The design presented in Figure 13.7a does not allow economical mass production as the costs for material and the manufacturing process are very high. Therefore, other types of basic constructions have been investigated for the cell design. A thin metallic, corrosion-resistant plate or foil serves as BiP. Frames,

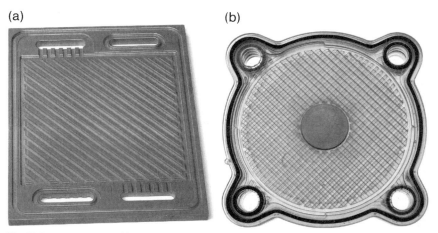

Fig. 13.7 (a) Conventional bipolar plate made of titanium with milled internal structures and (b) an advanced design of a BiP using injection-molded plastic plates with an internal Ti connector pin.

for example plastic molded, on both sides of the BiP fix the MEA and act as the current collectors, gaskets, and manifolds. Another approach is depicted in Figure 13.7b. A cheap, injection-molded and non-conductive separator plate provides the required mechanical structure of a BiP. Conductive connector pins or bridges are embedded in this separator plate to enable an electric current to flow between the current collectors of two adjacent cells. In particular, this concept is suited for smaller stacks, with their relatively small thermal loads.

In order to reduce the contact resistance between a bipolar plate and the adjacent current collector (due to formation of an oxide layer on the surface), often the bipolar plates are coated with precious metals such as gold or even platinum [27, 28].

Taking into account the high current densities of PEM electrolysis cells (see above) and the slim design of these cells, PEM electrolysis stacks offer very high power densities at similar voltage efficiencies ε_V compared with alkaline electrolysis stacks. At present, current densities from 0.5 to 2.0 A cm^{-2} and cell voltages from 1.7 to 2.1 V are usual. In most cases, PEM electrolysis stacks operate at temperatures between 323 and 363 K, and under high pressure in the range between 0.8 and 20.7 MPa [27]. With a well-designed electrolysis stack, the current efficiency ε_I can be up to 98%. According to Equation 13.9, or based on HHV, the power consumption at the stack level can be calculated to be 4.1 or 5.1 kWh Nm^{-3} hydrogen, respectively. In Figure 13.8, values for the power consumption of different commercial stacks and stacks under development are plotted against the nominal hydrogen production rate. The power consumption decreases only slightly for a higher production rate. By contrast, the power con-

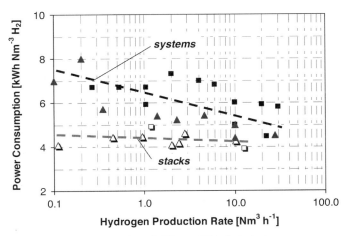

Fig. 13.8 Power consumption of different PEM electrolysis stacks and systems plotted against the nominal hydrogen production rate (■ commercial systems; ▲ systems under development; □ commercial stacks; △ stacks under development).

sumption at the system level clearly varies with the production rate, as discussed in the next section.

In terms of operating hours and lifetime, considerable progress has been made over the past decade. Due to the endothermic reaction, higher efficiencies and efficient cooling by the water recirculation, thermal management is facilitated compared with PEM fuel cells. Provided that a uniform current distribution over the electrodes can be assured, the risk of hot spots is lower than in fuel cells. However, the mechanical stress could be higher, as in most cases the stacks are operated under pressurized conditions, with pressure fluctuations caused by control and start/stop procedures. The most critical component of a PEM electrolysis cell or stack continues to be the membrane due to membrane thinning, with subsequent reduction of gas quality, and finally failure (leakage). Further typical degradation mechanisms within the stack are titanium embrittlement by hydrogen, corrosion/dissolution/agglomeration of the catalyst and its support, and deterioration of seals. Nevertheless, the lifetime of a well-designed and correctly operated PEM electrolysis stack can reach up to several tens of thousands of hours in industrial and military applications. As an example, Figure 13.9 presents long-term performance data for a PEM electrolysis stack 91E made by Hydrogenics, consisting of 13 cells with an active area of 90 cm^2.

It should be emphasized that water quality also has a strong influence on the degradation of the MEA and consequently plays an important role in determining the lifetime of the electrolysis cell. To achieve a long lifetime as reported above, deionized water with typical resistance values higher than 10 MΩ cm should be used.

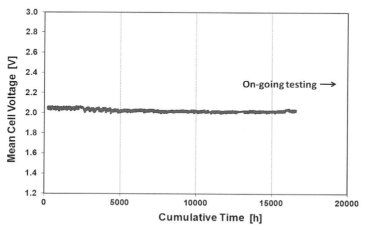

Fig. 13.9 Long-term performance data of a PEM electrolysis stack consisting of 13 cells with an active area of 90 cm^2. The stack operates at a current density $i_{cell} = 2.0$ A cm^{-2}. Data courtesy of Hydrogenics.

13.5
System Design

The basic configuration of a PEM electrolyzer is similar to that of an alkaline unit (see Figure 13.10). However, the system design is simpler due to the absence of lye as the liquid electrolyte. Apart from the electrolysis stack, the main components of the complete PEM electrolyzer are as follows:
- Power conditioning unit, consisting mainly of a transformer and rectifier to adapt the a.c. voltage to the required d.c. voltage of the stack (up to approximately 200 $V_{d.c.}$)
- Feed water pump, which feeds the water to the circulation loop on the oxygen side (at a given pressure) and – optionally – a purification stage
- (Low-pressure) circulation loop for the oxygen side, usually with a circulation pump, heat exchanger, water purification stage (ion-exchange resin), and gas-water separator. The main task of the circulation pump is to ensure stack cooling, rather than supplying water for the electrolysis cells. In most cases, the natural convection of the two-phase flow forced by the ascending gases is sufficient to supply the stack with water. For this reason, smaller systems can be designed without a circulation pump.
- (High-pressure) circulation loop for the hydrogen side with a gas-water separator
- Hydrogen conditioning unit with a demister to remove droplets and aerosols from the gas flow, heat exchanger, and condensate trap to reduce the dew point to approximately room temperature.
- Water drain to recapture the water transported via the membrane to the cathode (electroosmotic drag)
- Control and safety installation (not shown in Figure 13.10).

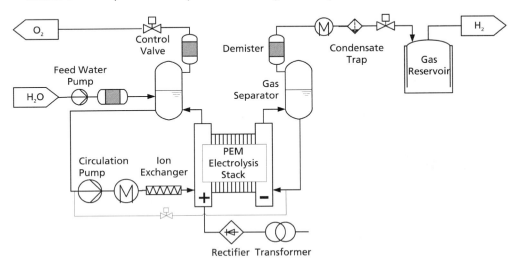

Fig. 13.10 Typical layout of a PEM electrolysis system. In this case, both sides operate under balanced pressure.

In general, a distinction in electrolyzer designs can be made between systems working at balanced pressure and those with differential pressure. In the former case, both sides of the electrolyzer are operated at the same pressure, which is regulated by an oxygen and hydrogen control valve. This concept is known from alkaline electrolyzers. In the latter case, the hydrogen side works under pressure whereas the oxygen side is not or only slightly pressurized. This approach demands a well-developed stack design, as the membrane has to withstand the differential pressure. However, costs can be reduced considerably, as components do not need to be compression proof. Most of the commercial system developments in PEM electrolysis apply this concept.

Note that Figure 13.10 does not contain further secondary subsystems of an electrolysis system, for example a water pretreatment stage if only tap water is available, the recooler, the hydrogen gas chiller (hydrogen drying), and a deoxidizer purification stage (reducing the amount of oxygen in the hydrogen).

Figure 13.8 shows that the power consumption of PEM electrolysis systems decreases strongly as the hydrogen production rate increases, whereas the power consumption of the stack decreases only slightly. Obviously, smaller systems require the same balance of plant as larger systems but the power demand of these auxiliary components and subsystems cannot be reduced linearly with the production rate. Smaller systems up to roughly $1 \, \text{Nm}^3 \, \text{h}^{-1}$ exhibit a power consumption of 6–8 kWh Nm^{-3} hydrogen. For larger systems $>10 \, \text{Nm}^3 \, \text{h}^{-1}$, the power consumption can be reduced to <6 kWh Nm^{-3} hydrogen.

So far, only a few manufacturers offer PEM electrolyzers commercially with a hydrogen production rate up to approximately $30 \, \text{N m}^3 \, \text{h}^{-1}$ or have systems available for demonstration projects and early-adopter partnering, for example Proton Energy Systems (United States), Hydrogen Technologies (StatoilHydro, Norway), Hydrogenics (Canada) and h-tec (Germany). Apart from gas generators for laboratory equipment (with production rates $\ll 1 \, \text{Nm}^3 \, \text{h}^{-1}$), the main fields of application are similar to those of small alkaline electrolyzers: production of semiconductor electronics, generator cooling in power plants, meteorological applications (registering balloons), and materials processing (coatings, metallurgy, etc.). In the United States, PEM electrolyzers are well established in niche markets for military and space applications, such as oxygen generator plants for nuclear submarines. Here, Hamilton Sundstrand in cooperation with Proton Energy Systems and Giner with Treadwell are advancing the development of this technology.

13.6
Conclusion

Water electrolysis is one of the simplest ways to produce hydrogen. In general, it is a clean and mature technology which is perfectly suited for coupling to electricity from renewable energy sources such as wind and the sun. Very pure hydrogen can be produced with small to very large production capacities. In particular, PEM water electrolyzers have been operated with high power densities at reasonable system efficiencies of approximately 4.0–5.0 kWh Nm^{-3} hydrogen produced. At the cell level, efficiencies higher than 90% based on HHV can be achieved at a current density of 1 A cm^{-2}. The PEM design permits high-pressure operation up to several MPa and very short response and start-up times, as are required for renewable energy sources. Stack operation with partial loads is possible in the range from 0 to 100% but is limited practically at a system level to 5–100%, due to the power consumption of auxiliary components. The lifetime of PEM electrolyzers was lengthened during the past decade and now approaches values comparable to those known from the alkaline technology. It can be concluded that PEM electrolysis is an established technology for laboratory, industrial, and military niche applications. Systems with a hydrogen production rate up to 30 $Nm^3 h^{-1}$ and a pressure of 3 MPa are commercially available. However, the cost must be reduced still further, which could be achieved by mass production (economy of scale). Efforts should be directed towards achieving further improvements in the catalysts and bipolar plates.

13.7
Symbols and Abbreviations

F Faraday constant = 96 485 C mol^{-1}
ΔG_R free energy of reaction [J mol^{-1}]
ΔH_R enthalpy of reaction [J mol^{-1}]
i_{cell} current density [A cm^{-2}]
$\dot{n}_{H_2}$ molar hydrogen flow [mol s^{-1}]
N number of units – here electrons and molecules – involved in the electro chemical reaction
P_{el} electric power [W]
$\dot{Q}_{diss}$ heat losses to the surroundings
ΔS_R entropy of reaction [J K^{-1} mol^{-1}]
T temperature [K]
V_{cell} cell voltage [V]
V_{th} thermoneutral cell voltage [V]
V_r reversible cell voltage [V]
ε_V voltage efficiency
BiP bipolar plate
HER hydrogen evolution reaction
HHV higher heating value
LHV lower heating value
OER oxygen evolution reaction

References

1 Smith, W. (2000) The role of fuel cells in energy storage. *J. Power Sources*, **86** (1–2), 74–83.
2 Smolinka, T. (2009) Hydrogen from water electrolysis. In (ed. J. Garche), *Encyclopedia of Electrochemical Power Sources*, Elsevier, Amsterdam, pp. 394–413.
3 Marshall, A., Børresen, B., Hagen, G., Tsypkin, M., and Tunold, R. (2007) Hydrogen production by advanced proton exchange membrane (PEM) water electrolysers – reduced energy consumption by improved electrocatalysis. *Energy*, **32** (4), 431–436.
4 Wittstadt, U., Wagner, E., and Jungmann, T. (2005) Membrane electrode assemblies for unitised regenerative polymer electrolyte fuel cells. *J. Power Sources*, **145** (2), 555–562.
5 Grigoriev, S. A., Millet, P., Korobtsev, S. V., Porembskiy, V. I., Pepic, M., Etievant C., Puyenchet, C., and Fateev, V. N. (2009) Hydrogen safety aspects related to high-pressure polymer electrolyte membrane water electrolysis. *International J. Hydrogen Energy*, **34** (14), 5986–5991.
6 Stucki, S., Scherer, G., Schlagowski, S., and Fischer, E. (1998) PEM water electrolysers: evidence for membrane failure in 100 kW demonstration plants. *J. Applied Electrochemistry*, **28** (10), 1041–1049.
7 McElroy, J. F. (1994) Recent advances in SPE® water electrolyzer. *J. Power Sources*, **47** (3), 369–375.
8 Antonucci, V., Di Blasi, A., Baglio, V., Ornelas, R., Matteucci, F., Ledesma-Garcia, J., Arriaga, L. G., and Aric, A. S. (2008) High temperature operation of a composite membrane-based solid polymer electrolyte water electrolyser. *Electrochimica Acta*, **53** (24), 7350–7356.
9 Baglio, V., Ornelas, R., Matteucci, F., Martina, F., Ciccarella, G., Zama, I., Arriaga, L. G., Antonucci, V., and Aric, A. S. (2009) Solid polymer electrolyte water electrolyser based on Nafion–TiO_2 composite membrane for high temperature operation. *Fuel Cells*, **9** (3), 247–252.
10 Cheng, J., Zhang, H., Chen, G., and Zhang, Y. (2009) Study of $Ir_xRu_{1-x}O_2$ oxides as anodic electrocatalysts for solid polymer electrolyte water electrolysis. *Electrochimica Acta*, **54** (26), 6250–6256.
11 Grigoriev, S. A., Porembsky, V. I., and Fateev, V. N. (2006) Pure hydrogen production by PEM electrolysis for hydrogen energy. *International J. Hydrogen Energy*, **31** (2), 171–175.
12 Rasten, E., Hagen, G., and Tunold, R. (2003) Electrocatalysis in water electrolysis with solid polymer electrolyte. *Electrochimica Acta*, **48** (25–26), 3945–3952.
13 Ahn, J. and Holze, R. (1992) Bifunctional electrodes for an integrated water-electrolysis and hydrogen–oxygen fuel cell with a solid polymer electrolyte. *J. Applied Electrochemistry*, **22** (12), 1167–1174.
14 Marshall, A., Sunde, S., Tsypkin, M., and Tunold, R. (2007) Performance of a PEM water electrolysis cell using $Ir_xRu_yTa_zO_2$ electrocatalysts for the oxygen evolution electrode. *International J. Hydrogen Energy*, **32** (13), 2320–2324.
15 Cheng, J., Zhang, H., Ma, H., Zhong, H., and Zou, Y. (2009) Preparation of $Ir_{0.4}Ru_{0.6}Mo_xO_y$ for oxygen evolution by modified Adams' fusion method. *International J. Hydrogen Energy*, **34** (16), 6609–6613.
16 Flores H. R. J. (2005) Optimization of membrane-electrode assemblies for SPE water electrolysis by means of design of experiments. PhD thesis, Albert-Ludwigs-Universität Freiburg.
17 Grigoriev, S. A., Millet, P., and Fateev, V. N. (2008) Evaluation of carbon-supported Pt and Pd nanoparticles for the hydrogen evolution reaction in PEM water electrolysers. *J. Power Sources*, **177** (2), 281–285.
18 Paunovic, P., Radev, I., Dimitrov, A. T., Popovski, O., Lefterova, E., Slavcheva, E., and Jordanov, S. H. (2009) New nanostructured and interactive supported composite electrocatalysts for hydrogen evolution with partially replaced platinum loading. *International J. Hydrogen Energy*, **34** (7), 2866–2873.

19 Pantani, O., Anxolabéhère-Mallart, E., Aukauloo, A., and Millet, P. (2007) Electroactivity of cobalt and nickel glyoximes with regard to the electro-reduction of protons into molecular hydrogen in acidic media. *Electrochemistry Communications*, **9** (1), 54–58.

20 Fuentes, R. E., García, B. L., and Weidner J. W. (2009) Effect of TiO_2 crystal structure on the activity of Pt–Ru towards electrochemical oxidation of methanol. *J. the Electrochemical Society*, in press.

21 Kondoh, M., Yokoyama, N., Inazumi, C., Maezawa, S., Fujiwara, N., Nishimura, Y., Oguro, K., and Takenaka, H. (2000) Development of solid polymer-electrolyte water electrolyser. *J. New Materials for Electrochemical Systems*, **3** (1), 61–65.

22 Marshall, A., Børresen, B., Hagen, G., Tsypkin, M., and Tunold, R. (2006) Electrochemical characterisation of $Ir_xSn_{1-x}O_2$ powders as oxygen evolution electrocatalysts. *Electrochimica Acta*, **51** (15), 3161–3167.

23 Song, S., Zhang, H., Ma, X., Shao, Z., Baker, R. T., and Yi, B. (2008) Electrochemical investigation of electrocatalysts for the oxygen evolution reaction in PEM water electrolyzers. *International J. Hydrogen Energy*, **33** (19), 4955–4961.

24 Smolinka, T. Grootjes, S., Mahlendorf, F., Hesselmann, J., and Makkus, R. (2006) Prototype of a reversible fuel cell system for autonomous power supplies. In *Proceedings of the 3rd European PV-Hybrid and Mini-Grid Conference*, CD-ROM: Aix-en-Provence, 11–12 May 2006, France. Regensburg: OTTI, 2006, pp. 131–136

25 Roen, L. M., Paik, C. H., and Jarvi, T. D. (2004) Electrocatalytic corrosion of carbon support in PEMFC cathodes. *Electrochemical and Solid-State Letters*, **7** (1), A19–A22.

26 Young, A. P., Stumper, J., and Gyenge, E. (2009) Characterizing the structural degradation in a PEMFC cathode catalyst layer: carbon corrosion. *J. Electrochemical Society*, **156** (8), B913–B922.

27 LaConti, A. B. and Swette, L. (2003) Special application using PEM-technology. In *Handbook of Fuel Cells – Fundamentals, Technology and Applications* (eds W. Vielstich, H. A. Gasteiger, and A. Lamm), John Wiley & Sons, Ltd, Chichester, Vol. 4, Chapter 55, pp. 745–761.

28 Jung, H.-Y., Huang, S.-Y., Ganesan, P., and Popov, B. (2009) Performance of gold-coated titanium bipolar plates in unitized regenerative fuel cell operation. *J. Power Sources*, **194** (2), 972–975.

14
Reforming and Gasification – Fossil Energy Carriers

Jens Rostrup-Nielsen

Abstract

This chapter summarizes the basic technologies available for the manufacture of hydrogen from fossil fuels followed by a description of important state-of-the-art flow schemes and potential alternatives. Finally, aspects of future applications for fuel cells and for synfuels are discussed.

Keywords: hydrogen production, fossil fuels, steam reforming, partial oxidation, coal conversion, fuel cells, carbon dioxide footprint

14.1
Introduction. The Need for H_2

Hydrogen is an important raw material for the chemical and the refinery industries, and it may play a future role in the energy sector [1–3]. The total hydrogen consumption can be estimated as about 250×10^9 Nm^3 per year, with the refinery industry accounting for the majority (>90%). About the same amount is used as mixtures of hydrogen and nitrogen for ammonia synthesis and mixtures of hydrogen and carbon oxides (synthesis gas) for methanol synthesis. With this and other uses in the chemical industry, the total hydrogen consumption amounted to about 50×10^6 Mt per year (630×10^9 Nm^3 per year) in 2002 [3, 4].

There is a rapidly growing need for more hydrogen production capacity in refineries [3, 4] where the use of hydrogen is increasing due to the need for deep desulfurization of fuels and for hydrocracking of heavy oil fractions aiming at an atomic ratio of H/C=2, being typical for transportation fuels. Traditionally, a major a part of the hydrogen consumption in refineries was covered by hydrogen produced as a by-product from other refinery processes, mainly catalytic reforming ("platforming"). However, as available oil resources become heavier, with higher contents of sulfur and metals, there is a need for additional hydrogen production capacity in refineries [3].

Hydrogen as an "energy vector" was discussed in the wake of the energy crisis in the 1970s [5]. Today, "hydrogen economy" is on the political agenda [4]. Hy-

Hydrogen Energy. Edited by Detlef Stolten
Copyright © 2010 WILEY-VCH Verlag GmbH & Co. KGaA, Weinheim
ISBN: 978-3-527-32711-9

drogen may replace hydrocarbons to provide a clean fuel with no carbon emissions for use in both stationary and mobile applications. Fuel cells will play a key role in both situations. Plants supplying hydrogen for the build-up of a future hydrogen infrastructure are faced with a dilemma when based on fossil fuels [6]. Centralized large-scale hydrogen production is penalized by significant costs of compression and transportation. Therefore, decentralized production at gas stations appears to be the optimum solution, but CO_2 sequestration appears feasible only with large-scale production. Without CO_2 sequestration, it may be better to use natural gas directly in the car.

With the increased interest in synfuels from coal, the supply of hydrogen plays an important role in determining the CO_2 footprint of the coal conversion.

14.2
Basic Technologies

14.2.1
Processes and Reactions

Hydrogen (and synthesis gas) can be produced from almost any carbon source, ranging from natural gas and oil products to coal and biomass. The main processes are listed in Table 14.1.

Today, natural gas and other hydrocarbons are the dominant feedstocks for the production of hydrogen because the investment required is about one-third of that for a coal-based plant. The main chemical reactions are listed in Table 14.2.

Table 14.1 Characteristics of processes for hydrogen from fossil fuels [2].

Process	Reactor	Feedstock	Temperature (°C)	Catalyst
Steam reforming	Heated reactor of many types	Light hydrocarbons	500–900	Ni/ceramic support
Prereforming	Adiabatic reactor	Natural gas, diesel	400–500	Ni/support
Water gas shift	Adiabatic reactor	Syngas	230–350	Fe_3O_4, $Cu/ZnO/Al_2O_3$
Autothermal reforming	Brick-lined reactor with burner and catalyst bed	Light hydrocarbons	–1050	Ni/ceramic support
Catalytic partial oxidation	Monolithic catalyst reactor	Light hydrocarbons, diesel	–950	Rh/ceramic support
Partial oxidation (gasification)	Brick-lined reactor with burner	Heavy hydrocarbons, coal, petcoke, biomass	1300	None

Table 14.2 Synthesis gas reactions.

Process	$-\Delta H^0_{298}$ (kJ mol^{-1})
Steam reforming:	
1. $CH_4 + H_2O \rightarrow CO + 3H_2$	−206
2. $C_nH_m + nH_2O \rightarrow nCO + (n+m/2)H_2$	−1175[a]
3. $CO + H_2O \rightarrow CO_2 + H_2$	41
Autothermal reforming (ATR):	
5. $CH_4 + 1.5 O_2 \rightarrow CO + 2H_2O$	520
6. $CH_4 + H_2O \rightarrow CO + 3H_2$	−206
7. $CO + H_2O \rightarrow CO_2 + H_2$	41
Catalytic partial oxidation (CPO):	
8. $CH_4 + 0.5 O_2 \rightarrow CO + 2H_2$	38
Water gas shift:	
9. $CO + H_2O \rightarrow CO + H_2$	41
Gasification:	
10. $C + 0.5 O_2 \rightarrow CO$	111[b]
11. $C + O_2 \rightarrow CO_2$	394[b]
12. $C + H_2O \rightarrow CO + H_2$	−131[b]

[a] For n-C_7H_{16}.
[b] For graphite.

14.2.2
Steam Reforming

Steam reforming of hydrocarbons is the preferred process for hydrogen and syngas today. It involves the conversion of steam and methane, and as these are two very stable molecules, the reaction (Table 14.2) requires the supply of heat. Industrial steam reforming is carried out in a fired reactor, the *tubular reformer* [1, 7], in which catalyst tubes are placed in a fired furnace supplying the heat for the reaction and the heat for arriving at the exit temperature required for the desired conversion. The catalyst tubes have a typical diameter of 10 cm and a length of about 10 m. They are made of high-alloy steel to withstand the high tube wall temperatures (typically close to 1000 °C) and the high thermal stresses caused by the high heat fluxes amounting to average fluxes of 0.1 MW m^{-2}. The industrial breakthrough for the process 50 years ago was the result of design for operation at high pressures (20–40 bar), thereby reducing the energy consumption for compression of the large volume of the syngas product. Today, tubular reformers are built for hydrogen production levels of more than 200 000 Nm^3 h^{-1}. The product gas leaves the reformer tubes close to thermodynamic equilibrium. This makes it easy to predict the operation and to calculate the need for heat supply from a simple enthalpy balance [1, 7].

The reformer tubes are filled with *catalyst*, which must show high mechanical stability because of exposure to high temperatures and steam partial pressures

[8]. With a typical activity of a nickel catalyst, the gas composition in the reformer tube quickly arrives close to the equilibrated gas [7, 9]. It can be shown that there is a huge surplus of catalyst activity in a tubular reformer. Still, there is a need for high catalyst activity because it means that the heat transfer for a given conversion can take place at a lower tube wall temperature [7].

In many situations when natural gas is not available, higher hydrocarbons become the preferred feedstock for the reforming process [10, 11]. Many refineries benefit from flexibility in feedstock, taking advantage of the surplus of various hydrocarbon streams in the refinery. Steam reforming of liquid hydrocarbons is also considered for hydrogen generation for fuel cells. This includes "logistic fuels" such as diesel and jet fuel [9].

The higher hydrocarbons are more reactive (per carbon atom) than methane, except for aromatic molecules. Benzene has a reactivity comparable to that of methane [8, 11]. This means that liquid hydrocarbons in principle can be easily converted by steam reforming, but in practice this is limited by the higher potential for poisoning by sulfur (being chemisorbed on the nickel surface) and by the higher risk of carbon formation.

Removal of hydrogen sulfide and lower thiols is easily accomplished over zinc oxide, whereas heavier sulfur components require hydrogenation (HDS) over CoMo catalysts [1].

The steam reforming reaction results in an $H_2:CO$ ratio close to 3. The $H_2:CO$ ratio can be varied over a wider range, as the reforming reactions are coupled to the shift reaction. In the manufacture of hydrogen, the reforming process is followed by a water gas shift reaction (Table 14.2) carried out in the presence of a copper catalyst at low temperatures (210–330 °C) to ensure complete conversion of carbon monoxide [1, 3].

The high temperatures required to achieve high equilibrium conversion of methane is in contrast to the potential of the catalyst, showing activity even below 400 °C [9]. Steam reforming at low temperatures would allow the use of cheaper construction materials and of low-temperature heat to drive the reaction. This has led to efforts to circumvent constraints by the use of a selective hydrogen membrane (Pd) [3, 12] installed in the catalyst bed with extraction of hydrogen, thus pushing the reforming equilibrium to higher conversion. Reactor simulations and experiments [12] have shown that the reformer exit temperature can indeed be reduced to below 700 °C while maintaining the same conversion. Membrane reforming results in hydrogen at low pressure and CO_2 at high pressure. The latter is good for CO_2 sequestration and a low supply pressure may be acceptable for fuel cells. However, there are still mechanical challenges to be overcome.

14.2.3
Partial Oxidation

Partial oxidation represents an alternative to steam reforming. The reaction heat is provided by the partial combustion of the hydrocarbon and hence there is no need for a complex heated reactor. On the other hand, the partial oxidation reactions result in low $H_2:CO$ ratios, meaning that larger reactors for the water gas shift reaction are required. Partial oxidation can be carried out in three ways [2, 13].

Noncatalytic partial oxidation (POX) requires high temperatures to ensure complete conversion of methane and to reduce soot formation. Some soot is normally formed and is removed in a separate scrubber system after the partial oxidation reactor. Gasification of heavy oil fractions may play an increasing role as these fractions are becoming more available because of falling demand. Finally, gasification of coal may become the route for hydrogen if CO_2 sequestration issues are resolved.

The autothermal reforming (ATR) process [1, 14] is a hybrid of partial oxidation and steam reforming using a burner and a fixed catalyst bed for equilibration of the gas. This design allows a decrease in the maximum temperature, and hence the oxygen consumption can be lowered. Soot formation can be eliminated by addition of an appropriate amount of steam to the feedstock and by a specific burner design. The autothermal reformer is a simple piece of equipment with a specifically designed burner and a fixed catalyst bed in a brick-lined reactor. The composition of the product gas will be determined by the thermodynamic equilibrium at the exit temperature and pressure, which in turn are determined by the adiabatic heat balance.

In catalytic partial oxidation (CPO) [2, 15], the reactants are premixed, and the conversion takes place in a catalytic reactor without a burner. The hydrocarbon feedstock and air (oxygen) are mixed and fed to the catalyst. The direct CPO reaction (reaction 8 in Table 14.2) and has a low heat of reaction (38 kJ mol^{-1}). However, in practice, the reaction is accompanied by the reforming and water gas shift reactions (reactions 1 and 3 in Table 14.2) and, at high conversions, the product gas will be close to thermodynamic equilibrium. CPO over noble metals is less sensitive to sulfur poisoning [16]. In the presence of oxygen, sulfur is oxidized to SO_2, which is not adsorbed on the catalyst. Nickel will be oxidized in presence of oxygen, which is not the case for rhodium, a typical catalyst for CPO. This means that rhodium stays active as long as oxygen is present. After depletion of oxygen, SO_2 will be reduced to H_2S, which will be chemisorbed on the catalyst and also on downstream catalysts and anodes.

Up to 40% of the costs of a syngas plant based on ATR (CPO or POX) are related to the oxygen plant [13]. Consequently, routes based on air and eliminating the cryogenic air separation plant have been suggested. The use of air in the process stream is possible only in once-through synthesis schemes, otherwise huge accumulations of nitrogen are unavoidable. Attempts to use air instead of oxygen result in large gas volumes and consequently in large feed/effluent heat exchangers and compressors. These are hardly feasible for large-scale plants. The use of air-blown autothermal reforming for power production (or catalytic partial oxidation) is considered, however, for large-scale manufacture of hydrogen combined with CO_2 sequestration [3]. Its implementation will depend strongly on imposed legislation.

Cheaper technology for oxygen manufacture may be one route to cost reduction in syngas manufacture. One attempt involves eliminating the oxygen plant and including a reactor concept with oxygen addition through a dense ceramic membrane for oxygen ion transportation [17, 18]. With the large driving force for transport across the membrane, there is no need for compression of air to the pressure of the process. The feasibility of this scheme is yet to be demonstrated [17].

14.2.4
Choice of Technology

For a given hydrocarbon feedstock, steam reforming remains the most economical and efficient technology for a wide range of capacities. For very large capacities, oxygen-blown technologies (autothermal reforming) become more economical because the economy of scale is more favorable for oxygen plants than for the tubular reformer [1, 17]. Parameters other than efficiency play a role for small units, such as simplicity, compactness, and (for automotive units) short start-up time. Air-blown CPO fulfils these requirements, in particular for fuel cell applications, where it is normally acceptable that the hydrogen stream contains nitrogen. A CPO plant has a simpler steam and heat recovery system than a steam reforming plant, but an air compressor is needed, which makes the technology less suited for high-pressure operation. If pure hydrogen is required, the costs of a small oxygen plant or a hydrogen-selective membrane need to be added, making CPO less favorable.

14.3
Process Schemes

14.3.1
Hydrogen by Steam Reforming of Hydrocarbons

From thermodynamics, hydrogen manufacture is favored by a high steam-to-carbon ratio, high exit temperature, and low pressure. The pressure should be close to the supply pressure (typically 25 bar). A high steam-to-carbon ratio means that a surplus of steam has to be heated in the reformer, which results in a larger reformer and high energy consumption. Therefore, modern hydrogen plants [3, 19] are normally designed for low steam-to-carbon ratios (gas 1.8 mol per C-atom). A low steam-to-carbon ratio reduces the mass flow through the plant and thus the size and costs of equipment. This means that less gas will be heated up in the reformer, with a resulting smaller number of tubes. The size of the reformer can be reduced further by increasing the inlet temperature. A high inlet temperature means a higher risk of carbon formation from higher hydrocarbons in the feed. This is eliminated by using an adiabatic prereformer in which higher hydrocarbons are converted. The conditions for a conventional and a modern hydrogen plant are compared in Table 14.3 [1, 19]. It is evident that the reformer is significantly smaller than for the conventional design.

Table 14.3 Typical conditions for hydrogen plants using steam reforming of methane (100 000 Nm^3 H$_2$ h^{-1}) [1]

Design	H$_2$O:CH$_4$	P (bar)	T$_{inlet}$ (°C)	T$_{exit}$ (°C)	H$_2$ (%)	CO (%)	CO$_2$ (%)	CH$_4$ (%)	Process duty (MW)
Conventional	4.5	30	500	850	75	11	11	3	97.8
Modern	1.8	30	650	920	71	18	4	7	85.4

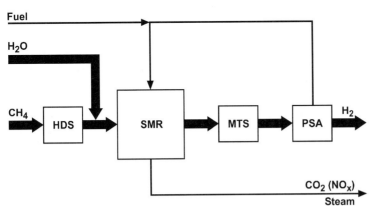

Fig. 14.1 Hydrogen plant (simplified scheme).

A low steam-to-carbon ratio increases the amount of unconverted methane from the reformer, but this is compensated for by increasing the reformer outlet temperature, typically to 900–950 °C. A flow diagram for a modern hydrogen plant is shown in Figure 14.1 [3, 19]. The carbon monoxide in the reformer product gas is converted partially in a shift converter and unconverted carbon monoxide and methane are recovered in a unit for pressure swing adsorption (PSA) for cleaning hydrogen to a purity of 99.9%. The PSA off-gas is then used as fuel for the reformer with additional natural gas. This means that the natural gas for fuel amounts to only about 5% of the total natural gas consumption. Therefore, a modern hydrogen plant is highly efficient [3]. The energy consumed in a methane-based plant (~12.6 GJ per 1000 Nm^3 H$_2$) is very close to the theoretical minimum (11.8 GJ per 1000 Nm^3 H$_2$) [3] when including credit for steam export.

Most hydrogen plants suffer from a surplus of steam production, which may not always be exported. Only about 50% of the heat fired in the tubular reformer is transferred through the tubes with the remaining 50% being recovered mainly as high-pressure steam [3]. In ammonia plants, the steam is used to drive the syngas compressor but in hydrogen plants there is little need for steam. By using convective reformers, the transferred duty to the process can be increased to about 80% with a resulting reduction of the steam export [19, 20]. In a convective reformer, both the flue gas and the product gas are cooled by heat exchange with the process gas flowing through the catalyst beds, so that they leave the reformer at about 600 °C [20]. Convective reforming is not economical for large-scale plants (because of lower heat transfer coefficients and hence the need for larger metal surfaces). Convective reforming can also improve the productivity of the fired reformer by utilizing the hot product gas for supplementary heat input to the process, that is, by "chemical recuperation" of the heat in the process gas instead of raising steam [19].

14.3.2
Hydrogen for Fuel Cells

Hydrogen manufacture for fuel cells differs from large-scale industrial hydrogen manufacture in a number of ways, such as: size, pressure, product quality, and system integration [20, 21].

The size is smaller, which allows for more compact designs that break the constraints of the tubular reformer, not the least the large surplus of catalyst. Microchannel reformers may take advantage of utilizing the catalyst more efficiently [9] either by catalyzed hardware on heat transfer surfaces [22] or by splitting the heat transfer and reaction in a series of adiabatic reactors and heat exchangers [23]. The catalyst beds may be monoliths with low pressure drop and the heat exchangers of compact microchannel design [23]. These types of fuel processors have been developed for a variety of fuels [24, 25], including natural gas [25] and diesel [26].

The low pressure makes pressure drop constraints more critical, often resulting in a risk of maldistribution of flow. Low pressure makes desulfurization of heavy feedstocks an issue that is difficult to resolve because HDS requires hydrogen at high pressure. This may result in selection of the less sulfur-sensitive CPO for the fuel processing, then leaving the sulfur removal problem to cleaning of the CPO product gas for H_2S.

Low pressure means that PSA is not directly applicable for purification of hydrogen. Instead, a series of catalytic steps should be used depending on the requirements on purity of the hydrogen product [20, 21, 24]. This includes low-temperature shift and methanation as used in conventional hydrogen plants or preferential oxidation of carbon monoxide (PROX). It is a challenge that most units must have simple start-up and shut-down procedures, allowing exposure of the hot catalysts to air. This has resulted in the development of noble metal catalysts for most of the steps. The layout of the fuel-producing system depends on the type of fuel cell. Low-temperature fuel cells (PEMFC and PAFC) require pure hydrogen, as carbon monoxide is a poison for the platinum anode. A typical flow scheme for a PEMFC unit is shown in Figure 14.2 [21].

For small-scale units (~ 50 kW) providing hydrogen for automotive fuel cells, the choice of technology is dictated by parameters such as simplicity and rapid response to transients.

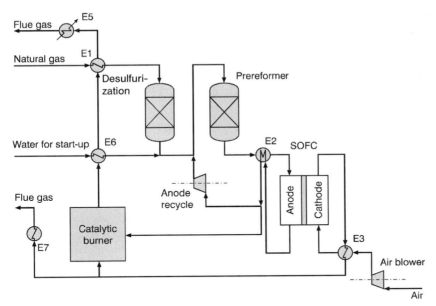

Fig. 14.2 Basic layout of the 250 kW$_e$ SOFC-based CHP system [21]. Fuel processing system. Reproduced with the permission of the author.

CPO may often be the preferred choice for fuel processing because of its compactness and higher flexibility towards sulfur, but the CPO route results in a lower efficiency than the steam reforming route. In principle, the steam reforming route to hydrogen and CO_2 leads to 4 mol H_2 per mol CH_4, equivalent to eight electrons, whereas the CPO route leads to 3 mol H_2 per mol CH_4 or six electrons (this figure should be corrected for the use of anode off-gas as fuel for the reformer). The reaction with oxygen results in an additional heat production which is highest for complete combustion and does not exist for steam reforming. This additional heat production can only be converted into electricity with the Carnot efficiency.

The combustion related to fuel processing can be eliminated in high-temperature fuel cells (SOFC, MCFC), for which it is possible to convert the free energy of methane (or hydrogen and other fuels) directly into electric energy by coupling the reforming reaction with the heat from the electrochemical reaction [20]. The system for direct fuel cells will still include combustion of the anode off-gas to provide heat for preheating and so on [21].

14.3.3
Alternative Routes for Less CO_2

Various attempts have been made to reduce the CO_2 footprint of the steam reforming process by using high-temperature heat for the reaction from nonfossil sources. The stoichiometric molar value is 0.25 $CO_2:H_2$ whereas the value for a modern natural gas-based plant is about 0.4. With an external heat supply, the value may be reduced to about 0.3.

The gas-cooled high-temperature nuclear reactor (HTGR) provides heat as hot helium at 900–950 °C and 40 bar. The hot helium was used as a heating source for a steam reformer (EVA) in the German ADAM-EVA system [27, 28] explored around 1980. The reformer product gas was meant to be transported over long distances and converted into superheated high-pressure steam by high-temperature methanation (ADAM). There is a renewed interest in the HTGR linked the manufacture of hydrogen [29].

Schemes coupling heat from concentrated solar power (CSP) with reforming were developed for energy transmission and storage (CETS) equivalent to the nuclear ADAM-EVA concept [30–32]. At first the aim was on central receivers and separate heat pipe reformers heated by sodium [33]. The preference was dry (CO_2) reforming for the distributed systems to avoid steam in the pipelines. CO_2 reforming requires a noble metal catalyst to avoid carbon formation [34, 35]. The process scheme was demonstrated at large scale [30, 36]. Later work moved away from central receivers and heat pipes and towards "volumetric receivers" with devices where the solar flux passes through a window and directly impinges a porous catalyst bed [37, 38]. A receiver reactor with a parabolic-shaped ceramic foam catalyst was tested in the so-called Caesar project [39]. The concept is still being studied [40–42].

The steam reforming process may be eliminated by using a cyclic process in which hydrogen is generated by reacting steam with a metal or a lower oxide. The resulting oxide is then reduced by reaction with methane, forming steam and CO_2 at a pressure well suited for sequestration. The addition of air is necessary to make the over all reaction thermoneutral [3, 43]:

$$CH_4 + 1.32\, H_2 + 0.34\, O_2 \rightarrow 3.32\, H_2 + CO_2 \tag{14.1}$$

This scheme was studied [43] using iron oxide:

$$\text{heat} + CH_4 + 4\, Fe_2O_3 \rightarrow 8\, FeO + CO_2 + 2\, H_2O \quad\quad (a)$$

$$8\, FeO + 4\, H_2O \rightarrow 4\, Fe_2O_3 + 4\, H_2 \quad\quad (b) \tag{14.2}$$

$$\text{heat} + CH_4 + 2\, H_2O \rightarrow CO_2 + 4\, H_2 \quad\quad (a+b)$$

The heat is supplied by burning some of the hydrogen with air. This scheme may have some advantage for CO_2 sequestration compared with steam reforming.

14.4 Hydrogen from Coal

14.4.1 Coal Conversion Schemes

Coal conversion was considered during the oil crisis in the 1970s with the scope of developing combined cycle power plants (IGCCs) and technologies for the manufacture of synfuels. Both schemes involve the manufacture of hydrogen. A number of gasification technologies were optimized and commercialized in the United States and Europe [44].

With the increasing use of coal, in particular in China, coal conversion has come into focus again. Clean coal conversion is an issue in the United States, which has large coal reserves, however with an increasing demand on CO_2 capture and storage (CCS).

In IGCCs, syngas is burned in a gas turbine followed by a steam turbine [45]. IGCC plants have CO_2 available at much higher pressures than envisaged with normal combustion, which makes carbon capture more feasible. With this, an IGCC plant becomes a hydrogen plant as illustrated in Figure 14.3.

High-temperature entrained flow gasifiers (GE, Shell, Siemens, etc.) [44] are preferred because of full conversion to syngas. The dust-free raw syngas is shifted to convert carbon monoxide to more hydrogen and CO_2. Sulfur-resistant shift catalysts are preferred to minimize the addition of steam [1]. CO_2 and H_2S are removed in a wash, leaving the hydrogen stream for the gas turbine. For ultrapure hydrogen, a final purification can be made in a PSA unit.

Coal conversion to liquid fuels (CTL) implies increasing the atomic ratio from 0.5–0.9 in coal (depending on its origin) to 2, typical for transport fuels. This may take place in two ways, as illustrated below.

Direct liquefaction by hydrogenation:

$$CH_x + (1 - x/2)H_2 \rightarrow CH_2 + \text{heat} \tag{14.3}$$

with external hydrogen supply, which may be manufactured by gasification of the heavy end of the product and unconverted feed or by steam reforming of light hydrocarbons or natural gas.

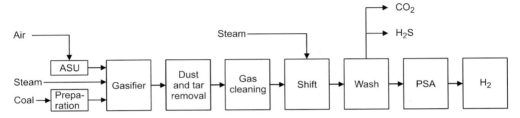

Fig. 14.3 Coal conversion to hydrogen: principal flowsheet.

Indirect liquefaction via gasification and shift followed by synthesis:

$$\text{"}2\,CH_x + 0.5(1-x)O_2 + (2-x)H_2O \rightarrow (CO + 2H_2) + CO_2 + \text{heat"} \quad (14.4)$$

$$\text{"}CO + 2\,H_2 \rightarrow CH_2 + H_2O + \text{heat"}$$

with internal manufacture of hydrogen. As shown, this requires gasification of more coal (2 CH_x per mol CH_2), resulting in CO_2.

The direct liquefaction (Bergius process) was practiced in Germany during World War 2. The start-up of the first modern commercial plant for direct conversion is now taking place in China (Majata, Inner Mongolia) [46]. The hydrogen is manufactured by coal gasification, resulting in a large side production of CO_2.

Indirect conversion of coal to fuels via the Fischer–Tropsch synthesis has been practiced for many years in South Africa (Sasolburg, Secunda). In China, a number of plants are in operation or under construction, including most synfuel syntheses (methanol, DME, diesel, gasoline, SNG).

14.4.2
The CO_2 Footprint

A comparison of the CO_2 footprint of the different process routes is given in Table 14.4 using carbon as the feed material for the stoichiometric conversion (i.e., disregarding the energy consumption for running the process).

It would be tempting to use steam gasification to reduce CO_2, but this endothermic reaction (reaction 12 in Table 14.2) consumes heat (131 kJ mol^{-1} or 18 GJ t^{-1} C). Moreover, high temperatures are required for the steam gasification depending on the reactivity of the coal. Therefore, the oxygen is necessary in conventional gasification to provide the heat required for the coal conversion by "internal" (autothermal) combustion. In principle, the heat from the autothermal combustion could be replaced by another energy source such as heat from CSP, high-temperature nuclear power (HTGR [47]), or concentrated solar power (CSP) [47], making steam gasification possible, but the interest in HTGR and coal is mainly linked to the manufacture of hydrogen [29, 49]. The hydrogen is added to the coal to liquids process to adjust the H_2:CO ratio, thereby

Table 14.4 CO_2 balance for hydrogen from fossil fuels: stoichiometric conversion (i.e., disregarding the energy consumption for running the process).

Reaction (mol WGS)	mol CO_2 per mol H_2	
Gasification	$C + 0.5\,O_2 + H_2O \rightarrow H_2 + CO_2$	1
Steam gasification	$C + 2\,H_2O \rightarrow 2\,H_2 + CO_2$	½
Steam reforming	$CH_4 + 2\,H_2O \rightarrow 4\,H_2 + CO_2$	¼

eliminating the water gas shift reaction and the related CO_2 emission. The hydrogen routes being considered are nuclear steam reforming (EVA type) of natural gas, high-temperature electrolysis, and thermochemical cycles. Solar reforming may be another solution.

14.4.3
Carbon Sequestration

CCS may well dictate the acceptance of coal conversion in the future [50]. CSS means that the efficiency of an IGGC power plant will decrease from 43 to 35% [6].

Capture of CO_2 by absorption using the wet scrubbing processes is well-known technology in the process industry. The capture costs are already a part of syngas-based plants. CO_2 capture is the most expensive part of CCS [51]. The cost of the absorption process is almost proportional to the partial pressure of CO_2.

The captured CO_2 must be compressed before sequestration, at best to a supercritical liquid (critical pressure of CO_2, 73.8 bar), which means a reduction in the volume to 0.3%. This requires a significant amount of energy. The minimum energy required (0.3 GJ t^{-1} CO_2) is determined by thermodynamics from the two conditions of state [1].

However, the CCS experience is limited and there is a need for more research and demonstration projects to create a solid basis for public acceptance.

14.5
Conclusion

Fossil fuels will remain the main source for hydrogen in the mid-term. Steam reforming of hydrocarbons is highly efficient. Membrane-driven reforming may have potential. Hydrogen from coal is used in clean coal power plants based on combined cycle (IGCC) and directly or indirectly in coal conversion to synfuels. Nuclear or solar-driven reforming might play a future role in reducing the CO_2 footprint. In the long term, the use of fossil fuels for hydrogen will be linked to the acceptance of CO_2 sequestration.

References

1 Rostrup-Nielsen, J.R. and Christiansen, L.J. (2010) *Concepts in Syngas Preparation*, Imperial College Press, London, to be published.
2 Rostrup-Nielsen, J.R. (2008) In *Encyclopedia of Catalysis* (ed. I.T. Hováth), Vol. 4, pp. 1–17
3 Rostrup-Nielsen, J.R. and Rostrup-Nielsen, T. (2002) *Cattech*, 6 (4), 150.
4 International Energy Agency (2005) *Prospects for Hydrogen and Fuel Cells*, IEA, Paris.

5 Strub, R.A. and Imarisio, G. (1980) *Hydrogen as an Energy Vector*, Reidel, Dordrecht.
6 Rostrup-Nielsen, J.R. (2004) *Catal. Rev. Sci. Eng.*, **46**, 246.
7 Rostrup-Nielsen, J.R. (1984) In *Catalysis, Science and Technology* (eds J.R. Anderson and M. Boudart), Vol. 5, Springer, Berlin, Chapter 1.
8 Rostrup-Nielsen, J.R. (2009) In *Handbook of Heterogeneous Catalysis* (eds G. Ertl, H. Knözinger, F. Schüth, and J. Weitkamp), Wiley-VCH Verlag GmbH, Chapter 13.11, pp. 1–24.
9 Rostrup-Nielsen, J.R., Sehested, J., and Nørskov, J.K. (2002) *Adv. Catal.*, **47**, 65.
10 Gøl, J.N. and Dybkjær, I. (1995) *HTI Q.*, (Summer), 27.
11 Rostrup-Nielsen, J.R., Dybkjær, I., and Christensen, T.S. (1998) *Stud. Surf. Sci. Catal.*, **113**, 2.
12 Aasberg-Petersen, K., Stub Nielsen, C., and Lægsgaard-Jørgensen, S. (1998) *Catal. Today*, **46**, 193.
13 Aasberg-Petersen, K., Bak Hansen, J.-H., Christensen, T.S., Dybkjær, I., Seier Christensen, P., Stub Nielsen, C., Winter Madsen, S.E.L., and Rostrup-Nielsen, J.R. (2001) *Appl.Catal. A: Gen.*, **22**, 379.
14 Christensen, T.S., Seier Christensen, P, Dybkjær, I., Bak Hansen, J-H., and Primdahl, I.I. (1998) *Stud. Surf. Sci. Catal.*, **119**, 883.
15 Enger, B.C., Lødeng, R., and Holmen, A. (2008) *Appl. Catal. A: Gen.*, **346**, 1.
16 Rostrup-Nielsen, J.R. (2007) *Stud. Surf. Sci. Catal.*, **167**, 153.
17 Rostrup-Nielsen, J.R. (2000) *Catal. Today*, **63**, 159.
18 Chen, C.M., Bennett, D.L., Carolan, M.F., Foster, E.P., Schinski, W.L., and Taylor, D.M. (2004) *Stud. Surf. Sci. Catal.*, **147**, 55.
19 Rostrup-Nielsen, J.R. and Winter-Madsen, S.E.L. (2008) *Prepr. Pap. Am. Chem. Soc. Div. Pet. Chem.*, **53** (1), 82.
20 Dybkjær, I. and Winter Madsen, S.E.L. (1997/98) *Int. J. Hydrocarbon Eng.*, **3** (19), 556.
21 Hansen, J.B. In *Proceedings of the 6th European SOFC Forum*.
22 Farrouto, R.J., Liu, Y., Ruettinger, W., Ilinich, O., Shore, L., and Giroux, T. (2007) *Catal. Rev.*, **49**, 141.
23 Seris, E.L.C., Abramowitz, G., Johnston, A.M., and Haynes, B.S. (2005) *Trans. I. Chem. E., Part A: Chem. Eng. Res. Dev.*, **83** (A6), 619.
24 Kolb, G. (2008) *Fuel Processing for Fuel Cells*, Wiley-VCH Verlag GmbH, Weinheim.
25 Mazanec, T. (2003) *Pet. Q.*, (Autumn), 149.
26 Irving, P. and Pickles, J. (2006) In *Abstracts of the 2006 Fuel Cell Seminar*, p. 242.
27 Kugeler, K., Niessen, H.F., Röth, M., Kamat, D., Böcker, B., Rüter, K., and Theis, A. (1975) *Nucl.Eng. Des.*, **34**, 65.
28 Fedders, H., Harth, R., and Höhlein, B. (1975) *Nucl. Eng. Des.*, **34**, 119.
29 Kuhr, R. (2008) *Nucl. Eng. Des.*, **238**, 3013.
30 Diver, R.B., Fish, J.D., Levitan, R., Levy, M., Meirovitch, E., Rosin, H., Paripatyadar, S.A., and Richardson, J.T. (1992) *Solar Energy*, **48**, 21.
31 Diver, R.B., Miller, J.E., Allendorf, M.D., Siegel, N.P., and Hogan, R.E. (2008) *J. Solar Energy Engl. Transl. ASME*, **130**, 0410011
32 Becker, M., Harms, R., and Müller, W.D. (1986) In *Proceedings of Intersoc. Energy Conference*, p. 920.
33 Richardson, J.T., Paripatyadar, S.A., and Shen, J.C. (1988) *AIChE J.*, **34**, 743.
34 Rostrup-Nielsen, J.R. and Bak Hansen, J.-H. (1993) *J. Catal.*, **144**, 38.
35 Richardson, J.T., Jung, J-K., Zhao, J. (2001) *Stud. Surf. Sci. Catal.*, **36**, 203.
36 Levy, M., Levitan, R., Rosin, H., and Rubin, R. (1993) *Solar Energy*, **50**, 179.
37 Diver, R.B. (1987) *Chemtech*, **17**, 606.
38 Hogan, R.E. Jr, Skocypec, R.D., Diver, R.B., Fish, J.D., Garrait, M., and Richardson, J.T. (1990) *Chem. Eng. Sci.*, **45**, 2751.
39 Buck, R., Muir, J.R., Hogan, R.E., and Skocypec, R.D. (1991) *Solar Energy Mater.*, **24**, 449.
40 Wörner, A. and Tamme, R. (1998) *Catal.Today*, **46**, 165.
41 Steinfeld, A. (2005) *Solar Energy*, **78**, 603.

42 Petrasch, J. and Steinfeld, A. (2007) *Chem. Eng. Sci.*, **62**, 4214.
43 Sanfilippo, D., Miracca, I., Cornaro, U., Mizia, F., Malandrino, A., Piccoli, V., and Rossini, S. (2004) *Stud. Surf. Sci. Catal.*, **147**, 91.
44 Higman, C. and van der Burgt, M. (2003) *Gasification*, Elsevier, Amsterdam.
45 Mills, G.A. and Rostrup-Nielsen, J.R. (1994) *Catal.Today*, **22**, 335.
46 (2009) *Chem. Eng.*, **6**, 11.
47 van Heek, H.H., Jürtgens, H., and Peter, W. (1973) *J. Inst. Fuel*, **46**, 249.
48 Z'Graggen, A. and Steinfeld, A. (2008) *Chem. Eng. Process.*, **47**, 655.
49 Greyvenstein, R., Correia, M., and Kriel, W. (2008) *Nucl. Eng. Des.*, **238**, 3031.
50 Azar, C., Lindgren, K., Larson, E., and Möllersten, K. (2006) *Climate Change*, **74**, 47.
51 Metz, B., Davidson, O., de Coninck, H., Loos, M., and Meyer, L. (2005) *IPCC Special Report on Carbon Dioxide Capture and Storage*, Intergovernmental Panel on Climate Change, Cambridge University Press, Cambridge.

15
Reforming and Gasification – Biomass

Achim Schaadt, Siegfried W. Rapp, and Christopher Hebling

Abstract

Gasification is an efficient method to convert both wet and dry biomass into a gaseous product with a usable heating value. Recently, the technology for the gasification of dry biomass has approached commercialization for some applications. In addition, a variety of processes for different products are currently tested in pilot plants. In future, these processes could provide a significant contribution for the sustainable production of biomass-derived energy carriers which could be used for mobile and stationary applications (e.g., fuel cells). In contrast to conventional gasification, hydrothermal processes can efficiently convert biomass with its natural water content. However, hydrothermal gasification still requires further research and development before large-scale pilot plants can be built.

Keywords: reforming, gasification, biomass

15.1
Introduction

In the near future, the global energy demand has to be met completely by renewable energy. Among the various sources of renewable energy, biomass is the type which provides the largest contribution to the worldwide energy supply today. In Germany, about 5% of the primary energy demand was provided by biomass in 2008 [1], whereas in less industrialized countries, this value can rise to 90%. Biomass can be converted to other forms of energy via either biological or thermochemical processes. In this chapter, we focus on thermochemical processes (e.g., gasification), which could produce a variety of fuels (e.g., hydrogen, methane) for different applications. This chapter summarizes the activities in current biomass gasification processes and the reforming of biomass-based components. Not only are pre-commercial processes presented, but also promising technologies that are not yet commercially feasible but have the potential to achieve this goal very soon are addressed.

Hydrogen Energy. Edited by Detlef Stolten
Copyright © 2010 WILEY-VCH Verlag GmbH & Co. KGaA, Weinheim
ISBN: 978-3-527-32711-9

15.2
Gasification of Biomass

15.2.1
Operating Conditions – Dry Biomass

Since lignocellulosic biomass is by far the most abundant biomass material with an annual worldwide yield of residues estimated at more than 220 billion tons, it is a very attractive and low-cost feedstock [2]. The most common thermochemical process is combustion, but gasification, which includes pyrolysis, has attracted increasing attention over the last 10–15 years due to increasing oil prices. At present, about 89% of the biomass which is used worldwide for energy supply is combusted by traditional firing and less than 0.8% is used in combined heat and power (CHP) plants [3]. The net efficiency for electricity generation based on biomass combustion is typically in the range of only 20–40% [4]. Another method for using biomass as a source of energy is co-firing in existing coal combustors. Since biomass leads to clogging of the coal feed systems, the fraction of biomass is limited to 5–10% of the total feedstock.

Gasification is a process which converts carbonaceous sources to a gaseous product with a useable heating value. Modern gasification processes conserve between 75 and 88% of the heating value of the original fuel. The commercial application of gasification began with the foundation of the London Gas, Light and Coke Company in 1812. This first process was carried out in the absence of oxygen (pyrolysis) and was based on coal in order to produce the widely used town gas. In the 20th century, town gas was gradually replaced by cheaper natural gas [5].

Gasification is mostly achieved by partial oxidation to form a gaseous mixture which consists mainly of hydrogen and carbon monoxide but also contains methane and carbon dioxide [6]. This mixture is called synthesis gas (syngas) because of its use in several fundamental synthesis reactions in the chemical industry.

A variety of gasification agents can be used: air, oxygen, steam, carbon dioxide (CO_2), and mixtures of these. Air, as it is very cheap, is widely used as the oxidant. However, a large amount of the inert component nitrogen is introduced into the process, so that the syngas contains more than 50% by volume of nitrogen. As a result, the heating value of the syngas produced is very low (3–6 MJ N^{-1} m^{-3}) [7, 8]. The application of oxygen instead of air as a gasification agent would increase the heating value of the syngas, but the additional oxygen generation would also increase the costs. Another approach to increase the heating value to 10–15 MJ N^{-1} m^{-3} is the application of steam as a gasification agent. The use of carbon dioxide as an oxidant is an interesting option, because it is always one of the product gases. Furthermore, it has been reported that CO_2 with a catalyst such as Ni/Al can convert char, tar, and methane into hydrogen and/or carbon monoxide [9–11].

When the heat for the endothermic gasification reaction is provided by the exothermic partial oxidation using oxygen or air, the process is called autothermal gasification. An example of directly heated gasification is the two-stage Viking process [12]. If carbon dioxide or steam is used as the oxidant, an indirect or external heat supply has to be provided. Examples of such an allothermal process are the fast internal circulating fluidized-bed (FICFB) gasification [13], Battelle gasification [14] and the biomass heatpipe reformer [15].

When using biomass as a source of energy, an important issue is the appropriate amount of biomass to be fed to the gasifier. The lower energy density of biomass (about 3.7 GJ m^{-3}) compared with fossil fuels (gasoline: 33 GJ m^{-3}) results in greater relative effort for transportation. For that reason, biomass plants should not exceed a capacity of approximately 40 MW for economic and ecological reasons. Holt and van der Burgt [16] proposed a two-stage concept in order to overcome this limitation in size and to benefit from the economies of scale. In the first step, flash pyrolysis is carried out, which increases the energy density to about 20 GJ m^{-3}. During flash pyrolysis, the biomass is heated at atmospheric pressure in the absence of oxygen for 1 s to a temperature of about 740 K, producing a liquid bio-oil as the main product (75% by weight). By-products are pyrolysis gas and coke, which can be used as an energy source to maintain the endothermic process. This flash pyrolysis can be built economically on a reasonably small scale and would be located very close to the source of biomass. The bio-oil produced can be transported to a central world-scale gasifier.

Although biomass pyrolysis is a fairly young technology, many small-scale commercial and demonstration plants are already in operation or in planning [5]. Within a European project, a set of properties for bio-oils was specified in order to guarantee at least a minimum quality for end users [17]. At the Forschungszentrum Karlsruhe (FZK) in Germany, initial gasification tests of biomass-based slurries under atmospheric conditions are being investigated in a 60 kW plant in order to gain a better understanding of gasification in an entrained flow reactor [18].

15.2.2
Reactor Types – Dry Biomass

Depending on their fluid mechanical properties, gasifiers are often classified into fixed-bed (sometimes called moving bed), fluidized-bed, and entrained flow processes. In moving-bed gasifiers, the biomass moves slowly downwards due to gravity during gasification. Either co-current or counter-current operation can be applied. Co-current flow means that the biomass and the oxidants flow in the same direction. In the counter-current configuration, a higher tar content is usually observed, since residues of pyrolysis products are entrained due to a shorter residence time for some of the biomass constituents.

Currently, fixed-bed biomass gasification is offered in combination with CHP plants generating an electrical output between 10 and 400 kW at a pre-commercial stage in Germany by only about 10 companies. The high potential of fixed bed gasifiers is related to their high thermal efficiency and the minimal pretreatment of biomass required. Fixed-bed gasifiers are economically feasible only on a scale smaller than 1 MW, whereas fluidized-bed gasifiers should be operated in plants larger than 10 MW for economic reasons.

In a fluidized-bed gasifier, stable fluidization is ensured by a bed material which may also act as a catalyst. The high flow rate of bed material provides optimal heat and mass transfer conditions. The fluidization is achieved by the gasification agent. In the case of a stationary or bubbling bed, the superficial velocities are in the range 1–2 m s^{-1}. When fast (circulating) fluidized-bed gasification is used, the superficial velocity is increased to 6 m s^{-1}, so that the bed material is entrained to the top of the gasifier and fed back to the reactor via a cyclone. In this case, the average size of the bed particles is in the range 0.2–0.4 mm. Table 15.1 summarizes the typical parameters used in fluidized-bed gasifiers.

At present, many teams are working on the three possible combinations of dual fluidized-bed gasifiers, namely two bubbling beds, one bubbling and one fast fluidized bed, or two fast fluidized beds [19]. The main reason is that this method allows the production of nitrogen-free syngas without the need for pure oxygen. This is implemented by using two chambers which can exchange hot heat-transport particles. In the first fluidized bed, the gasification is carried out by steam, whereas in the other bed combustion takes place to generate the heat for the endothermic gasification. Heat pipes have also been used to transport the heat between the two dual fluidized-bed systems [20]. The first proof of concept of the dual fluidized-bed technology has been demonstrated in Güssing (Austria). After the successful implementation, five other industrial plants with a biomass input ranging from 10 to 25 MW were constructed and are currently being tested [21].

In the third class of gasifiers, the entrained flow reactor, the biomass particles are moved through the reactor within seconds. Since the gasification temperature is very high (>1473 K) and the particle diameter is small, this short residence time is sufficient to achieve complete gasification. As a consequence of this high operating temperature, a low methane content (<1 vol.%) and a very low tar content (<10 mg N^{-1} m^{-3}) can be obtained. Although this almost tar-free gasification is very attractive, the handling of the molten slag, which is very aggressive at this high temperature, is very challenging. Furthermore, in order to

Table 15.1 Characteristic data for fluidized-bed gasifiers [19].

Temperature (K)	973–1173
Capacity (MW)	10–100
Fuel flexibility	High
Tar content (g Nm^{-3} dry gas)	1–20 (medium); <1 if catalytically active bed material is used
CH$_4$ content (vol.%)	10
Status	commercial for co-firing, CHP and IGCC (integrated gasification combined cycle)

15.2 Gasification of Biomass

obtain these small particles, fuel pretreatment (pyrolysis, grinding, etc.) is required. For these reasons, fluidized-bed technology is applied most often for biomass gasification.

15.2.3
Applications – Dry Biomass

The main obstacle to the commercialization of gasification plants is still the lack of sufficient experience and the challenge to prove a real economic advantage compared with combustion-based technology. Hence activities in research and development have to some extent moved from CHP applications towards syngas production in order to avoid competition with combustion technology.

However, the main advantages of gasification are its flexibility with respect to the type of biomass used and the variety of products obtained. The operating conditions and the reactor types presented in Sections 15.2.1 and 15.2.2, respectively, have a strong influence on the composition of the syngas. Depending on the process conditions, the following products are currently obtained:
- heat
- electricity
- syngas
- substitute natural gas (SNG)
- hydrogen
- Fischer–Tropsch fuel (FT fuel)
- methanol
- dimethyl ether (DME).

Figure 15.1 illustrates the processes and reactions that are necessary to obtain the different products. It becomes clear that the H_2/CO ratio is an important parameter and has to be adjusted within a certain range in order to obtain the desired product.

Starting from lignocellulosic biomass and applying a gasification process, raw syngas is generated, which can be combusted either to obtain heat for industrial processes or for use in a co-firing plant. In both cases, no gas cleaning is necessary. Since it is often desirable to generate both heat and electricity, more or less complex gas cleaning/tar reduction has to be carried out. Apart from most CHP applications, all downstream processes rely on heterogeneous catalysis, which implies high vulnerability to chlorine and sulfur poisoning. For long-term applications, the concentration of such undesired by-products should not exceed 50 ppbv [19].

As gasification technology permits the application of gas engines, gas turbines, and fuel cells, CHP biomass plants are expected to achieve market penetration very soon. Gasification-based CHP plants operated with gas engines are already in the pre-commercial phase or have at least been tested on a full plant scale (gas turbines) [22], whereas systems with solid oxide fuel cells (SOFCs) are at the very beginning of development [23, 24]. Fuel cells are attractive be-

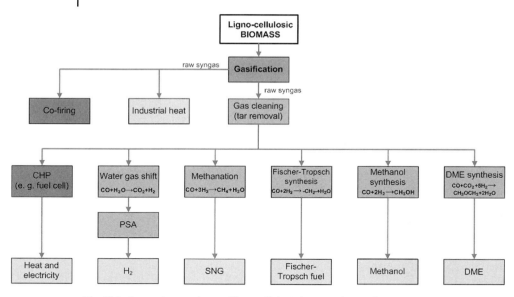

Fig. 15.1 Conversion products of lignocellulosic biomass by gasification.

cause their electrical efficiency, especially for partial loads, is much higher than that of gas engines or gas turbines. For the application discussed here, the SOFC is the preferred type of fuel cell, due to the low requirements on gas quality and the ability to convert carbonaceous fuels. Martini et al. [24] tested a SOFC in a demonstration plant (5 Nm^3 h^{-1}) with wood gas that had been previously purified. Although they could demonstrate general suitability, the power output and the mechanical reliability of the fuel cell were insufficient. Biollaz et al. [23] investigated the influence of diluted wood gas on the long-term performance of a SOFC. They found that tar does not influence the degradation of the fuel cell, provided that the tar content is below 5 g N^{-1} m^{-3}.

As a further conversion path, the generation of substituted natural gas made from syngas is considered. The first test of the demonstration plant for Bio-SNG in Güssing (Austria) was successful. The existing dual fluidized-bed gasification plant was coupled with a new 1 MW methanation reactor. In addition, a further gas cleaning step (sulfur trap) is included between gasification and methanation. Downstream of the methanation, the product is purified by CO_2 removal, drying, and H_2 separation. During 14 days of testing in June 2009, the gas quality achieved met the criteria to operate a gas station [19].

The dual fluidized-bed gasifier in Güssing was also used for steam gasification of biomass with *in situ* CO_2 absorption. This process is called absorption-enhanced reforming (AER), which yields a product gas with a higher hydrogen content than that obtained with other gasification processes [25]. This technology is based on the use of a reactive bed material (limestone) which absorbs/desorbs CO_2 by undergoing cyclic carbonation of CaO and calcination of $CaCO_3$.

Although this process had been demonstrated previously in a pilot reactor, this test in Güssing is the first one at an industrial scale (8 MW). The AER product gas obtained in Güssing consisted of approximately 50 vol.% (dry gas) of hydrogen instead of 40 vol.%, which would be a typical value in a standard gasification process. In addition, the gasification temperature (950 K) can be adjusted to be significantly lower than in other gasification processes (1150 K). Furthermore, the tar content in the raw product gas was reduced from 2–5 to 1 g Nm^{-3}. This can be explained by the catalytic activity of limestone with respect to tar reforming. Since the AER process already produces syngas with a high hydrogen content, it would be very suitable for the production of pure hydrogen to operate low-temperature fuel cells, which require a higher hydrogen quality than a SOFC. In order to increase the hydrogen content further, a water gas shift reaction has to be carried out after the AER process. In this state-of-the-art process, CO is converted into hydrogen using water (Figure 15.1). Depending on the type of fuel cell used, the product gas after the shift reactor is fed to a fuel cell (e.g., a high-temperature PEM fuel cell) or is further purified to a hydrogen concentration greater than 99.999% by a commercially available pressure swing adsorption (PSA) process. Hydrogen production from biomass gasification is considered to be the technology that could provide hydrogen from renewable sources at the lowest cost [26].

Despite the drawbacks of entrained-flow gasifiers mentioned in Section 15.2.2, this technology is used in CHOREN's Carbo-V® process for large-scale production of FT fuel [27]. In order to solve the previously mentioned problems, a three-stage process was developed. During preconditioning, the biomass with a water content of about 15% by weight is fed under pressure into a low-temperature gasifier (673–773 K) using oxygen and carbon dioxide as the oxidants and producing volatile gases and char. The gaseous components are converted by partial oxidation at temperatures above 1673 K (second stage) and the biocoke is separated, milled to a powder, and mixed with hot gases produced by the partial oxidation. This chemical quench reduces the temperature immediately to approximately 1173 K. The cold gas efficiency is reported [27] to be 81.4%. The technology was tested successfully in an (Alpha) plant with a 1 MW thermal input of biomass during more than 20 000 h of operation. In 2007–08, a (Beta) plant was erected close to the first one in Freiberg in Saxony, Germany, with a thermal power of 45 MW and an annual FT fuel production capacity of 15 000 tons, meeting the annual fuel demand of approximately 15 000 cars. Currently, the Beta plant is being tested and optimized within the European research project OPTFUEL, in order to identify the best process parameters for an even larger plant with an annual capacity of 200 000 tons of FT fuel. In the EU-funded Renew project, which was coordinated by Volkswagen AG, the manufacturing costs of FT fuel from biomass were calculated to be € 1.5 per liter.

Further liquid fuels that will be produced by biomass gasification in large quantities are dimethyl ether (DME) and methanol. In September 2009, the construction of a pilot plant for DME started in Sweden. The process is based on the technology of Chemrec, a company which specializes in reactor design for black liquor gasification. Black liquor is a by-product that is produced in large quantities in the pulp and paper industry. Since there is a large paper-producing industry in Sweden, it is estimated that half of all freight transportation in Sweden could be fueled by DME produced from black liquor. Within the EU's Seventh Framework Programme, the BioDME project will start DME production in a pilot plant with a capacity of approximately 1500 tons per year. In addition, the Swedish Energy R&D Board will fund a project to demonstrate DME and methanol production on an industrial scale (100 000 tons per year).

15.2.4
Hydrothermal Gasification – Wet Biomass

15.2.4.1 Overview

The gasification technologies presented up to now are not able to convert wet biomass (moisture content >70 wt%) economically because of their high energy consumption for the evaporation of water. Such wet biomass is very common and is produced by agriculture, bio-energy processes, biochemically based processes, and also by the food and beverage industry. Because the transportation of wet biomass would be very expensive and energy consuming, decentralized units are recommended [28]. On the one hand, a high dry matter content is useful for obtaining a high flow rate of product gas, but on the other hand, the feedstock has to be suitable for pumping. Depending on the type of biomass and the pretreatment applied, dry matter contents of up to 40 wt% have been used.

Typically, the water content in conventional gasification processes is in the range of only 10–15 wt%. Since natural drying takes a very long time, forced drying is often applied as the first process step in a gasification plant to ensure a feedstock with a uniform water content. In contrast to conventional gasification, a hydrothermal process is carried out, using an excess of water at increased elevated temperature and pressure. High pressure is applied in order to prevent water evaporation.

A distinction must be made between hydrothermal carbonization (main product: high-quality coke), hydrothermal liquefaction (main product: viscous oil) and hydrothermal gasification, which produces mainly hydrogen and methane. In this chapter, we focus on hydrothermal gasification. Hydrothermal gasification can be further divided into catalyzed aqueous-phase reforming to produce hydrogen (490–540 K), catalyzed near-critical gasification (at ~ 620 K) or catalyzed gasification (at ~ 670 K) in the supercritical state to produce methane, and supercritical water gasification (SCWG) with or without heterogeneous catalysis, to produce mainly hydrogen.

The product gas is obtained during hydrothermal gasification at high pressure, which avoids the need for further compression and simplifies the separation of CO_2. A gas with a high energy content is obtained. Hydrothermal processes open up the chance to convert biomass with its natural water content, that is, an extra drying procedure is avoided. Another advantage of hydrothermal gasification is a relatively low process temperature compared with other gasification processes. In hydrothermal processes, water acts as a solvent, as a catalyst or catalyst precursor, and as a reactant.

15.2.4.2 Aqueous Phase Reforming

In 2002, Cortright *et al.* published a fundamental paper on aqueous-phase reforming (APR) [29]. APR is carried out in the presence of a heterogeneous catalyst (e.g., Pt, but also Ni, Ru, Rh, Pd, Ir), with hydrogen being the main product. Since the temperature range used is also favorable for the water gas shift reaction, it is possible to produce hydrogen with very low amounts of CO in a single reactor. Depending on the reaction conditions, CO concentrations below 100 ppmv can be achieved. Compared with other hydrothermal gasification techniques, low pressure is applied (1–5 MPa).

Originally, the APR process was applied to convert diluted solutions of sugars (e.g., glucose) and polyols (e.g., glycerol, ethylene glycol, sorbitol) efficiently into hydrogen. The APR process was extended for feed streams containing 60 wt% of glucose in water at an efficiency higher than 55% [30]. Recently, the application of the APR technology to lignocellulosic biomass (southern pine sawdust) was reported for the first time [31].

15.2.4.3 Catalytic Hydrothermal Gasification for the Production of Methane

A thermodynamic analysis showed that a high methane yield could be obtained when conditions close to the critical point (T_c=647 K, p_c=22.1 MPa) are applied and the dry matter content is in the range of 10–20 wt.%. In order to benefit from this low temperature, a suitable catalyst (e.g., nickel) has been developed [32]. The application of a catalyst allowed the use of a cheaper reactor material, but the stability of the catalyst presented a problem that was hard to solve. The catalytic hydrothermal gasification of biomass has been investigated for more than 25 years. During this period, significant progress was made in the development of new catalyst formulations. As a result, continuous-flow processing has been successfully demonstrated in laboratory test rigs [33].

15.2.4.4 Supercritical Water Gasification

SCWG uses water above the critical point, which has completely different properties to water under ambient conditions. The density and the static relative dielectric constant decrease considerably. Under supercritical conditions, water is a fairly aggressive medium and behaves like a non-polar solvent, showing a high solubility in organic compounds and a low solubility of salts. The solvation of intermediates by hot compressed water, in particular in the supercritical state, significantly inhibits the formation of undesired products such as tar and coke [34].

SCWG is carried out at a comparatively high temperature ($\sim$870 K) in order to shift the equilibrium towards hydrogen. The dry matter content typically ranges from 10 to 20 wt%. Antal and co-workers developed fairly efficient carbon catalysts that were successfully applied to convert sewage sludge into a hydrogen-rich gas [35]. Although a catalyst increases the gas yield, there is also the risk that salts from the biomass clog the reactor.

The VERENA test facility, which is located in Karlsruhe (Germany), is the only larger SCWG plant known to the authors, with a throughput of 100 kg h^{-1}. In this pilot plant, different types of biomass (e.g., corn silage) have been successfully converted in the absence of a catalyst with a heat-transfer efficiency greater than 80%. The overall reported energy efficiencies for different SCWG processes are high but vary greatly, ranging from 44 to 65% [28]. Thermodynamic calculations clearly show that even small improvements of the heat exchanger have a strong effect on the overall efficiency. In addition to the heat exchange efficiency, a high heating rate is another aspect that still has to be improved in the field of SCWG.

In 2002, Matsumura [36] compared the SCWG with the alternative process of biomethanation (anaerobic digestion) with regard to efficiency and costs. Although the costs of methane production would have been 1.86 times higher than the price of Tokyo city gas at that time, it became clear that the SCWG process showed better results according to both criteria. In addition, Yoshida *et al.* [37], in an analysis of different biomass conversion technologies, found that SCWG was the most energy-efficient technology for biomass with a water content higher than 30 wt%.

15.2.5
Reforming of Biomass-Derived Energy Carriers

Biomass-derived fuels can be produced thermochemically, as shown in Section 15.2.3 or via biological processes. In the following, the reforming of different biofuels to a hydrogen-rich gas for fuel cell applications will be discussed:
- bioethanol and biomethanol
- biodiesel
- bioglycerol
- biogas.

Fuel cell systems based on ethanol are currently under development [38, 39] and it is assumed that they will play an important role in the near future for applications such as portable, mobile, and off-grid power supply, due to their high efficiency and low emissions. Although the technology of bioethanol production is mature and the worldwide annual production capacity is in the region of 6×10^7 m^3 (2007), there are currently many activities to use lignocellulosic residues instead of sugar-containing or starchy plants in order to avoid competition with food and to reduce the production costs [40]. In contrast to methanol, ethanol is a nontoxic compound and therefore much easier to handle. In addition, ethanol is characterized by a much higher energy density than methanol. Since the production costs of biomethanol are still too high, most of the methanol produced is based on nonrenewable sources. On the other hand, methanol reforming is carried out at a much lower temperature (methanol ~540 K, ethanol ~970 K) and less complex gas purification is required. Fuel cell systems based on methanol are already commercially available [41].

Biodiesel, which is a nontoxic and biodegradable fuel, is produced by transesterification of vegetable oil with methanol, with glycerol being obtained as an additional product. An example of activities in this field is the European BIOFEAT project, which is primarily aimed at developing a biodiesel-fueled, heat-integrated fuel processor for an electrical power output of 10 kW, to operate an auxiliary power unit (APU) for road vehicles. As a result, a high reformate quality with CO concentrations lower than 10 ppmv and a fuel processor efficiency of 87% were obtained. Further improvements are necessary in order to test the durability of the catalysts and to reduce the volume, the weight, and the start-up time of the system [42].

Glycerol, which is formed in huge amounts as a by-product of biodiesel production, is completely converted by conventional steam reforming at a temperature of only 400 °C [43]. In another study, glycerol was converted by steam reforming into synthesis gas, which was fed into an SOFC. The SOFC was operated with pure glycerol and with glycerol which was taken from a biodiesel plant, containing fatty matter, methanol, water, and ash as additional components. It was found that under the same reaction conditions, the yield decreased considerably from 100% of pure glycerol to 70% when by-product glycerol was used, and the deposition of carbon was observed [44]. Another interesting approach to convert glycerol into hydrogen could be aqueous-phase reforming [29].

Biogas, which contains mainly methane and carbon dioxide, is produced by anaerobic digestion. If biogas could be used in fuel cells instead of combustion in internal combustion engines, the energy conversion efficiency could be increased and the emissions could be reduced. Since biogas contains traces of contaminants, the use of high-temperature fuel cells such as SOFCs, which are more tolerant regarding impurities, is recommended. In addition, a SOFC would offer a better possibility for heat integration. A 1 kWe SOFC from Sulzer-Hexis has been successfully operated with biogas for 1 year [45].

A further interesting option in addition to steam reforming, autothermal reforming, and partial oxidation is the application of highly endothermic dry reforming or CO_2 reforming:

$$CH_4 + CO_2 \rightarrow 2CO + 2H_2 \quad \Delta H = 247 \text{ kJ mol}^{-1} \quad (15.1)$$

A disadvantage of dry reforming is the risk of carbon deposition by methane cracking and the Boudouard reaction. Since carbon deposition is thermodynamically possible up to 1270 K, it is necessary to develop a catalyst which kinetically inhibits carbon formation. In the work of Xu et al. [46], an Ni/Co bimetallic catalyst was developed which showed a high conversion performance (>93%) and low carbon deposition during a 290 h test.

References

1 Federal Ministry for the Environment, Nature Conservation and Nuclear Safety (2009), *Using Current Data of the Working Group on Renewable Energies – Statistics (AGEE-Stat)*, http://www.erneuerbare-energien.de/inhalt/39831/39882/ (accessed 28 October 2009).

2 Ren, N., Wang, A., Cao, G., Xu, J., and Gao, L. (2009) Bioconversion of lignocellulosic biomass to hydrogen: potential and challenges. *Biotechnol. Adv.*, **27**, 1051–1060.

3 Sterner, M. (2009) Wissenschaftlicher Beirat der Bundesregierung Globale Umweltveränderungen (WBGU): "Welt im Wandel: Zukunftsfähige Bioenergie und nachhaltige Landnutzung". Lecture given at Bio CHP Congress at Enertec, Leipzig, 27 January 2009, WBGU Expertise from 3 December 2008.

4 Caputo, A. C., Palumbo, M., Pelagagge, P. M., and Scacchia, F. (2005) Economics of biomass energy utilization in combustion and gasification plants: effects of logistic variables. *Biomass Bioenergy*, **28**, 35–51.

5 Higman, C. and van der Burgt, M. (2008) *Gasification*, 2nd edn, Elsevier, Amsterdam.

6 Knoef, H. (2005) *Handbook of Biomass Gasification*, Biomass Technology Group BTG, Enschede.

7 Gabra, M., Pettersson, E., Backman, R., and Kjellström, B. (2001) Evaluation of cyclone gasifier performance for gasification of sugar cane residue – Part 1: gasification of bagasse. *Biomass Bioenergy*, **21**, 351–369.

8 Zainal, Z. A., Rifau, A., Quadir, G. A., and Seetharamu, K. N. (2002) Experimental investigation of a downdraft biomass gasifier. *Biomass Bioenergy*, **23**, 283–289.

9 Devi, L., Ptasinski, K. J., and Janssen, F. J. J. G. (2003) A review of the primary measures for tar elimination in biomass gasification processes. *Biomass Bioenergy*, **24**, 125–140.

10 Garcia, L., Salvador, M. L., Arauzo, J., and Bilbao, R. (2001) CO_2 as a gasifying agent for gas production from pine sawdust at low temperatures using Ni/Al coprecipitated catalyst. *Fuel Process. Technol.*, **69**, 157–174.

11 Ollero, P., Serrera, A., Arjona, R., and Alcantarilla, S. (2003) The CO_2 gasification kinetics of olive residue. *Biomass Bioenergy*, **24**, 151–161.

12 Henriksen, U., Ahrenfeldt, J., Kvist Jensen, T., Gøbel, B., Dall Bentzen, J., Hindsgaul, C., and Holst Sørensen, L. (2006) The design, construction and operation of a 75 kW two-stage gasifier. *Energy*, **31**, 1542–1553.

13 Hofbauer, H., Rauch, R., Löffler, G., Kaiser, S., Fercher, E., and Tremmel, H. (2002) Six years experience with the FICFB-gasification process. In *Proceedings of the 12th European Conference on Biomass for Energy, Industry and Climate Protection*, Amsterdam, Netherlands.

14 Mansour, M. N., Chandran, R. R., and Rockvam, L. (2002) The evaluation of and advantages in steam reforming of black liquor. In *Fall Technical Conference and Trade Fair*, San Diego, CA, 8–11 September 2002, Session 63, TAPPI Press, Atlanta, GA.

15 Karellas, S., Karl, J., and Kakaras, E. (2008) An innovative biomass gasification process and its coupling with microturbine and fuel cell systems. *Energy*, **33**, 284–291.

16 Holt, N. A. and van der Burgt, M. J. (1997) Biomass conversion: prospects and context. Paper presented at 16th EPRI Gasification Technology Conference, San Francisco.

17 Oasmaa, A., Peacock, C., Gust, S., Meier, D., and McLellan, R. (2005) Norms and standards for pyrolysis liquids. End-user requirements and specifications. *Energy Fuels*, **19**, 2155–2163.

18 Santo, U., Seifert, H., Kolb, T., Krebs, L., Kuhn, D., Wiemer, H.-J., Pantouflas, E., and Zarzalis, N. (2007) Conversion of biomass based slurry in an entrained flow gasifier. *Chem. Eng. Technol.*, **30** (7), 967–969.

19 Hofbauer, H. (2009) Biomass gasification for synthesis gas – overview. In *Proceedings of the 17th European Biomass Confer-*

ence and Exhibition, 29 June–3 July 2009, Hamburg.

20. Karl, J. (2004) *Erzeugung von Synthesegas mit dem Biomass Heatpipe-Reformer – Betriebserfahrungen und Leistungsgrenzen*, Tagungsband 6, DGMK-Fachtagung Energie aus Biomasse, Velen, Germany.

21. Hofbauer, H. (2009) Necessary conditions for a successful market introduction of gasification technologies – example gasification/polygeneration plant (heat, power and bioSNG). Presented at the 3rd International Conference on Application of Biomass Gasification, CEP – CLEAN ENERGY POWER, Stuttgart.

22. Stahl, K., Neergaard, M., and Nieminen, J. (1999) Progress report: Varnamo biomass gasification plant. In *Proceedings of the 1999 Gasification Technology Conference*, 17–20 October 1999, San Francisco.

23. Biollaz, S.M.A., Hottinger, P., Pitta, C., and Karl, J. (2009) Results from a 1200 hour test of a tubular SOFC with woodgas. In Proceedings of the 17th European Biomass Conference and Exhibition, 29 June–3 July 2009, Hamburg.

24. Martini, S., Kleinhappl, M., and Hofbauer, H. (2009) Operation of a solid oxide fuel cell (SOFC) with producer gas (thermal gasification). In *Proceedings of the 17th European Biomass Conference and Exhibition*, 29 June–3 July 2009, Hamburg.

25. Koppatz, S., Pfeifer, C., Rauch, R., Hofbauer, H., Marquard-Möllenstedt, T., and Specht, M. (2009) H_2 rich product gas by steam gasification of biomass with in situ CO_2 absorption in a dual fluidized bed system of 8 MW fuel input. *Fuel Process. Technol.*, **90**, 914–921.

26. GermanHy (2009) *Woher kommt der Wasserstoff in Deutschland bis 2050?* Report on behalf of Federal Ministry of Transport, Building and Urban Development, Berlin.

27. Blades, T., Rudloff, M., and Schulze, O. (2005) Sustainable sunfuel from CHOREN's Carbo-V® process. Presented at ISAF XV, San Diego, CA.

28. Kruse, A. (2008) Review: supercritical water gasification, *Biofuels, Bioprod. Bioref.*, **2**, 415–437.

29. Cortright, R.D., Davda, R.R., and Dumesic, J.A. (2002) Hydrogen from catalytic reforming of biomass-derived hydrocarbons in liquid water. *Nature*, **418**, 964–967.

30. Rozmiarek, B. (2008) In *Hydrogen Generation from Biomass-Derived Carbohydrates via Aqueous Phase Reforming Process* (ed. J. Milliken), US Department of Energy, Washington, DC, pp. 1–6.

31. Valenzuela, M.B., Jones, C.W., and Agrawal, P.K. (2006) Batch aqueous-phase reforming of woody biomass. *Energy Fuels*, **20**, 1744–1752.

32. Elliot, D.C., Hart, T.R., and Neuenschwander, G.G. (2006) Chemical processing in high-pressure aqueous environments. 8. Improved catalysts for hydrothermal gasification. *Ind. Eng. Chem. Res.*, **45**, 3776–3781.

33. Elliott, D.C. (2008) Review: catalytic hydrothermal gasification of biomass. *Biofuels Bioprod. Bioref.*, **2**, 254–265.

34. Kruse, A. (2009) Hydrothermal biomass gasification. *J. Supercrit. Fluids*, **47**, 391–399.

35. Xu, X., Matsumura, Y., Stenberg, J., and Antal, M.J.J. (1996) Carbon-catalyzed gasification of organic feedstocks in supercritical water. *Ind. Eng. Chem. Res.*, **35**, 2522–2530.

36. Matsumura, Y. (2002) Evaluation of supercritical water gasification and biomethanation for wet biomass utilization in Japan. *Energy Conversion Manage.*, **43**, 1301–1310.

37. Yoshida, Y., Dowaki, K., Matsumura, Y., Matsuhashi, R., Li, D., Ishitani, H., and Komiyama, H. (2003) Comprehensive comparison of efficiency and CO_2 emissions between biomass energy conversion technologies – position of supercritical water gasification in biomass technologies. *Biomass Bioenergy*, **25**, 257–272.

38. Bowers, B.J., Zhao, J.L., Ruffo, M., Khan, R., Dattatraya, D., Dushman, N., Beziat, J.-C., and Boudjemaa, F. (2007) Onboard fuel processor for PEM fuel cell

vehicle. *Int. J. Hydrogen Energy*, **32**, 1437–1442.

39 Aicher, T., Full, J., and Schaadt, A. (2009) A portable fuel processor for hydrogen production from ethanol in a 250 W_{el} fuel cell system. *Int. J. Hydrogen Energy*, **34**, 8006–8015.

40 Ni, M., Leung, D.Y.C., and Leung, M.K.H. (2007) A review on reforming bio-ethanol for hydrogen production. *Int. J. Hydrogen Energy*, **32**, 3238–3247.

41 Ida Tech LLC (2009) http://www.idatech.com/uploadDocs/491014_IG_DS.pdf (accessed 4 December 2009).

42 Kraaij, G.J., Specchia, S., Bollito, G., Mutri, L., and Wails, D. (2009) Biodiesel fuel processor for APU applications. *Int. J. Hydrogen Energy*, **34**, 4495–4499.

43 Zhang, B., Tang, X., Li, Y., Xu, Y., and Shen, W. (2007) Hydrogen production from steam reforming of ethanol and glycerol over ceria-supported metal catalysts. *Int. J. Hydrogen Energy*, **32**, 2367–2373.

44 Slinn, M., Kendall, K., Mallon, C., and Andrews, J. (2008) Steam reforming of biodiesel by-product to make renewable hydrogen. *Bioresource Technol.*, **99**, 5851–5858.

45 Jenne, M., Zähringer, T., Schuler, A., Piskay, G., and Moos, D. (2002) Sulzer-Hexis SOFC systems for biogas and heating oil. In *Proceedings of the Fifth European Solid Oxide Fuel Cell Forum* (ed. U. Bossel), Lucerne, European Forum Secretariat, Oberrohrdorf, Switzerland, pp. 460–466.

46 Xu, J., Zhou, W., Li, Z., Wang, J., and Ma, J. (2009) Biogas reforming for hydrogen production over nickel and cobalt bimetallic catalysts. *Int. J. Hydrogen Energy*, **34**, 6646–6654.

16
State of the Art of Ceramic Membranes for Hydrogen Separation

Wilhelm-A. Meulenberg, Mariya E. Ivanova, Tim van Gestel, Martin Bram, Hans-Peter Buchkremer, Detlev Stöver, and José M. Serra

Abstract

Membrane technology can be integrated into many advanced system concepts for the production of liquid energy carriers and chemicals, for microfiltration, oxygen generation, low CO_2 emission power generation, hydrogen technology, and carbon dioxide capture. The separation of hydrogen by inorganic membranes for energy applications is a growing field with respect to the low efficiency losses of membrane-based processes. Inorganic hydrogen separation membranes find application in the separation of hydrogen from syngas in a fossil power plant and as an electrolyte in intermediate-temperature proton-conducting fuel cells or electrolyzers for hydrogen production. The separation can be performed by different membrane concepts. This chapter discusses the state of the art of (1) microporous ceramic membranes (amorphous and zeolitic) and (2) mixed proton-electron conducting ceramic membranes, focusing on their properties and current manufacturing and development status. The properties of each membrane type are discussed with respect to the foreseen range of applications. Finally, an outlook is given for R&D activities required in the next few years.

Keywords: hydrogen separation, microporous ceramic membranes, dense ceramic membranes, mixed proton-electron conductivity

16.1
Introduction

In a world of ever-increasing energy demands, alternative forms of energy are not only environmentally important, but also critical in replacing dwindling non-renewable energy sources. Hydrogen-based technologies are challenging solutions. There are many technological ways to produce hydrogen, having different efficiencies, for example electrolysis/thermolysis from water, coal gasification, and reforming of hydrocarbons, alcohols, and carbohydrates, each of them resulting in hydrogen with a different degree of purity. The impact of hydrogen

Hydrogen Energy. Edited by Detlef Stolten
Copyright © 2010 WILEY-VCH Verlag GmbH & Co. KGaA, Weinheim
ISBN: 978-3-527-32711-9

purity on the system performance can be very different depending on the type and concentration of the impurity species. In any case, it has detrimental effects under specific conditions. In order to power, for example, proton exchange membrane (PEM) fuel cell systems and to simplify the relevant design, hydrogen of excellent quality is a major demand. Hence membrane-based technologies and systems are directly relevant to the hydrogen separation and purification, thus taking their place amongst the world's most important technological strategies with high environmental impact and low efficiency losses.

Apart from H_2 purification needs, membranes are a key element in the separation of hydrogen isotopes, fuel cell applications, methane steam reforming, continuous separation of hydrogen from coal/biomass gasification, steam reforming gas, hydrocarbon reforming, and hydrogenation/dehydrogenation conversions, which make possible the process intensification in several chemical and petrochemical industries. In each of these applications, the corresponding membranes have to meet the relevant operational conditions together with prospects of reasonable costs in the long term. This leads to stringent requirements towards membranes in terms of efficiency, functional degradation, scalability, and safety of operation.

This chapter gives an overview of two different classes of ceramic membranes for hydrogen separation, namely microporous and dense membranes, covering the material diversity and relevant properties, together with some manufacturing aspects.

16.2
Microporous Membranes for H_2 Separation

This section presents a summary of the development of various microporous membranes for gas separation applications as reported in the literature, and in addition examples of membranes recently developed at Forschungszentrum Jülich. The focus is mainly on membranes reported for H_2–CO_2 separation, since there is currently immense interest in the application of such membranes in power plants. However, in a number of cases data are not available, so as an alternative data for other industrially relevant separations such as H_2–N_2 and CO_2–N_2 are provided.

16.2.1
Introduction

16.2.1.1 General Properties
A microporous ceramic separation membrane can be considered as a graded multilayer porous ceramic material – with macroporous, mesoporous, and microporous layers – in which the last membrane layer displays a pore size <2 nm. According to IUPAC notation, porous materials are classified into three kinds: microporous materials having pore diameters <2 nm, mesoporous mate-

rials have pore diameters between 2 and 50 nm, and macroporous materials have pore diameters >50 nm [1].

Microporous membranes are considered by many research groups as one of the candidates for separation problems involving small gas molecules (H_2, He, N_2, O_2, CO, CO_2, CH_4). Typically, microporous membranes are developed as an alternative to polymeric gas separation membranes for industrial applications in which polymeric membranes cannot perform well or do not have the required lifetime. Examples are applications including acid contaminants, steam, high temperatures, and high pressures. The main classes of microporous membranes, including amorphous SiO_2 and doped SiO_2 membranes, templated SiO_2 membranes, and zeolite membranes, are summarized in Table 16.1.

In this research field, amorphous SiO_2 membranes have been most extensively investigated, because of their cost-effective sol-gel preparation methods, their scalability, and their potential for combining high selectivity and permeability. Chemical vapor deposition (CVD) preparation methods have also frequently been used as an alternative to sol-gel coating, giving excellent results. These kinds of membranes will be discussed further in Section 16.2.2.

Another interesting approach reported by many researchers is the preparation of crystalline microporous SiO_2-based membranes (e.g., zeolites). Some of these membranes exhibit fascinating properties, including a well-defined ordered pore structure with a pore size of approximately 0.5 nm, and they are not susceptible to densification at higher temperatures or degradation in steam. Several groups have also succeeded in preparing zeolite membranes with good performance in certain gas and liquid separation applications, but the existence of intercrystalline pores is often reported as a factor that limits the separation efficiency for

Table 16.1 Properties of different classes of microporous membranes reported in the literature.

Microporous material	Preparation methods	Reported properties
Amorphous SiO_2	Sol-gel dip-coating or spin-coating CVD	High H_2 permeation Permeation $H_2 \gg CO_2 \geq N_2$ High H_2/N_2 selectivity Moderate to high H_2/CO_2 Low stability in water vapor
Doped amorphous SiO_2	Sol-gel dip-coating or spin-coating CVD	Improved stability towards water vapor
Templated/ordered SiO_2	Sol-gel dip-coating or spin-coating Hydrothermal treatment	Improved pore structure stability
Zeolite	Hydrothermal treatment	Low H_2/CO_2 or CO_2/H_2 selectivity
	Dip-coating or spin-coating	Moderate CO_2/N_2 selectivity High thermal stability

small gas molecules, including He and H_2. The potential of current zeolitic and SiO_2 membranes with an ordered pore structure for H_2 separation is discussed further in Section 16.2.3.

Due to the high chemical resistance of some non-silica materials in various membrane applications (e.g., TiO_2 and ZrO_2 membranes in ultra- and nanofiltration), research efforts have been devoted in the past few years to the potential application of non-silica membranes as an alternative to amorphous silica membranes in gas separation. In contrast to silica membranes, non-silica membranes with a connected network of micropores and selectivity for small gases have not yet been reported. However, significant progress has been achieved by a few groups in developing thin-film non-silica membranes with a pore size <2 nm (e.g., [1–4]). Depending on their X-ray diffraction (XRD) behavior, these membranes can also be subdivided into amorphous and crystalline membranes. They are both prepared using sol-gel methods and it appears that the trend in this research field is to prepare amorphous membranes. An example can be found in articles published by Puhlfürss and Voigt, who demonstrated the development of membranes with a pore size of approximately 1 nm [1, 2]. Further research and development have been focused on reducing the pore size, which could also lead to non-silica membranes with a potential for gas separation, but so far such membranes have not yet been achieved, to our knowledge.

16.2.1.2 Working principle

The working principle of a microporous membrane is often described in terms of three mechanisms: Knudsen diffusion, micropore diffusion, and surface diffusion. It is well known that for membranes with mesopores larger than 2 nm, gas transport occurs by a Knudsen diffusion mechanism. In this case, selectivity is proportional to the square root of the molecular weight ratio of the gases and is independent of temperature. The selectivities achieved for H_2–CO_2 and N_2–CO_2 are 4.67 and 0.8, and are therefore too low for most practical applications in gas separation units. In practice, however, this mechanism is also found for microporous membranes which show larger micropores, indicating that membranes with a pore size in the lower microporous region are required.

Micropore diffusion is often reported as a working mechanism in which the smaller gas molecule (He or H_2) permeates through the porous structure of the membrane, whereas the larger molecule is sieved out. The transport of the smaller molecule tends to be an activated process, with the rate increasing exponentially with temperature. This type of mechanism has been reported for amorphous SiO_2 membranes, for example in separation experiments with H_2 or He from N_2 or CO_2, and in practice these membranes show high permeabilities for He and H_2. Usually, a pore diameter of 0.3–0.5 nm is reported for such materials, but in fact it is not really clear whether the reported materials show "real" physical pores or whether a similar working principle occurs as in polymeric membranes, with the small gas molecule diffusing through structural spaces in the inorganic polymeric structure of the amorphous SiO_2 matrix.

Surface diffusion is usually reported as the transport mechanism for CO_2 separation membranes, for example for the separation of CO_2 from H_2, N_2, or CH_4. The mechanism involves the surface diffusion of adsorbed CO_2 along the pore walls and CO_2 selectivity is obtained when the surface transport of CO_2 outweighs the contribution from Knudsen diffusion. The mechanism is usually reported for membranes with larger pore diameters in comparison with amorphous SiO_2 membranes, including zeolite membranes and templated SiO_2 membranes, and also in the case of hybrid organic-ceramic membranes. In this case, the experimental test conditions are reported to play a major role. At lower temperatures, CO_2 adsorption is enhanced and the separation factor increases strongly. Further, the major effect of CO_2 adsorption on the separation factor is frequently confirmed by the difference in selectivity for single gas and mixed gas operation. However, in comparison with the previously mentioned micropore diffusion type of membranes, the separation factors achieved are – particularly in experiments with smaller gases – always much lower.

16.2.2
Amorphous SiO_2 Membranes

As mentioned in above, amorphous SiO_2 membranes have already been frequently investigated. With such membranes, separation factors for small gases (He, H_2) to larger gases (CO_2, N_2, CH_4) from several hundred to more than 1000 have been reported. According to different reports, the thickness of the membranes varies in the range 50–200 nm and is significantly smaller than any other kind of gas separation membrane, which is an important advantage for achieving the required permeability for commercial applications.

Amorphous SiO_2 membranes are basically synthesized by two methods: sol-gel coating and CVD. From a practical point of view, sol-gel coating appears to be a very interesting method, since the sol coating methods are the same as conventional powder suspension coating methods and exhibit the same advantages (inexpensive in terms of capital costs, simplicity of the equipment). There are a variety of coating methods used for making sol-gel membranes, of which spin-coating (flat membranes) and dip-coating (tubular membranes) are the most frequently used. A problem that is often reported for sol-gel coating is a lack of reproducibility, but this problem is probably often due to inaccurate working conditions and also the lack of clean room conditions.

CVD methods for preparing an SiO_2 membrane are usually divided into two types. In the first method, all precursors and reactants are provided on one side of the membrane, while the other side is usually under vacuum. In the second method, the reactants are supplied from the opposite sides of the substrate. CVD membranes frequently have higher selectivity for H_2, but they usually also have lower permeability. It should be noted, however, that literature data on the permeability of SiO_2 membranes show a large scatter, thus making a comparison of different types of membranes very difficult.

An important drawback frequently mentioned for amorphous SiO_2 membranes is, however, degradation of the membrane material in water-containing atmospheres. For this reason, a number of researchers have focused on developing alternative SiO_2 materials with improved stability. A variety of modified materials have been developed with dopants/additives incorporated, such as metals (Ni, Co), metal oxides (Al_2O_3, ZrO_2, TiO_2, Nb_2O_3), and organic groups.

Another drawback could be the limited thermal stability of a number of membranes reported in the literature. Frequently, these membranes are fired by a very low thermal treatment – sometimes just 300 °C – and/or in an inert atmosphere, which implies that the membrane can be sensitive to structural changes if they are subsequently applied at higher temperatures or in an oxidizing atmosphere.

Table 16.2 gives an overview of a few selected amorphous SiO_2 and doped SiO_2 sol-gel membranes. In this overview, membranes with special properties/compositions are summarized, rather than giving a comprehensive overview of a large number of more or less identical membranes. Further, Table 16.2 also gives an overview of membranes obtained in our laboratory, including amorphous SiO_2, NiO-doped SiO_2, Co_3O_4-doped SiO_2, and ZrO_2-doped SiO_2 membranes. Table 16.3 presents a few selected comparable CVD membranes, including SiO_2 and doped SiO_2 membranes and membranes made with precursors containing organic groups.

16.2.3
Zeolite Membranes and Templated SiO_2 Membranes

16.2.3.1 Zeolite Membranes

Zeolites are usually defined as microporous aluminosilicates with a very narrow pore size distribution. In contrast to the previously described amorphous SiO_2 membranes, they are crystalline and show an ordered pore structure. Zeolites are often regarded as candidate materials for gas separation membranes, since some of these materials can be made with a pore diameter in the lower microporous region. In addition, these crystalline materials have the required thermal stability and also some compositions show the required chemical stability.

Most zeolite membranes are synthesized by bringing a support material into contact with a zeolite seed solution and by bringing the system under controlled conditions of temperature and pressure so that the zeolite layer can nucleate and grow on the support. Alternatively, the zeolite layer can also be grown in one step on a support under hydrothermal conditions. Membranes of various zeolites, such as ZSM, silicalite, Y type, A type, and silicoaluminophosphate, have been synthesized on porous supports, including porous ceramic and stainless-steel supports.

Drawbacks frequently mentioned for zeolite membranes include the larger top layer thickness, which reduces the permeability achievable in comparison with thin-film SiO_2 membranes, the complicated synthesis routes, and the limited scale-up potential of the synthesis methods. The first point has been im-

Table 16.2 Properties of selected sol–gel SiO_2 and doped SiO_2 membranes reported in the literature and membranes prepared at Institute of Energy Research IEF-1.

Membrane and study	Preparation method	H_2 permeation	CO_2 permeation	N_2 permeation	Selectivity
SiO_2 de Vos [5]	Sol-gel with firing at 400 °C; Sol-gel with firing at 600 °C	4.7–7.3×10^{-7} mol m^{-2} Pa^{-1} s^{-1} (25 °C); 1.7–1.8×10^{-6} mol m^{-2} Pa^{-1} s^{-1} (200 °C); 3.9–4.1×10^{-7} mol m^{-2} Pa^{-1} s^{-1} (200 °C)	2–2.5×10^{-7} mol m^{-2} Pa^{-1} s^{-1} (25 °C); 2.3×10^{-7} mol m^{-2} Pa^{-1} s^{-1} (200 °C); 0.6–1.1×10^{-8} mol m^{-2} Pa^{-1} s^{-1} (200 °C)	0.9–1.0×10^{-8} mol m^{-2} Pa^{-1} s^{-1} (25 °C); 2.7×10^{-8} mol m^{-2} Pa^{-1} s^{-1} (200 °C); 0 mol m^{-2} Pa^{-1} s^{-1} (200 °C)	H_2/CO_2 2–3 (25 °C); H_2/CO_2 7.5 (200 °C) (single gas permeation); H_2/CO_2 37–66 (200 °C) (single gas permeation)
Methylated SiO_2 de Vos [6]	Sol-gel with methylated Si precursor and with firing at 400 °C	2–2.2×10^{-6} mol m^{-2} Pa^{-1} s^{-1} (200 °C)	3.8–4.1×10^{-7} mol m^{-2} Pa^{-1} s^{-1} (200 °C)	2.3–2.7×10^{-7} mol m^{-2} Pa^{-1} s^{-1} (200 °C)	H_2/CO_2 6 (200 °C); H_2/N_2 8 (200 °C) (single gas permeation)
SiO_2 Tsai [7]	Sol-gel with firing at 450 °C	3.4×10^{-7} mol m^{-2} Pa^{-1} s^{-1} (80 °C); 2×10^{-7} mol m^{-2} Pa^{-1} s^{-1} (50:50 H_2–N_2 gas mixture)	6.7×10^{-8} mol m^{-2} Pa^{-1} s^{-1} (80 °C)	6.7×10^{-10} – 1×10^{-9} mol m^{-2} Pa^{-1} s^{-1} (80 °C)	H_2/CO_2 5 (80 °C); CO_2/N_2 60 (80 °C) (80 °C, single gas permeation); H_2/N_2 270 (50:50 H_2–N_2 mixture)
NiO–SiO_2 Xomeritakis [8]	Sol-gel with 10–20% Ni(NO$_3$)$_2$·6H$_2$O, firing at 300–500 °C in vacuum		2.2–4.2×10^{-7} mol m^{-2} Pa^{-1} s^{-1} (10% NiO); 4.1×10^{-8} mol m^{-2} Pa^{-1} s^{-1} (20% NiO)		CO_2/N_2 16.4–21.9 (10% NiO); CO_2/N_2 92 (20% NiO) (25 °C, 10:90 CO_2–N_2 mixture, counter-current flow configuration, He sweep gas)

Table 16.2 (continued)

Membrane and study	Preparation method	H_2 permeation	CO_2 permeation	N_2 permeation	Selectivity
Ni–SiO$_2$ Kanezashi [9]	Sol-gel with firing in air at 550–650 °C or firing in steam at 550 °C (Si:Ni 2:1)	Firing at 650 °C in steam 4.4×10^{-7} mol m^{-2} Pa^{-1} s^{-1}		Firing at 650 °C in steam 8.8×10^{-10} mol m^{-2} Pa^{-1} s^{-1}	Firing at 550 °C H$_2$/N$_2$ 160 Firing at 650 °C H$_2$/N$_2$ 370 Firing at 650 °C in steam H$_2$/N$_2$ 400 (500 °C, single gas permeation)
Co–SiO$_2$ Battersby [10]	Sol-gel with 25% Co(NO$_3$)$_2$·6H$_2$O, firing 600 °C in air and 500 °C in H$_2$	5×10^{-10} mol m^{-2} Pa^{-1} s^{-1} (100 °C) 6×10^{-9} mol m^{-2} Pa^{-1} s^{-1} (250 °C)			H$_2$/CO$_2$ 70 (100 °C) H$_2$/CO$_2$ 500 (200 °C) H$_2$/CO$_2$ 1000 (250 °C) (single gas permeation)
Nb$_2$O$_5$–SiO$_2$ Boffa [11]	Sol-gel with firing at 500 °C in air	3.9×10^{-8} mol m^{-2} Pa^{-1} s^{-1} (200 °C)	8.5×10^{-10} mol m^{-2} Pa^{-1} s^{-1} (200 °C)	6.2×10^{-9} mol m^{-2} Pa^{-1} s^{-1} (200 °C)	H$_2$/CO$_2$ 46 H$_2$/N$_2$ 6.3 CO$_2$/N$_2$ 13 (200 °C, single gas permeation)
ZrO$_2$–SiO$_2$ Yoshida [12]	Sol-gel with firing at 570 °C in air or 570 °C in steam to enhance hydrothermal stability	10% ZrO$_2$ 2.6×10^{-7} mol m^{-2} Pa^{-1} s^{-1} (300 °C) 30% ZrO$_2$ 1.7×10^{-7} mol m^{-2} Pa^{-1} s^{-1} (300 °C) 50% ZrO$_2$ 3×10^{-8} mol m^{-2} Pa^{-1} s^{-1} (300 °C)	10% ZrO$_2$ 2.2×10^{-8} mol m^{-2} Pa^{-1} s^{-1} (300 °C) 30% ZrO$_2$ 8.8×10^{-8} mol m^{-2} Pa^{-1} s^{-1} (300 °C) 50% ZrO$_2$ 4.4×10^{-8} mol m^{-2} Pa^{-1} s^{-1} (300 °C)	10% ZrO$_2$ 8.8×10^{-8} mol m^{-2} Pa^{-1} s^{-1} (300 °C) 30% ZrO$_2$ 4×10^{-8} mol m^{-2} Pa^{-1} s^{-1} (300 °C) 50% ZrO$_2$ 8.8×10^{-8} mol m^{-2} Pa^{-1} s^{-1} (300 °C)	H$_2$/CO$_2$ 20 (30% ZrO$_2$) H$_2$/N$_2$ 45 (30% ZrO$_2$) (300 °C, single gas permeation)

Table 16.2 (continued)

Membrane and study	Preparation method	H_2 permeation	CO_2 permeation	N_2 permeation	Selectivity
SiO_2 IEF-1	Sol-gel with firing at 500 °C	5.5×10^{-8} mol m^{-2} Pa^{-1} s^{-1} (25 °C, 2.5 bar) 1.1×10^{-7} mol m^{-2} Pa^{-1} s^{-1} (100 °C, 2.5 bar) 1.6×10^{-7} mol m^{-2} Pa^{-1} s^{-1} (200 °C, 2.5 bar)	5.12×10^{-10} mol m^{-2} Pa^{-1} s^{-1} (25 °C, 2.5 bar) 3×10^{-10} mol m^{-2} Pa^{-1} s^{-1} (100 °C, 2.5 bar) 3.8×10^{-10} mol m^{-2} Pa^{-1} s^{-1} (200 °C, 2.5 bar)	0 mol m^{-2} Pa^{-1} s^{-1} (25 °C, 2.5 bar) 0 mol m^{-2} Pa^{-1} s^{-1} (100 °C, 2.5 bar) 0 mol m^{-2} Pa^{-1} s^{-1} (200 °C, 2.5 bar)	H_2/CO_2 107 (25 °C) H_2/CO_2 372 (100 °C) H_2/CO_2 410 (200 °C) H_2/CO_2 100 (1.5 bar) H_2/N_2 100 (25–200 °C) (2.5 bar, single gas permeation)
$NiO-SiO_2$ IEF-1	Sol-gel with 20% $Ni(NO_3)_2 \cdot 6H_2O$, firing at 500 °C	2.3×10^{-8} mol m^{-2} Pa^{-1} s^{-1} (25 °C, 2.5 bar) 6.2×10^{-7} mol m^{-2} Pa^{-1} s^{-1} (100 °C, 2.5 bar) 1.3×10^{-7} mol m^{-2} Pa^{-1} s^{-1} (200 °C, 2.5 bar)	4.7×10^{-10} mol m^{-2} Pa^{-1} s^{-1} (25 °C, 2.5 bar) 3.6×10^{-9} mol m^{-2} Pa^{-1} s^{-1} (100 °C, 2.5 bar) 2.4×10^{-9} mol m^{-2} Pa^{-1} s^{-1} (200 °C, 2.5 bar)	0 mol m^{-2} Pa^{-1} s^{-1} (25 °C, 2.5 bar) 0 mol m^{-2} Pa^{-1} s^{-1} (100 °C, 2.5 bar) 0 mol m^{-2} Pa^{-1} s^{-1} (200 °C, 2.5 bar)	H_2/CO_2 9.6 (25 °C) H_2/CO_2 17 (100 °C) H_2/CO_2 28 (200 °C) H_2/N_2 100 (25–200 °C) (2.5 bar, single gas permeation)
$Co_3O_4-SiO_2$ IEF-1	Sol-gel with 20% $Co(NO_3)_2 \cdot 6H_2O$, firing at 500 °C	1.3×10^{-8} mol m^{-2} Pa^{-1} s^{-1} (25 °C, 2.5 bar) 2.8×10^{-8} mol m^{-2} Pa^{-1} s^{-1} (100 °C, 2.5 bar) 6.6×10^{-8} mol m^{-2} Pa^{-1} s^{-1} (200 °C, 2.5 bar)	0 mol m^{-2} Pa^{-1} s^{-1} (25 °C, 2.5 bar) 4.8×10^{-10} mol m^{-2} Pa^{-1} s^{-1} (100 °C, 2.5 bar) 8.4×10^{-10} mol m^{-2} Pa^{-1} s^{-1} (200 °C, 2.5 bar)	0 mol m^{-2} Pa^{-1} s^{-1} (25 °C, 2.5 bar) 0 mol m^{-2} Pa^{-1} s^{-1} (100 °C, 2.5 bar) 0 mol m^{-2} Pa^{-1} s^{-1} (200 °C, 2.5 bar)	H_2/CO_2 100 (25 °C) H_2/CO_2 58 (100 °C) H_2/CO_2 78 (200 °C) H_2/N_2 100 (25–200 °C) (2.5 bar, single gas permeation)
ZrO_2-SiO_2 IEF-1	Sol-gel with firing at 500 °C (Zr:Si 1:2)	8.5×10^{-8} mol m^{-2} Pa^{-1} s^{-1} (200 °C, 2.5 bar)	5.7×10^{-9} mol m^{-2} Pa^{-1} s^{-1} (200 °C, 2.5 bar)	1.2×10^{-9} mol m^{-2} Pa^{-1} s^{-1} (200 °C, 2.5 bar)	H_2/CO_2 15 (200 °C) H_2/N_2 70 (200 °C) (2.5 bar, single gas permeation)

Table 16.3 Properties of selected CVD SiO_2 and doped SiO_2 membranes reported in the literature.

Membrane and study	Preparation method	H_2 permeation	CO_2 permeation	N_2 permeation	Selectivity
SiO_2 Gopala-krishnan [13]	CVD Counter-diffusion	5.1×10^{-7} mol m^{-2} Pa^{-1} s^{-1} (100 °C) 7×10^{-7} mol m^{-2} Pa^{-1} s^{-1} (400 °C)	9×10^{-8} mol m^{-2} Pa^{-1} s^{-1} (100 °C) 2×10^{-8} mol m^{-2} Pa^{-1} s^{-1} (400 °C)	3.5×10^{-8} mol m^{-2} Pa^{-1} s^{-1} (100 °C) 1.5×10^{-8} mol m^{-2} Pa^{-1} s^{-1} (400 °C)	H_2/CO_2 36 H_2/N_2 57 (400 °C, single gas permeation)
SiO_2 Oyama [14]	CVD	1.8×10^{-7} mol m^{-2} Pa^{-1} s^{-1} (600 °C)	8.1×10^{-11} mol m^{-2} Pa^{-1} s^{-1} (600 °C)		$H_2/CO_2 \gg 1000$ (600 °C, single gas permeation)
Al_2O_3–SiO_2 Gu [15]	CVD (0.02 ASB–TEOS)	1.6×10^{-7} mol m^{-2} Pa^{-1} s^{-1} (600 °C) 3×10^{-8} mol m^{-2} Pa^{-1} s^{-1} (180 °C)	2.5×10^{-10} mol m^{-2} Pa^{-1} s^{-1} (600 °C)		H_2/CO_2 590 (600 °C, single gas permeation)
SiO_2 Sea [16]	CVD with TEOS, PTES or DPDES	5×10^{-7} mol m^{-2} Pa^{-1} s^{-1} (200 °C, DPDES) 0.22×10^{-7} mol m^{-2} Pa^{-1} s^{-1} (200 °C, TEOS)	0.33×10^{-7} mol m^{-2} Pa^{-1} s^{-1} (200 °C, DPDES) 0.0033×10^{-7} mol m^{-2} Pa^{-1} s^{-1} (200 °C, TEOS)	0.079 mol m^{-2} Pa^{-1} s^{-1} (200 °C, DPDES) 0.0050×10^{-7} mol m^{-2} Pa^{-1} s^{-1} (200 °C, TEOS)	H_2/N_2 64 (200 °C) CO_2/N_2 4.2 (200 °C) CO_2/N_2 8.9 (30 °C) (single gas permeation) H_2/N_2 43 (200 °C) H_2/CO_2 66 (200 °C) (single gas permeation)
SiO_2 Araki [17]	HPCVD Counter-diffusion	5×10^{-8} mol m^{-2} Pa^{-1} s^{-1} (300 °C)			H_2/CO_2 1200 (300 °C, single gas permeation)

proved by developing thin zeolite layers, but these still show a significantly larger thickness than sol-gel membranes. The last two points are still under study and it is clear from the current literature that the difficulties in obtaining defect-free layers also lead to layers with a much lower separation factor than sol-gel membranes.

In the field of zeolite membrane formation, apart from a few notable examples, no membranes with molecular sieving properties, as described for SiO_2, have yet been reported. One remarkable example is the formation of a zeolite membrane by Nishiyama et al. [18], who simply dissolved a commercial zeolite powder in HCl and spin-coated the resultant sol-solution on to a porous glass substrate with 4 nm pores. As shown in Table 16.4, they obtained extremely high separation factors using this very simple preparation method. Usually, however, membranes are prepared by a hydrothermal growth process, which apparently results in membrane layers containing larger (meso)pores and probably also defects, thus preventing molecular sieving properties.

On the other hand, a number of membranes have been described which yield a significant separation for CO_2, especially when the measuring temperature approximates room temperature. This result is explained on the basis of the high CO_2 sorption capacity of the zeolitic membrane material and the occurrence of a surface diffusion mechanism. The occurrence of such a mechanism in the case of a CO_2–H_2 mixture was confirmed in a notable paper published by Hong et al. [19], who found that the larger CO_2 molecule permeates through the membrane whereas the smaller H_2 is retained. They obtained a CO_2/H_2 separation selectivity of 150 in a mixed gas experiment, whereas the selectivity for CO_2 and H_2 in single gas permeation experiments was around 2. Further, the selectivity was the highest at $-20\,°C$ and decreased strongly with increasing temperature, and at $35\,°C$ the selectivity was below 10, indicating that CO_2 adsorption inhibited the permeance of H_2 at lower temperatures in a similar way to that found for N_2 in CO_2–N_2 mixtures.

In order to improve the CO_2 adsorption of the membrane material and to enhance the gas selectivity of zeolite membranes, basic cations have been incorporated in the zeolitic structure. A method which has been described extensively includes immersing the membranes in solutions of alkali or alkaline earth metal ions. For example, zeolite Y membranes were prepared and subsequently treated by immersion in a solution of NaCl, LiCl, KCl, RbCl, CsCl, $MgCl_2$, $BaCl_2$, or $CaCl_2$, which improved the CO_2 separation (from N_2) significantly in comparison with the untreated membranes, suggesting improved CO_2 adsorption.

16.2.3.2
Templated SiO_2 Membranes

Templated SiO_2 membranes are related to sol-gel membranes and zeolite membranes. They are usually prepared using common sol-gel procedures, with the difference that a template is incorporated in the top layer and subsequently

Table 16.4 Properties of selected zeolite membranes reported in the literature.

Membrane and study	Preparation method	H_2 permeation	CO_2 permeation	N_2 permeation	Selectivity
NaA zeolite Nishiyama [18]	Dissolution of commercial Tosoh zeolite powder + coating resulting sol with nano-blocks (drying at 90 °C or drying + firing at 400 °C)	~ 5×10^{-8} mol m^{-2} Pa^{-1} s^{-1}		<3.6×10^{-11} mol m^{-2} Pa^{-1} s^{-1} (detection limit)	H_2/CO >660 H_2/N_2 >1600 H_2/CH_4 >1200 (25 °C, 50:50 mixture; Ar sweep gas)
SAPO-34 zeolite Hong [19]	In-situ crystallization	1–2×10^{-8} mol m^{-2} Pa^{-1} s^{-1} (−20 to −37 °C, single gas) 0.2–2.5×10^{-8} mol m^{-2} Pa^{-1} s^{-1} (−20 to −37 °C, mixed gas)	3.5–4×10^{-8} mol m^{-2} Pa^{-1} s^{-1} (35 °C, single gas) 2–3.5×10^{-8} mol m^{-2} Pa^{-1} s^{-1} (−20 to −37 °C, mixed gas)		CO_2/H_2 2 (20 °C, single gas) CO_2/H_2 ~ 150 (−20 °C, mixed gas) CO_2/H_2 ~ 25 (20 °C, mixed gas) CO_2/H_2 ~ 8 (37 °C, mixed gas) H_2/CH_4 20–30
ZSM-5 zeolite Bonhomme [20]	In situ crystallization	1.9–2.9×10^{-7} mol m^{-2} Pa^{-1} s^{-1} (25 °C, single gas)	2.9–5.9×10^{-7} mol m^{-2} Pa^{-1} s^{-1} (25 °C, single gas)	1.1–1.6×10^{-7} mol m^{-2} Pa^{-1} s^{-1} (25 °C, single gas)	CO_2/N_2 3–4 (25 °C, single gas permeation)
ZSM-5 zeolite Sebastian [21]	In situ crystallization		2.7×10^{-7} mol m^{-2} Pa^{-1} s^{-1}		CO_2/N_2 13 (25 °C, 50:50 CO_2–N_2 mixture)
ZSM-5 zeolite Shin [22]	Hydrothermal treatment + coating with TEOS sol and firing at 200 °C		4×10^{-8} mol m^{-2} Pa^{-1} s^{-1} (50:50 CO_2–N_2 mixture)	8×10^{-10} mol m^{-2} Pa^{-1} s^{-1} (50:50 CO_2–N_2 mixture)	CO_2/N_2 54 (25 °C) CO_2/N_2 15 (100 °C) (50:50 CO_2–N_2 mixture; He sweep gas)

removed to create an ordered pore structure. Also, in some cases the coated layer is treated hydrothermally in an analogous way to a zeolite membrane. Compounds which are frequently used include ionic surfactants [e.g., tetrapropylammonium bromide (TPA)], anionic surfactants (e.g., Brij) and Si precursors with an organic group attached, such as methyl, propyl, vinyl, and phenyl groups, and also aminated organic groups.

The advantage with respect to common zeolite membranes includes the simple sol-gel-like synthesis route and the ability to deposit membranes by simple dip-coating or spin-coating methods. Also, some authors have reported the advantage in comparison with normal sol-gel membranes of using a higher firing temperature without degradation of the microporous structure. Another advantage may be higher gas permeability, but this is not always clear from the gas permeation data in the literature, which show wide scatter. From literature results, it is also frequently unclear whether the membranes should be classified in the category of templated membranes – with the organic template removed – or hybrid membranes which still partly or completely contain the incorporated organic groups. According to West *et al.* [27], methyl groups are removed in the range 300–400 °C, but the complete removal of phenyl groups occurs only in the range 500–600 °C.

Table 16.5 gives an overview of a few selected membranes. Particularly for separation problems involving two larger gases, such as CO_2 and N_2, these membranes can be an alternative to common amorphous SiO_2 membranes, due to their somewhat larger pore radius. However, examples of H_2–CO_2 separation with such membranes can also be found in the literature.

16.3
Dense Ceramic Membranes for H_2 Separation

Dense ceramic membranes with both dominant proton and mixed proton-electron conductivity (MPEC) have great potential in a variety of applications, for example, as solid electrolytes for proton-conducting solid oxide fuel cells (PC-SOFCs) and in hydrogen separation systems and petrochemical technologies.

An innovative and challenging application field for MPEC dense ceramic membranes is the so-called *pre-combustion CO_2 capture concept*. This technology is an efficient way to produce high-purity hydrogen that can be further used for versatile applications requiring hydrogen, for example directly as a fuel in power plants or in vehicles. Along with hydrogen production, the pre-combustion concept also combines the simultaneous sequestration of the CO_2, which can be identified as a major benefit of great environmental relevance.

Generally, the pre-combustion approach consists of several stages which will be briefly described here. The partial oxidation of fossil fuel (methane, natural gas, or gasified coal) results in a syngas composed of hydrogen and carbon monoxide. Due to the water gas shift reaction that also takes place, the CO from

Table 16.5 Properties of selected templated SiO_2 membranes reported in the literature

Membrane and study	Preparation method	H_2 permeation	CO_2 permeation	N_2 permeation	Selectivity
SiO_2 Kim [23]	Sol-gel with 10 mol% methacryloxy-propyltrimethoxysilane Firing at 400–700 °C	2.4×10^{-7} mol m^{-2} Pa^{-1} s^{-1} (100 °C) 4×10^{-7} mol m^{-2} Pa^{-1} s^{-1} (200 °C) 6.8×10^{-7} mol m^{-2} Pa^{-1} s^{-1} (300 °C)	2×10^{-8} mol m^{-2} Pa^{-1} s^{-1} (100 °C) 1.8×10^{-8} mol m^{-2} Pa^{-1} s^{-1} (200 °C) 2×10^{-8} mol m^{-2} Pa^{-1} s^{-1} (300 °C) + H_2O: 2×10^{-9} mol m^{-2} Pa^{-1} s^{-1} (100 °C) 4–5×10^{-9} mol m^{-2} Pa^{-1} s^{-1} (200 °C) 6–7×10^{-9} mol m^{-2} Pa^{-1} s^{-1} (300 °C)	3.1×10^{-9} mol m^{-2} Pa^{-1} s^{-1} (100 °C) 4.5×10^{-9} mol m^{-2} Pa^{-1} s^{-1} (200 °C) 2.3×10^{-8} mol m^{-2} Pa^{-1} s^{-1} (300 °C)	H_2/CO_2: 12 (100 °C, single gas) 11 (100 °C, mixed gas) 22 (200 °C, single gas) 20 (200 °C, mixed gas) 34 (300 °C, single gas) 36 (300 °C, mixed gas) CO_2/N_2: 6.1 (100 °C, single gas) 13 (100 °C, mixed gas) 4.1 (200 °C, single gas) 6 (200 °C, mixed gas) 0.9 (300 °C, single gas) 1.6 (300 °C, mixed gas) (Ar sweep gas)
SiO_2 Kusakabe [24]	Sol-gel with 10 mol% octyltrimethoxysilane or dodecyltrimethoxysilane Firing at 600 °C	4–5×10^{-7} mol m^{-2} Pa^{-1} s^{-1} (−Si−C_8H_{17}, 100 °C) 5×10^{-7} mol m^{-2} Pa^{-1} s^{-1} (−Si−$C_{12}H_{35}$, 100 °C)	2–3×10^{-7} mol m^{-2} Pa^{-1} s^{-1} (−Si−C_8H_{17}, 100 °C) 2×10^{-7} mol m^{-2} Pa^{-1} s^{-1} (−Si−$C_{12}H_{35}$, 100 °C)	3×10^{-8} mol m^{-2} Pa^{-1} s^{-1} (−Si−C_8H_{17}, 100 °C) 4×10^{-8} mol m^{-2} Pa^{-1} s^{-1} (−Si−$C_{12}H_{35}$, 100 °C)	H_2/CO_2 2–3 (100 °C, single gas) CO_2/N_2 5–8 (100 °C, single gas) (Ar sweep gas)
SiO_2 Moon [25]	Sol-gel with 10 mol% methyltriethoxysilane (MTES) Firing at 550 °C + TEOS and firing at 300 °C		1–7×10^{-9} mol m^{-2} Pa^{-1} s^{-1} (25–300 °C, single gas)		CO_2/N_2 12–24 (100–170 °C; 50:50 gas mixture)

Table 16.5 (continued)

Membrane and study	Preparation method	H_2 permeation	CO_2 permeation	N_2 permeation	Selectivity
SiO_2 Xomeritakis [8]	Sol-gel with 8 mol% Brij 56 + TEOS SiO_2 sol Firing at 500°C Firing at 300–500°C in vacuum		1×10^{-7}–1.8×10^{-7} mol m^{-2} Pa^{-1} s^{-1} (25°C, 10:90 CO_2–N_2) 2.1×10^{-7} mol m^{-2} Pa^{-1} s^{-1} (25°C, 50:50 CO_2–N_2) 5.2×10^{-9} mol m^{-2} Pa^{-1} s^{-1} (70°C, 50:50 CO_2–N_2) 2.5×10^{-9} mol m^{-2} Pa^{-1} s^{-1} (100°C, 50:50 CO_2–N_2)	1.7×10^{-9} mol m^{-2} Pa^{-1} s^{-1} (25°C, 50:50 CO_2–N_2) 7×10^{-9} mol m^{-2} Pa^{-1} s^{-1} (70°C, 50:50 CO_2–N_2) 1.4×10^{-8} mol m^{-2} Pa^{-1} s^{-1} (100°C, 50:50 CO_2–N_2)	CO_2/N_2 44.5–78.4 (25°C, 10:90 CO_2–N_2 mixture; counter-current flow configuration, He sweep gas)
SiO_2 Osada [26]	Sol-gel with 25 mol% TPABr Firing at 300°C		7×10^{-8} mol m^{-2} Pa^{-1} s^{-1} (50°C, RH 0%) 4×10^{-9} mol m^{-2} Pa^{-1} s^{-1} (50°C, RH 3%) 8×10^{-10} mol m^{-2} Pa^{-1} s^{-1} (50°C, RH 6%) 3×10^{-8} mol m^{-2} Pa^{-1} s^{-1} (150°C, RH 0%) 2×10^{-8} mol m^{-2} Pa^{-1} s^{-1} (150°C, RH 3%) 2×10^{-8} mol m^{-2} Pa^{-1} s^{-1} (150°C, RH 6%)	1×10^{-8} mol m^{-2} Pa^{-1} s^{-1} (50°C, RH 0%) 2–3×10^{-10} mol m^{-2} Pa^{-1} s^{-1} (50°C, RH 3%) 1×10^{-11} mol m^{-2} Pa^{-1} s^{-1} (50°C, RH 6%) 6×10^{-9} mol m^{-2} Pa^{-1} s^{-1} (150°C, RH 0%) 5×10^{-9} mol m^{-2} Pa^{-1} s^{-1} (150°C, RH 3%) 4×10^{-9} mol m^{-2} Pa^{-1} s^{-1} (150°C, RH 6%)	CO_2/N_2: <5 (50°C, RH 0%) ~10 (50°C, RH 3%) 80 (50°C, RH 6%) ~4 (150°C, RH 0%) ~4 (150°C, RH 3%) ~4 (150°C, RH 6%) (10:90 CO_2–N_2 mixture)

syngas is transformed into CO_2 that is further separated from the hydrogen by means of the membrane.

One of the possible solutions for CO_2–H_2 separation is the integration of a *dense ceramic membrane with mixed proton-electron conductivity*. Among the advantages of this type of membrane are their lower price compared with precious metal-based membranes and their higher thermal, chemical, and mechanical resistance, which implies a longer lifetime without significant functional degradation.

The development and industrial implementation of dense ceramic mixed proton-electron conducting membranes faces several open issues that require further optimization. Among them can be highlighted: (i) the chemical stability of the membrane, in order to withstand the harsh operating conditions in a real power plant (still rather insufficient for several types of materials); (ii) the mixed proton-electron conductivity, which is determining for the hydrogen flux through the membrane (considerable conductivity and permeation values are required); and (iii) functional degradation issue under operating conditions (in the short and long term). These factors are the key features of the membrane for ensuring high performance and simultaneously having a great impact regarding the real implementation of the membrane in industrial applications. Hence an important prerequisite of R&D activities is close collaboration between the scientific and industrial communities.

In this section, an overview is presented focused on the variety of MPEC ceramic materials for manufacturing hydrogen-separating dense ceramic membranes. Several groups of materials with higher thermal tolerance are considered and available hydrogen permeability data are also presented.

16.3.1
Membrane Materials Overview

The ceramic materials for hydrogen separation membranes integrated in fossil power plants should have high proton and electron conductivity and low oxygen ion conductivity. Hydrogen permeation through a dense ceramic membrane is proportionally dependent on the ambipolar conductivity, expressed by $(\sigma_{H^+}\sigma_{el})/\sigma_{tot}$. The high-temperature proton conductors known so far exhibit rather poor electronic conductivity that determines the low values of the overall hydrogen flux through the membrane, which is their main drawback. In order to obtain appreciable electronic conduction, candidate materials need to contain reducible cations. The current state of the art also shows that the implementation of different doping schemes represents another possible approach to improve the hydrogen flux through the membrane. It is also worth mentioning that since the first report on high-temperature proton conduction in oxides by Stotz and Wagner [28], the defect chemistry and the proton conductivity of a variety of oxide materials have been intensively investigated [29–36]. In contrast, the ability of these materials to permeate hydrogen has received less attention. The kinetic and mechanistic description of hydrogen permeation behavior through mixed proton-electron conductors remains rather incomplete [37]. Undoubtedly, due to the practical benefits that the dense ceramic membranes offer when integrated on an industrial scale, the development of ceramic materials for membranes represents a challenging field in materials science.

16.3.2
Materials with Perovskite Structure

The proton conductivity in the perovskite type of oxide ceramics was first discovered by Iwahara et al. [38] in the early 1980s. The most extensively studied high-temperature proton-conducting ceramic materials with perovskite structure are $SrCeO_3$, $BaCeO_3$, and $SrZrO_3$, and also complex perovskites of the types $A_2B'B''O_6$ and $A_3B'B''O_9$ [39].

16.3.2.1
Cerates

The highest proton conductivity in hydrogen-containing atmospheres has been found for cerate-based materials, more particularly in $SrCe_{0.95}Yb_{0.05}O_{3-\delta}$ (SCYb) (0.7×10^{-2} S cm^{-1} at 900 °C in pure H_2) and $BaCe_{0.9}Nd_{0.1}O_{3-\delta}$ (BCNd) (2.2×10^{-2} S cm^{-1} at 900 °C in pure H_2). SCYb is essentially a non-oxygen ion conductor compared with BCNd, which exhibits oxygen ion conductivity of the same order of magnitude as its proton conductivity at temperatures higher than 800 °C. Similar behavior has been confirmed for $BaCe_{1-x}Y_xO_3$ (BCY), in which the oxygen ion conductivity is comparable to the proton conductivity of the material. $SrCe_{0.95}Y_{0.05}O_{3-\delta}$ (SCY), for example, has an electron conductivity that is too low for use as a membrane material and too high for use as an electrolyte in solid oxide fuel cells. Mixed conductivity of the cerate-based oxide materials, however, can be further improved by aliovalent doping on the Ce^{3+} site, and also by adding a metallic phase (10–40 vol.%) to the proton-conducting ceramic powder [39], and consequently producing a mixed conducting composite (cermet).

Doped proton conductors with the general formula $AB_{1-x}B'_xO_{3-\delta}$, where A is Ca, Sr, or Ba, B is Ce, Tb, Pr, or Th, and B' is Ti, V, Cr, Mn, Fe, or Co, were recently described in the patent literature. The best membranes of thickness 1–2 mm give hydrogen permeation fluxes in the range of $(8–16) \times 10^8$ mol cm^{-2} s^{-1} at 850–950 °C with the membrane exposed to oxygen and methane [40].

A study by Matsumoto et al. [41] on mixed proton-electron conduction and hydrogen permeation of $BaCe_{0.9-x}Y_{0.1}Ru_xO_{3-\delta}$ highlights the influence of Ru on acceptor-doped $BaCeO_3$. No hydrogen flux was detected when the membrane did not contain Ru or when the Ru concentration was $x=0.05$. For Ru concentrations of $x=0.075$ and 1, remarkable flux is measured with a maximum value of 6.47×10^{-8} mol cm^{-2} s^{-1}. It can be concluded that the hydrogen flux through an acceptor-doped $BaCeO_3$ membrane can be improved significantly by adding Ru as a dopant, which contributes mainly to the electron conduction of the material.

Zuo et al. [42] investigated the effect of Zr doping on hydrogen permeation of the cermet Ni–BaCe$_{0.8}$Y$_{0.2}$O$_{3-\delta}$ mixed proton-electron conductor. It was found that the hydrogen permeation fluxes increase with temperature from 600 to 900 °C for Ni–BaZr$_{0.8-x}$Ce$_x$Y$_{0.2}$O$_{3-\delta}$ membranes of all compositions. In the case of Zr-free membranes the flux reaches a maximum of 0.056 cm^3 min^{-1} cm^{-2} at 650 °C and then decreases slightly with increase in temperature. At each experimental temperature, the permeation flux decreases with increase in zirconium content from 10 to 40%. The highest flux of 0.056 cm^3 min^{-1} cm^{-2} is measured for Ni–BaZr$_{0.1}$Ce$_{0.7}$Y$_{0.2}$O$_{3-\delta}$ at 900 °C.

The preparation and hydrogen permeation of BaCe$_{0.95}$Nd$_{0.05}$O$_{3-\delta}$ is described in a paper by Cai et al. [43]. Membranes were prepared via the EDTA-citric acid complexing method. Figure 16.1 presents the hydrogen flux of a BaCe$_{0.95}$Nd$_{0.05}$O$_{3-\delta}$ membrane with a thickness of 0.7 mm as a function of temperature. The hydrogen flux increases with increase in hydrogen content and stream concentration at the feed side. For a stream concentration of 15 vol.%, the hydrogen permeation flux is 0.026 cm^3 min^{-1} cm^{-2} at 925 °C.

Hamakawa et al. [44] prepared thin layers of SrCe$_{0.95}$Yb$_{0.05}$O$_{3-\delta}$ and SrZr$_{0.95}$Y$_{0.05}$O$_{3-\delta}$ on porous SrZr$_{0.95}$Y$_{0.05}$O$_{3-\delta}$ substrates via the spin coating of colloidal suspension of powders prepared by the combustion method. Hydrogen permeation rates were measured for membranes with different thicknesses. The permeation rates through SrCe$_{0.95}$Yb$_{0.05}$O$_{3-\delta}$ films with thickness of 2–140 µm were as high as 500 times larger at 950 K than on 1 mm discs at the same temperature. For a film 2 µm thick, a permeation rate of 6.0×10^{-4} mol H$_2$ cm^{-2} min^{-1} was measured. H$_2$ permeation rates were proportional to the inverse of the membrane thickness, indicating that the permeation through SrCe$_{0.95}$Yb$_{0.05}$O$_{3-\delta}$ thin films

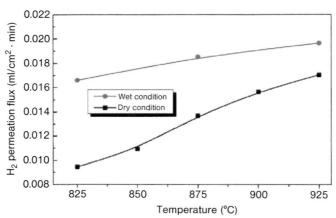

Fig. 16.1 H$_2$ permeation flux as a function of temperature under dry and wet conditions. Membrane thickness, 0.7 mm. Feed side: dry conditions (80 ml min^{-1} H$_2$ + 20 ml min^{-1} He); wet conditions (80 ml min^{-1} H$_2$ + 15 ml min^{-1} He + 5 ml min^{-1} steam). Sweep side: 29.5 ml min^{-1} Ar + 0.5 ml min^{-1} Ne [43].

even 2 µm thick is controlled by the bulk diffusion and not by the dissociative chemisorption of H_2 or by the boundary layer transport.

The conduction in SCYb ceramic becomes proton-electron-hole type as the oxygen partial pressure increases. In order to understand better the mixed conduction behavior of $SrCe_{0.95}Yb_{0.05}O_{3-\delta}$ in a hydrogen–oxygen cell, Hamakawa *et al.* [45] carried out a study on mixed conduction as a function of surrounding atmosphere. It was confirmed that Yb-doped Sr-cerate behaves as a proton-electron-hole mixed conductor when one side of the membrane is exposed to oxidizing conditions and the other side to 1% hydrogen at 1173 K. Under these conditions, H_2 permeation is possible since the membrane acts as an electrochemical diaphragm without direct current being applied to the cell.

The study by Li and Iglesia [37] was focused on the modeling and analysis of hydrogen permeation in mixed proton-electron conductors. They derived a mass transfer model for hydrogen permeation through proton-electron mixed conductors that has been developed on the basis of non-equilibrium thermodynamics. The model describes the concentrations of all charge carriers through equilibrium with the contacting gas phase at surfaces and through a pressure formalism within the membrane material. This model was applied to describe hydrogen permeation through dense $SrCe_{0.95}Yb_{0.05}O_{3-\delta}$ membranes. The rate-controlling role of electronic transport in H_2 permeation processes was confirmed, especially when the membrane is operated at very different hydrogen chemical potentials at both sides. Under these operating conditions, an internal region with a very low electron transfer number arises from a transition from n-type to p-type conduction as the hydrogen chemical potential within the membrane decreases from high values near the process side to much lower values at the permeate side. As a result, when the electronic mobility is high, either because of the nature of the membrane material or because of the presence of an external conducting circuit, large H_2 chemical potential gradients across the membrane exploit the full driving force without the adverse effects of the lower electronic transport rates associated with such chemical potential gradients. When electronic transport limits the H_2 permeation, the full effects of these large chemical potential gradients cannot be exploited, because of their negative effects on electron transport rates. Improvements in electronic (and H_2) transport rates require higher electron-hole diffusivities or more reducible membrane materials, which increase electron concentrations for a given chemical environment.

It is still an open issue to determine the effect on the proton conductivity of improving substantially the p-type conductivity, since the empirical observation is that the enhancement of the electronic conductivity reduces the proton conductivity.

Lin [46] presented a detailed review on the status and prospects of dense ceramic and microporous membranes. With respect to the ceramic membranes, the electron conductivity of the material was correlated with the ionization potential of the doping element. The electron conductivity of $SrCeO_3$ (2.82×10^{-4} S cm^{-1} at 900 °C) and $SrCe_{0.95}Yb_{0.05}O_{3-\delta}$ (8.50×10^{-3} S cm^{-1} at 900 °C) appeared to be inversely related to the ionization potential of the lanthanide element in these ceramics.

The ionization potential of the doping ions in proton-conducting $SrCeO_3$ (36.8 eV) and $SrCe_{0.95}Yb_{0.05}O_{3-\delta}$ (25.0 eV) were chosen to be reference values for the selection of Tb and Tm as doping elements having ionization potentials respectively higher (39.8 eV for Tb) and lower (23.7 eV for Tm) than Ce and Yb. Solid solutions of $SrCe_{0.95}Tb_{0.05}O_{3-\delta}$ (SCTb) and $SrCe_{0.95}Tm_{0.05}O_{3-\delta}$ (SCTm) were prepared by the citrate route described by Qi and Lin [47] and their electron conductivity was measured under different atmospheres by the four-point d.c. method. The results show that the electronic conductivity increases with decreasing ionization potential. SCTb with electron conductivity much lower than that of SCYb can be used as an electrolyte for fuel cell, and SCTm with a higher electron conductivity is a good membrane material. The hydrogen permeance of disc membranes with a thickness of 1–3 mm was measured with one membrane surface exposed to a 10% H_2–He mixture and the other to a 20% O_2–N_2 mixture. At 900 °C, the hydrogen permeance flux increases from 2.0×10^{-8} to 3.3×10^8 mol cm^{-2} s^{-1} as the membrane thickness decreases from 3 to 1 mm. The SCTm membranes could offer a higher hydrogen permeation flux if the thickness is further reduced.

An effective way to improve the hydrogen permeance of the mixed proton-electron-conducting membrane is to prepare the membrane as an asymmetric structure of a dense layer deposited on a porous substrate prepared from the same proton-conducting material. In the work of Cheng et al. [39], the dry pressing preparation procedure was reported as a method for manufacturing asymmetric membranes of $SrCe_{0.95}Tm_{0.05}O_{3-\delta}$ on a porous $SrCe_{0.95}Tm_{0.05}O_{3-\delta}$ substrate. It was found that the hydrogen permeation through the SCTm membrane is limited by the bulk diffusion of the charge carriers. An H_2 flux of 9.37×10^{-8} mol cm^{-2} s^{-1} was measured at 900 °C for a membrane with a thickness of 150 m when a mixture of 10% H_2–He was used at the upstream side and air was used at the downstream side, as shown in Figure 16.2.

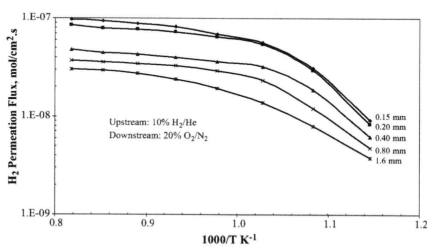

Fig. 16.2 H_2 permeation flux versus $1000/T$ for SCTm membranes of different thicknesses [39].

The flux value is about four times higher than that obtained for a 1.6 mm thick SCTm membrane. Moreover, the hydrogen flux increases when the downstream side is increased. This is a clear indication that the mixed conductivity of the SCTm membrane is of the proton-electron-hole type.

Mixed proton-electron conductivity of $SrCe_{1-x}Tb_xO_{3-\delta}$ ($x=0.025$, 0.05, and 0.10) under hydrogen-containing atmospheres was reported by Wei et al. [48]. The activation energy of the proton conduction is close to that of electron conduction for this material. It was found that an increase in downstream CO partial pressure from 0.001 to 0.1 atm leads to a small increase in the hydrogen flux from 1.4×10^{-2} to 1.6×10^{-2} cm^3 cm^{-2} min^{-1}. The hydrogen flux of the SCTb membrane also increases with increase in the upstream hydrogen partial pressure.

Zhan et al. [49] focused on the development and flux measurements of $SrCe_{0.95}Y_{0.05}O_{3-\delta}$ asymmetric membranes obtained by the dry pressing method. Membranes with a thickness of about 50 m were reported to yield hydrogen flux as high as 7.6×10^{-8} mol cm^{-2} s^{-1} at 950 °C (80% H_2–He). This value is about seven times higher than that of the symmetric membranes with a thickness of ∼ 620 m.

A hydrogen permeability study and also an estimation of the effect of microstructure on mixed proton-electron conductivity of Sr-cerate doped with 5% Eu were carried out by Song et al. [50]. The highest flux of 3.1×10^{-9} mol cm^{-2} s^{-1} was measured at 850 °C for a membrane sintered at 1773 K for 10 h (a membrane thickness 1.72 mm). In this case, the milling of the powder after the solid-state synthesis was carried out for 10 days. Permeation measurements showed that the H_2 flux decreases when the milling time is shorter and the sintering time is longer.

The hydrogen permeability of $SrCe_{0.95}Eu_{0.05}O_{3-\delta}$ and $SrCe_{0.95}Sm_{0.05}O_{3-\delta}$ membranes as a function of temperature, hydrogen partial pressure gradient, and water vapor pressure gradient was investigated in another study by Song et al. [51]. Under dry hydrogen conditions at 1123 K, the hydrogen permeation rates of dense membranes (1.72 mm thick) are 3.19×10^{-9} mol cm^{-2} s^{-1} for $SrCe_{0.95}Eu_{0.05}O_{3-\delta}$ and 2.33×10^{-9} mol cm^{-2} s^{-1} for $SrCe_{0.95}Sm_{0.05}O_{3-\delta}$. Under wet hydrogen conditions at 1123 K, the hydrogen permeation rates are 2.89×10^{-9} and 1.21×10^{-9} mol cm^{-2} s^{-1}, respectively.

16.3.2.2 Zirconates

The class of zirconate-based materials is also interesting for manufacturing hydrogen separation membranes, mainly due to the higher stability of zirconates under reducing atmospheres, in particular in CO_2 environments. However, the existing experimental data on the hydrogen flux through zirconate-based membranes are limited in comparison with the data available for the cerate class of membranes.

A mixed conducting hydrogen separation membrane, consisting of a proton-conductive oxide ($BaZr_{0.8}Y_{0.2}O_{3-\delta}$) and metallic palladium, was reported by Okada et al. [52]. In this case, a porous alumina tube was used as a support, and proton-conductive oxide particles were introduced into the microporous top layer of the support by impregnation. Palladium particles were deposited on the same porous layer by CVD. A hydrogen permeance (P_{H_2}) of 1.2×10^9 mol m^{-2} s^{-1} Pa^{-1} at 873 K was measured for the membrane. The selectivity for hydrogen (P_{H_2}/P_{N_2}) increased with operating temperature due to an increase in proton conductivity of the membrane, and $P_{H_2}/P_{N_2} = 5.7$ was attained at 873 K. The hydrogen separation mechanism of this membrane is considered to be based on the diffusion of dissolved protons through the cermet membrane.

Hamakawa et al. [44] reported a rather limited hydrogen flux through the $SrZr_{1-x}Yb_xO_{3-\delta}$ membrane with no oxygen in the sweep gas, due to its low electronic conductivity under non-oxidative atmospheres. This result indicates that the electronic conductivity limits the hydrogen permeance of proton-conductive membranes operating in a non-galvanic mode. As was outlined above, both proton and electron conductivities are required and determine the hydrogen flux through the dense ceramic membrane. Hence an effective route for improving the hydrogen permeability of the membrane is to increase the electron conductivity of the material. The measured hydrogen flux through a 1 mm thick $SrZr_{0.95}Y_{0.05}O_{3-\delta}$ disc at 700 °C and 200 kPa H_2 was lower than 10^{-8} mol cm^{-2} min^{-1}. Obviously the low hydrogen flux is due to the low conductivity of one of the required charge carriers (protons and electron–holes). Hibino et al. [53] attributed the lower hydrogen flux of Y-doped $SrZrO_3$ to the lower proton conductivity.

$CaZrO_3$-based oxides in which trivalent cations (In, Ga, Sc) partly replaced Zr were studied by Yajima et al. [54] with a focus on development and proton conductivity measurements. The electrochemical hydrogen permeation was carried out implementing these ceramic materials as a solid electrolyte. The charge carriers in the doped $CaZrO_3$ are protons and the proton transport number is almost unity in hydrogen atmospheres. The authors stated that using these oxides as a solid electrolyte, it may be possible to separate hydrogen from a mixed gas such as pyrolysis gas from hydrocarbons and to electrolyze steam to produce hydrogen.

16.3.3
Materials with a Defective Fluorite Structure (Ln_6WO_{12})

There have been only a few studies on the hydrogen permeation of the Ln_6WO_{12} class of materials with a defective fluorite structure.

The hydrogen flux through an Nd_6WO_{12} disc membrane with a thickness of 510 μm was investigated by Escolástico et al. [55]. At the feed side, a mixture of moist 20% H_2–He was used and at the sweep side argon (both sides at atmospheric pressure). In Figure 16.3, the hydrogen permeation flux of the Nd_6WO_{12} membrane is presented as a function of temperature. At moderate temperatures (below 900 °C), the electron conductivity seems to be the limiting factor, whereas at high temperatures (above 900 °C), p-type conductivity becomes pre-

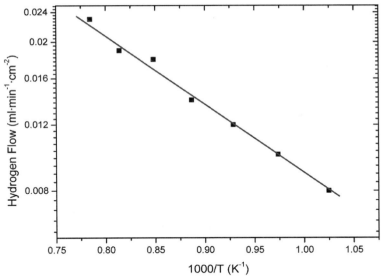

Fig. 16.3 Hydrogen permeation flux of an Nd_6WO_{12} membrane as a function of temperature [55].

dominant and the proton conductivity is the limiting step, likely due to the low concentration of protons in the oxide lattice, as indicated by thermogravimetric studies by Haugsrud and Kjølseth [56].

Permeability data prediction was reported by Norby and Haugsrud [57] for La_6WO_{12}, $SrCeO_3$, and Er_6WO_{12}. The flux was calculated considering 10 m thick layers with a feed side pressure of 10 atm H_2 and assuming that the flux was ruled by the Wagner equation, that is, through the bulk diffusion mechanism. The predicted hydrogen permeation for La_6WO_{12} at 800 °C is about 2.0 ml_n cm^{-2} min^{-1}, which is higher than for $SrCeO_3$ and Er_6WO_{12}, which yield a hydrogen flux at the same temperature of about 1.0 and < 0.1 ml_n cm^{-2} min^{-1}, respectively.

Haugsrud and Kjølseth [56] analyzed the feasibility of tungstates in membrane applications by comparing them with 5% Yb-doped $SrCeO_3$, which is one of the state-of-the-art mixed proton-electron conductors. Based on the ambipolar proton-electron conductivity, which is twice as high as the conductivity of perovskite at 1000 °C, it can be concluded that undoped La_6WO_{12} is a good candidate for membrane applications. Moreover, the behavior of Gd_6WO_{12} and Er_6WO_{12} was pointed out as interesting. The authors recognized the possibility of protons compensating electrons under partial reduction at low temperature.

16.3.4
Materials with Pyrochlore Structure

Pyrochlore-structured oxides containing Ti exhibit relatively high mixed oxygen ion–electron conductivities [58]. Pyrochlore-structured $La_2Zr_2O_7$ has been reported to show proton conductivity which has not been reported for the isostructural Ti-containing oxide [59]. Shimura et al. [60] found that $Y_2Ti_{1.8}M_{0.2}O_{7-\delta}$ with M = In or Mg did not show measurable proton transport. Based on empirical models to predict the proton conductivity in oxides, the class of A-site acceptor-doped $La_2Ti_2O_7$ may be considered to be a better candidate. The conductivity of undoped $La_2Ti_2O_7$ was previously measured at relatively high temperatures (800–1000 °C) and with no emphasis on possible effects of protons and proton transport [61].

Haugsrud and Norby addressed 2% Ca-doped $La_2Ti_2O_7$ [58]. The mixed proton-oxygen ion-electron conductivity of this material was studied in order to evaluate its potential as a candidate material for gas separation membranes. The conductivity was measured by the a.c. impedance spectroscopy as a function of the oxygen partial pressure and water vapor pressure in different atmospheres at temperatures between 300 and 1200 °C. The partial conductivities were determined by EMF measurements. The hydrogen flux across a Ca-doped $La_2Ti_2O_7$ film with a thickness of 10 μm was estimated on the basis of partial conductivity measurement. The authors assumed a pressure difference of 10 atm H_2 between the feed and the permeate sides and estimated that at 750 °C the membrane would yield a hydrogen flux in the order of 0.1 ml_N cm^{-2} min^{-1}. Comparing this value with the perovskites, it appeared that 5% Yb-doped $SrCeO_3$ would exhibit an H_2 flux higher by a factor of ~5 than $La_2Ti_2O_7$ under the corresponding reaction conditions.

16.4
Conclusion and Outlook

Sol-gel microporous membranes have been presented by many research groups as one of the candidates for hydrogen separation due to their ease of fabrication, low cost of production and scalability. Because of their pore size of approximately 0.3–0.5 nm, amorphous SiO_2 membranes have been most frequently investigated and high separation factors for small-molecule gases such as H_2 and N_2 or CO_2 have been accomplished.

Today, it is recognized, however, that these membranes suffer from serious stability problems on exposure to moisture and therefore much work is being done in searching for better materials. One approach is to improve the material stability by adding doping compounds, such as ZrO_2, TiO_2, Al_2O_3, and NiO. An extension of the doped silica membrane work is the synthesis of metal-doped membranes (e.g., Ni- and Co-doped). Another common method which could potentially be used to improve the material stability is the synthesis of hybrid organic-ceramic SiO_2 membranes.

Another interesting approach reported by many researchers is to prepare crystalline microporous SiO_2-based membranes (e.g., zeolites). Some of these membranes exhibit fascinating properties, including a well-defined pore structure with pore sizes of several ngströms and are not susceptible to densification at higher temperatures or degradation in steam. Several groups have succeeded in preparing zeolite membranes with good performance in gas and liquid separations, but the existence of larger pores and also intercrystalline pores often limits their molecular separation efficiency for small gases. Research is therefore focused on developing more effective synthesis methods and improving the separation properties of currently available membranes by surface modification techniques.

Due to the high chemical resistance of some non-silica materials, in particular TiO_2 and ZrO_2, much effort has also been devoted to the synthesis and potential application of supported non-silica membranes. Significant progress has been made in developing membranes with a pore size < 2 nm by sol-gel coating methods and research should continue in tailoring the pore size by developing more effective nanoparticle synthesis methods.

In addition to membrane processing research, one of the main challenges is to provide ideal module fabrication and sealing methods. The potential to fabricate the mentioned thin-film membranes on metallic carriers, which is currently under study by many groups, can, however, offer a solution for this problem.

In the field of dense ceramic proton-conducting membranes, significant progress has been made during the past few years, mainly motivated by the benefits that these membranes offer and the numerous applications in which they can be implemented.

The improvement of proton conductivity in terms of PC-SOFC application and of mixed proton-electron conductivity (the H_2 flux) in the case of H_2 separation membrane applications still remains amongst the most challenging materials requirements. Another crucial requirement for membrane implementation in targeted applications is the chemical stability, for example, in reduction-oxidation conditions, when the membrane acts as a solid electrolyte in the PC-SOFC, and in reducing atmospheres, for example, when H_2 separation through the membrane is considered.

Proton conductivity is rather sensitive to temperature due to the thermodynamic stability of proton defects in terms of proton content that contributes to the total conductivity of the material. Moreover, the stability and mobility of proton defects are rather individual for each kind of material. They are strongly dependent on the chemical nature and concentration of doping elements and governed by the thermodynamics of hydration. In this context, it should be mentioned that the proton defects have the highest stability in cerates, followed by the class of zirconate materials, but this trend follows the opposite tendency when the chemical stability of cerate and zirconate membrane materials is considered. In order to elucidate the basics of these phenomena and the complex dependences for different classes of materials, intensive research is strongly required. In any case, the proton-electron or dominant proton conductivity, the proton defects stability/mobility, and the chemical stability must be balanced with a definite degree of compromise in order to obtain the optimal properties of the membrane.

Acknowledgments

This work is supported by the Initiative and Networking Fund of the Helmholtz Association, contract HA-104 ("MEM-BRAIN") and by the Federal Ministry of Education and Research of Germany through the Northern European Innovative Energy Research Project, contract 03SF0330 ("N-INNER").

References

1 Puhlfürss, P., Voigt, A., Weber, R., and Morbé, M. (2000) Microporous TiO_2 membranes with a cut-off < 500 Da. *J. Membr. Sci.*, **174**, 123–133.
2 Voight, I., Fisher, G., Puhlfürss, P., Schleifenheimer, M., and Stahn, M. (2003) TiO_2 NF-membranes on capillary supports. *Sep. Purif. Technol.*, **32**, 87–91.
3 Sekulic, J., ten Elshof, J. E., and Blank, D. H. A. (2004) A microporous titania membrane for nanofiltration and pervaporation. *Adv. Mater.*, **72**, 49–57.
4 Tsuru, T., Hironka, D., Yoshioka, T., and Asaeda, M. (2001) Titania membranes for liquid phase separation: effect of surface charge on flux. *Sep. Purif. Technol.*, **25**, 307–314.
5 de Vos, R. M. and Verweij, H. (1998) Improved performance of silica membranes for gas separation. *J. Membr. Sci.*, **143**, 37.
6 de Vos, R. M., Maier, W. F., and Verweij, H. (1999) Hydrophobic silica membranes for gas separation. *J. Membr. Sci.*, **158**, 277–288.
7 Tsai, C. Y., Tam, S. Y., Lu, Y., and Brinker, C. J. (2000) Dual-layer asymmetric microporous silica membranes. *J. Membr. Sci.*, **169**, 255–268.
8 Xomeritakis, G., Tsai, C. Y., Jiang, Y. B., and Brinker, C. J. (2009) Tubular ceramic-supported sol-gel silica-based membranes for flue gas carbon dioxide capture and sequestration. *J. Membr. Sci.*, **341** (1–2), 30–36.
9 Kanezashi, M. and Asaeda, M. (2006) Hydrogen permeation characteristics and stability of Ni-doped silica membranes in steam at high temperature. *J. Membr. Sci.*, **271**, 86–93.
10 Battersby, S., Tasaki, T., Smart, S., Ladewig, B., Liu, S., Duke, M., Rudolpha, V., and Diniz da Costa, J. C. (2009) Performance of cobalt silica membranes in gas mixture separation. *J. Membr. Sci.*, **329**, 91–98.
11 Boffa, V., Blank, D. H. A., and ten Elshof, J. E. (2008) Hydrothermal stability of microporous silica and niobia–silica membranes. *J. Membr. Sci.*, **319** (1–2), 256–263.
12 Yoshida, K., Hirano, Y., Fujii, H., Tsuru, T., and Asaeda, M. (2001) Hydrothermal stability and performance of silica-zirconia membranes for hydrogen separation in hydrothermal conditions. *J. Chem. Eng. Jpn.*, **34** (4), 523–530.
13 Gopalakrishnan, S. and Diniz da Costa, J. C. (2008) Hydrogen gas mixture separation by CVD silica membrane. *J. Membr. Sci.*, **323**, 144–147.
14 Oyama, S. T., Lee, D., Hacarlioglu, P., and Saraf, R. F. (2004) Theory of hydro-

gen permeability in nonporous silica membranes. *J. Membr. Sci.*, **244**, 45–53.
15 Gu, Y., Harcarlioglu, P., and Oyama, S.T. (2008) Hydrothermally stable silica-alumina composite membranes for hydrogen separation. *J. Membr. Sci.*, **310**, 28–37.
16 Sea, B.K., Kusakabe, K., and Morooka, S. (1997) Pore size control and gas permeation kinetics of silica membranes by pyrolysis of phenyl-substituted ethoxysilanes with cross-flow through a porous support wall. *J. Membr. Sci.*, **130** (1–2), 41–52.
17 Araki, S., Mohri, N., and Yoshimitsu, Y. (2007) Synthesis, characterization and gas permeation properties of a silica membrane prepared by high-pressure chemical vapor deposition. *J. Membr. Sci.*, **290** (1–2), 138–145.
18 Nishiyama, N., Yamaguchi, M., Katayama, T., Hirota, Y., Miyamoto, M., Egashira, Y., Ueyama, K., Nakanishi, K., Ohta, T., Mizusawa, A., and Satoh, T. (2007) Hydrogen-permeable membranes composed of zeolite nano-blocks. *J. Membr. Sci.*, **306** (1–2), 349–354.
19 Hong, M., Li, S., Falconer, J., and Noble, R.D. (2008) Hydrogen purification using a SAPO-34 membrane. *J. Membr. Sci.*, **307** (2), 277–283.
20 Bonhomme, F., Welk, M., and Nenoff, T. (2003) CO_2 selectivity and lifetimes of high silica ZSM-5 membranes. *Microporous Mesoporous Mater.*, **66**, 181–188.
21 Sebastian, V., Kumakiri, I., Bredesen, R., and Menendez, M. (2007) Zeolite membrane for CO_2 removal: operating at high pressure. *J. Membr. Sci.*, **292**, 92–97.
22 Shin, D.W., Hyun, S.H., Cho, C.H., and Han Shin, M.H. (2005) Synthesis and CO_2/N_2 gas permeation characteristics of ZSM-5 zeolite membranes. *Microporous Mesoporous Mater.*, **85**, 313–323.
23 Kim, Y.S., Kusakabe, K., Morooka, S., and Yang, S.M. (2001) Preparation of microporous silica membranes for gas separation. *Korean J. Chem. Eng.*, **18** (1), 106–112.
24 Kusakabe, K. (1999) Pore structure of silica membranes formed by a sol-gel technique using tetraethoxysilane and alkyltriethoxysilanes. *Sep. Purif. Technol.*, **16** (2), 139–146.

25 Moon, J.H., Park, Y.J., Kim, M.B., Hyun, S.H., and Lee, C.H. (2005) Permeation and separation of a carbon dioxide/nitrogen mixture in a methyltriethoxysilane templating silica/γ-alumina composite membrane. *J. Membr. Sci.*, **250**, 195–205.
26 Osada, K., Ohnishi, T., Shin, Y., Yoshini, J., and Kuroishi, N. (1999) Development of inorganic membranes by sol-gel method for CO_2 separation, *Greenhouse Gas Control Technologies* (eds. Riemer, P., Eliasson, B., Wokaun, A.), Elsevier Science, pp. 43–45.
27 West, G.D., Diamond, G.G., Holland, D., Smith, M.E., and Lewis, M.H. (2002) Gas transport mechanisms through sol-gel derived templated membranes. *J. Membr. Sci.*, **203**, 53–69.
28 Stotz, S., and Wagner, C. (1966) Die Löslichkeit von Wasserdampf und Wasserstoff in festen Oxiden. *Ber. Bunsenges. Phys.Chem.*, **70** (8), 781–788.
29 Bonanos, N. (2001) Oxide-based protonic conductors: point defects and transport properties. *Solid State Ionics*, **145**, 265–274.
30 Kreuer, K.D. (2003) Proton conducting oxides. *Annu. Rev. Mater. Res.*, **33**, 333–359.
31 Kreuer, K.D. (1996) Proton conductivity: materials and applications. *Chem. Mater.*, **8**, 610–641.
32 Kreuer, K.D. (2000) On the complexity of proton conduction phenomena. *Solid State Ionics*, **136–137**, 149–160.
33 Iwahara, H. (1996) Proton conduct, ceramics and applications. *Solid State Ionics*, **86–88**, 9–15.
34 Nowick, A.S. and Du, Y. (1995) High-temperature protonic conductors with perovskite-related structures. *Solid State Ionics*, **77**, 137–146.
35 Norby, T. (1999) Solid-state protonic conductors: principles, properties, progress and prospects. *Solid State Ionics*, **125**, 1–11.
36 Phair, J.W. and Badwal, S.P.S. (2006) Materials for separation membranes in hydrogen and oxygen production and future power generation. *Sci. Technol. Adv. Mater.*, **7**, 792–805.
37 Li, L. and Iglesia, E. (2003) Modeling and analysis of hydrogen permeation

in mixed proton-electronic conductors. *Chem. Eng. Sci.*, **58**, 1977–1988.

38 Iwahara, H., Esaka, T., Uchida, H., and Maeda, N. (1984) Proton conduction in sintered oxides and its application to steam electrolysis for hydrogen production. *Solid State Ionics*, **3/4**, 359.

39 Cheng, S., Gupta, V., and Lin, Y. S. (2005) Synthesis and hydrogen permeation properties of asymmetric proton-conducting ceramic membranes. *Solid State Ionics*, **176**, 2653–2662.

40 White, J. H., Schwartz, M., and Sammells, A. F. (2000) Solid state proton and electron mediating membrane and use in catalytic membrane reactors. US Patent 6 037 514.

41 Matsumoto, H., Shimura, T., Higuchi, T., Tanaka, H., Katahira, K., Otake, T., Kudo, T., Yashiro, K., Kaimai, A., Kawada, T., and Mizusaki, J. (2005) Protonic-electronic mixed conduction and hydrogen permeation in $BaCe_{0.9-x}Y_{0.1}Ru_xO_{3-\delta}$. *J. Electrochem. Soc.*, **152** (3), A488–A492.

42 Zuo, C., Dorris, S. E., Balachandran, U., and Liu, M. (2006) Effect of Zr-doping on the chemical stability and hydrogen permeation of the $Ni–BaCe_{0.8}Y_{0.2}O_{3-\delta}$ mixed protonic-electronic conductor. *Chem. Mater.*, **18**, 4647–4650.

43 Cai, M., Liu, S., Efimov, K., Caro, J., Feldhoff, A., and Wang, H. (2009) Preparation and hydrogen permeation of $BaCe_{0.95}Nd_{0.05}O_{3-\delta}$ membranes, *J. Membr. Sci.*, **343**, 90–96.

44 Hamakawa, S., Li, L., Li, A., and Iglesia, E. (2002) Synthesis and hydrogen permeation properties of membranes based on dense $SrCe_{0.95}Yb_{0.05}O_{3-\alpha}$ thin films. *Solid State Ionics*, **48**, 71–81.

45 Hamakawa, S., Hibino, T., and Iwahara, H. (1994) Electrochemical H_2 permeation in a proton-hole mixed conductor and its application to a membrane reactor. *J. Electrochem. Soc.*, **141** (7), 1720–1725.

46 Lin Y. S. (2001) Microporous and dense inorganic membranes: current status and prospective. *Sep. Purif. Technol.*, **25**, 39–55.

47 Qi, X. and Lin, Y. S. (2000) Electrical conductivity and hydrogen permeation through mixed proton-electron conducting strontium cerate membranes. *Solid State Ionics*, **130**, 149–156.

48 Wei, X., Kniep, J., and Lin, Y. S. (2009) Hydrogen permeation through terbium doped strontium cerate membranes enabled by presence of reducing gas in the downstream. *J. Membr. Sci.*, **345**, 201–206.

49 Zhan, S., Zhu, X., Ji, B., Wang, W., Zhang, X., Wang, J., Yang, W., and Lin, L. (2009) Preparation and hydrogen permeation of $SrCe_{0.95}Y_{0.05}O_{3-\delta}$ asymmetrical membranes. *J. Membr. Sci.*, **340**, 241–248.

50 Song, S. J., Wachsman, E. D., Rhodes, J., Yoon, H. S., Lee, K. H., Zhang, G., Dorris, S. E., and Balachandran, U. (2005) Hydrogen permeability and effect of microstructure on mixed protonic-electronic conducting Eu-doped strontium cerate. *J. Mater. Sci.*, **40**, 4061–4066.

51 Song, S. J., Wachsman, E. D., Rhodes, J., Dorris, S. E., and Balachandran, U. (2004) Hydrogen permeability of $SrCe_{1-x}M_xO_{3-\delta}$ ($x=0.05$, M = Eu, Sm). *Solid State Ionics*, **167**, 99–105.

52 Okada, S., Mineshige, A., Kikuchi, T., Kobune, M., and Yazawa, T. (2007) Cermet-type hydrogen separation membrane obtained from fine particles of high temperature proton-conductive oxide and palladium. *Thin Solid Films*, **515**, 7342–7346.

53 Hibino, T., Mizutani, K., Yajima, T., and Iwahara H. (1992) Characterization of proton in Y-doped $SrZrO_3$ polycrystal by IR spectroscopy. *Solid State Ionics*, **58** (1–2), 85–88.

54 Yajima, T., Suzuki, H., Yogo, T., and Iwahara, H. (1992) Protonic conduction in $SrZrO_3$-based oxides. *Solid State Ionics*, **51** (1–2), 101–107.

55 Escolástico, S., Vert, V. B., and Serra, J. M. (2009) Preparation and characterization of nanocrystalline mixed proton-electronic conducting materials based on the system Ln_6WO_{12}. *Chem. Mater.*, **21**, 3079–3089.

56 Haugsrud, R. and Kjølseth, Ch. (2008) Effects of protons and acceptor substitution on the electrical conductivity of

La$_6$WO$_{12}$. *J. Phys. Chem. Solids*, **69**, 1758–1765.

57 Norby, T. and Haugsrud, R. (2007) High temperature proton conducting materials for H$_2$-separation. Invited Lecture at FZJ Workshop, 16 November 2007.

58 Haugsrud, R. and Norby, T. (2005) On the mixed ionic–electronic conductivity in Ca-doped La$_2$Ti$_2$O$_7$. Proceedings of the 26th Risø International Symposium on Materials Science. *Solid State Electrochemistry* (eds S. Linderoth, A. Smith, N. Bonanos, A. Hagen, L. Mikkelsen, K. Kammer, D. Lybye, P.V. Hendriksen, F.W. Poulsen, M. Mogensen, and W.G. Wang), Risø National Laboratory, Roskilde, Denmark.

59 Labrincha, J.A., Frade, J.R., and Marques, F.M.B. (1997) Proton conduction in La$_2$Zr$_2$O$_7$-based pyrochlore materials. *Solid State Ionics*, **99** (1–2), 33–40.

60 Shimura, T., Komori, M., and Iwahara, H. (1996) Ionic conduction in pyrochlore-type oxides containing rare earth elements at high temperature. *Solid State Ionics*, **86–88**, 685–689.

61 Balachandran, U. and Eror, N.G. (1982) Non-stoichiometric disorder in La$_2$Ti$_2$O$_7$ at elevated temperatures. *J. Less-Common Met.* **85**, 111–120.

17
Hydrogen System Assessment: Recent Trends and Insights
Joan M. Ogden

Abstract

The potential role of hydrogen in the world's future energy system has been analyzed in many assessments and remains a topic of vigorous, ongoing debate. A hydrogen economy involves not only new types of vehicles, but a new fuel infrastructure and development of low carbon primary resources. Given the number of possible configurations for a future hydrogen economy, systems analysis plays a key role in informing decision-makers in government and industry about the best hydrogen strategies and the prospects for hydrogen as compared to alternatives such as electricity, biofuels or fossil-derived liquid fuels.

While early assessments presented a vision of an "end-state" hydrogen economy, recent analyses have focused more on transitions: what is required to move toward widespread use of hydrogen energy starting from today's energy system. To understand the dynamics of transitions, analysts are using increasingly sophisticated tools: models of consumer choice and decision-making, infrastructure design models including spatial optimization and regional geographic data, and energy system modeling including interactions with the rest of the energy system. Lifecycle assessment is an important method for comparing the societal benefits of hydrogen to other fuels, considering the entire system, across multiple dimensions.

In this paper, I present an overview of recent trends in hydrogen system modeling and results from hydrogen systems assessments. In the last section, I propose a multi-attribute framework for evaluating hydrogen and other future fuels with respect to technical performance, cost, energy use, infrastructure development and greenhouse gas (GHG) emissions, while considering air pollution, energy security and reliability, water use, land use, and materials requirements. The goal is to illuminate hydrogen transition pathways that are viable across a wide range of economic and sustainability constraints, and to compare to other energy pathways.

Keywords: hydrogen assessment, systems analysis models, fuel cells, hydrogen vehicles, infrastructure, resources

Hydrogen Energy. Edited by Detlef Stolten
Copyright © 2010 WILEY-VCH Verlag GmbH & Co. KGaA, Weinheim
ISBN: 978-3-527-32711-9

17.1
Introduction

A large part of global primary energy use, greenhouse gas (GHG) emissions, and air pollution comes from direct combustion of fuels for transportation and heating. Reducing emissions and energy use from this multitude of dispersed sources (750 million vehicles, several billion households and millions of industrial sources worldwide) will mean replacing today's vehicles and heating systems with higher efficiency, low emission models and, ultimately, adopting new fuels that can be produced cleanly and efficiently from diverse sources. This is particularly crucial for transportation, where the number of vehicles could triple by 2050, and we are 97% dependent on a single primary energy source: crude oil.

In the near term, improving fuel economy is the best way to slow the growth of transportation GHG emissions and oil use, and should be aggressively pursued. However, even if we triple the fuel economy of today's fleet over a few decades and reduce reliance on cars through "smart growth," the growing number of vehicles around the world will constrain the rate of decline of oil use and CO_2 emissions. To make deeper reductions, we will need to couple high efficiency with a switch to low-carbon, non-petroleum fuels. Liquid fuels from biomass and fossil resources might play important roles, especially for applications such as air travel and heavy duty trucks. However, for the long term, two energy carriers stand out, offering high efficiency and zero emissions, and with a diverse, plentiful resource base: electricity and hydrogen. A third, long-term hope is to create a low-cost, easily stored liquid fuel directly from solar energy and CO_2 in the atmosphere, using genetically engineered organisms. If successful, this could obviate the need for electricity or hydrogen in transportation, both of which are difficult to store and transport, compared with liquid fuels. However, such methods are still in the basic research stage, whereas advances in battery and fuel cell technologies have brought electric and hydrogen vehicles to near-commercial status.

Depending on the application, hydrogen, electricity, or a liquid fuel may be preferred. Each new energy carrier faces technical uncertainties, market transition barriers, and costs. The potential role of hydrogen in the world's future energy system has been analyzed in many assessments [1–12] and remains a topic of vigorous, ongoing debate. A hydrogen economy involves not only new types of vehicles, but also a new fuel infrastructure and development of low-carbon primary resources. There are many ways to produce and deliver hydrogen to users (Figure 17.1). Given the number of possible configurations for a future hydrogen economy, systems analysis plays a key role in informing decision-makers in government and industry about the best hydrogen strategies and the prospects for hydrogen as compared with alternatives such as electricity, biofuels, or fossil-derived liquid fuels.

Whereas early assessments presented a vision of an "end-state" hydrogen economy [4, 7–9], recent analyses have focused more on transitions: what is

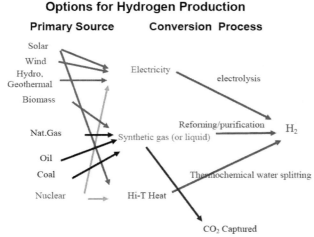

Fig. 17.1 Hydrogen production pathways.

required to move towards widespread use of hydrogen energy starting from today's energy system.

Some of the key transition questions are posed below:

- *What is required to initiate a hydrogen transition?* What will drive early markets for hydrogen vehicles? Who will build early infrastructure? Where will transition start? How many vehicles and stations are needed for an initial "roll-out"? How do we coordinate infrastructure and vehicle introduction? What can we learn from past transitions in the energy system?
- *How might the system evolve over time (and space)?* How is hydrogen market growth tied to infrastructure growth? What are the time constants for change? How would hydrogen supply pathways change over time? What is the "end state" (pipeline delivery, liquid hydrogen trucks?) Will hydrogen replicate the gasoline system? How do geographic factors impact regional infrastructure design?
- *How much will it cost?* What will it cost to build a hydrogen infrastructure and introduce hydrogen end-use systems such as fuel cell vehicles? When will hydrogen become cost competitive with incumbent technologies? How much time and investment will it take to get to this point? Are there low-cost strategies to build infrastructure?
- *What are the societal benefits of hydrogen compared with other options?* How does hydrogen compare in terms of GHG emissions, oil use, air pollution, and other factors? How soon could we realize these benefits?
- *What would a hydrogen transition mean for the rest of the energy system?* What is the role of hydrogen? What are the demands on primary energy resources? Are there synergies with electricity?
- *What is the role of policy?* Are technology-forcing policies required? Is a strong carbon policy sufficient to bring about large-scale use of hydrogen?

To understand the dynamics of transitions, analysts are using increasingly sophisticated tools: models of consumer choice and decision-making, infrastructure design models including spatial optimization and regional geographic data, and energy system modeling including interactions with the rest of the energy system. Interactions include both competition, between say hydrogen and electricity as vehicle fuels, and synergies, for example, using hydrogen as a form of electricity storage. Lifecycle assessment is an important way to compare the societal benefits of hydrogen with those of other fuels, considering the entire system, across multiple dimensions.

This chapter presents an overview of recent trends in hydrogen system modeling and results from hydrogen systems assessments. In the last section, a multi-attribute framework is proposed for evaluating hydrogen and other future fuels. The framework includes not only the commonly used metrics of technical performance, cost, energy use, infrastructure development, and GHG emissions, but also air pollution, energy security and reliability, water use, land use, and materials requirements. The goal is to illuminate hydrogen transition pathways that are viable across a wide range of economic and sustainability constraints, and to make comparisons with other energy pathways.

17.2
Survey of Hydrogen System Assessment Models: Recent Trends and Insights

Table 17.1 lists some current models that are used to assess hydrogen system design, cost, and environmental impacts. The models are grouped in several general categories.

17.2.1
Hydrogen Infrastructure Technology Components: Cost and Performance Data

Component-level models provide detailed engineering/economic information about the individual technologies that make up a hydrogen supply infrastructure, for example, compressed gas hydrogen storage cylinders, small-scale steam methane reformers, electrolyzers, hydrogen pipelines, biomass gasification plants, and liquefiers. Component-level models are the "building blocks" for more complex system models. For each technology, component models give an estimate of key performance indicators (conversion efficiency, energy input requirements, etc.), capital costs, operating and maintenance costs, and lifetime of equipment as a function of capacity. They also contain information about current status of the technology and projections for future improvements.

Recent Trends and Insights
Until the past few years, there was widespread disagreement in the literature about the costs and performance of hydrogen technologies. This has now been largely resolved with the adoption of authoritative and user-friendly spreadsheet

Table 17.1 Hydrogen system models.

- *Hydrogen Infrastructure Technology Components: Cost and Performance Data*
  - NRC [2, 10, 11]
  - H2A [13]
  - E3 Database [15]
- *Vehicle Technology Models*
  - PSAT (US DOE/Argonne National Laboratory) [16, 20]
  - MIT [17–19]
  - IEA (EPT) [3]
  - CONCAWE/EUCAR/JRC [21]
  - University of California, Davis [22–24]
- *Hydrogen Supply Chain Pathways Models*
  - NRC [2, 10, 11]
  - H2A Delivery Model (US DOE)/HDSAM[a] [14]
  - E3 Database [15]
  - HyPro[a] (Directed Technologies, Inc.) [27]
  - H2NowNPV model (TIAX) [28]
  - SSCHISM[a] (steady-state city infrastructure model, UC Davis) [25, 26]
  - Macro System Model MSM[a] (US DOE) [29]
- *Regional Infrastructure Development Case Studies*
  - HyDS[a] (NREL) [35, 36]
  - Hydra[a] (NREL) [37, 38]
  - HyDive[a] (NREL, System Dynamics) [29, 40]
  - US DOE Transition Analysis Scenario development [32, 33, 64]
  - UC Davis H2 Pathways Program/Sustainable Transportation Energy Pathways Program [41, 42]
    - ○ Early H2 FCV Rollout Strategies [43, 44]
    - ○ Hydrogen Infrastructure Transition model (HIT)[a] (dynamic programming for optimized H2 infrastructure build up; perfect foresight). Case studies in Beijing, California [48, 49]
    - ○ GIS Case Studies Coal → H_2 w/CCS[a], for Ohio, California (spatial optimization; limited foresight scenarios for time development) [45]
    - ○ GIS CA Biomass H2 from Rice Straw Study[a] (spatial optimization, time dependence and comparison to other biofuel pathways in future work) [46]
  - Imperial College/BP (dynamic programming) [50]
  - TIAX National Transition Model (2006) [28]
  - HyWays and HyNet Studies (European focus) [51–54]
- *Scenarios and Integrated Economic Systems (includes comparisons with other fuel/vehicle options)*
  - ETP (Global, IEA) [1, 3]
  - MARKAL (USEPA)[a] [59, 60]
  - NEMS (US DOE)[a] [57]
  - AMIGA (NETL, ANL) [58]
  - Pacific Northwest National Laboratory [62]
  - IIASA [63]
  - HyTrans (ORNL)[a] [32, 64]
  - H2CAS[a] (ANL agent-based) [65]

Table 17.1 (continued)

Well-to-Wheels Comparisons of Hydrogen and Other Fuels/Vehicles – GREET WTW Emissions (ANL) [67–69] – LEM (UC Davis) [70, 71] – E3 Database (LBST) [15]
Comparisons of Hydrogen and Other Fuels – CONCAWE [21]; DOE Multipath Study [20]; UC Davis STEPS [42]

a) Uses H2A component level data.

databases such as those developed for the United States National Academies [10, 11], the United States Department of Energy (US DOE)'s H2A models [13, 14] and the Ludwig Bolkow Systemtechnik (LBST) E3 model [15], which were well vetted with industry, and form a strong basis for more complex system models and analyses.

17.2.2
Hydrogen Vehicle Models

Several groups have developed vehicle models that allow comparison of hydrogen fuel cell vehicles with other types of internal combustion and battery vehicles [12, 16–23].

Recent Trends and Insights
There is broad agreement that hydrogen fuel cell vehicles will have about 2–2.5 times the fuel economy of a comparable gasoline internal combustion engine vehicle, and 1.5 times that of a gasoline hybrid. Fuel cell vehicles are currently much more expensive than internal combustion engine vehicles, but with mass production and technical progress, the "learned out" price premium is estimated to be $ 3600–6000 (15–30%) over a comparable gasoline vehicle [2, 16, 18–21].

17.2.3
Hydrogen Supply Chain Pathway Models

Several models have been developed that combine hydrogen components into complete "supply chain pathways" including either distributed or central hydrogen production, storage, delivery, and refueling stations (Figure 17.2). Supply chain models allow the user to compare the many ways of producing hydrogen and delivering it to users with respect to cost, emissions, and energy resource requirements.
- The United States National Academies [9] used an EXCEL-based model developed by Simbeck and Chang [10] to estimate the delivered cost of hydrogen to vehicles, for a variety of pathways, including centralized production from nat-

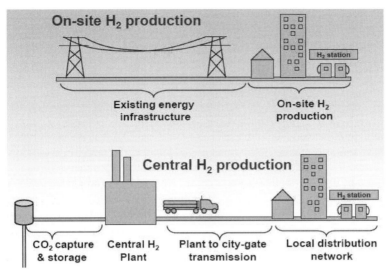

Fig. 17.2 Hydrogen supply chains [41].

ural gas, coal, biomass, nuclear energy, or electrolysis (with hydrogen delivery by truck or pipeline), and distributed production in small steam methane reformers or electrolyzers. In a subsequent assessment, the National Academies [2] adopted the H2A model's component costs and used a geographically specific supply chain model developed at the University of California, Davis [25, 26] that considered the effect of geographic factors such as differing local energy prices or density of demand on system design, to find the lowest cost hydrogen supply in different regions of the United States at different levels of demand.

- As part of the US DOE's H2A project, a team at Argonne National Laboratory developed a static EXCEL-based engineering/economic model for hydrogen delivery systems called HDSAM, including all the steps between the production plant and the vehicle [14]. Delivery is the part of a hydrogen system that is most variable with geography, energy prices, and demand, so HDSAM takes these effects into account.
- The US DOE has also supported the development of other supply chain models that incorporate dynamics. The HyPro Model developed by Directed Technologies, Inc. (DTI) analyzes options and tradeoffs of different choices for hydrogen infrastructure development. It uses the H2A database, and includes a variety of hydrogen supply pathways, sizing production and delivery systems to meet a specified level of demand (fraction of market penetration of hydrogen vehicles in the fleet) [27]. TIAX has developed several models for the US DOE for studying hydrogen steady-state pathways and build-out scenarios [28]. The H2NowNPV model evaluates the economics of several scenarios for the introduction of H_2 fuel cell vehicles (FCVs) in a series of "light-

house" cities, designs infrastructure, and estimates supply mix, energy use, emissions, and costs. The US DOE is also building a Macro System Model to link its various hydrogen infrastructure and well-to-wheels emissions models [29].
- The E3 Database was developed by LBST (Ludwig Bolkow Systemtechnik), an engineering/economics consulting firm and a major analyst of hydrogen systems and fuel cells for the European Union. This program allows the user to select a fuel/vehicle pathway for analysis, including primary energy source, fuel production, delivery, refueling, and use [15]. The E3 Database output gives costs and also provides emissions of GHGs and pollutants on a well-to-wheels basis. E3 considers a range of vehicles and fuels other than hydrogen. It was used in the GM Well-to-Wheels study [30], HyWays, and CONCAWE/JRC/UECAR studies [21], and allows a comprehensive comparison of hydrogen, biofuels, and conventionally fueled vehicles. The E3 Database is available as a commercial product and is widely used by industry and government analysts in Europe [15].

Recent Trends and Insights

Because there are many potential pathways for hydrogen production, storage, and delivery, supply chain comparisons are very useful for selecting among possible options. The lowest cost hydrogen pathway depends on the level of demand for hydrogen (which determines the scale of the system), and local feedstock and energy prices. As demand grows, the best choice for hydrogen supply changes. At first, hydrogen vehicles could be fueled by trucks or mobile refuelers fueled by the industrial merchant hydrogen system. Once hydrogen vehicles capture a few percent of the fleet in a localized area, onsite small steam methane reformers at refueling stations are preferred. As demand grows (once 25–50% of vehicles run on hydrogen), scale economies tend to favor a centralized hydrogen infrastructure with large hydrogen plants and truck or pipeline delivery [2, 25, 26, 31, 32]. Most studies indicate that in the long term hydrogen could be supplied to vehicles at $ 2–4 per kilogram [2, 9, 26, 31, 32] competitive with gasoline on a cents per mile basis.

17.2.4
Regional Infrastructure Development Case Studies

Hydrogen is generally more expensive to store and transport than liquid fuels, and like electricity, hydrogen could be made from diverse regionally available resources. This suggests that hydrogen might be made regionally, and that the infrastructure could vary significantly from one region to another. With the advent of widely available spatial data through geographic information systems (GIS), hydrogen energy analysts have carried out a number of realistic regional hydrogen infrastructure case studies that combine spatial analysis, engineering/economic data, and optimization. The design of a hydrogen energy system depends on geographic factors such as local energy prices, the location, size, and geo-

17.2 Survey of Hydrogen System Assessment Models: Recent Trends and Insights | 359

graphic density of hydrogen demand, and constraints imposed by existing infrastructure.
- Researchers at the US DOE the National Renewable Energy Laboratory (NREL) are developing tools to estimate regional hydrogen infrastructure costs [33, 34]. The HyDS-H2 model uses census data for American cities, NREL geographic databases, EIA energy information about energy costs, and hydrogen technologies costs from the H2A model to estimate infrastructure costs at different levels of demand [35, 36]. The Hydra model provides a set of GIS-based data and tools to permit spatial analysis of hydrogen infrastructure systems [37], including GIS databases on renewable resources, tools for estimating hydrogen demand using GIS data to screen for factors such as high population, income, existence of policies encouraging alternative fuels [38], and information on energy and transportation infrastructures. The HyDive dynamic model aims to understand the "chicken and egg" problem for initiating a hydrogen refueling infrastructure [39, 40]. HyDive uses systems dynamic methods to link the many decisions by consumers, fuel suppliers, and policymakers. A major feature is a consumer choice program (including a menu of many vehicle/fuel types) that uses system dynamics methods, and GIS data on consumer attributes to identify areas that could be early adopters of hydrogen. It further identifies optimal station locations to meet demand.
- Researchers in the University of California (UC) at Davis' Hydrogen Pathways and Sustainable Transportation Energy Pathways programs [41, 42] have developed a number of tools for geographic/spatial and engineering/economic analysis of regional hydrogen systems. Nicholas and Ogden used real city data on traffic flows, population, and locations of gasoline station, to find an optimal layout for early hydrogen refueling infrastructure in terms of metrics such as consumer travel time and proximity to stations. [43]. They explored [44] a "cluster strategy" for co-locating early hydrogen vehicles and infrastructure, finding that even with only a few hydrogen stations (1% of the total number of gasoline stations), strategic placement would allow good accessibility to fuel for early adopters with convenience similar to gasoline today.
UC Davis researchers have also conducted GIS case studies to design spatially optimized infrastructures. Specific case studies focus on hydrogen from fossil resources with carbon capture and storage in different American regions [45], hydrogen from biomass wastes in California [46], and impacts of hydrogen vehicles on California's energy system [47]. Lin and co-workers have investigated hydrogen transitions using a dynamic programming framework to design an optimal regional hydrogen infrastructure over time [48, 49]. An example of GIS-based infrastructure analysis is shown in Figure 17.3, depicting the build-up over time of an optimized coal-based hydrogen system with carbon capture and sequestration [45].
- Researchers at Imperial College London with BP have developed a dynamic programming model for regional hydrogen infrastructure build-up [50]. Hydrogen costs and CO_2 emissions were considered in building the optimal infrastructure over time. A variety of hydrogen production and delivery

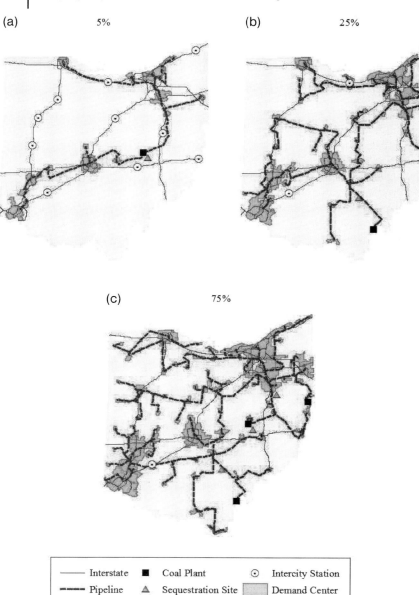

Fig. 17.3 Optimal infrastructure configuration at different market penetration levels for a coal-based hydrogen supply system with carbon capture and sequestration in Ohio, USA [45].

options were considered. The model was applied to building hydrogen infrastructure in Europe.
- The HyWays and HyNet studies have explored strategies for building hydrogen infrastructure in Europe. A broad consortium of government and industry partners contributed. LBST has coordinated these studies [51–53]. Detailed GIS studies were carried out for each member country. HyWays is a project to develop a coordinated European roadmap for hydrogen infrastructure development [54]. The HyNet project carried out a variety of engineering/economic studies, where geographic-specific scenarios for hydrogen infrastructure build-up were developed. Studies were largely steady state rather than dynamic.

Recent Trends and Insights
Combing spatial data with engineering/economic models is a powerful approach to gain insights into the design of hydrogen infrastructure in different regions. The optimal design is very sensitive to geographic factors such as spatial density of demand, local feedstock and energy prices, distances between primary resources for hydrogen production and demand, and location of existing infrastructure. For example, in regions with low-cost hydropower, the optimal infrastructure design will look very different to that in a region with low-cost coal with plentiful sites for carbon capture and sequestration. GIS analysis aids the design process by allowing the analyst to visualize the results on a map, and suggests low-cost strategies for building up a hydrogen infrastructure.

17.2.5
Scenario Studies and Integrated Energy/Economic System Models

Several recent scenario studies have looked at the implications of introducing hydrogen vehicles into the transportation system [2, 6, 19, 21, 31, 55, 56]. Although these scenarios are not optimizations and do not necessarily represent the lowest cost or lowest carbon system designs, they have the benefit of simplicity. Scenario studies suggest that it would take several decades before hydrogen would make a significant reduction in oil use or GHG emissions, but beyond this time frame, hydrogen could dramatically reduce GHG emissions and oil use from the transport sector.

Hydrogen pathways are included in many comprehensive energy/economic models that are used to model the future energy system. These models allow competition among energy carriers within a self-consistent representation of the economy, and seek an optimized, lowest cost solution. They are often used to explore the future makeup of the energy system, including the potential role of hydrogen under different economic, policy, and technology assumptions.
- The US DOE NEMS model is used to generate projections for American energy use and prices to 2030. A special version called NEMS-H2 has been developed to incorporate the latest costs and designs for hydrogen infrastructure development to 2050 [57].

- The AMIGA model has been used by researchers at NETL and ANL to investigate roles for hydrogen from coal in the United States transport sector [58].
- The United States Environmental Protection Agency (USEPA) uses a nine-region MARKAL model to estimate emissions from different energy system options. Demand for energy services is specified over a time period of up to 50 years. MARKAL is an optimization model with the objective of meeting these demands at the lowest cost from among energy supply and end-use technical descriptions [59]. An adaptation of this model was employed by Yeh *et al.* to examine different options for meeting policy goals for GHG reduction targets in the transportation sector [60].
- The International Energy Agency (IEA)'s ETP model is a comprehensive MARKAL-based model of the global energy system, built on the IEA's long experience with energy system modeling. It is a full model of the energy system, and competes a variety of energy technologies within the economy. It models technical changes, policies, costs, and market penetration of hydrogen versus other technologies in a variety of sectors of the economy [61]. Under some of the "Blue Map" scenarios, which require a 50% cut in carbon emissions by 2050, hydrogen plays a major role as a future fuel for light duty vehicles [3].
- Researchers at Pacific Northwest National Laboratory used the MiniCAM integrated assessment model to examine the potential for hydrogen in transportation [62]. The focus was on climate policy and technology options and the implications for hydrogen. A full slate of future transportation technologies were competed, including conventional fuels, biofuels, and hydrogen.
- IIASA researchers [63] used integrated assessment models to study a sustainable automobile transport scenario. Multiple sustainable development objectives are incorporated, including (i) continuing economic growth, with a moderate reduction in disparities in income between different world regions; (ii) maintaining a buffer of oil and gas resources to enhance security of energy supply, both globally and in vulnerable regions; (iii) abating GHG emissions to ensure that atmospheric CO_2 concentrations do not exceed double pre-industrial levels; and (iv) ensuring that global mobility demands are met, without resorting to assumptions about a large counter-trend shift to public transport or lower travel demand. It finds a growing role for biofuels and then hydrogen and fuel cells. Hydrogen plays a limited role until 2050, but becomes dominant in some scenarios by 2100.

Two recent economic models focus on the behavior of consumers and decision-makers in transitions.
- Researchers at Oak Ridge National Laboratory developed the HyTrans model to study hydrogen transitions in the United States [32, 64]. HyTrans is built around a consumer choice model for vehicles that selects among a variety of vehicle/fuel options, based on direct economic costs (vehicle price and fuel cost) and also attributes such as vehicle range, fuel availability, vehicle performance, and interior space.

- The H2CAS model being developed at Argonne National Laboratory uses an agent-based modeling approach to understand hydrogen transitions [65]. H2CAS simulates the behaviors and interactions of a large number of individuals (agents) and studies the macro-scale consequences of these interactions.

Recent Trends and Insights
Energy/economic optimization models such as MARKAL and NEMS generally found that rapid technology advancement was needed in addition to a strong carbon policy for hydrogen to capture a major market share. Moreover, it will take time for hydrogen technologies to penetrate the market and for building hydrogen supply. Therefore, hydrogen is likely to play a strong role in reducing GHG emissions beyond 2030. By 2050, hydrogen used in combination with other approaches such as reducing vehicle miles traveled and improving efficiency could lead to deep cuts in carbon emissions [2, 3, 9, 56]. Several assessments indicate that reaching societal goals of 50–80% reduction in GHG emissions by 2050 will require major adoption of electric-drive vehicles (hydrogen fuel cell or electric battery), especially in the light duty sector [3, 56].

17.2.6
Evaluating Societal Impacts: Well-to-Wheels Comparisons of Hydrogen and Other Fuels

A primary motivation for adopting hydrogen is to gain the societal benefits of reduced emissions and energy security. Several well-to-wheels analyses have compared the societal benefits of hydrogen vehicle/fuel pathways with other options (Figure 17.3):
- The US DOE Argonne National Laboratory's GREET model is a well-established and extensively documented EXCEL-based tool for estimating well-to-wheels energy use and air pollutant and GHG emissions for alternative fuel pathways and vehicle types. It has been widely used in a variety of studies [2,64,66]. It includes many hydrogen pathways and forms the basis for emissions calculations in HyTrans and other US DOE hydrogen program models [67–69].
- The LEM (Lifecycle Emissions Model) was developed at UC Davis by Dr Mark Delucchi to model lifecycle energy use and air pollutant and GHG emissions for a variety of fuel/vehicle pathways. This model includes several effects not included in GREET, such as impacts of land use. Several hydrogen pathways are included in LEM [70]. The AVCEM model, also developed at UC Davis by Dr. Mark Delucchi, estimates the costs of various externalities including air pollution and energy supply security. A societal lifecycle cost is estimated that includes both direct economic costs and externality costs [71].
- The LBST E3 model [15] also allows well-to-wheels comparisons of a wide range of fuel/vehicle pathways, including hydrogen from different sources, a variety of biofuels, and conventional fuels. In addition, it estimates infrastruc-

ture costs for these fuels using the H2Invest Model. This model was used in the CONCAWE/EUCAR/JRC study [21].

Recent Trends and Insights

Hydrogen offers the prospect of near-zero well-to-wheels emissions of GHGs and greatly reduced emissions of criteria air pollutants. The environmental impacts depend on the hydrogen pathway. Most studies find that the well-to-wheels emissions of GHGs are lower for hydrogen fuel cell vehicles using hydrogen derived from natural gas than for efficient gasoline hybrid vehicles. For hydrogen from renewables or fossil hydrogen with carbon capture and sequestration, the well-to-wheels emissions for hydrogen pathways are near zero.

17.2.7
Including Ancillary Costs and Benefits

Most lifecycle analyses to date have focused on emissions and energy use. Several authors have expanded their focus to estimate primary resource, land,

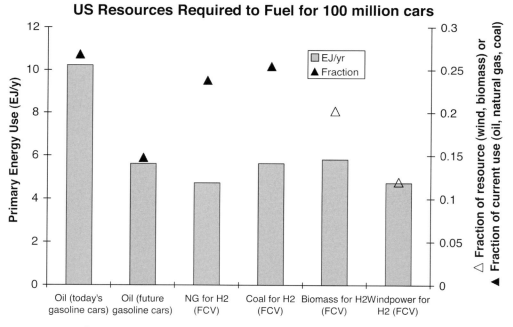

Fig. 17.4 Estimated primary energy use and resources needed to produce hydrogen for 100 million fuel cell vehicles in the United States. The biomass resource is assumed to be 800 million tonnes of biomass per year, and the wind resource is assumed to be 11 000 billion kWh of electricity per year.

water, and materials use associated with hydrogen energy systems, as compared with other fuels.

17.2.7.1 Primary Energy Resources

Many assessments have estimated the primary energy use needed for hydrogen compared with other fuels. A frequent question is "where will the hydrogen come from"?

Figure 17.4 shows a sample calculation of the amount of primary energy needed to make hydrogen for 100 million fuel cell cars in the United States (about 50% of the current American fleet or 33% of the projected fleet in 2050) [31]. The amount of primary energy required is given in exajoules (10^{18} J) per year on the left-hand y-axis. The fraction of the available annual resource (for biomass and wind) or the current use (for coal or natural gas) are shown on the right-hand y-axis. For reference, energy use for 100 million current gasoline vehicles and 100 million gasoline hybrids is shown. With hydrogen fuel cells, the amount of primary energy required is similar to that for gasoline hybrids, and considerably less than that for conventional gasoline cars. Ultimately, it is desirable to make hydrogen from zero or near-zero carbon sources. There are plentiful near-zero carbon resources for hydrogen production in the United States. For example, a mix of low-carbon resources including natural gas, coal (with carbon sequestration), biomass, and wind power could supply ample hydrogen for vehicles. With 20% of the biomass resource, plus 15% of the wind resource plus 25% added use of coal (with sequestration), 300 million hydrogen vehicles (approximately the entire American fleet projected in 2030) could be served with near-zero GHG emissions.

17.2.7.2 Land Requirements

The land requirements for producing hydrogen also depend on the particular pathway chosen. Table 17.2 shows the land requirements to produce hydrogen for a variety of renewable pathways. For comparison, the total United States land area is 9.1 million km^2 and the total cropland is 1.8 million km^2. The im-

Table 17.2 Land area required to produce renewable hydrogen.

	m^2/(GJ/year)	Land area to produce H_2 for 100 million FCV cars (km^2)
Electrolytic H_2		
Solar PV	1.89	5700
Solar thermal electric	5.71	17000
Wind	6.3–33	19000–99000
Hydropower	11–500	
H_2 via biomass gasification	50	150000

pacts of this level of land use have not been thoroughly examined, in terms of competing uses.

17.2.7.3 Water Requirements

King and Webber [72] estimated the water needed for different fuel/vehicle pathways. The water requirement for hydrogen depends sensitively on the pathway chosen. In general, hydrogen pathways relying on renewable electrolysis or steam methane reforming are estimated to use much less water than synthetic fuels from coal or biomass, and somewhat less water than gasoline. Water could become an important constraint on future energy production.

17.2.7.4 Materials Constraints

Materials availability could also become an issue for the widespread use of hydrogen. For example, hydrogen fuel cell vehicles are highly efficient, zero emission vehicles, which require the use of a platinum catalyst in the fuel cell. If fuel cell vehicles come into widespread use in the future, significant quantities of platinum catalyst would be needed [73]. Here we present a rough estimate of future platinum requirements in 2050.

According to the US DOE [74], a feasible 2015 goal for total platinum catalyst loading for automotive fuel cells is 0.19 g per kilowatt of fuel cell output power. Another recent assessment, by the California Air Resources Board [75], found a lower bound of 0.1 mg cm^{-2} of fuel cell area, which given a future fuel cell power density of 1 W cm^{-2}, translates into about 0.1 g Pt kW^{-1}.

Future light-duty fuel cell cars are likely to be hybrids (similar to the Toyota Prius) with some of the power coming from the fuel cell and some from a peak power battery [18]. This implies that a future light-duty FCV might have about a 40 kW fuel cell capacity. Hence the platinum requirements for a single fuel cell car would be

$$0.1\text{--}0.19 \text{ g kW} \times 40 \text{ kW FC per car} \approx 4\text{--}8 \text{ g of Pt per car} \tag{17.1}$$

In the IEA BLUE Map scenarios [3], the future number of FCVs could be as large as 50% of the total light-duty fleet or 1 billion FCVs by 2050. In this case, the total platinum in the FCV fleet would be

$$1 \text{ billion cars} \times 4\text{--}8 \text{ g of Pt per car} \approx 4000\text{--}8000 \text{ tonnes of Pt} \tag{17.2}$$

Assuming each car is replaced after 15 years, this would mean require the production of 270–540 tonnes per year of platinum. Current worldwide platinum production is 217 tonnes per year, but the total economically mineable platinum reserve is estimated to be 100 000 tonnes [76]. Providing platinum for FCVs would require a significant scale-up of production over time, but with recycling could be manageable in terms of the total resource.

Recent Trends and Insights

Hydrogen permits low-carbon, low-emission energy systems that could utilize diverse primary energy resources. However, there could be important constraints on land, water, and materials that are not well understood.

17.3
Towards a Comprehensive Framework for Hydrogen Systems Analysis

Systems assessments are moving towards incorporating multiple attributes when comparing fuels.

Usual metrics today include performance and cost of hydrogen systems, and emissions of GHG and air pollutants. Several studies [55, 71, 77–79] have estimated the societal costs of externalities, such as health damage from air pollution, climate impacts GHG emissions, and oil insecurity costs. When these costs are added to the direct economic cost, hydrogen and other low-emission options tend to be more competitive with conventional fuels.

Additional important attributes are water, land, and materials use. In addition, future energy systems need to be reliable and resilient to disruptions. These aspects have just begun to be explored for future fuels [47, 48, 80].

In the future, we may see energy system assessments based on diverse metrics, not all of which can be easily monetized. A possible set of metrics include cost, performance, reliability, and resilience to disruptions, GHG emissions, air pollutant emissions, water use, land use, compatibility with existing systems, and synergies with other parts of the energy system, especially electricity. Including these impacts in system assessments will give a more complete basis for decisions.

Increasingly, policymakers are concerned with the overall sustainability of future energy systems across many dimensions. Energy sustainability has been defined as "the ability of future generations to supply a set (or basket) of energy services to meet their demands without diminishing the potential for future environmental, economic, and social well-being." [81]. Several authors have suggested a developing a "sustainability index" incorporating these effects [82]. This work is still nascent, in part because it is difficult to value different attributes on the same scale. Certain hydrogen pathways appear to score well on many of the components of sustainability, but clearly more work is needed.

Systems analysis will continue to inform decisions and help invent our energy future. The toolkit for energy systems analysts is rapidly expanding, and yielding important new insights. With the availability of detailed spatial- and time-resolved data on energy use and the operation of energy systems, analysts can assess future designs with more depth and sophistication. Further, we can seek new configurations for using hydrogen within the energy system, for example, exploring its interactions with electricity [47,83]. As we look forward, there is ever more scope for creative thinking about the role of hydrogen within the future energy system.

References

1 Gielen, D. and Simbolotti, G. (2005) *Prospects for Hydrogen and Fuel Cells*, OECD/IEA, International Energy Agency Publications. Paris.

2 National Research Council of the National Academies, Committee on Assessment of Resource Needs for Fuel Cell and Hydrogen Technologies (2008) *Transitions to Alternative Transportation Technologies: a Focus on Hydrogen*, National Academies Press, Washington, DC; available at http://www.nap.edu/catalog.php?record_id=12222 (accessed 26 December 2009).

3 International Energy Agency (2008) *Energy Technology Perspectives 2008 in Support of the G8 Plan of Action: Scenarios and Strategies*, OECD/IEA, International Energy Agency Publications, Paris.

4 Winter, C.-J. and Nitsch, J. (1988) *Hydrogen as an Energy Carrier*, Springer, New York.

5 Ball, M. and Wietschel, M. (2009) *The Hydrogen Economy: Opportunities and Challenges*, Cambridge University Press, Cambridge.

6 McDowall, W. and Eames, M. (2006) Forecasts, scenarios, backcasts and roadmaps to the hydrogen economy: a review of hydrogen futures literature, *Energy Policy*, **34** (11), 1236–1250.

7 Ogden, J.M. and Williams, R.H. (1989) *Solar Hydrogen: Moving Beyond Fossil Fuels*, World Resources Institute, Washington, DC.

8 Ogden, J.M. and Nitsch, J. (1993) Solar hydrogen, in *Renewable Energy Sources for Fuels and Electricity* (eds T.B. Johansson, H. Kelly, A.K.N. Reddy and R.H. Williams), Island Press, Washington, DC, pp. 925–1009.

9 Jacobson, M.Z. and M.A. Delucchi. (2009) A path to sustainable energy by 2030, *Sci. Am.*, **301** (5), 58–65.

10 Simbeck, D.R. and Chang, E. (2002) *Hydrogen Supply: Cost Estimate for Hydrogen Pathways – Scoping Analysis*, Report to the National Renewable Energy Laboratory, Report NREL/SR-540-32525, SFA Pacific, Mountain View, CA.

11 National Research Council, National Academy of Engineering, Committee on Alternatives and Strategies for Future Hydrogen Production and Use (2004) *The Hydrogen Economy: Opportunities, Costs, Barriers and R&D Needs*, National Academies Press, Washington, DC; available at http://www.nap.edu/catalog.php?record_id=10922 (accessed 26 December 2009).

12 Plotkin, S. (2007) *Examining Hydrogen Transitions*, Report ANL-07/09, Energy Systems Division, Argonne National Laboratory, Argonne, IL.

13 US Department of Energy (2009) *H2A, Hydrogen Analysis Tool*, available at http://www.hydrogen.energy.gov/h2a_analysis.html (accessed 24 December 2009).

14 Mintz, M. (2007) III.2. Hydrogen delivery infrastructure analysis. In *United States Department of Energy Hydrogen Program FY 2008 Annual Progress Report*, US Department of Energy, Washington, DC, pp. 368–371. The H2A and HDSAM models are available through the H2A website, with full documentation (a users' manual), and has a user interface with a GUI that allows delivery option selection. The H2 delivery model is available at http://www.hydrogen.energy.gov/h2a_analysis.htm (accessed 24 December 2009).

15 Ludwig-Bölkow-SystemtechnikGmbH (2009) *E3 Database – A Tool for the Evaluation of Energy Chains*, L-B-Systemtechnik GmbH, Ottobrunn; presentation available at http://www.e3database.com/ (accessed 24 December 2009).

16 Rousseau, A. and Sharer, P. (2004) *Comparing Apples to Apples: Well-to-Wheel Analysis of Current ICE and Fuel Cell Vehicle Technologies*, Report 2004-01-1015, Argonne National Laboratory, Argonne, IL.

17 Weiss, M.A., Heywood, J.B., Schafer, A. and Natarajan, V.K. (2003) *Comparative Assessment of Fuel Cell Cars*, Report LFEE 2003-001 RP, Laboratory for Energy and the Environment, Massachusetts Institute of Technology, Cambridge, MA.

18 Kromer, M. A. and Heywood, J. B. (2007) *Electric Powertrains: Opportunities and Challenges in the U.S. Light-Duty Vehicle Fleet*, Report LFEE 2007-02 RP, Sloan Automotive Laboratory, Laboratory for Energy and the Environment, Massachusetts Institute of Technology, Cambridge, MA.

19 Bandivadekar, A., Bodek, K., Cheah, L., Evans, C., Groode, T., Heywood, J., Kasseris, E., Kromer, M., and Weiss, M. (2008) *On the Road in 2035: Reducing Transportation's Petroleum Consumption and GHG Emissions*, Report LFEE 2008-05 RP, Laboratory for Energy and the Environment, Massachusetts Institute of Technology, Cambridge, MA.

20 Plotkin, S. and Singh, M. (2009) Multipath transportation futures study: vehicle characterization and scenario analyses (draft), unpublished manuscript, Argonne National Laboratory, Argonne, IL.

21 CONCAWE/EUCAR/JRC (2007) *Well-To-Wheels Analysis of Future Automotive Fuels and Powertrains in the European Context, Well-To-Wheels Report Version 2c, March 2007*, WTW Report 010307.doc, European Commission, European Joint Research Center, Institute for Environment and Sustainability, Ispra; available at http://ies.jrc.ec.europa.eu/WTW (accessed 24 December 2009).

22 Burke, A. F. and Miller, M. (2009) *Simulated Performance of Alternative Hybrid-Electric Powertrains in Vehicles on Various Driving Cycles*, Research Report UCD-ITS-RR-09-09, Institute of Transportation Studies, University of California, Davis, Davis, CA.

23 Zhao, H. and Burke, A. F. (2009) *Optimum Performance of Direct Hydrogen Hybrid Fuel Cell Vehicles*, Research Report UCD-ITS-RR-09-12, Institute of Transportation Studies, University of California, Davis, Davis, CA.

24 Zhao, H. and Burke, A. F. (2008) *Modeling and Optimization of PEMFC Systems and its Application to Direct Hydrogen Fuel Cell Vehicles*, Research Report UCD-ITS-RR-08-30, Institute of Transportation Studies, University of California, Davis, Davis, CA.

25 Yang, C. and Ogden, J. (2009) *US Urban Hydrogen Infrastructure Costs Using the Steady State City Hydrogen Infrastructure System Model (SSCHISM)*; a beta copy of the model is available on the website of University of California, Davis, Institute of Transportation Studies, http://steps.ucdavis.edu/research/Thread_2/sschism/steady-state-city-hydrogen-infrastructure-sy stem-model-sschism/ (accessed 26 December 2009).

26 Yang, C. and Ogden, J. (2007) Determining the lowest-cost hydrogen delivery mode. *Int. J. Hydrogen Energy*, **32** (2), 268–286.

27 James, B. D., Perez, J., and Schmidt, P. (2007). VIII.1 using HyPro to evaluate competing hydrogen pathways. In *United States Department of Energy Hydrogen Program FY 2007 Annual Progress Report*, US Department of Energy, Washington, DC, pp. 1189–1193.

28 Unnasch, S., Lasher, S., and Chan, M. (2006) Hydrogen infrastructure –transition and cost. In *Proceedings of the 2006 Fuel Cell Seminar*, Oahu, HI, 13–17 November 2006.

29 Ruth, M. and Vanderveen, K. (2007) VIII.5 macro-system model. In *United States Department of Energy Hydrogen Program FY 2007 Annual Progress Report*, US Department of Energy, Washington, DC, pp. 1204–1207.

30 Gross, B. K., Sutherland, I., and Mooiewek, H. (2007) *Hydrogen Infrastructure Fueling Assessment*, R&D Report 11,065, Research and Development Center, General Motors Corp., Warren, MI.

31 Ogden, J. and Yang, C. (2009) Build-up of a hydrogen infrastructure in the US. In *The Hydrogen Economy: Opportunities and Challenges* (eds M. Ball and M. Wietschel), Cambridge University Press, Cambridge, pp. 454–482.

32 Greene, D. L., Leiby, P. N., James, B., Perez, J., Melendez, M., Milbrandt, A., Unnasch, S., and Hooks, M. (2008) *Analysis of the Transition to Hydrogen Fuel Cell Vehicles and the Potential Hydrogen Energy Infrastructure Requirements*, ORNL/TM-2008/30, Oak Ridge National Laboratory, Oak Ridge, TN.

33 Singh, M., Moore, J. and Shadis, W. (2005) *Hydrogen Demand, Production, and Cost by Region to 2050*, Report ANL/ESD/05-2, Argonne National Laboratory, Argonne, IL.

34 Gronich, S. (2006) Hydrogen and FCV implementation scenarios, 2010–2025. Presented at the US DOE Hydrogen Transition Analysis Workshop, Washington, DC, 9–10 August 2006; available at http://www1.eere.energy.gov/hydrogenandfuelcells/analysis/scenario_analysis_mtg.html (accessed 26 December, 2009).

35 Short, W. and Parks, K. (2005) III.4 energy systems analysis: hydrogen deployment system modeling. In *United States Department of Energy Hydrogen Program FY 2005 Annual Progress Report*, US Department of Energy, Washington, DC, pp. 35–39.

36 Parks, K. (2006). GIS-based infrastructure modeling. Presented at the US DOE Hydrogen Transition Analysis Workshop, Washington, DC, 9–10 August 2006; available at http://www1.eere.energy.gov/hydrogenandfuelcells/analysis/scenario_analysis_mtg.html (accessed 26 December 2009).

37 Levene, J. and Sparks, W. (2008) X.8 HyDRA: hydrogen demand and resource analysis tool. In *United States Department of Energy Hydrogen Program FY 2008 Annual Progress Report*, US Department of Energy, Washington, DC, pp. 1275–1278.

38 Melendez, M. (2007) Geographically based hydrogen demand and infrastructure rollout scenario analysis. Presented at the 2010–2025 Scenario Analysis for Hydrogen Fuel Cell Vehicles and Infrastructure, 31 January 2007, Washington, DC; available at www1.eere.energy.gov/hydrogenandfuelcells/analysis/pdfs/scenario_analysis_melendez1_07.pdf (accessed 26 December 2009).

39 Welch, C. (2006) *Lessons Learned from Alternative Transportation Fuels: Modeling Transition Dynamics*, NREL Technical Report NREL/TP-540-39446, National Renewable Energy Laboratory, Golden, CO.

40 Welch, C. (2007) Discrete choice analysis: H2FCV demand potential. Presented at the 2010–2025 US DOE Scenario Analysis Workshop for Hydrogen Fuel Cell Vehicles and Infrastructure, Washington, DC, 31 January 2007; available at www1.eere.energy.gov/hydrogenandfuelcells/analysis/pdfs/scenario_analysis_welch_07.pdf (accessed 26 December 2009).

41 Ogden, J. M. (2007). Lessons from the Hydrogen Pathways Program. Invited talk at the Transportation Research Board Meeting, Session on The Future of Energy in Transportation, Part 1: Breaking the Chains of Petroleum Dependence, 22 January 2007, Washington, DC.

42 Institute of Transportation Studies (2009) *Sustainable Transportation Energy Pathways Program*, Institute of Transportation Studies, University of California, Davis, Davis, CA; available at http://steps.its.ucdavis.edu/ (accessed 26 December 2009).

43 Nicholas, M. A. and Ogden J. M. (2007) Detailed analysis of urban station siting for California hydrogen highway network. *Transport. Res. Rec.*, 1983, 129–139.

44 Nicholas, M. A. and Ogden, J. M. (2010) *An Analysis of Near-Term Hydrogen Vehicle Rollout Scenarios for Southern California*, Research Report UCD-ITS-RR-10-xx, Institute of Transportation Studies, University of California, Davis, Davis, CA.

45 Johnson, N., Yang, C., and Ogden, J. (2008) A GIS-based assessment of coal-based hydrogen infrastructure deployment in the State of Ohio. *Int. J. Hydrogen Energy*, 33 (20), 5287–5303.

46 Parker, N. C., Fan, Y., and Ogden, J. (2010) From waste to hydrogen: an optimal design of energy roduction and distribution network. *Transport. Res. E Logistics*, in press.

47 McCarthy, R. W. (2009) *Assessing Vehicle Electricity Demand Impacts on California Electricity Supply*, Research Report UCD-ITS-RR-09-44, Institute of Transportation Studies, University of California, Davis, Davis, CA.

48 Lin, D. Z., Chen, C.-W., Ogden, J., and Fan, Y. (2008) The least-cost hydrogen infrastructure for Southern California. *Int. J. Hydrogen Energy*, 33 (12), 3009–3014.

49 Lin, D. Z., Ogden, J. M., Fan, Y., and Sperling, D. (2006) *The Hydrogen Infrastructure Transition (HIT) Model and Its Application in Optimizing a 50-year Hydrogen Infrastructure for Urban Beijing*, Research Report UCD-ITS-RR-06-05, Institute of Transportation Studies, University of California, Davis, Davis, CA.

50 Hugo, A., Rutter, P., Pistikopoulos, S., Amorelli, A., and Zoia, G. (2005) Hydrogen infrastructure strategic planning using multi-objective optimization. *Int. J. Hydrogen Energy*, **30**, 1523–1534.

51 Bünger, U. and Landinger, H. (2004) Considering a basic area-wide hydrogen fuelling station network in Europe, short version. Prepared for HyNet, 13 August 2004; available at http://www.hyways.de/hynet/HyNet_Basic_Area-Wide_HRS-Network_AUG2004.pdf (accessed 26 December 2009)

52 Bünger, U. (2004) Towards a European hydrogen energy roadmap. Preface to *HyWays – the European Hydrogen Energy Roadmap Integrated Project, Key Highlights Report*; available at http://www.hyways.de/hynet/HYNET-roadmap_Executive_Report_MAY2004.pdf (accessed 26 December 2009).

53 Bünger, U. (2005) The path to a hydrogen refueling infrastructure in Europe – HyNet and HyWays. Presented at the Hydrogen Pathways Workshop, Institute for Transportation Studies, University of California, Davis, Davis, CA, 28–29 June 2005; available at http://hydrogen.its.ucdavis.edu/workshops/Workshops/Casestudies/casestudies (accessed 26 December 2009).

54 Mulard, P. (2005) European HyWays and platform projects. Presented at the Hydrogen Pathways Workshop, Institute for Transportation Studies, University of California, Davis, Davis, CA, 28–29 June 2005; available at http://hydrogen.its.ucdavis.edu/workshops/Workshops/Casestudies/casestudies (accessed 26 December 2009).

55 Thomas, C. E. (2009) *Comparison of Transportation Options in a Carbon-constrained World: Hydrogen, Plug-in Hybrids and Biofuels*, National Hydrogen Association Report, 2009; available at http://www.hydrogen.energy.gov/pdfs/htac_july09_10_h2gen.pdf (accessed 24 December 2009).

56 McCollum, D. and Yang, C. (2009) Achieving deep reductions in US transport greenhouse gas emissions: scenario analysis and policy implications. *Energy Policy*, **37** (12), 5580–5596.

57 Wood, F. (2006) NEMS-H2. Presented at the US DOE H2 Scenario Analysis Workshop, Washington, DC, 10 January 2006; available at http://www1.eere.energy.gov/hydrogenandfuelcells/wkshp_h2_transition.html (accessed 26 December, 2009)

58 Balash, P., Hanson, D., Ruether, J., Schmalzer, D., Molburg, J., Keairns, D., Kern, K., Stirling, K., and Marano, J. (2005) *Use of Hydorgen for the Light Duty Transportation Fleet: technology and Economic Analysis*, unpublished report to the United States Department of Energy; available ar http://www.hydrogen.energy.gov/analysis_repository/project.cfm/PID=139 (accessed 26 December 2009).

59 Yeh, S., Loughlin, D. H., Shay, C., and Gage C. (2007) An integrated assessment of the impacts of hydrogen economy on transportation, energy use, and air emissions. *Proc. IEEE*, **94** (10), 1838–1851.

60 Yeh, S., Farrell, A. E., Plevin, R. J., Sanstad, A., and Weyant, J. (2009) Optimizing U.S. mitigation strategies for the light-duty transportation sector: what we learn from a bottom-up model. *Environ. Sci. Technol.*, **42** (22), 8202–8210.

61 Gielen, D. and Simbolotti, G. (2006) International Energy Agency, "Prospects for Hydrogen and Fuel Cells," Proceedings of the 2006 Transportation Research Board Meeting, 26 January 2006.

62 Geffen, C. A., Edmonds, J. A., and Kim, S. H. (2004).Transportation and climate change: the potential for hydrogen systems. In *Environmental Sustainability in the Mobility Industry: Technology and Business Challenges*, Special Publication SP-1865, SAE International, Warrendale, PA, pp. 13–20.

63 Turton, H. (2006) Sustainable global automobile transport in the 21st century:

an integrated scenario analysis. *Technological Forecasting and Social Change*, **73**, 607–629.

64 Greene, D., Leiby, P., and Bowman, D. (2007) *Integrated Analysis of Market Transformation Scenarios with HyTrans*, Report ORNL/TM-2007/094, Oak Ridge National Laboratory, Oak Ridge, TN.

65 Mintz, M. (2007) VIII.3 Analysis of the hydrogen production and delivery infrastructure as a complex adaptive system. In *United States Department of Energy Hydrogen Program FY 2007 Annual Progress Report*, US Department of Energy, Washington, DC, pp. 1197–1200; available at http://www.hydrogen.energy.gov/analysis_repository/project.cfm/PID=95 (accessed 26 December 2009).

66 Brinkman, N., Wang, M., Weber, T., and Darlington, T. (2005) *Well-to-Wheels Analysis of Advanced Fuel/Vehicle Systems – A North American Study of Energy Use, Greenhouse Gas Emissions, and Criteria Pollutant Emissions*, Research Report by General Motors Corporation and Argonne National Laboratory; available at http://www.transportation.anl.gov/pdfs/TA/339.pdf (accessed 24 December 2009).

67 Wang, M. (2002) Fuel choices for fuel-cell vehicles: well-to-wheels energy and emission impacts. *J. Power Sources*, **112**, 307–312.

68 Wang, M. and Elgowainy, A. (2009) Fuel-cycle analysis of fuel-cell vehicles and fuel-cell systems with the GREET model. In *2009 DOE Hydrogen Program Review*, 19 May 2009; available at http://www.hydrogen.energy.gov/pdfs/review09/an_12_wang.pdf (accessed 26 December 2009).

69 United States Department of Energy, Argonne National Laboratory, Transportation Technology R&D Center (2009) *GREET Model*; available at http://www.transportation.anl.gov/modeling_simulation/GREET/index.html (accessed 26 December 2009).

70 Delucchi, M.A. (2003) *A Lifecycle Emissions Model (LEM): Lifecycle Emissions from Transportation Fuels, Motor Vehicles, Transportation Modes, Electricity Use, Heating and Cooking Fuels, and Materials; Main Report*, Research Report UCD-ITS-RR–03-17-MAIN, Institute of Transportation Studies, University of California, Davis, Davis, CA; full documentation on this model is available at http://www.its.ucdavis.edu (accessed 24 December 2009).

71 Delucchi, M.A. (2005) *AVCEM: Advanced Vehicle Cost and Energy Use Model. Overview of AVCEM*, Research Report UCD-ITS-RR-05-17(1), Institute of Transportation Studies, University of California, Davis, Davis, CA.

72 King, C.W. and Webber, M.E. (2008) Water use and transportation. *Environ. Sci. Technol.*, **42** (21), 7866–7872.

73 Yang, C.-J. (2009) An Impending platinum crisis and its implications for the future of the automobile. *Energy Policy*, **37** (10), 1805–1808.

74 James, B.D. and Kalinoski, J. (2009) Mass production cost estimation for direct hydrogen PEM fuel cell systems for automotive applications. In *United States Department of Energy, Hydrogen Fuel Cells and Infrastructure Technologies Program Annual Merit Review, May 21, 2009*; available at http://www.hydrogen.energy.gov/pdfs/review09/fc_30_james.pdf (accessed 26 December 2009).

75 Kalhammer, F.R., Kopf, B.M., Swan, D.H., Roan, V.P., and Walsh, M.P. (2007) *Status and Prospects for Zero Emissions: Vehicle Technology Report of the ARB Independent Expert Panel 2007*. Prepared for the State of California Air Resources Board, Sacramento, CA; available at http://www.arb.ca.gov/msprog/zevprog/zevreview/zev_panel_report.pdf (accessed 26 December 2009).

76 US Geological Survey (2007) *Mineral Commodity Summaries, January 2007*, US Geological Survey, Reston, VA.

77 Ogden, J.M., Williams, R.H. and Larson, E.D. (2004) A societal lifecycle cost comparison of cars with alternative fuels/engines. *Energy Policy*, **32** (1), 7–27.

78 Bickel, P. and Friedrich, R. (eds) (2005) *ExternE, Externalities of Energy, Methodology 2005 Update*, Institut für Energiewirtschaft und Rationelle Energieanwendung – IER, Universität Stuttgart, Directorate-General for Research, Sustainable Energy Systems, EUR 21951, Office for

Official Publications of the European Communities, Luxembourg, 2005, ISBN 92-79-00423-9.

79 Jacobson, M. Z., Colella, W. G., and Golden, D. M.. (2005) Cleaning the air and improving health with hydrogen fuel-cell vehicles. *Science*, **308**, 1901–1908.

80 McCarthy, R. W., Ogden, J. M., and Sperling, D. (2007) Assessing reliability in energy supply systems. *Energy Policy*, **35** (4), 2151–2162.

81 Löschel, A., Johnston, J., Delucchi, M. A., Demayo, T. N., Gautier, D. L., Greene, D. L., Ogden, J., Rayner, S., and Worrell, E. (2009) Stocks, flows, and prospects of energy. In *Linkages of Sustainability* (eds T. E. Graedel and E. van der Voet), MIT Press, Cambridge, MA, pp. 389–418.

82 Zah, R., Böni, H., Gauch, M., Hischier, R., Lehmann, M. and Wäger, P. (2007) *Life Cycle Assessment of Energy Products: Environmental Assessment of Biofuels.* Written under a contract from the German Federal Office for Energy (BFE), the Federal Office for the Environment (BFE) and the Federal Office for Agriculture (BLW), Bern.

83 Yang, C. (2008) Hydrogen and electricity: parallels, interactions and convergence. *Int. J. Hydrogen Energy*, **33** (8), 1977–1994.

Storages

18
Physical Hydrogen Storage Technologies – a Current Overview
Bert Hobein and Roland Krüger

Keywords: hydrogen storage, hydrogen liquefaction, liquid hydrogen, fuel cell, refueling

18.1
Introduction

The shortage of fossil energy resources demands new energy carriers and energy carrier technologies in the future. Therefore, one of the major challenges currently in the automotive industry and research institutes worldwide is to develop and realize alternative fuel concepts for passenger cars. Physical hydrogen storage systems, that is, compressed and liquid storage, are currently the most mature technology to store hydrogen onboard road vehicles. A combination of compressed and liquid hydrogen storage is so-called cryo-compressed systems, which are currently under development. Since the first hydrogen vehicle generation demonstrated insufficient cruising ranges, new physical storage technologies, such as the 70 MPa technology, can increase vehicles' cruising ranges significantly. This chapter gives an overview of different physical hydrogen storage technologies with regard to their design and their operating and refueling capabilities.

18.2
General Overview

This chapter gives an overview of the three different physical hydrogen storage technologies, including fundamental hydrogen properties and individual safety aspects.

18.2.1
Properties of Hydrogen

Hydrogen is the most abundant element, comprising about three-quarters of the mass of the Universe. It is found in water, which covers 70% of the Earth's surface, and also in all organic matter. Hydrogen is the lightest of all elements, and in the gaseous state it diffuses immediately upwards into the air and does not pollute the ground or groundwater. It is colorless, odorless, and nontoxic. Hydrogen does not produce acid rain, deplete the ozone layer, or produce harmful emissions.

18.2.1.1 Hydrogen Storage Densities

The gravimetric energy density of hydrogen (33.33 kWh kg^{-1}) is approximately three times that of gasoline (12.7 kWh kg^{-1}). However, the low volumetric energy density of hydrogen (0.77 kWh l^{-1} at 35 MPa and 300 K) in comparison with gasoline (8.76 kWh l^{-1}) is the reason for the increased package volume required for a hydrogen storage system compared with conventional gasoline/diesel fuel systems (see Figure 18.1).

The fuel capacity of a hydrogen storage system obviously depends on the total tank volume, the maximum storage pressure, and the temperature of the gas stored. Nevertheless, since hydrogen does not show ideal gas behavior but rather real gas behavior, there is no linear correlation between hydrogen density

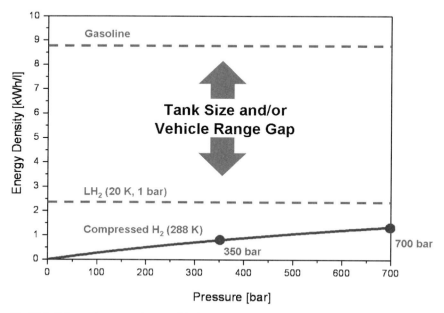

Fig. 18.1 Volumetric energy density of hydrogen compared with gasoline.

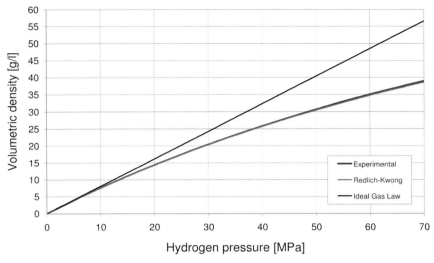

Fig. 18.2 Volumetric density of hydrogen at 300 K as a function of pressure (experimental data from [2]).

and gas pressure. Hence the amount of hydrogen stored at various conditions needs to be calculated on the basis of an appropriate equation of state, such as the Redlich-Kwong equation of state [1]:

$$p = \frac{RT}{V-b} - \frac{a}{\sqrt{T}V(V+b)} \qquad (18.1)$$

where $a = 0.42748 R^2 T_c^{2.5}/p_c$ and $b = 0.08664 RT_c/p_c$ are empirical constants, $R = 8.31451$ J mol^{-1} K^{-1} is the ideal gas constant, $p_c = 1.316 \times 10^6$ Pa is the critical pressure of H$_2$ [2], $T_c = 32.23$ K is the critical temperature of H$_2$ [2], p is pressure (Pa), V is molar volume (m^3 mol^{-1}), and T is absolute temperature (K).

Figure 18.2 shows a comparison of data calculated by the ideal gas law and by using the Redlich-Kwong equation of state with experimental data valid for hydrogen at 300 K. Although the ideal gas law data correlate well with the experimental data at low gas pressures below 10 MPa, they show a large error at 70 MPa of approximately 44%. However, the Redlich-Kwong equation of state provides a good correlation with the experimental data. At a gas pressure of 70 MPa it leads to an error of approximately 1% compared with the experimental data. Below 70 MPa, the error even decreases.

The advantage of liquid hydrogen (LH$_2$) over CGH$_2$ fuel is the high volumetric density at low pressure (see Figure 18.3) and favorable transportation characteristics, that is, it does not require transportation in high-pressure containers and can easily be delivered by truck.

At atmospheric pressure (0.1 MPa), LH$_2$ has a density of 70.9 g l^{-1}, which is 1.8 times the approximately 40 g l^{-1} of CGH$_2$ at 70 MPa and 288 K. According

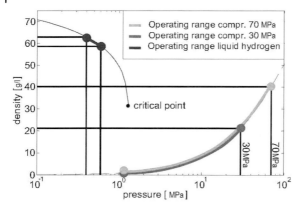

Fig. 18.3 Comparison of densities of CGH_2 and LH_2 [3].

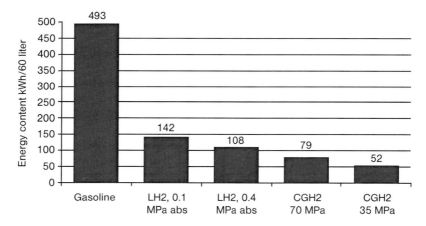

Fig. 18.4 Comparison of energy contents in a volume of 60 l [3].

to Krainz et al. [4], gaseous hydrogen would have to be compressed to about 200 MPa to achieve the same volumetric density as LH_2 at an operational pressure of 0.5 MPa. The volumetric energy density of LH_2 and CGH_2 fuels compared with gasoline for a completely filled virtual volume of 60 l is illustrated in Figure 18.4. The energy content of LH_2 fuel is four times lower than that of gasoline, whereas that of CGH_2 at 70 MPa is six times lower than that of gasoline. On the other hand, 1 l of LH_2 weighs only 0.07 kg.

It should be noted that the above-mentioned advantages of LH_2 over CGH_2 are based purely on fuel properties. When considering the performance of the system as a whole, other parameters should not be ignored, such as auxiliary systems (insulation, valves, safety devices) and gross volume required for the whole system. These additional weights and volumes have to be taken into account when considering gravimetric and volumetric densities of the complete fuel system.

18.2.1.2 Energy Required to Compress Hydrogen

The compression of hydrogen to its service pressure requires energy. The higher the service pressure, the greater is the energy required to compress the hydrogen gas. The compression work can be provided by a standard piston-type mechanical compressor. The theoretical work for isothermal compression is given by

$$\Delta G = RT \ln \left(\frac{P}{P_0} \right) \tag{18.2}$$

where T is the absolute temperature (K), P is the end pressure (Pa), and P_0 is the starting pressure (Pa).

In practice however, the ideal isothermal compression work cannot be realized. Instead, the adiabatic compression equation (Equation 18.3) seems to allow for more accurate calculations of the compression work, as shown below.

$$W = [\gamma/(\gamma - 1)] P_0 V_0 [(P/P_0)^{(\gamma-1)/\gamma} - 1] \tag{18.3}$$

where $V_0 = 11.11 \text{ m}^3 \text{ kg}^{-1}$ is the initial specific volume and $\gamma = 1.41$ is the adiabatic coefficient.

For a pressure range up to 80 MPa, the isothermal and adiabatic compression work is calculated using Equations 18.2 and 18.3, respectively, and is shown in Figure 18.5. The compression of hydrogen from 0.1 to 35 MPa consumes

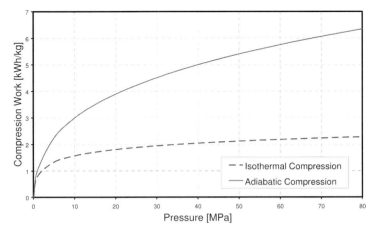

Fig. 18.5 Theoretical compression work for hydrogen.

2 kWh kg^{-1} according to the isothermal process, whereas the actual energy consumption according to the adiabatic process is 4.77 kWh kg^{-1} [14.31% of the lower heating value (LHV) of H_2], which is more than twice the isothermal equivalent. For the 0.1 to 70 MPa compression, the isothermal energy consumption is 2.24 kWh kg^{-1}, whereas the adiabatic compression energy consumption is 6.07 kWh kg^{-1} (18.21% of the LHV of H_2), almost three times the isothermal equivalent. Note that the LHV of hydrogen is 119.97 MJ kg^{-1} = 33.33 kWh kg^{-1} (see Section 18.2.1.1).

18.2.1.3 Energy Required to Liquefy Hydrogen

Liquefaction can be processed through the Joule–Thomson cycle (Linde cycle). In this cycle, the gas is first compressed and then cooled in a heat exchanger. It passes a throttle valve where it expands and therefore cools (Joule-Thomson effect). The cooled gas is fed back to expand again and cool further until it is liquefied. Here, hydrogen is usually precooled using liquid nitrogen (78 K) [5].

The theoretical energy (work) required to liquefy hydrogen from room temperature is 3.23 kWh kg^{-1}. However, the actual technical work required ranges from 27.8 kWh kg^{-1} for a small liquefaction plant (e.g., 10 kg h^{-1}) to a minimum of 11.1 kWh kg^{-1} (33.31% of the LHV of H_2) for a large plant (1000 kg h^{-1}). These values represent 83.42 and 33.31%, respectively, of the lower heating value of hydrogen.

18.3
Fuel System Design and Specifications

This section describes the design and specifications of different fuel systems, including system schematics and a description of the various components integrated into the systems.

18.3.1
Overview of Compressed Gaseous Hydrogen Storage System

Compressed hydrogen storage systems are designed and built according to a generic system schematic (see Figure 18.6). The vehicle is refueled at a standard fueling location that is equipped with a hydrogen refueling receptacle. The receptacle only connects with its counterpart at the station and allows refueling if the design pressure of the receptacle, that is, the vehicle, is equal to or higher than the design pressure of the refueling nozzle, that is,. the refueling station. A pressure regulator reduces the storage pressure to the supply pressure. The fuel is routed from the storage cylinder(s) to the powertrain of the vehicle.

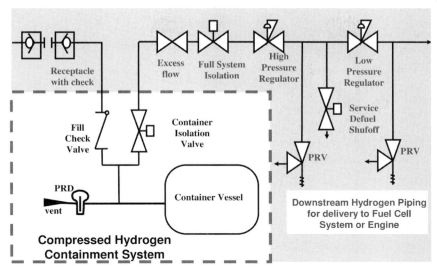

Fig. 18.6 Schematic of fuel system [6].

18.3.1.1 Components of CGH$_2$ Fuel System

One of the main system components, the hydrogen storage cylinder, is designed to optimize weight and function through the utilization of carbon fiber, similarly to CNG-type cylinders. Type 3 composite cylinders consist of a metallic liner fully wrapped with resin-impregnated continuous filament (carbon fiber). Type 4 composite cylinders include a liner made out of a polymer fully wrapped with resin-impregnated continuous filament (carbon fiber). All storage cylinders should be certified according to current national regulations. The filling pressure and temperature are limited to maxima of 87.5 MPa and 85 °C, respectively.

The in-tank isolation valve is activated by turning the key and applying 12 V d.c. Its integration makes it an essential part of the safety strategy of the vehicle. If hydrogen leakage is detected or during similar safety-critical events, it isolates the container from the rest of the powertrain to minimize potential risks to the operator. A direct connected pressure transducer measures the pressure in the extraction/refueling line. During refueling and extraction, the gas passes a filter to protect the valve and the downstream components from being damaged by larger particles. A manual shut-off valve enables the operator to isolate the container manually, overwriting the solenoid valve. During operation, an excess flow valve minimizes the gas flow in the event of a major gas leakage caused, for example, by rupture of a fuel line.

To prevent the tank from being over-pressurized, a directly connected pressure relief device (T-PRD) is activated and releases hydrogen into the hydrogen safety line of the vehicle. An in-tank temperature sensor measures the gas temperature necessary to calculate the actual fuel capacity of the cylinder. A bleed

Fig. 18.7 Hydrogen refueling and service receptacle [7].

port embedded parallel to the T-PRD outlet allows defueling of the tank even with a nonactivated solenoid for service and maintenance purposes.

A pressure regulator is located downstream of the in-tank isolation valve and reduces the outlet pressure of the cylinder. To protect the components downstream of the pressure regulator, for example in case of failure of a regulator, a pressure relief valve (PRV) is plumbed immediately after the regulator. In case the PRV opens, hydrogen is safely directed to the ambient atmosphere through the vehicle's vent line.

The refueling receptacle is installed under the tank flap and acts like a quick coupler with integrated check valve and filter (see Figure 18.7). Together with the corresponding refueling nozzle, it enables safe and fast refueling of the tank system with hydrogen up to 70 MPa. The service receptacle is integrated in the rear of the vehicle and connected with a T-fitting to the low-pressure fuel line downstream of the pressure regulator, and therefore allows defueling of the whole system at moderate pressures.

In addition to the main components outlined above, the hydrogen storage system includes additional components, such as check valves, manual isolation valves, and low-pressure sensors, which may differ for each system layout.

18.3.1.2 CGH$_2$ Refueling Strategy

To refuel 70 MPa hydrogen storage systems while guaranteeing short refueling times together with maximum fill percentage and state of charge (SOC), different strategies are currently under investigation. A first consensus was laid down in the SAE technical information report draft J2601. Here the station controls the refueling process with regard to the fueling process operating window (see Figure 18.8) and also the maximum fuel delivery temperature and pressure target as a function of ambient temperature.

A distinction is made between two main refueling modes: with and without communication between the vehicle and the refueling station. During a communication fill, the actual gas pressure, temperature inside the vehicle tank, and cylinder volume are communicated to the station via a wire connection or wireless IR connection. This permanent information flow guarantees a maximum SOC while minimizing the refueling time and keeping the tank status

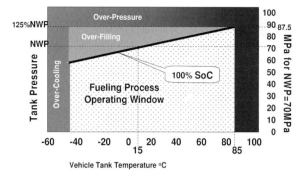

Fig. 18.8 Operating window of fueling process [8].

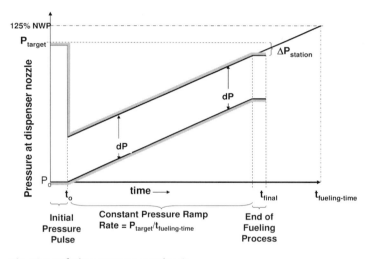

Fig. 18.9 Refueling pressure corridor [8].

within the fueling process window. For non-communication refueling, a different pressure target as a function of ambient temperature assures a safe refueling process, that is, the parameter tank pressure and tank gas temperature will remain within the fueling process operating window.

The fueling is conducted within the limits of a fueling pressure corridor illustrated in Figure 18.9. The mass flow rate is controlled by the station such that an average pressure increase is achieved. This ensures a refueling time of 3 min from 0 MPa to a maximum pressure of 87.5 MPa. Regardless of the cylinder size, the refueling is completed in the same time period. During refueling, the gas temperature inside the cylinder increases due to the heat of compression and the Joule-Thompson effect. A comparison of the magnitudes of these two phenomenon shows that the Joule-Thompson effect has an insignificant effect on the overall temperature rise when the thermodynamics of the entire process are considered. The temperature increase is dependent on, among other factors, the temperature of the inlet gas, the initial gas temperature inside the cylinder, the internal volume of the cylinder, the mass flow rate into the cylinder, and the heat transfer rate through the cylinder liner and laminate [9].

Different types of refueling station are currently in operation and/or under investigation. One common technology is based on a three-buffer system in which the gas is stored at 87.5 MPa. After connecting the refueling nozzle to the vehicle, the station determines the initial pressure in the vehicle's tank, for example by initiating a pressure pulse or, in the case of a communication fill, by gathering actual pressure data from the vehicle's tank sensor. The refueling process starts with overflowing from the first buffer storage. Under the aforementioned goal of maintaining a certain pressure increase, the first buffer is disabled after the flow rate decreases to a system-calculated limit, and the second and the third storage buffers are enabled successively. The refueling process ends when the target pressure is reached. For further refueling processes, the storage buffers are refilled from a general hydrogen source with the aid of a compressor. To allow short refueling times of less than 3 min, the temperature increase inside the vehicle's tank is compensated by precooling the gas to a minimum temperature of $-40\,°C$.

The additional energy necessary to cool the gas from, for example, 15 to $-40\,°C$ accounts for approximately 0.7% of the LHV for hydrogen (120 MJ kg^{-1}), disregarding the efficiencies related to the technical measures to cool, exchange the heat, and maintain the low temperature.

18.3.2
Liquid Hydrogen Storage System Overview

Typically, an on-board LH$_2$ storage system contains the following components: a container with support posts, a receptacle (or coupling), a pressure relief device, an automatic shut-off valve, a flexible or rigid fuel line, a hydrogen conversion system, and a boil-off management system. Figure 18.10 shows one such system.

18.3.2.1 The Container
Liquid hydrogen is stored in the container (0.6 MPa service pressure) at 20 K. This very low service temperature is in current applications maintained in double-walled cylindrical tanks, that is, there is an inner vessel (2) inside an outer vessel (1) and in the space between them there is a thermal insulation material in vacuum. Tanks should resist hydrogen embrittlement and show negligible hydrogen permeation. They are for these reasons made of either stainless steel or an aluminum alloy. Aluminum alloys are particularly well suited for tanks due to their inherent properties: low specific weight, high modulus and strength, high coefficients of thermal expansion, and good characteristics of

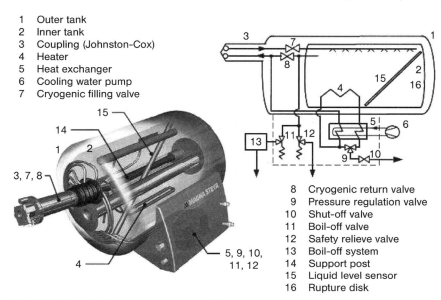

Fig. 18.10 Liquid hydrogen storage system [4].

1	Outer tank
2	Inner tank
3	Coupling (Johnston-Cox)
4	Heater
5	Heat exchanger
6	Cooling water pump
7	Cryogenic filling valve
8	Cryogenic return valve
9	Pressure regulation valve
10	Shut-off valve
11	Boil-off valve
12	Safety relieve valve
13	Boil-off system
14	Support post
15	Liquid level sensor
16	Rupture disk

thermal conductivity. In order to reduce heat entry by thermal radiation, the insulation material between tanks consists of several layers of multi-layer-insulation reflective aluminum or aluminized polymer foils separated by glass fiber spacers at a vacuum of about 10^{-3} Pa [3, 4].

The position of the inner vessel with respect to the outer vessel is maintained by the inner supports or suspension. They should resist not only crash forces without rupture but also acceleration during operation in addition to reducing the heat transfer to the inner vessel. Supports are typically made of coaxial tubes or loops of glass fiber- or carbon fiber-reinforced plastics with high mechanical strength and low thermal conductivity. The weight of a typical tank system for storing about 10 kg of hydrogen, including valves and heat exchanger, is about 150 kg [4].

One advantage of LH_2 over CGH_2 systems is that tank geometries no longer need to be restricted to a cylindrical shape due to the low pressures (0.6 MPa), therefore allowing wider design freedom.

18.3.2.2 Filling Procedure

The filling procedure is initiated when a special cryogenic nozzle is connected to the container's coupling (3). This connection must be gas-tight and thermally insulated. Start and stop of filling are controlled by an electrical signal communication between the fueling station and the tank control unit. This communication is also necessary for safety reasons [3]. During filling, the cryogenic filling valve (7) is opened, allowing LH_2 to flow from the filling station via the refueling line in the inner vessel due to the pressure difference between the two. The cryogenic return valve (8) is also opened during the process, allowing evaporated gaseous hydrogen to leave the tank and flow back to the filling station, while keeping the pressure in the tank at low levels. Both filling and return valves are closed after completion of the filling procedure.

18.3.2.3 Hydrogen Extraction

In fact, in cryogenic hydrogen storage systems, hydrogen is stored with equilibrium of its liquid and gaseous phases at very low temperatures. The density ratio between liquid and gas depends on the operating pressure of the system.

Hydrogen stored in the liquid state is supplied to the engine/fuel cell in the gaseous state at ambient temperature and appropriate flow rate and pressure levels. During extraction, the cryogenic return valve (8) opens and cold gaseous hydrogen flows from the inner vessel to the cooling water heat exchanger (5), where it heats up above ambient/room temperature. It then passes a shut-off valve (10) and runs through the fuel line towards the engine/fuel cell. Before entering the engine/fuel cell, the pressure level of the gas can be tuned by an additional pressure-control valve.

As gas extraction causes a pressure drop in the inner vessel, the latter should be heated such that further LH_2 evaporates, thus assuring the required operating pressure. In order to do this, some of the warmed gaseous hydrogen that has already left the tank is diverted (at the shut-off valve level) from the main stream and runs through an inner tank heater element (4).

18.3.2.4 Boil-Off Behavior

Boil-off-losses are closely connected to the driver's habits. For everyday users, the system losses are expected to tend towards zero [3]. However, when the vehicle is not being used, and both cryogenic valves are closed, heat enters the inner vessel via several components. The temperature increase inside the tank causes the pressure to increase to the so called boil-off pressure (a certain maximum overpressure with respect to atmospheric pressure to which the inner vessel is designed) and hydrogen has to be released to maintain the pressure below this level. The time required for fuel depletion is proportional to the hydrogen mass initially contained in the tank and to the heat leakage in the tank. According to Aceves et al. [10], a full 5 kg tank affected by a 1 W heat leak would lose all its hydrogen in approximately 3 weeks, whereas a tank under the same conditions, containing 1 kg, would be empty after approximately 4 days.

Hydrogen is expelled from the vehicle through the boil-off valve. As a redundancy measure, if the boil-off valve fails to operate, a safety relief valve(s) will open at its preset pressure. Also, losses of high vacuum could arise from a severe crash and lead to a very high heat input in the inner vessel, causing the boil-off rate to increase dramatically. In this case, the boil-off system management would no longer be capable of dealing with this rapid pressure increase and again the pressure relief valve(s) would be activated [3].

According to the ECE regulation draft for LH_2 storage systems, this boil-off gas must not simply be released into the atmosphere, but must be converted into harmless fluids/gases. This can be done by one of the following methods: catalytic burning to water within a boil-off management system, on-board thinning with air below ignition limits, or transformation into electric power in an on-board fuel cell-based auxiliary power unit [3].

18.3.3
Overview of Cryo-Compressed Hydrogen Storage Systems

In cryo-compressed systems, hydrogen is stored in cryogenic-capable insulated pressure vessels (cryo-compressed tanks) that can accept cryogenic LH_2, cryogenic CGH_2, or CGH_2 at ambient temperature. Cryo-compressed systems combine properties of both CGH_2 and LH_2 storage systems, and therefore system components and auxiliaries, filling procedure, and hydrogen extraction are similar to those already described.

18.3.3.1 The Container
The tank used in cryo-compressed storage systems must be capable of operating at high pressure (24 MPa or higher) and cryogenic temperatures (20 K). According to Aceves *et al.* [10], a type III CGH_2 tank design (aluminum-lined composite-wrapped) fulfils this requirement best as it combines moderate weight and affordable price with no hydrogen permeation. An example of a cryo-compressed tank is shown in Figure 18.11.

In order to reduce the heat transfer, an outer vacuum vessel and multi-layer vacuum insulation are used. The tank is instrumented for pressure, temperature, and liquid level and possesses safety devices to prevent failure if hydrogen leaks into the vacuum space. As mentioned above, the tank can be filled exclusively with LH_2, or it can be filled flexibly with LH_2, cryogenic CGH_2, or ambient-temperature CGH_2. According to Aceves *et al.* [10], the operation can be divided into two different modes: the LH_2 mode and the flexible mode.

18.3.3.2 Liquid Hydrogen Mode
In this mode, the system is operated with LH_2 at 20 K and 0.1 MPa. When operated in this mode, its great advantage over the conventional LH_2 system is the lower evaporative losses. Also, the fact that the tank is able to withstand higher pressures allows greater pressure increases before hydrogen has to be boiled off. Overall, this translates into an extension of the period of inactivity before the tank starts to have evaporative losses (boil-off), which will be referred to as dormancy.

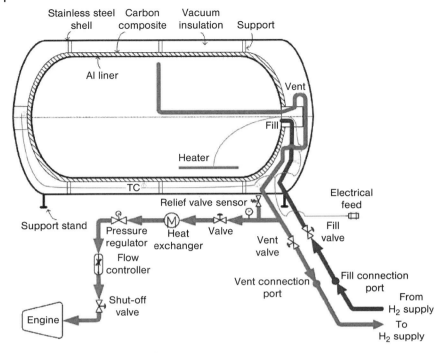

Fig. 18.11 Cryo-compressed tank system [11].

In order to assess/quantify the improvement in this respect, an experiment was conducted [10] in which a cryo-compressed tank with an inner volume of 84 l^{-1} was filled with LH$_2$ at 20 K and 0.1 MPa to 84% of its total volume and then parked until the hydrogen pressure in the tank reached its service pressure (20 MPa). The heat transfer rate into the tank was 1 W. The results of this experiment are shown in Figure 18.12. When the service pressure of the tank is reached, hydrogen has to be released from the tank to keep the pressure at an acceptable level. Dormancy increases as the tank's service pressure increases. For the service pressure of today's LH$_2$ tanks (0.6 MPa), the dormancy period is approximately 4–5 days, whereas a service pressure of 13 MPa corresponds already to a dormancy period of 2 weeks. The maximum shown service pressure of 20 MPa corresponds to a dormancy period of 17 days, a value which can be improved with higher service pressure tanks.

It should be noted that the results in Figure 18.12 were calculated from the first law of thermodynamics and the properties of hydrogen, neglecting the thermal capacity of the tank and also the conversion between the two different states of nuclear spin arrangement (*para-* and *ortho-*hydrogen) [10]. Both of these effects tend to increase the tank's dormancy, especially for temperatures above 80 K.

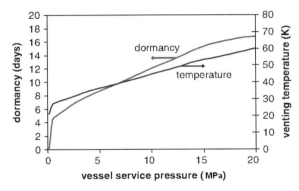

Fig. 18.12 Dormancy and hydrogen temperature at time of venting versus tank service pressure [10]

Figure 18.12 also shows the temperature of hydrogen at the time of venting and demonstrates that as the service pressure increases, hydrogen has to reach a higher temperature before venting is required. This means more heat transfer and consequently more time allowed before the boil-off point is reached.

Cryo-compressed tanks also promise a reduction in evaporative losses when the vehicle is being operated. An investigation [10] showed that an LH_2 tank loses hydrogen when a vehicle is driven on average less than 23 km a day (for a heat transfer rate of 1 W). On the other hand, under the same conditions, a cryo-compressed tank with a service pressure of 24 MPa only starts to lose hydrogen for an average daily driving distance of less than 5 km.

On top of this, the improved thermal endurance of cryo-compressed tanks allows the use of thinner insulation and consequently greater volumetric efficiency than LH_2 tanks. Moreover, cryo-compressed tanks can be filled with high-pressure LH_2, increasing their density advantage in comparison with LH_2 tanks. The density of LH_2 at 24 MPa is 86.5 g l^{-1}, which is substantially higher than the 70.9 g l^{-1} of saturated LH_2 at 0.1 MPa.

18.3.3.3 Flexible Mode

Cryo-compressed storage tanks can be filled with hydrogen at any state between 20 K LH_2 and ambient-temperature CGH_2. This takes advantage of the fact that these tanks are compatible with CGH_2 and LH_2 refueling infrastructures.

Filling the tank with CGH_2 instead of LH_2 is less energy demanding and is therefore expected to be considerably more economical. As seen in Section 18.3.2.2, the energy required to liquefy hydrogen is not expected to fall below 33% of hydrogen's LHV, whereas from Figure 18.5 it can be seen that the energy required to compress hydrogen from 0 to 25 MPa (4.22 kWh kg^{-1}) represents only 12.6% of hydrogen's LHV.

The concept described by Aceves *et al.* [10] that best exploits the potential of this storage technique is to use CGH_2 for normal/daily driving conditions (short

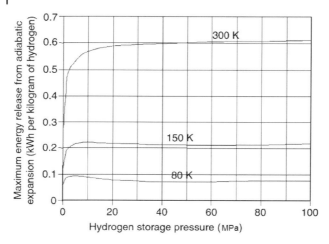

Fig. 18.13 Maximum mechanical energy per kilogram of hydrogen released upon expansion of hydrogen gas [11].

distances) and LH$_2$ exclusively for long distances. In this manner, important savings could be achieved by regularly filling the tank with ambient-temperature CGH$_2$ at lower cost while leaving the more expensive LH$_2$ for long trips. This concept also tends towards zero hydrogen losses as liquid filling would occur before direct hydrogen consumption. Hence a tank of 84 l inner volume (mounted in an efficient vehicle), filled with ambient-temperature CGH$_2$ at 24 MPa and 300 K, suffices for reaching a driving range of 200 km, which can be considered acceptable for average daily driving.

An important advantage of cryo-compressed systems operated with low-temperature hydrogen over conventional CGH$_2$ storage systems concerns safety in the event of a vessel rupture. Figure 18.13 shows the maximum mechanical energy released by sudden expansion (as in a vessel rupture) of 1 kg of high-pressure hydrogen gas at different temperatures.

As seen above, the energy release is mostly dependent on storage temperature, and pressure increases play a smaller role. At 300 K, a large pressure increase from 20 to 100 MPa would lead to a less than 10% increase in maximum released energy. On the other hand, a decrease in gas temperature from 300 to 80 K would lead to a more than 80% reduction in the released mechanical energy. It should be noted that the mechanical energy in Figure 18.13 is the maximum available work based on reversible adiabatic expansion from the pressure shown to 0.1 MPa, calculated from internal energy differences for hydrogen gas before and after isentropic expansion.

Finally, it could be stated that exclusively filling the cryo-compressed tank with LH$_2$ before long journeys would reduce the need for insulation material and hence allow for an increase in the tank's inner volume or for more compact tank designs. Also, the fact that tanks service pressures would be considerably lower than current CGH_2 system service pressures could allow for a reduction in the thickness of carbon fiber wrapping and consequently cost advantages [12].

18.4 Conclusion

Overall, the hydrogen storage systems described here are designed according to today's state-of-the-art hydrogen storage requirements and have been proven to provide acceptable fuel capacity and vehicle range, good functionality, and appropriate safety in particular with regard to North American and European standards and regulations.

References

1 Gardiner, M.R., Cunningham, J., and Moore, R.M. (2001) Compressed hydrogen storage for fuel cell vehicles, In *Future Transportation Technology Conference, Costa Mesa, California*, SAE Technical Paper Series SP-1635, SAE International, Warrendale, PA.
2 Vargaftik, N.B. (1975) *Tables on the Thermophysical Properties of Liquids and Gases*, John Wiley & Sons, Inc., New York.
3 Amaseder F. and Krainz G. (2006) *Liquid Hydrogen Storage Systems Developed and Manufactured for the First Time for Customer Cars*, SAE Technical Paper Series 2006-01-0432, SAE International, Warrendale, PA.
4 Krainz, G., Bartlok, G., Bodner, P., Casapicola, P., Doeller, Ch., Hoffmeister, F., Neubacher, E., and Zieger, A., *Development of Automotive Liquid Hydrogen Storage*, Magna Steyr Fahrzeugtechnik, Oberwaltersdorf, www.storhy.net.
5 Züttel, A. (2003) Materials for hydrogen storage, *Materials Today*, (September).
6 SAE International (2008) *Technical Information Report for Fuel Systems in Fuel Cell and Other Hydrogen Vehicles*, SAE TIR J2579, SAE International, Warrendale, PA.
7 WEH GmbH (2009), www.weh.com.
8 SAE International (2007) *Technical Information Report for Fueling Protocols for Gaseous Hydrogen Surface Vehicles*, SAE TIR J2601, unpublished draft, SAE International, Warrendale, PA.
9 Dicken, C.J.B. and Merida, W. (2007) *Journal of Power Sources*, **165**, 324–336.
10 Aceves, S.M., Berry, G.D., Martinez-Frias, J., and Espinosa-Loza, F. (2006) Vehicular storage of hydrogen in insulated pressure vessels, *International Journal of Hydrogen Energy*, **31**, 2274–2283.
11 Ahluwalia, R.K. and Peng, J.K. (2008) Dynamics of cryogenic hydrogen storage in insulated pressure vessels for automotive applications, *International Journal of Hydrogen Energy*, **33**, 4622–4633.
12 Aceves, S.M., Berry, G.D., Martinez-Frias, J., Espinosa-Loza, F., Ross, T., and Weisberg, A. (2007) *Storage of Hydrogen in Cryo-Compressed Vessels*, DOE Hydrogen Program – FY 2007, Annual Progress Report, US Department of Energy, Washington, DC.

19
Metal Hydrides

Etsuo Akiba

Abstract

The concept of hydrogen storage materials was proposed in late 1960's. Mg-based Mg_2Cu and Mg_2Ni are the first examples and $LaNi_5$ that works at room temperature was followed. In these early days, hydrogen storage materials are interstitial hydrides or hydrides of intermetallic compounds. Investigations on interstitial hydrides, a plenty of experimental results and empirical rules based on experiments have been reported. At the first part of this review, these achievements were briefly introduced because they are extremely suggestive for the research on non-interstitial hydrides as well as interstitial hydrides. In the later part, recent progress on interstitial hydrides and Mg based hydrides and their applications were briefly reviewed.

Keywords: metal hydrides, hydrogen storage, thermodynamics, kinetics

19.1
Introduction

Various metals and other elements form hydrides, some of which can be applied for hydrogen storage, transport, and delivery. In this chapter, properties of metal hydrides are described in two parts. Part I treats fundamentals of metal hydrides, including thermodynamic properties, crystal structures, kinetics, and properties required from an engineering standpoint. Part II discusses applications of metal hydrides including Ni–MH batteries and on-board hydrogen storage.

Hydrogen Energy. Edited by Detlef Stolten
Copyright © 2010 WILEY-VCH Verlag GmbH & Co. KGaA, Weinheim
ISBN: 978-3-527-32711-9

19.2
Part I: Fundamentals of Metal Hydrides for Hydrogen Storage

19.2.1
Elements Forming Hydrides

In the Periodic Table, shown in Table 19.1, there are various metallic elements, some of which form stable hydrides, which means that their hydrides are thermodynamically stable or the enthalpy change in hydride formation is a negative value. The reaction between a metal M and hydrogen is expressed by the following equation:

$$M + H_2 \rightleftharpoons MH_2 + Q \tag{19.1}$$

where Q is the heat of hydride formation.

Alkali metals forms ionic hydrides in which hydrogen has a negative charge. NaH is a typical case. Alkaline earth metals such as Mg form hydrides that have both ionic and covalent metal–hydrogen bonding. In general, ionic hydrides are

Table 19.1 The Periodic Table of the elements.

too stable to release hydrogen under the operating conditions of polymer electrolyte fuel cells (PEFCs), namely a maximum temperature of 80 °C and a minimum hydrogen pressure of a few bars.

Transition metals form interstitial hydrides. Hydrogen occupies interstitial sites formed by the metals in the crystal lattice. Grochala and Edwards published a review on non-interstitial hydrides [1]. Historically, interstitial hydrides have been intensively studied since early 1970s as the first materials that could be applied for hydrogen storage. LaNi$_5$ is a typical hydrogen storage material or hydrogen-absorbing alloy that was first reported in 1970 [2]. Transition metals except V cannot react reversibly with hydrogen under ambient conditions. Therefore, alloys that consist of more than two metallic elements are usually used for hydrogen storage purposes. In the late 1990s, other types of hydrogen storage materials were reported (these are introduced in another chapter). Hydrogen storage materials are now grouped into interstitial hydrides and non-interstitial hydrides.

Base metals such as Al form hydrides but not interstitial hydrides. Usually, the nature of base metal–hydrogen bond is covalent but it contains the nature of ionic bonding. In this chapter, metal hydrides such as MgH$_2$ and AlH$_3$ that are not interstitial hydrides will also be discussed.

The enthalpy change in hydride formation by transition metals, which is the same as the heat of hydride formation, depends on the position in the Periodic Table. If one compares Sc to Ni, from Sc to V the enthalpy change increases linearly (from a large, negative absolute value to a small absolute value), and from Cr to Ni it is positive but small in absolute value. This is shown schematically in Figure 19.1, which clearly demonstrates that there is a border region of stable–unstable hydrides between V and Cr in the Periodic Table for interstitial hydrides of transition metals. Other trends in the Periodic Table are similar but Pd is only exception that absorbs and desorbs hydrogen at room temperature under moderate hydrogen pressure.

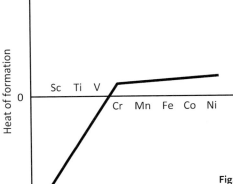

Fig. 19.1 Enthalpy change in hydride formation for the first row of the transition metals.

19.2.2
Thermodynamic Properties

19.2.2.1 Pressure–Composition Isotherms

The equilibrium relationship in a metal (alloy)–hydrogen system is illustrated by means of pressure–composition–temperature (p–c–t) isotherms such as shown in the example in Figure 19.2. When hydrogen gas is introduced into the system, it dissolves in the lattice of the metal and forms a solid solution. This takes place over the region AB in Figure 19.2. Hydrides formed from metal and hydrogen are chemical compounds and, therefore, the phase rule shown in Equation 19.2 can be applied:

$$f = c - p + 2 \tag{19.2}$$

The number of degrees of freedom of the system, f, is unity in the region where the metal (in practice the hydrogen solid solution phase) is in equilibrium with its hydride and gaseous hydrogen (in this case the number of components c is 2, and the number of phases p is 3). This means that, for a given temperature, the hydrogen equilibrium pressure is constant in the two-phase region (a solid solution phase and hydride in this case), shown as the plateau section, BC, in the isotherm in Figure 19.2. When all the solid solution phase has converted to the hydride phase, the number of degrees of freedom becomes two and the equilibrium pressure changes with the amount of solute hydrogen in the hydride phase (region CD).

Metal hydride formation is expressed by Equation 19.1. The change in Gibbs free energy, ΔG^0 of this reaction can be expressed by Equations 19.3 and 19.4:

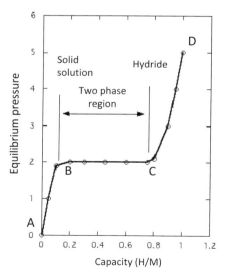

Fig. 19.2 Schematic diagram of pressure–composition isotherm.

$$\Delta G^0 = RT \ln P(H_2) \tag{19.3}$$

$$\Delta G^0 = \Delta H^0 - T\Delta S^0 \tag{19.4}$$

where R is the gas constant, T is the absolute temperature, $P(H_2)$ is the hydrogen equilibrium pressure, ΔH^0 is the enthalpy change of hydride formation and ΔS^0 is the entropy change of hydride formation. In Equation 19.3, the activity of solid states such as metal and hydride is set to 1. In Equation 19.1, Q is the heat of reaction (in this case hydride formation), which is equal to the enthalpy change of hydride formation, ΔH^0, in Equation 19.3. Here, we assume that reaction 19.2 is endothermic, which means that the hydride is more stable than the metal. This is the case in this chapter.

From Equations 19.3 and 19.4, equation 19.5 is obtained:

$$\ln P(H_2) = \frac{\Delta H^0}{RT} - \frac{\Delta S^0}{R} \tag{19.5}$$

The plot of the hydrogen equilibrium pressure versus the reciprocal of the absolute temperature, $1/T$, is called the van't Hoff plot. The slope and the intercept with the y-axis of the van't Hoff plot give the enthalpy change in hydride formation, ΔH^0, and the entropy change of hydride formation, ΔS^0, respectively.

In general, the difference in standard entropy between a metal and its hydride is very small and is of the order of 10 J K^{-1} mol^{-1}. The change in entropy with hydride formation, S^0, is mainly provided by the loss of the standard entropy of hydrogen gas (130.858 J K^{-1} mol^{-1}), which means that ΔS^0 can be assumed to be a constant and does not depend on the nature of materials that absorb hydrogen. Therefore, Equation 19.5 can be rewritten as

$$\ln P(H_2) = \Delta H^0/RT + C \tag{19.6}$$

where $C = -\Delta S^0/R$ is a constant and does not depend on the hydrogen storage material.

A sufficient numbers of sets of temperatures and equilibrium pressures are needed for calculation of the enthalpy of hydride formation when using the van't Hoff Equation 19.5. However, by using Equation 19.6, only a pair of the temperature and equilibrium pressure values is required to calculate the enthalpy change of hydride formation. This method is very convenient to use to obtain a first impression of the thermodynamic properties of a given metal-hydrogen system. Because the difference in standard entropy between metal and hydrogen is of the order of 10 J K^{-1} mol^{-1}, ΔS^0 calculated using the van't Hoff plot is usually around -110 J K^{-1} mol^{-1}.

Because only V as a metallic element reacts with hydrogen at ambient temperature and hydrogen pressure, alloys are usually used for hydrogen storage purposes. There are empirical rules that predict the stability of a given alloy–hydrogen system. The most popular and effective rule is Miedema *et al.*'s rule of reversed stability [3]. They proposed the following relationship between the hydride and alloy stabilities:

$$\Delta H(AB_nH_{2m}) = \Delta H(AH_m) + \Delta H(B_nH_m) - \Delta H(AB_n) \tag{19.7}$$

where A is hydride-forming metal and B is not. The heat of hydride formation, which is the equal to the enthalpy change of hydride formation, depends substantially on the element. However, that of B metals is not dependent on the element and is a small positive value. If the A atom is fixed, the sum of the first and second terms of the right-hand side of Equation 19.7 has almost the same value for a given class of alloy system. This means that if the third term on the right-hand side becomes more negative (equivalent to a more stable alloy), the left-hand side becomes more positive (equivalent to a more unstable hydride). In the case that the A atom is La, $LaNi_5$ and $LaCo_5$ are compared. The ΔH values of $LaNi_5$ and $LaCo_5$ are −168 and −80 kJ per mole of alloy, respectively. $LaCo_5$ hydrides are found to be more stable than those formed from $LaNi_5$. This clearly shows that these AB_5-type hydrides obey the rule of reversed stability. This empirical rule has been used effectively especially for the development of novel hydrogen-absorbing alloys.

19.2.3
Crystal Structures

The hydrogen atoms in the interstitial hydrides are located in interstitial sites among metal atoms, and it is very important to know the positions of these atoms. In solid-state compounds, atoms are generally fixed in certain crystallographic positions and, therefore, their properties are essentially determined by the arrangement of the component atoms, namely by their crystal structures. In practice, neutron diffraction is the only suitable method that can be used to determine the crystal structure of metal hydrides. Due to the large incoherent scattering cross-section of protium (hydrogen of atomic weight 1), it is necessary to use deuterides (deuterium compounds) of alloys when carrying out these neutron diffraction studies [4]. The structure difference between the protides (protium compounds) and the corresponding deuterides has not been reported for alloy–hydrogen systems [5], and therefore the crystal structure results obtained for deuterides are usually taken as also representing the structure of the hydrides in general. However, one should be careful to consider the crystal structure in the case when an isotope effect of hydrogen on the phase relation is reported, as for the V–H system.

Table 19.2 lists the crystal structures of classical alloy–hydrogen systems, and Table 19.3 lists those of selected metals and their hydrides (Na–NaH, Mg–MgH_2, Al–AlH_3, Ti–TiH_2, V–VH_2).

Table 19.2 Crystal structure parameters of selected binary hydrides.

Material	Space group	Lattice parameters (Å)
$LaNi_5$	$P6/mmm$	$a = 5.017$
		$c = 3.986$
$LaNi_5D_7$	$P6_3mc$	$a = 5.388$
		$c = 8.559$
$ZrMn_2$	$P6_3/mmc$	$a = 5.035$
		$c = 8.276$
$ZrMn_2D_3$	$P6_3/mmc$	$a = 5.391$
		$c = 8.748$
Mg_2Ni	$P6_222$	$a = 5.216$
Mg_2NiD_4	$C2/c$	$a = 14.342$
		$b = 6.403$
		$c = 6.483$
		$\beta = 113.52$
TiFe	$Pm\text{-}3m$	$a = 2.9789$
$TiFeD_{1.9}$	$Cmmm$	$a = 7.029$
		$b = 6.233$
		$c = 2.835$

Table 19.3 Crystal structure parameters of selected binary hydrides.

Material	Space group (structure type)	Lattice parameters (Å)
Na	$Im\text{-}3m$ (BCC)	$a = 4.235$
NaH	$Fm\text{-}3m$ (NaCl)	$a = 5.388$
Mg	$P6_3mc$ (HCP)	$a = 3.32439$
		$c = 5.25266$
MgH_2	$P4_2/mnm$ (rutile)	$a = 4.5025$
		$c = 3.0123$
Al	$Fm\text{-}3m$ (FCC)	$a = 2.870$
AlH_3	$R\text{-}3\ 2/c$	$a = 4.4493$
		$c = 11.8037$
Ti	$P6_3mc$ (HCP)	$a = 2.9064$
		$c = 4.6667$
TiH2	$Fm\text{-}3m$ (CaF_2)	$a = 4.440$
V	$Im\text{-}3m$ (BCC)	$a = 3.030$
VH2	$Fm\text{-}3m$ (CaF_2)	$a = 4.271$

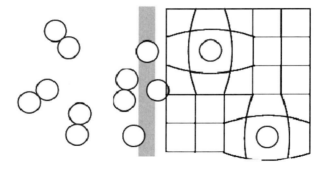

Hydrogen molecule (H$_2$)   Surface   Metal lattice

Fig. 19.3 Mechanism of hydrogen absorption to form interstitial hydride.

19.2.4
Kinetics

The direct reactions between the metal and gaseous hydrogen are solid–gas-phase reactions, which differ in some aspects from homogeneous reactions. Figure 19.3 shows schematically the reaction of hydrogen with hydrogen-absorbing alloys. The presence of boundaries influences the reaction kinetics and mechanism. Hydrogen gas exists only at the surface of the metallic materials. A hydrogen molecule (H$_2$) is first adsorbed on the surface, where it is decomposed into individual hydrogen atoms (H). These hydrogen atoms then diffuse into the bulk of the metallic material and finally form a metal hydride. The surface, which is one of the boundaries, plays an important role in hydride formation. A metal hydride is usually a fine powder. The hydrogenation of a fine powder proceeds independently inside each of the particles. A solid–gas-phase reaction generally starts in a particular part of the solid where reaction nuclei can readily form, such as grain boundaries, crystal defects, and impurities. Not all of the atoms of the solid contribute equally to the reaction. Therefore, in many cases, a simple kinetic equation cannot be used to describe the reaction. The method of preparation, activation procedure, and the past history of the metallic materials significantly influence the reaction kinetics, because they determine the particle size, the number of defects, and the size and nature of any precipitate species in the grain boundaries [6, 7].

In solid-state reactions, the fraction of a material that has reacted is used as a parameter for representing the change in the concentration of that substance in the original solid. If the parameter represents the ratio of the amount of reacted material to the total, that is, the original amount of material, the solid-state rate equation can be generally expressed by

$$F(\alpha) = kt \tag{19.8}$$

Table 19.4 Values of m used in Equation 19.10 and the corresponding rate equations.

Rate equation	m
$a^2 = kt$	0.62
$(1-a)\ln(1-a) + a = kt$	0.57
$[1-(1-a)^{1/3}]^2 = kt$	0.54
$1 - 2a/3 - (1-a)^{2/3} = kt$	0.57
$-\ln(1-a) = kt$	1.00
$1-(1-a)^{1/2} = kt$	1.11
$1-(1-a)^{1/3} = kt$	1.07
$a = kt$	1.24
$[-\ln(1-a)]^{1/2} = kt$	2.00
$[-\ln(1-a)]^{1/3} = kt$	3.00

where k and t are the rate constant and the reaction time, respectively.

The equation

$$a = 1 - \exp(-Bt^m) \tag{19.9}$$

where B and m are constants. Equation 19.9 is known as the Avrami–Erofeev equation. From Equation 19.9, the following can be derived:

$$-\ln[\ln(1-a)] = \ln B + m \ln t \tag{19.10}$$

Table 19.4 lists rate equations depending on the value of m for $0.15 < a < 0.50$, which means that from analysis using Equation 19.10 the mechanism of a solid–gas-phase reaction can be estimated. The original paper by Hancock and Sharp [8] gives the details of this very useful method.

19.2.5
Engineering Properties

As shown by Equation 19.1, hydride formation is usually endothermic. Therefore, heat should be removed for hydrogenation and be introduced for dehydrogenation. The heat required can be obtained from the enthalpy change of hydride formation, as mentioned earlier.

A metal hydride is obtained in the form of a fine powder and each powder attaches only by points. Therefore, the heat conductivity of a fine powder is significantly lower than its bulk form. To design the reaction vessel, heat conductivity is in many cases the most critical value to be considered. Table 19.5 gives the thermal conductivities of selected metal hydrides, including some of non-interstitial hydrides.

Table 19.5 Thermal conductivity of hydrogen storage materials and hydrides.

Material/hydride	Thermal conductivity/ W m^{-1} K^{-1}
Mg$_2$NiH$_x$	0.65
LaNi$_5$	0.2–0.8
LaNi$_5$H$_x$	0.1–0.75
MgH$_x$	1.2
NaAlH$_4$	0.46–0.75

19.3 Part II: Applications of Metal Hydrides

19.3.1 Types of Applications

Numerous metal hydrides have been found to be excellent hydrogen storage materials. Hydrogenation and dehydrogenation usually occur reversibly, as shown in Equation 19.1. Extensive application-oriented studies have been carried out in order to use metal hydrides for hydrogen storage vessels, secondary batteries, heat pumps, and so on.

19.3.2 Electrochemical Applications

Metal hydrides can also be formed by electrochemical reactions. Electrochemical hydrogen charging of steels and of other structural materials is used mainly in the research field of hydrogen embrittlement.

Since 1990, a more realistic electrochemical application of metal hydrides, the Ni metal hydride battery, has been commercialized [9, 10]. The overall reaction involved for battery application is

$$MH_x + NiOOH \rightleftharpoons MH_{x-1} + Ni(OH)_2 \qquad (19.11)$$

NiOOH is used for the cathode and metal hydride for the anode. The electrochemical reaction from the left to the right-hand side is the discharge reaction and the reverse reaction is the charge reaction.

AB$_5$-type hydrogen-absorbing alloys based on LaNi$_5$ have been used for this application since it was invented. To reduce the costs, mixed rare earth elements (mischmetal: Mm) are used for the commercially available products. To improve various performances, Co, Al, Mn, and other elements replace part of the Ni.

Kohno *et al.* reported (R,Mg)Ni$_{3+a}$ alloys for battery application [11]. The alloys reported had a layered structure consisting of AB$_2$ and AB$_5$ cells. It was reported that the alloy has more capacity and a higher charge/discharge rate than

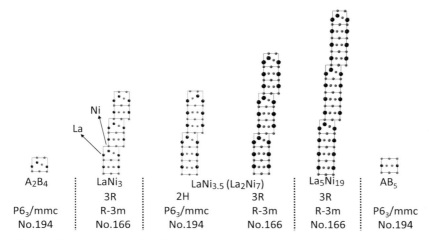

Fig. 19.4 Crystal structure of LaNi$_{3+x}$.

AB$_5$-type alloys as the electrode material. Figure 19.4 shows some of the crystal structures of this type of alloy. Kadir et al. [12] first reported this kind of alloy but the electrochemical properties, especially cycle life, were not sufficient for application. However, the alloy developed by Kohno et al. had sufficient cycle life and other significant performance properties for battery applications. Today, Sanyo, which is the largest secondary battery producer in the world, uses this type of alloy (named "super lattice alloy") for their best selling Ni–hydrogen battery "eneloop" [13].

Subsequently, a large number of papers to improve their application-oriented performance have been published, and the alloy has already been applied to commercially available products, including a hybrid vehicle made by a Japanese car manufacturer. However, the fundamentals of this type of alloy are not yet fully understood [14].

Investigation of the crystal structure of these alloys is not easy because the length of the c direction is 1–4 nm and usually they are mixture of more than two phases with very similar structures, as shown in Figure 19.3. Hayakawa et al. polished small pieces of the alloy to make single crystals [15]. From single-crystal structural analysis, it was found that Mg selectively occupies the AB$_2$ cell but not the AB$_5$ cell, as shown in Figure 19.5. Iwase et al. confirmed this hypothesis by crystal structure analysis of La$_2$Ni$_7$ that does not contain Mg [16]. La$_2$Ni$_7$ shows two plateaus in the p–c isotherm, whereas Mg-substituted alloys show a single, flat plateau. Hydrogen absorbed in the first hydride occupies the interstitial sites of the AB$_2$ cell and hydrogen occupies both AB$_2$ and AB$_5$ cells in the second hydride. If one compares the sizes of LaNi$_2$ and LaNi$_5$, the LaNi$_2$ cell is slightly larger. The atomic radii of La and Mg are 0.185 and 0.160 nm, respectively. By substituting Mg for La, the lattice size of the La(Mg)Ni$_2$ cell becomes small and fits the LaNi$_5$ cell.

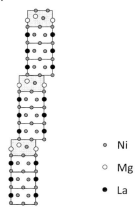

Fig. 19.5 Substitution sites of Mg in $LaNi_{3+x}$.

Historically, Ti- and/or Zr-based Laves phase alloys were extensively studied in the 1990s, but easy activation, a high rate capability and suitable cycle life were not simultaneously realized. Therefore, Laves phase alloys that were expected to have more capacity than AB_5-type alloys have never been commercialized.

Mg has more than double the electrical capacity per unit weight than a conventional $LaNi_5$-type alloy, but the issue of significant degradation of Mg-based alloy as the electrode material has not yet been resolved.

19.3.3
Hydrogen Storage, Transportation, and Delivery

19.3.3.1 Targets for On-Board Application

Hydrogen is widely accepted as the fuel for fuel cell vehicles. There have been various trials to look for an appropriate method of hydrogen storage on-board. At present, compressed highly pure hydrogen is chosen as the on-board hydrogen storage medium for most hydrogen-powered cars. A small number of car manufacturers use liquefied hydrogen for on-board storage.

For mobile application, especially for passenger vehicles, the requirements for on-board hydrogen storage are very severe. The United States Department of Energy set numerous system targets for on-board hydrogen storage. Table 19.6 lists a few of them that were revised in February 2009. Table 19.7 lists those before the revision. Table 19.8 lists those of the Japanese government.

Revision of the targets has been based on recent technical achievements in research and development in both the hydrogen and fuel cell fields. For example, progress has been made in the fuel (hydrogen) consumption of fuel cell vehicles. The most recent demonstration showed a cruising range of fuel cell vehicles of over 100 km using 1 kg of hydrogen. On-board storage of 5 kg of hydrogen seems to be sufficient to guarantee the same order of cruising range as for a, existing gasoline-powered vehicle.

Table 19.6 Hydrogen storage system targets in the United States after revision in February 2009.

Property	2010	2015	Ultimate
Gravimetric capacity (wt%)	4.5	5.5	7.5
Volumetric capacity (g l^{-1})	28	40	70
System cost	TBD[a]	TBD	TBD
Cycle life (cycles)	1000	1500	1500
Fill time for 5 kg of H$_2$ (min)	4.2	3.3	2.5

a) TBD, to be determined.

Table 19.7 Hydrogen storage system targets in the United States before revision.

Property	2010	2015
Gravimetric capacity (wt%)	6	9
Volumetric capacity (g l^{-1})	45	81
System cost ($ kg^{-1} H$_2$)	133	67
Cycle life (cycles)	1000	1500
Fill time for 5 kg of H$_2$ (min)	3	2.5

Table 19.8 Hydrogen storage targets in Japan for the hybrid system and materials (March 2008).

System	Property	Target
Hybrid tank system	Maximum pressure	35 MPa
	Weight capacity	3 wt%
	H$_2$ amount/volume/weight	5 kg/100 l/165 kg
	System cost	<200 000 yen per vessel
Materials	Weight capacity	>6.5 wt%
	Desorption temperature	<150 °C
	Cycle life	90% after 1000 cycles
	Material cost	1000 yen kg^{-1}

In general, volume density is more critical than weight density for passenger vehicles. Space for passengers and luggage is the most important factor for commercial vehicles to be sold to consumers.

19.3.3.2 Material Challenge to Reach the Targets

To reach the target, significant effort has been made in both material and system developments. In the following, typical examples of materials research in interstitial hydrides and binary hydrides (MgH$_2$ and AlH$_3$) are introduced. However, because of their weight, most metallic elements cannot reach the target mentioned earlier. If we assume that a metal forms a dihydride, MH_2, only Li, Be, Na, Mg, and Al among metals form hydrides with a higher hydrogen content than the target. In practice, Na forms NaH not NaH_2. The hydrogen content of NaH is about 4.3 mass%.

Novel Hydrides

The most successful progress in the investigation of novel interstitial hydrides has been made with Ti-based body-centered cubic (BCC) alloys [17]. Before finding these Ti-based BCC alloys, the highest hydrogen capacity at room temperature was less than 2 mass%. For example, $LaNi_5$, TiFe, and $Ti_{1.2}Mn_{1.8}$ have hydrogen capacities at room temperature of 1.4, 1.8, and 1.8 mass%, respectively. Ti-based BCC alloys that were named "Laves phase related BCC solid solution alloys" reversibly absorb hydrogen up to about 3 mass% at room temperature below 15 MPa [18].

Conventional metal hydrides such as $LaNi_5$ and $Ti_{1.2}Mn_{1.8}$ have the closest packing structures of $CaCu_5$ and $MgZn_2$ types, respectively. However, the BCC structure is not closest packing and, theretofore, it has more interstitial sites per metal than the closest packing structures, face-centered cubic (FCC) and hexagonal close packing (HCP).

The hydrogen capacities of TiH_2 and VH_2 are both around 4 mass%, which is less than the target set by some governments. Therefore, investigations of metal hydrides with lighter elements have been carried out. However, metallic elements that can absorb more than 5 mass% hydrogen if it forms MH_2 are very limited, as mentioned above. At present, significant effort has been made for Mg, Al and their related alloys.

Mg metal has a hydrogen capacity of 7.6 mass% and a ΔH value of -76 kJ mol^{-1} H_2. The hydrogen content of Mg hydride is above the target and Mg is rich as a resource, nontoxic and inexpensive. Using equation 19.6, the equilibrium hydrogen pressure of Mg hydride at 298 K is calculated to be of the order of 10^{-8} MPa. In practice, Mg can absorb hydrogen at room temperature but does not desorb hydrogen under the working conditions of PEFCs. Therefore, novel Mg-based alloys that have appropriate thermodynamic properties have been investigated. Using Equation 19.6, the window of ΔH suitable for on-board application is between -25 and -35 kJ mol^{-1} H_2 according to a report from one of major car manufacturers [19].

Unfortunately, Mg forms intermetallic compounds with a limited number of metallic elements such as rare earths, Ca, Ni, and Cu. Terashita *et al.* [20] and another Japanese group independently found that $(Mg,Ca)Ni_2$ absorbs and desorbs hydrogen under the same conditions as for $LaNi_5$. Of course, the ΔH of $(Mg,Ca)Ni_2$ is similar to that of $LaNi_5$. However, up to now no other metal hydride that contains Mg as the major component has been reported within the ΔH window mentioned above.

AlH_3 (alane) has a hydrogen capacity of over 10 wt% and a ΔH of -7.6 kJ mol^{-1} H_2 [21]. The hydrogen content of Al hydride is also above the target. Al is rich as a resource and inexpensive. The equilibrium hydrogen pressure of Al hydride is calculated to be of the order of 10^5 MPa at room temperature. In practice, because of surface protective layers, AlH_3 does not desorb hydrogen at room temperature [22]. On the other hand, it means that to form Al hydride it needs a significantly high hydrogen pressure. Saito *et al.* hydrogenated and dehydrogenated Al metal using an *in situ* high hydrogen pressure cell. The crystal structures were analyzed using energy-dispersive X-ray diffraction [23]. It was found that the equilibrium hydrogen pressure at 730 K is 6 GPa [23]. This pressure is different to that calculated using Equation 19.6. If we consider the critical temperature (13 K) and pressure (1.3 MPa) of hydrogen, hydrogen at this temperature and pressure is a supercritical fluid and has significant reaction activity. It is strongly expected that Al-based hydrogen-absorbing materials with a ΔH window from 25 to -35 kJ mol^{-1} H_2 can exist, but they have not been reported so far.

Improvement of Kinetics

Kinetics is also a key aspect for on-board application because the refueling time should be as short as possible. The target refueling time shown in the tables above is less than 5–10 min. In many cases, the surface of the metal hydride plays a critical role with respect to the reaction rate. Hydrogen molecules are adsorbed on the surface and decompose to atoms on the surface. However, the hydrogen-absorbing alloys to form interstitial hydrides are usually treated in air, which makes the surface oxidized. An oxidized surface likely prevents hydrogen adsorption and decomposition or at least reduces the activity of the surface for reactions.

In order to resolve these issues and accelerate the reaction, a catalyst is usually used. To improve the hydrogen reaction rate of Mg, in the early 1980s Bogdanovic *et al.* added organic materials [24] and Ono *et al.* added a small amount of Ni [25]. After their pioneering work, major efforts have been made to accelerate the reaction rate of Mg-based alloys. In the 1990s, the ball-milling technique became popular in the field of metal hydrides. Recently, the most significant progress in catalysts using ball-milling has been made by the DKSS group in Germany [26]. They investigated a large numbers of materials, both metals and oxides. Nb_2O_5 addition apparently gave the best results. As shown Figure 19.6, hydrogen absorption is completed in less than 30 s at 573 K and under 0.84 MPa of hydrogen [26].

To improve reaction kinetics of AlH_3, Sandrock *et al.* reported that an alkali metal works as a effective catalyst [22]. However, the reported decomposition temperature of AlH_3 is still slightly above working the temperature of on-board PEMFCs.

Destabilization of Hydrides

In order to destabilize a given hydride, a second compound is introduced. This idea for Mg-based alloys was proposed by Vajo *et al.* [27] (Figure 19.7). The ΔH of the hydride is reduced by the enthalpy of formation of the second compound from the initial phase such as a metallic element and hydrogen gas. In the case of MgSi, it is reported that the reaction did not proceed.

The above reaction needs diffusion of atoms during hydrogenation and dehydrogenation, which differs completely from conventional interstitial hydrides in which hydrogen simply occupies the interstitial sites in the metal sub-lattice.

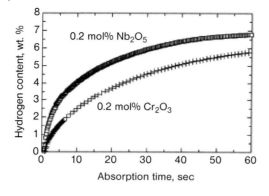

Fig. 19.6 Kinetics of hydrogen absorption by Nb_2O_5-doped Mg.

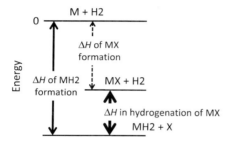

Fig. 19.7 Concept of destabilization of stable hydrides.

However, under the experimental conditions of the reaction with Mg_2Si, diffusion of atoms is not sufficient to complete the reaction in a certain time.

The idea that Vajo et al. [27] and Dronheim and co-workers [28] independently proposed is mostly applied to complex hydrides, especially borohydrides. Details are considered in another chapter.

19.3.3.3 Development of Storage System

Mori et al. demonstrated a high-pressure metal hydride tank system for on-board hydrogen storage [29]. Ti-based Laves phase or BCC alloys set inside a heat exchanger were put into a high-pressure (35 MPa) cylinder. The metal hydride is usually a fine powder and, therefore, the materials can occupy only around half of the volume. The remaining half can be filled by high-pressure hydrogen at 35 MPa. The working pressure range of metal hydrides extends form a few MPa to 35 MPa, which means that hydrides with an enthalpy change in hydride formation of around -25 kJ mol^{-1} H_2 are most suitable for this application. This makes two significant improvements for on-board hydrogen storage. One is that the heat of hydride formation during hydrogen refueling that must be removed as rapidly as possible becomes small enough to be cooled using an on-board radiator. The other is that hydrogen can be released at $-30\,°C$ to start the fuel cell on cold mornings in the major car markets. The maximum pressure of the tank is reported to be 35 MPa. Before that, the concept of the combination of high pressure and metal hydrides had been proposed by Takeichi *et al.* [30]. They termed this combination a "hybrid tank," and hybrid tank is now the most popular term to express this type of on-board hydrogen storage system. One of the drawbacks of a high-pressure cylinder is the use of carbon fiber, which is extremely expensive and cannot be recycled at present. As the second generation of "hybrid tank," Mori *et al.* reported a shell and tube type of tank using metallic materials. Because it is made of metallic components, the pressure range is reported to be up to 15–20 MPa [31]. In order to cover the same temperature range as the first-generation tank, they used two alloys with different equilibrium pressures.

From fundamental analysis to demonstration, a wide range of techniques for the fabrication of hydride tanks have been developed by various manufacturers, including United Technology and Sandia National Laboratory in the United States, CNRS at Grenoble in France and a company in Australia. In most cases, interstitial hydrides or Mg-based alloys are applied because they are technically available and can be produced on a large scale.

19.3.3.4
Other Applications

When introducing renewable energy to the electricity grid, fluctuating power should be stored to stabilize the grid and balance the supply and demand. Electricity can be effectively converted to hydrogen using an electrolyzer and the hydrogen formed is favorably stored by metal hydrides. In stationary applications, weight density is not a serious issue, but volume density and the cost are more critical than in mobile applications. Metal hydrides readily react at ambient temperature and pressure. In addition, the volume density of metal hydrides is better than that of compressed gas and liquefied hydrogen. Energy storage will be one of the most promising application fields of metal hydrides in near future.

As shown in Equation 19.1, hydride formation is a chemical reaction and the heat of hydride formation arises to a certain extent. If two metal hydrides are used, thermodynamic systems can be designed, based on a heat pump and a refrigerator. Two different levels of heat drive thermodynamic systems using metal hydrides. In thermodynamic systems using metal hydrides, electricity is not needed. Because two different heat sources are not available everywhere, the application of thermodynamic systems using metal hydrides is limited.

19.4
Conclusion

Interstitial hydrides have many favorable properties for various applications. They show rapid reaction rates, long cycle life, recyclability, significant volume density of hydrogen, and so on. Only the weight density is an issue. At present,

Ti-based BCC alloys have the maximum hydrogen capacity of slightly less than 3 mass% at room temperature.

To meet the capacity target set for mobile applications, Mg- and Al-based materials have been intensively investigated. As mentioned earlier, the enthalpy changes of hydride formation of both Mg and Al hydrides are far beyond the window applicable for combination with PEMFCs. Much effort has been expended to destabilize the hydrides while maintaining their capacity and this should remain one of the major tasks in the field of metal hydrides.

Metal hydrides have sufficient kinetics for applications. Therefore, even though the capacity of the material does not reach to the target, metal hydrides are used for engineering purposes. Materials research and engineering development should be conducted side-by-side. Metal hydrides will be used widely for research and development of hydrogen storage materials.

References

1 Grochala, W. and Edwards, P. P. (2004) *Chem. Rev.*, **104**, 1283–1315.
2 van Vucht, J. H. N., Kuijpers, F. A., and Brunimg, H. C. A. M. (1970) *Philips Res. Rep.*, **25**, 133.
3 van Mal, H. H., Buschow, K. H. J., and Miedema, A. R. (1974) *J. Less-Common Met.*, **35**, 65.
4 Yvon, K. (1984) *J. Less-Common Met.*, **103**, 53.
5 Yvon, K. and Fischer, P. (1988) In *Hydrogen in Intermetallic Compounds I* (ed. L. Schlapbach), Springer, Berlin, pp. 87–138.
6 Boulet, J. M. and Gerard, N. (1983) *J. Less-Common Met.*, **89**, 151.
7 Mintz, M. H. and Bloch, J. (1985) *Prog. Solid State Chem.*, **16**, 163.
8 Hancock, J. D. and Sharp, J. H. (1972) *J. Am. Chem. Soc.*, **55**, 74.
9 Sakai, T., Yuasa, A., Ishikawa, H., Miyamura, H., and Kuriyama, N. (1991) *J. Less-Common Met.*, **172–174**, 1194.
10 Ishikawa, H. (1991) *J. Less-Common Met.*, **172–174**, A31.
11 Kohno, T., Yoshida, H., Kawashima, F., Inaba, T., Sakai, I., Yamamoto, M., and Kanda, M. (2000) *J. Alloys Compd.*, **311**, L5.
12 Kadir, K., Sakai, T., and Uehara, I. (1997) *J. Alloys Compd.*, **257**, 115.
13 Sanyo Electric Co., Ltd. (2010) Sanyo eneloop Global Site, http://sanyo.com/eneloop/ (accessed 20 February 2010).
14 Yarty's, V. A., Riabov, A. B., Denys, R. V., Sato, M., and Delaplane, R. G. (2006) *J. Alloys Compd.*, **408–412**, 273.
15 Hayakawa, H., Akiba, E., Gotoh, M., and Kohno, T. (2005) *Mater. Trans.*, **46**, 1393; Akiba, E., Hayakawa, H., and Kohno, T. (2006) *J. Alloys Compd.*, **408–412**, 280.
16 Iwase, K., Sakaki, K., Nakamura, Y., and Akiba, E. *J. Phys. Chem. C*, submitted.
17 Akiba, E. and Iba, H. (1998) *Intermetallics*, **6**, 461.
18 Okada, M., Kuriiwa, T., Tamura, T., Takamura, H., and Kamegawa, A. (2001) *Met. Mater. Korea*, **7**, 67.
19 Mori, D., Hirose, K., Haraikawa, N., Takiguchi, T., Sinozawa, T., Matsunaga, T., Toh, K., Fujita, K., Kumano, A., and Kudo, H. (2007) SAE (Society of Automotive Engineers) Technical Paper 2007-01-2011.
20 Terashita, N., Kobayashi, K., Sasai, T., and Akiba, E. (2001) *J. Alloys Compd.*, **327**, 275.
21 Sinke, G. C., Walker, L. C., Oetting, F. L., and Stull, D. R. (1967) *J. Chem. Phys.*, **47**, 2759.
22 Sandrock, G., Reilly, J., Graetz, J., Zhou, W.-M., Johnson, J., and Wegrzyn, J. (2005) *Appl. Phys. A*, **80**, 687; Graetz, J. and Reilly, J. J. (2006) *J. Alloys Compd.*, **424**, 262.
23 Saito, H., Machida, A., Katayama, Y., and Aoki, K. (2008) *Appl. Phys. Lett.*, **93**, 151918.

24 Bogdanovic, B., Liao, S.-T., Schwickardi, M., Sikorsky, P., and Spliethoff, B. (1980) *Angew. Chem. Int. Ed. Engl.*, **19**, 818; Bogdanovic, B. (1985) *Angew. Chem. Int. Ed. Engl.*, **24**, 262.

25 Ono, S., Akiba, E., and Imanari, K. (1981) In *Proceedings of the Miami International Symposium on Metal–Hydrogen Systems*, 13–15 April 1981, Miami Beach, FL, p. 467.

26 Barkhordarian, G., Klassen, T., and Bormann, R. (2003) *Scr. Mater.*, **49**, 213.

27 Vajo, J.J., Mertens, F., Ahn, C.C., Bowman, R.C. Jr, and Fultz, B. (2004) *J. Phys. Chem. B*, **108**, 13977; Vajo, J.J., Skeith, S.L., and Mertensohn, F. (2005) *J. Phys. Chem. B*, **109**, 3719.

28 Barkhordarian, G., Klassen, T., Dronheim, M., and Bormann, R. (2007) *J. Alloys Compd.*, **440**, L18.

29 Mori, D., Kobayashi, N., Shinozawa, T., Matsunaga, T., Kubo, H., Toh, K., and Tsuzuki, M. (2005) *J. Jpn. Inst. Met.*, **69**, 308 (in Japanese).

30 Takeichi, N., Senoh, H., Yokota, T., Tsuruta, H., Hamada, K., Takeshita, H.T., Tanaka, H., Kiyobayashi, T., Takano, T., and Kuriyama, N. (2003) *Int. J. Hydrogen Energy*, **28**, 1121.

31 Mori, D., Hirose, K., Komiya, K., Ishikiriyama, M., Haraikawa, N., Toh, K., Fujita, K., Watanabe, S., Miyahara, M., Mikuriya, S., and Tsukuhara, M. (2008) *Abstracts of International Symposium on Metal–Hydrogen Systems, Iceland, 2008*.

20
Complex Hydrides

Andreas Borgschulte, Robin Gremaud, Oliver Friedrichs, Philippe Mauron, Arndt Remhof, and Andreas Züttel

Abstract

The complex hydrides, particularly borohydrides, are currently under discussion as potential solid hydrogen storage materials. Their gravimetric hydrogen density exceeds that of transition metal hydrides by one order of magnitude. The electronic structure of complex hydrides such as $LiBH_4$ differs significantly from that of metallic hydrides, in which the hydrogen atoms occupy the interstitial sites of the metal host lattice. This has severe consequences for their physical and chemical properties such as the crystalline and electronic structure of the compound and ultimately their applicability as hydrogen storage materials. In a broader sense, certain organic hydrides have properties similar to those of complex hydrides and therefore are possibly suitable as hydrogen storage materials. In this chapter, we give an overview on the various materials and hydrogen storage solution based on them.

Keywords: complex hydrides, borohydrides, hydrogen storage, metal hydrides, organic hydrides, solid storage

20.1
Introduction

Intermetallic compounds and alloys based on transition metals are used in specific applications as storage materials (see Table 20.1). These metal hydrides are well established and serve as fast and reliable hydrogen storage materials in (stationary) hydrogen reservoirs (Schlapbach and Züttel, 2001; Schlapbach, 2009) and as hydride electrode materials in rechargeable nickel metal hydride batteries (Willems, 1984). Hydrogen absorption from the gas phase and also from the electrolyte is possible. The maximum storage capacity in these alloys is typically less than 2 mass%. In metallic hydrides, atomic hydrogen occupies the interstitial sites of the metal host lattice. The covalent character of the hydrogen–metal bond is small (Smithson *et al.*, 2002) and therefore hydrogen can diffuse on interstitial sites. Accordingly, the diffusion of hydrogen in some inter-

Hydrogen Energy. Edited by Detlef Stolten
Copyright © 2010 WILEY-VCH Verlag GmbH & Co. KGaA, Weinheim
ISBN: 978-3-527-32711-9

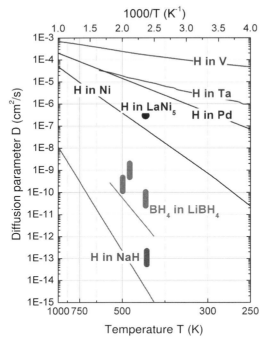

Fig. 20.1 Diffusion of hydrogen in various metal hydrides. V, Ta, Ni, and Pd, Wipf (1997); LaNi$_5$, Schoenfeld et al. (1994); NaH, line DFT calculations, Singh and Ejit (2008), dots tracer diffusion, Borgschulte et al. (2009); LiBH$_4$, dots tracer diffusion, Borgschulte et al., (2008 a, b), line NMR, Corey et al. (2008).

metallic compounds is fast with small activation energies (Fukai, 1993) (Figure 20.1). Together with the catalytically active surface for the hydrogen dissociation, the hydrogen sorption kinetics in such compounds is fast. The hydrogen atom contributes with its electron to the band structure of the metal, and therefore the binding energy of hydrogen can be tailored by partial substitution of the constituent metal atoms.

Complex hydrides exceed the gravimetric hydrogen density of transition metal hydrides by one order of magnitude (see Table 20.1). However, hydrogen in complex hydrides is covalently bound and arranged in subunits such as [AlH$_4$]$^-$ and [BH$_4$]$^-$ ("complexes"; Züttel et al., 2007) with a fixed stoichiometry. It is not known whether hydrogen can be removed from such a subunit without degradation of the whole compound, that is, diffusion and subsequent desorption of hydrogen would then require the movement of the whole subunit and/or degradation of it (Du, 2006). Still, in 1997, Bogdanovic and Schwickardi (1997) demonstrated reversible sorption behavior of catalyzed sodium alanate resulting in isotherms similar to those known from classical metal hydrides. The sorption reaction is only reversible under technically applicable conditions, if the material is doped with transition metal compounds, most efficiently with titanium compounds (Lindsay and Ronge, 1967; Bogdanovic and Schwickardi, 1997, Anton, 2004).

A multitude of experimental and theoretical methods have been applied to unravel the mechanism of the formation and decomposition process of NaAlH$_4$, but so far without a conclusive model on an atomic level. Furthermore, similar kinetically effective additives for other complex hydrides, for example the borohydrides, have not yet been found. The mostly slow kinetics of complex hydrides can be understood as being due to two main effects: (i) the covalent bond and (ii) the decomposition into several (at least two) solid phases. That means that in addition to slow hydrogen diffusion, the essential but very slow metal atom diffusion hinders the formation and decomposition of complex hydrides (Schueth et al., 2004; see also the illustration in Figure 20.2). It has been observed that most non-doped complex hydrides do not decompose at a substantial rate before reaching the melting point (Dymova et al., 1974, 1975; Züttel et al., 2003). Above the melting point, the crystalline order of the ionic constituents is reduced and the cation (e.g., Li$^+$) and the anion (e.g., [BH$_4$]$^-$) units exhibit a significantly increased mobility (Matsuo et al., 2007). Some complex hydrides are already liquid at room temperature, for example Al[BH$_4$]$_3$, and are, therefore, potentially easy to handle hydrogen storage materials. Commercially

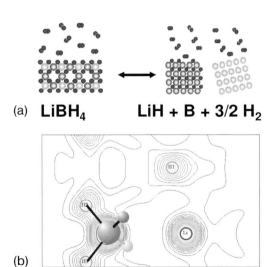

Fig. 20.2 The two most important characteristics of complex hydrides depicted using the example of LiBH$_4$: the "complex" reaction (a), the special electronic structure of a covalently bound pseudo-ion, BH$_4^-$, coordinated with the counterion, Li$^+$, forming an ionic crystal as depicted by an electron distribution plot through a crystal plane including two hydrogens (white), boron (gray) and Li (b) (F. Buchter, to be published).

available borohydrides are synthesized by wet chemical metathesis reactions in organic solvents such as diethyl ether or isopropylamine (Brown et al., 1981), using the high mobility of complex ions in solution (Schlesinger et al., 1953). Furthermore, it has been shown that some solvents enable H–D exchange, that is, they have catalytic properties (Smith et al., 1977). Interestingly, it was found that the formation of certain alanates from the elements is facilitated when performed in solution (e.g., $LiAlH_4$ in THF; Graetz et al., 2008). An alternative to inorganic compounds is the storage of hydrogen in liquid organic compounds, for example, hydrogen absorption by benzene to form cyclohexane. This reaction takes place reversibly for several chemical systems at moderate temperatures with a relatively high storage capacity (see Table 20.1).

Below we discuss the physics, chemistry, and the applicability of complex hydrides and similar covalently bound hydrides. We focus on the following:
- the structure of complex hydrides
- thermodynamics of complex hydrides

Table 20.1 Key parameters of various metal hydrides.

Hydride	Practical gravimetric H density (mass%)	Melting point (°C)	Desorption temperature at 1 bar (°C)	Enthalpy of formation, ΔH (kJ mol^{-1} H)	References
$LaNi_5H_6$	1.25	–	9	−14.75	Percheron et al., 1985
MgH_2	7.6	–	275	−37	Stampfer et al., 1960
Ti-doped $NaAlH_4$	5.6	180	∼190	−35.2, −18.5	Dymova et al., 1974
Ti-doped $NaAlH_4$	∼4	180	∼25	>−35.2, >−18.5	Streukens, 2006
$LiBH_4$	13.8	270	∼300	−37	Züttel et al., 2003 Mauron et al., 2008
$Ca(BH_4)_2$	8.6	295	∼320	−16	Mikheeva, 1979 Miwa et al., 2006
$Al(BH_4)_3$	16.9	−64.5	20	−3	Schlesinger et al., 1940 Miwa et al., 2006
$LiNH_2$	6.5	–	∼260	−40.5	Chen et al., 2002
Methyl-cyclohexane	6	−126	250–400	−35.9	Taube et al., 1983

- organic hydrides as hydrogen storage materials
- hydrogen storage systems using complex and organic hydrides.

20.2
The Structure of Complex Hydrides

The lightweight alkali and alkaline earth metal borohydrides and alanates are crystalline solids, in which the negative $[BH_4]^-$ and $[AlH_4]^-$ ions are ionically bound to a metal cation. The four hydrogen atoms surrounding the central atom form a regular tetrahedron which can be envisioned as being embedded inside a cube. Thereby, each vertex of the tetrahedron is a vertex of the cube, and each edge is a diagonal of one of the cube's faces. Two sets of high-symmetry axes are connected with this structure. First, there are three twofold axes, called c2, which are normal to the cube's face, and four threefold axes, called c3, which coincide with the body diagonals of the cube. Both symmetry operations map the tetrahedron to it. There are two possibilities to match the four vertices of a tetrahedron to the eight vertices of the cube. These two possible arrangements may be converted into one another by a 90° rotation around a c2 axis, while a rotation around c3 preserves the tetrahedron's orientation.

Within the anion, the negative charge is carried by the hydrogen. Apart from the charge transfer from the metal, there is an additional electron transfer from the central atom to the H. In $LiBH_4$ for example, this results in an effective charge of about 1.5 electrons per H atom (F. Buchter, to be published). The attractive Coulomb interaction between the ions would favor a tridentate configuration, in which three H atoms face the neighboring metal ion. This configuration, however, cannot be achieved simultaneously for all ions, and the system is frustrated. Depending on the size and the valence of the metal ion, different crystallographic structures are realized as a compromise between the frustration, the Coulomb interaction, and the entropy which originates from the disorder of the anions' arrangements. Most of the materials undergo structural phase transitions. For some of them, even different structural allotypes are known to coexist at room temperature.

Under ambient conditions, the borohydrides of the monovalent alkali metals crystallize in a cubic NaCl-like structure. In this case, the symmetry axes of the BH_4 tetrahedron coincide with the low-index crystallographic axes. The c_3 axes are thereby aligned with the [100] axes, whereas the c3 axes lie along the [111] axis, which are the body diagonals of the cubic lattice. $LiBH_4$ is an exception to this rule. At room temperature, it crystallizes in an orthorhombic structure. Heat capacity studies of the MBH_4 series suggest phase transitions at 190 K for $NaBH_4$, 76 K for KBH_4, 44 K for $RbBH_4$, and 27 K for $CsBH_4$ (Stevenson et al., 1955). The low-temperature phases of $NaBH_4$ and KBH_4 have been determined to be ordered tetragonal structures (Fischer and Züttel, 2004; Renaudin et al., 2004). In the case of $RbBH_4$ and $CsBH_4$, no diffraction evidence for a lowering of structural symmetry has been found (Renaudin et al., 2004).

LiBH$_4$ undergoes a structural phase transition to a hexagonal high-temperature (HT) structure at around 380 K (Soulie et al., 2002). Diffraction studies brought to the fore anomalously high hydrogen thermal displacement amplitudes in the HT phase (Hartman et al., 2007, Buchter et al., 2008), which were attributed to dynamic disorder in this phase (Hagemann et al., 2004). The disorder also strongly influences the phonon spectrum of the crystal. The phonon density of states (PDOS) of the low-temperature (LT) phase depends quadratically on the phonon energy, as expected from the Debye theory for acoustic vibrations (harmonic oscillator) in crystalline solids, whereas the HT phase shows a linear dependence of the PDOS as a function of the phonon energy. The excess of density of states of the HT phase reveals a high lattice anharmonicity and it is a characteristic feature of glasses and disordered systems (Buchter et al., 2008). The dynamics of LiBH$_4$ have also been studied by NMR spectroscopy. Recent studies by Matsuo et al. (2007) and Corey et al. (2008) revealed a sudden line narrowing of the ^{7}Li resonance at the phase transition point, indicative of rapid lithium diffusion in the HT phase. Thereby, hopping rates of 10^9 s^{-1} have been observed. The NMR-determined hopping is in agreement with large ionic electrical conductivity (Matsuo et al., 2007).

Apart from the ambient pressure phases, several high-pressure phases have recently been revealed for LiBH$_4$ (Filinchuk et al., 2007) and NaBH$_4$ (Filinchuk et al., 2008). All phase transitions in the monovalent borohydrides were reported to be fully reversible.

The divalent alkaline earth metal borohydrides crystallize in more complex crystal structures with unit cells comprising up to 330 atoms (Ozolins et al., 2008). Depending on the sample preparation and treatment, several phases have been found to coexist at room temperature (Her et al., 2007; Buchter et al., 2009). Even though several phase transitions have been observed in the past, the complete phase diagrams are still to be determined. Analogous to the behavior observed for the alkali metal borohydrides, reorientational motions of the [BH$_4$]$^-$ ion have been observed (Fichtner et al., 2008).

20.3
Thermodynamics of Complex Hydrides

In contrast to the interstitial hydrides, where the metal lattice hosts the hydrogen atoms on interstitial sites, desorption of the hydrogen from the complex hydride leads to decomposition of the complex hydride and a mixture of at least two phases in addition to hydrogen is formed. For alkali metal borohydrides and alanates, the decomposition reaction is described by the following equations:

$$A(BH_4) \rightarrow (AH + \text{``}BH_3\text{''} \rightarrow) AH + B + \tfrac{3}{2}H_2 \tag{20.1}$$

and

$$A(AlH_4) \rightarrow (AH + \text{"}AlH_3\text{"} \rightarrow) \; ^1\!/_3 A_3AlH_6 + {^2\!/_3} AH$$
$$\rightarrow {^1\!/_3} A_3AlH_6 + {^2\!/_3} Al + H_2 \rightarrow AH + Al + {^3\!/_2} H_2 \tag{20.2}$$

In the case of the alanates, the hexahydride phase as an intermediate product has been identified experimentally; however, the existence of an intermediate phase in the case of the borohydrides and in particular its structure is still under discussion (Ohba et al., 2006; Orimo et al., 2006; Züttel et al., 2007; Hwang et al., 2008; Friedrichs et al., 2008).

Amides release ammonia during decomposition:

$$2A(NH_2) \rightarrow A_2NH + NH_3 \tag{20.3}$$

If mixed with the corresponding hydride, the following reaction:

$$AH + A(NH_2) \rightarrow A_2NH + H_2 \tag{20.4}$$

is reversible and formation of ammonia is suppressed. The existence of similar fluent intermediates such as B_2H_6 and AlH_3 during decomposition of borohydrides and alanates, respectively, is discussed later. In hydrogen storage in metal hydrides, we make use of the fact that the chemical potential of gaseous hydrogen is equal to that of hydrogen dissolved in the metal and to hydrogen as the metal hydride for hydrogen concentration in the two-phase regime. This leads to a constant equilibrium pressure of hydrogen at constant temperatures ("plateau pressure"). The measurement of pressure–composition isotherms (pcTs) was only recently achieved for complex hydrides, revealing the existence of a two-phase regime for the decomposition and formation of complex hydrides ($NaAlH_4$, Dymova et al., 1975; Bogdanovic and Schwickardi, 1997; $LiBH_4$: Mauron, 2008; $LiNH_2$, David et al., 2007). The reason is their rather slow kinetics, thus requiring elevated temperatures and/or catalysts in order to reach quasi-equilibrium conditions.

The explanation for the plateaux is similar to that for metal hydrides: the chemical potentials of hydrogen in the various phases remain equal, and only the amount of the corresponding phases change. Thus, from isotherms determined at various temperatures, van't Hoff plots of the corresponding plateau pressures reveal the full set of thermodynamic parameters, that is, enthalpy of reaction and entropy of reaction. Table 20.1 lists the enthalpy of desorption ΔH of selected complex hydrides.

The thermodynamic characterization by means of isotherm measurements has been reported for only a few complex hydrides. An empirical model has been developed to describe the stability of complex hydrides. The stability of metal borohydrides is in relation to the localization of the charge on the anion. Nakamori et al. (2006) calculated the enthalpy of formation of a series of borohydrides by means of density functional theory and found a linear relationship between the enthalpy of formation and the electronegativity. Furthermore, the experimentally measured decomposition temperature of the borohydrides decreases with increasing electronegativity (Nakamori et al., 2006), and Nickels et al. (2008) demonstrated that mixed metal borohydrides may be synthesized with decomposition properties that approach the average of those of the single metal borohydrides. The decomposition temperature is the temperature at which hydrogen gas with a pressure of 1 bar is in equilibrium with the solid hydride ($T_{dec} = \Delta H/\Delta S$). The alanates are very much less stable than the corresponding borohydrides but show a similar correlation with the electronegativity of the cation. Furthermore, the difference between the stabilities of the alanates and borohydrides is explained by the different Pauling electronegativities of B and Al, namely 2.04 and 1.61, respectively. For metallic hydrides, the enthalpy of formation, that is, the enthalpy difference between the hydride and the intermetallic compound, is equivalent to the enthalpy for the hydrogen desorption reaction besides inelastic lattice effects. For complex hydrides, the enthalpy of formation, that is, the enthalpy difference between the hydride and the constituent elements, is significantly different from the enthalpy of hydrogen desorption, because of the formation of the hydride of the cation or another intermediate hydride.

In order to desorb the hydrogen at ambient temperature and 1 bar pressure, the enthalpy of desorption of the hydride should be $\Delta H \approx -40$ kJ mol^{-1} H$_2$ (see Table 20.1). New quaternary borohydrides may be tailored by the change of the composition to the appropriate thermodynamics. A different approach is the utilization of composites made of existing binary and complex hydrides, which react with each other upon hydrogen desorption and form a new compound or alloy and thereby alter the desorption enthalpy (Reilly and Wiswall, 1967). Furthermore, the LiNH$_2$–LiH system is a system that is reversible only as a composite (Equation 20.4). This system is still too stable, which results in a too low plateau pressure at ambient operating temperatures (compare Figure 20.3, Table 20.1). The plateau pressure of the Mg-substituted amide system LiNH$_2$–MgH$_2$ is about 30 bar at 200 °C and hence it is less stable with respect to the pure Li-based system (Luo, 2004). However, the hydrogen storage density is limited to 4.5 mass%. Composites made of borohydrides with binary hydrides, the so-called reactive hydride composites (RHCs) (Barkhordarian et al., 2004; Vajo et al., 2004) have been described. RHCs show not only the expected thermodynamic destabilization, but also an improvement of absorption kinetics. Referring to the composite MgH$_2$ + 2LiBH$_4$, during the endothermic dehydrogenation reaction the exothermic formation of MgB$_2$ proceeds and thereby the total reaction enthalpy is lowered. Absorption proceeds again from MgB$_2$ and LiH (Boesenberg et al., 2007). In spite of the lowered value of the reaction enthalpy in LiBH$_4$ + MgH$_2$, rehydrogenation is possible at a pressure of 50 bar hydrogen and temperatures below 300 °C, which are much more moderate conditions than in the pure LiBH$_4$ system (Friedrichs, 2008). Several other systems show similar behavior (see, e.g., Barkhordarian et al., 2007, Siegel et al., 2007, Remhof et al., 2009).

Although significant progress has been made in enhancing the hydrogen sorption kinetics, the ambitious goal of room temperature hydrogen storage in

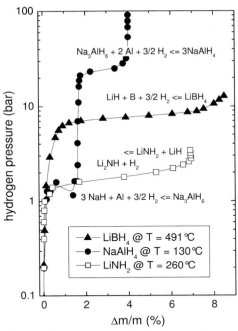

Fig. 20.3 Pressure composition isotherms of various complex hydrides: Ce-doped NaAlH$_4$ (Lohstroh and Fichtner, 2007), LiBH$_4$ (Mauron et al., 2008), and LiNH$_2$ + LiH (David et al., 2007).

complex metal hydrides has not been achieved so far. There are two main routes for further improving hydrogen sorption kinetics: nano-structuring and catalytic additives. The first route is in principle straightforward: the impact of the low diffusivity of hydrogen and metal atoms is reduced if the transport pathways are as short as possible. However, the experimental implementation is challenging. In most cases, comminution by, for example, ball-milling has no great effect, if the initial morphology can change during cycling (Fichtner, 2005). During the phase formation of hydrides and metals, the corresponding grain sizes equilibrate to the values they would reach without artificial nano-structuring (Fichtner, 2005). To overcome this problem, nano-scaffolds have been proposed, hindering the separation of phases. Recent research on LiBH$_4$ confined in activated carbon or mesoporous silica (Kostka et al., 2007; Vajo and Olson, 2007) or NaAlH$_4$ in mesoporous carbon aerogel (Balde et al., 2008) or silica (Zheng et al., 2008) showed promising lowering of the desorption temperature and faster desorption of hydrogen. The second route towards room temperature storage is the use of additives (catalysts). Hydrogen storage in NaAlH$_4$, for example, is only reversible at technically interesting reaction rates if the system is doped with transition metal compounds, most efficiently with titanium compounds. (Lindsay and Rouge, 1967; Bogdanovic and Schwickardi, 1997; Anton, 2004) Although many empirical studies have confirmed the extraordinarily

high activity of these Ti compounds and kinetics of the catalyzed systems reaching technical needs have been achieved, the origin of the high activity of the catalysts still remains elusive. This may also explain why no similarly active catalyst for borohydrides has been found yet. The catalyst may affect all elementary steps involved, namely by:
- supporting dissociation/recombination of hydrogen (Bellosta von Colbe et al., 2006; Borgschulte et al., 2008b)
- supporting formation/splitting of intermediates such as alane (AlH_3) (Chaudhuri et al., 2006; Fu et al., 2006) or borane species (BH_x) (Friedrichs et al., 2008)
- supporting diffusion of hydrogen in the corresponding phases (Palumbo et al., 2005; Singh et al., 2007)
- supporting diffusion of metal atoms (Bogdanovic et al., 2007)
- reducing nucleation barriers of new phases (Leon et al., 2006)
- interfering with the anion-cation charge transfer (Li^+–BH_4^-) (Wellons et al., 2009).

Indications for each of the effects have been found, but without finding a conclusive model of the mechanism(s). There is common belief that reactions of hydrogen with boron or aluminum are in principle slow and therefore the formation of B–H and Al–H bonds is the rate-limiting step for the formation of borohydrides. This implies that once the B–H and Al–H bonds have been formed, the reaction proceeds without major constraints. Indeed, nearly 60 years ago Schlesinger et al. (1953) showed the formation of $LiBH_4$ from the reaction of LiH with diborane (B_2H_6) as a suspension in diethyl ether at room temperature. Recent research reproduced this result also for the solid synthesis of $LiBH_4$ from B_2H_6 and LiH (Friedrichs et al., 2009). Similarly, alane (AlH_3) clusters, once formed, rapidly react with NaH to form $NaAlH_6$ and $NaAlH_4$ phases (Chaudhuri et al., 2006). There is strong evidence that Ti on Al surfaces facilitates the formation of alane, and thereby catalyzes the formation of $NaAlH_4$. Unfortunately, a similarly effective catalyst for B–H has not yet been found. Furthermore, in contrast to the thermal desorption of hydrogen from alanates, where pure hydrogen gas is observed, the desorption product from borohydrides is often a mixture of hydrogen, diborane, and higher boranes (Kostka et al., 2007). A catalyst for borohydride should therefore promote the formation of the B–H bond during the formation process and the splitting during desorption.

20.4
Organic Hydrides for Hydrogen Storage

Storing hydrogen in the form of a liquid organic hydride is a relatively new approach to storing hydrogen, which offers some unique advantages. In the past, the use of liquid organic hydrides did not attract much attention as the hydrogen liberation step (dehydrogenation) from organic molecules is strongly endothermic (Taube et al., 1983), thereby requiring much higher reaction temperatures (600–700 K) than the operating conditions of fuel cells (Coughlan and Keane, 1990). Recently, the investigation of organic molecules with heteroatom(s) for hydrogen storage opened up the possibility of thermodynamic tailoring (Pez et al., 2006, 2008). One of the prototype compounds as the liquid organic hydride is 9-ethylcarbazole with a hydrogen capacity of 5.7 mass%. It was found that the incorporation of nitrogen and fused rings into this organic structure can drastically decrease the enthalpy of dehydrogenation, giving a method to tailor the thermodynamics for reversible hydrogen storage (Clot et al., 2007). Various organic hydrogen storage systems with potentially commercially applicable properties are currently under discussion, in detail:

- decalin ($C_{10}H_{18}$) $\rightleftharpoons$ naphthalene ($C_{10}H_8$) + 5 H_2, $\Delta H = 59$ kJ mol^{-1} H_2, 7.3 mass% (Hodoshima et al., 2005);
- n-heptane (C_7H_{16}) $\rightleftharpoons$ toluene (C_7H_8) + 4 H_2, $\Delta H = 63$ kJ mol^{-1} H_2, 8 mass% (Newson et al., 1998);
- methylcyclohexane (C_7H_{14}) $\rightleftharpoons$ toluene (C_7H_8) + 3 H_2, $\Delta H = 72$ kJ mol^{-1} H_2, 6 mass% (Taube et al., 1983);
- perhydroethycarbazole $\rightleftharpoons$ ethylcarbazole + 6 H_2, $\Delta H = 25$ kJ mol^{-1} H_2, 5.7 mass% (Pez et al., 2006, 2008; Morawa et al., 2009);
- formic acid (HCOOH) $\rightleftharpoons$ CO_2 + H_2, $\Delta H = 32$ kJ mol^{-1} H_2 (Morrison and Boyd, 1972), 4.3 mass% (Fellay et al., 2008).

Similarly to complex hydrides, the catalysis of formation/decomposition of the C–H bond is the main challenge due to the strong covalent nature of the hydrogen bond (Taube et al., 1983, Züttel et al., 2008; Lu et al., 2009). On the other hand, the hydrogenation of unsaturated hydrides is a widely used process in the petrochemical industry, thus providing the technology for such systems (e.g., Ho, 2004).

20.5
Hydrogen Storage Systems Using Complex and Organic Hydrides

20.5.1
Hydrogen Storage in Solid Complex Hydrides

The functioning of hydrogen storage tanks based on classical transition metal-based alloys (e.g., AB5 alloys) or doped sodium alanate has been extensively demonstrated in the past. The huge amount of heat released during fueling of a tank large enough to drive a car (up to 500 kW; Yang et al., 2009) requires a sophisticated cooling–heating system with corresponding costs and energy losses (Lozano et al., 2009). The challenge lies in the principle of the on-board hydrogen exchange in hydrogen storage materials, which is basically a chemical reaction accompanied by heat exchange. As the heat of formation is related to the equilibrium pressure of a hydrogen storage material, its optimum is fixed at

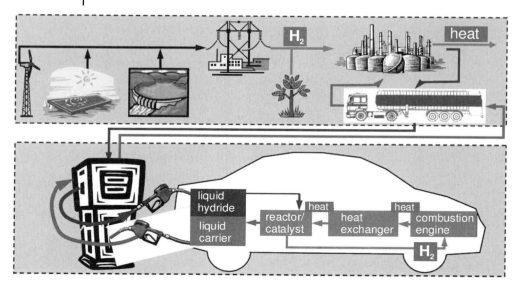

Fig. 20.4 Hydrogen storage system using in liquid hydrides and liquid carriers. From von Wild et al. (2008).

$-40 \, \text{kJ mol}^{-1}$ H_2, and also after optimization towards this value the heat exchange problem remains severe (Yang et al., 2009).

20.5.2
Hydrogen Storage in Liquid Hydrides

The use of liquid or fluid hydrogen carrier allows fueling to be separated from formation of the hydride. In 1984, a 16 ton demonstration truck was powered by hydrogen produced from the liquid hydrogen carrier methylcyclohexane on board the truck and directly coupled to the combustion engine (Taube et al., 1983), and the truck was refueled in a conventional manner. More recently, the idea has been elaborated into a system using renewable hydrogen stored and transported as liquid hydrides (see Figure 20.4) (see also Toseland, 2009). Hydrogen absorption will then be carried out in centralized locations, and "hydrogen fueling" will take place in a manner similar to fueling liquid gasoline. Hydrogen desorption from the liquid hydride will take place in on-board reactors. The spent fuel will be exchanged with regenerated liquid carrier during fueling. The main advantages of this liquid hydride system are:
- Only slight changes to the existing infrastructure are needed, in particular the use of existing fuel stations.
- There are no severe requirements on liquid hydride/carrier tank, in particular no waste heat problem during fueling.

- Heat evolved during hydrogen absorption is processed on a large scale and can therefore be used in households/industry.
- Heat needed for desorption taking place in a reactor is taken from the waste heat of hydrogen combustion in combustion engines or fuel cells.

References

Anton, D. L. (2004) *J. Alloys Compd.*, **356**, 400–404.

Balde, C. P., Hereijgers, B. P. C., Bitter, J. H., and de Jong, K. P. (2008) *J. Am. Chem. Soc.*, **130**, 6761–6765.

Barkhordarian, G., Klassen, T., Bormann, R. (2004) International Patent Application WO 2006/063627 (priority date 2004).

Barkhordarian, G., Klassen, T., Dornheim, M., and Bormann, R. J. (2007) *J. Alloys Compd.*, **440** (1–2), L18–L21.

Bellosta von Colbe, J. M., Schmidt, W., Felderhoff, M., Bogdanovic, B., and Schueth, F., (2006) *Angew. Chem. Int. Ed.*, **45** (22), 3663–3665.

Bösenberg, U., Doppiu, S., Mosegaard, L., Barkhordarian, G., Eigen, N., Borgschulte, A., Jensen, T. R., Cerenius, Y., Gutfleisch, O., Klassen, T., Dornheim, M., and Bormann, R. (2007) *Acta Mater.*, **55** 3951–3958.

Bogdanovic, B. and Schwickardi, M. J. (1997) *J. Alloys Compd.*, **253**, 1.

Bogdanovic, B., Eberle, U., Felderhoff, M., and Schueth F. (2007) *Scr. Mater.*, **56**, 813.

Borgschulte, A., Hug, P., Züttel, A., Racu, A., and Schoenes, J. (2008a) *J. Phys. Chem. A*, **112**, 4749–4753.

Borgschulte, A., Züttel, A., Hug, P., Barkhordarian, G., Eigen, N., Dornheim, M., Bormann, R., and Ramirez-Cuesta, A. J. (2008b) *Phys. Chem. Chem. Phys.*, **10**, 4045–4055.

Borgschulte, A., Pendolino, F., Gremaud, R., and Züttel, A. (2009) *Appl. Phys. Lett.*, **94**, 111907.

Brown, H. .C., Choi, Y. M., and Narasimhan, S. (1981) *Inorg. Chem.*, **20**, 4454–4456.

Buchter, F., Łodziana, Z., Mauron, Ph., Remhof, A., Friedrichs, O., Borgschulte, A., Züttel, A., Sheptyakov, D., Strässle, T., and Ramirez-Cuesta, A. J. (2008) *Phys. Rev. B*, **78**, 094302.

Buchter, F., Łodziana, Z., Remhof, A., Friedrichs, O., Borgschulte, A., Mauron, Ph., Züttel, A., Sheptyakov, D., Palatinus, L., Ch^{3}opek, K., Fichtner, M., Barkhordarian, G., Bormann, R., and Hauback, B. C. (2009) *J. Phys. Chem. C*, **113**, 17223.

Chaudhuri, S., Graetz, J., .Ignatov, A., Reilly, J. J., and Muckerman, J. T. (2006) *J. Am. Chem. Soc.*, **128**, 11404.

Chen, P., Xiong, Z., Luo, J., Lin, J., and Tan, L. (2002) *Nature*, **420**, 302.

Clot, E., Eisensteinand, O., and Crabtree, R. H. (2007) *Chem. Commun.*, **22**, 2231–2233.

Corey, R. L., Shane, D. T., Bowman, R. C. Jr, and Conradi, M. S. (2008) *J. Phys. Chem. C*, **112** (47), 18706.

Coughlan, B. and Keane, M. A. (1990) *Catal. Lett.*, **5**, 89–100.

David, W. I. F., Jones, M. O., Gregory, D. H., Jewell, C. M., Johnson, S. R., Walton, A., and Edwards, P. P. (2007) *J. Am. Chem. Soc.*, **129**, 1594–1601.

Du, A. J., Smith, S. C., and Lu, G. Q. (2006), *Phys. Rev. B*, **74**, 193405.

Dymova, T. N., Eliseeva, N. G., Bakum, S. I., and Dergachev, Yu. M. (1974) *Dokl. Akad. Nauk SSSR*, **215**, 1369; Engl., 256.

Dymova, T. N., Dergachev, Yu. M., Sokolov, V. A., and Grechanaya, N. A. (1975) *Dokl. Akad. Nauk SSSR*, **224**, 591; Engl., 556.

Fellay, C., Dyson, P. J., and Laurenczy, G. (2008) *Angew. Chem. Int. Ed.*, **47**, 3966.

Fichtner, M. (2005) *Adv. Eng. Mater.*, **7**, 443.

Fichtner, M., Chlopek, K., Longhini, M., and Hagemann, H. (2008) *J. Phys. Chem. C*, **112**, 11575.

Filinchuk, Y., Talyzin, A., Chernyshov, D., and Dmitriev, V. (2007) *Phys. Rev. B*, **76**, 092104.

Filinchuk, Y., Chernyshov, D., Nevidomskyy, A., and Dmitriev, V. (2008) *Angew. Chem. Int. Ed.*, **47**, 529.

Fischer, P. and Züttel, A. (2004) *Mater. Sci. Forum*, **443–444**, 287.

Friedrichs, O., Buchter, F., Borgschulte, A., Remhof, A., Zwicky, C. N., Mauron, Ph., Bielmann, M., and Zütte, A. (2008) *Acta Mater.*, **56**, 949–954.

Friedrichs, O., Borgschulte, A., Kato, S., Buchter, F., Gremaud, R., Remhof, A., and Züttel, A. (2009) *Chem. Eur. J.*, **15**, 5531.

Fu, Q. J., Ramirez-Cuesta, A. J., and Chi, S. (2006) *J. Phys. Chem. B*, **110**, 711–715.

Fukai, Y. (1993) *The Metal-Hydrogen System, Basic Bulk Properties*, Springer Series in Materials Science, Vol. 21, Springer, Berlin.

Graetz, J., Wegrzyn, J., and Reilly, J. J. (2008) *J. Am. Chem. Soc.*, **130** (52), 17790–17794.

Hagemann, H., Gomes, S., Renaudin, G., and Yvon, K. (2004) *J. Alloys Compd.*, **363**, 129.

Hartman, M., Rush, J., Udovic, T., Bowman, R. Jr, and Hwang, S.-J., (2007) *J. Solid State Chem.*, **180**, 1298.

Her, J.-H., Stephens, P. W., Gao, Y., Soloveichik, G. L., Rijssenbeek, J., Andrus, M., and Zhao, J.-C. (2007) *Acta Crystallogr., Sect. B*, **63**, 561.

Ho, T. C. (2004) *Catal. Today*, **98**, 3–18.

Hodoshima, S., Takaiwa, S., Shono, A., Satoh, S., and Saito, Y. (2005) *Appl. Catal. A*, **283**, 235.

Hwang, S.-J., Bowman, R. C., Reiter, J. W., Rijssenbeek, J., Soloveichik, G. L., Zhao, J.-C., Kabbour, H., and Ahn, C. C. (2008) *J. Phys. Chem. C*, **112**, 3165.

Johnson, R., David, W. I. F., Jones, M. O., and Edwards, P. P., (2009) *Chem. Asian J.*, **4** (6), 849–854.

Kostka, J., Lohstroh, W., Fichtner, M., and Hahn, H. (2007) *J. Phys. Chem. C*, **111**, 14026.

Leon, A., Kircher, O., Fichtner, M., Rothe, J., and Schild, D. (2006) *J. Phys. Chem. B*, **110**, 1192.

László, H., and Tabor, M. (2002) *Appl. Catal. A*, **226**, 319–322.

Lindsay, K. L. and Rouge, B. (1967) Preparation of alkali metal hydrides, US Patent 3 505 036.

Lohstroh, W., and Fichtner, M. (2007) *Phys. Rev. B*, **75**, 184106.

Lozano, G., Eigen, N., Keller, C., Dornheim, M., and Bormann, R. (2009) *Int. J. Hydrogen Energy*, **34**, 1896.

Lu, R.-F., Boethius, G., Wen, S. H., Su, Y., and Deng, W.-Q. (2009) *Chem. Commun.*, 1751.

Luo, W. (2004) *J. Alloys Compd.*, **381**, 284; **385**, 316.

Matsuo, M., Nakamori, Y., Orimo, S.-I., Maekawa, H., and Takamura, H. (2007) *Appl. Phys. Lett.*, **91**, 224103.

Mauron, Ph., Buchter, F., Friedrichs, O., Remhof, A., Bielmann, M., Zwicky, C. N., and Züttel, A. (2008) Stability and reversibility of LiBH$_4$, *J. Phys. Chem. C*, **112**, 906.

Mitcheeva, V. J., Maltsera, N. N., and Kedrova, N. S. (1979) *Russian Journal of Inorganic Chemistry*, **24**, 225.

Miwa, K., Aoki, M., Noritake, T., Ohba, N., Nakamori, Y., Towata, S., Züttel, A., and Orimo, S. (2006) *Phys. Rev. B*, **74**, 155122.

Morawa, K., Rentsch, D. L., Friedrichs, O., Züttel, A., Ramirez-Cuesta, A. J. T., and Tsang, S. C. (2009) *Int. J. Hydrogen Energy*, in press.

Morrison, R. T. and Boyd, R. N. (1972) *Organic Chemistry*, 2nd edn, Allyn and Bacon, Boston, p. 596.

Nakamori, Y., Miwa, K., Ninomiya, A., Li, H., Ohba, N., Towata, S. I., Züttel, A., and Orimo, S. I. (2006) *Phys. Rev. B*, **74**, 045126.

Newson, E., Haueter, Th., Hottinger, P., Von Roth, F., Scherer, G. W. H., and Schucan, T. H. (1998) *J. Hydrogen Energy*, **23**, 905.

Nickels, E. A., Jones, M. O., David, W. I. F., Johnson, S. R., Lowton, R. L., Sommariva, M., and Edwards, P. P. (2008) *Angew. Chem. Int. Ed.*, **47**, 1–4.

Ohba, N., Miwa, K., Aoki, M., Noritake, T., and Towata, S.-I. (2006) *Phys. Rev. B*, **74**, 075110.

Orimo, S.-I., Nakamori, Y., Ohba, N., Miwa, K., Aoki, M., Towata, S.-I., and Züttel, A. (2006) *Appl. Phys. Lett.*, **89**, 021920.

Ozolins, V., Majzoub, E. H., and Wolverton, C. (2008) *Phys. Rev. Lett.*, **100**, 135501.

Palumbo, O., Cantelli, R., Paolone, A., Jensen, C. M., and Srinivasan, S. S. (2005) *J. Phys. Chem. B*, **109**, 1168.

Percheron-Guégan, A., Lartigue, C., and Achard, J. C. (1985) *J. Less-Common Met.*, **109**, 287.

Pez, G. P., Scott, A. R., Cooper, A. C., and Cheng, H. (2006) US Patent 7101530.

Pez, G. P., Scott, A. R., Cooper, A. C., Cheng, H., Wilhel, F. C., and Habdourazak, A. (2008) US Patent 7351395.

Reilly, J. J. and Wiswall, R. H. (1967) *Inorg. Chem.*, **6**, 2220.

Remhof, A., Friedrichs, O., Buchter, F., Mauron, Ph., Kim, J. W., Oh, K. H., Buchsteiner, A., Wallacher, D., and Züttel, A. (2009) *J. Alloys Compd.*, **484**, 654–659.

Renaudin, G., Gomes, S., Hagemann, H., Keller, L., and Yvon, K. (2004) *J. Alloys Compd.*, **375**, 98.

Schlapbach, L. (2009) *Nature*, **460**, 809–811.

Schlapbach, L. and Züttel, A. (2001) *Nature*, **414**, 353.

Schlesinger, H. I., Sanderson, R. T., and Burg, A. B. (1940) *J. Am. Chem. Soc.*, **62**, 3421.

Schlesinger, H. I., Brown, H. C., Abraham, B., Bond, A. C., Davidson, N., Finholt, A. E., Gilbreath, J. R., Hoekstra, H., Horvitz, L., Hyde, E. K., Katz, J. J., Knight, J., Lad, R. A., Mayfield, D. L., Rapp, L., Ritter, D. M., Schwartz, A. M., Sheft, I., Tuck, L. D., and Walker, A. O. (1953) *J. Am. Chem. Soc.*, **75**, 186.

Schönfeld, C., Hempelmann, R., Richter, D., Springer, T., Dianoux, A. J., Rush, J. J., Udovic, T. J., and Bennington, S. M. (1994) *Phys. Rev. B*, **50**, 853.

Schüth, F., Bogdanović, B., and Felderhoff, M. (2004) *Chem. Commun.*, 2249.

Siegel, D. J., Wolverton, C., and Ozolins, V. (2007) *Phys. Rev. B*, **76**, 134102.

Singh, S. and Eijt, S. W. H. (2008) *Phys. Rev. B*, **78**, 224110.

Singh, S., Eijt, S. W. H., Huot, J., Kockelmann, W. A., Wagemaker, M., and Mulder, F. M. (2007) *Acta Mater.*, **55**, 5549–5557.

Smith, E., James, B. D., and Peachey, R. M. (1977) *Inorg. Chem.*, **16**, 2057.

Smithson, H., Marianetti, A., Morgan, D., Van der Ven, A., Predith, A., and Ceder, G. (2002) *Phys. Rev. B*, **66**, 144107.

Soulié, J.-Ph., Renaudin, G., Cerny, R., and Yvon, K. (2002) *J. Alloys Compd.*, **346**, 200.

Stampfer, J. F., Holley, C. E., and Suttle, J. F. (1960) The magnesium–hydrogen system, *J. Am. Chem. Soc.*, **82**, 3504.

Stephenson, C. C., Rice, D. W., and Stockmayer, W. H. (1955) *J. Chem. Phys.*, **23**, 1960.

Streukens, G., Bogdanovi, B., Felderhoff, M., and Schueth, F. (2006) *Phy. Chem. Chem. Phys.*, **8**, 2889.

Taube, M., Rippin, D. W. T., Cresswell, D. L., and Knecht, W. (1983) *Int. J. Hydrogen Energy*, **8**, 213.

Toseland, B. (2009) *APCI, Reversible Liquid Carriers for an Integrated Production, Storage and Delivery of Hydrogen, DOE Reviews*, http://www.hydrogen.energy.gov/pdfs/review09/pd_38_toseland.pdf (accessed 09/2009).

Vajo, J. J. and Olson, G. L. (2007) *Scr. Mater.*, **56**, 829–834.

Vajo, J. J., Mertens, F., Ahn, C. C., Bowman, R. C. Jr, and Fultz, B. (2004) *J. Phys. Chem. B*, **108**, 13977.

von Wild, J., Freymann, R., and Zenner, M. (2008) Potentiale von alternativen Wasserstoff-Speicherungstechnologien, *VDI Tagung, „Innovative Antriebe"*, Dresden, 2008.

Wellons, M. S., Berseth, P. A., and Zidan, R. (2009) *Nanotechnology*, **20**, 204022.

Willems, J. J. G. (1984) *Philips Res.*, 39.

Wipf, H. (1997) Diffusion of hydrogen in metals, *Top. Appl. Phys.*, **73**, 52

Yang, J., Sudik, A., Wolverton, C., and Siegel, D. J. (2010) *Chem. Soc. Rev.*, **39**, 656–675.

Zheng, S., Fang, F., Zhou, G., Chen, G., Ouyang, L., Zhu, M., and Sun, D. (2008) *Chem. Mater.*, **20**, 3954–3958.

Züttel, A., Wenger, P., Rensch, S., Sudan, P., Mauron, Ph., and Emmenegger, C. (2003) *J. Power Sources*, **118**, 1.

Züttel, A., Borgschulte, A., and Orimo, S.-I. (2007) *Scr. Mater.*, **56**, 823.

Borgschulte A., Goetze, S., and Züttel, A. (2008) *Hydrogen as a Future Energy Carrier*, Wiley-VCH Verlag GmbH, Weinheim, p. 241.

21
Adsorption Technologies

Barbara Schmitz and Michael Hirscher

Abstract

One promising possibility to store hydrogen is cryosorption of hydrogen molecules on nanoporous materials possessing large internal surfaces. The physisorption process is fast and fully reversible and, therefore, short refuelling times can be realized. This chapter introduces the basics of hydrogen adsorption and various classes of nanoporous adsorbents including the most advanced materials. The influence of the specific surface area, the pore size and the chemical composition on the hydrogen storage properties are discussed. Finally, the relevant parameters for technical applications and the requirements to possible storage materials are highlighted.

Keywords: adsorption, hydrogen adsoprtion, hydrogen storage, physisorption

21.1
Adsorption

Adsorption occurs when two non-polar atoms or molecules approach each other. The two particles feel a long-range non-dispersive attraction, called van der Waals interaction, which is caused by fluctuations in the electron shell. Oncoming closer a short-range repulsion force dominates the interaction, which is caused by an overlap of the eigenfunctions of these particles. If a molecule or atom is approaching the surface of a material it will remain at the equilibrium distance at which the attractive and repulsive forces between the surface atoms and the particle compensate each other. This phenomenon is called (physical) adsorption or physisorption as no chemical binding is involved. The free particle is called adsorptive, the material on which it is adsorbed is the adsorbent, and the adsorbed particles form the adsorbate. The number of particles that can be adsorbed on a defined specific surface area (SSA) depends on the adsorptive, the adsorbent, temperature, and pressure. For materials possessing large SSAs, adsorption can be technologically applied to store gases in solids. This storage

Hydrogen Energy. Edited by Detlef Stolten
Copyright © 2010 WILEY-VCH Verlag GmbH & Co. KGaA, Weinheim
ISBN: 978-3-527-32711-9

technique is fully reversible and possesses fast kinetics that can be controlled by pressure and temperature variation.

21.2
History of Adsorption

The ancient Egyptians, Greeks, and Romans used the effect of adsorption to process liquids, for example, applying clay, sand, and charcoal for desalination of water, clarification of oil and fat, and medical applications [1]. But the first reports of gas adsorption in charcoal date from the 1770s by Scheele, Priestley, and Abbé Fontane. In 1814, de Saussure described the exothermic reaction of gas adsorption and 40 years later the heat of adsorption for various gases in charcoal was measured by Favre. By this time he spoke of "wetting of solids by gases" as the term "adsorption" was first introduced in 1881 by Kayser. In that year, Chappuis and Kayser investigated the pressure dependence of the amount of gas adsorbed. An important step was made by Freundlich in 1907, who found a mathematical equation for the pressure-dependent adsorption which is nowadays still used, the Freundlich equation. In 1909, for the first time hydrogen storage by adsorption was investigated, when McBain described the uptake of hydrogen in carbon materials. But it took until the early 20th century when quantitative studies on adsorption of gases on various solids were performed and Zsigmondy, Polanyi, and Langmuir developed a theory to describe this phenomenon of adsorption. By that time, Langmuir found that the gas was adsorbed as a monolayer and therefore the surface area of the material can be calculated by using the maximum amount adsorbed. This model is still used for determining the specific surface area of microporous materials, even though Langmuir pointed out that this model was for flat surfaces and not applicable to small pores. In 1938, Brunauer, Emmett, and Teller noticed that, at temperatures close to the boiling point of the respective gas, adsorption occurred and it was not only adsorbed as a monolayer as described by Langmuir, but also in multilayers. The Brunauer-Emmett-Teller (BET) model offers an alternative way to determine the specific surface area of a material. One of the first who suggested inexpensive hydrogen storage on activated carbons at cryogenic temperatures were Carpetis and Peschka [2, 3] in 1980. Over the last 30 years, many new microporous materials for adsorption of gases have been synthesized, for example novel carbon structures, zeolites, organic polymers, and coordination polymers.

21.3
Hydrogen Adsorption

The heat of adsorption which is involved in physical adsorption is for hydrogen typically around 1–10 kJ mol^{-1} [4]. At low pressures, gases are adsorbed in a monolayer, but on going to higher pressures and lower temperatures close to the boiling temperature, multilayer adsorption can also occur. For hydrogen, the heat of condensation is rather low (0.9 kJ mol^{-1}), which is reflected by the low boiling point (20.4 K) and critical point (33.0 K). Therefore, at typical adsorption temperatures between 77 K and room temperature, hydrogen will be adsorbed as a monolayer and no multilayers will be formed. This is also reflected in the adsorption isotherm, which is for hydrogen at cryogenic temperatures a typical type I isotherm according to the IUPAC classification. At low pressures, the amount of hydrogen adsorbed on a surface increases strongly with the applied pressure. When the pressure is increased further, the isotherm reaches a saturation plateau which reflects the coverage of the surface with a monolayer of adsorbed hydrogen. The maximum amount of hydrogen which can be adsorbed on a solid therefore depends on the surface area of the material. This was first pointed out by Züttel *et al.*, who assumed a monolayer of liquid hydrogen on a surface and calculated a proportionality factor of 2.28 wt% per 1000 m^2 surface area [5]. Consequently, high surface area materials are needed for the effective storage of hydrogen.

21.4
Materials

The most obvious way to increase the specific surface area of a material is to decrease the particle size. However, this increase in outer surface area is limited as particles below a certain diameter are not stable and, for example, start to agglomerate in order to reduce their surface energy. The only way to increase the surface area further is to use materials possessing inner surfaces, that is, porous materials which are accessible to guest species. Depending on their pore size, these materials are classified as nano- or microporous (pores of less than 20 Å in diameter), mesoporous (pore diameter 2–50 nm), and macroporous (pore diameter greater than 50 nm).

21.4.1
Carbon Materials

High surface area carbon materials include non-crystalline activated carbons and nanostructured materials such as graphite nanofibers, and single- and multi-walled carbon nanotubes. Activated carbons are non-crystalline materials which can be produced easily, inexpensively, and in high quantity. They have been used as gas absorbers for many years and the first publication on hydrogen adsorption in high surface area carbon was in 1967 by Kidnay and Hiza [6]. One of the great advantages which led to the wide use of activated carbons is that carbon can be produced very cheaply from any carbonaceous precursors, mineralogical as well as organic materials. To increase the porosity of these carbons, they can be activated by either thermal or chemical treatment. Activated carbons with very different characteristics of surface area and pore size are formed. Most exhibit specific surface areas of 700–1800 m² g⁻¹ but some highly porous materials have extremely large specific surface areas, as for example AX-21 with 3000 m² g⁻¹ [7]. One of the main drawbacks of activated carbons is that they have a wide pore size distribution. Carbon materials with controlled pore size can be synthesized by using inorganic porous templates as zeolites or mesoporous silica where the carbon precursor is deposited and carbonized. Subsequently, the template is removed and a highly porous carbon material with well-defined pore size is left as replica of the porous template [8–11].

In order to produce carbons with well-defined structural properties, many nanostructured adsorbents have been synthesized, such as graphite nanofibers (GNFs), single-walled carbon nanotubes (SWCNTs) and multi-walled carbon nanotubes (MWCNTs). Graphite nanofibers have a diameter of approximately 3–100 nm and a length of up to 1 mm. They have been known for a long time as a nuisance occurring during the catalytic conversion of carbon-containing gases. Nowadays this "nuisance" finds wide application in electronics, for example, as polymer additives and in catalysis. They are formed by thermal decomposition of hydrocarbons on catalysts [12] and the graphene layers stack either parallel or as so-called fishbones. These two forms of GNFs expose exclusively either basal graphite planes or edge planes. In additon to straight fibers, also bi-directional, twisted, helical, and branched fibers were observed [13]. The specific surface area of GNFs range mostly between 100 and 300 m² g⁻¹. GNFs are actually close relatives of carbon nanotubes (CNTs). These were discovered by Iijima [14] and are rolled graphene sheets with an inner diameter of about 1 nm and length of 10–100 μm. SWCNTs consist of one layer of graphene rolled and closed at both ends by fullerene-like caps. Owing to the rolling direction of the graphene sheet, SWCNTs appear in three types: armchair, zigzag, and chiral [15]. Depending on the type, the SWCNTs are either semiconducting or metallic. Usually SWCNTs form bundles due to the van der Waals forces between individual tubes. Typically, purified SWCNTs possess an SSA of about 1000 m² g⁻¹. MWCNTs are an arrangement of concentric nanotubes with an inner diameter of approximately 1.5–15 nm and an outer of about 2.5–30 nm. The interlayer distance between the individual tubes is 0.34–0.36 nm, which is close to the interlayer distance in graphite. Therefore, the SSA of MWCNTs is much lower than that of SWCNTs.

21.4.2
Zeolites

Zeolites are highly crystalline aluminosilicate materials which occur in nature and have been widely applied in industry for a long time. As natural zeolites are not very pure and their number is limited to about 40 different kinds, the synthesis of zeolites increased the possible applications for this class of materials. Zeolites are built of TO_4 tetrahedrons sharing all four corners, where T indicates typically Si^{4+} and Al^{3+}, and have the general formula $M^{n+}_{m/n}[(SiO_2)_p(AlO_2)_m] \cdot xH_2O$, where M is a cation which neutralizes the negative charge on the aluminosilicate framework [16]. As the T–O–T bond is very flexible, the tetrahedral units can yield a huge variety of framework topologies. To maintain the electroneutrality of the structure, for every Si^{4+} substituted with an Al^{3+} there is an additional extra-framework metal ion (M) adsorbed in the framework. The additional M cations are usually alkali or alkaline earth metals. The ion-exchange capacity of these materials depends on the size of the accommodating pores and also on the charge of the cations. These extra-framework metal ions produce a strong electrostatic field. The presence of these strong electrostatic forces inside the framework can produce strong polarizing sites in the free volume of the zeolites. The non-porous building units of zeolites result in a dead volume for guest species which contributes to the weight of the material. Therefore, the maximum SSA which can be achieved with zeolites is about 1000 $m^2\ g^{-1}$ [17].

21.4.3
Organic Polymers

Porous organic polymers combine high surface areas due to microporosity with the synthetic diversity of organic chemistry [18]. Even though most organic polymers are so flexible that they fill the space very efficiently and therefore the material exhibits no porosity, there are new types of organic polymers, polymers of intrinsic microporosity (PIM) [18], which are more rigid because they are composed of fused-ring components and show good porosity [19, 20]. The advantage of these organic polymers is that they contain only very light elements such as carbon, hydrogen, nitrogen, and oxygen. Furthermore, functionalities can be easily introduced by a large number of synthetic routes. PIMs have specific surface areas of up to 1000 $m^2\ g^{-1}$ [18]. Besides the PIMs, another class of porous organic polymers are hyper-crosslinked polymers (HCPs) [21], which were first reported in the 1970s [22]. The porosity of HCPs is produced by swelling a precursor in a solvent and than fixing the structure by crosslinking of the precursor molecules [23]. The resulting HCPs offer specific surface areas of up to 2000 $m^2\ g^{-1}$ [24].

21.4.4
Metal-Organic Frameworks

Metal-organic hybrid coordination polymers, often called metal-organic frameworks (MOFs), are a new class of crystalline microporous high-surface-area materials. They consist of metal ions or coordinated metal clusters which are linked to each other by organic ligands. Therefore, MOFs exhibit a very open structure with well-defined pore size and shape. After removal of solvents, very high porosity and SSA, as for example MOF-177 with an SSA of 5640 $m^2\ g^{-1}$ [25], are observed, making MOFs the crystalline materials with the lowest specific density. For the metal clusters mainly transition metals are used, but recently also some coordination polymers based on lighter materials such as Al^{3+} or Mg^{2+} [26–28], which lowers the weight of the framework. Additionally, there is a huge variety of rigid organic ligands which have mainly two or three binding sites to the metal clusters. Therefore, MOFs offer a giant playground for synthesizing new materials with tailored properties such as pore size, surface area, and chemical composition. Because of the framework structure, MOFs possess only accessible surface area and no dead volume. Their well-defined structure makes them ideal candidates for the fundamental study of the microscopic nature of hydrogen adsorption. MOFs can be produced industrially in large amounts, which permits their technical application [29].

21.5
Hydrogen Storage

21.5.1
Specific Surface Area

About 10 years ago, an intensive study of carbon nanostructures as hydrogen storage media started due to reports of exceptionally high hydrogen storage at room temperature [30, 31]. However, even after several years of intensive investigation, these values could not be reproduced [32–34]. SWCNTs, MWCNTs, and GNFs do not exhibit any extraordinary hydrogen uptake. At room temperature, a maximum of 0.6 wt% was found [35–39] and up to 2.7 wt% only at an extremely high pressure of 500 bar [39]. Even at 77 K the hydrogen uptake of SWCNTs is only about 2 wt%. By a model, based on the assumption of adsorption in a monolayer with a density of liquid hydrogen, Züttel et al. [5] showed a clear correlation of the hydrogen storage capacity of nanostructured carbons with their SSA. The maximum hydrogen storage capacity calculated by this model is 2.28 wt% per 1000 m^2 of surface area. Since these novel carbon nanostructures typically have SSAs below 1000 $m^2\,g^{-1}$, their hydrogen uptake is rather low. In contrast, some activated carbons possess much higher SSAs. The best example of high SSAs in carbons is AX-21 with 3000 $m^2\,g^{-1}$ [7]. This material stores up to 5 wt% at 77 K and 20 atm, which corresponds to 1.6 wt% per 1000 m^2, and is therefore also beyond the calculated maximum capacity.

For all zeolites, the adsorption capacity varies widely not only with the framework type but also with chemical compositions and operating conditions. The maximum hydrogen uptake for zeolites at 77 K is 2.19 wt% at 15 bar for CaX zeolite [40]. Vitillon et al. [16] calculated theoretically an upper limit for the hydrogen uptake by estimating the maximum amount of molecular hydrogen which could be placed in the pore system of several zeolite types, regardless of the mechanism of adsorption or encapsulation. For FAU and RHO zeolite, a maximum amount of 2.86 wt% was predicted even under extreme conditions (low temperature and high pressure) or in the presence of ideal binding sites (negligible volume and high binding energy).

For PIMs, the maximum amount of hydrogen adsorbed is so far 2.7 wt% at 77 K and 10 bar [41, 42]. HCPs prepared by self-condensation of bischloromethyl monomers with a surface area of 1900 m² g⁻¹ show a maximum hydrogen uptake of 3.7 wt% at 77 K and 15 bar [43].

The first measurements of hydrogen storage in MOFs claimed storage capacities of up to 1 wt% at room temperature and 20 bar and 4.5 wt% at 0.8 bar and 77 K for MOF-5 [44]. These very promising results could not be reproduced independently and storage capacities of approximately 0.15 wt% at room temperature and 20 bar are more realistic. Three different laboratories measured independently an uptake of 5 wt% for MOF-5 at 77 K [25, 45, 46]. Higher excess adsorption capacities of approximately 7 wt% at 77 K were found for MOF-177 [25, 47] and MOF-5 prepared completely in an inert atmosphere [48].

Experimentally, it was shown in various publications that the maximum excess hydrogen uptake at 77 K of carbon materials depends almost linearly on the specific surface area [35, 49, 50] as calculated by Züttel *et al.* [5]. The experimentally found values are in the region of 2×10^{-3} wt% m⁻², which is often referred to as Chahine's rule [51], that is, approximately 1 wt% per 500 m². The relation also holds for zeolites [40] and PIMs [18]. Panella *et al.* showed for the first time that MOFs behave similarly, independent of their chemical composition [46], as later shown for a even larger variety of MOFs by other groups [25, 52, 53].

Figure 21.1 summarizes various experimental results for the maximum excess hydrogen uptake as a function of the SSA.

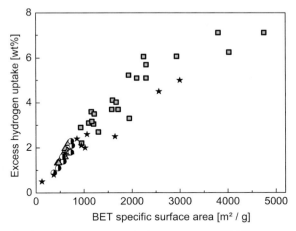

Fig. 21.1 Maximum excess hydrogen storage of MOFs (squares) [54, 55], zeolites (triangles) [40], Prussian Blue analogs (half-filled circles) [56], and carbon materials (stars) [7, 35] at 77 K correlated with their BET specific surface area.

21.5.2
Pore Size

The simplest model for a pore is a slit pore. When the pore is wide enough, the adsorption potential is the same as for a flat surface. With decreasing pore width, the potentials of both walls start to overlap and the heat of adsorption increases. Similarly, the adsorption potential increases when the surface is not flat but exhibits some curvature (e.g., in pores) and the overlapping potential of neighboring atoms enhances the adsorption. Therefore, hydrogen is adsorbed preferentially in small pores before being adsorbed in larger pores with lower interaction potential.

For carbide-derived carbons with well-defined and tunable pore size, Yushin *et al.* [57] found at low pressures no correlation between hydrogen storage and surface area. Instead, they observed with decreasing pore size an increase in the uptake at low pressures, indicating a higher heat of adsorption for small pores.

SWCNTs arrange in a two-dimensional triangular lattice when forming bundles. The effect of overlapping van der Waals potentials in these bundles was calculated theoretically for different adsorption sites. The heat of adsorption increases depending on the adsorption site from the outer surface, to the pore in open tubes, to the grooves between two tubes and finally the interstitial sites between three tubes [58]. This was shown experimentally by low-temperature thermal desorption spectroscopy [59], inelastic neutron scattering [60], low-temperature hydrogen adsorption [61], and Raman spectroscopy [62, 63].

For several nanoporous aluminophosphates (AlPOs) which all have the same average chemical composition ($AlPO_4$), a clear correlation between decreasing pore diameter and increasing heat of adsorption was shown by Jhung *et al.* [64]. The heat of adsorption was calculated from hydrogen adsorption measurements at 77 and 87 K, which also showed higher hydrogen uptake at low pressures for AlPOs with smaller pores. The same group investigated a series of different zeolites and showed that for zeolites with similar aluminum content the heat of adsorption is determined by the pore size [65].

For PIMs, the hydrogen uptake at low pressures depends strongly on the pore size, as materials with more ultramicropores show better hydrogen uptake at low pressures [41]. For different HCPs, the highest heat of adsorption of 7.25 kJ mol^{-1} was measured for the material with the smallest pores [43].

Several MOF structures possessing different pore sizes have been studied concerning the heat of adsorption of hydrogen by different techniques. Low-temperature thermal desorption spectroscopy (TDS) shows a higher desorption temperature for MOFs with smaller pores, indicating a higher heat of adsorption with decreasing pore diameter [54]. Temperature-dependent measurements of adsorption isotherms for different MOFs showed that for technologically relevant hydrogen uptake the heat of adsorption is higher for adsorbents with smaller pores [53]. For Cu-BTC with a bimodal pore structure, TDS shows two desorption maxima, indicating a different heat of adsorption for each pore. Similar results have been obtained by neutron powder diffraction in Cu-BTC [66]. For increasing deuterium uptake, progressive filling of the smaller pores (5 Å) followed by filling of the larger pores (9 Å) is observed. Theoretical calculations for MOFs by several groups [67–70] found a similar contribution of the metal clusters and organic linkers to the physisorption of hydrogen in the range of significant hydrogen uptake. Furthermore, materials with smaller pores generally show a higher heat of adsorption. Therefore, the pore size is the most important factor in order to tune the heat of adsorption in the range of technologically relevant hydrogen storage.

21.5.3
Chemical Composition

The binding strength of hydrogen on the surface of porous materials is mainly determined by two different effects, van der Waals and electrostatic potentials. Van der Waals interaction increases the heat of adsorption with decreasing pore size caused by overlap of the van der Waals potentials of the individual surface atoms. In porous materials with polarizing metal sites, the electrostatic potential also needs to be considered. In zeolites, for example, cations offer strong electrostatic adsorption sites [71, 72]. For zeolites with varying concentration of aluminum cations, Jhung et al. showed a clear correlation of the heat of adsorption with the concentration of aluminum cations [65]. A very similar situation with strong electrostatic potential can be found for MOFs with unsaturated metal sites at the metal clusters. Forster et al. showed for a MOF with unsaturated Ni sites that hydrogen is preferentially adsorbed close to the metal clusters [73]. However, the number of adsorption sites at unsaturated metal centres is limited and, therefore, for MOFs the heat of adsorption at very low hydrogen uptake is determined only by these metal sites. With increasing hydrogen concentration, the van der Waals interaction in the pores and therefore the pore diameter dominates the heat of adsorption [53].

Even though the heat of adsorption can be increased for a fraction of the hydrogen molecules by introducing open metal sites, this does not increase the maximum storage capacity, which is determined by the surface area.

21.6
Total Storage Capacity

The relevant parameter for technological application is the gravimetric storage capacity which can be achieved in a tank system. Figure 21.2 shows a two-dimensional schematic representation of the density of gas adsorbed in a pore. Close to the surface the gas is adsorbed and therefore the density is high, but with increasing distance from the surface the density decreases to the density of the free gas, which is determined by the applied pressure. For the characterization of adsorption materials, mostly the excess hydrogen uptake is reported, which is the amount adsorbed by a material in comparison with a non-adsorb-

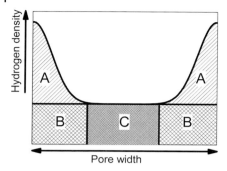

Fig. 21.2 Density of gas in a pore. A denotes the excess, A+B the absolute, and A+B+C the total storage capacity.

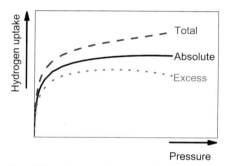

Fig. 21.3 Isotherms of excess, absolute, and total hydrogen uptake.

ing material of the same volume surrounded by the gas. In Figure 21.2, the excess uptake is given by area A whereas the free gas is represented by B and C. With increasing pressure, the density of the adsorbed layer A increases until a saturation coverage is reached. However, the density of the free gas is constantly rising with increasing pressure. Therefore, the excess adsorption isotherm will show a maximum (Figure 21.3), in contrast to a type I behavior. A type I isotherm represents the absolute adsorption, which includes the free gas in the region of the adsorbed layer, that is, the areas A and B in Figure 21.2. It is not possible to measure experimentally the absolute amount adsorbed, therefore it is calculated from the excess amount adsorbed by assuming an average density for the adsorbed layer. As shown, for example, by Züttel et al., a good assumption of the density of the adsorbed layer is the density of the liquid. However, this slightly overestimates the amount adsorbed on the surface of the material. For technical applications the total amount of hydrogen in a porous material is important. Even if no adsorption were to occur, gas could be stored in the pores. For an adsorbing material, total adsorption is then given by the excess adsorption plus the free hydrogen gas in the pore volume if no adsorption would have occurred, that is, in Figure 21.2 areas A, B, and C. The total uptake therefore increases monotonically with increasing pressure and the amount is higher than for excess or absolute adsorption (Figure 21.3). Therefore, for technical applications, the total uptake, which is higher than the excess uptake used to characterize the material, at the operating pressure is the relevant material parameter. In contrast, in hydrides, the adsorbed amount and the total uptake are identical since these are solid materials without pores.

The total storage capacity of the material can only be transferred to the tank system if the tank vessel is completely filled with the storage material without any voids. However, in a real tank system, the material is usually filled in as a powder or small pellets, which will result in inter-particle void spaces. Therefore, for microporous materials and for hydrides, the total storage capacity is decreased depending on the packing density of the powder or pellets.

21.7
Conclusion

Adsorption technologies making use of the physisorption of hydrogen in high surface area microporous materials are a promising route towards high-density hydrogen storage. The adsorption process possesses fast kinetics with a low evolution of heat and is totally reversible. At high pressures, the maximum hydrogen uptake is linearly correlated with the SSA and is independent of the chemical composition of the material. Furthermore, at technologically relevant hydrogen concentrations, the pore diameter of the cavity determines the heat of adsorption. For microporous materials containing metal sites, a higher heat of adsorption is only observed at low hydrogen concentration. So far, the most advanced materials are activated carbons and MOFs and for both material classes cost-efficient routes for up-scaling to high quantities exist. The best gravimetric storage capacity was recorded for MOF-177 with about 11 wt% on a materials basis at 77 K and 70 bar [47]. The maximum volumetric hydrogen uptake of 59.1 g l^{-1} was measured for a Cu-based MOF, PCN-11, at 30 K and 3.5 bar [74].

References

1 Rouquerol, J., Rodriguez-Reinaso, F., Sing, K. S. W. (1994) *Characterization of Porous Solids III*. Elsevier Science, Amsterdam, p. 109.
2 Carpetis, C. and Peschka, W. (1980) A study on hydrogen storage by use of cryoadsorbents. *Int. J. Hydrogen Energy*, **5** (5), 539–554.
3 Peschka, W. and Carpetis, C. (1980) Cryogenic hydrogen storage and refueling for automobiles. *Int. J. Hydrogen Energy*, **5** (6), 619–625.
4 Züttel, A. (2004). Hydrogen storage methods. *Naturwiss.*, **91**, 157–172.
5 Züttel, A., Sudan, P., Mauron, P., and Wenger P. (2004) Model for the hydrogen adsorption on carbon nanostructures. *Applied Physics A – Mat. Sci. Proc.*, **78** (7), 941–946.
6 Kidnay, A. J. and Hiza, M. J. (1966) High pressure adsorption isotherm of neon, hydrogen and helium at 76 K. In *Advanced in Cryogenic Engineering* (ed.

Timmerhaus, K. D.), Vol. 11, Plenum, New York, p. 338.

7 Chahine, R. and Bose, T. K. (1994) Low-pressure adsorption storage of hydrogen. *Int. J. Hydrogen Energy*, **19** (2), 161–164.

8 Kim, J. Y., Kim, S. H., Lee, H.-H., Lee, K., Ma, W., Gong, X., and Heeger, A. J. (2006) New architecture for high-efficiency polymer photovoltaic cells using solution-based titanium oxide as an optical spacer. *Advanced Materials*, **18** (5), 572–576.

9 Kim, T.-W., Park, I.-S., and Ryoo, R. (2003) A synthetic route to ordered mesoporous carbon materials with graphitic pore walls. *Angew. Chem. Int. Ed.*, **42** (36), 4375–4379.

10 Kyotani, T., Ma, Z., and Tomita, A. (2003) Template synthesis of novel porous carbons using various types of zeolites. *Carbon*, **41**, 1451–1459.

11 Ryoo, R., Joo, S. H., and Jun, S. (1999) Synthesis of highly ordered carbon molecular sieves via template-mediated structural transformation. *J. Phys. Chem. B*, **103** (37), 7743–7746.

12 De Jong, K. P. and Geus, J. W. (2000) Carbon nanofibers: catalytic synthesis and applications. *Catalysis Reviews: Science and Engineering*, **42** (4), 481–510.

13 Rodriguez, N. M. (1993) A review of catalytically grown carbon nanofibers. *J. Mat. Res.*, **8** (12), 3233–3250.

14 Iijima, S. (1991) Helical microtubules of graphitic carbon. *Nature*, **354**, 56–58.

15 Dresselhaus, M. S., Dresselhaus, G., and Saito, R. (1995) Physics of carbon nanotubes. *Carbon*, **33** (7), 883–891.

16 Vitillo, J. G., Ricchiardi, G., Spoto, G., and Zecchina, A. (2005) Theoretical maximal storage of hydrogen in zeolitic frameworks. *Phys. Chem. Chem. Phys.*, **7** (23), 3948–3954.

17 Morris, R. E. and Wheatley, P. S. (2008) Gas storage in nanoporous materials. *Angew. Chem.-Int. Ed.*, **47** (27), 4966–4981.

18 McKeown, N. B., Budd, P. M., and Book, D. (2007) Microporous polymers as potential hydrogen storage materials. *Macromol. Rapid Comm.*, **28** (9), 995–1002.

19 Budd, P. M., McKeown, N. B., and Fritsch, D. (2005) Free volume and intrinsic microporosity in polymers. *J. Mat. Chem.*, **15** (20), 1977–1986.

20 McKeown, N. B., Budd, P. M., Msayib, K. J., Ghanem, B. S., Kingston, H. J., Tattershall, C. E., Makhseed, S., Reynolds, K. J., and Fritsch, D. (2005) Polymers of intrinsic microporosity (PIMs). *Chemistry – Europ. J.*, 11(9), 2610–2620.

21 Germain, J., Frechet, J. M. J., and Svec, F. (2007) Hypercrosslinked polyanilines with nanoporous structure and high surface area: potential adsorbents for hydrogen storage. *J. Mat. Chem.*, **17** (47), 4989–4997.

22 Davankov, V. A., Rogozhin, S. V., and Tsjurupa, M. P. (1973) Factors determining degree of swelling of crosslinked polymers. *Angewandte Makromolekulare Chemie*, **32**, 145–151.

23 Svec, F., Germain, J., and Frechet, J. M. J. (2009) Nanoporous polymers for hydrogen storage. *Small*, **5** (10), 1098–1111.

24 Ahn, J. H., Jang, J. E., Oh, C. G., Ihm, S. K., Cortez, J., and Sherrington, D. C. (2006) Rapid generation and control of microporosity, bimodal pore size distribution, and surface area in Davankov-type hyper-cross-linked resins. *Macromol.*, **39** (2), 627–632.

25 Wong-Foy, A. G., Matzger, A. J., and Yaghi, O. M. (2006) Exceptional H_2 saturation uptake in microporous metal-organic frameworks. *J. Am. Chemi. Soc.*, **128** (11), 3494–3495.

26 Dinca, M. and Long, J. R. (2005) Strong H_2 binding and selective gas adsorption within the microporous coordination solid $Mg_3(O_2C-C_{10}H_6-CO_2)_3$. *J. Am. Chem. Soc.*, **127** (26), 9376–9377.

27 Dietzel, P. D. C., Blom, R., and Fjellvag, H. (2008) Base-induced formation of two magnesium metal-organic framework compounds with a bifunctional tetratopic ligand. *Eur. J. Inorg. Chem.*, **23**, 3624–3632.

28 Volkringer, C., Loiseau, T., Marrot, J., and Ferey, G. (2009) A MOF-type magnesium benzene-1,3,5-tribenzoate with two-fold interpenetrated ReO_3 nets. *Cryst. Eng. Comm.*, **11** (1), 58–60.

29 Czaja, A. U., Trukhan, N., and Müller, U. (2009) Industrial applications of metal-

organic frameworks. *Chem. Soc. Rev.*, **38**, 1284–1293.

30 Chambers, A., Park, C., Baker, R.T.K., and Rodriguez, N.M. (1998) Hydrogen storage in graphite nanofibers. *J. Phys. Chem. B*, **102** (22), 4253–4256.

31 Dillon, A.C., Jones, K.M., Bekkedahl, T.A., Kiang, C.H., Bethune, D.S., and Heben, M.J. (1997) Storage of hydrogen in single-walled carbon nanotubes. *Nature*, **386**, 377–379.

32 Hirscher, M., Becher, M., Haluska, M., Dettlaff-Weglikowska, U., Quintel, A., Duesberg, G.S., Choi, Y.-M., Downes, P., Hulman, M., Roth, S., Stepanek, I., and Bernier, P. (2001) Hydrogen storage in sonicated carbon materials. *App. Phys. A*, **72**, 129–132.

33 Tibbetts, G.G., Meisner, G.P., and Olk, C.H. (2001) Hydrogen storage capacity of carbon nanotubes, filaments, and vapor-grown fibers. *Carbon*, **39** (15), 2291–2301.

34 Hirscher, M. and Becher, M. (2003) Hydrogen storage in carbon nanotubes. *J. Nanosci. Nanotech.*, **3** (1–2), 3–17.

35 Panella, B., Hirscher, M., and Roth, S. (2005) Hydrogen adsorption in different carbon nanostructures. *Carbon*, **43** (10), 2209–2214.

36 Ritschel, M., Uhlemann, M., Gutfleisch, O., Leonhardt, A., Graff, A., Taschner, C., and Fink, J. (2002) Hydrogen storage in different carbon nanostructures. *App. Phys. Lett.*, **80** (16), 2985–2987.

37 Schimmel, H.G., Nijkamp, G., Kearley, G.J., Rivera, A., de Jong, K.P., and Mulder, F.M. (2004) Hydrogen adsorption in carbon nanostructures compared. *Mat. Sci. Eng. B – Solid State Materials for Advanced Technology*, **108** (1–2), 124–129.

38 Rzepka, M., Lamp, P., and de la Casa-Lillo, M.A. (1998) Physisorption of hydrogen on microporous carbon and carbon nanotubes. *J. Phys. Chem. B*, **102** (52), 10894–10898.

39 Jorda-Beneyto, M., Suarez-Garcia, F., Lozano-Castello, D., Cazorla-Amoros, D., and Linares-Solano, A. (2007) Hydrogen storage on chemically activated carbons and carbon nanomaterials at high pressures. *Carbon*, **45** (2), 293–303.

40 Langmi, H.W., Book, D., Walton, A., Johnson, S.R., Al-Mamouri, M.M., Speight, J.D., Edwards, P.P., Harris, I.R., and Anderson, P.A. (2005) Hydrogen storage in ion-exchanged zeolites. *J. All. Comp.*, **404–406**, 637–642.

41 Budd, P.M., Butler, A., Selbie, J., Mahmood, K., McKeown, N.B., Ghanem, B., Msayib, K., Book, D., and Walton, A. (2007) The potential of organic polymer-based hydrogen storage materials. *Phys. Chem. Chem. Phys.*, **9** (15), 1802–1808.

42 Ghanem, B.S., Msayib, K.J., McKeown, N.B., Harris, K.D.M., Pan, Z., Budd, P.M., Butler, A., Selbie, J., Book, D., and Walton, A. (2007) A triptycene-based polymer of intrinsic miciposity that displays enhanced surface area and hydrogen adsorption. *Chem. Comm.*, **1**, 67–69.

43 Wood, C.D., Tan, B., Trewin, A., Niu, H., Bradshaw, D., Rosseinsky, M.J., Khimyak, Y.Z., Campbell, N.L., Kirk, R., Stoeckel, E., and Cooper, A.I. (2007) Hydrogen storage in microporous hypercrosslinked organic polymer networks. *Chem. Mat.*, **19** (8), 2034–2048.

44 Rosi, N.L., Eckert, J., Eddaoudi, M., Vodak, D.T., Kirn, J., O'Keeffe, M., and Yaghi, O.M. (2003) Hydrogen storage in microporous metal-organic frameworks. *Science*, **300** (5622), 1127.

45 Dailly, A., Vajo, J.J., and Ahn, C.C. (2006) Saturation of hydrogen sorption in Zn benzenedicarboxylate and Zn naphthalenedicarboxylate. *J. Phys. Chem. B*, **110** (3), 1099–1101.

46 Panella, B., Hirscher, M., Pütter, H., and Müller, U. (2006) Hydrogen adsorption in metal-organic frameworks: Cu-MoFs and Zn-MoFs compared. *Ad. Funct. Mat.*, **16** (4), 520–524.

47 Furukawa, H., Miller, M.A., and Yaghi, O.M. (2007) Independent verification of the saturation hydrogen uptake in MoF-177 and establishment of a benchmark for hydrogen adsorption in metal-organic frameworks. *J. Mat. Chem.*, **17** (30), 3197–3204.

48 Kaye, S.S., Dailly, A., Yaghi, O.M., and Long, J.R. (2007) Impact of preparation and handling on the hydrogen storage

properties of Zn$_4$O(1,4-benzenedicarboxylate)$_3$ (MoF-5). *J. Am. Chem. Soc.*, **129** (45), 14176–14177.

49 Nijkamp, M.G., Raaymakers, J.E.M.J., van Dillen, A.J., and de Jong, K.P. (2001) Hydrogen storage using physisorption – materials demands. *App. Phys. A – Mat. Sci. Proc.*, **72** (5), 619–623.

50 Texier-Mandoki, N., Dentzer, J., Piquero, T., Saadallah, S., David, P., and Vix-Guterl, C. (2004) Hydrogen storage in activated carbon materials: role of the nanoporous texture. *Carbon*, **42** (12–13), 2744–2747.

51 Chahine, R. and Bose, T.K. (1996) Characterization and optimization of adsorbents for hydrogen storage. In *Hydrogen Energy Progress XI*. (eds. T.N. Veziroglu, C.J. Winter, J.P. Baselt, and G. Kreysa), published on behalf of the International Association for Hydrogen Energy, Stuttgart, Germany, vols 1–3, pp. 1259–1263.

52 Hirscher, M. and Panella, B. (2007) Hydrogen storage in metal-organic frameworks. *Scripta Materialia*, **56** (10), 809–812.

53 Schmitz, B., Müller, U., Trukhan, N., Schubert, M., Frey, G., and Hirscher, M. (2008) Heat of adsorption for hydrogen in microporous high-surface-area materials. *Chem. Phys. Chem.*, **9** (15), 2181–2184.

54 Panella, B., Hönes, K., Müller, U., Trukhan, N., Schubert, M., Pütter, H., and Hirscher, M. (2008) Desorption studies of hydrogen in metal-organic frameworks. *Angew. Chemie, Int. Ed.*, **47** (11), 2138–2142.

55 Murray, L.J., Dinca, M., and Long, J.R. (2009) Hydrogen storage in metal-organic frameworks. *Chem. Soc. Rev.*, **38** (5), 1294–1314.

56 Kaye S.S. and Long, J.R. (2005) The role of vacancies in the hydrogen storage properties of Prussian blue analogues. *Catalysis Today*, **120** (3–4), 311.

57 Yushin, G., Dash, R., Jagiello, J., Fischer, J.E, and Gogotsi, Y. (2006) Carbide-derived carbons: effect of pore size on hydrogen uptake and heat of adsorption. *Ad. Funct. Mat.*, **16** (17), 2288–2293.

58 Pradhan, B.K., Sumanasekera, G.U., Adu, K.W., Romero, H.E., Williams, K.A., and Eklund, P.C. (2002) Experimental probes of the molecular hydrogen-carbon nanotube interaction. *Phys. B – Cond. Matter*, **323** (1–4), 115–121.

59 Panella B., Hirscher, M., and Ludescher, B. (2007) Low-temperature thermal-desorption mass spectroscopy applied to investigate the hydrogen adsorption on porous materials. *Microporous and Mesoporous Mat.*, **103** (1–3), 230–234.

60 Georgiev, P.A., Ross, D.K., De Monte, A., Montaretto-Marullo, U., Edwards, R.A.H., Ramirez-Cuesta, A.J., Adams, M.A., and Colognesi, D. (2005) In situ inelastic neutron scattering studies of the rotational and translational dynamics of molecular hydrogen adsorbed in single-wall carbon nanotubes (SWNTs). *Carbon*, **43** (5), 895–906.

61 Wilson, T., Tyburski, A., DePies, M.R., Vilches, O.E., Becquet, D., and Bienfait, M. (2002) Adsorption of H$_2$ and D$_2$ on carbon nanotube bundles. *J. Low Temp. Phys.*, **126** (1–2), 403–408.

62 Panella, B. and Hirscher, M. (2008) Raman studies of hydrogen adsorbed on nanostructured porous materials. *Phys. Chem. Chem. Phys.*, **10** (20), 2910–2917.

63 Williams, K.A., Pradhan, B.K., Eklund, P.C., Kostov, M.K., and Cole, M.W. (2002) Raman spectroscopic investigation of H$_2$, HD, and D$_2$ physisorption on ropes of single-walled, carbon nanotubes. *Phys. Rev. Lett.*, **88** (16), 165502/1–4.

64 Jhung, S.H., Kim, H.K., Yoon, J.W., and Chang, J.S. (2006) Low-temperature adsorption of hydrogen on nanoporous aluminophosphates: effect of pore size. *J. Phys. Chem. B*, **110** (19), 9371–9374.

65 Jhung, S.H., Yoon, J.W., Lee, J.S., and Chang, J.-S. (2007) Low-temperature adsorption/storage of hydrogen on FAU, MFI, and MOR zeolites with various Si/Al ratios: effect of electrostatic fields and pore structures. *Chemistry – Eur. J.*, **13** (22), 6502–6507.

66 Peterson, V.K., Liu, Y., Brown, C.M., and Kepert, C.J. (2006) Neutron powder diffraction study of D$_2$ sorption in Cu$_3$(1,3,5-benzenetricarboxylate)$_2$. *J. Am. Chem. Soc.*, **128** (49), 15578–15579.

67 Frost, H. and Snurr, R. Q. (2007) Design requirements for metal-organic frameworks as hydrogen storage materials. *J. Phys. Chem. C*, **111** (50), 18794–18803.

68 Klontzas, E., Mavrandonakis, A., Froudakis, G. E., Carissan, Y., and Kopper W. (2007) Molecular hydrogen interaction with IRMOF-1: a multiscale theoretical study. *J. Phys. Chem. C*, **111** (36), 13635–13640.

69 Kuc, A., Heine, T., Seifert, G., and Duarte, H. A. (2008) On the nature of the interaction between H_2 and metal-organic frameworks. *Theoretical Chem. Accounts*, **120** (4–6), 543–550.

70 Sillar, K., Hofmann, A., and Sauer, J. (2009) Ab initio study of hydrogen adsorption in MOF-5. *J. Am. Chem. Soc.*, **131** (11), 4143–4150.

71 Ramirez-Cuesta, A. J., Mitchell, P. C. H., Ross, D. K., Georgiev, P. A., Anderson, P. A., Langmi, H. W., and Book, D. (2007) Dihydrogen in cation-substituted zeolites X – an inelastic neutron scattering study. *J. Mat. Chem.*, **17** (24), 2533–2539.

72 Palomino, G. T., Carayol, M. R. L., and Arean, C. O. (2006) Hydrogen adsorption on magnesium-exchanged zeolites. *J. Mat. Chem.*, **16** (28), 2884–2885.

73 Forster, P. M., Eckert, J., Heiken, B. D., Parise, J. B., Yoon, J. W., Jhung, S. H., Chang, J.-S., and Cheetham, A. K. (2006) Adsorption of molecular hydrogen on coordinatively unsaturated Ni(II) sites in a nanoporous hybrid material. *J. Am. Chem. Soc.*, **128** (51), 16846–16850.

74 Wang, X. S., Ma, S. Q., Rauch, K., Simmons, J. M., Yuan, D. Q., Wang, X. P., Yildirim, T., Cole, W. C., Lopez, J. J., de Meijere, A., and Zhou, H. C. (2008) Metal-organic frameworks based on double-bond-coupled di-isophthalate linkers with high hydrogen and methane uptakes. *Chem. Mat.*, **20**, 3145–3152.

Policy Perspectives, Initiatives and Cooperations

22
National Strategies and Programs

Jörg Schindler

Abstract

The world faces an imminent transition to a postfossil era. The supply of crude oil has reached a plateau since 2005 and is expected to decline significantly in the coming decades. Natural gas, coal and nuclear energies will not be able to compensate this decline. The future of energy will have to be renewable. This will lead to a growing role of hydrogen as an energy carrier, especially in transport.

National strategies and programs for hydrogen are driven by the motives of securing energy supply and providing clean fuel alternatives.

The paper describes the past evolution and the present focus of these programs in Europe, North America and Asia.

Keywords: national fuel strategies, fossil fuels, hydrogen-fueled vehicles, fuel cell vehicles

22.1
The Imminent Transition to a Postfossil Energy World

22.1.1
The Future of Fossil Energies

According to the International Energy Agency (IEA), in the future the world will see an ever-increasing role of fossil energy sources. In its World Energy Outlook 2008, the IEA describes a reference scenario for global energy demand and supply for fossil, nuclear, and renewable primary energy sources [1].

This scenario is shown in Figure 22.1. As can be seen, crude oil is the most important energy source, followed by coal and natural gas. The energy content of all energy sources is expressed in *million tons of oil equivalent (Mtoe)* to make them comparable. The reference scenario up to 2030 sees no change in the ranking and assumes continued growth of all fossil energies. Nuclear and renewable energies will grow slightly, but the dominance of fossil fuels remains practically unchanged. This scenario indicates that the future will be more or

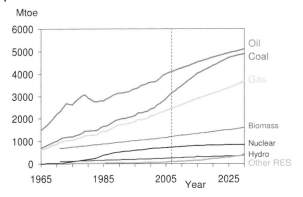

Fig. 22.1 World energy outlook by the IEA – reference scenario. Sources: historical data, *BP Statistical Review of World Energy* [2]; Projection, IEA [1].

less like the past, just more of everything. Accordingly, the message is that *business as usual (BAU)* can continue for at least another two decades.

However, in the report there is also the warning of an imminent oil crunch if a number of preconditions for a growing oil supply are not met; the following quotations are taken from the Executive Summary [1]:

- The world's energy system is at crossroads. Current global trends in energy supply and consumption are patently unsustainable – environmentally, economically, socially. But that can – and must – be altered; there's still time to change the road we're on.
- Preventing catastrophic and irreversible damage to the global climate ultimately requires a major decarbonization of the world energy sources.
- For all the uncertainties highlighted in this report, we can be certain that the energy world will look a lot different in 2030 than it does today.

One should note that the finiteness of fossil energies, especially of oil, is not mentioned. The message is blatantly contradictory: on the one hand, the IEA with its reference scenario declares BAU as being possible; on the other hand, the energy world in 2030 is supposed to be completely different. To get a more reliable picture of the future, we have to leave conventional wisdom behind.

22.1.1.1 Crude Oil

The analysis of the future availability of crude oil up to 2030 is based mainly on a study for the Energy Watch Group (EWG) in 2007 [3]. Figure 22.2 shows the annual oil discoveries in terms of proved and probable reserves since 1920 and also the annual production rates [4]. The units are gigabarrels (Gb) per year.

The peak of discoveries took place in the 1960s. Since the 1980s, the yearly oil production has exceeded the volume of new discoveries and the discrepancy is growing over time.

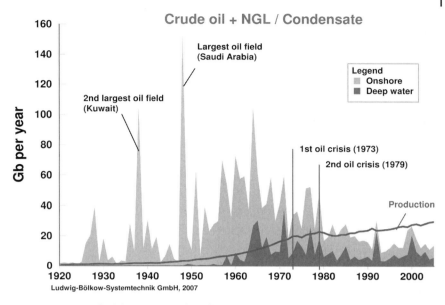

Fig. 22.2 History of oil discoveries and production. Source: HIS Energy [4].

Oil production in regions having passed their peak can be predicted with some certainty for the following years. In these cases, remaining reserves are no longer the decisive parameter for projecting future production rates.

In 2008, the oil production of all oil-producing countries past the peak plus countries whose production is regarded as being more or less on a plateau accounted for approximately 50 Mb per day. In the future, the aggregate oil production of this group of countries will decline each year.

If world oil production is to stay constant in the coming years, or is even to grow, then only the oil-producing countries in the Middle East (ME) remain eventually to increase their production in order to compensate for the decline in the rest of the world. This is not regarded as being possible.

Also for the ME countries a peak is projected in the near future, followed by a gradual decline. The main reasons are that the EWG thinks that ME reserves are grossly overstated by about 300 Gb and that the bulk of the current production comes from a very few, very old oil fields which already have problems maintaining production levels.

Based on the data and assessments sketched previously, the EWG study describes a scenario of the possible future oil supply up to 2030 (Figure 22.3). The key findings of the EWG study are:

- "Peak oil is now" (the scenario has 2006 as the date of peak oil).
- The most important result of the study is the steep decline in oil production after the peak.

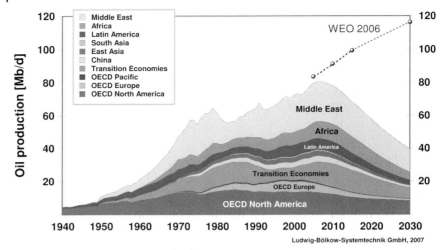

Fig. 22.3 The EWG scenario – "Peak Oil is now".

Recent data compiled by the US Energy Information Administration (EIA) show that the world's conventional oil production has been on a plateau since mid-2004 (actually the year with peak production was 2005) [5]. In spite of surging oil prices from 2004 to mid-2008, supply could not follow demand any longer.

The onset of the ultimate decline is imminent. This will signal the beginning of the end of the era of fossil fuels.

22.1.1.2 Natural Gas

A scenario by Ludwig-Bölkow-Systemtechnik GmbH (LBST) shows that according to reported reserves, global natural gas production can grow for another 5–15 years by approximately 25% until production peaks and the decline starts. Whether this projected growth will actually take place is an open question at the moment.

22.1.1.3 Coal

When discussing the future availability of fossil energy resources, conventional wisdom has it that globally there is an abundance of coal which allows for increasing coal consumption far into the future. This is either regarded as being a good thing, as coal can be a possible substitute for the declining crude oil and natural gas supplies, or as a horror scenario leading to catastrophic consequences for the world's climate. However, the discussion rarely focuses on the question: how much coal is there really?

This question was addressed by an EWG study on coal [6]. One important finding of this study is that the quality of the data on coal reserves and resources is poor, at both global and national levels. Even though the quality of reserve data is poor, an analysis of possible future production profiles based on these data is still deemed meaningful.

According to past experience, it is very likely that the available statistics are biased on the high side and therefore projections based on these data will give an upper boundary of possible future development. Accordingly, future production profiles have been developed using logistic fitting to past production.

The result of the study is that global coal production may still increase over the next 10–15 years by about 30%, mainly driven by Australia, China, the Former Soviet Union countries (Russia, Ukraine, Kazakhstan), South Africa, and the United States. Production will then reach a plateau and will eventually decline thereafter.

22.1.2
The Fossil and Nuclear Supply Outlook

The previously described supply scenarios for crude oil, natural gas, and coal can be integrated into a scenario of the future availability of all fossil and nuclear energy sources (Figure 22.4).

The scenario should be read as a qualitative statement with the numbers just indicating the likely magnitudes of possible contributions of individual fossil and nuclear energy sources at specific dates in the future. Although the exact numbers naturally will be uncertain (and will remain so), the qualitative de-

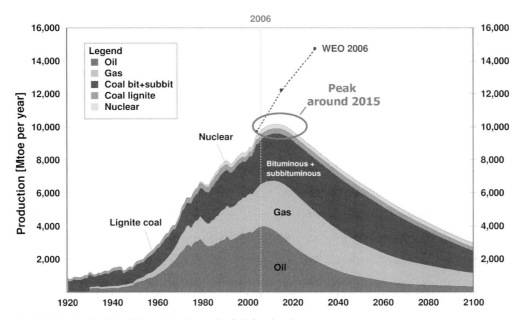

Fig. 22.4 Peak oil will be followed by the peak of all fossil and nuclear energies. Source: Schindler and Zittel [7].

scription of the future supply outlook has a very high likelihood of representing the possible fossil and nuclear energy availability in the coming decades.

The message of this scenario is dramatic (and possibly surprising to many observers):

- The current advent of peak oil will lead to the subsequent peaking of all fossil (and nuclear) energy supplies in the very near future.

Even though natural gas and coal are expected to peak one and two decades, respectively, later than oil, the imminent decline of oil production will have as a consequence the peaking of all fossil and nuclear energy sources in about 5 years time – around 2015.

The 21st century will see the transition to a postfossil energy world.

22.1.3
The Energy Future is Renewable

Fossil fuel supply will eventually decrease, for reasons of production constraints and probably also due to an increasingly rigorous climate policy. Renewable energies can be derived from biomass and by all technologies that render electricity (hydropower, wind, solar). The potential for energy from biomass is limited, but the electricity-generating potential is huge.

During recent years, renewable energies have been growing faster than expected by many (among the most prominent also the IEA). Especially the share of renewable electricity has increased significantly in certain regions during the last decade. Renewable energy technologies allow a fast ramp-up of power generation capacity. Because of their modularity, wind power plants and photovoltaic systems can be built with small lead times and have short construction times. In Europe (especially Germany and Spain), in the United States, and recently in China, wind power is booming. Photovoltaics are also seeing an exponential rise.

Climate policy was the main driver in the past and justification for subsidies, but also the huge potential and the cost reduction potential through economies of scale. In view of the large future markets for renewable energies, industry policy aspects are increasingly important.

Costs are already coming down. It is expected that renewable electricity will be competitive within the next decade.

22.2
The Role of Secondary Energy Carriers

In a fossil context, *electricity* is a secondary energy carrier. Is there a need for an additional secondary energy carrier? Electricity is now mainly a secondary energy carrier derived from the conversion of fossil and nuclear primary energies. With the increasing share of renewable energies, renewable electricity will eventually become the new and dominant primary energy. However, the complexities associated with storing electricity require improved storage concepts and capacity. Due to the fluctuating nature of most renewable energy sources, the need for energy storage will increase.

Hydrogen is a secondary energy carrier that can be produced from all primary energy sources and electricity. In the transition to a postfossil energy world, the dominant energy source will be renewable electricity. The rationale for producing hydrogen is the fact that it can be stored more easily than electricity. The technologies are either storage of gaseous hydrogen in pressure vessels or liquefaction at very low temperatures.

Large-scale underground hydrogen storage is a proven technology and can provide an excellent storage medium for electricity. Electrolytic hydrogen stored in underground caverns and re-electrified via combined cycle turbine power plants or via fuel cells on board vehicles (cycle efficiency 30–40%) incurs higher losses than competing technologies such as pumped hydro (cycle efficiency 80%) and adiabatic compressed air (cycle efficiency 70%). However, the potential of these alternatives is limited regarding possible applications and storage capacity. Hydrogen utilizes the storage capacity of caverns 60 times better than advanced compressed air storage.

22.3
Hydrogen in Transport

The most promising applications for hydrogen are to be found in the transport sector. There are two reasons for this: (1) the transport sector, which now relies nearly completely on fuels made from oil, has to shift progressively to renewable energies; and (2) hydrogen can be stored more easily than electricity. This is especially true for the energy storage on board of vehicles.

The currently emerging trend of the increasing electrification of the drivetrain of road vehicles will clearly benefit from the growing share of renewable electricity. Available technologies start with hybrid powertrains, leading to battery electric vehicles, range-extended vehicles, plug-in hybrids, and fuel cell hybrid vehicles (FCHVs). The automotive industry is about to start the transition from internal combustion engines (ICEs) to batteries and fuel cells. However, at the moment it is not clear whether there will be a significant market presence before 2015. Mass-manufactured battery electric and plug-in hybrid vehicles presumably will remain expensive and most likely will not provide a cost benefit compared with FCHVs.

FCHVs are the only foreseeable alternative technology for cost-effective longer range (>300 km) passenger cars, duty vehicles, and buses meeting customer requirements. Battery electric powertrains within present technology (but even with advances also in view of limits set by physics) will be suitable only for short-range vehicles.

22.4
National Strategies and Programs

22.4.1
Challenges

Most industrialized countries/regions (Canada, Europe, Korea, Japan, United States) and also emerging economies (Brazil, China, India) have established national programs to support the development and market introduction of hydrogen and fuel cells (H2&FCs). H2&FCs are key technologies for the transition from fossil fuels to renewable energies.

The programs are all driven to a varying extent by the three motives of: (1) security of energy supply mainly for transport and also the electrification of remote areas and grid stability; (2) reduction of local pollutants and greenhouse gas emissions in transport; and (3) industry policy goals, that is, the advancement of technological leadership and business opportunities.

With regard to the future oil supply, all programs were designed to prepare for an eventually diminishing fossil energy supply as described in the first part of this chapter, but most probably assuming time horizons lying two, three, or even more decades in the future. However, the developments in the oil markets in recent years (rising prices, price volatility, stagnating supply) have led to unprecedented downturns and crises in the automotive industry, first in the United States, and then a few years later in practically all countries except China. Because of peak oil being *now*, there is no lead time left any longer to prepare for the transition to renewable transport fuels. This causes an unexpected urgency for governments and industry.

Recently, a growing awareness by international institutions of the future challenges can be observed (see, for example, the unprecedented warnings by the IEA of an imminent oil crunch). This growing awareness is also reflected by the growing number of regional networks and projects in the area of hydrogen and fuel cells.

22.4.2
Europe

On a *European* level, there have been research activities in the general field of hydrogen technologies funded by the European Commission going on since the late 1980s. However, in the early years there was no explicit strategy to support these activities. The process to formulate an explicit European hydrogen strategy was mainly motivated by the American and Japanese hydrogen and fuel cell programs and really began with the formation of a body named the High Level Group on H_2 and FC. Members of this group were high-level representatives from industry and research institutions. The group worked under the auspices of the Directorate-General for Research. This group published in 2003 the report *Hydrogen Energy and Fuel Cells – a Vision of Our Future*, which constituted a major step in formulating a European strategy [8].

This was followed in 2004 by the initiation of the EU Hydrogen & Fuel Cell Technology Platform (HFP), in which all the relevant players cooperated. This body was responsible for the more detailed strategy formulation which in 2005 led to the publications *Strategic Research Agenda* [8] and *Deployment Strategy* [9]. The ideas presented in these publications were then discussed comprehensively and eventually in 2006 the HFP General Assembly finalized the *Implementation Plan* [10].

The strategic projects described above were building on important earlier projects and were also influenced by a series of parallel projects addressing specific strategic areas. The following projects should be considered in this context:

- EQHHPP (1989–1999): The Euro-Quebec Hydro-Hydrogen Pilot Project investigating the renewable production of hydrogen, its transport, and final use in a variety of applications [12].
- EIHP (1998–2004): This project aimed at creating the basis for the harmonization of necessary legislation for the approval of hydrogen vehicles and refueling infrastructures in Europe [13].
- HyNet (2001–2004): Major tasks of the HyNet – hydrogen network – were the development of a well-balanced European roadmap for a hydrogen energy infrastructure, an RTD strategy, and the assessment of the socio-economic and political issues associated with the move towards a hydrogen-based energy future. HyNet was a precursor of the HyWays project.
- HyWays (2004–2007): For the timeframes 2020, 2030, and 2050, an EU Hydrogen Energy Roadmap was formulated. For the participating regions, goals for greenhouse gas emissions and preferred hydrogen production and infrastructure technologies were integrated into the proposed roadmap for the build-up of supply infrastructures and end-use technologies [14].
- HyApproval (2005–2007): The aim of this project was to prepare a "handbook for approval of hydrogen refueling stations" which can be used to certify public hydrogen filling stations in Europe [15].
- HyLights (2006–2008): This project was a Coordination Action to accelerate the commercialization of hydrogen and fuel cells in the field of transport in Europe. HyLights assisted all stakeholders in the preparation of the next important phase for the transition to hydrogen as a fuel and energy vector to be produced in the long-term by renewable energies [16].

These projects contributed basic inputs for the formulation of a comprehensive European strategy.

The most recent step was the foundation of the FCH Joint Undertaking, an institution founded in accordance with Article 171 of the European Contract. The intention is to provide a long-term institutional framework for a public-private partnership under the auspices of the DG Research. This construction has the purpose of taking care of the needs of industry. The aim is to enhance and accelerate the development and commercialization of fuel cell and hydrogen technologies in the timeframe 2010–2020.

The activities of the FCH Joint Undertaking are now funded by the European Commission in a 7 year program (in the framework of FP7 until 2013) with M€470. Industry has to contribute at least the same amount. Individual projects applying for funding are subject to a public tender and are evaluated by an independent body. About one-third of the budget is envisaged for applications in transport and another third for stationary applications [17].

In *Germany*, the formulation of strategies and programs also evolved over a long time. Early industry-driven research activities in the 1980s (storing hydrogen in metal hydrides in cars by Daimler) and ICE cars fueled with liquid hydrogen (BMW) since 1980 followed by the first prototypes of fuel cell vehicles in the 1990s (Daimler), led in 1998 to the formation of an industry grouping of automotive companies and energy companies to address the long-term perspectives of alternative fuels and propulsion systems. This grouping, with the name Verkehrswirtschaftliche Energiestrategie (VES), or in English Transport Energy Strategy (TES), is still active and is the industrial partner for the government in the process of formulating fuel strategies and especially a strategy regarding hydrogen applications [18, 19].

Emerging from the VES, a major step forward was the founding of the Clean Energy Partnership (CEP) in 2003. The goal of the CEP is to prove that hydrogen can be used by regular customers in road transport and that renewable energies can be used for hydrogen production. Possible obstacles on the way to commercialization after 2015 will be identified and removed. CEP is operating three hydrogen refueling stations (HRSs) in Berlin, which is being extended to Hamburg with one large HRS in 2010 and four to follow by 2014, and will be connected with a highway HRS half way between the two cities.

The described activities led to a German public–private partnership program named NIP-NOW [20]. NIP stands for National Innovation Program for Hydrogen and Fuel Cell Technology, drawn up jointly by the Ministries responsible for transport, economy, research, and the environment. The operating agent for this program is NOW, the National Organization for Hydrogen and Fuel Cell Technology.

Technology validation and initial infrastructure clusters for fuel cell mobility and stationary applications are a main focus of these programs. The largest part of the budget is assigned to mobile applications (42%), followed by industrial applications (26%) and residential applications (14%). The program funding extends over 10 years with a total budget of B€1.4, cost shared with at least 50% from industry.

Denmark has a substantial funding of hydrogen fuel cell mobility of approximately M€45 per year and is gradually building up a hydrogen refueling infrastructure. Hydrogen vehicles, like battery vehicles, are exempted from a 180% luxury sales tax levied on conventional vehicles.

Norway has already built a rudimentary hydrogen refueling infrastructure and has started the first limited fleet vehicle operations [21].

22.4.3
North America

California has a long-standing clean air policy implemented by the Californian Air Resources Board (CARB). An important target of this policy is the reduction of emissions of local pollutants caused by road traffic. To achieve this goal, the Zero Emission Vehicle (ZEV) requirement has been mandatory in California since 1990. Until March 2008, this regulation required large-volume car manufacturers (selling >60 000 vehicles per year in California) to provide a certain share as ZEVs, of which at least 50% had to be fueled by hydrogen. This was the main driver for car companies with high-volume sales in California to develop fuel cell vehicles (FCVs), namely Ford, General Motors, Daimler-Chrysler, Honda, Nissan, and Toyota. With this ZEV regulation, California had a decisive influence on the technological development goals of the car industry worldwide: only "big sellers" in California started to develop fuel cell drive systems. Other major European and Japanese car companies with a lesser presence in California ignored the FC technology. Hence it is a fair observation that this Californian policy has had a greater influence on the technological orientation and progress of the global car industry than any other national program before or since.

Since April 2008, the required number of vehicles for the timeframe 2012–2014 was reduced to at least 7500 (while in the period 2012–2017 between 27 500 and 75 000 vehicles will come into use, depending on the ZEV type). However, now that part of the ZEV requirement formerly attributed explicitly to FCVs can also be met by battery electric vehicles (BEVs). Car manufacturers now have the choice to produce either FCVs or BEVs. Some observers suspected that this might be a major blow for FCVs. It is therefore worth noting that this change in the ZEV regulation has not yet changed the policy of companies developing FCVs. Several car manufacturers have emphasized that they will maintain their FCV development and commercialization goals for 2015, namely Daimler, Honda and Toyota.

In April 2004, CARB started the California Hydrogen Highway Network (CaH2Net) as a public-private partnership [22]. The mission of the CaH2Net is to promote a rapid transition to a clean hydrogen transportation economy in California. The main focus is the build-up of a hydrogen refueling infrastructure. In addition to energy security and air quality, the reduction of greenhouse gases is for the first time also stated as a goal.

On a federal level, the *United States* has had significant hydrogen and fuel cell programs only since 2002. A major driver to start such a program was the shock of "9–11." In November 2001, the DOE held a conference (National Hydrogen Vision Meeting) to discuss policies to enhance American energy security. The outcome was a vision paper highlighting the potential role of hydrogen [23] and as a consequence a national program for hydrogen. FCVs using hydrogen – which eventually could be produced domestically – opened the perspective of greater energy security and independence (this is also a reason why FCVs with onboard reforming of gasoline did not appear as an attractive option). Also worth noting is the Energy Policy Act of 2005 [24]. This law has a separate chapter on hydrogen and envisages possible budgets for this topic amounting to several hundred million dollars for the coming years (however, this law is very controversial for its bias towards fossil energies). Hydrogen from nuclear energy is regarded as being an option; there is no exclusive focus on renewable energies.

Since 2002, the United States has spent on average some US$250 million annually. The new Obama administration is not backing hydrogen technologies. However, the funding for hydrogen and fuel cell technologies (DOE's EERE program) was recently restored in Congress to at least US$153 million for the fiscal year 2010, after an initial attempt by the US administration to cut the budget significantly. Before that, California (with the project CA H2 Highway) together with South Carolina had made it clear that they will continue with their hydrogen and fuel cell demonstration and infrastructure activities. There are also numerous other American states that have FCV and hydrogen infrastructure demonstration projects.

To sum up, in the United States we can observe two different political motivations for political programs in the area of hydrogen and fuel cells: a focus mainly on clean air in California and a focus mainly on energy security at the federal level. Climate protection until now was not an issue at the federal level, and in California only in recent years.

In *Canada*, awareness of the strategic potential of hydrogen goes back to the late 1980s. This expressed itself in the remarkable report *Hydrogen – National Mission for Canada* prepared by the Advisory Group on Hydrogen Opportunities in June 1987 for the Ministry of State for Science and Technology and for the Ministry of Energy, Mines, and Resources. This is perhaps the first comprehensive attempt to describe a hydrogen strategy worldwide.

A major milestone for hydrogen in Canada was the Euro-Quebec Hydro-Hydrogen Pilot Project (EQHHPP), which started in 1989 and ended after 10 years in 1999. The original driving idea was to find a way to use the huge underutilized hydro-power potential of Quebec by exporting the energy in the form of hydrogen. In this project were involved from the Canadian side the province of Quebec, the utility Hydro-Quebec, and industry partners. This project, because of its size (about 60 participating companies and research organizations), its comprehensive scope, its duration, and international cooperation, had a great influence on hydrogen activities worldwide.

In Canada today, in addition to basic R&D in research departments, several demonstration and market preparation activities are under way. Examples are the Hydrogen Highway in British Columbia, the Hydrogen Village in Ontario and the Vancouver FCV program. The Whistler Winter Olympics will be served by 20 fuel cell transit buses which will afterwards be operated for another 5 years at least.

22.4.4
Asia

Japan is a country which nowadays lacks domestic fossil energy resources. Also, until recently in Japan there was a prevalent perception that the country has hardly any renewable energy resources. The idea of hydrogen as an energy vector with the potential to import renewable energies at a large scale from other parts of the world therefore seemed attractive. Such a vision of a global hydrogen network was the subject of the so-called WE-NET project (standing for World Energy Network Using Hydrogen), which was launched in the early 1990s. NEDO and MITI (later METI) coordinated and funded the project. This vision was certainly inspired by the EQHHPP which had started a few years earlier in the late 1980s.

The WE-NET project (which was mainly focused on fuel supply) was in later years gradually complemented and substituted by the Japan Hydrogen and Fuel Cell (JHFC) demonstration project. This new focus on FCV applications was most probably a reaction to the development of FCVs by Daimler which started in 1994. This was reinforced when Daimler announced the market introduction of FCVs for the early years in the new century. This technological competition together with the Californian ZEV requirement was the main motivation for several Japanese car manufacturers to intensify their FCV development and commercialization efforts.

The Japanese car industry is preparing the production of FCVs and has announced and reaffirmed this intention during recent months. Honda announced the production of a prototype series of 200 vehicles until 2011. Toyota plans to take up mass volume manufacturing of FCVs in 2015 and Honda in 2018.

In Japan, stationary fuel cells in the 1 kWe class are about to being commercialized. The targeted production volumes are at least 10 000 units up to 2010 and 60 000–100 000 units by 2015.

China has concentrated its hydrogen activities in two programs: the 863 program and the 973 program, which both provide funding for electric vehicles and battery and fuel cell vehicles [25]. The 863 project also provides funds for the development and trials of hydrogen refueling stations [26].

There are 13 electric mobility metro regions in China which are aiming to put 60 000 alternative fuel vehicles on the road by 2012. The incentives provided by the Ministry of Finance under this program for FCVs are in the order of RMB 200 000 or about € 21 000 per vehicle (about € 6500 per BEV). The aim of the program is to reduce air pollution in megacities and to improve fuel supply diversity.

Korea has a very ambitious program for hydrogen and fuel cells. The plan is to have about 3000 fuel cell cars and 200 fuel cell buses on the road by 2012, and also by 2012, the installation of 10 000 small residential fuel cell power generators (<3 kWe), 2000 medium-sized building integrated fuel cell generators (5–50 kWe), and 300 large distributed power generators (0.25–1 MWe) [27].

India, in its National Hydrogen Energy Roadmap up to 2020, plans to have one million vehicles fueled by hydrogen, vehicles with internal combustion engines and also with fuel cells. Also, the installation of an electricity generation capacity using fuel cells of 1000 MW is envisaged. A budget of B$5.2 (B€3.7) is estimated for the whole program up to 2020 [28].

22.5
Conclusion

The evolution of national strategies and programs shows no clear trend. In the early days, Europe together with Canada (EQHHPP) and Japan (WE-NET) were leading the way by elaborating ambitious visions for the global transport of energy using hydrogen as a carrier. However, these visions were not really followed up and were not imitated in other parts of the world. In later years, the focus increasingly shifted to hydrogen as a fuel for transport with the development of fuel cells for cars. Japan is a good example.

In 2002, the United States at a federal level developed a comprehensive vision for hydrogen which did not take off in the following years, and activities are now low key. The initiative is with California and other states and regions.

On a European level, in this decade strategy formulation has shown direction and continuity. Also in Germany, the national strategy has gathered pace based on a decade-long commitment by the automotive industry.

Climate protection was always a major issue in Europe regarding hydrogen. This is now also moving up in priority in California.

China is rapidly following the development of industrialized countries with a steep rise in car ownership, albeit still at a low level. The more successful this development is, the sooner this success will undermine its preconditions – in China and globally. China is well aware of the limits to oil supply and therefore sees the need for hydrogen as a future automotive fuel [29].

Peak oil will mark the end of business as usual. This will certainly bring a new urgency to the further development of hydrogen and fuel cell technologies. Perhaps it is worthwhile to revisit some of the visions developed in the past at various times in various regions of the world.

Acknowledgment

The author thanks Reinhold Wurster of Ludwig-Bölkow-Systemtechnik for his valuable inputs to this chapter.

References

1. International Energy Agency (IEA) (2008) *World Energy Outlook 2008*, IEA, Paris.
2. BP (2009) *BP Statistical Review of World Energy*, various editions 2009 and earlier, BP, London.
3. Energy Watch Group, Schindler, J., and Zittel, W. (2008) *Crude Oil – The Supply Outlook*, 2007, revised edition 2008, http://www.energywatchgroup.org (accessed 12 February 2010).
4. IHS Energy (2006), *Petroleum Economics and Policy Solutions (PEPS)*, IHS Energy, Geneva and London.
5. Koppelaar, R. (2009) *Oilwatch Monthly*, October 2009, ASPO Netherlands, Amsterdam, http://www.peakoil.nl (accessed 12 February 2010).
6. Energy Watch Group, Zittel, W., and Schindler, J. (2007) *Coal: Resources and Future Production*, http://www.energy-watchgroup.org (accessed 12 February 2010).
7. Schindler, J. and Zittel, W. (2007) Alternative world energy outlook 2006. A possible path towards a sustainable future. In *Advances in Solar Energy*, Vol. 17 (ed. Y. Goswami), earthscan, London, pp. 1–44.
8. DG Research (2003) *Hydrogen Energy and Fuel Cells – a Vision of our Future*, http://ec.europa.eu/research/energy/pdf/hlg_vision_report_en.pdf (accessed 12 February 2010)
9. European Commission (2005) *European Hydrogen and Fuel Cell Technology Platform, Strategic Research Agenda*, July 2005, European Commission, Brussels.
10. European Commission (2005) *European Hydrogen and Fuel Cell Technology Platform, Deployment Strategy*, August 2005, and *Deployment Strategy – Progress Report 2005*, October 2005, European Commission, Brussels.
11. European Commission (2007) *European Hydrogen and Fuel Cell Technology Platform, Implementation Plan – Status 2006*, March 2007, European Commission, Brussels.
12. Hydrogeit, HydroGeit.de, http://www.wasserstoff-autos.info/eqhhpp.htm (accessed 12 February 2010).
13. L-B-Systemtechnik, EIHP European Integrated Hydrogen Project, http://www.eihp.org (accessed 12 February 2010).
14. Ludwig-Bölkow-Systemtechnik, HyWays.de, http://www.hyways.de/hynet/welcome.html (accessed 12 February 2010).
15. Ludwig-Bölkow-Systemtechnik, HyApproval.org, http://www.hyapproval.org (accessed 12 February 2010).
16. Ludwig-Bölkow-Systemtechnik, HyLights.eu, http://www.hylights.eu (accessed 12 February 2010).
17. European Industry Grouping for a Fuel Cell and Hydrogen Joint Technology Initiative (NEW-IG), http://www.fchindustry-jti.eu/ (accessed 12 February 2010).
18. Bundesministerium für Verkehr, Bau und Stadtentwicklung, VES – Kraftstoff der Zukunft, http://www.bmvbs.de/Klima_-Umwelt-Energie/Mobilitaet-Verkehr-,2997/VES-Kraftstoff-der-Zukunft.htm (accessed 12 February 2010).
19. Bundesministerium für Verkehr, Bau und Stadtentwicklung, Klima, Umwelt & Energie, http://www.bmvbs.de/artikel-,302.22711/Mobil-mit-Wasserstoff-Clean-En.htm (accessed 12 February 2010).
20. NOW Nationale Organisation Wasserstoff- und Brennstoffzellentechnologie, http://www.now-gmbh.de/ (accessed 12 February 2010).
21. HYNOR, hynor.no, http://www.hynor.no/hynor–1/view?set_language=en (accessed 12 February 2010).
22. State of California, California Hydrogen Highway, http://www.hydrogenhighway.ca.gov/ (accessed 12 February 2010).

23 United States Department of Energy (2002) *A National Vision of America's Transition to a Hydrogen Economy – to 2030 and Beyond*, US DOE, Washington, DC.

24 109th Congress, Public Law 109-58 (2005) *ENERGY POLICY ACT OF 2005*, http://frwebgate.access.gpo.gov/cgi-bin/getdoc.cgi?dbname=109_cong_public_laws&docid=f:publ05 8.109 (accessed 12 February 2010).

25 Ming, P.-W., Lun, J.-G., and. Mytelka, L. (2007) *Hydrogen and Fuel-Cell Activities in China, 2007*, http://www.idrc.ca/en/ev-132182-201-1-DO_TOPIC.html (accessed 12 February 2010).

26 Ma, J.-X. (2007) *R&D Progress of FCVs and Hydrogen Infrastructure in Shanghai*, presentation given on 12 July 2007, Paris.

27 Kim, C.-S. (2009) *R&D Status and Prospects on Fuel Cells in Korea*, presentation given at the F-Cell Conference in Stuttgart, September 2009, http://www.f-cell.de (accessed 12 February 2010).

28 (a) Malhotra, R.K. (2009) *Hydrogen Energy Roadmap of India*, Indian Oil Corporation, NHA Annual Conference 2009; (b) Pal, N.K., Malhotra, R.K., and Kumar, A. (2009) *Hydrogen Initiatives in India*, Indian Oil Corporation, NHA Annual Conference 2009

29 Anon (2006) Oil import cap may force fuel cell vehicle leadership on Chinese market, *Automotive News*, 27 November 2006 [see interview "Bis 2020 soll jedes vierte Auto in China mit Wasserstoff fahren", *Die Welt*, 24 November 2006].

23
Renewable Hydrogen Production

Alan C. Lloyd, Ed Pike, and Anil Baral

Abstract

The need for renewable energy in an environmentally constrained world leads us to examine the potential role of hydrogen produced from renewable sources. Rationales for renewable hydrogen include climate, local air quality, and sustainability goals. Hydrogen can be produced from a variety of current and potential new future renewable resources for applications including transportation and electricity generation. We also examine the effectivenss of a number of policies that have been implemented or adopted. These policies range from specific standards for renewable hydrogen production to broader policies to increase renewable energy or reduce GHG more broadly.

Keywords: renewable hydrogen, hydrogen production, transportation, electricity generation, fuel cells

23.1
Introduction

This chapter provides some background on the need for renewable energy in an environmentally constrained world, and the role of renewable hydrogen. This includes a brief discussion of hydrogen use in transportation and in electricity generation, and hydrogen production from a variety of resources. This chapter also addresses the rationales for renewable hydrogen to address climate, local air quality, and sustainability goals, and describes a number of policies that have been implemented or adopted that will directly or indirectly promote renewable hydrogen production and utilization.

23.2
Rationale for Renewable Hydrogen

The world is facing unprecedented challenges on a number of fronts: demand for energy; the dramatic increase in the number of megacities globally; and the need to reduce greenhouse gases dramatically to constrain atmospheric temperature

Hydrogen Energy. Edited by Detlef Stolten
Copyright © 2010 WILEY-VCH Verlag GmbH & Co. KGaA, Weinheim
ISBN: 978-3-527-32711-9

increases to 2 °C or less. While each of these issues is distinct, they are linked by the need to identify and deploy zero and near-zero emissions technologies. Although the developed world has accounted for the majority of world energy use, the United States Energy Information Administration predicts that demand for energy will increase by 70% from 2006 to 2030 in the developing world [1] due to substantial economic development with the concomitant increases in energy demand in terms of electricity and fuels. Increasing standards of living are likely to lead to more cars, larger homes, greater use of air conditioning and other electrical appliances, increased air travel, and other demands for industrial, commercial, and residential energy use [2]. Local impacts include degraded air quality, while global impacts include climate change. A substantial increase in energy efficiency in electricity generation, in buildings, and in other uses is the first option, together with greater fuel economy for vehicles, for both the developed and the developing world. However, this strategy is necessary but not sufficient to meet the energy, environmental, and climate challenges mentioned above [3, 4].

Renewable energy has been recognized as playing a key role in our energy future, with a diversity of renewable energy sources which would vary geographically [5]. There are many benefits, but a new set of policy initiatives is needed to address a number of challenges and capture fully the potential of renewable energy. Nearly two decades later, we are seeing some of these policy initiatives [6] in place, but the deployment of renewable electricity generation has been slower than the environmental, energy, and climate challenges require.

The use of renewable energy to generate hydrogen offers the potential for a truly zero emission technology for a variety of applications described below. Renewable energy and other sources of hydrogen are shown in Figure 23.1, which highlights one of the attractive features of hydrogen as an energy currency, namely that it can be produced from a variety of sources. In addition, hydrogen is seen as an attractive energy carrier [7, 8] for a wide variety of applications, from small to large scale, including vehicles, ships, and aircraft. However, with increasing concern about the buildup of greenhouse gases in the atmosphere, it is recognized that the potential of hydrogen for addressing all the issues cited above can only be met completely if hydrogen is produced from renewable sources. These include wind, solar, geothermal, wave power, biomass, and other potential renewable energy precursors to electricity production.

In addition, many renewable energy hydrogen production technologies offer the potential virtually to eliminate local and regional pollutants contributing to negative public health effects (some energy use would still be embedded in the materials used to manufacture these technologies). For instance, over 122 million people in the United States live in areas that do not meet national ozone standards [9]. One study predicts over two million mortalities globally from ozone levels above pre-industrial levels by 2050 [10]. Although long-term future effects of air pollution are difficult to predict, this study shows the long-term importance of this issue.

Developing renewable energy to produce electricity and hydrogen can occur in parallel with the more efficient production and use of fossil fuels, potentially

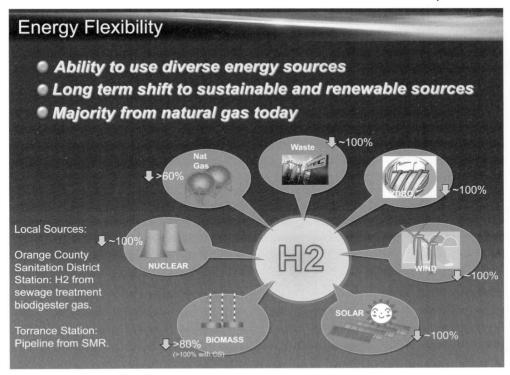

Fig. 23.1 Energy flexibility and GHG reduction for hydrogen (source: Honda presentation to US Department of Energy Hydrogen and Fuel Cell Advisory Committee November 2009 meeting).

combined with carbon capture and sequestration, as fossil fuels are not likely to be eliminated in the near or intermediate term. However, hydrogen and especially renewable hydrogen offer alternatives to limited fossil resources for transportation and other applications.

Renewable hydrogen offers the economic opportunities and energy security benefits of creating fuels domestically for many nations that currently rely on imported petroleum and other fossil fuels, and also the potential for new forms of economic development. Scott [11] identified both hydrogen and electricity as energy currencies that are synergistic and used the word "hydricity" for a joint energy currency of hydrogen and electricity. Later sections of this chapter explain some of the potential synergies between renewable electricity and renewable hydrogen.

The utilization of renewable energy with hydrogen is a component of an intriguing concept advanced by Rifkin as part of a potential Third Industrial Revolution (J. Rifkin, personal communication, 2009). Rifkin identified previous synergies between energy usage, such as the coal-powered steam energy and later the internal combustion engine, and advances in communications starting with the printing process and later with communications enabled by the development of electronics. Rifkin envisioned a Third Industrial Revolution (which is being explored in several areas of Europe and to a lesser degree in the United States):

> We are now on the cusp of a Third Industrial Revolution. Today, the same design principles and smart technologies that made possible the internet and vast distributed global communication networks, are just beginning to be used to reconfigure the world's power grids so that people can produce renewable energy and share it peer-to-peer, just like they now produce and share information, creating a new, decentralized form of energy use. We need to envision a future in which millions of individuals can collect and produce locally generated renewable energy in their homes, offices, factories, and vehicles, store that energy in the form of hydrogen and share their energy with each other across a continent-wide intelligent intergrid.

23.2.1
Overview of Transportation Applications

By far the major application of hydrogen in the transportation sector is through its application in the fuel cell using proton exchange membrane (PEM) technology. Fuel cell technology vehicles typically are 2–3 times more efficient than today's internal combustion engine. Both of these technologies will improve in future years, along with light weighting of vehicles that will improve the efficiency of both types of technologies while decreasing the amount of fuel storage needed on the vehicle to achieve a desired range. Fuel cells are poised for commercialization beginning in 2015 if further technical development addresses technical challenges such as reducing fuel cell stack costs, on-board hydrogen storage, and hydrogen infrastructure. Table 23.1 illustrates potential future fuel cell vehicle costs. The application of hydrogen in the transportation sector (both light-duty vehicles and buses) has been widely covered in many publications [12–16].

The greenhouse gas (GHG) reduction benefits of hydrogen fuel cell vehicles are dependent on the source of hydrogen, as shown in Figure 23.2. It is increasingly recognized that electric drive vehicles (powered by renewable energy) and the potential expanded use of biofuels will be necessary to achieve a dramatic reduction in greenhouse gases as required by 2050. For example, California will require a dramatic per capita reduction in GHG emissions as shown in Figure 23.3. (It should be noted, however, that fuel cell vehicles powered by hydrogen produced from natural gas steam reforming will lead to a reduction in GHGs by up about 30% compared with current internal combustion engines [18].) Zero tailpipe vehicle technologies powered by non-emitting renewable energy will also achieve significant reductions in pollutants causing local and regional air pollution.

Table 23.1 Estimated incremental OEM costs (US$) for vehicle technologies compared with the 2030 NA-SI (Naturally aspiration spark ignition); the impact of optimistic battery projection (based on $150 per kWh for a high-energy battery) and conservative fuel cell projection (based on $75 per kW) is reflected in parentheses.

Component	Turbo	Diesel	HEV	PHEV-10	PHEV-30	PHEV-60	FCV	BEV
Drive train								
Motor/controller	–	–	600	800	800	800	1400	1400
Engine/transmission	500	700	200	100	100	100	–3500	–3500
Fuel cell	–	–	–	–	–	–	3000 (4500)	–
Energy storage								
Battery	–	–	900 (700)	1500 (1200)	2800 (2200)	4600 (3700)	1000	12000 (8600)
H_2 storage (150 l)	–	–	–	–	–	–	1800	–
Miscellaneous								
Exhaust	0	500	0	0	0	0	–300	–300
Wiring, etc.	–	–	200	200	200	200	200	200
Charger	–	–	–	400	400	400	–	400
Total	500	1200	1900 (1700)	3000 (2700)	4300 (3700)	6100 (5200)	3600 (5100)	10200 (6900)

Source: Kromer and Heywood, 2007 [17].

Hydrogen has also been used in internal combustion engines, both for light-duty vehicles and in buses. Both Mazda and BMW have developed vehicles to run in a hydrogen internal combustion engine. However, BMW has recently announced that it would discontinue its program utilizing hydrogen, Mazda has not yet reached the demonstration phase, and the inefficiencies for this application seem to preclude large-scale deployment of this technology.

Hydrogen has been employed blended with natural gas in heavy-duty vehicle applications. Hydrogen compressed natural gas (HCNG) engines offer the benefit of reduced nitrogen oxide emissions compared with pure CNG, attaining low emissions of both greenhouse and criteria pollutants [19]. The trend towards increasing use of natural gas in the transportation sector in various regions of the world, particularly India, could see greater use of hydrogen for this application.

23.2.2
Other Applications of Hydrogen

This section covers three additional applications of hydrogen: use in forklifts for material and goods handling; use in backup power for telecommunication towers; and as a storage medium to facilitate the intermittent nature of several types of renewable energy.

470 23 Rerewable Hydrogen Production

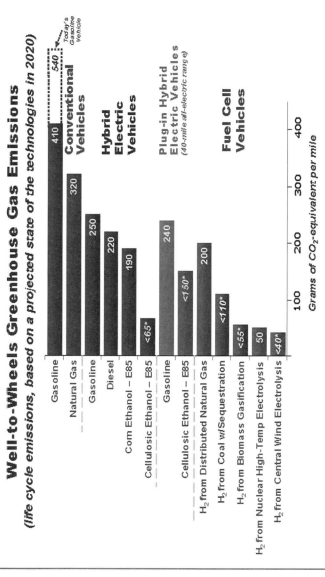

Fig. 23.2 Transportation technologies to reduce GHG emissions.

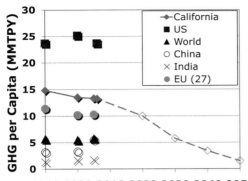

Fig. 23.3 GHG per capita and California goals.

23.2.2.1 Fork Lifts

There are a very large number of forklifts in application throughout the world for material handling, goods movement, and port applications. Forklifts used in enclosed spaces can use batteries when zero exhaust emissions are required. However, recharging the batteries can take a significant amount of time once the batteries are discharged. The application of hydrogen fuel cells for this use provides not only zero emissions, but also a much faster recharging using hydrogen and thus higher productivity.

23.2.2.2 Backup Power Applications

Hydrogen fuel cells are increasingly used to provide power for various backup power applications. Examples of extended run, backup power applications are shown in Table 23.2. All of these examples can be used in conjunction with batteries or as replacement for batteries. They have the benefit of operating for a significant period of time and can use renewable hydrogen in a clean and quiet operation. In some cases, fuel cells will replace diesel generators, which are both noisy and highly polluting sources of ozone-forming pollutants and particulates.

One of the greatest growth areas for the application of hydrogen fuel cells is to provide backup power for telecommunication towers. As the use of cell phones has increased dramatically throughout the world, there is a great need to improve the reliability of the telecommunication towers. Towers placed in re-

Table 23.2 Examples of extended backup power applications.

Utility	Telecommunications; highway/railway signaling and communications; pumping
Security	Surveillance and sensing
Commercial	Uninterruptible power supply

mote areas, and in developing areas without reliable electricity supply, will require more reliable power. Hence telecommunication operators are looking for reliable backup power to ensure uninterrupted cell phone operations. Hydrogen fuel cells have been increasingly used to meet these needs in the last few years. As a result, an increasing market has developed in the United States, Europe, India, Indonesia, and Latin America for fuel cells in the 1–5 kW range which can operate on hydrogen (the hydrogen can be delivered or generated on-site from renewable resources or from reforming methanol or natural gas). The use of renewable hydrogen in this application is ideal for reducing air pollution and reducing or eliminating GHGs from this energy source.

23.2.2.3 Hydrogen as a Storage Medium for Renewable Energy

Renewable energy resources are often located away from electric demand centers. Although transmission lines can be built or upgraded, this infrastructure is extremely costly. Hydrogen production can help make use of "stranded" renewable resources when access to electrical transmission is not available, although a method of local use or transportation would be necessary. Some examples in Argentina, Canada, and the United States produce between 30 and 260 kg per day (K. Harris, personal communication, 23 December 2009).

As stated earlier, the need to reduce GHGs dramatically will lead to increasing uses of renewable energy. Wind and solar energy are two common forms of intermittent renewable energy for replacing fossil fuels. Intermittent resources, especially wind power, would benefit from some form of energy storage to match energy demand better. As noted by Rifkin and described later in the renewable hydrogen pathways and policy drivers sections (Sections 23.3 and 23.4), hydrogen is a form of energy storage that could help overcome this potential barrier to national and regional renewable energy goals.

Several challenges remain before hydrogen and associated technologies can be widely deployed for the applications described in this chapter, particularly in the transportation sector. These include technical, economic, and social issues. How effectively and sustainably these are addressed will largely dictate how significant a role hydrogen will play in addressing environmental, climate, and energy challenges.

More in-depth discussion of potential renewable hydrogen sources for these potential applications and some of the policy drivers to encourage its use are provided in the following two sections.

23.3
Renewable Hydrogen Pathways

This section provides a brief overview of a number of potential renewable hydrogen production pathways. These pathways range from commercially available electrolysis using renewable electricity to potential future pathways using

23.3.1
Electrolysis Using Renewable Electricity

Electrolysis can create renewable hydrogen using electricity from a variety of centralized or distributed renewable resources. In electrolysis, water is split into hydrogen and oxygen by passing electric current through ionized water. Hydrogen gas is collected at the cathode and oxygen gas at the anode (equation 23.1). An advantage of using renewable zero carbon electricity is that it lowers the carbon intensity of hydrogen to near zero. Electrolysis requires a theoretical minimum of 39 kW h per kg of hydrogen produced, and in practice a range of 54–67 kW h kg^{-1} [20], so this production pathway is very dependent on electricity pricing (and also electrolyzer cost). An efficiency of 58% has been obtained for a proton exchange membrane (PEM) electrolyzer with a production capacity of 100 kg per day [21]. Studies by the United States National Renewable Energy Laboratory (NREL) have found that the United States has sufficient resources to replace gasoline usage with renewably generated hydrogen- if and when scientific and technical barriers are overcome [22].

$$2H_2O \rightarrow 2H_2 + O_2 \qquad (23.1)$$

Examples of centralized electricity generation sources include wind (large-scale), solar thermal, geothermal, and biomass (whether hydro is considered renewable depends on the type of hydro resources, such as run of the river versus storage behind large dams, and how renewable energy is defined). Renewable electricity can be used to generate hydrogen via electrolysis on-site at these central locations, avoiding energy losses of electricity transmission and distribution and costs needed for any additional electrical transmission system infrastructure. Electrical transmission costs and development timeframes can be especially important when renewable resources are located away from demand centers. A downside of centralized hydrogen generation is the cost and energy needed to distribute the hydrogen to fueling stations through options such as truck, rail, or pipeline if renewable resources are not located close to hydrogen demand centers. Thus, central electricity generation may be used for either central or distributed electrolysis generation of hydrogen based on availability of infrastructure for moving these resources to customers and the relative cost of any infrastructure upgrades needed for these two options.

Another option is local hydrogen generation via electrolysis using renewable energy generated near the demand location. Small-scale local electrolysis (100–500 kg per day) is particularly attractive when the number of hydrogen-powered vehicles is small. It can also catalyze the growth of hydrogen use. Photovoltaic solar and wind are two potential approaches for local generation, and can be used either with or without grid access. Grid access, where available, offers several advantages for distributed hydrogen production even when renewable electricity can be generated locally. First, a "net metering" arrangement could allow export of excess power during peak electrical production and balancing imports when demand is low. This would allow electrolyzer sizing based on average production on an annual basis, rather than over-sizing the electrolyzer to capture peak production levels of the local renewable resource. Second, if 100% renewable energy is not required, then local generation could be blended with grid power, which may facilitate incremental hydrogen production scale-up that may be needed due to gradually increasing hydrogen demand (e.g., fuel cell vehicle fleets).

Local hydrogen production would avoid the need for transportation and distribution, but may be limited by available land area, and the best areas for renewables may not be located at the same sites with maximum demand. Distributed electricity generation (on-site or near hydrogen stations) by itself will not necessarily be sufficient to serve demand entirely, if hydrogen completely displaced gasoline in the United States; in most cases however, local renewables combined with other in-state resources would still suffice [22].

23.3.2
Costs of Hydrogen Production from Electrolysis

The United States Department of Energy (DOE) has set the goal of achieving the cost of delivered hydrogen at $ 2.0–3.0 per gallon of gasoline equivalent (gge) based on fuel energy content before tax by 2015, independent of hydrogen pathways. One gge is approximately equal to 1 kg of hydrogen on an energy content basis. When used in a fuel cell vehicle, however, each BTU from hydrogen is expected to provide two to three times the useful work to the wheels as gasoline burned in an internal combustion engine, which is not reflected in these cost estimates.

Costs of hydrogen vary depending on technologies, scale of production, and modes of delivery. Normally, hydrogen produced from a central plant has a lower cost than hydrogen produced from decentralized and small units at fueling stations, as the economies of scale and lower industrial rates for power and feedstocks more than compensate for handling and delivery costs [23].

According to Nicholas and Ogden [24], the cost of decentralized electrolysis vary from $ 3.5 per gge for using non-renewable electricity from the grid to $ 4.1–6.1 per gge for using 100% renewable electricity from the grid to $ 19.4 per gge for using electricity from photovoltaic cells. According to the California Air Resources Board (CARB) [25], costs of hydrogen today including transport are $ 7.2, $ 4.4, and $ 11.30 per gge for distributed electrolysis, central wind electrolysis, and central solar electrolysis, respectively. Although estimates may vary based on specific scenarios and assumptions, renewable hydrogen costs are significantly higher than the $ 2.6 per gge cost for hydrogen from central natural gas reforming. Electrolysis costs are expected to drop significantly in the future [26], and increased efficiency of a fuel cell vehicle would help bridge the gap between the cost of renewable hydrogen and current petroleum prices. In addition, other alternatives including those listed below are less well established but have the potential for lower renewable hydrogen costs in the future.

23.3.3
Renewable Hydrogen from Biomass

Renewable hydrogen can be produced from biomass in a number of different ways. Several pathways follow technologies that are well established but have not yet been widely deployed for hydrogen production using biomass (as noted earlier, conventional electrolysis and renewable electricity from biomass are both well established). This sub-section describes (1) biomass gasification, (2) biomethane production followed by steam methane reforming, and (3) pyrolysis. Several less developed potential biomass technologies are described later.

23.3.3.1 Biomass Gasification
Steps involved in hydrogen production from biomass gasification are shown Figure 23.4. In gasification, the prepared biomass feedstock reacts with the limited amount of oxygen under high temperatures and pressures (1150–1425 °C at 400–1200 psig). This produces syngas which consists of hydrogen and carbon monoxide:

$$C_aH_b + a/2\,O_2 \rightarrow b/2\,H_2 + aCO \tag{23.2}$$

The yield of hydrogen can further be improved by including a water gas shift reaction step in which carbon monoxide reacts with water in the presence of catalysts to produce hydrogen and carbon dioxide:

$$CO + H_2O \rightarrow H_2 + CO_2 \tag{23.3}$$

Hydrogen produced in reactions 23.2 and 23.3 can be purified using multi-bed pressure swing adsorption (PSA) to remove methane, water, CO, N_2, and CO_2 and either compressed or liquefied for delivery. It is also possible to remove CO_2 by chemical adsorption.

Hydrogen production from biomass gasification alone has not been demonstrated on a commercial scale so far. However, biomass has been co-processed along with coal in a 25 to 75% ratio (by mass) to produce hydrogen via gasification on a commercial scale. Biomass gasification facilities in the range of 100 MW and under have been constructed in Finland, Germany, The Netherlands, Portugal, and Sweden, producing syngas [27].

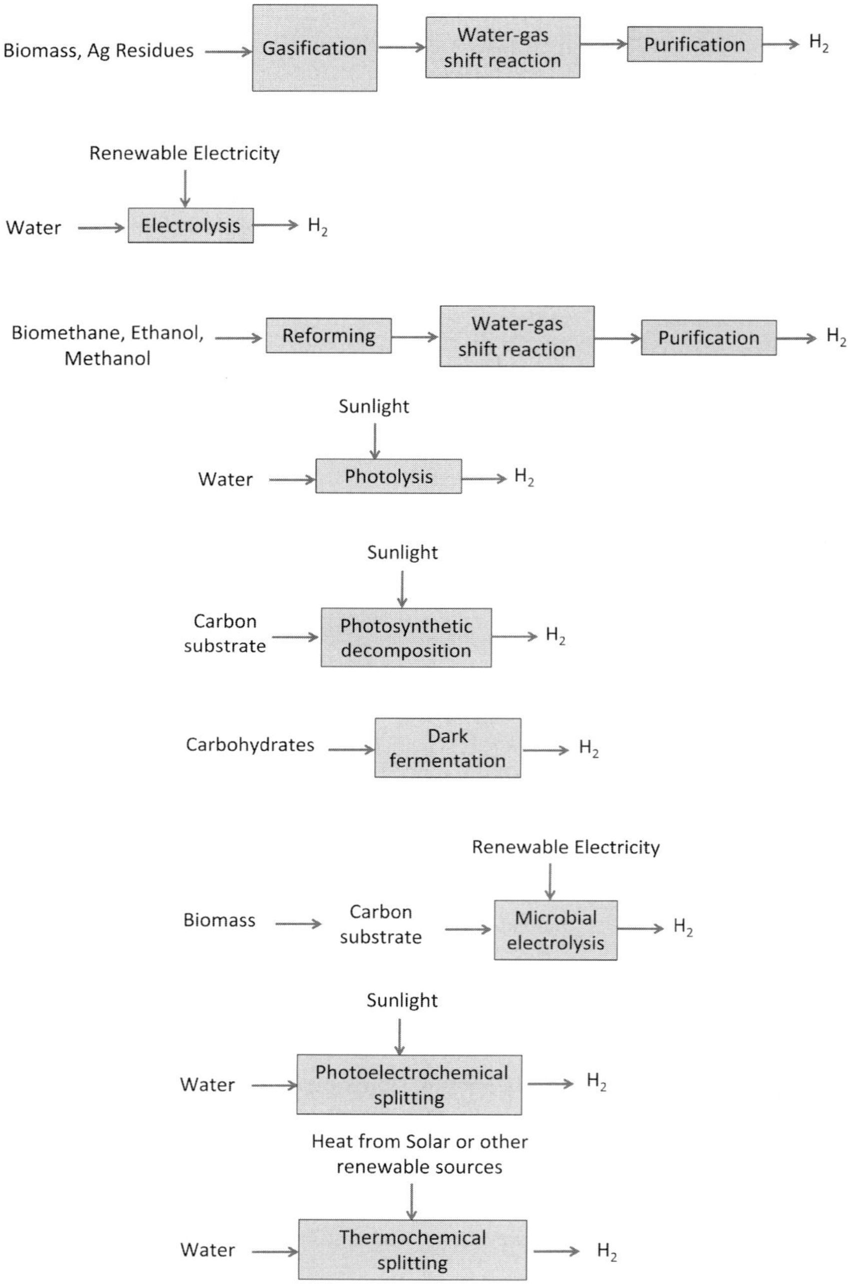

Fig. 23.4 Renewable hydrogen production pathways.

23.3.3.2 Biomethane Steam Reforming

Steam reforming is a well-established technology for hydrogen generation from natural gas (CH_4), including hydrogen used at refineries for desulfurization and denitrification of conventional fuels. The same process would apply to methane (CH_4) created from biological sources. First, biomethane is generated from biomass, agriculture residues, manure, and Municipal Solid Waste (MSW) either using anaerobic digesters or in landfills. Biomethane is subjected to reforming in which biomethane reacts with steam at 750–800 °C to produce syngas in the presence of catalysts. To avoid catalytic poisoning, biomethane is pretreated to remove any sulfur and chlorine from the biomethane stream.

$$CH_4 + H_2O \rightleftharpoons 3H_2 + CO \tag{23.4}$$

Syngas is further subjected to a water gas shift reaction to convert CO and H_2O to hydrogen and CO_2 according to equation 23.3.

In addition to biomethane, organic compounds derived from biomass such as ethanol and methanol can also used to generate hydrogen via steam reforming. For thermally unstable organic compounds such as sugars and polyols, aqueous phase reforming can be used; this is carried out at low temperature (225–265 °C) and pressure (35 bar).

23.3.3.3 Biomass Pyrolysis

As an alternative to biomass gasification, biomass can be converted to bio-oil using pyrolysis at high temperatures. Bio-oil can then be transported to fuel stations where it is converted to hydrogen and carbon dioxide. Bio-oil can be volatilized using ultrasonic atomization and can be subjected to partial oxidation to obtain syngas containing significant amounts of CO and hydrogen [28]. Carbon monoxide reacts with water formed during partial oxidation to produce hydrogen and CO_2 via the water shift reaction.

23.3.4
Application of Carbon Capture and Sequestration to Renewable Hydrogen Production

Carbon capture and sequestration (CCS) efforts globally have been focused on fossil-generating technologies, which can potentially improve the GHG footprint of fossil-based hydrogen production. For instance, a project in Norway has sequestered over 10^7 tonnes of CO_2 since 1996 [29]. These technologies could improve the sustainability of fossil-generated hydrogen and could also be applied to renewable bioenergy sources used to generate hydrogen, although there are a number of challenges that would need to be overcome [30].

For instance, there are a number of fossil integrated gasification combined cycle (IGGC) projects proposed with CCS. For instance, the "Green Gen" project in China's Tianjin City has begun construction with the first phase (250 MW) expected to come online in 2011 and expanding to 650 MW by 2016. Another planned IGGC project is a project to demonstrate 90% CCS in Central Valley in California with construction beginning in 2011 and operation in 2015; and a 900 MW project at Hatfield Colliery in the United Kingdom. These plants will burn the produced hydrogen for power, although the technology could also be used to produce hydrogen for transportation or other uses. Although these projects do not involve biomass, they would set a precedent for potential IGGC plants that blend in biomass feedstocks.

In addition to gasification, the DOE has recently funded two CCS demonstrations for methane steam reforming. Each project is expected to capture and sequester over 10^6 US tons of CO_2 per year. Methane steam reforming is a good potential candidate for CCS because concentrated CO_2 is released as a result of the reaction creating hydrogen and is already captured in a number of existing facilities at refineries (although not necessarily for permanent storage). These natural gas projects would reduce the carbon footprint of hydrogen production, although they would not utilize a renewable resource.

Application of CCS to biomethane steam reforming would reduce carbon emissions, potentially below zero, while utilizing a renewable resource. This application of CCS to biomethane may depend on location-specific availability of economical access to permanent storage options such as local geological storage and/or transportation to storage locations. These options may be more challenging for smaller biomethane projects that are not sited next to existing resources due to significant fixed costs for geological storage and/or pipeline transportation of CO_2.

23.3.5
Renewable Hydrogen in Early Research and Development

This section provides a brief overview of some potential emerging technologies. Biological hydrogen production, photoelectrochemical hydrogen production, and thermochemical production may emerge as promising technologies in the future.

23.3.5.1 Biological Hydrogen Production

This pathway takes advantage of microbial activities to produce hydrogen. There are four potential biological hydrogen technologies: photolytic (direct water splitting), photosynthetic (solar-aided organic decomposition), fermentative (organic decomposition), and microbial electrolysis. Of these, the last also utilizes electricity to enhance the microbial activity.

The photolytic process uses green algae and cyanobacteria and sunlight to split water directly into hydrogen and oxygen ions. A hydrogenase enzyme is involved in producing hydrogen gas from hydrogen ions. The other process using solar energy, the photosynthetic process, relies on purple non-sulfur bacteria to decompose organic compounds, preferably organic acids into hydrogen. Under nitrogen-deficient conditions, these bacteria capture sunlight near the infrared region and decompose acids with the help of nitrogenase enzymes.

In the fermentative process (also called dark fermentation), anaerobic bacteria act on carbohydrate-rich substrates to produce hydrogen. In microbial electrolysis, bacteria grown in the anodic chamber decompose carbon substrates such as acetic acid and cellulose, releasing electrons, protons, and carbon dioxide when a small amount of electricity is applied to an electrolysis cell. The protons migrate to the cathode, where they are discharged as hydrogen gas. Feedstocks used in microbial electrolysis can be obtained from biomass using hydrolysis and fermentation processes. The microbial electrolysis is an efficient process and the reported H_2 production yields are in the range 54–91% for various feedstocks [31]. To enhance the hydrogen production yield, it may be desirable to develop an integrated pathway that combines two or more of the technologies described above, such as the use of dark fermentation waste as a feedstock for microbial electrolysis.

23.3.5.2 Photoelectrochemical Hydrogen Production

A photoelectric pathway uses sunlight to split water into hydrogen and oxygen with the help of photoactive semiconductor materials. This production pathway may be suitable for centralized hydrogen production. A number of materials have been explored for use in photoelectrodes. For semiconductor materials that are stable in aqueous condition, the DOE's target is to achieve an efficiency of greater than 16%, with current levels well short of that goal [26].

23.3.5.3 Thermochemical Hydrogen Production

A potential thermochemical pathway uses heat from concentrated solar thermal energy and recyclable chemicals such as cadmium oxide to split water into hydrogen and oxygen. Thermochemical reactions often require temperatures higher than 1000 °C, thus requiring a match with concentrating solar thermal resources capable of achieving such high heat levels. A typical example is cadmium oxide thermochemical cycle. Cadmium oxide decomposes into cadmium and oxygen at 1450 °C. Elemental cadmium then reacts with water to produce cadmium oxide and hydrogen, and cadmium oxide is recycled back. So far, more than 300 such thermochemical cycles have been theoretically identified [26].

In areas with good solar potential, this technology offers potential efficiencies much higher than in converting solar energy to electricity and then using electrolysis to generate hydrogen. However, significant challenges remain for selecting the optimum thermochemical reaction cycle(s) and matching them with potential solar concentrating thermal technologies [32]. Although solar thermal technologies are not yet fully commercialized, broad deployment for power production (for instance, capacity over 4800 MW is under review in California [33]) would likely provide spill-over benefits for this potential hydrogen production pathway.

23.3.6
Sustainability Criteria

Sustainability criteria are becoming important considerations in low carbon fuel policies to encourage the production and use of energy in a sustainable way. For transportation fuels important sustainability criteria include GHG emissions, criteria pollutants, water and land use, and social welfare factors. For example, the European Fuel Quality Directive and Renewable Energy Directive have an elaborate framework for reporting and/or complying with environmental and social sustainability criteria. For hydrogen to play a major role in transportation and energy, it needs to have favorable environmental and ecological footprints.

Hydrogen generated from most renewable sources has low or near-zero life cycle GHG emissions and can play a major role in climate change mitigation, as noted elsewhere in this chapter. For hydrogen produced from dedicated energy crops, however, GHG emissions from land use changes and energy inputs can be a cause of concern, especially if feedstocks are grown in existing croplands and cause food crop displacement, or if carbon-rich forests and peatland are brought under cultivation. However, hydrogen produced from agriculture residues, waste grease and oils and other organic waste does not have land-use GHG emissions. Similarly, hydrogen produced via electrolysis, direct photolysis, thermochemical, and photo-electrochemical pathways is less likely to compete with cropland or result in the destruction of carbon-rich biological storage areas.

Water consumption is another important sustainability consideration. Water usage for hydrogen production via electrolysis will not necessarily exceed that for petroleum and ethanol production [34], but availability and quality may be local issues for any form of energy production.

Air quality is another important potential sustainability criterion. Many pathways to renewable hydrogen displace fossil fuels, resulting in air emissions with cleaner alternatives. In some cases, renewable energy (such as burning biomass for electricity production) will also require review for local and regional air quality concerns.

23.4
Renewable Hydrogen Policy Drivers

The purpose of this section is to provide an overview of direct and indirect policy drivers for renewable hydrogen production and utilization at the regional, national, and sub-national levels. Technologies that have not yet been commercialized face a variety of barriers to get through the demonstration and early commercialization phase [35]. Examples of potential barriers that apply to renewable hydrogen production include up-front costs, the cost and risks of demonstrating a new technology, sunk costs in existing infrastructure, information

23.4.1
Direct Renewable Hydrogen Policies

California statute SB1505 creates a standard requiring that transportation hydrogen funded by the State of California is produced from at least 33% renewable resources. The standard also requires a 50% reduction in nitrogen oxides and reactive organic gases compared with gasoline on a well-to-tank basis and no increase in toxic emissions [36]. Since these standards do not account for the zero tailpipe emissions from a fuel cell vehicle, the well-to-wheels emission reduction would be much larger. These standards will also apply to all hydrogen used in transportation in California once production levels reach 3500 tonnes per year state-wide. CARB staff have determined that these standards would lead to significant reductions in pollutants such as GHGs and NOx and increase the use of in-state resources [37]. (Note that this statute does not apply to electricity used for battery electric vehicle recharging, which is addressed below.)

CARB has recently announced $ 14.9 million in funding for hydrogen stations with a total H_2 production capacity of 660 kg per day, with $ 7.6 million in funding for three stations producing 220 kg H_2 per day from hydrogen from 100% renewable resources, as shown in Table 23.3 (note: 1 kg of $H_2 \approx 1$ gallon of gasoline on an energy content basis, as noted earlier). Two of the projects use renewable electricity to produce hydrogen via electrolysis while the Fountain Valley project utilizes a fuel cell to produce hydrogen, electricity, and heat from digester biogas. The aggregate costs of renewable hydrogen production is higher at this demonstration stage than methane steam reforming, although the cost of photovoltaic cells has been falling [38] and other renewable technology costs are likely to decrease if they are developed and deployed broadly.

Table 23.3 Recent California state funding for renewable hydrogen stations.

Location	State funding ($ million)	Capacity (kg per day)	Renewable resource(s)	Production technology
Orange County Sanitation District, Fountain Valley, CA	2.7	100	Digester biogas (methane)	Fuel cell
CalState University, Los Angeles, CA	2.7	60	On-site solar and wind power and off-site renewables	Electrolysis
Alameda-Contra Costa Transit, Emeryville, CA	2.7	60	On- and off-site photovoltaic solar	Electrolysis

Iceland is also promoting renewable hydrogen based on abundant local geothermal renewable energy, including vehicles and marine use to supply on-board electricity (propulsion is provided by a diesel motor) [39]. Iceland is currently operating a number of fuel cell passenger vehicles using hydrogen, thus displacing imported petroleum. Although Iceland's situation is unique, it does provide an interesting example for regions with abundant renewable resources of some type. Additionally, this example shows an option for reducing dependence on petroleum in the transportation sector – a potential policy driver in specific locations.

In Germany, the National Innovation Program for Hydrogen and Fuel Cell Technology has brought together industry, scientists, and the German federal government to accelerate development of this technology. Funding priorities include consistency with Germany's climate protection goals and Renewable Energy Sources Act [40].

23.4.2
North American Renewable Portfolio Standards, the European Renewable Energy Directive, and China's Renewable Energy Law

Renewable portfolio standards (RPS) established by American states and Canadian provinces in North America require a minimum level of renewable electricity, typically as a percentage of usage, although sometimes on a megawatt basis. In the United States, 28 states have state-wide requirements for renewable electricity and they generally range from 15 to 33% renewable content with a deadline of 2012–2020 [41] (with a few variations), and several Canadian provinces also have RPS. The United States Congress is also considering a national RPS. The European Union has also established a goal of 20% renewable energy by 2020, including both renewable electricity and renewable transport fuels (which are addressed further below). A number of individual countries also have national programs, including Germany's feed-in tariffs offering favorable prices for renewable electricity generation. In addition, China has established renewable energy goals of 10% by 2010 and 15% by 2020 [42].

These standards can facilitate renewable hydrogen in several ways. First, 21 of the states in the United States provide credit for fuel cells powered by renewable resources [43]. Biogas is similarly listed in Article 2 of the EU Directive as an "energy from renewable sources," indicating that electricity generation from biomethane in a fuel cell would be eligible for the broader 2020 goal [44]. Hence these are examples of standards that create a direct incentive for the use of biogas converted to hydrogen and then used to generate electricity.

Second, these standards are intended to increase the amount of renewable energy on the grid in China, Europe, and North America and would therefore also increase the amount used for any hydrogen produced via grid-tied electrolysis, both systems with purely grid electricity and systems using local renewable resources along with grid electricity to fill in gaps.

Third, hydrogen is one option for energy storage to facilitate renewable electricity targets. In California, for instance, the Economic and Technology Advancement Advisory Committee (ETAAC) has identified energy storage as a key challenge for meeting the state's 33% RPS by helping match the availability of renewable resources such as wind, geothermal, and to some extent also solar with demand [45]. Energy storage via hydrogen is one of the several options that is recommended for further consideration, competing with existing methods such as pumped hydro energy storage and new technologies such as battery storage in the future. Similar issues with resources fluctuations arose in a study of several northern European countries with hydrogen as a potential solution [46], and intermittent renewable resources have also been identified as a challenge for meeting renewable energy goals in China [47]. Small-scale demonstration projects that use hydrogen to store energy from intermittent renewable resources are currently operating in Europe, North America, and Asia [48].

Hydrogen can also help overcome some types of transmission bottlenecks. In some areas, power can be transmitted during off-peak periods for local storage in the form of hydrogen for use via distributed generation to support local electricity demand (in renewable resources areas where little or no transmission capacity exists, this would not be an option). For instance, the ETAAC has noted that transmission bottlenecks are a major constraint to renewable energy development [49]. Electricity produced from hydrogen fuel cell vehicles when not in use is one option for storing off-peak electricity and then creating electricity during peak demand [50].

23.4.3
Low-Carbon Transportation Fuel-Related Standards

The European Renewable Energy Directive also requires that at least 10% of energy used by the transportation sector should come from renewable sources by 2020. Therefore, renewable transportation hydrogen produced from biomass or biogas or renewable electricity (including solar, thermal, wind, and biomass) can qualify as one of the renewable fuels to meet the required target, creating a direct incentive for renewable hydrogen.

In North America, the California Low Carbon Fuel Standard (LCFS) requires that the transportation sector reduce GHG emissions intensity by 10% by 2020. Any fuels that have lower carbon intensity than diesel and gasoline can qualify as low-carbon fuels. Since renewable hydrogen can have substantially lower carbon intensity than gasoline and diesel, it can be an important fuel in meeting a 10% reduction target. For example, hydrogen produced from renewable biomass via on-site reforming has a carbon intensity of 33 g CO_2 MJ^{-1} as compared with 96 g CO_2 MJ^{-1} of gasoline [51]. The carbon intensity of hydrogen from electrolysis powered by renewable resources such as wind or solar should be effectively zero. A regional consortium of 11 northeastern and mid-Atlantic states are planning to develop low-carbon fuel standards similar to that of California's LCFS and thus create a similar incentive for hydrogen production and use including renewable hydrogen (and renewable electricity for transportation use). The member states have agreed to sign a Memorandum of Understanding. In Canada, British Columbia and Ontario also signed a Memorandum of Understanding with California in 2007 to implement standards similar to LCFS to reduce GHG emissions from the transportation sector by 10% by 2020.

At the United States federal level, the proposed Renewable Fuel Standard (RFS2) does not specifically mention renewable hydrogen, although biogas can qualify as renewable fuel. The RFS2 definition of a "renewable fuel" includes any fuel from biomass material including grain, starch, fats, greases, oils, and biogas. Therefore, it can be argued that renewable hydrogen should qualify as a renewable fuel under the RFS2 if it comes from biomass. However, it is not clear that hydrogen produced via electrolysis from renewable resources such as wind, solar, or hydro can qualify under the RFS2, hence the RFS2 is a less clear incentive for renewable hydrogen than the LCFS described above and EU Renewable Energy Directive.

23.4.4
Pricing Greenhouse Gas Through Fees or a Cap and Trade System

A number of regions have implemented a carbon or GHG fee (for instance, Finland, the United Kingdom, and British Columbia), or a cap and trade system (the EU and northeastern American states, and proposed for a number of western American states and Canadian provinces) [52] to establish a price on carbon or GHGs. Some of these systems are already intended to cover transportation emissions (Finland, British Columbia, and proposed western American states and Canadian provinces). The EU system will be extended to aviation and could include the transportation sector more broadly in the future, and proposed United States cap and trade legislation would also include transportation. This raises the question of what effect these policies would have on incentives for renewable hydrogen.

Figure 23.5 shows the influence of a carbon price on hydrogen from partially and fully renewable zero GHG fuels compared with steam methane reforming and petroleum refined to meet California air quality standards. Based on a GHG price of 20 (US$ 29) per tonne of CO_2, an incentive of approximately 0.27 (US$ 0.39) per kilogram of H_2 would be created for renewable hydrogen compared with conventional petroleum. Hence a price signal of this level would not be sufficient by itself to overcome existing price gaps between current renewable hydrogen technologies and fossil-based fuels such as traditional petroleum fuels and hydrogen from methane steam reforming. It could potentially supplement other efforts.

23.5
Conclusion

Hydrogen will potentially play a large role in meeting long-term global GHG reduction goals and local air quality goals, particularly in the transportation sector.

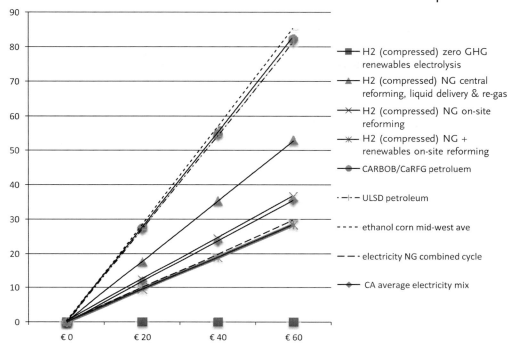

Fig. 23.5 Influence of carbon price on price of hydrogen and other fuels in euro cents per kilogram H_2 or equivalent.

While deployment of technologies such as fuel cell vehicles with hydrogen from natural gas would represent a significant advance over current vehicle technologies, hydrogen from renewable resources offers a sustainable pathway to the very deep reductions that are likely to be needed for long-term climate stabilization and local air quality goals.

When domestic renewable resources are available, hydrogen can also increase energy security and economic development by displacing imported fossil fuels. Current renewable resources such as wind, solar, geothermal, and sustainable biomass will likely play an important part, while other new renewable hydrogen production technologies may also emerge. Although policies for renewable hydrogen will have the greatest impact in the long term, forward-thinking policies for the further development and deployment of renewable hydrogen pathways will be needed in the meanwhile in order to achieve long-term success.

Acknowledgment

The authors would like to acknowledge the generous financial support of the Climate Works Foundation, the Energy Foundation, and the William and Flora Hewlett Foundation.

References

1 US Energy Information Administration (2009) *International Energy Outlook*, http://www.eia.doe.gov/oiaf/ieo/ieorefcase.html (accessed 30 December 2009).

2 Economist (2010) *The Economist Pocket World in Figures*, 2010 edn, Profile Books, London.

3 Monbiot, G. (2007) *Heat: How to Stop the Planet Burning*, Penguin Books, London.

4 Stern, N. (2009) *A Blueprint for a Safer Planet*, The Bodley Head, London.

5 Johansson, T.B., Kelly, H., Reddy, A.K.N., and Williams, R.H. (eds) (1993) *Renewable Energy: Sources for Fuels and Electricity*, Island Press, Washington, DC.

6 Jacobson, M.Z. and Delucchi, M.A. (2009) A pathway to sustainable energy by 2030. *Sci. Am.*, (November), 58–65.

7 Winter, C.-J. and Nitsch J. (eds) (1998) *Hydrogen as an Energy Carrier: Technologies Systems Economy*, Springer, Berlin.

8 Winter, C.-J. (2009) Hydrogen energy abundant, efficient, clean: a debate over the energy-system-of-change. *Int. J. Hydrogen Energy*, 34, S1–S22.

9 US EPA (2009) *8-Hour Ozone Area Summary*, 13 November 2009, http://www.epa.gov/air/oaqps/greenbook/gnsum.html (accessed 30 December 2009).

10 Delin, N.E., Wu, S., Nam, K.M., Reily, J.M., Paltsev, S., Prinn, R.G., and Webseter, M.D. (2009) Global health and economic impacts of future ozone pollution, *Environ. Res. Lett.*, 4, 044014 (9 pp.).

11 Scott, D.S. (2007) *Smelling Land: The Hydrogen Defense Against Climate Catastrophe*, Canadian Hydrogen Association, Université du Québec, Trois Rivières.

12 Sperling, D. and Gordon, D. (2009) *Two Billion Cars*, Oxford University Press, Oxford.

13 California Fuel Cell Partnership (2009) Various Reports, California Fuel Cell Partnership, West Sacramento, CA.

14 Thomas C.E.S. (2009) Transportation options in a carbon constrained world: hybrids, plug-in hybrids, biofuels, fuel cell electric vehicles, and battery electric vehicles. *Int. J. Hydrogen Energy*, 34, 9279–9296.

15 Lloyd, A.C. (2000) Hydrogen and clean air regulations in California. In *On Energies-of-Change. The Hydrogen Solution* (ed. C.J. Winter), Gerling Akademie Verlag, Munich.

16 Ogden, J. and Yang, C.(2009) Build up of a hydrogen infrastructure in the US. In *The Hydrogen Economy: Opportunities and Challenges* (eds. Ball, M., Wietschel, M.), Cambridge University Press, New York, pp. 454–482.

17 Kromer, M.A. and Heywood, J.B. (2007) *Electric Powertrains: Opportunities and Challenges in the U.S. Light-Duty Vehicle Fleet*, Massachusetts Institute of Technology, Cambridge, MA, May 2007, Table 53; underlying assumptions are summarized in Tables 51 and 52.

18 Ogden, J. (2008) In *Transitions to Alternative Transportation Technologies: a Focus on Hydrogen*, National Academy of Engineering, Washington, DC, Chapter 6.

19 Collier, K. (2002) Hydrogen/natural gas blends for heavy and light duty applications. In *Proceedings of the 2002 Department of Energy Hydrogen Program Review*, NREL/CP-610-32405, US Department of Energy, Washington, DC.

20 Levene, J.L., Mann, M.K., Marholis, R., and Milbrandt, A. (2005) An analysis of hydrogen production from renewable electricity sources, NREL. Prepared for the ISES 2005 Solar World Congress.

21 Porter, S. (2008) *Hydrogen Generation from Electrolysis, 100 kgh/day Trade Study*. 2008 Annual Progress Report, DOE; available at http://www.hydrogen.energy.gov/pdfs/progress08/ii_b_2_porter.pdf (accessed 29 December 2009).

22 Milbrandt, A. and Mann, M. (2007) NREL, http://www.afdc.energy.gov/afdc/pdfs/41134.pdf (accessed 29 December 2009).

23 Simbeck, D. and Chang, E. (2002) *Hydrogen Supply: Cost Estimate for*

Hydrogen Pathways¢Scoping Analysis, NREL/SR-540-32525, National Renewable Energy Laboratory, Golden, CO.

24. Nicholas, M. and Ogden, J. (2009) Analysis of a cluster strategy for near term hydrogen infrastructure rollout in southern California, PowerPoint presentation, Institute of Transportation Studies, UC Davis, Davis, CA.

25. California Air Resources Board (CARB) (2009) *Proposed Economic Analysis for the Low Carbon Fuel Standard*, California Environmental Protection Aging, Air Resources Board, Sacramento, CA.

26. FreedomCAR and Fuel Partnership (2009) *Hydrogen Production Roadmap: Technology Roadmap to the Future*, Appendix C-1; available at http://www1.eere.energy.gov/hydrogenandfuelcells/pdfs/h2_production_roadmap.pdf (accessed 29 December 2009).

27. Gasification Technologies Council (2009) *On-Line Database*, http://www.gasification.org/database1/search.aspx (accessed 29 December 2009).

28. Marda, J.R., Dibenedetto, J., Mckibben, S., Evans, R.J., Czernik, S., French, R.J., and Dean, A.M. (2009) Non-catalytic partial oxidation of bio-oil to synthesis gas for distributed hydrogen production. *Int. J. Hydrogen Prod.*, **34** (20), 8519–8534.

29. EnergyPedia News (2009) Norway: StatoilHydro's Sleipner carbon capture and storage project proceeding successfully, http://www.energy-pedia.com/article.aspx?articleid=134204, March 2009 (accessed 29 December 2009).

30. UK Department for Business Innovation and Skils (2009) *CCS Demonstration Competition*, http://webarchive.nationalarchives.gov.uk/+/http://www.berr.gov.uk//whatwedo/energy/sources/sustainable/ccs/ccs-demo/page40961.html (accessed 22 December 2009).

31. Shaoan, C. and Logan, B.E. (2007) Sustainable and efficient biohydrogen production via electrohydrogenesis. *Proc. Natl. Acad. Sci. USA*, **104** (47), 18871–18873.

32. Turner, J., Sverdrup, G., Mann, M.K., Maness, P.-C., Kroposki, B., Ghirardi, M., Evans, R.J., and Blake, D. (2008) Renewable hydrogen production. *Int. J. Energy Res.*, **32**, 379–407.

33. California Energy Commission (2009) *Status of All Projects*, spreadsheet available at http://www.energy.ca.gov/sitingcases/all_projects.html (accessed 29 December 2009).

34. Wu, M., Mintz, M., Wand, M., and Arora, S. (2009) *Consumptive Water Use in the Production of Ethanol and Petroleum Gasoline*, Argonne National Laboratory, Argonne, IL, Executive Summary, January 2009.

35. Brown, M., Chandler, J., Lapsa, M., and Sovacool, B. (2008) *Carbon Lock-in: Barriers to Deploying Climate Change Mitigation Technologies*, Oak Ridge National Laboratories, Oak Ridge, TN, revised 2008; *Focus for Success: a New Approach to Commercializing Low Carbon Technologies*, UK Carbon Trust, Witney, 2009; ETAAC, *Advanced Technology to Meet California's Climate Goals: Opportunities, Challenges, and Barriers*, Section 1, 14 December 2009.

36. California Air Resources Board (2008) *Concept Paper, Environmental and Energy Standards for Hydrogen Production*, Senate Bill 1505, CARB, 3 April 2008; available at http://www.arb.ca.gov/msprog/hydprod/1505concepts.pdf (accessed 29 December 2009).

37. CaH2Net (2009) *2008 Report to the Legislature, January 2009*, http://www.hydrogenhighway.ca.gov/update/cah2net2008reporttoleg.pdf (accessed 29 December 2009).

38. ETAAC (2009) *Advanced Technology to Meet California's Climate Goals: Opportunities, Challenges, and Barriers*, Section 4.2, 14 December 2009; available at http://www.arb.ca.gov/cc/etaac/etaac.htm (accessed 17 December 2009).

39. Skúlason, J.B. and Maack, M. (2007) *Sustainable Marine and Road Transport, H2 in Iceland*, Icelandic New Energy, Reykjavík, http://www.newenergy.is/newenergy/upload/files/project_papers/sustainable_marine_and_road_transporth2_in.pdf (accessed 29 December 2009).

40. NOW (2009) *Project Funding Criteria*, http://www.now-gmbh.de/index.php?id=44&L=1#c113

41 Renwable Portfolio Standards Fast Sheet, April 2009, http://www.epa.gov/CHP/state-policy/renewable_fs.html (accessed 17 December 2009).
42 National Development and Reform Commission (2007) *China's Long Term Renewable Energy Development Plan*, www.sdpc.gov.cn/zcft/zcfbtr/2007tongzhi/t20070904_157352.htm (accessed 9 March 2010).
43 North Carolina Solar Center and Interstate Renewable Energy Council (2010) *Database of State Incentives for Renewables and Efficiency*, http://www.dsireusa.org/incentives/incentive.cfm?Incentive_Code=NJ05Ree=1 (accessed 17 December 2009).
44 European Commission (2009) *Directive 2009/28/EC of the European Parliament and of the Council of 23 April 2009 on the Promotion of the Use of Energy from Renewable Sources and Amending and Subsequently Repealing Directives 2001/77/EC and 2003/30/EC*; available at http://eur-lex.europa.eu/LexUriServ/LexUriServ.do?uri=CJ:L:2009:140:0016:0062:en:pdf.
45 ETAAC (2008) *Recommendations of the Economic and Technology Advancement Advisory Committee*, 11 February 2008, Sections 5-F and 5-G; ETAAC (2009) *Advanced Technology to Meet California's Climate Goals: Opportunities, Challenges, and Barriers*, Section 4.2, 14 December 2009; available at http://www.arb.ca.gov/cc/etaac/etaac.htm (accessed 17 December 2009).
46 Sorensen, B. (2009) A renewable energy and hydrogen scenario for northern Europe. *Int. J. Energy Res.*, **32**, 471–500.
47 Oster, S. (2009) China law forces clean-energy use, *The Wall Street Journal Asia*, Hong Kong, 28 December, 1.
48 Schoenung, S. and Keller, J. (2009) International experience in fuel cells and hydrogen for electric power applications. Presentation to HTAC, 5 November 2009; available at http://hydrogen.energy.gov/pdfs/htac_nov09_14_international_experience.pdf (accessed 18 December 2009); *Stuart Island Energy Initiative*, http://www.siei.org/systemoverview.html (accessed 18 December 2009).
49 ETAAC (2009) Report, p. 4-2.
50 ETAAC (2008) Report, p. 5-15.
51 California Air Resources Board (CARB) (2009) *Proposed Regulation to Implement the Low Carbon Fuel Standard. Volume I Staff Report: Initial Statement of Reasons*, California Environmental Protection Agency, Air Resources Board, Sacramento, CA.
52 ICCT (2009) *Existing Experience on Allocating of Value of Greenhouse Gas Allowances/Fees/Taxes*. Prepared for April 2009 Allocations Workshop Backgrounder; available at http://www.next10.org/next10/pdf/allocations/ab_32_objectives.pdf (accessed 17 December 2009).
53 ETAAC (2009) *Advanced Technology to Meet California's Climate Goals: Opportunities, Challenges, and Barriers*, pp. 4-12 and 4-13, 14 December 2009.

24
Environmental Impact of Hydrogen Technologies
Ibrahim Dincer and T. Nejat Veziroglu

Abstract

This chapter discusses the role of hydrogen technologies as a potential solution for current and future environmental problems, and to provide a better environment and sustainable development. It also assesses the hydrogen production methods for sustainable hydrogen production by evaluating their greenhouse gas emissions and air pollutants. Two case studies are presented to discuss the environmental impact and energy utilization aspects, and to highlight the importance of the topic and show that these can help achieve a better environment and sustainability.

Keywords: energy, exergy, fuel cell, hydrogen, life cycle assessment, environment, economics, sustainable development

24.1
Introduction

Energy is a key element of the interactions between nature and society, and is considered a key input for the environment and sustainable development. Environmental and sustainability issues span a continuously growing range of pollutants, hazards, and eco-system degradation factors that affect areas ranging from local through regional to global. Some of these concerns arise from observable, chronic effects on, for instance, human health, whereas others stem from actual or perceived environmental risks such as possible accidental releases of hazardous materials. Many environmental issues are caused by or relate to the production, transformation, and use of energy, for example, acid rain, stratospheric ozone depletion, and global climate change. Recently, a variety of potential solutions to current environmental problems associated with harmful pollutant emissions have evolved. Hydrogen energy systems appear to be the one of the most effective solutions and can play a significant role in providing a better environment and sustainability (Dincer, 2002).

In the literature, there have been limited studies on sustainability aspects of hydrogen energy systems (including fuel cell systems) undertaken by several researchers (e.g., Afgan et al., 1998; Hart, 2000; Baretto et al., 2003; Pehnt, 2003; Afgan and Carvalho, 2004; Kwak et al., 2004; Hopwood et al., 2005; Midilli et al., 2005 a, b). Of these, Afgan and Carvalho (2004) give an overview of the potential of multi-criteria assessment of hydrogen systems. With suitable selection of the criteria comprising performance, environment, market, and social indicators, the assessment procedure is adapted for the assessment of selected options of the hydrogen energy systems, and their comparison with new and renewable energy systems. Hopwood et al. (2005) pointed out that sustainable development, although a widely used phrase and idea, has many different meanings and therefore provokes many different responses. In broad terms, the concept of sustainable development is an attempt to combine growing concerns about a range of environmental issues with socio-economic aspects. Sustainable development implies a smooth transition to more effective technologies from the point of view of environmental impact and energy efficiency. As was pointed by Midilli et al. (2005 a, b), increasing concerns about urban air pollution, energy security, and climate change will expedite the transition to a "hydrogen economy". Kwak et al. (2004) indicated that new hydrogen-powered fuel cell technologies in both their high- and low-temperature derivatives are more effective and cleaner than conventional energy technologies, and can be considered one of the pillars of a future sustainable energy system. Barreto et al. (2003) examined future perspectives for fuel cells and developed a long-term hydrogen-based scenario of the global energy system. Their scenario illustrates the key role of hydrogen in a long-term transition towards a clean and sustainable energy future. Hart (2000) stated that hydrogen from renewables, coupled with fuel cell generation on demand, provides an elegant and complementary solution to this problem. It is suggested, therefore, that fuel cells are not only a future economically competitive option for sustainable energy conversion, but are also a complementary option in the sustainable energy systems of the future.

The primary objective of this chapter is to discuss the role of hydrogen and fuel cell systems for a sustainable future, to highlight the importance of exergy in achieving this, and to develop some new models for environmental impact and sustainability aspects of hydrogen and fuel cell systems and applications. In addition, two case studies on the life cycle assessment of fuel cell vehicles and hydrogen production systems from energy, environmental, and sustainability points of views are also presented.

24.2
Sustainable Development

Sustainable development requires a sustainable supply of clean and affordable energy resources that do not cause negative societal impacts (e.g., McGowan, 1990; OECD, 1995; Hui, 1997; Afgan et al., 1998; Dincer and Rosen, 1998; 2005; Hammond, 2004). Supplies of energy resources such as fossil fuels and uranium are finite. Energy sources such as sunlight, wind, and falling water are generally considered renewable and therefore sustainable over the relatively long term. Wastes and biomass fuels are also usually viewed as sustainable energy sources. Wastes are convertible to useful energy forms through technologies such as waste-to-energy incineration facilities.

Much environmental impact is associated with energy-resource utilization. Ideally, a society seeking sustainable development utilizes only energy resources that release no or minimal emissions to the environment and thus cause no or little environmental impact. However, since all energy resources may somehow lead to some environmental impact, increased efficiency can somewhat alleviate the concerns regarding environmental emissions and their negative impacts. For the same services or products, less resource utilization and pollution are normally associated with increased efficiency.

Sustainability often leads local and national authorities to incorporate environmental considerations into energy planning. The need to satisfy basic human needs and aspirations, combined with the increasing world population, will make the need for successful implementation of sustainable development increasingly apparent. Various hydrogen energy-related criteria that are essential to achieving sustainable development in a society include the following:

- information about and public awareness of the benefits of sustainability investments
- environmental and sustainability education and training
- appropriate energy and exergy strategies for better efficiency
- promoting environmentally benign technologies
- clean hydrogen production technologies
- development of sustainable hydrogen economy infrastructure
- commercially viable and reliable hydrogen energy systems, including fuel cells
- availability and utilization of renewable energy resources
- use of cleaner technologies for production, transportation, distribution, storage and use
- a reasonable supply of financing and incentives
- academia-industry-government partnership programs
- policy development for sustainable energy programs
- appropriate monitoring and evaluation tools
- road maps for future implementation.

Environmental concerns are significantly linked to sustainable development. Activities which continually degrade the environment are not sustainable. For example, the cumulative impact on the environment of such activities often leads over time to a variety of health, ecological, and other problems.

Clearly, a strong relation exists between efficiency and environmental impact since, for the same services or products, less resource utilization and pollution are normally associated with increased efficiency. Note that improved energy efficiency leads to reduced energy losses. Most efficiency improvements produce direct environmental benefits in two ways: (i) operating energy input requirements are reduced per unit output, and pollutants generated are correspondingly reduced, and (ii) consideration of the entire life cycle for energy resources and technologies, suggests that improved efficiency reduces environmental impact during most stages of the life cycle. That is why assessing the future hydrogen technologies, such as fuel cells, over their entire life cycle is essential to obtain correct information on energy consumption and emissions during various life cycle stages, to determine competitive advantages over conventional technologies, and to develop future scenarios for better sustainability.

In recent years, the increased acknowledgment of humans' interdependence with the environment has been embraced in the concept of sustainable development. With energy constituting a basic necessity for maintaining and improving standards of living throughout the world, the widespread use of fossil fuels may have impacted the planet in ways far more significant than first thought. In addition to the manageable impacts of mining and drilling for fossil fuels and discharging wastes from processing and refining operations, the "greenhouse" gases created by burning these fuels are regarded as a major contributor to a global warming threat. Global warming and large-scale climate change have implications for food chain disruption, flooding, and severe weather events, such as hurricanes.

It is obvious that utilization of hydrogen and fuel cell technologies can help reduce environmental damage and achieve sustainability. Such technologies essentially do not consume fuel, contribute to global warming, or generate substantial waste provided that hydrogen is produced through clean and renewable energy resources. In this respect, hydrogen and fuel cell technologies can provide more efficient, effective, environmentally benign, and sustainable alternatives to conventional energy technologies, particularly fossil fuel-driven ones.

Hydrogen and fuel cell technologies have a crucial role to play in meeting future energy needs in both rural and urban areas. The development and utilization of such technologies should be given a high priority, especially in the light of increased awareness of the adverse environmental impacts and political consequences of fossil-based generation. The need for sustainable energy development is increasing rapidly in the world. In fact, widespread use of these technologies is important for achieving sustainability in the energy sectors in both developing and industrialized countries. These technologies are a key component of sustainable development for four main reasons:

- They have numerous advantages, such as being energy efficient and compatible with renewable energy sources and carriers for future energy security, economic growth, and sustainable development.
- They generally cause much less environmental impact than other conventional energy sources and technologies. The variety of hydrogen and fuel cell technologies provides a flexible array of options for their use in various applications.
- Hydrogen cannot be depleted since the basic source is water. If used carefully in appropriate applications, it can provide a fully reliable and sustainable sup-

ply of energy almost indefinitely. In contrast, fossil fuel and uranium resources are diminished by extraction and consumption.
- These technologies favor system decentralization, and local and individual solutions that are somewhat independent of the national network, thus enhancing the flexibility of the system and providing economic and environmental benefits to small isolated populations. Also, the small scale of the equipment often reduces the time required from initial design to operation, providing greater adaptability in responding to unpredictable growth and/or changes in energy demand.

It is important to note that if we produce hydrogen through conventional technologies using fossil fuels, this will not make hydrogen inherently clean in that they may cause some burden on the environment in terms of pollutant emissions, solid wastes, resource extraction, or other environmental disruptions. Nevertheless, the overall use of these technologies almost certainly can provide a cleaner and more sustainable energy system than increased controls on conventional energy systems. This is in fact clearly shown in the case studies.

To overcome obstacles in initial implementation, programs should be designed to stimulate a hydrogen energy market, so that options can be exploited by industries as soon as they become cost-effective. Financial incentives should be provided to reduce up-front investment commitments and infrastructure costs for production, transportation, distribution, storage, and use, and to encourage design innovation, in addition to research and development activities along with commercialization practices.

24.3
Sustainable Development and Thermodynamic Principles

24.3.1
Interdisciplinary Triangle

As mentioned earlier, energy is a key element of the interactions between nature and society, and is considered a key input for economic development and sustainable development. Energy use is very much governed by thermodynamic principles and, therefore, an understanding of thermodynamic aspects of energy can help us understand pathways to sustainable development (Dincer and Rosen, 2005). The impact of energy resource utilization on the environment and the achievement of increased resource-utilization efficiency are best addressed by considering exergy. The exergy of an energy form or a substance is a measure of its usefulness or quality or potential to cause change and provide the basis for an effective measure of the potential of a substance or energy form to impact the environment. It is important to mention that in practice a thorough understanding of exergy and the insights it can provide into the efficiency, environmental impact, and sustainability of energy systems is required for the

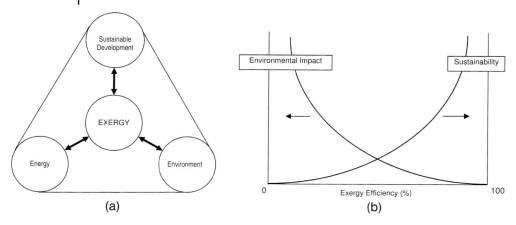

Fig. 24.1 (a) The interdisciplinary triangle of exergy. (b) Qualitative illustration of the relation between the environmental impact and sustainability of a process and its exergy efficiency.

engineer or scientist working in the area of energy systems and the environment. During the past decade, the need to understand the linkages between exergy, energy and environmental impact has become increasingly significant (Dincer, 1998; Dincer and Rosen, 1999). Dincer and Rosen (2004) considered exergy as the confluence of energy, environment and sustainable development and illustrated this in a triangle (Figure 24.1a). The basis for this treatment is the interdisciplinary character of exergy and its relation to each of these disciplines.

When we look at the general energy efficiency (η) and exergy efficiency (ψ) definitions:

$$\eta = \frac{\text{energy in product outpouts}}{\text{energy in inputs}} = 1 - \frac{\text{energy loss}}{\text{energy in inputs}} \quad (24.1)$$

and

$$\psi = \frac{\text{exergy in product outpouts}}{\text{exergy in inputs}} = 1 - \frac{\text{exergy loss} + \text{exergy consumption}}{\text{exergy in inputs}} \quad (24.2)$$

it is obvious that reducing losses will increase the efficiency. The relation between exergy efficiency, sustainability, and environmental impact is illustrated in Figure 24.1b. There, sustainability is seen to increase and environmental impact to decrease as the exergy efficiency of a process increases. The two limiting efficiency cases in Figure 24.1b appear to be significant:
- As the exergy efficiency approaches 100%, the environmental impact associated with process operation approaches zero, since exergy is only converted from one form to another without loss (either through internal consumption

or losses). Also, sustainability approaches infinity because the process approaches reversibility.
- As the exergy efficiency approaches 0%, sustainability approaches zero because exergy-containing resources are used but nothing is accomplished. Also, the environmental impact approaches infinity because, to provide a fixed service, an ever-increasing quantity of resources must be used and a correspondingly increasing amount of exergy-containing wastes is emitted.

Although this chapter discusses the benefits of using thermodynamic principles, especially for exergy, to assess the sustainability and environmental impact of energy systems, this area of work, particularly for hydrogen and fuel cell systems, is relatively new. Further research is, of course, needed to gain a better understanding of the potential role of exergy in such a comprehensive perspective. This includes the need for research to (i) define better the role of exergy in environmental impact and design, (ii) identify how exergy can be better used as an indicator of potential environmental impact, and (iii) develop holistic exergy-based methods that simultaneously account for technical, economic, environmental, sustainability, and other factors.

Nevertheless, hydrogen appears to be one of the most promising energy carriers for the future. It is considered an energy-efficient, non-polluting fuel. When hydrogen is used in a fuel cell to generate electricity or is combusted with air, the only products are water and a small amount of NO_x, depending on the source of hydrogen and its impurity. Hydrogen that is produced from renewable resources and used in fuel cells can provide sustainable energy to power fuel cell vehicles. The total system, including distribution, refueling, and on-board storage of hydrogen, may prove superior to batteries recharged with grid power. A hydrogen-powered fuel cell vehicle may offer a market entry for hydrogen and renewable resources in transportation. Attractive transitional applications of hydrogen include use in combustion engine vehicles and production from natural gas. In either case, the environmental and energy policy consequences are significantly less than with the continued use of oil-derived fuels in conventional combustion engine vehicles. Fuel cells, which employ hydrogen to produce electricity, particularly proton exchange membrane (PEM) fuel cells, can be used to power a wide variety of applications. This is especially true in transportation, where there are several options for providing hydrogen for the fuel cells.

Recently, there has been increased interest in hydrogen energy and fuel cell applications for both stationary and mobile power generation. This interest has been motivated by the fuel cells' high efficiency, even in small-scale installations, and their low waste emissions. Recent legislative initiatives in California, USA, aimed at mandating the introduction of zero-emission vehicles, and the failings of other technologies (e.g., the limited range and long refueling times of battery-powered vehicles) have further promoted the investigation of fuel cells in mobile applications.

Thermodynamic principles can be used to assess, design, and improve energy and other systems, and to understand environmental impact and sustainability issues better. For the broadest understanding, all thermodynamic principles must be used, not just those pertaining to energy. Hence, many researchers feel that an understanding and appreciation of exergy, as defined earlier (see Figure 24.1a), is essential to discussions of sustainable development.

Beyond individual behavior, we should think collectively about how society meets its energy needs, including decisions about energy resource selection, efficiency, and the role of hydrogen and fuel cell technologies.

An inexpensive and stable energy supply is a prerequisite for social and economic development, both in households and at the national level. Indeed, energy is essential to human welfare and quality of life. However, energy production and consumption generate significant environmental problems (at global, regional, and local levels) that can have serious consequences and even put at risk the long-term sustainability of the planet's ecosystems. The relationship between energy consumption and production and sustainability is, therefore, complex, as shown earlier by Dincer and Rosen (2004).

We consider sustainable development here to involve four key factors in terms of environmental, economic, social, and resource/energy sustainability under global sustainability, as shown in Figure 24.2. It is clearly seen that all these factors are interrelated.

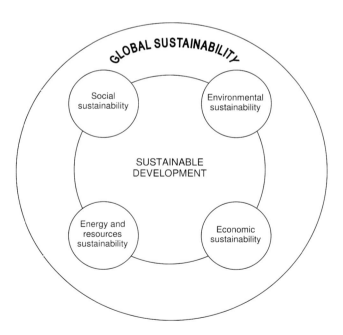

Fig. 24.2 Four key factors of sustainable development under global sustainability.

24.4
Hydrogen Versus Fossil Fuels

Global primary power consumption today is about 13 TW [1 TW (terawatt) is equal to 10^{12} W], most of which is fossil based (Hoffert et al., 2002; US DOE, 2004; Muradov and Veziroglu, 2008) The demand for energy continues to rise because of two main reasons: (a) the continuing increase in world population and (b) the growing demand by developing countries to improve their living standards. Hence there has been a continuous increase in consumption of fossil fuels, but it is expected that the world fluid fossil fuel production will peak soon, and will then begin to decrease (Elliot and Turner, 1972; Root and Attanasi, 1978; Parent, 1979; Bockris and Veziroglu, 1983). Figure 24.3 shows estimates of the production rates of the fluid fossil fuels. It can be seen that the peak is expected to occur in about 20–30 years from now.

Since, at the consumer end, a large proportion of the energy is consumed in the form of a fluid fuel and since the natural (fossil) fluid fuel resources are being depleted rap[idly, new (synthetic) fuels are being considered to close the gap between the demand and the production of fluid fossil fuels. They include synthetic gasoline (or Syn-Gas), synthetic gas (or synthetic natural gas, SNG), liquid hydrogen, gaseous hydrogen, ethanol, and methanol. These will be compared by taking into account production costs, utilization efficiencies, and environmental effects, in addition to factors such as resource conservation and transportation.

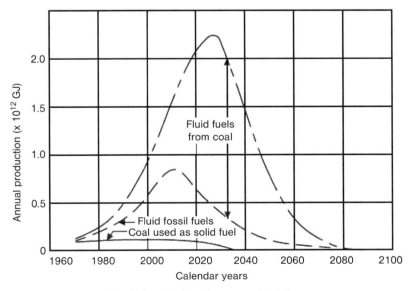

Fig. 24.3 Estimates of world fossil fuel production. Modified from Awad and Veziroglu (1984).

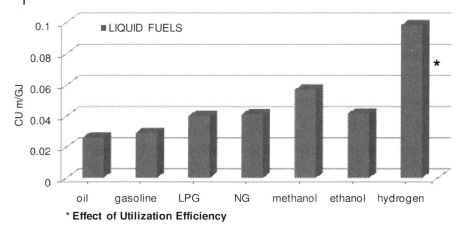

Fig. 24.4 Volume per unit energy (HHV) generated for liquid fuels (Veziroglu and Barbir, 1992).

In the case of surface vehicles and airplanes, they must carry their fuel for a certain distance before replenishing their fuel supply. In the case of space transportation, a space vehicle must carry its fuel, and also the oxidant, necessary for its scheduled range. Therefore, it is important that the transportation fuel be as light as possible and also take up as little space as possible. Figure 24.4 presents the energy per unit volume for the various fuels that are currently used and/or under consideration for future use. In this figure, liquid fuels are compared, on an energy per unit volume basis; oil and gasoline are on the lower side, and hydrogen occupies more space than any other fuel for a given amount of energy. If the utilization efficiency advantage of hydrogen is taken into account, there is about a 36% reduction in the volume of hydrogen.

In comparing the fuels, it is important to compare the utilization efficiencies at the user end. For utilization by the user, fuels are converted into various energy forms, such as mechanical, electrical, and thermal. Studies show that in almost every instance of utilization, hydrogen can be converted to the desired energy form more efficiently than the fossil fuels (or the synthetic fossil fuels). In other words, hydrogen would result in energy conservation due to its higher utilization efficiencies.

Hydrogen can be converted to electricity in fuel cells with much greater efficiency (Lavi and Trimble, 1979) than is possible in thermal power plants using fossil fuels. Whereas the conversion efficiencies for the latter are in the range 35–38%, the practical efficiencies in hydrogen fuel cells are 65–70%. In advanced hydrogen fuel cells being developed now, it is expected that the efficiencies will rise to 80–90%. This is an important unique property of hydrogen, which can also increase the conversion efficiencies in transport vehicles. Even if the end use is mechanical power (such as in automobiles, buses, or trucks), hydrogen fuel cell-electric motor combination would yield far greater conversion efficiencies than an internal combustion engine running on fossil fuels. Hydrogen-powered vehicles with internal combustion engines have been proved to be more efficient than gasoline vehicles. Hydrogen can be considered more thermally efficient than gasoline, primarily because it burns better in excess air and permits the use of a higher compression ratio. Data from engine tests indicate that hydrogen combustion is 15–50% more thermally efficient. The overall fuel efficiency of a hydrogen vehicle, which takes into account both the thermal efficiency and the weight of the vehicle, is also better when compared with a gasoline vehicle. On average, hydrogen vehicles are 22% more efficient. As hydrogen can burn both in lean fuel–air mixtures and in rich mixtures, it can cause large improvements in fuel use efficiencies in the stop–start type of city driving. Investigations show that for a given number of passengers and a given payload, a subsonic jet passenger airplane would use 19% less energy if it were to use liquid hydrogen instead of fossil-based jet fuel. In the case of a supersonic jet plane, the efficiency advantage of hydrogen is even greater: it is 38% better than jet fuel. In some industrial, commercial, and residential applications, such as in heating and cooling, fuels are converted to thermal energy.

Experiments (Billings, 1979; Hydrogen Update, 1981) show that hydrogen can be converted to thermal energy 24% more efficiently than fossil fuels. Gas turbine electric power plants using liquid hydrogen may have favorable efficiency benefits if the cryogenic energy of LH_2 is converted to useful work (Tsujikawa and Sawada, 1982). The above-discussed hydrogen utilization efficiencies are summarized in Figure 24.5. The "utilization efficiency factor" is defined as the ratio of the fuel utilization efficiency to the hydrogen utilization efficiency for a given application. The terms η_H and η_F are the utilization efficiencies of hydrogen and fossil fuels, which are 1 and 0.72, respectively, defined as the weighted

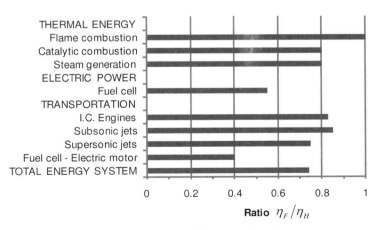

Fig. 24.5 Ratios of fossil fuel utilization efficiency to that of hydrogen for various applications. Modified from Veziroglu and Barbir (1992).

mean efficiency with which a given fuel can be converted to the desired energy form at the user end.

Since the utilization of fuels can adversely affect the environment, it is important to examine the environmental compatibility of fuels. Throughout the process of fossil fuel consumption (extraction, transportation, processing, and particularly their end use – combustion), there are harmful impacts on the environment, which cause direct and indirect negative effects on the economy. Excavation of coal devastates the land, which has to be reclaimed, and is out of use for several years. During the extraction, transportation, and storage of oil and gas, spills and leakages occur, which cause water and air pollution. Refining processes also have a negative environmental impact. Most of the environmental impact of fossil fuels occurs during the end use. Their end use involves combustion, irrespective of the final purpose (i.e., heating, electricity production, or motive power for transportation). The main constituents of fossil fuels are carbon and hydrogen, but also some other ingredients, which are originally in the fuel (e.g., sulfur) or are added during refining (e.g., lead, alcohols). Combustion of the fossil fuels produces various gases (CO_x, SO_x, NO_x, hydrocarbons), soot and ash, droplets of tar, and other organic compounds, which are all released into the atmosphere and produce air pollution.

Mainly CO_2, but also CO, CH_4, hydrocarbons, NO_x, chlorofluorocarbons, and so on, in the atmosphere allow the sun's ultraviolet and visible radiation to penetrate and warm the Earth, but then absorb some of the infrared energy that the Earth radiates back into the atmosphere. By blocking the escape of this radiation, these gases effectively form a thermal blanket around the Earth. To rebalance the incoming and outgoing radiation, the Earth's temperature must increase. For over a century, scientists throughout the world have expressed concern over the increase in atmospheric CO_2 concentration, which could affect the global climate by increasing the temperature of the Earth (Arrhenius, 1896; Callender, 1939; Wilier, 1950; Plass, 1956a, b; Chamberlin et al., 1982). Resulting primarily from the use of fossil fuels, CO_2 emissions may alter the radiative balance of the Earth, increasing global temperatures and dramatically changing the Earth's climate.

In order to obtain manageable mathematical relationships to predict the mean Earth temperature, a thermal model of the Earth has been developed. A schematic diagram of the model is presented in Figure 24.6. As can be seen, it is assumed that the atmosphere is enclosed between the Earth and a spherical "effective" homogeneous cloud cover at the mean cloud height. The radiative properties of the cloud cover are selected such that from the point of view of radiation heat transfer, the effect is the same as if it were not continuous but piecemeal as it is in Nature. Only the atmosphere between the mean cloud height and the Earth is taken into account. The atmosphere outside the mean cloud cover is neglected. This is reasonable since the density of the atmosphere outside the mean cloud height (3 km) is very low and most of the CO_2, the primary variable in the problem, accumulates below the mean cloud height because of its higher density. In the model, the mean solar constant just outside

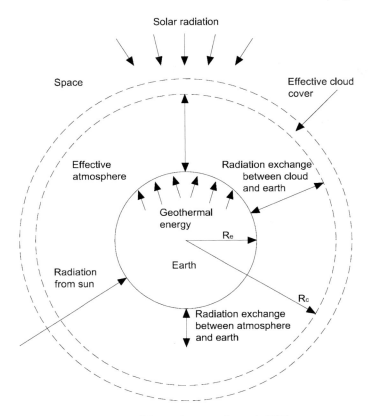

Fig. 24.6 Thermal model of the Earth (Veziroglu et al., 1989).

the effective cloud cover layer is considered, in order to avoid taking the Sun itself into account. The mathematical relationship of the thermal model of the Earth has been studied in more detail (Veziroglu et al., 1989).

Higher global temperatures are just one aspect of CO_2-induced changes to the natural environment. Human activities are also responsible for releasing other greenhouse gases (CO, CH_4, hydrocarbons, NO_x, chlorofluorocarbons, aerosols, etc.) into the atmosphere, and there have been studies to understand the climatic effects of these gases (Lacis et al., 1981; Wang and Pinto, 1981; Chamberlin et al., 1982; McDonald, 1982; NASA, 1983). However, CO_2 is found to be the most critical in producing the greenhouse effect. It was estimated that to satisfy world's growing appetite for energy, support economic growth, and stabilize atmospheric CO_2 levels at an acceptable level, at least 10 TW of carbon-neutral power has to be generated by the middle of the present century (Argonne National Laboratory, 2005; Muradov and Veziroglu, 2008). According to Hoffert et al. (2002), the stabilization of atmospheric CO_2 concentrations at 550, 450 and 350 ppm levels would require the generation of 15, 25 and

>30 TW, respectively, of carbon-free power by 2050 (Muradov and Veziroglu, 2008).

Veziroglu and Basar (1974) have recommended replacing the present fossil fuel system by the hydrogen energy system, in order to reduce atmospheric CO_2, pollution and acid rain, and improve the quality of life. In this energy system, the by-product of combustion is water vapor, not CO_2 or the other pollutants such as those produced by the combustion of fossil fuels. Their model predicts that if hydrogen replaces fossil fuels, the atmospheric pollution would be reduced by 90% and the quality of life would be 10 times better. Veziroglu et al. (1976) also investigated the effect of replacing part of the fuel demand with hydrogen on atmospheric pollution. The results showed that replacing 25% of the demand by hydrogen would reduce the pollution by 10%, replacing 50% of the demand by hydrogen would reduce the pollution by 25%, replacing 75% of the demand by hydrogen would reduce the pollution by 60%, and replacing all of the demand by H_2 would reduce the pollution by 90%. Awad and Veziroglu (1984) studied the effect of increasing atmospheric CO_2, and also the effects of pollution and acid rain, on the environment. Their estimate showed that on a worldwide basis, the damage due to rising oceans caused by the greenhouse effect alone is US$ 0.29 per gigajoule of fossil fuels consumed. The total damage due to air pollution, acid rain, and greenhouse effect added up to $7.64 per gigajoule of fossil fuel energy used.

In addition, acid rain is damaging the historical monuments, modern buildings and structures. Oil spills and leaks pollute drinking water sources and damage beaches. Such effects will now be considered according to the recipients of the damage.

The consequences of fossil fuel consumption, land devastation, air and water pollution, acid rain, the greenhouse effect, and climatic changes have already caused severe economic damage. However, although some ecosystems do not have a direct economic value, it does not mean that they are worthless and that they should be ignored. It is clear that air and water pollution, acid rain, and climatic changes will alter competitive outcomes and destabilize natural ecosystems in unpredictable ways. In many cases, the indirect effects, such as changes in habitat, in food availability, and in ecological chains, may have a greater impact than the direct physiological effects (Smith and Tirpak, 1988). Therefore, as a conservative estimate, the damage is estimated to be twice as great as the measurable and already calculated damage to the various ecosystems (plants, forests, animals, aquatic ecosystems, etc.).

Fossil fuels in general consist of coal, petroleum, and natural gas. Their contributions to environmental damage are different (e.g., they are not equal to their energy contents). Hence, in order to apportion a given type of damage amongst the fossil fuels, a pollution factor (P_{in}) will be used. It can be defined as a factor that is proportional to the environmental damage produced for a given fuel, i, and for a given type of damage, n. In the absence of more definitive data, in general CO_2 emissions will be used to obtain the pollution factors, that is, the pollution factor for each fuel will be assumed to be proportional to

Table 24.3 Pollution factors.

	CO_2 emissions (kg GJ^{-1})		
	Coal 85.50	Petroleum 69.40	Natural gas 52.00
Damage type (n)	Pollution factor (P_{in})		
	Coal $i=1$	Petroleum $i=2$	Natural gas $i=3$
Damage due to emissions	1.00	0.81	0.60
Damage due to oil spills, leakages	0.00	1.00	0.00
Damage due to strip mining	1.00	0.00	0.00

Source: Barbir et al. (1990).

CO_2 emissions per unit energy. In the case of oil spills and leakages, the pollution factors for coal and natural gas should be equal to zero, since they do not contribute to the damage in question. Similarly, in the case of strip mining, the pollution factors for petroleum and natural gas are zero. Estimated pollution factors are presented in Table 24.1.

The following relationship can be used to obtain the environmental damage for a given fuel and for a given type of damage, C_{in} (Barbir et al., 1990):

$$C_{in} = D_n \{\text{factor to apportion } D_n \text{ to fossil fuel } i\}/F_i \tag{24.3}$$

$$C_{in} = D_n \frac{F_i P_{in}}{\sum (F_i P_{in})} \frac{1}{F_i} \tag{24.4}$$

or

$$C_{in} = C_n P_{in} \tag{24.5}$$

where C is the damage per unit of modified energy consumption, D is the estimated annual damage, F is the fuel consumption, and the subscript n refers to the type of damage.

Using the appropriate data and the above equations, the unit damage for coal (C_{1n}), petroleum (C_{2n}), and natural gas (C_{3n}) can be calculated for each type of environmental damage for the desired year. Values of annual damage, modified fossil fuel consumption, damage per unit energy, the year of estimate, and the applicable references for the types of the environmental damage have been discussed in more detail (Barbir et al., 1990).

In calculating the cost of fuels to the society, their environmental effects and damage must certainly be considered. The cost of the environmental damage due to fossil fuel utilization is not included in the market price of fossil fuels, and can be considered as external costs. These costs are paid by society, and/or eventually will be paid by society, since in the long term any disturbed ecosystem will affect human society, its environment, and its economy.

In order to compare the overall environmental effects of various fuels, it will be better to compare the energy systems based on these fuels. We shall therefore consider the following three possible scenarios: (1) fossil fuel energy system; (2) coal and coal-based synthetic fuels system; and (3) hydrogen system based on renewable energy sources. Figure 24.7 presents the CO_2 (a) and other

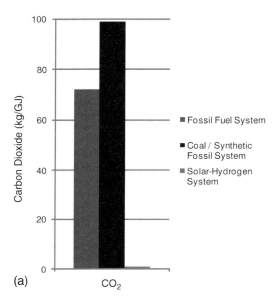

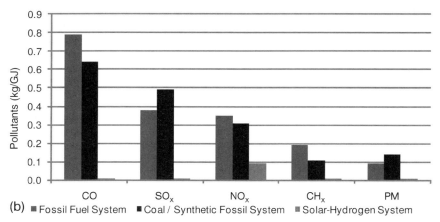

Fig. 24.7 (a) Carbon dioxide produced by three energy systems and (b) pollutants produced by three energy systems. Modified from Veziroglu and Barbir (1992).

pollutants (b) for the three energy systems in bar-chart format. It can be seen that the coal–synthetic fossil system is the worst from the environmental point of view, and the solar–hydrogen energy system is the best. The solar–hydrogen system will not produce any CO_2, CO, SO_x, hydrocarbons, or particulates, except some NO_x. However, the solar–hydrogen-produced NO_x is much less than those produced by the other energy systems. Finally, it is expected that achieving a sustainable development in the energy sector will be possible only if fossil fuels are replaced by clean and renewable energy sources, of which the solar–hydrogen system is a possibility that is full of advantages.

24.5
Future Energy Systems

The economic growth of modern industrialized society has been based mainly on the utilization of energy locked in fossil fuels. A simplified version of the fossil fuel energy system can be summarized as follows: fossil fuels are used for transportation, for heat generation in residential, commercial, and industrial sectors, and for electric power generation. For transportation, mostly petroleum products are used (gasoline, diesel fuel, jet fuel, etc.). Heat generation includes space heating, domestic water heating, cooking, steam generation, and direct heating and/or drying in various industrial processes. All three forms of fossil fuels are used for these purposes. In electric power generation, coal is used mainly for the base load generation, and natural gas and heating oil for peak load. Part of the electric power is produced by hydroelectric and nuclear power. All this use is rapidly resulting in critical environmental problems throughout the world, as we explained in detail earlier when considering hydrogen versus fossil fuel.

Worldwide conversion from fossil fuels to hydrogen would eliminate many problems and their ramifications. The optimal endpoint for conversion to a hydrogen economy is the substitution of clean hydrogen for the present fossil fuels. The production of hydrogen from non-polluting sources (such as solar energy) is the ideal way (Zweig, 1992; Momirlan and Veziroglu, 2005). Many researchers and scientists agree that the solution to these critical environmental problems would be to replace the existing fossil fuel system by a hydrogen energy system. Hydrogen is a very efficient and clean fuel. Its combustion will produce no greenhouse gases, no ozone layer-depleting chemicals, little or no acid rain ingredients, and no pollution. Hydrogen, produced from renewable energy (e.g., solar) sources, would result in a permanent energy system, which we would never have to change.

However, other energy systems have been proposed for the post-petroleum era, such as a synthetic fossil fuel system. In this system, synthetic gasoline and synthetic natural gas will be produced using abundant deposits of coal (Veziroglu and Sahin, 2008).

In a coal and synthetic fossil fuel system, it is assumed that the present fossil fuel system will be continued by substitution with synthetic fuels derived from coal wherever convenient and/or necessary. Patterns of energy consumption are also assumed to remain unchanged. Coal will be used extensively for thermal power generation and for electric power generation, because it is much cheaper than synthetic fuels. However, some end-uses require fluid fuels. Therefore, it has been assumed that synthetic natural gas (SNG) will be used for thermal energy generation (primarily in the residential sector) and also as a fuel for surface transportation, where it will share the market with synthetic gasoline. Synthetic jet fuel will be used in air transportation. In cases where synthetic fuels will be produced from coal, more than 1 GJ of coal will be used for each gigajoule of a synthetic fuel manufactured and consumed. Consequently, the environmental damage caused by 1 GJ of a synthetic fuel produced from coal is greater than the estimated damage caused by coal consumption. This can be expressed as follows:

$$E_{syn} = E_{coal} P_s \tag{24.6}$$

where E_{syn} is the environmental damage due to use of 1 GJ of a synthetic fuel (including hydrogen) produced from coal, is the environmental damage due to coal, E_{coal} and is the synthetic fuel pollution factor. The total CO_2 generated per unit energy has been taken as a measure of the pollution factor. Table 24.2 presents the pollution factors for various synthetic fuels compared with coal.

In the solar-hydrogen energy system, it is assumed that the conversion to hydrogen energy will take place, and one-third of the hydrogen needed will be produced from hydropower and two-thirds by direct and indirect (other than hydropower) solar energy forms. It will be assumed that half of the thermal energy will be achieved by flame combustion, one-quarter by steam generation with hydrogen–oxygen steam generators, and the last quarter by catalytic combustion; electric power will be generated by fuel cells; half of the surface transportation will use gaseous hydrogen-burning internal combustion engines and the other half will use fuel cells. In air transportation, both subsonic and supersonic,

Table 24.2 CO_2 emissions and pollution factors for coal-derived synthetic fuels.

Fuel	CO_2 emissions (kg GJ^{-1})	Pollution factor
Coal	85.5	1
Coal GH$_2$	116.3	1.36
Hydro GH$_2$	0	0
Solar GH$_2$	0	0
SNG	116.3	1.36
Coal LH$_2$	145.4	1.70
Hydro LH$_2$	0	0
Solar LH$_2$	0	0
Syn-Gasoline	131.6	1.54
Syn-Jet	131.6	1.54

Source: data taken from Barbir and Veziroglu (1992).

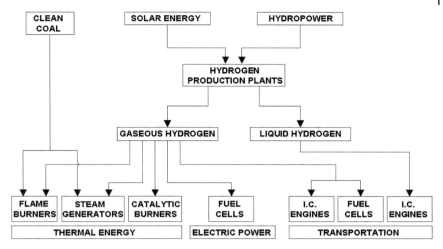

Fig. 24.8 Combination of clean coal and renewable hydrogen energy system as the least-cost energy system. Modified from Barbir and Veziroglu (1992).

liquid hydrogen will be used. Figure 24.8 shows the proposed solar-hydrogen energy system.

An energy system based on coal and coal-derived synthetic fuels would be the most costly alternative. Of course, conversion to a new energy system is not possible in a short time period. For this reason, the hydrogen energy system may become cost competitive in the coming decades, but the transition should start as soon as possible, because only with adequate technology and market developments can the projected costs of hydrogen be achieved.

For the transition period, clean coal technologies could be used for thermal energy generation, satisfying approximately 30% of the total energy needs. Such a system, which employs coal for thermal energy generation and hydrogen from renewable energy sources in the electricity generation and transportation sectors, would be the least-cost energy system. The least-cost energy system, that is, a combination of coal and a renewable hydrogen energy system, is shown in Figure 24.8. Also in this transition period the solar-hydrogen energy system could help to save our economy and our planet.

24.6
Case Study I

Sustainable development of the world will require hydrogen energy (e.g., Dincer, 2007; Veziroglu, 2007). Hydrogen usage would reduce greenhouse gas emissions that pollute the atmosphere and contribute to climate change. Therefore, hydrogen energy systems can play a significant role in providing a better envi-

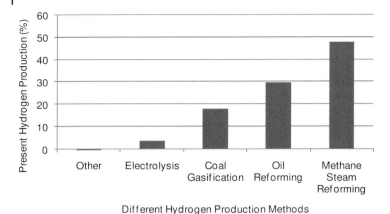

Fig. 24.9 Feedstock used in the present global hydrogen production. Data from Ewan and Allen (2005).

ronment and sustainability. The worldwide demand for hydrogen is growing rapidly, but there are few options for affordable hydrogen production free of greenhouse gas emissions.

Several methods have been identified for producing hydrogen. Among these, steam methane reforming of natural gas (SMR) is currently the most widely used. Greenhouse gases are currently emitted in the present primary methods for the production of hydrogen, for example by SMR. As can be seen in Figure 24.9, current hydrogen production methods are still based on processes that mainly extract hydrogen from fossil fuel feedstock (Barretto et al., 2003; Ewan and Allen, 2005; Kothari et al., 2008). However, a hydrogen economy requires hydrogen to be produced with sustainable, environmentally benign, and non-fossil sources. Clearly, we need to develop new methods to produce hydrogen in cleaner ways.

Hydrogen can also be produced as a clean fuel from the world's sustainable non-fossil primary energy sources such as solar energy, wind energy, hydropower, biomass, and geothermal resources. When considering the use of renewable energy for hydrogen production, geothermal resources seem to be an attractive option for production of hydrogen, owing to its relatively low cost and ready availability.

Eco-friendly hydrogen production via renewable energy is very important to save the environment as it does not emit any greenhouse gases during operation. Renewable-based hydrogen production methods are still in the developmental stage, but they have great potential for hydrogen production and in the hydrogen economy.

In a case study, the sustainability of a renewable-hydrogen system through exergy efficiency was investigated. A comparison of the various processes associated with the renewable-hydrogen production in terms of exergy efficiency and sustainability index has also been done.

24.6.1
Analysis

As a case study, we analyzed four low-temperature thermochemical and hybrid cycles, namely (a) Cu–Cl, (b) Li–NO$_3$, (c) Mg–Cl (d) H$_2$SO$_4$ and High Temperature Steam Electrolysis (HTSE) systems in terms of energy and exergy efficiencies. The energy and exergy efficiencies of HTSE and thermochemical and hybrid cycles of Balta et al. (2009a,b, 2010) were considered to evaluate the sustainability index of the each system.

Some parameters that can help researchers and design engineers to develop a sustainable hydrogen production system that uses renewable sources of energy are energetic and exergetic efficiencies, recycling ratio, sustainability index, improvement potential, and environmental impact factor. These parameters are discussed in the following subsections.

24.6.1.1 Energy Efficiency

For a steady-state process, the general energy balance can be expressed as the total energy input equal to total energy output:

$$\sum \dot{E}_{in} = \sum \dot{E}_{out} \tag{24.7}$$

The energy efficiency of the overall cycle, based on first law of thermodynamics, can be defined as

$$\eta_{overall} = \frac{(1-r)\dot{m}_{H_2} LHV_{H_2}}{\dot{W}_{in} + \sum \dot{Q}_{in}} \tag{24.8}$$

where is the lower heating value per kilomole of hydrogen and is taken as 239.92 MJ kmol^{-1} H$_2$, and r is the total recycling ratio.

24.6.1.2 Exergy Efficiency

Exergy efficiency, which is also called second-law efficiency, is usually defined as the outlet exergy divided by the inlet exergy of streams. In this context, an exergy balance is used in formulating an exergy efficiency for the studied systems. By considering a steady-state steady-flow process for all, the rate at which exergy enters the cycles equals the rate at which exergy exits plus the rate at which exergy is destroyed within the system. We also need to assume a heat rejection rate and consider that about 30% of the total heat transfer rate is lost to the surroundings for exergy loss calculations. The exergy efficiency of the overall system is defined as the ratio of the exergy of the hydrogen to the total exergy input, which can be determined from

$$\psi_{overall} = \frac{(1-r)\dot{E}x_{H_2}}{\dot{W}_{in} + \sum \dot{E}x_{in}} \tag{24.9}$$

24.6.1.3 Recycling Ratio

The recycling ratio r is an important parameter for hydrogen production processes and is defined as the ratio of the number of kilomoles of unreacted substances to the number of kilomoles of reactants supplied. The recycling ratio affects both the energy and exergy efficiencies. However, there is not enough information about the recommended recycling ratios of the analyzed systems in the open literature. In this context, we performed a parametric study to investigate the system's performance for a range of practical recycling ratios.

24.6.1.4 Exergy Efficiency and Sustainability Index

Sustainable development requires not only that a sustainable supply of clean and affordable energy resources be used, but also that the resources should be used efficiently. Exergy methods are very useful tools for improving efficiency, which maximize the benefits and usage of resources and also minimize undesired effects (such as environmental damage). Exergy analysis can be used to improve the efficiency and sustainability (Cornelissen, 1997).

Rosen et al. (2008) defined a relation between exergy efficiency (ψ) and the sustainability index (SI) as

$$\psi = 1 - \frac{1}{SI} \qquad (24.10)$$

This relation shows how sustainability increases with the exergy efficiency of a process increases.

24.6.1.5 Exergy and Environmental Impact Factor

Many researchers have suggested that the most suitable method to reduce the environmental impact is through exergy, because it is a measure of the departure of the state of a system from that of the environment (Rosen, 1986; Cornelissen, 1997; Rosen and Dincer, 1997, 1999). Thus, exergy has an important role to play in providing a better environment. The environmental impact can be reduced by increasing the energy and exergy efficiency. Increased efficiency also reduces the exergy losses.

24.6.1.6 Environmental Impact Ratio

The concentrations of most of the greenhouse gases have increased with industrial expansion. One of the most important greenhouse gases is CO_2, and the emissions of CO_2 play a crucial role in climate change. The greenhouse effect increases with increase in the amount of CO_2 in the atmosphere. Actually, all resource use leads to some degree of environmental impact. In this context, we defined a ratio which is based on CO_2 emission by renewable and non-renewable technologies:

$$EIR = \frac{CO_{2es} \, (g \, kW \, h^{-1})}{CO_{2coal} \, (g \, kW \, h^{-1})} \tag{24.11}$$

The *environmental impact ratio* (*EIR*) can be defined as the ratio of CO_2 emission by a particular renewable/non-renewable technology to CO_2 emission by coal-based technology. In other words, the environmental impact ratio is the fraction of CO_2 emitted by a renewable/non-renewable technology as compared with coal-based technology.

24.6.2
Results and Discussion

Energy and exergy efficiencies of the process cycles are calculated as shown in Figure 24.10 versus various recycling ratios from 0 to 0.9. The energy efficiencies of each cycle are determined to be about 50%, based on the complete reaction and lower heating value (*LHV*). In this calculation, the auxiliary works are not considered.

The highest energy efficiency, such as 51%, is obtained by Cu–Cl and H_2SO_4 cycles. However, the exergy efficiency of Cu–Cl cycle is better than that of H_2SO_4 cycle: 65% versus 57%. Also, it can be seen in Figure 24.10b that the cycle exergy efficiencies are slightly higher than their corresponding energy efficiencies which are calculated based on LHV_{H_2}. This difference may be caused by the effect of inlet heat and inlet exergy. In Equations 24.7 and 24.8, the *LHV* of hydrogen is very close to its exergy concept since inlet exergy also includes the chemical exergy of compounds, while inlet works are the same in both equations. However, the inlet heat is higher than inlet exergy since inlet exergy also includes the exergetic temperature factor. Hence exergy efficiency is higher than energy efficiency.

Renewable-based hydrogen production using thermochemical cycles appears to be a promising solution for the future hydrogen economy to be run on a renewable basis, and it will help reduce the environmental impact (greenhouse gas emissions) and hence be beneficial for sustainable development. Therefore, the copper-chlorine cycle has been identified as a highly promising option among all the cycles considered for thermochemical hydrogen production.

Figure 24.11 shows the sustainability index for the copper-chlorine cycle, which varies from 1.07 to 2.89, while the recycling ratio increases from 0 to 0.9.

The energy and exergy efficiencies of high-temperature electrolysis were found to be 87 and 86%, respectively (Balta *et al.*, 2009 a, b). One of the main factors affecting the hydrogen production cost is the temperature of the electrolyzer. Figure 24.12 illustrates the effects of the electrolyze temperature on the energy and exergy efficiencies. Energy efficiency values for the overall system vary between 80 and 87%, whereas the corresponding exergy efficiency values range from 79 to 86%, respectively. If we compare the hydrogen production through thermochemical cycles, the HTSE electrolyzer unit efficiency is higher than that of thermochemical cycles.

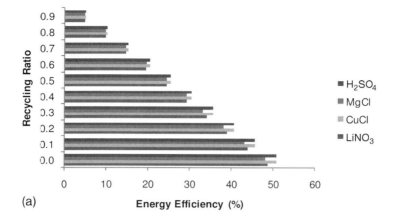

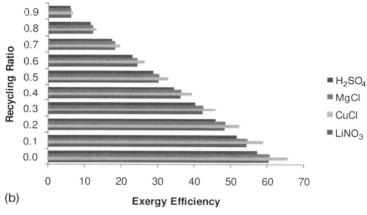

Fig. 24.10 (a) Energy efficiencies and (b) exergy efficiencies of low-temperature cycles versus recycling ratio.

Exergy is evaluated with respect to a reference environment and is used to standardize the quantification of exergy. The reference state temperature is a state of a system in which it is at equilibrium with its surroundings. Figure 24.13a shows the effects of reference temperature on the sustainability index of the system. It can be clearly seen that the sustainability index of the HTSE increases nonlinearly with the reference environment temperature. In this regard, at higher reference state temperatures, less energy is consumed and hence the corresponding sustainability increases.

The sustainability index for operating temperatures of the HTSE system can be seen in Figure 24.13b. The sustainability index for the HTSE process of Balta et al. (2009a, b) varies from 4.76 to 7.58 with respect to operating temperatures varying between 473 and 1173 K. Note that the overall electrolyzer system

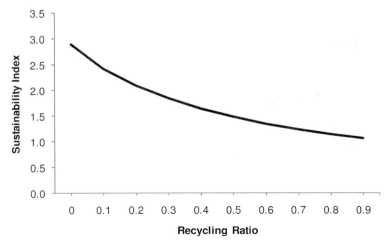

Fig. 24.11 Variation of sustainability index with recycling ratio of Cu–Cl cycle.

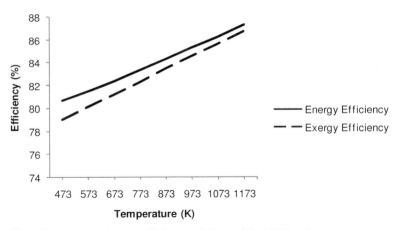

Fig. 24.12 Energy and exergy efficiency variations of the HTSE system at various operating temperatures.

efficiency is always less than the electrolysis process efficiency due to various irreversibilities and losses taking place in different components, including coupling pumps and turbines. Some energy losses may occur if such devices are used.

Greenhouse gas emissions are one of the key parameters that define the sustainability of energy usage, generation, and transportation. As mentioned previously, one of the most important greenhouse gases is CO_2. We calculated the environmental impact ratio of some energy technologies based on CO_2 emissions. Figure 24.14 shows that the environmental impact ratio of the various en-

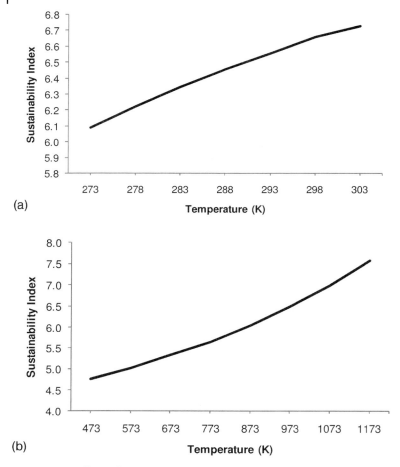

Fig. 24.13 (a) Effects of reference state temperature on the sustainability index of HTSE. (b) Corresponding sustainability index for the HTSE system at various operating temperatures.

ergy sources (data taken from Armannsson et al., 2005 and IAEA, 1997). It is found that nuclear power is the most sustainable option, but the technology suffers from the nuclear waste disposal problem and the technology is also costly. Compared with nuclear power, wind power produces much less CO_2 during its operation and hence poses better sustainability. Since it involves revolving parts, for example wind mills, maintenance is frequently required. Biomass can also be a very good option and it also emits comparatively less CO_2 into the atmosphere, but collection of biomass may be problematic. Unlike wind, solar power does not involve any moving parts and can be used as an option, but at present it suffers from low conversion efficiencies and high costs.

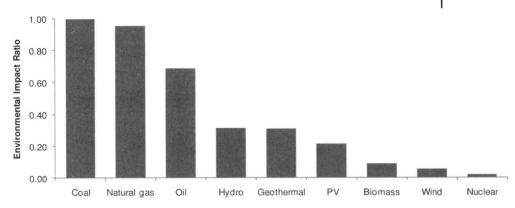

Fig. 24.14 Environmental impact ratio of the various energy sources.

Geothermal steam plants emit typically only 5–30% of the CO_2 emitted by a coal or natural gas plant and about 20–40% of the CO_2 from an oil plant, per kilowatt hour. Note that geothermal emissions are most significantly impacted by technology choices (DiPippo, 1991). The CO_2 emission of a geothermal source may change region by region.

We have therefore investigated the potential role of renewable energy for sustainable hydrogen production. In this regard, renewable-based hydrogen production through an HTSE and some thermochemical and hybrid cycles were studied in terms of energy and exergy efficiencies and sustainability index. The environmental impact ratio shows that the renewable sources are much more environmentally benign than non-renewables, such as fossil fuels (coal, natural gas, and oil).

24.7
Case Study II

In this case study, three methods were considered for life cycle assessment (LCA) of hydrogen production as shown in Figure 24.15, namely (i) by natural gas reforming, (ii) by water electrolysis using solar energy, and (iii) by water electrolysis using wind energy, and compared with conventional gasoline production in terms of GHG emissions and air pollutants. Environmental, sustainability, and economic aspects were investigated for the above three methods for comparison purposes. The results were used for evaluation of the changes that result from hydrogen implementation as a fuel in fuel cell vehicles instead of gasoline. The environmental and economic indicators of the gasoline used in conventional transportation were estimated for the standard processes of crude oil processing and distillation (for details, see Granovskii et al., 2006).

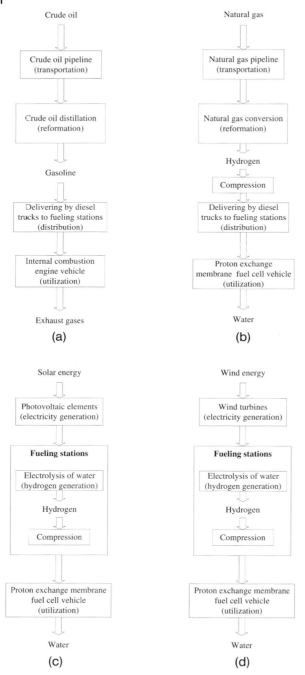

Fig. 24.15 Essential steps in utilizing (a) crude oil, (b) natural gas, (c) solar energy, and (d) wind energy in transportation.

24.7.1
Analysis

In the analysis, hydrogen and gasoline represent the final products. The conventional production methods are reforming of natural gas for hydrogen and crude oil distillation for gasoline. Figure 24.15a–d present the flow charts for gasoline and hydrogen production and use in transportation vehicles. In general, each step in the figure is accompanied by total fossil fuel energy use E, which can be expressed as

$$\Delta E = \Delta E_{\text{dir}} + \Delta E_{\text{ind}} \tag{24.12}$$

where ΔE_{dir} is the direct use of fossil fuel energy to perform a step and ΔE_{ind} is the indirect use of fossil fuel energy used in construction materials and equipment, and also for installation, operation, maintenance, decommissioning, and so on.

Note that fuel energy represents the fuel's lower heating value (LHV) or, in other words, the heat released by the complete burning of all fuel components to CO_2 and H_2O (steam). It should be noted that construction materials are also produced from mineral sources (ores, limestone, etc.), which, like fossil fuels, have value. To account for this, the energy equivalent of construction materials and devices (EEQ) is introduced and the following procedure proposed to calculate indirect energy ΔE_{ind} as follows:

$$\Delta E_{\text{ind}} = \frac{\sum EEQ + EOP}{LFT} \tag{24.13}$$

where ΣEEQ is the summation of the energy equivalents of construction materials and devices related to a given technological operation. EOP is the operation energy (i.e., the fossil fuel energy required for installation, construction, operation, maintenance, decommissioning, etc. of the equipment). LFT is the lifetime of the unit performing any step as outlined in operation. The intensity data of embodied energies and the respective cost data are taken from (EIOLCA, 2005; White, 1998, EIA, 1997).

The direct and indirect fossil fuel energy consumptions lead to different kinds of pollution. greenhouse gases (e.g., with CO_2, CH_4 and NO_2 constituting the greatest part). The global warming potentials of these substances, according to the International Panel on Climate Change, are 1, 23 and 310, respectively (Granovskii et al., 2006).

24.7.2
Results and Discussion

The two fossil fuel technologies as outlined in Figure 24.15 are characterized by their pipeline transportation efficiencies. To evaluate and compare the energy consumption and environmental impact of natural gas and crude oil pipeline transportation, equal lengths of pipelines (e.g., 1000 km) were considered. The mechanical work or electricity required to perform pipeline transportation is assumed to be produced by a gas turbine unit with an average efficiency $\eta_{gt}=0.33$ (Cleveland, 2004).

The energy losses in the process of natural gas reforming comprise ~14% of the total energy of the initial flows of methane (Rosen, 1996). The energy efficiency and environmental impact to produce 1 MJ of gasoline were estimated according to the energy consumption and carbon dioxide emission data for all petroleum refineries in the United States in 1996 (IEA, 1999; EIA, 2005). The total intensities of direct energy consumption and direct greenhouse gas emissions for natural gas and crude oil transportation and reforming (e.g., distillation) processes per megajoules per second of fuels produced are as follows:

- $\Delta E_{dir} = 0.273$ MJ s^{-1}, CO_2 equivalent emission = 75.7 g s^{-1} for hydrogen at $\eta = 78.5\%$
- $\Delta E_{dir} = 0.173$ MJ s^{-1}, CO_2 equivalent emission = 12.1 g s^{-1} for gasoline at $\eta = 85.3\%$.

Here, the indirect energy evaluation for the natural gas reforming process is based on the hydrogen plant material requirements data given by Spath and Mann (2001) and embodied energy magnitudes taken from White (1998): 1.4 GJ t^{-1} for concrete for an amount of 10,242 t; 34.4 GJ t^{-1} for steel for an amount of 3272 t; 201.4 GJ t^{-1} for aluminum for an amount of 27 t; and 23.5 GJ t^{-1} for iron for an amount of 40 t. It is assumed that the operation energy to install, maintain, and operate the equipment is equal to the embodied energy which is consumed to produce it. Comparing the values of direct and indirect energies and their emissions, it can be observed that indirect energy (ΔE_{ind}) is more than 10 times less than direct energy (ΔE_{dir}) and the indirect greenhouse gas emissions are only 0.5% of the direct emissions. No data are available to calculate the indirect energy consumption for the crude oil refinery process. However, as seen in Lange and Tijm (1996), the capital cost of crude oil distillation is lower than that for natural gas reforming and therefore we can assume that for the case of natural gas reforming, the indirect energy use and indirect greenhouse gas emissions are negligible in comparison with direct emissions.

We now consider the indirect energy consumption and indirect greenhouse gas emissions of a photovoltaic system which converts solar energy into direct current (d.c.) electricity, as transformed by inverters to alternating current (a.c.) electricity and transmitted to the power grid. At fueling stations, a.c. electricity is used to electrolyze water to produce hydrogen as shown in Figure 24.15. The data for a 1231 kW building-integrated photovoltaic system at Silverthorne, CO, USA (Meier, 2002), which utilizes thin-film amorphous silicon technology, are considered here. We calculate the amount of total embodied energy in a solar cell block, with a surface area of 157.2 m^2 and a lifetime of 30 years, to be 688.44 GJ per unit. For the entire photovoltaic system, consisting essentially of solar cell block, inverters and wiring, we can summarize the total indirect energy flow and CO_2 equivalent greenhouse gas emission per second of lifetime for a 1231 kW thin film photovoltaic system as 973.77 J s^{-1} and 0.0132 g s^{-1}, respectively. In addition, a procedure similar to that used for a wind power plant is applied to evaluate the indirect energy flows and greenhouse gas emissions associated with electrolysis. Taking into account the efficiency of electrolysis and transmission losses, the 1231 kW photovoltaic system combined with water electrolysis can produce 823.5 J s^{-1} of hydrogen energy. Therefore, we end up with the following results:

- $\Delta E_{ind} = 973.77$ J s^{-1}, CO_2 equivalent emission = 0.0132 g s^{-1} for a 1231 kW photovoltaic system
- $\Delta E_{ind} = 64.4$ J s^{-1}, CO_2 equivalent emission = 0.0007 g s^{-1} for electrolysis
- Total $\Delta E_{ind} = 1038.1$ J s^{-1}, total CO_2 equivalent emissio = 0.0139 g s^{-1}.

Note that the fossil fuel energy consumption ratio becomes 0.79 (1 MJ of hydrogen produced relates to a E_{ind} of 1.261 MJ s^{-1} and a CO_2 equivalent emission of 16.88 g s^{-1}), meaning that the fossil fuel energy consumed (embodied in materials, equipment, etc.) is 0.79 times less that the energy of the hydrogen produced.

Basically, wind energy is converted to mechanical work by wind turbines and then transformed by an alternator to a.c. electricity, which is transmitted to the power grid as shown in Figure 24.15. The efficiency of a wind turbines depends on several factors, such as wind velocity. Applications of wind energy normally make sense only in areas with high wind activity. Here we use the data for a 6 MW wind power generation plant taken from White and Kulcinski (2000) for analysis and comparison purposes. Based on data of Spath and Mann (2004) for electrolysis to produce hydrogen with an efficiency of 72% (on an LHV basis), the indirect energy and greenhouse gas emissions are 6.61 and 5.64%, respectively, of those for a wind power generation plant. Accounting for ~7% electricity loss during transmission, the efficiency of hydrogen production is 66.9%. Hence a 6 MW wind power plant combined with electrolysis of water could produce 4.01 MJ s^{-1} of hydrogen.

Here we calculate the indirect energy use and indirect greenhouse gas emissions for hydrogen production via a wind power plant and water electrolysis and end up with the following results:

- $\Delta E_{ind} = 1.087$ MJ s^{-1}, CO_2 equivalent emission = 26.06 g s^{-1} for a 6 MW wind power generation plant
- $\Delta E_{ind} = 0.072$ MJ s^{-1}, CO_2 equivalent emission = 1.47 g s^{-1} for electrolysis
- Total $\Delta E_{ind} = 1.159$ MJ s^{-1}, total CO_2 equivalent emission = 27.53 g s^{-1}.

Note that the fossil fuel energy consumption ratio η becomes 3.46 (1 MJ of produced hydrogen relates to a ΔE_{ind} of 0.289 MJ s^{-1} and a CO_2 equivalent emission of 6.85 g s^{-1}), meaning that the fossil fuel energy consumed (embodied in materials, equipment, etc.) is 3.46 times less that the energy of the hydrogen produced. This is due to the fact that the wind energy is considered "free" and is not included into the expression for η. This value should not be confused with the energy efficiency of wind power generation plants, which is about 12–

25% and usually calculated as the ratio of electricity produced to the sum of all sources of input energy (mainly kinetic energy of wind).

The density of hydrogen under standard conditions is low. The direct energy consumed and direct greenhouse gas emissions to compress hydrogen were evaluated assuming an isothermal compression efficiency η_{cmp} of 0.65 and a typical gas-turbine power plant efficiency η_{gt} of 0.33. A maximum pressure $p_{max} = 200$ atm (Amos, 1998) is considered. Minimum pressures before compression of $p_{min} = 1$ and 20 atm were taken from Spath and Mann (2001) for hydrogen production through electrolysis and natural gas reforming, respectively.

Hydrogen distribution is replaced by electricity distribution when using wind and solar energy (see Figure 24.15). The distribution of compressed hydrogen after its production via natural gas reforming is similar to that for liquid gasoline, but compressed hydrogen is characterized by a lower volumetric energy capacity and higher material requirements for a hydrogen tank. According to the 1997 Vehicle Inventory and Use Survey, the average "heavy–heavy" truck in the United Sates ran 6.1 miles per gallon of diesel fuel (CRA, 2004). The direct fuel (i.e., diesel) energy consumption, assuming that a distance of 300 km is traveled before refueling for a truck with a 50 m^3 tank, was evaluated. The direct energy consumption and greenhouse gas emissions were associated with the combustion of the diesel fuel.

Environmental indicators characterize the environmental impacts of technologies. Here, the implementation of hydrogen as a fuel for Proton Exchange Membrane Fuel Cell (PEMFC) vehicles is considered from the point of view of changes in air pollution and greenhouse gas emissions. In a PEMFC stack, electricity (which is converted into mechanical work in electrical motors with efficiency higher than 90%) is generated via the following electrochemical reactions:

Anode: $2H_2 \rightarrow 4H^+ + 4e^-$ (24.14a)

Cathode: $O_2 + 4H^+ + 4e^- \rightarrow 2H_2O$ (24.14b)

These reactions generally occur at low temperature (<100 °C) and involve the separation of oxygen from air at the cathode. Under these conditions, the formation of harmful nitrogen oxides is inhibited and only water is produced during power generation. Hence the utilization of hydrogen in PEMFC vehicles can be considered as ecologically benign, with respect to direct vehicle emissions. Any associated emissions of pollutants and greenhouse gases are associated with hydrogen production.

In internal combustion engine (ICE) vehicles, gasoline (a mixture of hydrocarbons) is combusted in air:

$C_nH_m + [(n+m)/4]O_2 \rightarrow nCO_2 + (m/2)H_2O + Q$ (24.15) where Q is in part converted to mechanical work. According to the Carnot principle, the higher the temperature of fuel combustion, the more mechanical work can be extracted theoretically. The average temperature of the combusting mixture of gasoline and air is about 1300 °C. At such high temperatures, the formation of NO_x is promoted.

Evaporation of gasoline and incomplete combustion lead to emissions of volatile organic compounds and CO. Air pollution and greenhouse gas emissions are associated with gasoline production and its utilization in ICE vehicles. Table 24.3 gives emissions of greenhouse gases and air pollutants during the life cycle from production of hydrogen and gasoline to their utilization in vehicles. The quantities of air pollutants were evaluated assuming that direct and indirect greenhouse gas emissions (during the production of construction materials and equipment) are from natural gas combustion for all life cycle stages, excluding crude oil distillation. In the latter case, air pollution was evaluated assuming propane combustion. For the stage of gasoline and hydrogen distribution by trucks, air pollution was estimated assuming diesel fuel combustion. The last line in Table 24.3 represents the emissions which are obtained by using GREET 1.6 software (e.g., GRT, 2005). They are consistent with our data for ICE vehicles powered by gasoline. However, application of this model to PEMFC vehicles powered by hydrogen leads to unsatisfactory results, because in this model the efficiency of a fuel cell vehicle was taken to be about three times higher than that of an ICE vehicle, and energy consumption and emissions connected to the gaseous hydrogen compression were not taken into account.

Figure 24.16 present the normalized emissions of greenhouse gases (GHG) and air pollution (AP) for different methods for hydrogen production as a function of the change in the fraction of hydrogen. These emissions decrease with the transition from gasoline to hydrogen for any production technology. The first derivative corresponds to the intensity of the economic criterion decline with hydrogen fraction. A negative sign implies a decrease in the effectiveness of capital investments. The first derivatives and taken with the opposite sign indicate the intensities of air pollution and greenhouse gas emission reductions, respectively. These are clearly presented in Table 24.4.

To account for the interaction between environmental and economic criteria as a result of implementation of a new technology, sustainability indexes are introduced. They are obtained by the summation of the corresponding economic and environmental intensities. Thus, the sustainability index for air pollution reduction becomes

$$SI_{AP} = \frac{d\gamma}{dX_{H_2}} + \frac{dAP}{dX_{H_2}} \tag{24.16}$$

and the sustainability index for greenhouse gas emissions is

$$SI_{GHG} = \frac{d\gamma}{dX_{H_2}} + \frac{dGHG}{dX_{H_2}} \tag{24.17}$$

Table 24.3 Greenhouse gas emissions and air pollution during the life cycle of hydrogen and gasoline production and utilization (per MJ of LHV).

Sub-cycle	GHG (g)	CO (g)	NO$_x$ (g)	VOCS[a] (g)
Hydrogen from natural gas				
Natural gas pipeline transportation and reforming to produce hydrogen	75.7	0.0217	0.0259	0.0544
Hydrogen compression	5.94	0.00413	0.00492	0.000383
Hydrogen delivery to fueling stations	3.13	0.00722	0.0453	0.00135
Total	84.8	0.0330	0.0761	0.0561
Hydrogen from wind energy				
Electricity generation and hydrogen production through electrolysis at fueling stations	6.85	0.00468	0.00558	0.000435
Hydrogen compression	13.7	0.00949	0.0113	0.000881
Total	20.55	0.0142	0.0169	0.00132
Hydrogen from solar energy				
Electricity generation and hydrogen production through electrolysis at fueling stations	16.9	0.0115	0.01375	0.00107
Hydrogen compression	13.7	0.00949	0.0113	0.000881
Total	30.6	0.0210	0.0251	0.00195
Gasoline from crude oil				
Crude oil pipeline transportation and distillation to produce gasoline	12.1	0.0120	0.0610	0.0237
Gasoline delivery to fueling stations	0.19	0.00044	0.00276	0.0000826
Gasoline utilization in ice vehicle[b]	71.7	0.864	0.0508	0.146
Total	84.0	0.876	0.115	0.170
GREET 1.6 (total)	75.3	0.827	0.072	0.139

a) Methane emissions are included in VOCs.
b) Source: Walwijk et al. (1999).

Table 24.4 Calculated intensities and sustainability indexes of different method of hydrogen production for PEMFC vehicles.

Hydrogen production technology	Reduction in air pollution (AP)	Reduction in GHG emissions	Sustainability index for air pollution reduction	Sustainability index for GHG emissions reduction
Hydrogen-natural gas	0.663	0.279	0.113	−0.271
Hydrogen-wind	0.946	0.825	0.081	−0.036
Hydrogen-solar	0.920	0.739	0.069	−0.25

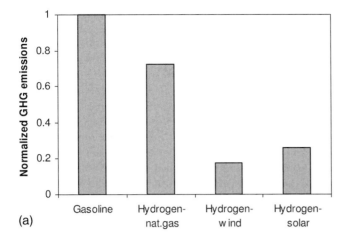

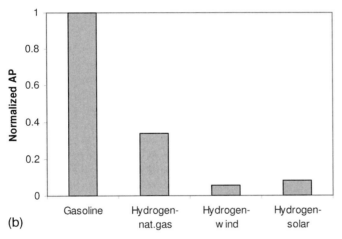

Fig. 24.16 (a) Normalized GHG emissions for gasoline and hydrogen use as a fuel for ICE and PEMFC vehicles and (b) normalized AP index for gasoline and hydrogen use as a fuel for ICE and PEMFC vehicles.

The resulting sustainability indexes are presented in Table 24.4. The negative indexes indicate that the degree of emissions reduction (environmental criterion) is lower than the degree of effectiveness of capital investments (economic criterion) decline. The positive indexes imply that the reduction of the emissions exceeds the decrease in the effectiveness of capital investments. An analysis of sustainable indexes permits one to choose an optimal strategy for an implementation of competing environmentally benign technologies. For instance, it can be concluded from the values presented (e.g., Table 24.4) that the production of hydrogen from wind energy and its application in PEMFC vehicles reduce greenhouse gas emissions more effectively but the traditional technology of natural gas reforming is more favorable for reducing air pollution.

24.8 Conclusion

This chapter has discussed the key environmental and sustainability issues for the present and future, and identified hydrogen as a potential solution. The benefits of hydrogen production technologies have been highlighted by using the principles of thermodynamics (particularly exergy) and life cycle assessment to evaluate their key roles in sustainable development.

For society to attain or try to attain sustainable development, efforts should be devoted to developing hydrogen technologies. Renewable energy utilization in hydrogen production can provide a potential solution to current environmental problems. Advanced hydrogen and fuel cell technologies can provide environmentally responsible alternatives to conventional energy systems, in addition to greater flexibility and decentralization.

To realize the energy, exergy, economic, and environmental benefits of hydrogen and fuel cell technologies, an integrated set of activities should be conducted including research and development, technology assessment, standards development, and technology transfer. These can be aimed at improving efficiency, facilitating the substitution of these technologies and other environmentally benign energy currencies for more harmful ones, and improving the performance and implementation characteristics of these technologies.

Acknowledgments

The support of the Natural Sciences and Engineering Research Council of Canada is gratefully acknowledged. The authors also acknowledge the assistance of Mr M. Tolga Balta.

Nomenclature

AP air pollution
$CH2$ cost of hydrogen
CNG cost of natural gas
EEQ energy equivalent, MJ
E intensity of energy flows, MJ s^{-1}
$\dot{E}$ energy rate, kW
ΔE intensity of energy consumption to perform a technology, MJ s^{-1}
E_{syn} environmental damage due to synthetic fuel

E_{coal}	environmental damage due to coal
EIR	environmental impact ratio
EMB	energy embodied in construction materials, MJ
EOP	operation energy, MJ
$\dot{E}x$	exergy rate, kW
e^-	electron
GHG	greenhouse gas
H_2	hydrogen
H^+	hydrogen cation
IEE	intensity of the energy equivalent, $ MJ^{-1}
IPM	investments to produce construction materials, $
i, j	indexes
LFT	lifetime of unit, s
LHV	lower heating value, kJ
M	mass of the air pollutant, g/MJ
$\dot{m}$	mass flow rate, kg/s
P_s	synthetic fuel pollution factor
Q	heat released, kJ
r	recycling ratio
SI	sustainability index
VOC	volatile organic compound
W	capacity of hydrogen energy production, MJ s^{-1}
w	weighting coefficient of an air pollutant
$\dot{W}$	power rate, kW
X_{H_2}	fraction of hydrogen

Greek Symbols

γ	capital investments effectiveness
η	energy efficiency
η_H	utilization efficiency of hydrogen
η_F	utilization efficiency of fossil
ψ	exergy efficiency

Subscripts

AP	air pollution
dir	direct
es	energy source
g	gasoline
in	input
ind	indirect
ng	natural gas
out	output

References

Afgan, N. H. and Carvalho, M. G. (2004) Sustainability assessment of hydrogen energy systems. International Journal of Hydrogen Energy, 29, 1327–1342.

Afgan, N. H., Al-Gobaisi, D., Carvalho, M. G., and Cumoc, M. (1998) Sustainable energy development. Renewable and Sustainable Energy Reviews, 2, 235–286.

Amos, W. (1998) *Costs of Storing and Transporting Hydrogen*, National Renewable Energy Laboratory, Report NREL/TP-570-25106, US Department of Energy, Washington, DC.

Argonne National Laboratory (2005) *Basic Research Needs for Solar Energy Utilization*, Report of the US Department of Energy Basic Energy Sciences Workshop on Solar Energy Utilization.

Armannsson, H., Fridriksson, T., and Kristjansson, B. R. (2005) CO_2 emissions from geothermal power plants and natural geothermal activity in Iceland. Geothermics, 34, 286–296.

Arrhenius, S., (1896) On the influence of carbonic acid in the air upon the temperature of the ground, Philosophical Magazine, 41, 237.

Awad, A. H. and Veziroglu, T. N. (1984) Hydrogen versus synthetic fossil fuels, Int. Journal of Hydrogen Energy, 9, 355–366.

Balta, M. T., Dincer, I., and Hepbasli, A. (2009a) Thermodynamic assessment of geothermal energy use in production through high-temperature steam electrolysis. In *Proceedings of the Fourth International Exergy, Energy and Environment Symposium*, 19–23 April 2009, AUS, Sharjah, UAE, Paper No. 52-19598.

Balta, M. T., Dincer, I., and Hepbasli, A. (2009b) Energy and exergy efficiency analyses of geothermal based hydrogen production. In *Proceedings of the International Conference on Hydrogen Production*, May 2009, Oshawa, Canada, Paper No. ICH2P-883-6, pp. 225–242.

Balta, M. T., Dincer, I., and Hepbasli, A. (2010) Potential methods for geothermal-based hydrogen production. International Journal of Energy Research, doi: 10.1002/er.1589.

Barbir, F. and Veziroglu, T. N. (1992) Effective costs of the future energy systems. International Journal of Hydrogen Energy, 17 (4), 299–308.

Barbir F., Veziroglu T. N., Plass H. J. (1990) Environmental damage due to fossil fuels use, International Journal of Hydrogen Energy, 15 (10), 739–749.

Baretto, L., Makihira, A., and Riahi, K. (2003) The hydrogen economy in the 21st century: a sustainable development scenario. International Journal of Hydrogen Energy, 28, 267–284.

Billings R. E. (1979) Hydrogen homestead. In *Hydrogen Energy System* (eds T. N. Veziroglu and W. Seifritz), Vol. 4, Pergamon Press, Oxford, pp. 1709–1730.

Bockris, J.O'M. and Veziroglu, T. N. (1983) A solar hydrogen economy for U.S.A. International Journal of Hydrogen Energy, 8, 323.

Callender, G. (1939) The artificial production of carbon dioxide and its influence on temperature. Quarterly Journal of the Royal Meteorological Society, 64, 223.

Chamberlin, J., Foley, H., McDonald, G., and Ruderman, M. (1982) Climate effects of minor atmospheric constituents. In *Carbon Dioxide Review: 1982* (ed. W. C. Clarke), Clarendon Press, Oxford.

Cleveland C. J. (ed.) (2004) *Encyclopedia of Energy*, Elsevier, New York.

Cornelissen, R. L. (1997) Thermodynamics and sustainable development. PhD Thesis, University of Twente, The Netherlands.

CRA (2004) *Diesel Technology and the American Economy*, Report D032378-00, Charles River Associates, Washington, DC.

Dincer, I. (1998) Thermodynamics, exergy and environmental *impact, In Proceedings* of the ISTP-11, the Eleventh International Symposium on Transport Phenomena, 29 November–3 December 1998, Hsinchu, Taiwan, pp. 121–125.

Dincer, I. (2002) Technical, environmental and exergetic aspects of hydrogen energy systems. International Journal of Hydrogen Energy, 27, 265–285.

Dincer I. (2007) Environmental and sustainability aspects of hydrogen and fuel cell systems. International Journal of Energy Research, 31, 29–55.

Dincer, I. and Rosen, M.A. (1998) A worldwide perspective on energy, environment and sustainable development. International Journal of Energy Research, 22 (15), 1305–1321.

Dincer, I. and Rosen, M.A. (1999) the intimate connection between exergy and the environment. In *Thermodynamic Optimization of Complex Energy System* (eds A. Bejan and E. Mamut), Kluwer Academic, Dordrecht, pp. 221–230.

Dincer, I. and Rosen, M. (2004) Exergy as a drive for achieving sustainability. International Journal of Green Energy, 1 (1), 1–19.

Dincer, I. and Rosen, M.A. (2005) Thermodynamic aspects of renewables and sustainable development. Renewable and Sustainable Energy Reviews, 9 (2), 169–189.

DiPippo, R. (1991) Electricity generation and environmental impact. Renewables series – geothermal energy. Energy Policy, 798–807.

EIA (1997) *US DOE: Emissions of Greenhouse Gases in the United States 1996*, DOE/EIA-0573(96), Energy Information Administration, Washington, DC

EIA (2005) *Official Energy Statistics from the US Government*. Energy Information Administration, Washington, DC, http://www.tonto.eia.doe.gov/steo_query/app/pricepage.htm (accessed 22 October 2009).

EIOLCA (2005) *Economic Input–Output Life Cycle Assessment. Green Design Initiative*. Carnegie Mellon University, Pittsburgh, PA, http://www.eiolca.net (accessed 22 October 2009).

Elliot, M.A. and Turner, N.C. (1972) Estimating the future rate of production of the world's fossil fuels. Presented at the American Chemical Society 163rd National Meeting, Division of Fuel Chemistry Symposium on Non-Fossil Chemical Fuels, Boston.

Ewan, B.C.R. and Allen, R.W.K. (2005) A figure of merit assessment of the routes to hydrogen. International Journal of Hydrogen Energy, 30, 809–819.

Granovskii, M., Dincer, I., and Rosen, M.A. (2006) Life cycle assessment of hydrogen fuel cell and gasoline vehicles. International Journal of Hydrogen Energy, 31, 337–352.

GRT (2005) *Greenhouse gases, Regulated Emissions, and Energy use in Transportation model GREET (1.6)*, Argonnes Transportation Technology Research and Development Center, Argonne, IL, http://www.transportation.anl.gov/ttrdc/greet (accessed 22 October 2009).

Hammond, G.P. (2004) Towards sustainability: energy efficiency, thermodynamic analysis, and the two cultures. Energy Policy, 32, 1789–1798.

Hart, D. (2000) Sustainable energy conversion: fuel cells-the competitive options? Journal of Power Sources, 86, 23–27.

Hoffert, M., Caldeira, K., Benford, G., Criswell, D., Green, C., and Herzog, H. (2002) Advanced technology paths to global climate stability: energy for the greenhouse planet. Science, 298, 981–987.

Hopwood, B., Mellor, M., and OBrien, G. (2005) Sustainable development: mapping different approaches. Sustainable Development, 13, 38–52.

Hui, S.C.M. (1997) From renewable energy to sustainability: the challenge for Hong Kong. In *Proceedings of the POLMET 97 Conference*, 25–27 November 1997, Hong Kong Institution of Engineers, Hong Kong, pp. 351–358.

Hydrogen Update (1981) Hydrogen update – Tappan, *Hydrogen Progress Magazine*, 6 (2), 7, Billings Corporation, Independence, MO.

IAEA (1997) *Sustainable Development and Nuclear Power*, International Atomic Energy Agency, Vienna.

IEA (1999) *Automotive Fuels for the Future: the Search for Alternatives*, International Energy Agency IEA-AFIS, Paris.

Kothari, R., Buddhi, D., and Sawhney, R.L. (2008) Comparison of environmental and economic aspects of various hydrogen production methods. Renewable and Sustainable Energy Reviews, 12, 553–563.

Kwak, H.-Y., Lee, H.-S., Jung, J.-Y., Jeon, J.-S., and Park, D.-R. (2004) Exergetic and thermoeconomic analysis of a 200-kW phosphoric acid fuel cell plant. Fuel, 83 (14–15), 2087–2094

Lacis, A., Hansen, J., Lee, P., Mitchell, T., and Lebedeff, S. (1981) Greenhouse effect of trace gases, 1970–80, Geophysical Research Letters, 8, 1035.

Lange, J. P. and Tijm, P. J. A. (1996) Processes for converting methane to liquid fuels: economic screening through energy management. Chemical Engineering Science, 51, 2379–2387.

Lavi, A. and Trimble L. C. (1979) OTEC for hydrogen production. In *Hydrogen Energy System* (eds T. N. Veziroglu and W. Seifritz) Vol. 1, Pergamon Press, Oxford, pp. 142–167.

McDonald, G.(1982) *The Long Term Impacts of Increasing Atmospheric Carbon Dioxide Levels*, Ballinger Publishing Co., Cambridge, MA.

McGowan, J. G. (1990) Large-scale solar/wind electrical production systems-predictions for the 21st century. In *Energy and the Environment in the 21st Century* (eds J. W. Tester, D. O. Wood, and N. A. Ferrari), Massachusetts Institute of Technology, Boston, MA, pp. 256–277.

Meier, P. (2002) Life-cycle assessment of electricity generation systems and applications for climate change policy analysis, PhD Thesis, University of Wisconsin, Madison, WI.

Midilli, A., Ay, M., Dincer, I., and Rosen, M. A. (2005 a) On hydrogen and hydrogen energy strategies – I: current status and needs. Renewable and Sustainable Energy Reviews, 9 (3), 291–307.

Midilli, A., Ay, M., Dincer, I., and Rosen, M. A. (2005 b) On hydrogen and hydrogen energy strategies – II: future projections affecting global stability and unrest. Renewable and Sustainable Energy Reviews, 9 (3), 309–323.

Momirlan, M. and Veziroglu T. N. (2005) The properties of hydrogen as fuel tomorrow in sustainable energy system for a cleaner planet. International Journal of Hydrogen Energy, 30, 795–802.

Muradov, N. Z. and Veziroglu T. N. (2008) "Green" path from fossil-based to hydrogen economy: an overview of carbon-neutral technologies. International Journal of Hydrogen Energy, 33, 6804–6839.

NASA (1983) *The Climate Impact of Increasing Atmospheric CO_2 with Emphasis on Water Availability and Hydrology in the United States*, Draft, prepared for the US Environmental Protection Agency by the Goddard Institute for Space Studies, New York.

OECD (1995) *Urban Energy Handbook*, Organization for Economic Cooperation and Development (OECD), Paris.

Parent, J. D. (1979) *A Survey of United States and Total World Production, Proved Reserves, and Remaining Recoverable Resources of Fossil Fuels and Uranium as of December 31, 1977*, Institute of Gas Technology, Chicago.

Pehnt, M., (2003) Assessing future energy and transport systems: the case of fuel cells. Part I: methodological aspects. International Journal of Life Cycle Assessment, 8 (5), 283–289.

Plass, G. N. (1956a) The carbon dioxide theory of climatic change. Tellus, 8, 1411.

Plass, G. N. (1956b) Effect of carbon dioxide variations on climate. American Journal of Physics, 24, 376.

Root, D. and Attanasi, E. (1978) *American Association of Petroleum Geologists Bulletin*.

Rosen, M. A. (1986) The development and application of a process analysis methodology and code based on exergy, cost, energy and mass, PhD Thesis, University of Toronto.

Rosen, M. A. (1996) Comparative assessment of thermodynamic efficiencies and losses for natural gas-based production processes for hydrogen, ammonia and methanol. Energy Conversion Management, 37, 359–367.

Rosen, M. A. and Dincer, I. (1997) On exergy and environmental impact. International Journal of Energy Research, 21, 643–654.

Rosen, M. A. and Dincer, I. (1999) Exergy analysis of waste emissions. International Journal of Energy Research, 23, 1153–1163.

Rosen, M. A., Dincer, I., and Kanoglu, M. (2008) Role of exergy in increasing efficiency and sustainability and reducing environmental impact. Energy Policy, 36, 128–137.

Smith, B. J. and Tirpak, D. A. (eds) (1988) *The Potential Effects of Global Climate Change on the United States* (draft), US Environmental Protection Agency, Office of Policy, Planning and Evaluation and Office of Research and Development, Washington, DC

Spath, P. L. and Mann, M. K. (2001) *Life Cycle Assessment of Hydrogen Production via Natural Gas Steam Reforming*, Report NREL/TP-570-27637, National Renewable Energy Laboratory, US Department of Energy Laboratory, Washington, DC.

Spath, P. L. and Mann, M. K. (2004) *Life Cycle Assessment of Renewable Hydrogen Production via Wind/Electrolysis*, Report NREL/MP-560-35404, National Renewable Energy Laboratory, US Department of Energy Laboratory, Washington, DC.

Tsujikawa, Y. and Sawada, T. (1982) Analysis of a precooled gas turbine cycle combined with an auxiliary cycle with liquefied hydrogen as fuel. In *Hydrogen Energy Progress IV* (eds T. N. Vezirogllu, W. D. Van Vorst, and J. H. Kelley), Proceedings of the Fourth World Hydrogen Energy Conference, Vol. 3, Pergamon Press, New York, pp. 951–960.

US DOE (2004) *Annual Energy Review 2003*. US Department of the Environment, Energy Information Administration, Washington, DC.

Veziroglu, T. N. (2007) 21st century's energy: hydrogen energy system. International Scientific Journal for Alternative Energy and Ecology, 4, 29–39.

Veziroglu, T. N. and Barbir, F. (1992) Hydrogen: the wonder fuel. International Journal of Hydrogen Energy, 17, 391–404.

Veziroglu, T. N. and Basar, O. (1974) *Dynamics of Universal Hydrogen Fuel System*, Hydrogen Economy Miami Energy (THEME) Conference, Clean Energy Research Institute, University of Miami, S15:93–S15:110.

Veziroglu, T. N. and Sahin S. (2008) 21st century's energy: hydrogen energy system. Energy Conversion and Management, 49, 1820–1831.

Veziroglu, T. N., Kakac, S., Basar, O., and Forouzanmehr, N. (1976) Fossil/hydrogen energy mix and population control. International Journal of Hydrogen Energy, 1, 205–217.

Veziroglu, T. N., Gurkan, I., and Padki, M. M. (1989) Remediation of greenhouse problem through replacement of fossil fuels by hydrogen. International Journal of Hydrogen Energy, 14 (4), 257–266.

Walwijk, M., Buckman, M., Troelstra, W. P., and Elam, N. (1999) *Automotive Fuels for the Future*, International Energy Agency, Paris.

Wang, W. and Pinto, J. P. (1981) Climatic effects due to halogenated compounds in the Earth's atmosphere, Journal of Atmospheric Science, 7, 333–338.

White, S. (1998) Net energy payback and CO_2 emission from ^{3}He fusion and wind electrical power plants, PhD Thesis, University of Wisconsin, Madison, WI.

White, S. and Kulcinski, G. (2000),Birth to death analysis of the energy payback ratio and CO_2 gas emission rates from coal, fission, wind, and DT-fusion electrical power plants. Fusion Engineering and Design, 48, 473–481.

Wilier, H. C. (1950) Temperature trends of the past century. *Centenary Proceedings of the Royal Meteorological Society*, 195.

Zweig, R. M. (1992) The "Hydrogen Economy" – Phase I, In *Proceedings of the Ninth World Hydrogen Energy Conference*, Paris, pp. 1995–1997.

Strategic Analyses

25
Research and Development Targets and Priorities
Clemens Alexander Trudewind and Hermann-Josef Wagner

Abstract

Coming from the state of energy distribution and conversion technologies there are many alternatives for transforming the energy system to a hydrogen economy. Many techniques could be introduced for the same purpose but level of development and benefit differ from time to time. Therefore the paper highlights targets of a sustainable energy system, political frameworks, scenarios of infrastructural developments and the state of the art for several hydrogen technologies. The most relevant research fields which were identified concern the reduction of expenses for efficient catalysis by reducing material inputs as well as manufacturing costs of fuel cells and the development of large scale (HT-)electrolysis for adapting to regenerative electricity.

Keywords: hydrogen energy, hydrogen production, hydrogen storage, hydrogen transportation, research and development targets, R&D

25.1
Introduction

Nearly every vision of a sustainable energy supply by hydrogen (H_2) is based on the use of regenerative energy carriers. For development towards a sustainable hydrogen economy, it is important to build up a cost-efficient infrastructure which relies on a secure hydrogen demand. In addition, much research and development (R&D) work has to be done. In order to identify R&D targets in the field of hydrogen technologies, it is necessary to determine the current status, to identify future potential, to uncover weaknesses, and to classify urgent R&D needs.

Hydrogen Energy. Edited by Detlef Stolten
Copyright © 2010 WILEY-VCH Verlag GmbH & Co. KGaA, Weinheim
ISBN: 978-3-527-32711-9

25.2
Procedure

The current stage of development was determined from literature data and interviews with experts. Scenarios were investigated as they analyze the future development of technologies and provide information to the rate of market shares.

In order to classify R&D targets, criteria have to be defined. For this classification, the current stage of development, the requirements in the future, and the relevance of a technology in the time line could be used. The discrepancy between the stage of development and the future requirements reveals the necessary R&D targets while the relevance of technologies is used to identify the urgency of the R&D needs.

The current stage of development distinguishes between "basic research," which is associated with fundamental research in the laboratory, technologies in the "demonstration" phase, and technologies which are available commercially and thus "state of the art." The relevance describes on the one hand the development in time and on the other the market mindshare. For simplification, the relevance distinguishes between an "increasing," a "constant," and a "decreasing" relevance. As a result of both criteria, priorities can be obtained as "high," "medium," or "low."

25.3
Scenarios

Looking at scenarios in order to identify the relevance of different techniques, two kinds of scenarios have to be distinguished. The first tries to estimate the economic development of technologies and compare it with technologies which are already established in the market. Such scenarios are able not only to calculate the economic development but also to compare technical and ecological parameters. These parameters finally give evidence on the stage of development, requirements, and goals of the considered technologies.

The other kind of scenarios provides information on ecological impacts and market shares of developing technologies in an energy system. The basis of these scenarios is parameters of the first kind of scenario and also assumptions on political will and surrounding conditions. The results therefore serve as recommendations for actions. It is obvious that different assumptions in the scenarios will lead to different results.

Investigations of different scenarios, which focused on building up a hydrogen economy in European Union countries (EU25), showed that penetration rates and market shares are dependent on the development and differ for different technologies. For example, in the EU25 the energy demand for hydrogen is expected to be between 14.5 and 48 EJ in 2050 and the production mix varies with oil and gas prices (Ball *et al.*, 2007; Tzimas *et al.*, 2007).

However, there are also similarities in the development of the scenarios. The initial production of hydrogen takes place in centralized production plants that use fossil energy carriers. By-products from the industry are also important at the beginning. Independent of policy goals, gasification of coal will enter the market due to price advantages of coal over other fossil energy carriers such as methane and oil. Another advantage of coal gasification might be the possibility of power plants with integrated gasification combined cycles (IGCC) to produce either electricity or hydrogen depending on the actual electricity tariff.

Renewable energies for hydrogen production enter the market only in the long term. These technologies especially seem to play a role in remote areas because regenerative energies are used for the generation of electricity there. To store and make uniform the fluctuating renewable energies therefore argues for hydrogen production by electrolyzers.

In the future, the supply of stationary and mobile applications is dependent on the development of the hydrogen economy. However, by 2050, the installation of a hydrogen pipeline network will have been started in all scenarios. On-site reforming and the transportation of liquid hydrogen from centralized produced hydrogen by trailers will play a major role at the beginning. Whereas on-site reforming will become less important and will not play a role in the long term, the relevance of trailers will be affected by the development. The more slowly the hydrogen economy develops, the more trailers will be needed for transporting hydrogen to consumers. The reason can be seen in a late but then high demand of hydrogen which cannot be provided by pipelines alone due to late installation and a lack of pipelines. Therefore, trailers will distribute the hydrogen. This development will then result in a delay in building up a hydrogen pipeline network. However, in every scenario pipelines will continue until 2050, so that pipelines will play a significant role in the distribution of hydrogen.

The results of the scenario analysis are summarized in Table 25.1.

Table 25.1 Future market mindshare of technologies for produced and distributed hydrogen.

Technology	Future market mindshare of produced and distributed hydrogen
Industrial by-products	Constant
Centralized reforming	Increasing
On-site reforming	Important during the introduction on the market, then becoming less important
Gasification	Increasing with increasing oil and methane prices
Electrolysis	Increasing in the mid- and long term (H_2 production from renewable energies)
Transportation by pipelines	Increasing (start up dependent on development)
Transportation by trailer	Important during introduction in the market, then dependent on development increasing or decreasing but plays a major role in every scenario

25.4
Investigation of Technologies

The stage of development and the identification of R&D targets were investigated for each technique separately. As the hydrogen economy involves many market sectors and technologies, it is useful to divide it up. Following the hydrogen chain, three different parts could be identified. The first part accounts for the production, conditioning, and purification, the second for storage, transportation, and distribution, and the third for applications of hydrogen.

25.4.1
Production, Conditioning, and Purification of Hydrogen

In the field of hydrogen production, there are many techniques at different stages of development. Investigated hydrogen production techniques were electrolysis, reforming, gasification, photobiological, and photochemical hydrogen production. These techniques are discussed separately and then Table 25.2 summarizes the stage of development, market shares, R&D needs, and priorities.

25.4.1.1 Electrolysis
Electrolysis can generally be differentiated into alkaline, polymer electrolyte membrane (PEM), and high-temperature (HT) electrolysis. They each offer the opportunity to store and to make uniform renewable energies, but the different types of electrolysis differ in their stage of development.

Alkaline electrolysis is state of the art and is operated in the industry. The PEM electrolysis is also available but there are only a few applications so that the technology is seen to be in a demonstration phase. HT electrolysis features the lowest stage of development and some would consider it to be at the stage of fundamental research.

An important goal for each kind of electrolysis is the development of large-scale electrolyzers which are able to shut down quickly, as they offer the use of fluctuating, renewable energies and reserves of power plants.

Industrial alkaline electrolyzers are sized up to 150 MW_{el}, have an efficiency of 67–85% [on the basis of the higher heating value (HHV)] and feature a lifetime of 30 years. They operate at temperatures up to 80 °C and pressures up to 30 bar. The purity of the hydrogen produced reaches 99.9%. For hydrogen production, a process energy input of 4.7 kWh Nm^{-3} H_2 is required. This energy input could be reduced to 4.1 kWh Nm^{-3} H_2 by coating the electrolytes using a vacuum plasma spraying method (Brinner and Huh, 2002). Actual research and development work on alkaline electrolyzers focus on the development of catalysts with hafnium in order to reduce overvoltages and to increase the efficiency (Stoji et al., 2006). When increasing temperatures to reduce the ion conductivity were investigated, problems arose from a more corrosive milieu which corrodes electrodes and decreases their lifetime (NAE, 2004). Furthermore, requirements for stabilizing diaphragms were increased (Emonts, 2000). In order to prevent or to reduce subsequent compression energies, investigations are also carried out to increase the operating pressure (BMWA, 2005).

In addition to alkaline electrolyzers, PEM electrolyzers also operate at 80 °C. Usually the pressure is 15 bar. Mitsubishi had developed a PEM electrolyzer which produces 2.5 $Nm^3 h^{-1}$ of H_2 at an operating pressure of 350 bar. Although the efficiency of 47–53% is lower than that of alkaline electrolyzers, a higher H_2 purity of 99.999% and a higher gas pressure compensate for this disadvantage. The main focus of R&D is on the development of the catalyst and the reduction of membrane's ion resistance by humidification, a reduction in thickness, and higher process temperatures. The optimization of current conductors by designing the porosity structure promises to reduce costs and replacement of expensive materials by small catalytic structures (BMWA, 2005; Bello and Junker, 2006; Millet, 2006; NAE, 2004; Ni et al., 2006 a).

When talking about huge amounts of hydrogen produced by electrolysis, it is generally related to HT electrolysis. High temperatures of 800–1000 °C reduce costs because overvoltages are reduced and nearly 35% of the necessary energy could be provided by cheap process heat. Therefore, it is expected that higher efficiencies will be achieved compared with alkaline and PEM electrolysis. In order to reduce ion conductivity, investigations are focused on the replacement of the yttrium-stabilized zirconium electrolyte by oxidized bismuth, scandium-stabilized zirconium, and lanthanum strontium gallium manganite (LSGM). However, extensive research is still needed. Another research topic relates to the introduction of heat. Allothermal electrolysis can be differentiated from autothermal and the natural gas-assisted steam (NGAS) electrolysis. As allothermal electrolysis introduces process heat of 800–900 °C into the electrolyzer, autothermal electrolysis introduces heat with temperatures much lower than 800 °C. To reach operational temperatures, further heating is necessary in the autothermal process. Autothermal HT electrolysis is favored over alkaline or PEM electrolysis provided that the process heat is higher than 240 °C. NGAS electrolysis introduces heat by oxidation of natural gas within the electrolyzer. The disadvantages of this process relate to material problems such as the conductivity of ions and the development of homogeneous dense ceramic membranes (Pham et al., 2000; Martinez-Frias et al., 2001; Gallet and Grastien, 2004–2005; BMWA, 2005; Bello and Junker, 2006; Ni et al., 2006 b; Sigurvinsson et al., 2007).

Since alkaline and PEM electrolysis are generally available, R&D targets belong to process optimization. The urgency of developments is therefore low so that priority is classified as medium. As HT electrolysis is intended to produce huge amounts of hydrogen, although the stage of development is low and much research still has to be done, the urgency of research and therefore the priority are high.

25.4.1.2 Reforming

Reforming of hydrocarbons, especially methane, could generally be seen as state of the art. Commercial systems reach efficiencies of 70–75% when they are operated at temperatures of 800–900 °C and pressures of 20 bar. The purity of hydrogen after purification (by pressure swing adsorption) is 99.6–99.95%. The development of small-scale reformers is investigated intensively to enable small-scale applications to have access to this technology. The R&D needs are of medium priority because the technology is generally available so that R&D is mainly focused on optimization (Dreier and Wagner, 2000; Spath and Mann, 2001; Klett *et al.*, 2002).

25.4.1.3 Gasification

In addition to the reforming process, the gasification of fuels could also be seen as state of the art. This is especially true for the gasification of coal. In contrast, the gasification of biomass affords opportunities to optimize process conditions and to detect applicable materials. The processes operate at temperatures of around 850 °C and pressures of 35 bar. After purification the hydrogen purity is 99.9%. With reference to relevance in the near and medium term, the R&D targets are of medium priority.

25.4.1.4 Photobiological Hydrogen Production

Photobiological hydrogen production is considered as basic research. Whereas biophotolysis is based on photosynthesis to produce hydrogen, photofermentation relies on the splitting of organic compounds. Conversion efficiencies from light energy to hydrogen of both processes are low. Because efficiency determines the scaling of photobioreactors and affects investment costs, increasing the conversion efficiency is of the utmost importance. For this reason, the biological components of different organisms have to be characterized. After that, the most efficient components have to be identified for designing an optimized organism. At the same time, photobioreactors have to be scaled and designed and process parameters have to be identified. In order to produce hydrogen without CO_2 emissions, photobiological hydrogen production possesses high potential which has to be exploited. The goal is to increase the conversion efficiency up to $\sim 10\%$. The priority of this development is very low with reference to the early build-up of a hydrogen infrastructure (Reith *et al.*, 2003).

25.4.1.5 Photochemical

In addition to photobiological hydrogen production, production via a photochemical process is considered as basic research. Conversion efficiencies of 2.5–8% can be reached today, but the lifetime is short due to oxidation of the anode within a few hours or days. The improvement of the conversion efficiency of photochemical cells is therefore an important goal of research as it influences the scale of cells. Research has to be carried out to understand the applicability of electrolyte materials. Furthermore, the lifetime of photochemical cells has to be increased significantly by preventing oxidation (NAE, 2004; Kelly and Gibson, 2006).

In order to produce hydrogen without CO_2 emissions, photochemical hydrogen production possesses high potential which has to be exploited. With reference to the scenarios, the priority of these development targets is very low.

25.4.1.6 Gas Conditioning

The purification of hydrogen by pressure swing adsorption (PSA) is technically mature and state of the art. PSA is operated at 9–30 bar and reaches purities of 99.999% hydrogen. The CO concentration thereby is lower than 10 ppm. The energy demand for this process was calculated to be 0.11–0.15 kWh Nm^{-3} H_2. For this technology, no R&D need could be identified, but it has to be considered whether alternative gas conditioning processes such as membrane technologies are able to simplify the process conditions. Especially when carbon capture and storage (CCS) processes are considered, membranes would be an interesting opportunity for simplifying the process (Hamelinck and Faaij, 2001; Myers et al., 2002; Spath et al., 2003; Wang and Delucchi, 2005).

The techniques discussed for the production and purification of hydrogen, which have been considered earlier, are depicted with their stage of development, market shares, and R&D needs and priorities in Table 25.2.

Table 25.2 Summary of the stage of development, future market mindshares, and R&D needs for production and purification processes, and classification of the priorities of the identified R&D needs.

Process	Stage of development	Future market mindshare	R&D needs	Priority
Alkaline electrolyzer	State of the art	Increasing	Integration of renewable energies Development of large-scale electrolyzers Reduction of overvoltages by coating technologies for catalysts Increasing efficiency by hafnium Increasing the end pressure to avoid later compression Increasing the stability of diaphragms	Medium
PEM electrolyzer	Demonstration	Increasing	Integration of renewable energies Development of large-scale electrolyzers Reduction of ion resistance by humidification and reduction of thickness and temperature Structure of porosities of the current collectors	Medium

Table 25.2 (continued)

Process	Stage of development	Future market mindshare	R&D needs	Priority
HT electrolyzer	Basic research	Increasing	Integration of renewable energies Development of large-scale electrolyzers Integration of process heat Decrease of the operating temperature Decrease of the layer thickness of the electrolyte Improvement of ion conductivity Development of homogeneous dense ceramic membranes Mass production of 3–5 MW electrolyzers	High
Reforming	State of the art	Increasing	Improvement of process conditions by improving catalysts Optimization of small reformers	Medium
Gasification	State of the art (coal) Demonstration (biomass)	Increasing	Improvement of process conditions Gas conditioning Detecting of applicable materials	Medium
Photobiological	Basic research	Constant	Characterization of components Design of a replicable organism with high efficiency from light to hydrogen Design of photobioreactors for efficient introduction of light into the bioreactor Optimization of process variables	Low
Photochemical	Basic research	Constant	Improvement of conversion efficiency Understanding of material selection Decrease of oxidation Increase of durability	Low
Pressure swing adsorption	State of the art	Increasing	Alternative gas conditioning processes	Medium

25.4.2
Storage, Transportation and Distribution of Hydrogen

After production and purification, hydrogen has to be stored and distributed to consumers, who might also store the hydrogen. As hydrogen is intended to be used as fuel for vehicles, many different storage systems have been investigated as they promise to fulfill high mass-based storage densities. Due to the number of basic approaches, a general survey is given and some detailed descriptions have been omitted. To obtain a general idea, the techniques were differentiated into physical and chemical storage approaches.

25.4.2.1 Physical Storage of Hydrogen

Physical storage of hydrogen can be divided into the storage of compressed gas and the storage of liquefied hydrogen. A combination of both aggregate states will be called cryogenic.

The compression of hydrogen has been used in the chemical industry for decades and is thus state of the art. However, compression processes to end pressures of over 550 bar hydrogen deviate from the properties of ideal gases. Therefore, the real behavior of hydrogen becomes more important and the energy for compression increases. For the exact calculation of the compression work, no equations of state are available which cover a wide range of pressure and temperature (Robinson and Handrock, 1994). For this reason, such an equation of state might be an important R&D target. However, as simplified equations exist, the urgency of this R&D need is low and thus of medium priority. Efficiencies of compression processes are 65–85%, depending on the type of compressor. The lifetime of compressors is 10–30 years and end pressures of 200–300 bar are state of the art. Pressures of 750–800 bar have been demonstrated with storage tanks for mobile applications, so pressure storage vessels are available in principle (BMWA, 2005; Petrovic, 2005). However, insulation and manufacturing have to be improved further in order to realize storage densities of 9% by 2015 (Stenzel, 2006; European Commission, 2007). Underground storage with operating pressures of 80–160 bar allows a future hydrogen economy for storing huge amounts of hydrogen. Although no underground hydrogen storage has been demonstrated, it could generally be seen as available as underground storage for methane exists. This accounts for the realization of such an underground storage in order to collect practical experience (Hoffmann, 1994). Micro glass beads were investigated as they promise high storage densities for mobile applications. However, they are still in the laboratory stage and thus refer to basic research. Micro glass beads are investigated with respect to shaping and increase in charging and storage pressure in addition to the use of new catalysts (Robinson and Handrock, 1994).

Liquefaction of hydrogen was first used in the aerospace industry in the 1950s. Hence liquefaction by the Claude process and the Sterling process is state of the art. As the energy demand for liquefaction is up to 13 kWh kg^{-1}, alternative liquefaction processes such as magnetocaloric liquefaction have to be considered (Winter, 1989; Hoffmann, 1994). Fundamental research is needed for this purpose (Peschka, 1984). However, liquefaction techniques are available and although liquefaction will become more important in the future, the R&D needs are of medium priority. Further R&D needs are seen for available liquid storage tanks. Boil-off losses could accumulate to 1% of the storage volume per day (Winter, 1989). Hence materials research for isolation in order to minimize boil-off losses is needed. In addition, improvements are desirable for cooling cryogenic high-pressure storage and for increasing mass-based storage density to 12% by 2015 (European Commission, 2007). The priority of R&D needs is seen as medium due to the important role of liquid hydrogen in an upcoming hydrogen economy. Table 25.3 summarizes the results for physical hydrogen storage techniques.

Table 25.3 Summary of the stage of development, future market mindshares, and R&D needs for physical hydrogen storage techniques, and classification of the priorities of the identified R&D needs.

Technique	Stage of development	Future market mindshare	R&D needs	Priority
Compression	State of the art		Equation of state for a wide range of pressures and temperatures	Medium
Pressure vessel	State of the art (450 bar) Demonstration (700–825 bar)	Increasing	Insulation and manufacturing of storage facilities Increasing the pressure to increase the mass-based storage density from 4 to 9% Safety	High
Underground storage	State of the art	Constant	Realization	Medium
Micro glass beads	Basic research	Constant	Shaping of glass beads Utilization of catalysts Increasing the mass-based storage density to 9%	Low
Liquefaction	State of the art (Claude process, Sterling-process)	Increasing	Decreasing the energy demand by alternative liquefaction technologies (magneto caloric liquefaction)	Medium
Liquid hydrogen storage	State of the art	Constant	Decreasing boil-off losses Materials research for isolation	Medium
Cryogenic storage	Demonstration	Increasing	Decreasing boil-off losses Materials research for isolation Long-term tests for aluminum composite storage tanks Increasing the mass-based storage density to 12% by 2015	Medium

25.4.2.2 Chemical Storage of Hydrogen

Hydrocarbons, ammonia, water, and iron sponges are chemical compounds from which hydrogen can be extracted. However, storage in these cases will not be discussed here. Rather. hydrogen storage based on sorption effects will be analyzed. This type of storage is still in laboratory stage and hence considered as basic research. The most important group are metal hydrides. Carbon-based materials, also called nanotubes, and other inorganic nano materials are also being investigated.

In the case of metal hydrides, subgroups could be identified. Some metal hydrides are able to store hydrogen reversibly, so they are classified as "reversible metal hydrides." The other kind of metal hydrides are called "chemical metal hydrides" as they release hydrogen during reactions with water or during heat supply. The chemical metal hydrides have to be regenerated before they can store hydrogen again.

As mentioned earlier, the reversible metal hydrides are at a low development stag. In most cases, hydrogen is stored during an exothermic reaction while desorption takes place when heat is supplied. The properties of metal hydrides could be modified by using multicomponent metal hydrides. Investigations try to associate single elements in order to create favorable characteristics. R&D targets aim for high storage capacities, good reaction kinetics, and low desorption temperatures. As a huge number of possible alloys have to be considered, fundamental reaction mechanisms are often unknown (Fichtner et al., 2005). Furthermore, research has to clarify the action of catalysts (Janot, 2005). As the stage of development is low and high storage densities are promised but have not yet been reached, R&D targets are of medium priority.

Chemical metal hydrides are also in the stage of basic research. Borohydrides, slurries, and film hydrides are grouped as chemical metal hydrides. Their success is dependent on the development of regeneration processes. Borohydrides could be made from borax. During desorption, storage densities of 8–11% were obtained. Key technologies of borohydrides are the regeneration processes as they are difficult and the reaction time is slow. Moreover, with an increasing number of adsorption and desorption cycles, the storage capacity decreases significantly. To understand the reactions, more fundamental research is needed. Investigations are focused on simplifying these processes. If regeneration processes could be realized under moderate conditions (pressure and temperature), borohydrides could be an interesting option for hydrogen storage because of their high storage densities.

Slurries are made from light metal hydrides and oil, which provides a reaction with water and facilitates pumping. When hydrogen is needed, a controlled water reaction releases pure hydrogen, which could be used directly in fuel cells. The exact control of the exothermic metal hydride water reactions is of particular importance in order to be able to use a hydride slurry as a powerful source of hydrogen. Problems exist in the control of the reaction rate and the on–off switching. These problems have to be resolved in order to be able to use slurries in the future.

Film hydrides are also at an early stage of development and hence are only a few specifications are known. Physical techniques for the separation of vapor and the combination of plasma or ion radiation allow thermodynamically unstable materials to be made. These materials are characterized by defects, impurities, and deformations. However, adsorption and desorption processes could not be explained by thermodynamic functions alone. Thus the reaction mechanisms are not fully understood, and fundamental research is necessary (Gregoire, 2000; McClaine et al., 2000; Vielstich, 2003; Fakiouglu et al., 2004; Laversenne and Bonnetot, 2005; Milcius et al., 2005; Au and Jurgensen, 2006). With reference to high storage densities of 9% and the low stage of development, R & D needs are of high priority because chemical metal hydrides could be of great importance for mobile applications.

Carbon is known to be a good adsorber of gases due to its porosity and interconnections between the atomic carbon and the molecular gas. The sorption process hence could be controlled by the surrounding conditions. In contrast, descrption needs heat supply of 180–325 °C. As the physicochemical properties of hydrogen storage in carbon-based materials are not fully understood, optimization of materials with respect to increased storage capacities are complicated. Moreover, there are no exact definitions so that structural modifications of nanotubes and nanofibers are no one could proof investigations could not be verified. These findings necessitate the development of standardized measuring methods and further fundamental research. Table 25.4 summarizes the results for chemical hydrogen storage techniques.

25.4.2.3 Transportation and Distribution

Transportation and distribution can be effected by pipelines or trailers. As a result of small existing hydrogen pipelines which have operated for decades, hydrogen pipelines could be seen as state of the art. They are usually operated at pressures below 100 bar and could be used as storage devices. In a developing future, one might also consider the transition from natural gas pipelines to hydrogen pipelines, which is possible in principle. Hence important R & D needs are not seen for pipelines or trailers and they are of low priority (Table 25.5).

25.4.3
Applications

For portable fuel cells, R & D needs are cost reductions, increased lifetime, development of catalysts and development of appropriate designs. Furthermore, membrane electrode assemblies (MEAs), the reduction of cross-over losses, and the development of catalysts for cost reductions and of cooling devices for active and passive cooling are being investigated. The priorities of these research topics are high because portable fuel cells are expected to penetrate the market soon.

Combustion engines and turbines are already operated with hydrogen. Hence they are at the demonstration stage of development. During market penetration,

Table 25.4 Summary of the stage of development, future market mindshares, and R&D needs for chemical hydrogen storage techniques, and classification of the priorities of the identified R&D needs.

Technique	Stage of development	Future market mindshare	R&D needs	Priority
Reversible hydride metal storage	Basic research	Constant	Increasing the mass-based storage density to 2% Improvements of kinetics Decreasing the desorption temperature Development of catalytic hydride systems Structural analysis Research into reaction mechanisms	Medium
Chemical metal hydride	Basic research	Increasing	Regeneration processes for metal hydrides Improvement of process conditions (pressure and temperature) Improving the stability of cyclic adsorption and desorption and the reaction kinetics of $LiBH_4$ Control of reaction rates of metal hydride–water reactions Especially on–off control Understanding the adsorption and desorption characteristics of film hydrides Increasing the mass-based storage density to 9%	High
Carbon-based materials	Basic research	Increasing	Understanding of physio-chemical properties Development of standardized measurement methods	High

combustion engines for mobile applications could be used. R&D needs are adaptation to process conditions and the increased efficiency.

In the chemical industry, ammonia is an important preliminary product and is used in multiple applications such as a raw material for fertilizers, a cooling agent, in pharmaceuticals, and as a solvent in the textile industry. The synthesis of ammonia and the production of methanol are the main users of hydrogen, with market shares of 75% and 8%, respectively.

The synthesis of ammonia has been a routine operation for many years, so the use of hydrogen is seen as state of the art. The chemical reactions are generally well known but efficient catalysts are still under development. Elemental reaction steps and the development of catalytic materials are the fields of R&D,

Table 25.5 Summary of the stage of development, future market mindshares, and R&D needs for transportation and distribution techniques, and classification of the priorities of the identified R&D needs.

Technique	Stage of development	Future market mindshare	R&D needs	Priority
H_2 pipelines	State of the art	Increasing	None	Low
H_2 trailer	State of the art	During market penetration important, afterwards dependent on future development increasing or decreasing, in either case important shares	None	Low

but it has to be concluded that the synthesis and splitting of ammonia are well understood (Schlögl, 2003).

In addition to the synthesis of methanol from methane, methanol production from carbon dioxide and hydrogen is state of the art. This conclusion is supported by a demonstrated process (Pasel *et al.*, 2000). In the long term, large-scale systems have to be tested in order to gather work experience. Further R&D needs include a better understanding of the reaction steps in the catalyzer. One important goal is the understanding of the splitting of hydrogen into two hydrogen atoms in suitable catalyzers and the binding of carbon dioxide (Seidel *et al.*, 2003). Since the general reaction process is understood, the priority of the R&D needs is low.

In the field of combined heat and power generation, R&D needs are in the cost reduction of fuel cells. The applicability of such combined heat and power fuel cell plants in industry has to be proved in each particular case. As CHP fuel cell power plants are expected to have a high potential for operation, cost reductions of fuel cells are of high priority. Table 25.6 summarizes the results for hydrogen applications.

25.5
Conclusion

Investigations show that urgent R&D targets exist in all parts of a hydrogen economy, in production, storage, and application. In general, catalysts are an important research and development field as the materials involve major costs. As far as R&D needs are seen for sophisticated techniques such as transportation and the production of methanol or ammonia, these R&D needs are of low priority.

In the field of hydrogen production, the development of large-scale HT electrolyzers, which are able to shutdown quickly and to utilize renewable energies and process heat, is of high priority.

Table 25.6 As all hydrogen applications are commercially available and future market mindshare increases with time, R&D needs differ from technique to technique, resulting in different levels of priorities.

Technique	Stage of development	Future market mindshare	R&D needs	Priority
Portable fuel cells	Demonstration and state of the art	Increasing	Increasing efficiency and adaptation to process conditions	High
Combustion	State of the art	Increasing	Increasing efficiency and adaptation to process conditions	Medium
Synthesis of ammonia	State of the art	Increasing	Improvement of reaction kinetics by developing catalysts Cost reduction by developing catalysts	Low
Methanol production	State of the art	Increasing	Realization of large-scale processes Understanding of elemental chemical reaction mechanisms	Low
CHP fuel cell plant	Demonstration	Increasing	Cost reduction Estimation of potentials Increasing lifetime of plant	High

Although pressure storage of hydrogen could be seen as state of the art, one development target of high priority is to increase the storage pressure to increase the associated storage densities to 9%. Therefore, insulation and manufacturing of storage vessels have to be developed and safety systems have to be implemented.

As chemical metal hydrides promise to reach the highest mass-based storage densities but relevant research has not yet succeeded, development targets of high priority are an increase in storage densities and the improvement of process conditions.

The use of fuel cell technologies is closely connected to hydrogen. The main goal for fuel cells is reduction of costs. Therefore, R&D targets of high priority are the development of catalysts, the extension of lifetimes, the reduction of cross-over losses, and investigations of direct and indirect cooling systems for fuel cells.

Acknowledgments

We thank the Forschungszentrum Jülich within the Helmholtz-Gemeinschaft for financial support. In addition, we thank Jörg Brandenberg, Jan Haussmann, Dominik Möllenbrink, Andreas Möllney, and Dipl.-Ing. Peter Stenzel for help and discussions.

References

Au, M. and Jurgensen, A. (2006) Modified lithium borohydrides for reversible hydrogen storage. *J. Phys. Chem. B*, **110** (13), 7062–7067.

Ball, M., Wietschel, M., and Rentz, O. (2007) Integration of hydrogen economy into the German energy system: an optimising modelling approach. *Int. J. Hydrogen Energy*, **32** (10–11), 1355–1368.

Bello, B. and Junker, M. (2006) Large scale electrolysers. In *Proceedings of the 16th World Hydrogen Energy Conference*, Lyon, 13–16 June 2006, International Association for Hydrogen Energy (IAHE), CD-ROM, Ref.-No. 215.

BMWA (2005) *Strategiepapier zum Forschungsbedarf in der Wasserstoff-Energietechnologie*, Strategiekreis Wasserstoff des Bundesministeriums für Wirtschaft und Arbeit, Bundesministerium für Wirtschaft und Arbeit, Munich.

Brinner, A. and Hug, W. (2002) Dezentrale Herstellung von Wasserstoff durch Elektrolyse. In *f-cell 2002: Dokumentation f-cell 2002*, f-cell, Stuttgart, 14–15 October 2002, Wirtschaftsförderung Region Stuttgart, Stuttgart, 2002, www.dlr.de/gk/desktopdefault.aspx/tabid-2830/4377_read_6417 (accessed 12 February 2010).

Dreier, T. and Wagner, U. (2000) Perspektiven einer Wasserstoff-Energiewirtschaft. Teil 1: Techniken und Systeme zur Wasserstoffherstellung. *Brennst. Wärme Kraft (BWK)*, **52** (12), 41–46.

Emonts, B. (2000) *Teststand zur Qualifizierung von Diaphragmen für die alkalische Wasserelektrolyse bei hohem Druck*, Projekt-Nr. 261 107 99, Arbeitsgemeinschaft Solar NRW, Jülich.

European Commission (2007) *European Hydrogen and Fuel Cell Technology Platform: Implementation Plan – Status 2006*, LBST GmbH, Ottobrunn.

Fakiouglu, E., Yürüm, Y., Veziroglu, A., et al. (2004) A review of hydrogen storage systems based on boron and its compounds. *Int. J. Hydrogen Energy*, **29** (13), 1371–1376.

Fichtner, M.O., Fuhr, O., Kircher, O., and Léon, A. (2005) Wasserstoffspeicherung in Nanomaterialien. *Nachr. Forschungszentrum Karlsruhe*, **37** (1), 52–58.

Gallet, D. and Grastien, R. (2004–2005) High temperature electrolysis. *Clefs CEA*, (50/51) 40–41.

Grégoire Padró, C.E., and Lau, F. (2000) Advances in Hydrogen Energy (Proceedings of an American Chemical Society Symposium on Hydrogen Production, Storage, and Utilization, held 22–26 August 1999, in New Orleans, LA), Kluwer Academic/Plenum Publishers, New York, ISBN 0-306-46429-2.

Hamelinck, C. and Faaij, A. (2001) *Future Prospects for Production of Methanol and Hydrogen from Biomass – System Analysis of Advanced Conversion Concepts by ASPEN-Plus Flowsheet Modelling*, Report No. NWS-E-2001-49, University of Utrecht, Department of Science Technology and Society, Utrecht.

Hoffmann, V.U. (1994) *Wasserstoff: Energie mit Zukunft*, Teubner, Stuttgart.

Janot, R. (2005) Metallic amides and imides as new materials for reversible hydrogen storage. *Ann. Chim. Sci. Matér.*, **30** (5), 505–517.

Kelly, N.A. and Gibson, T.L. (2006) Design and characterization of a robust photo-electrochemical device to generate hydrogen using solar water splitting. *Int. J. Hydrogen Energy*, **31** (12), 1658–1673.

Klett, M.G., White, J.S., Schoff, R.L., et al. (2002) *Hydrogen Production Facilities Plant Performance and Cost Comparison*

Laversenne, L. and Bonnetot, B. (2005) Hydrogen storage using borohydrides. In: *Ann. Chim. Sci. Matér.*, **30** (5), 495–503.

Martinez-Frias, J., Pham, A.-Q., and Aceves, S.M. (2001) *Analysis of a High-Efficiency Natural Gas-Assisted Steam Electrolyzer for Hydrogen Production*, UCRL-JC-144368, US Department of Energy, Washington, DC.

McClaine, A.W., Breault, R.W., Larsen, C., et al. (2000) Hydrogen transmission/storage with metal hydride–organic slurry and advanced chemical hydride/hydrogen for PEMFC vehicles. In *Proceedings of the Year 2000, Hydrogen Program Review*, US Department of Energy, Washington, DC.

Milcius, D., Pranevicius, L.L., and Templier, C. (2005) Hydrogen storage in the bubbles formed by high-flux ion implantation in thin Al films. *J. Alloys Compd.*, **398** (1), 203–207.

Millet P. (2006) GenHyPEM: an EC-supported Strep program on high pressure PEM water electrolysis. In *Proceedings of the 16th World Hydrogen Energy Conference*, Lyon, 13–16 June 2006, International Association for Hydrogen Energy (IAHE).

Myers, D. B., Ariff, G. D., James, B. D., et al. (2002) *Cost and Performance Comparison of Stationary Hydrogen Fueling Appliances (Task 2 Report)*. US Department of Energy, Arlington, VA.

NAE (2004) National Academy of Engineering (NAE), Board on Energy and Environmental Systems (BEES), and Department of Engineering and Physical Science (DEPS), *The Hydrogen Economy: Opportunities, Costs, Barriers, and R&D Needs*, National Academies Press, Washington, DC.

Ni, M., Leung, M. K. H., and Leung, D. Y. C. (2006a) Electrochemistry modelling of proton exchange membrane (PEM) water electrolysis for hydrogen production. In *Proceedings of the 16th World Hydrogen Energy Conference*, Lyon, 13–16 June 2006, International Association for Hydrogen Energy (IAHE), CD-ROM, Ref.-No. 018.

Ni, M., Leung, M. K. H., and Leung, D. Y. C. (2006b) Prospect of solid oxide steam electrolysis for hydrogen production. In *Proceedings of the 16th World Hydrogen Energy Conference*, Lyon, 13–16 June 2006, International Association for Hydrogen Energy (IAHE), CD-ROM, Ref.-No. 212.

Pasel, J., Peters, R., and Specht, M. (2000) Methanol – Herstellung. und Einsatz als Energieträger für Brennstoffzellen. In *Forschungsverbund Sonnenenergie (FSV): Themen 1999/2000*, Forschungsverbund Sonnenenergie (FSV), Berlin.

Peschka, W. (1984) *Flüssiger Wasserstoff als Energieträger Technologie und Anwendungen*, Springer, Vienna.

Petrovic, T. (2005) *Photobiologische Wasserstofferzeugung durch Mikroalgen – Beschreibung konkurrierender Systeme zur H_2-Erzeugung*. Endbericht zum Forschungsforhaben 85.65.69-T-170, LEE-39, Ruhr-Universität Bochum, Lehrstuhl für Energiesysteme und Energiewirtschaft, Bochum.

Pham, A.-Q., Haslam, J. J., Wallmann, H., DiCarlo, J., Glass, and R. S. (2000) *Natural-Gas-Assisted Steam Electrolysis for Distributed Hydrogen Production*. US Department of Energy, Washington, DC; available at www.osti.gov/bridge/purl.cover.jsp?purl=/15004105-W9EDC8/native/ (accessed 12 February 2010).

Reith, J. H., Wijffels, R. H., and Barten, H. (2003) *Bio-methane and Bio-hydrogen – Status and Perspectives of Biological Methane and Hydrogen Production*. Dutch Biological Hydrogen Foundation, Petten.

Robinson, S. and Handrock, J. (1994) *Hydrogen Storage for Vehicular Applications: Technology Status and Key Development Areas*. Sandia Laboratories, Albuquerque, NM.

Schlögl, R. (2003) Katalytische Ammoniaksynthese – eine "unendliche Geschichte"? *Angew. Chem.*, **115**, 2050–2055.

Seidel, R., Muhler, M., and Grünert, W. (2003) Chemisch entzaubert: Zinkoxid steuert Katalyse. In RUBIN (Special Edition Chemie RUBIN), 44–49, ISSN 0942-6639.

Sigurvinsson, J., Mansilla, C., Lovera, P., and Werkoff, F. (2007) Can high temperature steam electrolysis function with geothermal heat? *Int. J. Hydrogen Energy*, **32** (9), 1174–1182.

Spath, P. L. and Mann, M. K. (2002) *Life Cycle Assessment of Hydrogen Production via Natural Gas Steam Reforming*, National Renewable Energy Laboratory (NREL), Golden, CO.

Spath, P. L., Mann, M. K., and Amos, W. A. (2003) *Update of Hydrogen from Biomass – Determination of Delivered Cost of Hydrogen*, Milestone Completion Report NREL/MP-510-33112, National Renewable Energy Laboratory (NREL), Golden, CO.

Stenzel, P. (2006) *Wasserstoffspeicherung – Literaturrecherche zum aktuellen Sachstand und Ausblick auf neue Optionen*, Projektarbeit LEE-D-57, Ruhr-Universität Bochum, Lehrstuhl für Energiesysteme und Energiewirtschaft (LEE), Bochum.

Stojic, D. L., Cekic, B. D., Vasil J., et al. (2006) Alkaline electrolysis – Hf based intermetallics as cathode materials. In *Proceedings of the 16th World Hydrogen Energy Conference*, Lyon, 13–16 June 2006, International Association for Hydrogen Energy (IAHE), CD-ROM, Ref.-No. 190.

Tzimas, E., Castello, P., and Peteves, S. (2007) The evolution, size and cost of a hydrogen delivery infrastructure in Europe in the medium and long term. *Int. J. Hydrogen Energy*, **32** (10–11), 1369–1380.

Vielstich, W. (ed.) (2003) *Handbook of Fuel Cells: Fundamentals, Technology and Applications. Volume 3: Fuel Cell Technology and Application, Part 1*. John Wiley & Sons, Inc., New York.

Wang, G. and Delucchi, M. (2005) *Appendix K: Input–Output Data for Hydrogen from Coal and Hydrogen from Biomass Conversion Processes. A Lifecycle Emissions Model (LEM): Lifecycle Emissions from Transportation Fuels, Motor Vehices, Transportation Modes, Electricity Use, Heating and Cooking Fuels, and Materials*, UCD-ITS-RR-03-04K, University of Califonria, Institute of Transportation Studies, Davis, CA.

Winter, C. J. (1989) *Wasserstoff als Energieträger: Technik, Systeme, Wirtschaft*. Springer, Berlin.

26
Life Cycle Analysis and Economic Impact

Ulrich Wagner, Michael Beer, Jochen Habermann, and Philipp Pfeifroth

Abstract

This chapter presents a life cycle analysis of different potential fuels and power trains for passenger cars. The focus of the study is electric vehicles powered by batteries and fuel cells. First the cumulative energy demand (*CED*) is introduced as an instrument to compare the different technologies. Subsequently, several process chains for transportation are shown in a holistic approach. Finally, an economic and ecological comparison of the different drive technologies is used to work out the constraints and the necessary future development for hydrogen and electric power trains. The results are taken from several reports for the Bavarian Hydrogen Initiative (wiba), which is coordinated by the authors.

Keywords: life cycle analysis, fuel cells, electric vehicles, hydrogen power train, electric power train, cumulative energy demand

26.1
Introduction

In all industrialized and developing countries, a steady increase in motor-driven traffic is taking place. Between 1965 and today, traffic has almost tripled. During the same period, significant progress in automotive engineering has been achieved, especially in terms of fuel consumption. However, these advances were accompanied by higher motor power, the need for a multitude of supplementary equipment and greater vehicle weight. Consequently, only a slight decrease in average fuel consumption has taken place.

At present, the largest part of the energy demand for transportation applications is covered by petroleum products. Considering the finite limited fossil energy sources and an estimated significant increase in global energy demand, the search for alternative energy carriers and power train concepts is increasing. Discussions about greenhouse gas emissions and global warming are further forcing the need for innovative and sustainable energy systems.

Hydrogen Energy. Edited by Detlef Stolten
Copyright © 2010 WILEY-VCH Verlag GmbH & Co. KGaA, Weinheim
ISBN: 978-3-527-32711-9

Prior to a full-scale introduction of alternative energy carriers and power trains, the ecological advantages and disadvantages in comparison with established, conventional process chains must be evaluated. In terms of a full life cycle assessment, environmental impacts such as primary energy (PE) demand, space requirements, greenhouse gas emissions, and the appearance of ecotoxic and human toxicological substances are of special interest. Furthermore, economic and social facets and the availability of resources have to be considered [1].

26.2
Definitions and Methodology

For the production and use of a certain fuel, a multitude of different processes must be analyzed. These are interlinked with each other in so-called process chains, sometimes in a complex manner. For each process, an energetic input-output analysis based on the lower heating value (LHV) was performed and for each process chain an individual analysis was carried out, yielding the cumulative energy demand (CED) and cumulative emissions for a specific product or provision of services.

In general, the CED is defined as the sum of the PE demand evolving from the production, use, and disposal of an economic good or object, respectively, the PE demand which can be causally assigned to this economic good or object (both, products and services). Thus, the CED represents the sum of the CEDs for the production (CED_P), use (CED_U), and disposal (CED_D) of an object [2, 3].

Dreier [4] defined characteristic parameters and introduced a methodology which allows for a sophisticated analysis of different energy systems and a well-balanced assessment of the individual technologies. According to this methodological approach, regenerative and non-regenerative energy demands are investigated separately by using the following definitions:

1. *Cumulative regenerative energy demand (CRED)*: the sum of the regenerative energy demands for a product or provision of services is defined as $CRED$. All of the further considerations in this study assume that regenerative energies are inexhaustibly available.
2. *Cumulative non-regenerative energy demand (CNRED):* the sum of non-regenerative energy demands for a product or provision of services is defined as $CNRED$.

Thus, the CED can be split into the two subcomponents, $CRED$ and $CNRED$:

$$CED = CED_P + CED_U + CED_D = CRED + CNRED \tag{26.1}$$

To assess the input of regenerative energies in process chain analyses, the regenerative PE input was set equal to the amount of useful energy produced according to the practice of the United Nations Department of Economic and

Social Affairs Statistics Division [5] and the Statistical Office of the European Communities (SOEC) [6]. In doing so, energy converters using regenerative energy sources are in fact evaluated in terms of their energetic benefit rather than in terms of their technical capacities and efficiencies. This procedure offers the possibility to set up and quantify the energy flows and energy balance within the individual system boundaries without overweighing the regenerative energy inputs.

For the assessment of non-regenerative energy inputs (fossil and nuclear) on a PE basis, the overall efficiency of supply (OES) was used. In the case of fuels and other materials with an LHV, OES_f is defined as the ratio of the LHV and the cumulative energy demand for the supply of the fuel (CED_s) [2]:

$$OES_f = \frac{LHV}{CED_s} \tag{26.2}$$

Thus, the energetic efforts for the supply as well as for the production and the disposal of the plants, all valued as primary energy, must be considered for the calculation of CED_s.

Accordingly, the overall efficiency of electricity supply (OES_{el}) includes the overall efficiencies of generation and distribution to the end-user and is defined as the ratio of electrical energy W_{el} and CED_s [2]:

$$OES_{el} = \frac{W_{el}}{CED_s} \tag{26.3}$$

26.3
Extraction, Conversion, and Distribution of Fuels

Within the scope of this study, state of the art and future ways of fuel supply as depicted in Figure 26.1 are considered. A PE carrier has to be supplied and converted into a fuel, which is transported and distributed to the final consumer. Utilization of different fuels can take place either in a vehicle powered by an internal combustion engine (ICE), in a battery electric car, or in a fuel cell (FC) vehicle.

Subsequently, the CRED and CNRED for process chains of selected fuels will be evaluated aggregated according to the methodology outlined above and depicted as kWh PE per kW h final energy (FE), which is equivalent to the reciprocal OES. The energy demand values cover supply of energy carriers, the conversion process to fuels, transport, and distribution to the final consumer, that is, to the vehicle tank (well-to-tank analysis).

The following fuel supply chains for Germany were investigated:
- Diesel fuel and gasoline are produced from crude oil in German refineries or imported from other countries. In process chain analyses, the imported shares from different countries of origin are taken into account by considering the associated routes of transport [8].

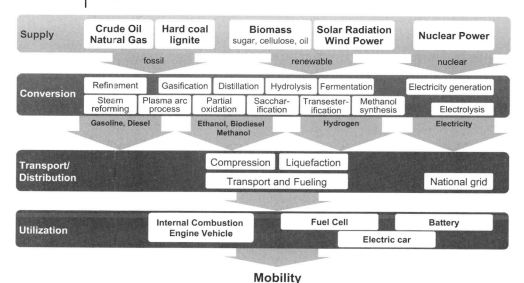

Fig. 26.1 Potential power trains and means of fuel supply for road traffic [1, 7].

- Compressed natural gas (CNG) is supplied to the consumer at fast-fill service stations where natural gas is drawn from the public distribution system. At these service stations, natural gas is compressed and stored at pressures up to 280 bar and filled directly into the vehicle tank on demand [8].
- Liquefied petroleum gas (LPG) consists of propane and butane gas in different mixing ratios according to the application area and region. It must not be confused with liquefied natural gas (LNG). LPG is a distillation product of crude oil. In an LPG separation plant, fractions of methane and ethane are extracted and propane and butane separated. Butane has a 30% higher LHV than propane, but its boiling point of about 0 °C (under normal pressure) causes difficulties, especially in Northern Europe. Hence in colder regions, mixtures with a higher propane fraction are used. In warmer regions, more butane is adopted – the advantage is a lower gas pressure at high temperatures.
- For the supply of liquefied hydrogen (LH$_2$), several process chains have been analyzed. Gaseous hydrogen can be produced by steam reforming of natural gas, gasification of biomass (e.g., wood), or electrolysis. A further differentiation was applied within process chains including electrolysis subsystems: electricity can either be generated by photovoltaic systems and wind power in Germany or through solar thermal power plants, for example in North Africa, with subsequent high-voltage direct current power transmission to Germany. After liquefaction, LH$_2$ is delivered to the petrol station in cryogenic containers and filled directly into super-insulated vehicle tanks [9].
- Methanol (MeOH) is produced from a CO-, CO_2-, and H_2-containing synthesis gas. Two techniques were chosen for the production of the synthesis gas:

combined steam reforming of natural gas [8] and gasification of biomass (residual forest wood in particular) [10]. Distribution of methanol to filling stations and vehicle fueling takes place in the same way as with conventional liquid fuels.
- Rape methyl ester (RME, biodiesel) is produced from raw rape oil by refining or transesterification and supplied at petrol stations like diesel fuel. All of the co-products (rape straw, shred, glycerine) along the process chain are used [4].
- Electric vehicles (EVs) are passenger cars powered by an electric motor in connection with an accumulator. The calculation is based on the electricity supply by the German electricity mix, that is, a primary energy adoption of about 50% coal, 32% nuclear power, 11% natural gas, and 7% others (including renewable energy carriers) for 2005. The accumulator's efficiency is assumed to be 80%.

26.4
Results of Process Chain Analyses

26.4.1
Provision of Hydrogen

26.4.1.1 Aquatic Electrolysis

For aquatic electrolysis, water is fractionated into its components hydrogen (H_2) and oxygen (O_2) using a d.c. voltage. H_2 is collected at the cathode and O_2 at the anode. A semipermeable membrane separates the emerging gases and permits charge equalization.

For lower capacities up to $20\ m^3\ h^{-1}$, proton exchange membrane (PEM) (also known as polymer electrolyte membrane) electrolyzers are adopted [11]. In addition, alkaline aquatic electrolysis is one of the most usual processes. In this kind of electrolysis, the ions needed for the ion exchange are typically provided by a caustic potash solution, but the underlying type of process is the same [12].

Figure 26.2 shows a flow diagram for the electrolytic provision of hydrogen using electricity from the German electricity mix in 2005. According to this, for provision of hydrogen a four times higher PE adoption is needed compared with its energy content. The major losses are caused by the supply of electricity, but the electrolysis itself and the compression of the hydrogen to 300 up to 700 bar pressure also cause losses.

In addition to the reflection of the electricity mix, often pure regenerative electricity production is considered. This "ecological" current is, however, generally fed into the electricity grid and therefore part of the electricity mix.

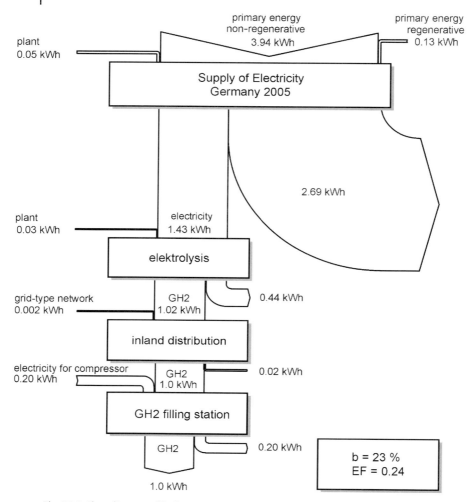

Fig. 26.2 Flow diagram of hydrogen provision by electrolysis [3, 12].

26.4.1.2 Steam Reforming of Natural Gas

At present, most hydrogen is produced by steam reforming of natural gas. This process has an efficiency of 70–80% and is therefore one of the most efficient methods. An unfavorable circumstance of natural gas reforming is that the same amount of CO_2 is emitted as would have been released if combusted [12].

Figure 26.3 presents an energy flow diagram for the whole chain of steam natural gas reforming. In contrast to the electrolysis in Figure 26.2, here the cryogen liquefaction of hydrogen is illustrated. For liquefaction a temperature below −253 °C is needed, which is the reason for the high energy demand of this process. According to that, the whole supply chain has a PE efficiency of 46%. The provision of gaseous hydrogen can be accomplished with an efficiency of ∼57%.

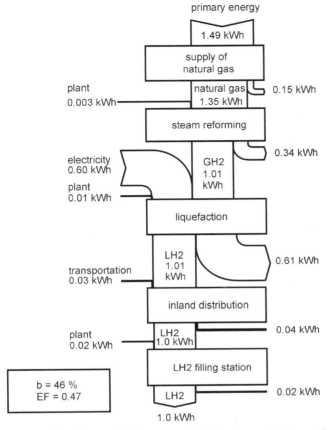

Fig. 26.3 Flow diagram of hydrogen provision by steam natural gas reforming [3, 12].

Steam reforming of natural gas is typically used for the central hydrogen provision. Certainly it is possible to produce hydrogen locally where needed for the FC. Mostly autothermic reforming is adopted – a combination of steam reforming and partial oxidation. As a first step towards a hydrogen economy, this would be an opportunity to make hydrogen accessible in areas which could not be supplied centrally.

26.4.2
Cumulative Energy Demand

The average useful life of a passenger car is assumed to be 10 years and 150 000 km. Furthermore, indirect expenditures are included in the calculation, that is, car maintenance and infrastructure investments such as construction and maintenance of the road network, which are taken into account proportionately. After 10 years, the age-related maintenance of the vehicle increases sub-

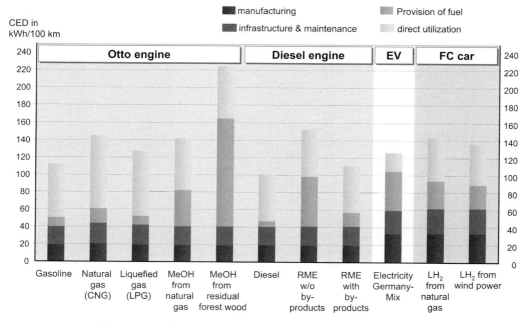

Fig. 26.4 Cumulative energy demand (CED) of the considered technologies.

stantially. Further, the consumption and emission values are no longer state of the technology [3].

Considering the total CED of the analyzed vehicle and power train variants, the PE consumption per 100 km leads to the results in Figure 26.4. The lowest input is required by the diesel engine, owing to its high development status and the highest efficiency of energy supply of all energy sources considered.

Due to the larger energy demand for manufacturing of the EV – batteries and electric motors – the advantages of the electric power train's high efficiency are compensated in most instances.

The CED of the FC power train during current development is significantly higher than that of the gasoline variants. There is much room for improvement, however, as hydrogen could be provided by a variety of production alternatives. By using renewable energy sources, a considerable lower CED could be obtained in the future.

In general, the part of renewable energy assumed to be used for the production of fuels could also be used in sectors other than transportation, such as the direct use of electricity from wind or solar power in the residential or industrial sector. Furthermore, the probability of using renewable energies in transportation is much lower than that in other sectors because a low-cost gasoline infrastructure is already established. Any alternative has to compete with the existing fuel supply networks. An analysis of the optimal use of regenerative energies either for transportation or for any other sector, however, is beyond the scope of the results shown. In the future, several factors may promote an increased use of renewable fuels for road traffic, above all increasing air pollution and greenhouse gas emissions and also a politically and economically desired diversification of primary energy sources. As CNRED represents the use of fossil resources and thus net generation of CO_2 emissions, CNRED associated with the use of renewable fuels must be lower than that for conventional fuels to achieve a positive effect. Reducing the holistic consideration of the CED on the non-regenerative part will yield the CNRED as shown in Figure 26.5.

As the variant MeOH from residual forest wood comes out badly in primary energy consumption (see Figure 26.4), Figure 26.5 shows the advantages of the adoption of MeOH, resulting from the high fraction of regenerative energy for fuel production. In contrast, the values for the conventional fuels do not change substantially, caused by their fossil origin with a negligible fraction of renewable energies.

The two RME variants do not differ in Figure 26.5, as the by-products of RME production (mostly coarse rape grist) are normally used for animal food and therefore substitute only regenerative energy. Hence the by-products are not offset to determine the CNRED. Overall, RME-powered vehicles are economical in terms of non-regenerative energy consumption, with a level of about 50% compared with CNG and MeOH and also LH_2 from natural gas.

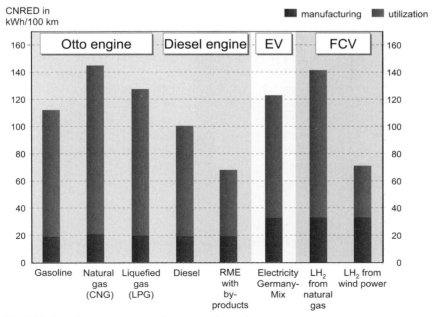

Fig. 26.5 Cumulative non-regenerative energy demand (CNRED) of the considered technologies.

The main results of these process chain analyses can be summarized as follows:
- Regarding fossil energy carriers, techniques for the conversion of natural gas into MeOH and LH_2 are associated with significantly higher CED and $CNRED$ than the supply of the conventional fuels (diesel, gasoline, and CNG).
- Alternative fuels such as MeOH and LH_2 have energetic advantages compared with conventional fuels if they are produced from regenerative energy carriers.
- Due to the high input of regenerative energy, most renewable fuels show a higher CED than fossil fuels. In contrast, the $CNRED$ of renewable fuels is much lower in most cases.
- Hydrogen is unfavorable in terms of CED because of the high energy demands for liquefaction [9]. In this case, the type of electricity generation system for liquefaction has a crucial impact. In Figures 26.4 and 26.5, the use of electricity from the German grid was implied for liquefaction of hydrogen. Substitution of electrical power from the German grid by hydro or wind power would result in a reduction of the $CNRED$ by two-thirds.

Figure 26.6 shows four Sankey diagrams of the analyzed vehicles' process chains. In each case, the data are normalized for a run of 100 km. On average, 13.5 kW h of the effective energy is required to drive this distance. Depending on technology, losses occur at different stages in the process chain. The CED for building the infrastructure and maintenance is similar for all types of vehicles. The energy demand for manufacturing the vehicle itself is split up into drive train and chassis for battery- and FC-powered vehicles. The CED_P for the whole car is then about one-third higher than the CED_P for conventional vehicles.

As shown, the highest losses result from the combustion processes in the conventional vehicles analyzed. The refining process and natural gas supply have a fairly high efficiency. In the process chains of electric cars, the supply of electricity and hydrogen has a major impact on the overall CED. The utilization ratio of the electric power train is very high at about 79% in both cases. Losses due to energy storage have to be taken into account when considering batteries and storage of liquid hydrogen.

26.4.3
Cost Analysis

Today and in the coming years, the reference basis for costs for new technologies will be the conventional fossil-fueled cars. Figure 26.7 illustrates the capital, fixed, and fuel costs for a 1 km drive for different power trains. Compared with diesel cars, the costs per kilometer of H_2-FC vehicles are still about 10–15% higher. Examining the different proportions of expenses, it becomes clear that this does not result only from the higher capital costs of the capital-intensive FC power train.

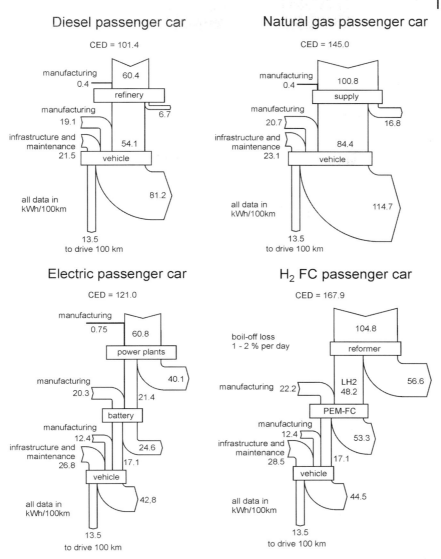

Fig. 26.6 Process chain analyses for considered vehicles, according to [13].

The assessed fuel costs of hydrogen from biomass are approximately 40% higher than those of hydrogen from steam reforming and are generally higher than those of conventional fuels. This is caused by the higher logistical costs for the preparation and distribution of biomass. Conflicts and bottlenecks in the use of biomass in addition to rising crude oil and natural gases prices can increase fuel costs for both conventional and H_2-FC vehicles.

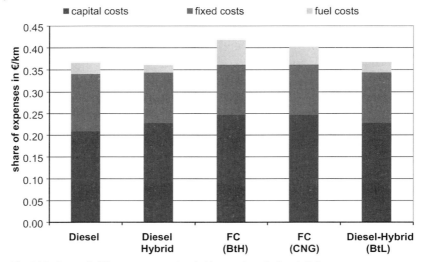

Fig. 26.7 Costs of different power trains (without mineral oil tax) [14].

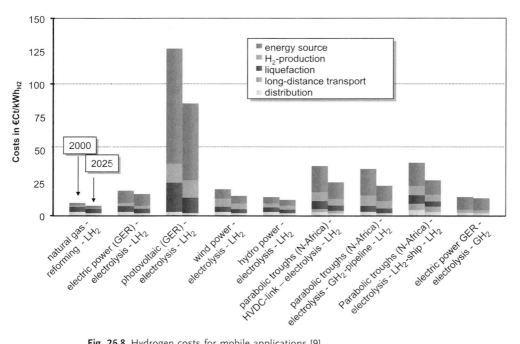

Fig. 26.8 Hydrogen costs for mobile applications [9].

In Figure 26.8, the hydrogen costs for mobile applications are listed in detail for different production processes. It shows the higher fuel costs that have to be considered when considering H_2-FC vehicles as a CO_2-free alternative to conventional fossil-fueled cars because of using, for example, hydrogen produced by photovoltaic-powered electrolysis.

However, even the more cost-competitive production processes such as natural gas steam reforming still offer a major opportunity to face the actual and immediate emission problems of the transport sector.

26.5 Conclusion

The results obtained in this study demonstrate, from a cumulative point of view, that battery-powered electric vehicles can reach overall efficiencies near those of conventional cars. They are more efficient than diesel-powered cars if the utilization ratio of the electricity supply exceeds 35%. H_2-FC vehicles have a similar efficiency to natural gas- or diesel-fueled cars provided that the overall efficiency of hydrogen supply is about 65–70%. To achieve this aim, further investigations are needed.

Alternative fuels such as RME, MeOH from biomass or electricity, and hydrogen from regenerative energy sources show significantly lower $CNRED$ along their process chains than do conventional fuels. With respect to preserving fossil energy sources and reducing emissions, low $CNRED$ is crucial.

Both technologies – H_2-powered FC and electric vehicles – have their advantages as they are emission free at the site of use. This includes the possibility of CO_2 sequestration in more efficient central facilities. Therefore, fossil fuels could be used with reduced impact on climate change. This leads to a huge variety of primary energy carriers that can be utilized for transportation.

Even though electric vehicles – battery and fuel cell powered – show a significantly higher CED_P than gasoline and diesel ICE vehicles, the opportunity to use renewable energy carriers is yielding the lowest specific $CNRED$ and CO_2 emissions over the lifetime of the vehicle. In the case of fossil fuels, $CNRED$ and CO_2 emissions are dominated by the individual fuel supply chains and direct consumption, not by vehicle production, maintenance, and disposal. Due to the inherent $CNRED$ and CO_2 neutrality of regenerative fuels, $CNRED$ and CO_2 emissions during utilization are virtually zero. As long as no hydrogen infrastructure is provided, on-board storage of hydrogen and also electricity remains a crucial problem, and the use of RME in an ICE vehicle or MeOH from biomass in a PEFC vehicle represents a reasonable transitional solution.

By comparing different power train concepts, namely ICE and electric systems, one has to bear in mind that these represent strongly unequal evolutionary states. Whereas ICE vehicles have long been mass produced, FC and electric vehicles are still at the research and development stage. Consequently, the database for the propulsion concepts under consideration also varies strongly in quality and quantity. In the future, both technologies will undergo significant improvements and other technologies will also have to be taken into account, most notably hybrid systems. Thus, a continuous assessment of different power train systems with the associated fuel supply chains is essential in order to fill data gaps and evaluate the environmental impacts of each technology. In addition to technological issues, social and economic factors such as consumer acceptance and availability of fuels are of equal importance.

The most important criterion for decision making in the cases considered in this chapter are the costs, both for the propulsion system and for the supply and distribution of fuels. Even though fuel cells and electric vehicles can be regarded as cleaner and more efficient automotive engines, many challenges still have to be met before the successful commercialization of fuel cell vehicles can be achieved [15, 16]. Consequently, the design requirements of the automotive industry in the future will inevitably be connected with the design and configuration of a low-cost fuel infrastructure. Given the considerable business risk involved in creating an automotive fuel infrastructure based on regenerative fuels, a supportive political framework and diversification of the risk with individual corporate strategies will be essential.

In the near future, storage of electricity, produced by fluctuating renewable energy sources, could be a new field of business for energy suppliers. Electric vehicles, powered by batteries and also by electrolytic hydrogen, may contribute as additional electricity consumers. Intelligent control of timed electricity supply and demand will provide the ability to meet future challenges for the energy economy.

References

1 Wagner, U., Eckl, R., and Tzscheutschler, P. (2005) Energetic life cycle assessment of fuel cell powertrain systems and alternative fuels in Germany. In *Science Direct Energy*, Elsevier, Amsterdam, Volume 31, 3062–3075.
2 German Association of Engineers – Society for Energy Technology (VDI-GET) (1997) *Cumulative Energy Demand – Terms, Definitions, Methods of Calculation*, VDI Guideline No. 4600: 1997. Beuth Verlag, Berlin.
3 Wagner, U., Geiger, B., and Schäfer, H. (1998) Energy life cycle analysis of hydrogen systems. *Int. J. Hydrogen Energy*, **23** (1), 1–6.
4 Dreier, T. (2000) *Cumulative System Analysis and Potentials of Biofuels*, E&M Energie und Management, Herrsching.
5 United Nations Department of Economic and Social Affairs Statistics Division (2002) *2000 Energy Statistics Yearbook*, United Nations Publications, New York.
6 Carvalho Neto, J., Deimezis, N., Lecloux, M., Karadelogou, P., and Bailey, R. A. (1991) *Energy in Europe – Annual Energy Review*, Commission of the European Communities, Brussels.
7 Richter, S. and Wagner, U. (2006) Hydrogen Production. In *Landolt-Börnstein Numerical Data and Functional Relationships*, Vol. VIII/3, Springer, Berlin.
8 Forschungsvereinigung Verbrennungskraftmaschine (FVV) (1998) *Fuel Cell Study Final Report*, Project No. 686:1998, FVV, Frankfurt.

9 Angerloher, J. and Dreier, Th. (2000) *Techniken und Systeme zur Wasserstoff-bereitstellung – Perspektiven einer Wasserstoff-Energiewirtschaft (Teil 1)*, Bavarian Hydrogen Initiative, Forschungsstelle für Energiewirtschaft, Munich.

10 Dreier T. (1999) *Biofuels – Energetic, Environmental and Economic Analysis*, E&M Energie und Management, Herrsching.

11 TÜV Süddeutschland (2002) *Energiewelt Wasserstoff – Wissen, Technologie, Perspektive*, TÜV Süddeutschland, Munich.

12 Beer, M. and Fieger, C. (2008) *Perspektiven der Brennstoffzellen- und Wasserstofftechnologie – Einsatz innovativer KWK-Technologien in der Hausenergieversorgung*, Forschungsstelle für Energiewirtschaft, Munich.

13 Corradini, R. and Krimmer, A. (2003) *Perspektivenstudie 4 – Systemvergleich alternativer Antriebstechnologien*, Forschungsstelle für Energiewirtschaft, Munich.

14 Baitsch, M. and Richter, S. (2007) *Energieeffizienz alternativer Kraftstoffe aus Biomasse und Erdgas im Vergleich mit konventionellen Kraftstoffen*, Bavarian Hydrogen Initiative, Forschungsstelle für Energiewirtschaft, Munich.

15 Beer, M., Gobmaier, T., Hauptmann, F., Mauch, W., Podhajsky, R., Steck, M., and von Roon, S. (2007) *Ganzheitliche dynamische Bewertung der KWK mit Brennstoffzellentechnologie – Forschungsvorhaben im Forschungsverbund EduaR&D*, Forschungsstelle für Energiewirtschaft, Munich.

16 Wagner, U., Richter, S., and von Roon, S. (2005) *Strategiepapier zum Forschungsbedarf in der Wasserstoff-Energietechnologie*, Forschungsstelle für Energiewirtschaft, Munich.

27
Strategic and Socioeconomic Studies in Hydrogen Energy
David Hart

Abstract

The introduction of hydrogen as an energy carrier is not only a technology and science challenge, also requiring the engagement of many socioeconomic disciplines. Analysing the conventional economic implications of different potential manifestations of a hydrogen economy is important, but so is the inclusion of externality assessment. New business models and financing schemes need to be developed to help us understand the balance between short and long term costs and rewards, and between private and social impacts. The behaviour and perceptions of different actors in any possible transition and future have a strong bearing on whether or not it can ever reach fruition, and methods such as actor and transition analysis can assist in their evaluation. All of these disciplines and more must be linked with the science, technology and engineering aspects of hydrogen energy deployment in order to assess its potential and derive appropriate methods of support.

Keywords: hydrogen energy, socioeconomics, economic analysis, social behavior, finance

27.1
Introduction

> *"The difficulty lies, not in the new ideas, but in escaping from the old ones ..."*   John Maynard Keynes

Energy is a complex and multi-disciplinary subject. To understand fully its advantages, drawbacks, and potential requires deep and careful study of fundamental science, engineering, resources, infrastructure, economics, policy, and many other areas. Hydrogen energy is no less complex. The potential introduction of hydrogen as an energy carrier into the global energy system (the so-called Hydrogen Economy) could augment and, eventually, potentially supplant current energy carriers, which have disadvantages in terms of, primarily, environmental characteristics and resource availability and distribution. These disadvantages are generally poorly valued in terms of the final price of conventional energy. Inevitably, hydrogen also has disadvantages, primarily linked to its fundamental nature. Two key aspects are the requirement to produce hydrogen from other, primary, resources and the difficulty of storing and transporting hydrogen in meaningful amounts.

However, the interest in introducing hydrogen is primarily predicated on its benefits: the potential for zero local emissions; dramatically reduced or zero greenhouse gas emissions; energy security through the use of local resources and through distributed energy; and potentially even the emergence of new industry. These benefits are overwhelmingly social in nature – so-called "public goods," but the cost of switching to hydrogen is more likely to be private, and will only become public with careful and well-considered policy intervention. Designing such policy interventions is complex, as many well-intentioned policies have unintended consequences. Many of these complexities are primarily socioeconomic in nature, that is, they have wider implications for society and the economy. However, this part of the Hydrogen Economy is probably also the least well understood. Considerable research has been conducted and is ongoing into the techniques of production, storage, and end-use of hydrogen, into new materials and scientific methods, and in demonstration projects. Alongside this, fewer groups are conducting socioeconomic research, although it will be of equal importance in understanding the best ways of introducing hydrogen energy in the future.

27.2
Defining Socioeconomics

According to Wikipedia (http://en.wikipedia.org/wiki/Socioeconomics), "Socioeconomics is the study of the social and economic impacts of any product or service offering, market intervention or other activity on an economy as a whole and on the companies, organisations and individuals who are its main economic actors. These effects can usually be measured in economic and statistical terms, such as growth in the size of the economy, the number of jobs created (or destroyed), or levels of home ownership or Internet penetration; and in measurable social terms such as life expectancy or levels of education."

It is this combination of economic and social factors that is fundamentally important and which requires careful analysis – not only of supply, demand, and price, but also of changes in quality of life due to other factors. These might include, *inter alia*, the physical environment, political stability, or income distribution. Socioeconomics is therefore not an economic theory, although socioeconomic analysis may use economic theories to understand potential impacts of activities.

Table 27.1 An introductory nomenclature for socioeconomics.

Economic analysis:	Social behavior:
• Macroeconomics	• Public acceptance
• Microeconomics	• Pressure group behavior
• Externality evaluation	• Social structures
• Energy service evaluation	• Psychology
Visions and futures:	**Finance:**
• Forecasting and backcasting	• Financing structures
• Scenario development	• Markets
Drivers and barriers:	• Institutions
• Actor analysis	• Risk
• Transition analysis	**Business:**
	• Business structures
	• Organizations

27.3 Examples

The boundaries of socioeconomic analysis cannot be simply drawn. Table 27.1, although far from exhaustive, provides some examples of relevant areas in the form of a simple nomenclature.

Clearly, many of these disciplines and analyses can be interlinked. In addition, it is essential that they are considered in conjunction with the underlying technical development and status of hydrogen energy technologies. However, a better understanding of these areas also allows a better understanding of what technology developments may succeed, and helps to focus specific research. Each is considered in turn in the subsequent discussion.

27.4 Economic Analysis

Most of the economic analyses carried out relevant to hydrogen energy are related to costs – studies have investigated the cost of individual technologies, of infrastructures, of cost evolution over time, and increased production volume. These have typically used engineering models for predicting the potential for different hydrogen technologies such as fuel cells and fuel processors. Such analyses demonstrate that hydrogen technologies, although expensive in the near term due to limited development and use, are competitive with conventional technology once they are mature. They also show the difficulty of switching from the status quo to a new technology. Although the stable end state of such an introduction is economically competitive, the transition itself is costly and requires careful management. Government support and cooperation between different industry sectors is required to make the change as smooth as possible.

Some much larger studies have looked at the introduction of hydrogen energy in a wider economic context, using input–output models to develop an understanding of the potential changes in such things as a country's GDP due to a switch from petroleum products to hydrogen, or to employment in different sectors. Economies that rely heavily on conventional car manufacturing, for example, could view hydrogen as a threat to the status quo or an opportunity for new technology introduction and business revitalization.

The study of externalities also falls under the broader heading of economics. In economic terms, an externality is a consequence of an action that is not reflected (internalized) in the price or cost of that action. A typical example would be the cost of providing health care for those who suffer from pollution due to emissions from internal combustion engine vehicles running on petroleum products. The price of the petroleum product itself to the consumer does not include this cost, which is normally borne by society as a whole, in the form of taxation or contributions to healthcare. Evaluating these externalities is fraught with difficulty and controversy, as it can require economic valuation of such immeasurables as a human life. However, by assessing these externalities it is possible to produce more reasoned arguments about the value to society of switching to alternative fuels such as hydrogen.

Climate change has perhaps emerged as the greatest and one of the most difficult externalities to value and attribute. Changing weather patterns and water levels do nevertheless result in significant costs, and the potential contribution to their reduction is an area in which hydrogen energy could play a major role. Recently, however, the majority of analyses have focused not on the value of the possible reduction in climate change that might be attributable to changing our energy systems, but simply to potential reductions in greenhouse gas emissions themselves.

27.5
Visions and Futures

In attempting to understand the potential for hydrogen energy, it is useful to use tools and methods that enable different possible futures to be examined and pathways to be analyzed. Forecasting is typically represented by the projection forward of different possible hydrogen futures – scenarios – based on different assumptions of boundary conditions. These might include the speed of technology development, prices for different resources such as oil, or aggressive environmental taxes. Using these boundary conditions, it is then possible to build internally consistent pictures of what the future could look like, and draw conclusions from them, such as whether the future scenario meets anticipated future environmental requirements. Backcasting adopts a different approach – a scenario of the future is still developed, but it is used as a basis to work backwards from that point to the present day. By doing this, conclusions can be drawn about the speed of change required, possible break points and path dependency between the current and future points, and whether or not such change is realistic.

Scenario development (such as that undertaken by Shell) also permits the analysis of potential futures in terms of – or due to – their impacts. For example, scenarios can be constructed around a future posited requirement to reduce greenhouse gas concentrations in the atmosphere, which could be achieved by different mixes of technologies, fuels, and procedures, some of which may include hydrogen energy. Different scenarios with different amounts of hydrogen in the energy mix can point to different requirements for technology and policy development.

27.6
Social Behavior

Social behavior, and particularly the attitudes of the general public, can be fundamental in determining the success or failure of a new technology. Equally important are the attitudes of those developing and introducing the technology. Both genetically modified organisms and nuclear power have a long history of consumer protest, in part due to their fundamental natures. Research has shown, however, that the attitude of scientists, multinationals, and governments has played a crucial role in these protests. Where people are given information and left to make up their own minds, they are overwhelmingly more positive about change and new technologies than when they feel it has been forced upon them.

Any successful introduction of hydrogen energy will depend significantly on the general public and the way they are approached regarding hydrogen energy. Studies to date around the world have shown that the current perception of a hydrogen economy and hydrogen technologies is broadly benign. A very few of the older generation associate hydrogen with the Hindenburg airship or (erroneously) with nuclear bombs, but more associate it with clean energy, and the majority make no association, negative or positive. Analyses of public attitudes to hydrogen vehicles, particularly buses, in their own towns and cities have shown overwhelming support for the technology, although people are more wary of the fueling infrastructure. In well-studied cases in London and Washington, DC, both BP and Shell experienced significant resistance to the construction of public hydrogen fueling stations. As it transpired, this resistance was not due to the fact that hydrogen was involved, but to the way in which the local residents were consulted about the process.

Ongoing analyses and studies repeated in the same areas over time suggest that as more information becomes available, and as trials and tests continue to run, attitudes become more positive. These studies have also given valuable evidence of the best ways of providing information and engaging with local stakeholders to help ensure that public attitudes do not turn negative due to poor procedure.

27.7
Drivers and Barriers

An alternative way of looking at future energy options is through the specific examination of drivers and barriers – physical, regulatory, financial, or other influences on the development of different technologies and pathways. Although this is implicit, and may be explicit, in some of the analyses above, the focus here is the drivers and barriers themselves and in how they may be reinforced or overcome as appropriate.

Two specific methodologies that may be applied in this case are actor analysis and transition analysis. The former is usually applied to specific examples where the different actors who can influence the case under investigation can be readily identified, and examines their possible influence. Actors may include such diverse groups as technology developers and adopters from research organizations to major corporations, legislators, technology operators such as bus drivers, and members of the general public. The implications of their actions on the case in point can be directly investigated.

Transition analysis is in some ways similar to the examination of path dependency in backcasting approaches, but focuses more directly on the actual transitions. Transitions in energy systems have historically varied from the relatively smooth (the gradual uptake of natural gas and gas turbines in some countries) to the more disruptive (the forced replacement, due to legislative enforcement, of conventional two-stroke engines and fuel with compressed natural gas over a very short timeframe in Delhi). Typically, disruptive transitions entail higher transition costs as existing infrastructure and other investments may not have been amortized, and major modifications may be required.

27.8
Finance

Although fundamentally important and examined in great depth for conventional energy, the financial aspects of hydrogen energy have been less well considered to date. The economic analysis that has been conducted, as outlined above, has focused primarily on aspects such as total costs of infrastructure, cost reduction for new technologies through learning by doing, and comparatively straightforward discounted cashflow analysis. Financing the technology development and infrastructure roll-out, understanding which entities bear what risk, and understanding the role of markets is more traditionally a financial analysis. For example, economic analysis can tell us that billions of dollars are required to roll out sufficient hydrogen fueling stations to enable a million vehicles to operate in a given geographical area, but it does not necessarily tell us if the risk/reward profile of the investment would be attractive to a private company, or how the investment may be phased to ensure that positive cashflow can be achieved in the shortest time.

The aspects above depend significantly on a wide range of financial conditions, including the flexibility and openness of local markets, the institutions involved, and the willingness and ability of government to underwrite risk. For example, liberalized energy markets have been described as having potentially counterbalanced effects on investment in new technologies: (i) the competitive nature of the markets may inspire new technology development and uptake so that firms can maintain r gain market share versus their rivals; and (ii) the focus on the bottom line can act to minimize investment in research and development and prevent companies from taking risks – for example with new technologies. The reality is between these two extremes, but the costs of developing and implementing a technology must normally be recoverable within the investment timeframe of any organization if they are to go ahead with an investment.

However, actors in the financial markets are able to bring together very large finance packages and enable investment in major infrastructure projects, as is clear from the financing available for such things as offshore wind farms. If the terms and potential rewards are appealing, then the simple cost of a project is, as it long as it remains within reasonable boundaries, irrelevant. Understanding and evaluating risk are an essential part of understanding the attractiveness of an investment, as the financial calculations will include uncertainty related to technology, policy and regulation, business and market stability, and many other factors. For example, a project that is financially appealing due to a government incentive such as a feed-in tariff could founder if the tariff were removed – which may happen due to a change of government or a change in government attitude.

27.9
Business

Business analysis, although linked in many ways to finance, requires further tools and approaches. Although the economics and the financial conditions of a given investment may look appealing, they may not link to a business's strategic goals, or the investment may simply be out of the company's reach. Equally, different stages of development of a technology may require different business structures. Consider a major automotive original equipment manufacturer (OEM) developing a breakthrough technology such as fuel cell vehicles. The new development is very high risk and of uncertain reward, particularly as its success depends not only on the successful implementation of the fuel cell vehicle itself, but also on the provision of hydrogen at enough fueling stations to give customers comfort, on the acceptance of both hydrogen and fuel cell vehicles within society, and on other competing technologies. Meanwhile, the research and development funds required are significant and compete within the organization against other, more incremental, projects such as the improvement of conventional internal combustion engines. One solution is to "ring-fence" the funds for research and development of the fuel cell technology and put the entire fuel cell R&D organization into a separate business unit where it faces different requirements in terms of time to market and return on investment. Introducing radical new technologies typically does not lead to major returns in the short term, and unless they are given sufficient time to mature and pay back, investments may not be recovered. This can conflict with shareholder expectations, for example, where payback times of months or very few years may be expected.

27.10
Conclusion

The field of socioeconomic analysis is broad and fundamentally important to understanding the potential for the development, uptake, and success of new technologies, including those relating to hydrogen energy. Analysis not only of economics and cost structures, but also financing mechanisms, company behavior, future scenarios, and public attitudes must be brought together to allow coherent interpretation of relevant opportunities, weaknesses, and required developments. These analyses must be undertaken in conjunction with the work being done on hydrogen energy topics in fundamental science, technology development, and engineering. Each set of disciplines is important in enabling the introduction of hydrogen energy, and equally important is the communication between them.

Further Reading

Barreto, L., Makihira, A., and Riahi, K. (2003) The hydrogen economy in the 21st century: a sustainable development scenario. *Int. J. Hydrogen Energy*, **28** (3), 267–284.

Bleischwitz, R., Bader, N., and Trümper, S.C. (2009) The socio-economic transition towards a hydrogen economy. *Energy Policy*, in press.

Chernyavs'ka, L. and Gullì, F. (2009) Measuring the environmental benefits of hydrogen transportation fuel cycles under uncertainty about external costs. *Energy Policy*, in press.

Doll, C. and Wietschel, M. (2008) Externalities of the transport sector and the role of hydrogen in a sustainable transport vision. *Energy Policy*, **36** (11), 4069–4078.

Hart, D. (2004), Hydrogen, end uses and economics. In *Encyclopedia of Energy* (ed. Cutler, J.C.), Elsevier, New York, pp. 231–239.

Hart, D., Anghel, A.T., Huijsmans, J., and Vuille, F. (2009) A quasi-Delphi study on technological barriers to the uptake of hydrogen as a fuel for transport applications – production, storage and fuel cell drivetrain considerations. *Journal of Power Sources*, **193** (1), 298–307.

Hisschemöller, M., Bode, R., and van de Kerkhof, M. (2006) What governs the transition to a sustainable hydrogen economy? Articulating the relationship between technologies and political institutions. *Energy Policy*, **34** (11), 1227–1235.

Hodson, M., Marvin, S., and Hewitson, A. (2008) Constructing a typology of H2 in cities and regions: an international review. *Int. J. Hydrogen Energy*, **33** (6), 1619–1629.

Houghton, T. and Cruden, A. (2009) An investment-led approach to analysing the hydrogen energy economy in the UK. *Int. J. Hydrogen Energy*, **34** (10), 4454–4462.

Hugh, M. J., Yetano Roche, M., and Bennett, S. J. (2007) A structured and qualitative systems approach to analysing hydrogen transitions: key changes and actor mapping. *Int. J. Hydrogen Energy*, **32** (10–11), 1314–1323.

Madsen, A. N. and Andersen, P. D. (2009) Innovative regions and industrial clusters in hydrogen and fuel cell technology. *Energy Policy*, in press.

Marbán, G. and Valdés-Solís, T. (2007) Towards the hydrogen economy? *Int. J. Hydrogen Energy*, **32** (12), 1625–1637.

McDowall, W. and Eames, M. (2006) Forecasts, scenarios, visions, backcasts and roadmaps to the hydrogen economy: a review of the hydrogen futures literature. *Energy Policy*, **34** (11), 1236–1250.

McDowall, W. and Eames, M. (2007) Towards a sustainable hydrogen economy: a multi-criteria sustainability appraisal of competing hydrogen futures. *Int. J. Hydrogen Energy*, **32** (18), 4611–4626.

Moriarty, P. and Honnery, D. (2009) Hydrogen's role in an uncertain energy future. *Int. J. Hydrogen Energy*, **34** (1), 31–39.

Mumford, J. and Gray, D. (2009) Consumer engagement in alternative energy – can the regulators and suppliers be trusted? *Energy Policy*, in press.

O'Garra, T., Mourato, S., Garrity, L., Schmidt, P., Beerenwinkel, A., Altmann, M., Hart, D., Graesel, C., and Whitehouse, S. (2007) Is the public willing to pay for hydrogen buses? A comparative study of preferences in four cities. *Energy Policy*, **35** (7), 3630–3642.

O'Garra, T., Mourato, S., and Pearson, P. (2008), Investigating attitudes to hydrogen refuelling facilities and the social cost to local residents. *Energy Policy*, **36** (6), 2074–2085.

Ricci, M., Bellaby, P., and Flynn, R. (2008) What do we know about public perceptions and acceptance of hydrogen? A critical review and new case study evidence. *Int. J. Hydrogen Energy*, **33** (21), 5868–5880.

Schulte, I., Hart, D., and van der Vorst, R. (2004) Issues affecting the acceptance of hydrogen fuel. *Int. J. Hydrogen Energy*, **29** (7), 677–685.

Shayegan, S., Pearson, P. J. G., and Hart, D. (2009) Hydrogen for buses in London: a scenario analysis of changes over time in refuelling infrastructure costs. *Int. J. Hydrogen Energy*, **34** (19), 8415–8427.

Sherry-Brennan, F., Devine-Wright, H., and Devine-Wright, P. (2009) Public understanding of hydrogen energy: A theoretical approach. *Energy Policy*, in press.

Stephens, J. C., Wilson, E. J., and Peterson, T. R. (2008) Socio-political evaluation of energy deployment (SPEED): an integrated research framework analyzing energy technology deployment. *Technol. Forecast. Social Change*, **75** (8), 1224–1246.

Suurs, R. A. A., Hekkert, M. P., and Smits, R. E. H. M. (2009) Understanding the build-up of a technological innovation system around hydrogen and fuel cell technologies. *Int. J. Hydrogen Energy*, **34** (24), 9639–9654.

Yetano Roche, M., Mourato, S., Fischedick, M., Pietzner, K., and Viebahn, P. (2009) Public attitudes towards and demand for hydrogen and fuel cell vehicles: a review of the evidence and methodological implications. *Energy Policy*, in press.

28
Market Introduction for Hydrogen and Fuel Cell Technologies
Marianne Haug and Hanns-Joachim Neef

Abstract

This article takes stock of the visions, roadmaps and the status quo of market introduction for hydrogen and fuel cell technologies. Based on innovation theory concepts, the study examines whether the framework conditions for a smooth transition form RD&D to market introduction exist. International, regional and national partnerships have started to develop supportive advocacy coalitions. Niche markets for value-driven, but limited product range like toys, portable, auxiliary and back-up applications are being established. But, market formation/creation, development of a competitive industry and rules and regulations for the prime applications are still in their infancy. Instead, a patch work of country-, region- and product specific policies has emerged to allow "supported commercialization" beyond the demonstration stage. The study concludes that without more attention to the specifics of co-evolution of technologies and policies for each major market segment, the ongoing market introduction efforts of fuel cell and hydrogen technologies is likely to falter or create eternal niche markets.

Keywords: hydrogen energy, fuel cell, market introduction

28.1
Introduction

The 21st century began with renewed interest in fuel cell technologies and hydrogen as a promising energy carrier to meet energy and carbon savings objectives. Ambitious roadmaps for market introduction and penetration within 15–20 years were developed. New international collaborations emerged to address research, development, and demonstration (RD&D), investment, policy, and regulatory barriers. Substantial financing in the order of about US$ 1 billion (public) and US$ 3–4 billion (private) per year [1] was mobilized. Governments and industry forecasts saw market introduction and cost reductions within reach, starting with portable micro fuel cell applications, stationary power units, and, shortly thereafter, the introduction of fuel cell/hydrogen buses and automobiles.

Hydrogen Energy. Edited by Detlef Stolten
Copyright © 2010 WILEY-VCH Verlag GmbH & Co. KGaA, Weinheim
ISBN: 978-3-527-32711-9

Successful market introduction of any product requires a high-quality product technology and appropriate infrastructure; competitive cost and consumer acceptance. Consumers will only buy a product which serves their purposes, is reliable, is easy to operate, and at acceptable cost.

Market introduction is hardly a linear process: after a mostly long RD&D phase, it involves successive and overlapping steps:
- deployment and evaluation steps with feedback into RD&D to prove technological readiness;
- development of production technologies for volume manufacturing;
- development, adoption, and adjustments to incentives and regulatory measures;
- campaigns to achieve consumer acceptance followed by sales and marketing efforts; and
- industrial investment in production facilities, infrastructure and service.

The innovation literature depicts this process through a five-step S-form diffusion curve [2]. Innovations achieve cost/performance competitiveness through a combination of learning – whether learning "through research", learning "by doing" or "by using" – and the "scale effect" of increasing volumes through adoption, commercialization, and mass production. During the RD&D and early adoption phases, large differences exist between social and private returns to investment due to the knowledge spill-over and lock-in effects. These differences are particularly pronounced for energy technologies. Public support in the form of technology push or market pull incentives help overcome these barriers and address the "valley of death," the lack of financing to bring applied research to market.

The public/private RD&D programs for hydrogen and fuel cell technologies have begun to deal with these well-identified innovation hurdles in a determined manner. They fall short in two aspects:
- First, they do not deal with the differences between social and private returns to investment that occur during the market introduction/commercialization phase. Such differences continue to exist because of market failures and multiple externalities: the cost of carbon, energy security, and the infrastructure-related network externalities have not been internalized.
- Second, fuel cell and hydrogen technologies are "radical innovations." They require a technological paradigm and system shift towards sustainable development, a societal and institutional undertaking that goes far beyond the standard introduction of a new product in the market place.

"Studies of how firms bring radical new technologies to market have shown how early stage development is dominated by efforts to get the technology to work and often ill defined or wildly wrong assumptions about markets." [3]

This finding from the literature was confirmed over and over again for fuel cells and hydrogen technologies. In the face of media hype and huge public expectations, governments, industry, and market researchers had to revise their wildly optimistic market forecasts year after year for every single application. Profit goals were not met [4] and targeted units not shipped. Now, public and media attention has shifted to the next low-carbon technology hype, electro mobility with battery cars.

A large body of empirical case studies trace the delays, hurdles and missed promises of fuel cell and hydrogen technologies and provide insights for policy makers and industry. They inform, *inter alia*, about specific hydrogen and fuel cell applications [5–8], assess policies and institutions [9–10], and review different stages of the diffusion process [4, 11] and cross-county experiences [12–14].

This chapter first summarizes the conceptual insights in market introduction from the innovation literature, then provides a brief overview of the most important roadmaps, the international cooperation initiatives, and the status quo for the markets of major fuel cell and hydrogen technologies, and finally draws lessons for the future.

28.2
Market Introduction of Radical Innovations: What Do We Know from the Literature?

Beyond the classic technology diffusion model, evolutionary economics and innovation system theories develop conceptual and explanatory approaches to guide the understanding of innovation processes [15,16]. Recent co-evolutionary studies incorporate multi-level micro-meso-macro analyses to understand the transition from technological niches to the transformation of socio-technical regimes and society as a whole [17–19] – approaches that are directly applicable to the transition of our present fossil-based to a low-carbon energy system.

Several authors advocate a new approach to "niche markets" in the case of "radical innovations," such as hydrogen and fuel cells. Kemp *et al.* [20] suggest "strategic niche market management" to provide market participants with a protected space to innovate and be able to try out new products. Countries and industries that foster product and market development in such niche markets accelerate market introduction and cost competitiveness, and may also achieve first mover advantage, develop dominant designs, and become "lead markets" for global diffusion of new technologies. Strategic niche market management in support of low-carbon technologies and climate/energy security goals can be justified from an economic point of view as they address well-known market failures. In practice, the results – and the pitfalls – are the same as policy interventions in support of industrial, employment or export development policies.

Grubb *et al.* [21] advocate market pull support during two phases of the market introduction of new low carbon technologies: first, during a pre-commercial stage where established firms adopt the technology and/or new firms emerge, and second, during a "market accumulation phase" where the use of the technology increases, niches develop, and protected or subsidized markets allow scale expansion. They are silent on the length of such "supported commercialization." While the multiple externalities may well justify public policy support during market introduction and commercialization, the question is: how much support and for how long? Who will carry the costs if the technology never develops beyond a niche market?

Geels [18], Satorius [22], and Foxon [23] stress, therefore, the importance of co-evolution of technology and policy regimes. There is a need for "transition management" and time strategies. "System failures," whether ineffective policies, changing technologies, or unforeseeable circumstances, will inevitably arise at each stage of the market introduction process. They need to be identified and addressed by market participants flexibly and in a timely manner.

Jacobsson and Bergek [19] pinpoint four framework conditions for the adoption of low carbon energy technologies. They require (i) a stable process of market formation/creation, (ii) the entry of new firms and the development of a competitive industry, (iii) institutional change with rules and regulations that accommodate the new technology, and (iv) the development of supportive "technology-specific advocacy coalitions."

How far has this process advanced in the case of hydrogen and fuel cell technologies? What differences exist between countries? How effective are the national public/private partnerships and international collaboration as "technology-specific advocacy coalitions" to solve technology and regulatory barriers and accelerate market introduction? Is the public support and private investment effort sufficient to meet the announced targets or will hydrogen and fuel cell technologies linger in perpetual niche markets?

Who will arbitrate support levels and instruments between competing clean technologies [24]? The literature is still silent on competition among carbon-neutral or energy-saving technologies, and the appropriate policy response during market introduction. A McKinsey study warns that "markets will migrate" as support for a given sustainable technology increases or vanes [25]. What will this mean for hydrogen and fuel cell technologies? Will the public support and attention to electro mobility with battery cars scale down investments and incentives for hydrogen and fuel cells in the transport sector and push back market introduction? Will the type of support for combined heat and power (CHP) and decentralized power from renewables retard the market introduction of fuel cells for stationary power in Germany compared with Japan?

Economists, eco-innovation researchers, and policy makers have written extensively and universally to embrace carbon pricing to accelerate market uptake of low-carbon technologies and support a global public good [26]. The literature is silent on internalizing energy security costs/externalities. This is an unfortunate gap for energy efficiency innovations such as fuel cells. Energy security externalities vary among countries as they depend on both the energy resources endowment and past energy policies of the specific country. Internalizing energy security costs can call for and justify country-specific policy support measures.

This chapter will not be able to answer all these questions. It will draw on the concepts highlighted above, examine the status quo of co-evolution of technology and policies during market introduction, and point to lessons for policy makers and industry.

28.3
The Fuel Cell and Hydrogen Road Maps: from Visions to Public/Private Coalitions

"Roadmaps communicate visions, attract resources from business and government, stimulate investigations, and monitor progress. They become the inventory of possibilities for a particular field, thus stimulating earlier, more targeted investigations. They facilitate more interdisciplinary networking and teamed pursuits." Robin Graves (Motorola)

The fuel cell and hydrogen roadmaps drawn up during the last decade have been valuable to clarify and monitor goals, technological pathways, investment priorities, and requirements for institutional change. They helped mobilize additional private and public funds, created national and international partnerships and synergies, raised public awareness of these technologies, and, thus, built the beginnings of a "technology-specific advocacy coalition."

Such roadmaps, also called visions, national programs, implementation plans, development plans, and so on, are partly formulated along political goals and estimate budget requirements for RD&D and market preparation. Some roadmaps provide numerical values for market expectations in the future and mostly contain technological milestones to be achieved in the coming years. Roadmaps underlie a constant revision influenced by political and administration changes, regular progress meetings, budget revisions, changing industry commitment, and international communication on public and industry level.

A few highlights from fuel cell/hydrogen roadmaps worldwide are sketched below:
- The Japanese government and industries have pursued a substantial collaborative RD&D program on hydrogen and fuel cell technologies since the 1980s. The latest roadmap includes the launch of "supported commercialization" for residential CHP systems with 1 kW polymer electrolyte membrane (PEM) fuel cells in fiscal year (FY) 2009, with investment subsidies of up to 50% of the system costs [27]. For fuel cell vehicles and hydrogen stations, the industrial association called "Fuel Cell Commercialization Conference" [28] developed a scenario which sees commercialization beginning in 2015 and full commercialization during the period 2020–2030.
- The Hydrogen Posture Plan issued in 2006 by the United States Departments of Energy and Transport and the Hydrogen, Fuel Cells and Infrastructure Technologies Program [29] describe the goals to be reached by about 2015 with a strong government role in this RD&D phase. When the performance and durability of the technologies are validated, the government may consider

becoming an early technology adopter, and could enact policies to nurture the development of an industry capable of delivering significant quantities of hydrogen to the market place. Industry's role would become increasingly dominant as market penetration increases in the years after 2015. Several partnerships support the implementation of the roadmap such as the Hydrogen Utility Group, the FreedomCAR and Fuel Partnership, State/local governments such as the California Fuel Cell Partnership, the Federal Interagency Partnerships, and industry associations.

- China's fuel cell and hydrogen roadmap and program [30] is part of the seven key areas of the National Mid-to-Long Term Sci-Tech Plan (2006–2020). Priority areas are mechanism and materials in hydrogen production (from renewables and from coal with carbon capture and storage), hydrogen storage, and fuel cell development and systems. Not much is known about market perspectives. General statements from officials indicate that the Chinese market is big enough for the mass production of fuel cell vehicles [31].
- Korea adopted a new ambitious program and roadmap on hydrogen and fuel cells. The vision for 2040 [32] foresees that 15% of final energy is provided through hydrogen; that the GDP portion of fuel cell industry is 5%; that 60% of hydrogen is produced from renewables; that 22% of the power plants will be replaced by fuel cell generation; and that 23% of residential power is replaced by fuel cell generation.
- The roadmap of the European Hydrogen and Fuel Cell Technology Platform [33] was formulated with broad participation of the European industry and science community. Based on the preparatory work of the Platform, an industry-led public–private partnership was created in 2009 to implement the European roadmap together with important country-specific national and regional programs. The Fuel Cell and Hydrogen Joint Undertaking [34] brings together the European industry (currently 59 companies from 15 countries) and science community with the European Community, represented by the European Commission. The RD&D program aims at creating a technology basis for commercialization in the time frame 2015–2020. In line with the European program are country- and region-specific programs of the Member States, bringing together government, industry, and the science community.

In summary, the current visions, roadmaps, and public/private programs for hydrogen and fuel cell technologies are focused on RD&D and market preparation. They have been successful in forming "technology-specific coalitions," attracting businesses, and defining common goals. Preoccupied with technological readiness, attention to the dynamic of market introduction has remained limited.

28.4
International Cooperation: Value Added During Market Introduction?

Several organizations and initiatives are promoting international cooperation for hydrogen and fuel cell technologies. The International Energy Agency (IEA) has a long history of collaboration in energy technologies as part of its recently renovated "IEA Framework for International Energy Technology Cooperation" [35]. The focus of cooperation has been traditionally on pre-competitive R&D, with demonstration and market introduction activities having less priority. The IEA Hydrogen and the IEA Advanced Fuel Cell Implementing Agreements are the main vehicles of this R&D collaboration.

In 2003, the IEA Hydrogen Coordination Group was established. It completed a review of national R&D programs on hydrogen and fuel cells [1] and their market prospects [36]. The analysis concluded that hydrogen is likely to conquer a significant market share *only* if effective policies for CO_2 mitigation and energy security are in place and combined with considerable reductions of hydrogen and fuel cell costs. Only then will hydrogen emerge as a player in the future transport sector. Under less optimistic assumptions for technology development and policy measures, hydrogen and fuel cells are unlikely to reach the critical mass that is needed for successful market uptake.

The International Partnership for the Hydrogen Economy (IPHE) was established in 2003 and provides a forum for advancing policies and common codes and standards that can accelerate the cost-effective transition to a global hydrogen economy to enhance energy security and environmental protection [37].

Based on a series of international workshops on infrastructure development for hydrogen in the energy economy, the IPHE together with the IEA concluded in 2009 [38] that the advancement of a hydrogen economy will continue to pose numerous challenges and opportunities. Diverse hydrogen-related technologies will need to be developed, and willing, informed investors will be needed to fund them. Successful implementation cannot be achieved by a single plan, because infrastructure planning and design possibilities must consider regional, economic, and political contexts. Rather, a wide range of stakeholders must be involved in the planning process, including local, state, and national government agencies, automakers, hydrogen equipment manufacturers, financial and insurance institutions, industry, utilities, and policy makers.

The development of regulations, codes, and standards (RCS) for hydrogen and fuel cell technologies is an integral part of the process towards commercialization. Several international organizations such as ISO, IEC, and CEN/CENELEC, but also the establishment of Global Technical Regulations at the UN level, are contributing to a safe performance of hydrogen and fuel cell products and systems. In response to the call for coordination and covering the complete range of RCS across hydrogen production, delivery, storage, and end use, the IPHE [39] has started, inter alia, an activity to analyze and develop recommendations on common RCS, identify and promulgate good practices to advance the safe use of hydrogen and fuel cell technology, and coordinate with other international organizations

Whereas the value of international RD&D collaboration is widely acknowledged, effective international collaboration during market introduction/commercialization is less tested. Such collaboration beyond information sharing raises issues of competitiveness for countries and companies. RCS can be used to acquire competitive advantages in export markets, establish dominant designs, or restrict imports. Nevertheless, the global objective of bringing to market energy-efficient, carbon-neutral technologies justifies international collaboration for fuel cell and hydrogen during market introduction if it addresses potential market failures. For example, the focus of such collaboration may be to analyze and *avoid* distortions in RCS or share experiences for effective hydrogen infrastructure development.

28.5
Market Introduction: The Status Quo

Fuel cells and hydrogen technologies face a myriad of markets for each application, different conventional competitors, and diverse incentives and regulations. For the purpose of this chapter, three groups of applications will be discussed:
- Portable, niche transport, and material handlings applications. Several products in these market segments are either already cost/performance competitive or are beginning to build a market based on superior product characteristics.
- Stationary power/heat systems for domestic and commercial applications. Market introduction requires competitive cost/performance in comparison not only with conventional heat/power technology, but also with other low-carbon/energy-saving technologies.
- Transportation with buses, ships, and passenger cars, and the associated fuel supply and infrastructure. These market segments are in the RD&D phase. However, infrastructure development has started and will be expanded in line with the number of hydrogen cars on the road.

No typical business model for the production and marketing of fuel cell products has yet emerged for any of these applications. Some companies are developing and producing all critical elements of the supply chain from fuel cells, stacks, and modules to systems. Other companies are concentrating on the production of innovative elements and cooperate with system integrators, original equipment manufacturers (OEMs), or large users such as utilities. The OEMs have the knowledge of comparable conventional systems, established distribution chains, and market access. Industry participants along the supply chain and the end-use industries are currently looking for cooperation, joint ventures, and mergers, or for outsourcing. While there is a trend to source fuel cells and stacks globally, manufacture/assembly of systems has remained close to the major expected markets.

Two factors impact the developing industry and supply chain structure. On the one hand, the OEMs and their experience with global supply chain management push component supply towards global competition. On the other, the procurement restrictions of military demand and the many country-specific partnerships and incentives foster national or regional supply chains. Typically, "shake out" starts with market introduction, but may be delayed if a long period of "supported commercialization" impacts sourcing decisions.

To achieve performance competitiveness in the larger volume markets, the durability of hydrogen and fuel cell systems must improve so that reliability can be guaranteed. To achieve cost competitiveness, two aspects need attention: first, the classic competitive cost reduction through a combination of further technology development, more efficient supply chain management and the scale effect of large production volumes; and second, appropriate policies and regulations that provide a level playing field for fuel cell and hydrogen in the important market segments compared with conventional and other low-carbon/energy-saving technologies.

28.5.1
Early Markets for Portable, Niche Transport, and Materials Handling Applications

For more than a decade, fuel cells for portable, niche transport, and materials handling applications have been considered the obvious candidates for market introduction and commercialization. While earlier announcements by Casio, Toshiba, NEC, and others [5] for consumer electronics did not materialize, some companies such as SFC Smart Fuel Cell (see Box 28.1) entered the commercialization stage with a range of portable fuel cell products.

Market assessments for portable applications vary depending on fuel cell sizes and user characteristics: Pike Research [41] included in its report *Fuel Cells for Portable Power Applications* the entire sub-7 kW fuel cell industry with an estimated turnover of US$ 122 million in 2008 and forecasts a market size of US$ 2.7 billion in 10 years. Fuel Cell Today [42] reports consistent market growth for units up to 1.5 kW from about 5000 units shipped in 2005 to 9000 in 2008. The present consumer electronics market is led by manufacturers in Asia. Firms in China and Taiwan manufacture commercially fuel cell toys and educational units for export. Firms in Korea and Japan focus on demonstration and early adoption of either integrated DMFC mobile phone/laptop products or external DMFC battery chargers for mobile phones or laptops. Toshiba announced in October 2009 the commercial launch of its "Dynario" DMFC battery charging unit [43]. The increased power demands for mobile phone features such as Internet, GPS and MP3 players or for laptops should, in principle, generate additional consumer demand. However, not all users of mobile phones, digital cameras, or laptops look for these additional power-hungry services and are willing to pay more. Agnolucci [5] projects that fuel cells may capture 5% of the mobile phone market over a 5 year period after market introduction, 2% of the camcorder and digital camera market, and as much as 30% of the laptop market.

> **Box 28.1**
> **Market deployment at a commercial stage: SFC Smart Fuel Cell in Germany**
> SFC Smart Fuel Cell [40] is a pioneer and market leader in fuel cell technologies for mobile and off-grid power applications – serving the leisure, industrial, mobility, and defense markets. SFC was the first company worldwide with commercially available, proven DMFC (direct methanol fuel cell) products. Founded in 2000, the company has already sold over 16 000 systems by September 2009, and logged millions of operating hours in end-user environments. The company has alliances with leading companies in a wide range of industries and created a convenient fuel cartridge supply infrastructure. SFC is headquartered in Brunnthal/Munich, Germany, and has operations in the United States.
>
> The success of SFC is based on the continuous development of the DMFC technology, including fuel cartridges, and its unique hybrid approach which combines fuel cells with other sources of power, including traditional batteries and solar panels.
>
> In 2006, the EFOY fuel cell product family was introduced to the market. Today, customers can choose from a range of fuel cell generators with a charging capacity ranging from 600 to 6000 W h per day. These environmentally friendly, highly efficient, zero pollution generators have widespread applications in portable, vehicle-based, and stationary off-grid markets. At present, 50 vehicle manufacturers (OEMs) already offer EFOY power supplies as an auxiliary power unit (APU).
>
> SFC is participating in several projects of the German National Innovation Program Hydrogen and Fuel Cell Technology (NIP). Topics covered are continued cost reduction, product adaptation for higher-power markets, and field trials.

Consumer demand and market introduction can only follow the removal of regulatory barriers. The International Civil Aviation Organization (ICAO) recommended to its members in 2008 to allow the carriage of fuel cells and a limited number of methanol refueling cartridges for electronic devices on aircraft. Many countries have adopted such regulations. This is a crucial prerequisite for the market introduction of fuel cell-powered mobile phones or laptops.

The defense sector offers an important RD&D and demand stimulus for electronic devices, auxiliary power units (APUs), and unmanned vehicles. North American and European firms are the major suppliers of this market at present. Fuel Cell Today [42] reports the shipment of hundreds of units by companies such as SFC Smart Fuel Cell, Neah Power, and Jadoo Power alone in 2008. Nevertheless, the entire range of military applications is still in the stage of pre-commercial demonstration and deployment. The "rugged" or highly sophisticated units may not be the type of products the civilian market demands.

Material handling equipment (forklifts and trucks) is a sizable market with millions of vehicles worldwide and offers fuel cells a potential beyond limited niche markets. Most developers such as Hydrogenics, Nuvera, and Plug Power are developing PEM units, while Jülich/Jungheinrich, and Oorja Protonics focus on DMFC. United States government support was responsible for the 2008–2009 market growth for fuel cell material handling equipment. Over and above the fuel cell investment tax credit of US$ 3000 kW^{-1} or 30% of the unit price, whichever is less, in 2009 the United States mobilized US$ 10.8 million for the deployment of fuel cell forklift/truck systems under the American Recovery and Reinvestment Act of 2009 (ARRA), funded the demonstrations of about 100 forklifts at the Defense Logistics Agency, and supported further demonstrations under the US DOE Fuel Cell Market Transformation Initiative.

The early markets described above offer the fuel cell/hydrogen industry an array of niche markets where the performance characteristics of fuel cells have a comparative advantage and/or where price elasticity of demand is low. Two market trajectories are possible under these circumstances. Either fuel cells will be adopted and commercialized only in specialized niche markets, or these early markets, in particular fuel cells for material handling equipment and electronics, allow sizeable cost reductions through volume production, supply chain, and technology improvements in addition to the adoption of supportive regulations and industry-wide codes and standards.

The United States supports these segments through sizable investment tax credits until 2016 and strong defense sector demand. This is in addition to RD&D support and development of codes and standards by the national bodies. While most portable applications should need no policy support for market introduction, governments and industry may want to identify and correct early possible gaps, market failures, or policy inconsistencies in national and international markets.

28.5.2
The Market for Stationary Fuel Cell Power and Heat Systems

The most important market segments for these applications are the following:
- Small stationary systems for either residential CHP or UPS systems in the order of 1–10 kW, using low- and high-temperature PEM fuel cell and solid oxide fuel cell (SOFC) technology. The recent Fuel Cell Today survey [42] reports global, cumulative installations of about 11 000 units by end 2008. PEM units have become dominant, accounting for 95% of the 4000 units shipped in 2008.
- Larger stationary combined heat and power systems in the range from 10 kW to >2 MW, using primarily molten carbonate fuel cells (MCFCs) and phosphoric acid fuel cells (PAFCs) and to a lesser extent SOFC or PEM fuel cell technology. Cumulative MW$_{el}$ installed had reached 180 MW$_{el}$ by 2008 with an average of 50 units installed per year and an average size of 1 MW. Market demand focuses on three categories of applications: the 10 kW units, the 250–400 kW units for commercial buildings, and the >2 MW units for power plants [42].

In the aftermath of Hurricane Katrina, new requirements for extended runtime back-up power units emerged in the United States for the telecommunications sector. North American and European firms, led by IdaTech and Hydrogenics, accelerated development and market introduction of uninterrupted power supply/back-up systems (UPS). UPS systems are now ready for commercialization with orders placed not only in the United States, but also in India, Africa, and other countries. Additional firms are concentrating on this emerging market, such as Plug Power, Hydra Fuel Cells, CellKraft, Dantherm, and P21. In parallel, developers are expanding their product range by offering UPS for the residential market. This is the classic example of both performance characteristic- and government regulation-induced demand that can open niche markets and set new standards worldwide.

In 2005, the New Energy Foundation (NEF), a coalition of Japanese electrical utilities, fuel companies, and fuel cell manufacturers, was tasked with the objective of deploying 1 kW residential units. The program was supported through subsidies to fuel suppliers (gas and oil companies). Manufacturers were required to produce a minimum of 30 units with a subsidy of 6 million Japanese yen per unit when the program started. The stepwise reduction of the subsidy to 2.2 million yen in FY2008 followed the reduction of production cost over time; 3307 units were installed between FY2005 and FY2008.

In 2009, the ENE-FARM program was launched in Japan that offers a suite of residential CHP systems sold by several different manufacturers but marketed under one brand. Sales of 4000 units are targeted for FY2009 thanks to Government subsidies of 50% of the purchase price, up to 1.4 million yen per unit. Up to September 2009, additional 1500 units were installed under the new support scheme [27]. Subsidies will be decreased in the coming years according to expected lower costs and prices. As detailed in Box 28.2, Toshiba as one of the participating companies is ready for rapid commercialization of its system.

With the ENE-FARM program, Japan switched the focus of public subsidies from technology providers to consumers to ensure the creation of demand and markets. Further, the government relies on a critical mass of firms – such as ENEOS CellTech, Panasonic, Toshiba, and Toyota – that must compete for market shares in this "supported commercialization" effort. If domestic market growth for 1 kW co-generation units leads to early cost reduction and performance improvements, Japan may emerge as the lead market for residential co-generation units worldwide and establish its product range as dominant designs.

According to reports [42], starting from 2010 Korea will provide an 80% subsidy to cover the cost of purchasing and installing fuel cells in homes, with the subsidy falling to 50% between 2013 and 2016 and to 30% from 2017 and 2020. Manufacturers can anticipate 100–200 installations per year from 2010, increasing to around 1000 per year by 2013.

The potential European market for residential power and heat systems is attracting manufacturers from outside Europe. Japanese producers are eying export, especially to Europe [44], a natural progression from its present market

> **Box 28.2**
> **Market deployment of residential cogeneration systems: Toshiba in Japan**
> Toshiba Fuel Cell Power Systems Corporation is a Japan-based company for PEM technology. For 1 kW residential co-generation systems, Toshiba claims to have started the initial commercial stage already. All Japanese manufacturers (Toshiba, Panasonic, ENEOS Cell Tech, Toyota, and up to 2008 Ebara-Ballard) have demonstrated more than 3300 systems during FY2005–2008; 748 units were delivered from Toshiba. Toshiba will produce several thousand units from FY2009 onwards with the target of more than 10 000 units per year in FY2012.
> Like Toshiba, Panasonic aims at 3000–5000 unit sales in 2010 and production of 10 000 units of 1 kW stationary PEM fuel cells for the period 2008–2010. ENEOS CellTech aims at the production of 10 000 units of 1 kW stationary PEM fuel cells in 2010 [44]. With a view to consumer motivation, it is worth noting that the competing Japanese manufacturers are selling their fuel cell systems under the common brand name ENE-FARM and are using TV advertisements for promotion of their products.
> The present Toshiba ENE-FARM unit is based on 20 000 h of operation time, high reliability, and high efficiency. Cost reduction to one-eighth of FY2004 costs have been realized for FY2009 due to performance improvements of components, system simplification, manufacturing process improvement, and most recently mass production. Further cost reductions are expected.

position. Other companies are investing in European subsidiaries to produce or assemble residential systems assuming supportive policies for market introduction in the future (see Box 28.3).

Market introduction of large stationary systems, mostly MCFCs and PACFCs, continues to depend on sizable market introduction subsidies. For example, in 2000 California started the Self Generation Initiative Program (SGIP), which subsidizes the uptake of renewable and fuel cell power generation technologies. Starting in 2008, SGIP new single fuel cell installations of maximum 3 MW can apply for an investment subsidy, with a sliding scale from 100% of the subsidy for the first MW to 25% for the final MW. The subsidy is US$ 4.50 per watt for installations using renewable fuel, and US$ 2.50 per watt using fossil fuels (except diesel). The current funding round for this program extends to 2012. The effect of this program has been to see California exhibit the fastest growth of large stationary fuel cells anywhere in the world [42].

As the examples of Japan, California, and the UPS systems illustrate, very specific regulations and industry cooperations have been instrumental in the few markets where stationary fuel cell power and heat systems have reached market introduction. These are not universal, but should improve cost competitiveness for the companies and products involved. The further market trajectory

> **Box 28.3**
> **Market deployment at an early stage: CFCL in North-Rhine Westphalia (Germany)**
> CFCL is an Australian-based company for solid oxide fuel cell (SOFC) technology which has constructed a volume production facility in North-Rhine Westphalia (Germany) [45]. Here the European market can be reached, the supply chain is available, and the public support for industrial settlements is strong. The present modular generator for residential co-generation uses the SOFC module, brand-named GENNEX, which is the result of several product iterations since 2003. In the phase up to December 2011, fuel cells are delivered by German suppliers. For example, the ceramic manufacturer HC Starck has made investments to increase its capacity to 700 000 SOFCs per year [46]. Know-how of CFCL and its German suppliers combines to a delivery chain for mass production of efficient fuel cells. In Heinsberg, North-Rhine Westphalia, CFCL will integrate the fuel cells into stacks with its technological know-how and use the SOFC stack as core component in a 2 kW system for producing power and heat at very high efficiencies for residential use. CFCL is cooperating with French, English, and German manufacturers for heating systems to use their know-how in conventional and innovative system design. CFCL is aiming at "green" customers first, which are utilities, organizations, and institutions willing and able to cover part or all of the extra cost of the first demonstration projects. With increased production capacity, feedback from the demonstration and continued RD&D, cost is expected to fall to a level which is commercially acceptable. Taking into account the externalities for the use of this technology – the environmental and energy security externalities – it is expected that incentives for use will be provided by political measures as long as production costs are higher than prices for alternative residential heating or cogeneration systems.

for stationary heat and power systems will depend not only on cost reductions but also on policy support in place for competing renewable power/heat systems and the emissions trading schemes or environmental taxes and their effect on the cost of conventional systems. Regulatory bias can slow or accelerate market introduction.

28.5.3
Hydrogen and Fuel Cell Technologies for Transport

From a perspective of technical maturity much has been achieved in the past decade:
- About 120 fuel cell buses have been or are being tested under various demonstration support programs – in Europe (CUTE, ECTOS), Australia (STEP), China (UNDP GEF, "863" plan), Japan (JHF I and II), South Korea, United

States (SunLine Transit, Connecticut Transit), and Brazil (UNDP GEF), to list just the most important programs.
- More than 1000 prototype, concept, and demonstration cars are being tested with broad involvement of all major car und stack manufacturers under the various national and regional RD&D programs. Japanese, American and European carmakers lead the manufacturing segment with an array of different models, such as the Honda FCX Clarity, the GM Equinox, and the Daimler BlueZERO F-Cell. The original market driver for fuel cell cars has been and will remain the California Zero Emission Vehicle Mandate (ZEV).

However, costs for fuel cell vehicles are still exorbitant. The current cost of a fuel cell bus averages US$ 1.5 million, and passenger cars cost well above US$ 100 000. Extrapolating the current development status of a fuel cell stack to a production of 500 000 cars per year, the US DOE [29] reports a value of US$ 61 kW^{-1} and compares this value with the target of US$ 30 kW^{-1} for 2015. Around this year, "technology readiness" should be reached and infrastructure and markets are supposed to be built up within 20 years after 2015 and would reach saturation 10 years later.

In Japan, earlier targets of, for example, 50 000 fuel cell vehicles by 2010 and 5 million vehicles by 2020 have been replaced by more cautious statements such as, "It is the general consensus among Japanese energy providers and automakers that fuel cell vehicles will be introduced into the public market in 2015."

Daimler reported in 2009 [47] that their fuel cell fleet consists of 99 vehicles: 60 "Generation 2" cars, 36 buses, and three light duty cars. Until 2010, the number of cars ("Generation 3") will be around 200. Mass production of "Generation 5" vehicles is expected around 2020. This market introduction path is in line with statements by other car manufacturers.

International car manufacturers and energy companies have signed agreements on producing fuel cell vehicles and building up a hydrogen infrastructure with possible expansions into neighboring countries (see Box 28.4).

Less than 100 hydrogen fueling stations exist, from Singapore, Shanghai, New York, and Berlin to Reykjavik. Regional clusters in California and Europe have taken on the pioneering role. For example, the California Hydrogen Highway Program developed clusters around Los Angeles and the Bay area that may be connected to the British Columbia Transit fleet/stations after the Winter Olympics. The Scandinavian Hydrogen Highway Partnership (SHHP) is designed to link the stations in Norway (HyNor), Sweden, and Denmark. Germany continues to develop its hydrogen infrastructure under its National Innovation Program (NIP), the Hamburg–Berlin and North-Rhine Westphalian clusters, and a fuel station network in southern Germany. In Japan, in Phase II of the Hydrogen and Fuel Cell Demonstration Project (JHFC II), several hydrogen stations will be upgraded to 700 bar. The Research Association of Hydrogen Supply/Utilization Technology (HySUT) of 13 oil and gas utilities, and other industries, was established in July 2009 to install and operate a hydrogen infrastructure [27].

> **Box 28.4**
> **Corporate collaboration on the development and market introduction of fuel cell vehicles and hydrogen infrastructure**
> A "Letter of Understanding on the Development and Market Introduction of Fuel Cell Vehicles" [48] was signed in September 2009 by Daimler, Ford, GM/Opel, Honda, Hyundai/KIA, Renault/Nissan, and Toyota with the anticipation that from 2015 onwards a significant number of fuel cell vehicles could be commercialized. This number is aimed at a few hundred thousand units over the life cycle on a worldwide basis. The signing OEM "strongly support the idea of building up a hydrogen infrastructure in Europe, with Germany as starting point and at the same time developing similar concepts for market penetration of hydrogen infrastructure in other regions of the world, with one USA market, Japan and Korea as further starting points." In parallel, a "Memorandum of Understanding on H2 Mobility" [49] was signed by Daimler, EnBW, Linde, NOW, OMV, Shell, Total, and Vattenfall, agreeing on a build-up plan for hydrogen infrastructure, flanking expected serial production of fuel cell vehicles starting in 2015. In a first phase until 2011, the focus will be on standardization of hydrogen fueling stations, and a joint business plan for area-wide roll-out in Germany. In phase two, the respective action plan will be implemented.

Further, regions can be early adopters by using existing infrastructure. For example, the industrial by-product hydrogen and the hydrogen pipeline of 230 km in North-Rhine Westphalia could be used for fuel cell systems during the relatively slow growth of hydrogen consumption in the early market phases. It has been calculated that ~ 350 million N m^3 per year can be regarded as by-product hydrogen, which is sufficient to operate $\sim 260\,000$ cars. The hydrogen pipeline is the basis for the 200 million project "NRW Hydrogen HyWay" to establish a hydrogen infrastructure within the region [50].

Will fuel cell passenger cars reach commercialization along the projected time horizons around 2015, or will earlier retail sale of electric battery or hybrid cars corner the low-carbon transport market? The earlier market experiences of battery cars through "supported commercialization" will be an enormous benefit for the development of fuel cell vehicles. Such a benefit will include battery technology, learning about marketing for new technologies for road transport, and experience with public support measures in regional, national, and international markets.

28.6
Conclusion: Co-evolution of Technology and Policy

There are basically four elements to reduce cost of hydrogen and fuel cell systems:
- technological progress achieved through continuous RD&D and feedback from demonstration and commercialization into improved system development;
- a competitive industry structure, with integration of large, small, and medium enterprises;
- stepwise production increase with the goal of mass production; and
- stepwise development of clean fuel supply and the associated infrastructure.

The existing roadmaps and public/private partnerships focus primarily on maturing the technology through RD&D. The industry and policy approach to market introduction has been selective and eclectic and varies greatly among countries, as the examples in this chapter highlight. Important, competitive niche markets are developing because of favorable value propositions for speciality applications such as APUs or military applications, because of regulatory mandates and new standards for UPS in the telecommunication sector, or through effective consumer demand for toys or other portable application. Although important and helpful, they are unlikely to produce the volumes or cost reductions needed for competitive market introduction of hydrogen and fuel cell technologies in the power/CHP or transport sectors.

The main drivers for the high-volume fuel cell/hydrogen markets in the power/heat and transport sectors are the potential to contribute significantly to CO_2 mitigation and energy security in the medium and long term. Market introduction cannot be seen in technology-specific isolation or only in relation to conventional systems: stationary fuel cell power/CHP systems compete with renewable power/CHP. Transport will have to rely on efficient powertrains with either hydrogen or electricity from carbon-free sources. As country-specific market incentives exist for these competing technologies and systems to address industrial development, energy security, or climate concerns, competition among regulatory approaches is unavoidable, will segment markets, and will tend to limit diffusion.

Policy makers, industry, and researchers need to pay more attention to the co-evolution of technology and policies during the market introduction stage. "Supported commercialization" may be justified if private and social returns on investment still differ during market introduction/commercialization, and if fuel cell and hydrogen technologies offer the best low-carbon investment option in a given market segment. Industry needs to carry the business risk of market introduction, but carefully calibrated and consistent policies are needed to induce consumer demand and ensure competitive markets.

References

1 International Energy Agency (2003) *Hydrogen and Fuel Cells, Review of National R&D Programs*, OECD/IEA.
2 Rodgers, M.E. (1962) The Diffusion of Innovations, *The Free Press*, New York.
3 Hendry, C., Harborne, P., and Brown, J. (2007) Niche entry as a route to mainstream innovation: learning from the phosphoric acid fuel cell in stationary power. *Technology Analysis and Strategic Management*, **19** (4), 403–425.
4 PriceWaterhouseCoopers (2007) The Promise of Clean Power? *2007 Survey of Public Fuel Cell Companies*. Energy, Utilities and Mining, Renewables Series. PriceWaterhouseCoopers.
5 Agnolucci, P. (2007) Economics and market prospects of portable fuel cells. *Int. J. Hydrogen Energy*, **32**, 4319–4328.
6 Agnolucci, P. (2007) Hydrogen infrastructure for the transport sector. *Int. J. Hydrogen Energy*, **32**, 3526–3544.
7 Markard, J. and Truffer, B. (2008) Actor-oriented analysis of inn, ovation systems: exploring micro-meso linkages in the case of stationary fuel cells. *Technology Analysis and Strategic Management*, **20** (4), 443–464.
8 Stiller, C., Seydel, P., Bünger, U., and Wietschel, M. (2008) Early hydrogen user centers and corridors as part of the European hydrogen energy roadmap (HyWays). *Int. J. Hydrogen Energy*, **33**, 4193–4208.
9 Jochem, E. (ed.) (2009) *Improving the Efficiency of R&D and the Market Diffusion of Energy Technologies*, Physica-Verlag, Heidelberg.
10 Brand, M., Frey, G., Gross, B., Horst, J., Kimmerle, K., and Leprich, U. (2006) *Analyse und Bewertung von Instrumenten zur Markteinführung stationärer Brennstoffzellen*, Sachverständigenauftrag im Auftrag des Bundesministeriums für Wirtschaft und Technologie, Projekt Nr. 85/05, IZES (Institut für ZukunftsEnergieSysteme gGmbH, Saarbrücken), Saarbrücken, 15 September 2006, http://www.izes.de/cms/upload/pdf/markteinf_brezel_lang.pdf (accessed 13 February 2010).
11 Harborne, P., Hendry, C., and Brown, J.E. (2007) The development and diffusion of radical technological innovation: the role of bus demonstration projects in commercializing fuel cell technology. *Technology Analysis and Strategic Management*, **19** (2), 167–187.
12 Brown, J.E., Hendry, C., and Harborne, P. (2007) An emerging market in fuel cells? Residential combined heat and power in four countries. *Energy Policy*, **35**, 2174–2186.
13 Vasudeva, G. (2009) How national institutions influence technology policies and firms' knowledge-building strategies: a study of fuel cell innovation across industrialized countries. *Research Policy*, **38**, 1248–1259.
14 Yazici, M.S. (2009) Hydrogen and fuel cell activities at UNIDO-ICHET. *Int. J. Hydrogen Energy*, available online 2 June 2009.
15 Carlsson, B. (2006) Internationalization of innovation systems. A survey of the literature. *Research Policy*, **35** (1), 56–67.
16 Kemp, R. and Volpi, M. (2008) The diffusion of clean technologies: a review with suggestions for future diffusion analysis. *J. Cleaner Production*, **16** (1) S1, S14–S21.
17 Geels, F.W. (2002) Technological transition as evolutionary reconfiguration processes: a multi-level perspective and a case-study. *Research Policy*, **31** (8/9), 1257–1274.
18 Geels, F.W. (2005) *Technological Transitions and System Innovations: a Co-evolutionary and Sociotechnical Analysis*, Edward Elgar, Cheltenham.
19 Jacobsson, S. and Bergek, A. (2004) Transforming the energy sector: the evolution of technological systems in renewable energy technology. *Industrial and Corporate Change*, **13** (5), 815–849.
20 Kemp, R., Schot, J., and Hoogma, R. (1998) Regime shifts to sustainability through processes of niche formation: the approach of strategic niche management. *Technology Analysis and Strategic Management*, **10** (2), 175–195
21 Grubb, M., Haj-Hasan, N., and Newbery, D. (2008) Accelerating innovation and

strategic deployment in UK electricity: application to renewable energy. In *Delivering a Low-Carbon Electricity System* (eds M. Grubb, T. Jamasb, and M. Pollitt), Cambridge University Press, Cambridge, pp. 333–360.
22. Satorius, C. and Zundel, S. (eds) (2005) *Time Strategies, Innovation and Environmental Policy*, Edward Elgar, Cheltenham.
23. Foxon, I. and Pearson, P. (2008) Overcoming barriers to innovation and diffusion of cleaner technologies: some features of a sustainable innovation policy. *J. Cleaner Production*, **16** (1), S148–S161.
24. Nemet, G.F. (2009), Demand-pull, technology-push, and government-led incentives for non-incremental technical change. *Research Policy*, **38**, 700–709.
25. Hensley, R., Knupfer, S., and Pinner, D. (2009) Electrifying cars: how three industries will evolve. *McKinsey Quarterly*, June 2009. www.mckinseyquarterly.com (accessed 15 Nov. 1009)
26. Stern, N.H. (2007) *The Economics of Climate Change*, Cambridge University Press, Cambridge.
27. Aki, A. (2009) Research, development, and deployment of fuel cells and hydrogen in Japan. Presented at the FCH JU Stakeholders General Assembly, Brussels, 26–27 October 2009, http://ec.europa.eu/research/fch/index_en.cfm?pg=sga2009_presentations (accessed 18 November 2009).
28. Fuel Cell Commercialization Conference of Japan (2008) Commercialization of Fuel Cell Vehicles and Hydrogen Stations to Commence in 2015. FCCJ, 4 July 2008, http://fccj.jp/pdf/20080704sks1e.pdf (accessed 18 November 2009).
29. Satyapal, S. (2009) DOE fuel cell technologies program. Presented at the FCH JU Stakeholders General Assembly, Brussels, 26–27 October 2009, http://ec.europa.eu/research/fch/index_en.cfm?pg=sga2009_presentations (accessed 18 November 2009).
30. Sun, H.-H. (2008) Policy, Programme and Activities in H2/FC Sector in China. Presented at the International Partnership for the Hydrogen Economy Steering Committee Meeting, Moscow, 22 April 2008, http://www.iphe.net/docs/Resources/Policy_China_4–2008.pdf (accessed 18 November 2009).
31. AMFI Newsletter 2/2007 (2007) China's hydrogen fuel cell vehicles. *IEA Advanced Motor Fuels Implementing Agreement, 2007*, http://virtual.vtt.fi/virtual/amf/news/amfinewsletter2007_2april.pdf (accessed 18 November 2009).
32. Jun, S.-H. (2007) Member State Korea. Presented at the International Partnership for the Hydrogen Economy Steering Committee Meeting, Sao Paulo, 24–25 April 2007, http://www.iphe.net/partners/korea.html (accessed 13 February 2010).
33. European Hydrogen and Fuel Cell Technology Platform (2007) *Implementation Plan – Status 2006*, http://ec.europa.eu/research/fch/pdf/hfp_ip06_final_20apr2007.pdf#view=fit&pagemode=none (accessed 18 November 2009).
34. Fuel Cells and Hydrogen Joint Undertaking (2009) http://ec.europa.eu/research/fch (accessed 18 November 2009).
35. International Energy Agency (2003) *Implementing Agreement Highlights*, 2002–2003 edn, OECD/IEA.
36. International Energy Agency (2005) *Prospects for Hydrogen and Fuel Cells*, OECD/IEA 2003.
37. IPHE (2006) *Terms of Reference*, http://www.iphe.net/docs/terms_of_reference.pdf (accessed 18 November 2009).
38. IPHE (2009) International Partnership for the Hydrogen Economy, *Building the Hydrogen Economy: Enabling Infrastructure Development, Summary of Workshops*, http://www.iphe.net/docs/Resources/IPHE-IEA_Workshops_Summary_Report_FINAL.pdf (accessed 18 November 2009).
39. IPHE (2006) *Terms of Reference for the "IPHE Regulations, Codes and Standards Working Group"*, Internal Document, International Partnership for the Hydrogen Economy.
40. SFC Smart Fuel Cell (2009) *SFC Smart Fuel Cell – Company History*, http://www.sfc.com/en/sfc-history.html (accessed 18 November 2009).

41 Pike Research (2009) *Fuel Cells for Portable Power Applications*, http://www.pikeresearch.com/research/fuel-cells-for-portable-power-applications (accessed 18 November 2009).

42 Fuel Cell Today (2008–2009) www.fuelcelltoday.com (accessed 18 November 2009). Reports can be located with the search option on the Fuel Cell Today homepage:
Calaghan Jerram, L. (2008) *2008 Bus Survey*, http://www.fuelcelltoday.com/online/survey?survey=2008-12%2F2008-Bus-Survey.
Adamson, K.-A. (2008) *2008 Large Stationary Survey*, http://www.fuelcelltoday.com/online/survey?survey=2008-08%2F2008-Large-Stationar y-Survey.
Calaghan Jerram, L. and Dehamma, A. (2009) *2009 Hydrogen Infrastructure Survey*, http://www.fuelcelltoday.com/online/survey?survey=2009-06%2F2009-Infrastructure- Survey-Free.
Butler, J. (2009) *Portable Fuel Cell Survey 2009*, http://www.fuelcelltoday.com/online/survey?survey=2009-04%2F2009-Portable-Survey.
Callaghan Jerrram, L. and Adamson, K.-A. (2009) *2009 Niche Transport Survey*, http://www.fuelcelltoday.com/online/survey?survey=2009-08%2F2009-Niche-Transport.
Adamson, K.-A. (2009) *Small Stationary Survey 2009*, http://www.fuelcelltoday.com/online/survey?survey=2009-03%2FSmall-Stationary–2009.
Callaghan Jerram, L. (2009) *2009 Light Duty Vehicle Survey*, http://www.fuelcelltoday.com/online/survey?survey=2009-05%2F2009-Light-Duty-Vehi cle-Free.
Butler, J., Callaghan Jerram, L., and Dehamma, A. (2009) *2009 Q3 Legislation Review*, http://www.fuelcelltoday.com/online/survey?survey=2009-09/2009-Q3-Legislation.

43 Fuel Cell Today (2009) *Fortnightly Newsletter 11–04-2009*, http://www.fuelcelltoday.com/online/newsletters/view?newsletter=Fortnightly%2F20 09-11-04-Fortnightly-Newslette (accessed 18 November 2009).

44 Kato, K. (2009) Fuel cell marketing in Japan. In *Proceedings of the f-cell Symposium, 28–29 September 2009, Stuttgart*, Peter Sauber Agentur, Stuttgart, www.f-cell.de/files/order_form_documentation_2009_english_1. pdf (accessed 13 February 2010).

45 Ceramic Fuel Cells Limited (2009) *Heinsberg Factory Opening 2 October 2009*, http://www.cfcl.com.au/Heinsberg_Opening/ (accessed 18 November 2009)

46 BoerseGo.de (2008) *Ceramic Fuel Cells kooperiert mit H.C. Starck, um Großlieferungen von Brennstoffzellenkomponenten zu sichern*, 29 January 2008, http://www.boerse-go.de/nachricht/Ceramic-Fuel-Cells-kooperiert-mit-HC-Starck-um-Groslieferungen-von-Brennstoffzellenkom,a772559.html (accessed 18 November 2009).

47 Wind, J. (2009) Elektrifizierung des Automobils. In *Proceedings of the f-cell Symposium, 28–29 September 2009, Stuttgart*, Peter Sauber Agentur, Stuttgart, www.f-cell.de/files/order_form_documentation_2009_english_1.pdf (accessed 13 February 2010).

48 Daimler, Ford, GM/Opel, Honda Hyundai/KIA, the Alliance Renault/Nissan, Toyota (2009) *Letter of Understanding on the Development and Market Introduction of Fuel Cell Vehicles*, 8 September 2009, http://www.hydrogenlink.net/download/LoU-fuelcell-cars.pdf (accessed 18 November 2009).

49 Daimler, EnBW, Linde, OMV, Shell, Total, Vattenfall, NOW (2009) *Initiative "H_2 Mobility" – Major Companies Sign Up to Hydrogen Infrastructure Built-up Plan in Germany*, 10 September 2009, http://www.daimler.com/dccom/0-5-658451-1-1236356-1-0-0-0-0-0-12080-7165-0-0-0-0 -0-0-0.html (accessed 18 November 2009).

50 EnergieRegion.NRW (2009) *Hydrogen – Key to Worldwide Sustainable Energy Economy. Examples from North-Rhine Westphalia from Production to Application*. EnergieAgentur. NRW, Düsseldorf.

29
Hydrogen and Fuel Cells around the Corner – the Role of Regions and Municipalities Towards Commercialization

Andreas Ziolek, Marieke Reijalt, and Thomas Kattenstein

Abstract

New clean energy technologies such as hydrogen and fuel cells are seen as a potential solution for the world wide future energy and transportation needs. Many development and funding programs exist worldwide which vary from region to region. The need for a better coordination of these programs and the preparation of a long-term commercialization strategy is crucial for the effective support and the effective integration for these technologies at a local level. Regional and local organizations have to be involved at an early stage in the development of new funding frameworks and relevant programs for hydrogen and fuel cell technologies, such as HyRaMP.

Keywords: hydrogen, fuel cells, commercialization, regional development, municipal development

29.1
Introduction

There are no doubts about the serious and difficult challenges for the development of hydrogen and fuel cell technologies that have arisen for several reasons over the last few years. The road to commercialization of this emerging technology has faced many ups and downs during this period [1–4]. For example, the EU made a major step forward and launched its first public–private partnership in 2008, the Joint Undertaking for Fuel Cells and Hydrogen (FCH JU), to develop the market for these new energy technologies. At the same time, the new United States government for some reasons cut its support for these technologies. After intensive discussion, this decision has been reversed recently with the strong support of the United States House of Representatives.

The continuing, from a technical point of view dispensable, discussions on the role of electric mobility and in particular on the competition between battery storage and fuel cell systems have caused serious uncertainties in particular regarding policy and customer perspectives. Recently announced multi-billion

Hydrogen Energy. Edited by Detlef Stolten
Copyright © 2010 WILEY-VCH Verlag GmbH & Co. KGaA, Weinheim
ISBN: 978-3-527-32711-9

government support programs in response to the world economic crisis have favored green car development in Europe and the United States, putting a clear focus, however, on battery-powered cars. As a response, the world's largest car manufacturers sent out a communication in September 2009 that they are preparing the market role for fuel cell vehicles starting in 2015, as part of their development of electric vehicles in addition to battery and hybrid cars (in that communication, they also requested a clear commitment of the hydrogen production and distribution industry to build a sustainable hydrogen refueling infrastructure) [5]. In the meantime, a broad common understanding has been achieved that battery electric vehicles and fuel cell electric vehicles are comprehensive technologies with many synergies that need a coordinated and joint development. Industry has already aligned development and funding programs for both technologies in addition to many public funding programs. From today's perspective, clean transport programs including battery and fuel cell technologies appear even stronger and clearer than before this discussion.

However, the still lasting international economic crisis affecting in particular development and investment programs for new energy technologies has also caused serious difficulties. Thus, several stakeholders have reduced or postponed their financial involvement in future energy technology developments such as hydrogen and fuel cell technology [6–10]. Fortunately, recent announcements and activities clearly indicate that several key partners are already increasing their efforts again or are planning to do so soon, and are still committed to this technology. Hydrogen and fuel cell technologies are already approaching commercialization in several other applications such as material handling and uninterrupted power supply technologies and thousands of fuel cell systems were shipped in 2009 [11, 12].

Even if the hydrogen and fuel cell technology has "survived" recent challenges and seems to be back on the right track towards commercialization, another key challenge still needs to be tackled, namely the immediate need for a better coordination of development and funding programs worldwide and in particular in Europe, and the preparation of a long-term commercialization strategy. In this context, the role of regions and municipalities has become more and more important for many reasons. The increasing number of regional and local hydrogen and fuel cell development and demonstration initiatives, an extract of which is presented in Session SA.7 on Regional Activities, points to the crucial role of regions and municipalities in facilitating the development of these first markets, involving local industries and providing a first consumer base, as acknowledged, for instance, by the European Commission [13]. The several initiatives around the world show the need to establish more immediate coordination to ensure leveraging of different funding levels, effective joint procurement schemes, and facilitating regulatory frameworks to support an environmentally and economically sound development of hydrogen as a clean energy carrier at a local level. Concrete action is needed in the following areas:

1. in-depth reviews of current and potential research and development budgets at a local level to leverage EU, national, and local funding;

2. increased efforts to synchronize multi-level budget cycles and topics;
3. development of comprehensive joint procurement schemes at national and international levels that facilitate access of local actors to global markets;
4. adaptation and harmonization of local authorization requirements for hydrogen applications;
5. identification of key local stakeholders that could be the frontrunners of new industrial value chains;
6. integrated approach to environmentally and economically sustainable development of electric transport at a local level with regard to ensuring the efficient use of primary energy to power battery vehicles or produce hydrogen for use in fuel cell vehicles.

Those actions will only be effective if regional and local organizations active in clean energy technology deployment are involved at an early stage in the development of new funding frameworks and relevant programs at national, EU, and international level. This chapter describes the potential roles of regional and local activities in conjunction with national or multi-national programs and also possible tools and mechanisms which may be used or developed to facilitate regional and local strategies towards the commercialization of hydrogen and fuel cell technologies. The intended "multi-level" collaboration towards a unique strategy for commercialization of fuel cells and hydrogen technologies is still at a very early stage worldwide and several obstacles need to be removed and existing challenges need to be tackled. In 2008, a promising approach to facilitate this kind of collaboration was made with the establishment of the European Regions and Municipalities Partnership for Hydrogen and Fuel Cells (HyRaMP) This initiative is described here as an example of a catalyst of regional collaboration demonstrating the huge potential of involving these strategic partners in the overall approach.

29.2
The Role of Regional and Local Activities

Regions and municipalities are facing different jurisdictions and responsibilities in terms of the development and the commercialization of new technologies. Due to diverse motivations or political or economic objectives and depending on the current development status of the particular technology, local or regional stakeholders are able to implement or deploy adequate mechanisms and tools to support and to foster the development and market introduction of new technologies such as hydrogen and fuel cells. Local or regional stakeholders have access to different instruments and tools to pave the way for new technology initiatives, such as

- local or regional financed *research and development programs* – often designed and drafted in close collaboration with national or multi-national programs in a comprehensive manner;

- *public procurement programs* with special focus on a support of particular new technologies;
- design of a comprehensive *legal framework* to support the deployment of a new technology in certain applications or in a particular area;
- inauguration of *market incentives* to support financially additional costs for new technologies – although this tool is organized and designed at national or at least multi-regional level.

The first three instruments, especially relevant for regions, are described in detail in the following.

29.2.1
Research and Development Programs

Research and development programs by nature support the development process of new technologies in an early stage and look at a mid- to long-term timeframe. They are especially dedicated to support local partners and stakeholders during of the development and to impact the technical development and to include particular interests or topics. Local or regional financed research and development programs are usually designed in a comprehensive concept with national or other regional framework programs and address a particular interest or certain strength of a region or city in conjunction with the overall development.

In Germany, for instance, most federal states have developed their own research and development programs for hydrogen and fuel cell technology [14]. In order to facilitate close collaboration and intensive communication between these different programs and to use public financial support as efficiently as possible, a Federal State Working Group on Hydrogen and Fuel Cell Technology has been established. The Chair of this working group is also member of the Board of the National Program Committee. Thus, an effective and close coordination between the various German federal state programs and the German National Innovation Program for Hydrogen and Fuel Cell Technology is assured. The total annual budget of all German federal state programs is approximately 35 million and therefore corresponds to half of the annual budget of the German federal program [14].

Also in Spain there is close collaboration between engaged Spanish regions such as Aragon and Catalonia and the National Hydrogen and Fuel Cell Association [15]. In other cases, a common approach and close collaboration of different regions or municipalities is intended to compensate for the lack of a national program. This is valid, for instance, in Italy, where intensive efforts towards the establishment of a national program for hydrogen and fuel cell technology are in place but a concrete program is still lacking [16]. However, northern Italian regions such as Piedmont, Lombardy and Bolzano have a very close collaboration. For example, the region of Piedmont intends to invest nearly 100 million in hydrogen and fuel cell technology from 2007 to 2013 [16].

A similar engagement is expected from other Italian industry clusters such as Lombardy and Bolzano.

In France, significant efforts and programs towards hydrogen and fuel cell technology are also in place. So far, the national program looks like a research program with some industrial aspects relating to market development and demonstration [17]. More industry-oriented programs can be found in some French departments such as Rhône-Alpes and Isère [18]. Also, significant growth of several regional activities at state level in the United States and at province level in Canada has been remarkable within the last few years. Even the relation between federal and regional programs and financial support has changed significantly towards a much stronger role of states and provinces. Again, mainly the industrialized regions are the driving forces. As examples, Ontario in Canada and Ohio, Pennsylvania, and South Carolina in the United States should be mentioned as representing many other initiatives in North America [19–22].

Today, a good proportion of the public financial support for hydrogen and fuel cell development is still allocated and spent on research and development activities and programs worldwide. This correctly reflects the current state of development of the technology. For instance, in Europe there is almost no comprehensive program in place approaching the next step towards commercialization. However, as the technology is making serious progress many stakeholders are starting to look for programs beyond the research and development phase.

However, local or regional research and development programs are designed to promote or to involve local partners and to strengthen the local research community and local industry. The key objective is to support local stakeholders to find their role within the international competition and to secure the participation of local partners in emerging markets and business opportunities. That is why regional public funding programs often accompany a necessary change in important industry sectors (in the case of hydrogen and fuel cell technology, the relevant industry branches are the energy sectors and the automotive industry, including all involved suppliers and component manufacturers). Regional research and development programs are often initiated and operated by highly industrialized economic hotspots facing serious changes to the the existing economy. Thereby, the financial budgets of some European regional programs are even higher than those of smaller nations such as Denmark and The Netherlands.

29.2.2
Regional and Local Procurement Programs

Regional and local public procurement programs with a special focus on new technology are often a very effective tool to promote new technologies towards market introduction. They are usually limited in terms of numbers to a small set of demonstrators. However, this phase could be considered as steps one (deployment and evaluation steps with feedback into research, development, and demonstration to prove technological readiness) and two (development of production technologies for volume manufacturing) within the "five-step S-form diffusion curve" of technology commercialization [23]. According to this theory, procurement programs follow an intensive research and development phase. In practice, both phases are often mixed up and sometimes it is a challenge to distinguish between the two steps.

At an early stage of market introduction, intensive tests and trials are needed to prove the state of development and maturity of a new technology to be commercialized. As problems and mistakes are very likely and almost unavoidable, this step needs to be done in a "protected area." Public entities acting as first customers and being prepared to handle such problems and difficulties could play an important role in this context. As such public projects for the demonstration of new technologies used to be prepared with the close involvement of local public authorities, and success of the installation or application of new equipment is in their own interest, the need for patience and continuous support can be recognized. This valuable practical support of local or regional authorities, and their openness towards new technologies based on a clear idea of the role of new technologies within a regional development concept, are most important for a successful demonstration in addition to significant financial support. The concerted procurement of a limited number of pre-serial prototypes and a trial program within an every-day operation in close collaboration with the manufacturer in a protected and manageable framework is therefore considered to be a nearly "perfect" opportunity to prepare a new technology for the next step of commercialization.

The European HyFleet: CUTE project is recognized worldwide as an excellent example of successful public procurement initiatives and in addition for the successful utilization of regional, national, and multi-national funds. In particular, the unique and strong engagement of the public transport agency in Hamburg, Hamburger Hochbahn AG, should be mentioned. In addition to the significant financial involvement regarding investment into a fleet of fuel cell transit buses with the support of the German Federal Government and the European Commission, the comprehensive and intensive involvement of Hamburger Hochbahn in the maintenance and service program of the buses over the last few years is most important and probably a guarantee for the successful operation. Today, Hamburger Hochbahn is still running the largest fleet of fuel cell buses worldwide and even for a much longer duration than predicted at the beginning of the project [24]. Many other municipalities and cities, such as London, Oslo, and Bolzano, used the Hamburg approach as an archetype for their own projects and programs. The current activities of the Danish capital city Copenhagen with regard to the support and promotion of hydrogen and fuel cell technology is certainly an interesting example of a targeting public procurement program.

Driven by the overall objective of Copenhagen to become the world's first carbon-neutral capital city by 2025 [25], the Danish capital has initiated a first public procurement program for eight fuel cell-powered vehicles to be used by public authorities apart from a fleet of battery electric vehicles. The new hydrogen vehicles and the filling station were delivered and put into operation in early December 2009. Further vehicles will follow soon as the whole public fleet in Copenhagen, from a bicycle to a waste collection truck, should be operated emission free by 2025. With this approach, Copenhagen is setting a clear framework for the upcoming public procurement and is taking responsibility regarding the financial support of new technologies.

In a similar manner, procurement programs for stationary fuel cell systems have been established in particular in Japan, for instance in Fukuoka Prefecture [26]. This, in conjunction with the Japanese central government designed and financed program called the Hydrogen City Fukuoka Project, is part of a national approach to commercializing stationary fuel cells as decentralized combined heat and power (CHP) units across Japan and foresees the installation of a total of 150 units in the Fukuoka City by 2015. The CHP systems are already or will soon be installed in both public and private buildings at costs similar to those of commercial heating systems.

Recent experience indicates that public procurement programs for advanced or new technologies are often applied by communities or regions that are keen to promote themselves as a technology-oriented habitat rather than seeking the support of local technology manufacturers. Hence public procurement programs used to (and maybe need to) involve technology from several suppliers around the world. Public procurement is therefore not necessarily an adequate instrument to support local industry or technology. Possible motivations to apply such tools include climate protection objectives like the example that Copenhagen is showing or an intention to promote a certain location with advanced technology in conjunction with large public sport or culture events such as the 2006 FIFA World Soccer Championship in Germany, the 2008 Olympic Games in Beijing, the 2008 World Expo in Zaragoza, the 2010 Winter Olympic Games in Vancouver and the 2010 World Expo in Shanghai. All these events have or will display and promote hydrogen and fuel cell technologies. For example the public transport company in British Columbia (BC Transit) will launch a fleet of 20 new fuel cell buses in Whistler [27] and during the 2010 World Expo in Shanghai six fuel cell buses, 90 fuel cell passenger cars and 100 fuel cell sightseeing cars will be deployed [28]. Many programs or projects that have been initiated or launched in conjunction with major public events are still running or have been followed by new projects after the events.

29.2.3
Basic Regulatory Instruments

Basic regulatory instruments such as access restriction by congestion charge and environmental zones with priority for climate-protecting vehicles and with fees for polluting are also options to promote new efficient and environmentally friendly technologies. Well-known examples are the Zero Emission Vehicle Regulation in California in the United States and the Low Emission Zone Regulation in the London area in the United Kingdom [29, 30].

These instruments are very efficient with regard to the support of clean technologies. The government of California made a globally noteworthy decision to force the entry of climate-protecting vehicles in California, which led to significant global development efforts by the automotive industry. Various experts think that the environmental legislation in California is ultimately responsible for the beginning of the market development of fuel cell cars.

In addition to the legal framework, a progressive approval policy is also a very effective option to promote the distribution of new technologies. A positive standpoint of the approval authorities towards hydrogen cars and a hydrogen infrastructure, for example, helps considerably the development and market launch of these techniques. Many global regions and municipalities have already gained know-how and adjusted their basic regulatory framework to the new technologies. The same applies to the fiscal assessment of hydrogen as a fuel or the general classification of hydrogen as a fuel with the consequences for handling and usage.

Japan is a positive example where structured and comprehensive work has been carried out and experience gained in terms of standardization and approval at different locations with a strong cooperation between the respective prefectures.

29.3
HyRaMP – Organizing Local and Regional Drivers in Europe

The SA.7 session on Regional Activities at the upcoming 18th World Hydrogen Energy Conference 2010 in Essen, Germany, will present a small selection of significant initiatives and players around the world acting on behalf of municipalities, agglomerations, and regions. There is no doubt that these powerful partners need to be involved in the upcoming challenges and tasks to commercialize hydrogen and fuel cell technologies.

However, doing so is as challenging as commercializing new technologies. As an example, there are more than 200 regions and municipalities across Europe. How do they communicate or even contribute to a common process? How can joint planning or joint programming be managed?

The European Municipality and Regions Partnership for Hydrogen and Fuel Cells (HyRaMP) was established in April 2008 by 22 European Regions with the support of the European Commission to promote the regional development and deployment of fuel cell and hydrogen technologies in close collaboration and alignment with the EC Joint Undertaking on Hydrogen and Fuel Cells (FCH JU) and to become a powerful platform to manage the communication and to organize the involvement of local and regional strength into the European approach. Europe's public–private partnership FCH JU believes that the regions have a significant role in large-scale demonstration projects to show the economic viability of these technologies. To support the EU and FCH JU, the regions need to work together to develop supply chains that can compete with the best in the world.

HyRaMP's objective is to foster the adoption of fuel cell and hydrogen technologies in Europe. HyRaMP works with its members and partners in the EC and FCH JU to contribute to reductions in carbon dioxide emissions and improvements in environmental protection in addition to economic and jobs growth. It will also work to introduce novel technologies, to harmonize initiatives/activities, and to establish new opportunities in an environmentally friendly and sustainable manner.

HyRaMP's mission is to provide the European regions and municipalities with a representative body that is coherent, distinguishable, and influential. It will work with the European FCH JU, relevant stakeholders, and decision makers from the public and private sectors to make the European fuel cells and hydrogen vision a reality. The Partnership allows its members to play a key role in the development and implementation of strategies to assist the uptake of these new technologies.

Today, more than 30 European regions and municipalities are members of HyRaMP. They have founded a legal entity to act on their behalf and represent their interest towards the European stakeholders. HyRaMP members are already making ambitious efforts in hydrogen and fuel cell development and have clearly understood their role and responsibilities. Together they are managing a total annual R&D budget of about 35 million that will increase to 70 million by 2015. This is equal to the average EC contribution to the FCH JU program. Apart from significant financial resources, HyRaMP board members are proactively preparing HyRaMP's contribution to a joint planning of the FCH JU annual program.

Furthermore, HyRaMP is already acting as a project partner within the FCH JU program on behalf of its members. For instance, together with the international Hydrogen Bus Alliance (HBA) and several other partners, HyRaMP is actively contributing towards a long-term strategy to deploy and operate hydrogen buses across Europe. The HyRaMP members involved will contribute to finance the procurement of buses and infrastructure and maintain the bus fleets by their local service staff. However, even more important – HyRaMP is facilitating a step-by-step approach to map the next steps for bus deployment across Europe.

29.4
Conclusion

Recent political and industrial developments indicate that there will not be just one quick resolution to our future energy and transport needs, but rather a set of comprehensive technology solutions. The EU is currently setting up six new public–private partnerships, such as the FCH JU, called European Industrial Initiatives under the EU Strategic Energy Technology Plan (SET Plan), to support the market development of bio-energy, wind, solar, carbon capture, and sequestration and fission energy, and smart grid applications to achieve the EU's decarbonization objectives by 2050. In addition, it is creating a Smart Cities initiative to support local frontrunners to implement these technologies in an accelerated manner and become models for followers. The United States government has approved significant budget increases for renewable energy and smart grids. China is increasingly looking at Carbon Capture and Sequestration (CCS) and is collaborating with the EU to advance rapid implementation further by adapting regulatory frameworks. However, in all initiatives and future strategy plans, the role of hydrogen and fuel cells is clear – hydrogen is seen as a brilliant tool for large-scale and mid- to long-term storage of surplus renewable energy, and fuel cells are the most efficient energy conversion technology. Both components will play a crucial role in future energy systems.

The implementation of all these technologies is facing similar challenges to today's hydrogen and fuel cell initiatives that have been outlined in this chapter. Therefore, multi-level coordination is needed not only to ensure effective support for specific technologies but also to facilitate the effective integration of all clean energy technologies at a local level. The involvement and input of local decision makers at an early stage while defining future energy and transport policy are crucial to allow an accelerated uptake of these technologies by leveraging research and development funding, broadening joint procurement schemes and lowering regulatory barriers.

The development of a new transport strategy for Europe for 2010–2020 as outlined in the EU's recent communication on Sustainable Transport for Europe is another opportunity to include the need for more coordination in the development of the local recharging and refilling infrastructure for electric transport applications, including fuel cell vehicles [31]. At a global level, innovative new technologies such as hydrogen and fuel cells need to become part of the development of a comprehensive technology transfer framework, as discussed at the last UN Climate Change Conference, to enable local industries to broaden their markets to important emerging economies [32].

With regard to hydrogen and fuel cell developments in Europe, well-established organizations such as HyRaMP can offer insight into and support of the development of effective new funding facilities of financial institutions such as the European Investment Bank, and also the work programs of the FCH JU.

References

1 Kurpjuweit, K. (2009) *Versuch mit Wasserstoffbussen verpufft. Das Wasserstoffprojekt bei der BVG steht vor dem Scheitern*, art 270, 2746512, http://www.tagesspiegel.de/berlin/BVG-Wasserstoffbusse (accessed 22 February 2010).

2 Markhoff, J. (2009) New battery for electric cars may go the extra distance, *New York Times*, 21 September 2009, p. 4.

3 Madslien, J. (2009) Car firms disagree about electric future, http://news.bbc.co.uk/go/pr/fr/-/2/hi/business/8260722.stm (accessed 22 February 2010).

4 Wilson, J. (2009) California plans no exit from hydrogen highway, http://www.grist.org/article/index/2009-05-14-california-hydrogen-fuel-cell/PALL/, 14 May 2009.

5 Nationale Organisation Wasserstoff- und Brennzellentechnologie (NOW GmbH) (2009) Initiative "H2 Mobility" – Major companies sign up to hydrogen infrastructure built-up plan for Germany, http://www.now-gmbh.de/index.php?id=209&L=1 (accessed 2 February 2010).

6 Green Car Congress (2009) $26.4B DOE FY 2010 budget request cuts funding for hydrogen fuel cell vehicles; with recovery act funding boosts support for PHEVs, biomass and biorefineries, http://www.ecosilly.com/2009/05/07/264b-doe-fy-2010-budget-request-cuts-funding-for-hydrogen-fuel-cell-vehicles-with-recovery-act-funding-boosts-support-for-phevs-biomass-and-biorefineries/ (accessed 22 February 2010).

7 Shepardson, D. (2009) House approves energy bill that boosts vehicle research funding, http://www.detnews.com/article/20090717/AUTO01/907170437/1148/rss25, Detroit News Washington Bureau, 17 July 2009.

8 van der Zee, B. (2009) Has electric killed the hydrogen car? Have manufacturers looking for a green option decided that hydrogen is just too niche and that electricity is a better bet?, http://www.guardian.co.uk/environment/green-living-blog/2009/sep/16/electric-killed-hydrogen-car, 18 September 2009.

9 Bergi, T. (2009) Shell goes cold on wind, solar, hydrogen energy, Thomson Reuters, http://www.reuters.com/article/environmentNews/idUSTRE52G4SU20090317 (accessed 22 February 2010).

10 HyWeb (2009) Turmoil over proposed US budget cuts for hydrogen and fuel cells – newly appointed US Energy Secretary, Steven Chu, entered the DOE command bridge with much verve, announcing drastic budget cuts for hydrogen and fuel cell (H2/FC) technologies, 20 May 2009.

11 Satyapal, S. (2009) Overview of hydrogen and fuel cell activities. Presented at the Fuel Cells and Hydrogen Joint Undertaking Stakeholders General Assembly, Brussels, 27 October 2009.

12 Myers, C. (2009) Overview of U.S. hydrogen and fuel cell activities. Presented at the 9th Annual Meeting of the Fuel Cell and Hydrogen Network North Rhine-Westphalia, Düsseldorf, 10 December 2009.

13 Barbaso, F. (2009) The politics of clean energy: fuel cells and hydrogen on the road to Copenhagen, Keynote Speech, Fuel Cells and Hydrogen Joint Undertaking Stakeholders General Assembly, Brussels, 27 October 2009; http://ec.europa.eu/research/fch/pdf/barbaso.pdf (accessed 22 February 2010).

14 Baues, H. (2009) Hydrogen and fuel cell activities of the German Federal States – a survey on recent activities. Presented at the 2nd HyRaMP General Assembly, Brussels, October 2009.

15 Correas, L. (2008) Experiences in Activities to raise Awareness: Expo 2008 in Zaragoza. Presented at the 8th Annual Meeting of the Fuel Cell and Hydrogen Network North Rhine-Westphalia, Düsseldorf, 21 November 2008.

16 Damosso, D. (2009) Hydrogen and fuel cell activities in Italy: national and regional perspectives. Presented at the 9th Annual Meeting of the Fuel Cell and Hydrogen Network North Rhine-Westphalia, Düsseldorf, 10 December 2009.

17 Frois, B. (2009) Hydrogen and fuel cell activities in France. Presented at the Fuel Cells and Hydrogen Joint Undertaking Stakeholders General Assembly, Brussels, 27 October 2009.

18 Biasotto, J., Maio, P., Gravelier, M., Altmann, M., Gaertner, S., and Wurster, R. (2009) Development and implementation of the French Region Rhône-Alpes's hydrogen and fuel cell strategy. Abstract for 18th World Hydrogen Energy Conference 2010, Essen, May 2010.

19 Ontario Ministry of Research & Innovation (2007) *The Ontario Fuel Cell Innovation Program*, http://www.mri.gov.on.ca/english/news/Green082307_bd1.asp (accessed 22 February 2010).

20 Department of Development, Ohio State Government (2009) *Overview on Ohio Fuel Cell Program*, http://www.emtec.org/ae/fuelcells/hydrogen/index.php (accessed 22 February 2010).

21 Wast Penn Power Sustainable Energy Fund (WPPSEF) (2005) *Fuel Cells*, http://www.wppsef.org/fuelcells.html (accessed 22 February 2010).
22 Baxter-Clemmons, S. and Daetwyler, C. (2009) Hydrogen and fuel cells in South Carolina, USA, Abstract for 18th World Hydrogen Energy Conference 2010, Essen May 2010.
23 Rodgers, M.E. (1962) The Diffusion of Innovations, The Free Press, New York.
24 Klingenberg, H. (2009) Improving cooperation between regional programs and the FCH Joint Undertaking: Hamburg, Presented at the Fuel Cells and Hydrogen Joint Undertaking Stakeholders General Assembly, Brussels, 27 October 2009.
25 Williams, A. (2009) Copenhagen aims to be carbon neutral by 2025 (APF), http://hostednews/afp/article/ALeqM5jt1eqr-AWQqO3_2R2D5PMrP2c50g, 19 March 2009.
26 Watanabe, S. (2009) Hy-Life Project. Fukuoka's challenges towards a hydrogen society, Presented at the 9th Annual Meeting of the Fuel Cell and Hydrogen Network North Rhine-Westphalia, Düsseldorf, 10 December 2009.
27 Burelle, C. (2009) Country update Canada. Presented at International Partnership for the Hydrogen Economy, Joint SC/ILC Meeting, Washington, DC, 2 December 2009.
28 Cai, J.N. and Pan, X.M. (2009) Chinese hydrogen update. Presented at International Partnership for the Hydrogen Economy, Joint SC/ILC Meeting, Washington, DC, USA. 2 December 2009.
29 Union of Concerned Scientists (2009) California's Zero Emission Vehicle (ZEV) Program, http://www.ucsusa.org/clean_vehicles/solutions/advanced_vehicles_and_fuels/californias-zero-emission–2.html (accessed 30 January 2009).
30 Transport for London (2009) *The Low Emission Zone is Now in Operation*, information leaflet, http://www.tfl.gov.uk/assets/downloads/roadusers/lez/LEZ/lez-information-leaflet.pdf (accessed 22 February 2010).
31 European commission – Mobility & Transport (2009) European strategies, future of transport, http://ec.europa.eu/transport/strategies/2009_future_of_transport_en.htm (accessed 22 February 2010).
32 United Nations Framework Convention on Climate Change (UNFCCC) (2010) Technology Transfer Clearing House, Technology in Copenhagen, http://unfccc.int/ttclear/jsp/index.jsp (accessed 22 February 2010).

30
Zero Regio: Recent Experience with Hydrogen Vehicles and Refueling Infrastructure

Heinrich Lienkamp and Ashok Rastogi

Abstract

The project Zero Regio, co-financed by the European Commission within the 6th Framework Program, aims at demonstrating hydrogen infrastructure and fuel cell passenger vehicles in European cities. Demonstration activities have taken place in Germany and Italy. Hydrogen from two different sources – chemical by-product in Germany and natural gas reforming in Italy – was employed in fleet demonstration. Five Mercedes-Benz A-Class F-CELL vehicles, including one vehicle with 700 bar storage, have been tested in Frankfurt and three fuel cell vehicles (Panda) from Fiat have been demonstrated in Italy. Both fleets have gone through real-life driving cycles over a long period (3 years ending in November 2009). Experience with this demonstration is presented. Evaluation of data collected and analyzed during the demonstration is presented, providing important information on the performance, availability, maintenance requirements, and consumption of the FCVs tested in comparison with conventional vehicles. Some socioeconomic aspects of the fleet demonstration and the dissemination activities are also presented. Experience with refueling the FCVs is presented. At both sites, Frankfurt in Germany and Mantova in Italy, public multi-energy service stations have been built within the project. Significant characteristics of hydrogen infrastructure at these stations, such as high-pressure pipeline transport, on site production, compression schemes, precooling of hydrogen, dispenser designs, and communication between vehicle and dispenser, with regard to refueling of compressed hydrogen (at 350 and 700 bar) is briefly described. Based on the demonstration experience, areas in both fuel cell vehicles and hydrogen infrastructure where further development is necessary are highlighted. Future developments regarding the filling station and the growing demand for hydrogen and filling stations are also discussed.

Keywords: Zero Regio, by-product hydrogen, hydrogen infrastructure, 700 bar refueling, infrared communication, fuel cell vehicles, socioeconomics

Hydrogen Energy. Edited by Detlef Stolten
Copyright © 2010 WILEY-VCH Verlag GmbH & Co. KGaA, Weinheim
ISBN: 978-3-527-32711-9

30.1
Introduction

A number of demonstration projects on hydrogen and fuel cell technology in urban transport are in progress worldwide. In Europe, the project Zero Regio, partly sponsored by the European Commission, deals with low-duty fuel cell vehicles and the relevant refueling technology in two cities, Frankfurt in Germany and Mantova in Italy (Figure 30.1). The project has the overall objective of developing and testing low-emission urban transport systems based on hydrogen as an alternative motor fuel. The project addresses the complete well-to-wheel chain of hydrogen at the two demonstration sites. In Frankfurt, hydrogen produced as a by-product in a chemical plant is used, whereas in Italy, hydrogen is produced from natural gas in an on-site reformer. Infrastructure systems for transport, compression, storage, and distribution of hydrogen have been developed at both sites. Pioneering fuel cell vehicle fleets (A-Class from Mercedes-Benz in Germany and Panda from Fiat in Italy) have been operated over a period of 3 years under real-life conditions. An associated research program to evaluate the data and experience gathered from the demonstration of the infrastruc-

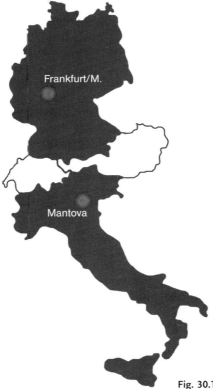

Fig. 30.1 Demonstration sites Frankfurt and Mantova.

ture and the vehicle fleets has been carried out. The Zero Regio project consortium consists of 16 partners from four European countries. The project is coordinated and managed by the present authors.

This chapter focuses on the work performed and the results obtained at the two sites of the Zero Regio project. Hydrogen infrastructure is described in the next two sections. Significant components of the hydrogen infrastructure, namely production, storage, compression, transport to refueling station, and the refueling dispensers, are described briefly. Fuel cell vehicle fleets, their demonstration, and the evaluation of results are described in the subsequent section. Some comments are provided on socioeconomic studies performed in the project, followed by the dissemination activities undertaken. The chapter concludes with a discussion of future prospects.

30.2
Hydrogen Production and Quality

Three different pathways have been employed in Zero Regio to obtain hydrogen for refueling the passenger vehicles, and these are described. The quality of the hydrogen obtained is compared with the quality standards required by the fuel cell vehicles.

30.2.1
By-Product Hydrogen

Figure 30.2 shows the facility at Infraserv receiving the by-product hydrogen, collecting it in a gasometer of $10\,000\,\mathrm{m}^3$ volume, and compressing it to two pressures, 7 and 254 bar. By-product hydrogen is produced in a chlorine-soda electrolysis plant electrolyzing sodium chloride on the basis of the amalgam process using Hg as cathode. The hydrogen stream is treated in washers, driers, and active coal filters in the chlorine plant before feeding it to the gasometer. Hydrogen is further conditioned at Infraserv as shown in Figure 30.2 before feeding it to medium- and high-pressure networks. At the location marked with Q in Figure 30.2, the gas quality is monitored continuously. After the 254 bar compressor and the particle filter, hydrogen is stored in bottles and buffer storage and is available for trailer truck transport. This gas is transported via a high-pressure pipeline, as described in the next section, to the hydrogen refueling station for fuel cell passenger vehicles, constructed within Zero Regio.

Table 30.1 shows the hydrogen quality measured at the trailer station. Listed also are values for safe operation of fuel cell vehicles specified in SAE J 2719 [1]. All values measured at the trailer station meet the vehicle requirements, which are in fact specific to polymer electrolyte membrane (PEM) fuel cell operation. Hydrogen quality with particular reference to components such as Hg and moisture content is also relevant for the safety considerations of other components in the infrastructure (valves, storage, pipeline, etc.) and vehicle (storage tank).

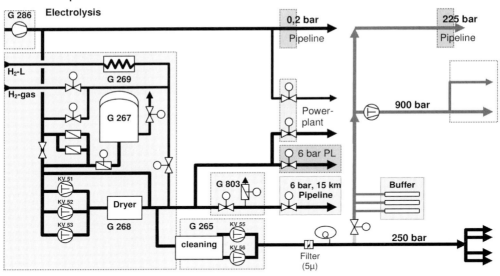

Fig. 30.2 Hydrogen center and hydrogen network at Infraserv Höchst (© Infraserv GmbH & Co. Höchst KG).

Table 30.1 Hydrogen quality standards and measured values at the demonstration sites [a] (source: Infraserv).

No.	Component	Measured values			SAE J2719 specification	OEM (car) specification
		At Infraserv (by-product)	At Sapio (industrial H_2)	At Eni (on-site reformer)		
	Grade	>99.99%	99.999%	99.995%	>99.99%	>99.98%
1	CO	0.5 ppm	<0.1 ppm	<0.1 ppm	<0.2 ppm	<1 ppm
2	CO_2	0.44 ppm	<0.1 ppm	<0.07 ppm	<1 ppm	<1 ppm
3	S compound	n.p.	n.p.	<0.004 ppm	<0.004 ppm	<0.01 ppm
4	THC (C_nH_m)	0.3 ppm	<0.1 ppm	<0.02 ppm (CH_4)	<2 ppm	<1 ppm
5	O_2	0.02 ppm	n.p.	n.m.	<5 ppm	<5 ppm
6	NH_3	n.p.	n.p.	n.p.	<0.1 ppm	<6 ppm
7	N_2, Ar, He	N_2 67 ppm	<5, 0, 15 ppm	<10, 0.1, 35 ppm	<100 ppm	<200 ppm
8	H_2O (gas and liquid)	0.74 ppm	<5 ppm	n.m.	<5 ppm	<5 ppm
9	Na^+	<0.01 ppm	n.p.	n.p.	<0.05 ppm	<0.05 ppm
10	K^+	<0.01 ppm	n.p.	n.p.	<0.05 ppm	<0.08 ppm
11	Formaldehyde (HCHO)	n.p.	n.p.	n.p.	<0.01 ppm	n.s.
12	Formic acid (HCOOH)	n.p.	n.p.	n.p.	<0.2 ppm	n.s.
13	Halogenates	n.p.	n.p.	n.p.	<0.05 ppm	n.s.
14	Hg	<0.007 ppb	n.p.	n.p.	n.s.	n.s.
15	Particle	n.m.	n.m.	n.m.	<1 µg m l^{-1}	<10 ppmw
16	Particle size	5 µm filter	n.m.	n.m.	<10 µm	<10 µm

a) ppm=ppm (vol.), ppb=ppb (vol.); all values under normal conditions (NTP); n.p., not present; n.m., not measured; n.s., not specified.

30.2.2
Industrial Hydrogen

During the initial phase of fleet demonstration in Italy, hydrogen produced in an industrial steam reforming plant of Sapio was transported by truck in cylinder packs (200 bar) to the refueling station. A storage room has been built at the service station in Mantova to store this hydrogen. Under normal circumstances, hydrogen produced on-site (see the next section) is used. Industrial hydrogen represents a back-up source for emergency situations or when the on-site plant is not in operation. Table 30.1 shows the quality of industrial hydrogen supplied by Sapio.

30.2.3
Hydrogen Produced On-Site

The on-site hydrogen production plant is based on the Short Contact Time – Catalytic Partial Oxidation (SCT-CPO) technology of Eni for the conversion of natural gas, together with air and steam, into syngas. Syngas is then further processed inside a high-temperature water gas shift reactor (HTS). Hydrogen is purified from the syngas stream inside a pressure swing adsorption unit (PSA). The plant scheme is shown in Figure 30.3.

The plant has been skid mounted (Figure 30.4) and the main skids are located inside a bunker construction. Each process skid is equipped with wheels and the whole assembly is mounted over tracks in order to have the possibility of extracting it from the bunker for maintenance. The compact design of the whole assembly (7 m length, 1.9 m width and 2.7 m height) represents one of the main features of the hydrogen production plant. The plant is designed to

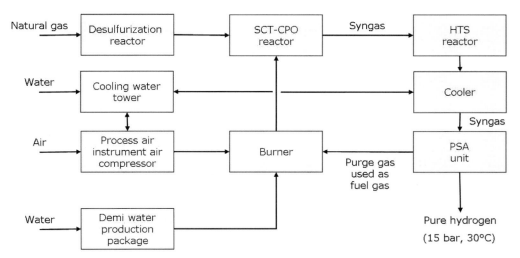

Fig. 30.3 On-site reformer scheme (source: Eni S.p.A.).

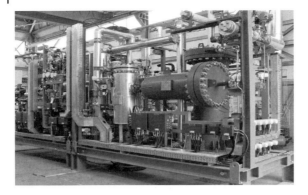

Fig. 30.4 On-site reformer plant on skid (source: Eni S.p.A.).

produce 20 $Nm^3\ h^{-1}$ of hydrogen flow with a possibility of a 60% turndown ratio. In line with the demand for hydrogen by the demonstration fleet, the plant has been optimized also to be operated intermittently in batch mode. The high quality of the hydrogen produced is shown in Table 30.1.

30.3
Refueling Infrastructure

Matching the above hydrogen sources and the already existing infrastructure at the two demonstration sites, different refueling systems have been developed. These are described here, including compression equipment and the transport infrastructure.

30.3.1
Multi-Fuel Public Service Stations

Both in Frankfurt and Mantova, public service stations have been built selling the usual fossil transport fuels in addition to hydrogen. Hydrogen refueling systems have been integrated in the stations. Figures 30.5 and 30.6 provide an overall impression of these service stations.

30.3.2
Refueling CGH2

In Germany, at the hydrogen center of Infraserv Höchst (see also Figure 30.2), an ionic liquid compressor developed and engineered by Linde is employed to compress hydrogen from 225 to 900 bar. Hydrogen is transported to the Agip service station via a high-pressure pipeline designed for 1000 bar. The pipeline design and layout are shown schematically in Figure 30.7.

Fig. 30.5 Agip station in Frankfurt (© Infraserv GmbH & Co. Höchst KG).

Fig. 30.6 Agip station in Mantova (source: Mantova Town Hall).

This pipeline feeds two Linde refueling dispensers, one for 350 bar and the other for 700 bar vehicle storage, at the service station. The dispensing scheme can be seen in Figure 30.8. The feed line to the 700 bar dispenser goes through an ultra-low cold fill to precool hydrogen to −40 °C in order to obtain a fast filling rate in accordance with SAE standards [2]. Figure 30.9 shows the two dispensers.

In Italy, a hydrogen gas refueling system with one dispenser is built for 350 bar vehicles. Hydrogen from cylinder pack storage or the on-site production facility is compressed with a two-stage membrane compressor and stored in a high-pressure storage bank. The dispenser (see Figure 30.10) is equipped with two nozzles, one for cars and the other for buses. Each nozzle has a 3 m long hose and a breakaway fitting. Prior to the dispensing system at the service station in Mantova, refueling has been performed with a mobile refueling station supplied by Air Products and operated by Sapio.

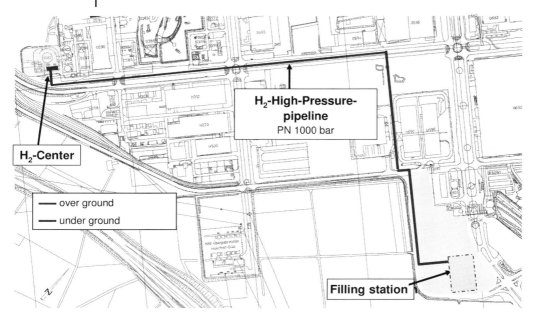

Fig. 30.7 Pipeline layout at industrial park, Höchst (© Infraserv GmbH & Co. Höchst KG).

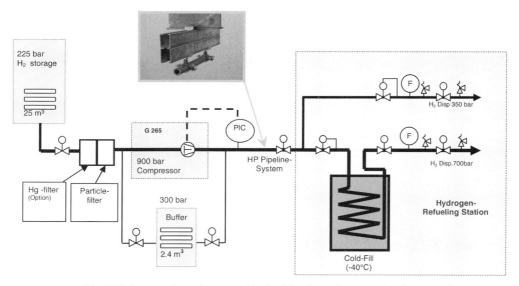

Fig. 30.8 Compression scheme used in Frankfurt (© Infraserv GmbH & Co. Höchst KG).

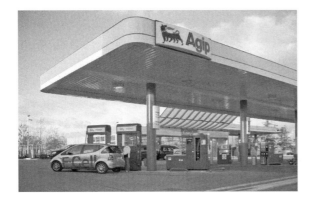

Fig. 30.9 Hydrogen dispensers in Frankfurt (© Infraserv GmbH & Co. Höchst KG).

Fig. 30.10 Hydrogen dispenser in Mantova (source: Sapio).

30.3.3
Refueling LH$_2$

At the Agip service station in Höchst, there is also a distribution and refueling system engineered and implemented by Linde for sub-cooled liquid hydrogen, truck delivered by Linde as well. The LH$_2$ refueling system (see Figure 30.11) consists of a 10 000 l liquid hydrogen storage tank, LH$_2$ dispenser, LH$_2$ transfer pump and a pressure regulation station for control of auxiliary gases such as nitrogen. This facility has been employed for refueling BMW Series 7 hydrogen cars, but outside the Zero Regio project contract.

Fig. 30.11 (a) LH$_2$ storage tank at Frankfurt station; (b) LH$_2$ dispenser in Frankfurt (source: Linde AG).

30.3.4
Approval of Hydrogen Refueling Stations

Considerable time and effort have been devoted to obtaining approval certificates for the hydrogen refueling facilities at the two sites. Different practices are used by the authorities in Germany and Italy. There have been severe delays in starting up the system in Italy due to time-consuming approval formalities.

In Germany, a two step procedure has been established. An approval for the building works (station building including all the filling stations) is required. Second, approval for operating each of the filling stations is required. Correspondingly procedures in accordance with certain federal laws are followed. All the safety-relevant documents such as hazop investigations are required for the operating permissions. The building approval was obtained in 10 weeks and the operating approvals for all the three dispensers were obtained in 4.5 months. Overall the approvals were obtained in good time and the refueling station could be opened in November 2006 as planned.

In Italy, the procedure was unclear to start with. An application was filed for a conventional service station, which was approved. Hydrogen facilities were then added to this application in accordance with so-called "single authorization act" in Italy. In spite of all the detailed safety-related documentation, the approval formalities took over 10 months for building permission for a hydrogen-integrated public service station. This resulted in a delay in infrastructure construction and the service station was opened in September 2007 instead of the planned November 2006. Based on experience in Mantova, Italian ministry of interior released an official law in 2006 (DM 31/08/2006) for hydrogen refueling station buildings.

Details of the approval formalities practiced at the two sites are described in a report [3]. Based on the safety analysis and experience with the hydrogen refueling stations in Zero Regio, national guidelines have been developed and published by VdTÜV [4]. These guidelines have the status of being a national standard on the requirements for hydrogen refueling stations in Germany.

30.3.5
Operational Experience

Experience with all the dispensers has been satisfactory. Apart from some minor incidents, such as faults in control electronics and valves, which could be repaired within a day, the reliability and availability of the dispensers have been high. Dispensers can be used by car drivers in the usual manner after initial instructions. During the first year of operation in Germany there was an accident at the 700 bar dispenser, which was hit by a station customer (driving a conventional car) while reversing his vehicle. This has been examined in detail. The conclusion was that it was entirely due to the negligence of the customer. The dispenser was damaged but no hydrogen leakage occurred. The dispenser was replaced by a new one. Such incidents unfortunately repeated twice more during the second year of operation at different dispensers in Germany, but without serious damage or hydrogen leakages. Pipeline operation in Germany has been without incidents except in the first start-up phase when the pipeline was not sufficiently purged with hydrogen initially.

Experience with the ionic compressor in Germany has been satisfactory to some extent. The first year of operation did not show any serious malfunction apart from some unexpected incidents. In the first year the compressor had an availability of 88%. During the second year of operation there was a malfunction in the cooling water circuit and as a result the ionic compressor was severely overheated. After this incident the whole compressor system had to be reworked (hydraulic system, high pressure cylinders, valve, replacing sealing etc.). In this year the availability of the system decreased to 70% because of the long still standing period for overhauling the total system. During this time the dispensers could be only supplied with 200 bar hydrogen (back up). After the reworking of the compressor the system has worked without any major incident. It has low energy consumption and needs low preventive maintenance. At this point it should be stated that the ionic compressor employed in Zero Regio was the first prototype world-wide operating at 900 bar pressure.

Experience with the refueling system for liquid hydrogen has been positive throughout the demonstration phase. Over 3700 kg of liquid hydrogen was filled in 860 fillings over a time period of 27 months.

30.4
FCV Fleets and Demonstration

Fuel cell passenger vehicles have been demonstrated at both sites over a period of 3 years under realistic urban conditions and different applications. The refueling infrastructure described in Section 30.3 has been used to refuel these cars. Demonstration details and results obtained are presented in this section.

30.4.1
Fleet Description

Figure 30.12 shows the vehicles demonstrated. Five A-Class F-CELL from Daimler were used in Germany, one of which had a 700 bar storage tank and fuel cell system. In Italy, three FC-Panda cars from Fiat, all with 350 bar storage, have been demonstrated. Whereas the vehicles in Germany use a hybrid configuration (fuel-cell and a NiMH battery), the Italian vehicles tested are powered entirely by fuel cells. Some specifications of the vehicles are given in Figure 30.12.

The vehicles were equipped with a data acquisition system which stores all the vehicle parameters to be used later in data analysis. The acquisition system developed by CRF within the project is shown in Figure 30.13. Data are transferred from time to time to a base station installed at a central location. The Daimler vehicles were already installed with a data acquisition system. The base station was installed at Fraport where the cars were stationed. The data from the base station was retrieved via the Internet to the central server of Daimler. In addition to on-board data acquisition, some data on mileage and refueling was noted by the drivers in the vehicle log-book. In case of emergency drivers could contact the responsible people at Daimler or CRF/Fiat. A number of drivers were trained at both sites. Minor repairs and maintenance could be done

Range: 177 km (@350 bar)　　　　　Range: 300 km (@350 bar)
Maximum speed: 140 km/h　　　　Maximum speed: 130 km/h (electr. limited)
Ballard PEM 72 kW　　　　　　　　Nuvera PEM 70 kW
5 cars, one with 700 bar storage　　　3 cars

Fig. 30.12 (a) F-CELL vehicle tested in Germany (source: Daimler AG); (b) fuel cell Panda tested in Italy (source: Fiat).

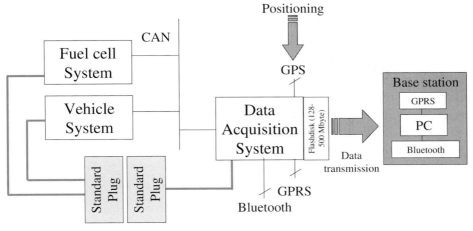

Fig. 30.13 Data acquisition system employed in Panda vehicles (source: Fiat).

on-site and for major repairs or transmission-related problems the vehicles were taken to the OEMs.

30.4.2
Homologation Formalities

The fleet in Germany was leased to the project and each vehicle was homologated as an individual approval in accordance with the German law (§ 21 StVZO) before delivery to the project. The situation was different in Italy. The three cars were acquired by Regione Lombardia and had to undergo a series of tests for homologation after delivery. Like infrastructure, this process turned out to be very time consuming and tedious. The Ministry of Transportation and the Ministry of the Interior in Italy were involved and required many tests, such as an endurance test for 2000 km on test tracks with different driving cycles and tests of the hydrogen sensors and so on. The three cars were finally homologated as single prototypes with certain conditions, namely that cars may be driven only with a 200 bar refill pressure and they require permission from the local authorities for circulating on the public roads [5]. It has not been possible to relax these restrictions during the lifetime of the project.

30.4.3
Operational Experience

Following the fleet delivery plan, two cars have been operated during the first year by Fraport, Germany, for ground transportation services. About 11 500 km have been driven and about 120 kg of hydrogen has been refilled. The cars have shown good performance with almost 100% availability. The only time lost due to vehicles was right at the beginning when the fuel cell stacks of both cars showed low cells (voltage drop), which was due to the high nitrogen content in the hydrogen being refilled. This was in turn, as mentioned earlier, due to the fact that the transport pipeline was not purged long enough with hydrogen before start-up and had some residual nitrogen left. The next three cars were delivered in the second year and it was decided to put them to different applications by different users. The 700 bar vehicle was used by Daimler and one 350 bar F-CELL was operated by Infraserv in the industrial park Höchst and the neighboring area. Fraport operated 3 cars throughout the second year.

During the second year, about 18 500 km were accumulated by the five cars. The 700 bar car joined the demonstration a few months later. The car at Infraserv was damaged in an accident caused by the driver at the parking lot. Otherwise the fleet availability remained high. During the third year, one car was transferred from Fraport to TÜV Hessen, also a project partner. During the third year, the fleet covered about 32 000 km. Overall the fleet demonstrated in Germany performed very well with an availability of 95%.

The fleet in Italy functioned equally well, with high availability. It has been used by Mantova Town Hall for different office work in the city of Mantova and the neighborhood. Due to the homologation problems stated above, only one car was used during the first year. The second car was delivered towards the end of the first year and the third car towards the end of the second year. The vehicles not officially registered were used for dissemination purposes by CRF personnel. The three cars together covered 7200 km in year two. In Italy, fuel cell stacks in each car were repaired on average once per year. During the third and the last demonstration year, all three cars have been operated with official number plates and have covered about 23 000 km. Cars in Italy had an average consumption of 1.06 kg per 100 km.

30.4.4
Data Analysis and Assessment

Data acquired by the on-board data acquisition system on each car are transferred via CD-ROMs to JRC, Ispra, at regular time intervals and are analyzed in detail in order to calculate the fuel consumption and analyze the energetic performance of the cars. The amount of data acquired by each vehicle is huge – 19 parameters are written at a frequency of 5 Hz.

This data are appropriately filtered and analyzed for certain predefined driving events or runs. Details of the energy distribution from tank to wheel are analyzed. This provides a good insight into the energy efficiency at each stage of transmission. For hybrid transmission (A-Class F-CELL of Daimler) the state of charge of the battery also plays a role [6]. Results are then assessed based on monthly statistics.

In addition to the above-mentioned data analysis, one car from each fleet is tested at the VELA (Vehicle Environmental Laboratory) facility via chassis dynamometer testing with simulated driving cycles such as NEDC and constant speeds. This testing was performed at the beginning and at the end of the demonstration to analyze the difference in performance with time.

30.5
Socioeconomic Investigations

The technical assessment of hydrogen infrastructure and fuel cell vehicles was accompanied by socioeconomic studies, covering acceptance, economic impact, and policy issues.

Regarding acceptance, results are presented from interviews with vehicle drivers and surveys of high school students participating in demonstration events. Acceptance is generally found to be high. Users are found to be very satisfied with the vehicles, except for their limited driving range. Reservations are observed with regard to societal costs and their funding, the energy sources used to produce hydrogen fuel, the user experience at the refueling stations, and road safety problems due to low vehicle noise. Safety is not otherwise found to cause acceptance problems.

The contributions of introducing hydrogen as a transport fuel to the security of transport fuel supply and to eco-efficient transport in Europe depend on the primary energy feedstock used for hydrogen production. Hydrogen based on non-fossil and non-combustible primary energy has the advantage over natural gas-based hydrogen of being tapped from European sources that are emission free. Regarding economic impact, the future costs of hydrogen as a transport fuel have been compared with the expected future costs of conventional petroleum-based fuels. The study shows that hydrogen fuel can be competitive at the expected future oil prices (above $ 100 per barrel) without preferential tax treatment, provided that high conversion efficiency in hydrogen production is achieved. At the oil price where hydrogen based on natural gas becomes competitive, hydrogen based on electrolysis can be expected to be more competitive.

As regards policy and regulatory issues, the influence of the introduction of standards on the cost of hydrogen energy is evaluated. Through the scale methodology (learning curves), the technological trajectories and the costs of this energy source are appraised (life cycle assessment approach) and compared with the current values resulting from the project and from the utilization of the DOE model.

30.6
Dissemination

Considerable effort has been spent on dissemination activities during the project. In the initial phase of the project, these were promotion activities aimed at making the project known via a dedicated website and presenting the concept at different conferences and fairs. After the construction phase, the service stations were inaugurated officially with high visibility inviting prominence from politics, science, technology and industry. A CD-ROM was produced and distributed to spread information about the project widely. After the first year of demonstration, a mid-term workshop was held at the WHTC 2007 in Montecatini, Italy.

Fig. 30.14 Honda and Mazda refueling hydrogen at Frankfurt station (source: Infraserv).

All 16 project partners contribute to dissemination actions. Over 20 articles have been published in high-diffusion news media such as *Frankfurter Allgemeine Zeitung*, local papers in Mantova and Lombardia, and CHEManager. After the completion of the service stations, information about them was broadcast several times on radio and TV. Many school visits and other visible actions have been taken at both refueling stations. The project team at Mantova has also visited many schools in the Lombardia region and presented the project and the fuel cell vehicles to the students.

The project has been presented via posters, papers, and presentations at over 80 conferences, workshops, and meetings and also at over 25 industrial fairs such as the Hannover Fair and Ecomondo. The news about Zero Regio has been spread very widely with the result that reporter teams from abroad have visited the Höchst industrial park and have made films, for example TV Vanguarda Globo Brazil in August 2008.

The results obtained and the know-how developed regarding Zero Regio can be exploited further. The service stations in Höchst and Mantova can be used to refill other hydrogen vehicles. At the station in Höchst, many vehicles have refilled hydrogen, as shown in Figure 30.14. The Honda FCX Clarity and Mazda RX8 have been refilling on a regular basis. Refilling of the next-generation B-Class FCV from Daimler is being tested extensively. Know-how and experience with approval formalities for hydrogen infrastructure and vehicles in Europe, pipeline transport of hydrogen, use of by-product hydrogen as a transport fuel, using the latest 700 bar refueling technology, and ionic liquid compression can be exploited further.

30.7
Conclusion

This chapter has presented a short résumé of the Zero Regio project, the first demonstration project on fuel cell passenger vehicles co-financed by the European Commission. The objectives set at the project beginning in 2004 have been realized. Infrastructure for production, transport, storage, and distribution has been developed following different pathways in two European cities. The infrastructure developed has been operated and demonstrated successfully for fuel cell vehicle fleets. Two dedicated small FCV fleets have been operated and tested under real-life conditions over a period of 3 years. Experience with the fleet performance has been to full customer satisfaction and it can be expected that the next generation of the FCVs (e.g. B-Class F-CELL) will perform even better and will make another step forward to real-life application. Some important lessons regarding approval formalities have been learnt in Italy which will help in planning future activities in this area in Italy and also other European countries. This experience must be taken as a strong incentive to streamline and harmonize the regulatory framework for hydrogen infrastructure and vehicles in Europe.

The hydrogen infrastructure built, tested, and integrated in public service stations in Frankfurt and Mantova can be employed for much larger fleets and are therefore obvious sites for future larger demonstrations. The experience gained with their operation should be exploited in building a network of similar infrastructure in other urban areas. In the initial phase, by-product hydrogen should be exploited much more than is done at present.

Acknowledgments

Zero Regio has been co-financed by the European Commission within the 6th Framework Program. The authors, on behalf of the project consortium, acknowledge the financial and organizational help from the Commission, in particular the project officers Mr. Sabater and Dr. Leiner. The work presented in this chapter was performed by the project consortium, and the authors as the project coordinators would like to thank all the consortium members for their contributions to the project and providing their results for this publication. Work of the following colleagues is acknowledged in particular: on Infrastructure development – Mr. Boening, Infraserv; Mr. Winkler, Linde; Dr. Grimolizzi, Dr. Ogliari, Sapio; Dr. Basini, Eni; Mr. Irmler, Eni Deutschland; on fleet preparation and demonstration – Mr. Weber, Fraport; Mr. Grasman, Mr. Klein, Mrs. Ewers, Daimler; Dr. Sutti, Dr. Volpi, Dr. Trevisani, Mr. Oneda, Mantova Town Hall; Mr. Melzi, CRF; on data evaluation – Dr. Perujo, Mr. Nauwelaers, JRC at Ispra; on socioeconomic studies – Dr. Dorigoni, Universitá Bocconi; Mr. Helby, University of Lund; Dr. Hansen, University of Roskilde; on safety and approval matters – Mr. Schork, Mr. Komrowski, TÜV Hessen; Dr. Fiaccadori, Mantova Town Hall; and on project management in Italy – Mrs. Sarno, Dr. Rota, Mr. Canobio, Regione Lombardia.

References

1 SAE (2005) *Information Report on the Development of a Hydrogen Quality Guideline for Fuel Cell Vehicles*, SAE Paper J2719, SAE International, Warrendale, PA.
2 SAE (2006) *Hydrogen Surface Vehicle Refueling System*, Draft SAE Paper J2601, SAE International, Warrendale, PA.
3 Zero Regio (2006) *Licensing Procedures and Experience with Obtaining Approval from Authorities at Both Sites (Höchst in Germany and Mantova in Italy)*, Zero Regio Report D 4.7.
4 VdTÜV (2009) *Requirements of Hydrogen Filling Stations*, VdTÜV Merkblatt 514, www.vdtuev.de/publikationen/merkblaetter.
5 Zero Regio (2007) *Status Report on Safety and Other Regulations for FCVs in Italy Before Bringing Them on Road*, Zero Regio Report D 8.6.
6 Zero Regio (2008) *Report with Initial Evaluation for DaimlerChrysler Car Fleet*, Zero Regio Report D 6.4.

Safety Issues

31
Safety Analysis of Hydrogen Vehicles and Infrastructure
Thomas Jordan and Wolfgang Breitung

Abstract

This chapter summarizes the state-of-the-art related to safety analysis of hydrogen vehicles and infrastructure. Many aspects have been treated in more detail in the reports of the European Commission-supported HySafe Network (www.hysafe.net). The potential use of hydrogen as an energy carrier and in particular its application as a fuel in vehicles require new operational parameters which are very different from those currently used in the chemical and petrochemical industries. This is why the industrial experience gained so far is applicable only partially. Safety analyses require the definition of acceptable risk levels, which are implicitly or explicitly contained in standards or regulations or might be derived from established technologies. A quantitative measure of risk is the product of a probabilistic factor representing the frequency of occurrence of hazardous events and of the associated damage. As these frequencies of failures are largely unknown and difficult to predict, it is recommended to focus on the deterministic evaluation of the damage and of the effectiveness of mitigation measures. For the deterministic consequences, evaluation of the distribution, ignition, and combustion phenomena are the key elements of the event sequence. The distribution of hydrogen is well understood as long as phase transition or near-critical conditions might be excluded. The latter phenomena are linked mainly to liquid hydrogen applications. The knowledge and modeling capabilities regarding the ignition phenomena are still incomplete. The classification of ignition sources into weak and strong igniters and the degree of conservatism related to assumptions about timing and location are still under development. Transitional behavior of flames plays an important role. The principal mechanisms are well captured, but the transfer to real accidental scenarios and the new physical domains is not yet accomplished. Commonly agreed, harmonized performance-based standardization on an international level relies on an appropriate understanding of the relevant mechanisms. This understanding will be further developed by internationally coordinated pre-normative research.

Keywords: hydrogen vehicles, safety, infrastructure, accident scenarios

Hydrogen Energy. Edited by Detlef Stolten
Copyright © 2010 WILEY-VCH Verlag GmbH & Co. KGaA, Weinheim
ISBN: 978-3-527-32711-9

31.1
Motivation of Hydrogen-Specific Safety Investigations

Although hydrogen has been safely produced, distributed, and used in the chemical industry for many decades, the developed safety procedures and technologies provide only limited guidance for the applications that are foreseen for its energetic use in a future hydrogen economy.

In the case of hydrogen-powered vehicles, hydrogen will be used in a decentralized infrastructure in relatively small amounts (several kilograms per user) by a large population without special training in the safety of combustible gases. The transition of vehicle fuels from liquid hydrocarbons to gaseous or cryogenic hydrogen requires the adaptation of automobile design and safety technology to the special properties of hydrogen.

As absolute safety measures are difficult to communicate, the public will only accept relatively safe technologies. Therefore, a safety level equal to or better than that of current fuels has to be achieved. This requires an objective and systematic investigation of the behavior of hydrogen under normal operating conditions, component malfunctions, vehicle collisions, and service/repair conditions. Unacceptable risks should be identified early in time and prevented by design or safety management measures before the acceptance of hydrogen technologies is adversely affected.

For the safe handling of hydrogen and for a safe design of the related technologies one has to understand the basic safety-relevant properties and phenomena. Figure 31.1 compares several important safety-relevant physical and chemical properties of hydrogen and gaseous hydrocarbons such as density and buoyancy. As can be seen, significant differences exist in the properties of hydrogen and hydrocarbons under standard conditions in stoichiometric mixtures. Hydrogen exhibits a very low density and ignition energy compared with gaseous hydrocarbons, whereas it has high values of buoyancy, diffusion in air, flammability range, specific heat of combustion, laminar burning velocity, and detonation sensitivity. These differences will be less pronounced for hydrogen originating from liquid spills at cryogenic temperatures. However, as the propagation and final outcome of an accidental hydrogen release depend very sensitively on the spatial configuration, initial and boundary conditions of actual properties, and thermodynamic state of the materials involved, one has to know this input and the influence on mixing, ignition, and combustion as accurately as possible.

In the following sections, the expected effects of the specific properties of hydrogen will be discussed in more detail. Indications are given of how the phenomena influence the consequences of accidents.

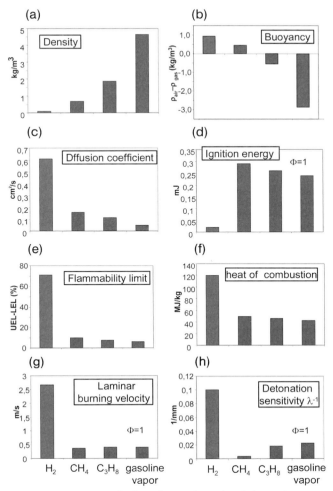

Fig. 31.1 Comparison of safety-relevant physical and chemical properties of hydrogen, methane, propane, and gasoline vapor.

31.2
Phenomena

In a phenomena identification and ranking exercise (PIRT) performed in the European Commission-funded Network of Excellence (NoE) HySafe, the safety-relevant phenomena have been selected by expert judgment for further investigation. The study concluded that all kinds of releases in confined or at least partially confined spaces and associated mitigation measures are the most critical issues. The following sections illustrate some important aspects of the related phenomena.

31.2.1
Buoyancy

Figure 31.1b displays the buoyancy of hydrogen in air. The high buoyancy of hydrogen compared with gaseous hydrocarbons suggests that if space is limited in the vertical direction, a partial premixed layer will develop beneath the roof. In a free environment, hydrogen experiences an initial vertical acceleration of 13g, rises, and mixes rapidly with air, which dramatically reduces the possible consequences of a hydrogen vehicle accident. The latter statement further relies on limited inventories – of the order of several kilograms – and on a moderate spatial congestion at the release point.

A somewhat different behavior is to be expected from liquid tank releases or cold hydrogen clouds evaporating from liquid spills. The density of very cold hydrogen close to the boiling temperature is close to or even higher than the density of air.

31.2.2
Diffusion

Figure 31.1c compares the diffusion coefficients of hydrogen and hydrocarbons. The large value of $0.6 \text{ cm}^2 \text{ s}^{-1}$ for hydrogen indicates that hydrogen distribution may occur noticeably also in the downward direction. Distances of a few centimeters will be affected by H_2 diffusion within a few seconds. As a result of the high diffusivity of hydrogen, the maximum burning velocity for a hydrogen-air premixed flame occurs at an equivalence ratio $\varphi = 1.8$, whereas for hydrocarbon-air flames it occurs at around $\varphi = 1.1$ [1]. In most real settings, the diffusion and all related effects such as flame propagation will be enhanced considerably by turbulent transport.

31.2.3
Ignition

Figure 31.1d shows that the minimum ignition energy of hydrogen-air mixtures at stoichiometry (29.5 vol.% H_2 in air) is exceptionally low, 0.019 mJ, which is about one order of magnitude lower than those of most hydrocarbons [2]. Weaker sparks or other small energy deposition mechanisms may therefore trigger an ignition in hydrogen rather than in hydrocarbons. On the other hand, it is frequently stated that most sparks from static electricity, switches, relays, and electrical motors have energies well above 1 mJ. For such strong ignition sources, the difference in ignition energy does not matter. Other ignition mechanisms, such as diffusion ignition, gas heating by shock focusing, or reflection, are very relevant to hydrogen, but not to hydrocarbons. Overall, a larger variety of potential ignition mechanisms exists for hydrogen than for hydrocarbons [3]. Some progress in the modeling of diffusion ignition has been made recently [4, 5]. Although often referred to, the heating up due to the Joule-Thomson effect at standard temperature is far too small to ignite hydrogen jet releases. Moreover, the cooling of the gases due to the acceleration often prevails in real expansion processes.

Figure 31.1e displays the concentration range of flammable compositions in air for different gases, depicted as the difference between the upper and lower flammability limits. Under standard laboratory conditions, the combustible H_2-air mixtures range from about 4 to 75 vol.% of H_2. This large range of combustible H_2-air compositions is of high relevance to safety. If a hydrogen leak occurs, a much larger part of the H_2-air cloud can find an ignition source and burn out compared with hydrocarbons. The flammability limits of hydrogen expand with temperature; for example, the lower flammability limit drops from 4% at NTP to 3% at 100 °C. On the other hand the lower flammability limit increases to 7.2% for horizontally propagating flames and to 8.5–9.5% for downward and spherically propagating flames [1].

It should be noted that hydrogen-rich regions of the cloud (> 29.5 vol.% H_2) cannot react completely due to the local oxygen deficiency. However, the partial hydrogen burning generally creates sufficient overpressure and expansion flow to mix the residual hot mixture of H_2, steam, and N_2 with fresh air. Hence the remaining hydrogen is consumed in a secondary burning.

31.2.4
Combustion

The heat of combustion of hydrogen on a mass basis (120 MJ kg^{-1}) is significantly higher than that of hydrocarbons due to the low molar weight (Figure 31.1f). Burning of 1 kg of hydrogen in an accident correspondingly causes more energy to be released into the environment, which may result in higher temperature or pressure loads. The radiation spectrum of a hydrogen fire contains little IR but high UV fractions. This reduces the likelihood of secondary fires, but on the other hand it makes small hydrogen fires difficult to detect, particularly in bright daylight. Without entrainment of other material such as dust, the color of the flame will remain blue. The spectra of cryogenic hydrogen burns have been analyzed by Eckl et al. [6].

Figure 31.1g shows that the laminar burning velocity (S_L) of stoichiometric H_2-air mixtures, 1.91 m s^{-1}, is significantly larger than that of hydrocarbon-air mixtures. S_L characterizes the general reaction kinetics of the combustion process and it is a fundamental property of a combustible mixture. The high chemical reactivity of H_2-air mixtures leads to fast flames in particular in confined and highly obstructed areas. The influence of confinement has been studied in comparison with methane- and propane-air mixtures [7]. Even lean H_2-air mixtures with only 15% of hydrogen can burn faster than stoichiometric hydrocarbon-air mixtures.

The combustion reaction in premixed systems usually start after a weak ignition with a deflagration, which might – after acceleration – transit into a detonation (DDT). With a strong ignition, for example with chemical explosives, a direct initiation of a detonation might be reached. However, in usual accident scenarios strong igniters are not present.

Deflagrations are combustion waves in which the reaction is controlled by diffusion of heat and radicals from the burned gas to the unburned gas. Deflagrations with flame speeds lower than or equal to the velocity of sound in the unburned mixture are called "slow deflagrations," and those with higher flame speeds are referred to as "fast deflagrations." The criteria for flame acceleration requires that the expansion ratio, defined by the mixture composition and the thermodynamic state, is larger than the critical value of 3.75 (Figure 31.2).

The accelerated flame might even turn into a detonation. Detonations are supersonic combustion waves with the highest overpressures in the range of 20 times the initial pressure. In detonations, the unburned gas is ignited by rapid adiabatic compression to a temperature that is higher than the self- or auto-ignition temperature. Detonation waves are three-dimensional phenomena which show a cellular structure created by superposition of different shock wave fronts. This so-called detonation cell size λ decreases with increasing chemical reactivity of the mixture, so that λ^{-1} can serve as a measure of detonation sensitivity. The detonation cell size of stoichiometric H_2-air at standard conditions is approximately 15 mm. The higher reactivity of H_2-air and in particular H_2-O_2 systems results in a higher tendency to undergo spontaneous DDT compared with hydrocarbons (Figure 31.1 h).

The actual consequences of hydrogen combustions are thermal and mechanical loads on the environment. Because of the high flame speed and the radia-

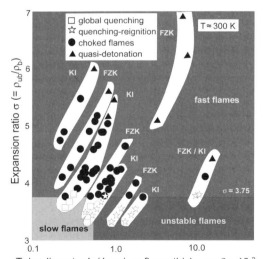

Fig. 31.2 Criterion for flame acceleration in H_2-air mixtures.

tion characteristics, the pressure effect hazard prevails when compared with other chemical energy carriers. Depending on the actual duration, peak overpressures of 2–5 kPa are able to destroy windows and 20–50 kPa might destroy masonry or even concrete buildings. The human organs which are most sensible to pressure waves are the ear drums and lungs [8, 9]. In the case of waves with a positive phase duration longer than 10 ms and a peak overpressure above 0.8 bar, for example, lung damage must be expected. Shorter waves need higher overpressures to produce similar damage.

The injury thresholds for the ear drums were reviewed extensively by Richmond et al. [8]. Their statistical treatment of the experimental data resulted in the two bands for 1 and 50% probability of ear drum rupture. The width of the bands represents the scattering due to parameters such as age or gender. With increasing Δp^+, the event of an ear drum rupture leads to increasing grades of ear injury, affecting also the inner ear bones. These data are conservative and appropriate for confined local explosions in small rooms such as a garage.

The thermal loads of a jet flame have been reported [10], and an experimental comparison of the thermal effects on the passengers in vehicle fires has been provided [11].

31.3
Safety Analysis Procedures

The assessment of safety requires determining acceptable risk limits and the actual risk. Risk is defined as the cumulated product of probabilities and associated adverse consequences and damage. Tolerated or acceptable risk limits are either explicitly prescribed by law or rely on comparison with established and accepted alternative technologies, a concept referred to as comparable risk analysis. Generally acceptable limits might differ for the general public and for professionals, working at a refueling station, for instance.

The basic step is identifying potential hazards with the help of any so-called HAZID methodology. This is usually supported by an expert team including designers, operators, and other experienced persons. Examples are system operation-driven procedures such as HAZOP and/or component failure-driven, empirical consequence estimation methods such as FMEA [12, 13].

The poor experience base regarding the energetic use of hydrogen can be improved with two database activities, which have been set up for recording hydrogen-related near misses and accidents [see the HySafe database HIAD (www.hysafe.net/hiad) [14], and the United States Department of Energy (DOE) initiative (http://h2incidents.org/)]. Another way to generate probabilistic data for estimating frequencies consists in extrapolating experience with compressed natural gas [15]. Some collections of hydrogen-related incidents can be found in the literature. In one report [16], 287 accidents with gaseous and liquid hydrogen were analyzed and classified in a systematic manner. Based solely on this report, it might be concluded that liquid hydrogen (LH2) represents a reduced hazard potential.

An older review of hydrogen fire and explosion incidents can be found elsewhere [17]. The frequent updates on the HySafe website indicate that hydrogen accidents are not only a topic of the past, but also of the present.

Compared with damage determination, the uncertainties in the probabilities and frequencies are orders of magnitude higher. Therefore, it is often recommended to focus on consequence analysis to exclude any serious hazard and thereby reduce the risk by appropriate mitigation measures (Figure 31.3).

The first level of mitigation measures addresses the design of the system including sensor equipment. Once the design has reached the maximum possible degree of safety, the next line of defense is to limit the sources releasing hydrogen into the environment. The next step to limit consequences is to exclude ignition sources as far as possible. If this cannot be achieved completely, the likelihood of fast flames should be decreased by reducing confinement or by avoiding heavily obstructed geometries. In addition, favorable conditions for detonation onset should be avoided to the largest possible extent. Finally, as a last resort, a sufficiently strong shelter may be used to protect against pressure and temperature effects.

In addition to sensors and ventilation, autocatalytic recombiners, water sprays, and fire and detonation arrestors can be considered for mitigation. Passive tech-

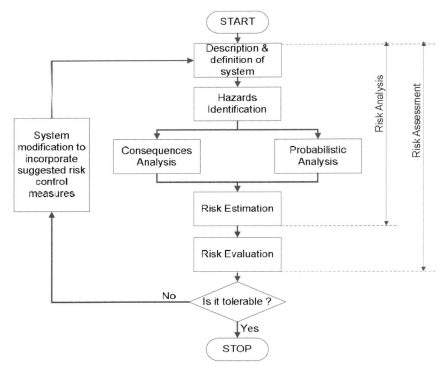

Fig. 31.3 Design iterations for reaching tolerable risk.

nologies should be preferred to active systems requiring additional energy supply. Generally a countermeasure is most efficient when it influences accident progression in an early stage. The principle of "defense-in-depth" has proven to be very successful in other areas of energy technology. It tries to include in the design a series of safety barriers which are completely independent and rely on different physical principles. The combination of several of the measures described above may represent a robust "defense-in-depth" approach to a safe hydrogen-powered system.

31.4 Scenarios

An extensive scenario selection and ranking exercise has been conducted via a PIRT analysis by the NoE HySafe. In the following, an extract will be given with the focus on vehicles and infrastructure. One conclusion for storage infrastructure and transportation was that
- Release from storage tanks (through faulty or leaking connections, or, in the case of refueling stations, at the level of the dispenser hose) into confined or partially confined atmospheres have received a high priority vote.

The other relevant application area addressed in the PIRT was commercial and passenger cars. Here the conclusion was to investigate
- Safety of H_2 vehicles in confined environments such as tunnels, public or private car parks, and maintenance workshops. Damage to systems or components including the tank (because of accidents or external causes such as fire) could lead to releases of H_2 and the formation of confined potentially explosive clouds. For private cars with smaller quantities of H_2 involved, high release rate issues have been ranked high.
- The performance and reliability of systems and components, including tanks: in some cases (PRD), even nominal behavior (i.e., the device is functioning as intended) can have dangerous consequences, for example if the release happens in a confined environment (e.g., a garage).
- The performance of the H_2 tanks under mechanical or thermal loads.
- Failure to follow "good practices" [for car mechanics in maintenance activities (purging of systems), or for emergency crews on scenes of accidents].

In summary, the most critical scenarios are represented by H_2 vehicles in garages, during refueling, in repair shops, and in tunnels.

In any case, a hazard develops only if hydrogen is released in liquid or gaseous form from vehicle components. By mixing with air, combustible H_2-air clouds can be created and then pose a threat to the passengers, and also to the external environment including persons such as first responders. The consequences of such a loss of hydrogen confinement depend on many design details of the car.

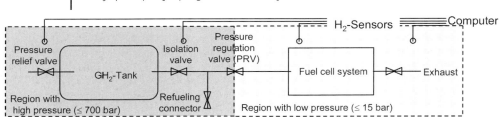

Fig. 31.4 Simplified layout of a hydrogen car using compressed gaseous hydrogen and a fuel cell.

Figure 31.4 displays a simplified layout of a hydrogen car equipped with compressed gaseous hydrogen and a fuel cell. The hydrogen system can generally be divided into a high-pressure part and a low-pressure part. The two parts are separated by the pressure regulation valve (PRV). All the components of such a system are carefully selected and tested for their performance and compatibility with hydrogen. The test objectives vary from non-destructive to destructive, from single to endurance tests at the level of single components, subsystems, or the entire system. Generally recommended design guidelines for hydrogen cars are listed in reports [18–20] and in the public documentation of the EIHP project (www.eihp.org).

Testing and following good practice, standards, and regulations will reduce the risk but may not rule out even catastrophic component failure.

31.4.1
Investigation of Garage Scenarios

The safety performance of hydrogen vehicles in garages has been addressed in several studies. In the HySafe InsHyde project, the permeation limitations for hydrogen vehicles in garages were discussed. The distribution of a characteristic release in a garage-like environment has been studied [21, 22]. The effectiveness of active and passive ventilation measures were also included in combustion simulations [23]. Initiated by distribution tests performed by INERIS with an unclear impact of residual turbulence, the special device HyGarage was built by CEA, which allows detailed optical measurements of the flow properties.

The initiating events causing hazardous situations include stress cracking or delamination of diffusion barriers, aging of sealing material, faulty valves, malfunction of the pressure relief device (PRD), mechanically destroyed piping, purging of the fuel cell stack, and boil-off in LH2-driven vehicles.

For example, the effects of different release rates of cold boil-off gas from an LH2 tank into initially stagnant air at ambient temperature in a simple rectangular garage have been investigated [24]. It is assumed that the cryogenic LH2 tank leads to 170 g of hydrogen boil-off gas per day and that this amount is released in five venting intervals of 34 g each. The first scenario considers this amount to be released in 10 s, whereas the release duration in the second scenario is 100 s.

If the amount of 34 g of H_2 were to be homogeneously distributed in the garage volume, then the resulting H_2 concentration would be only 0.6 vol.%. This value is far below the ignition limit of 4 vol.% of H_2. Does such a hydrogen release represent any risk? The GASFLOW program [25] has been applied to simulate hydrogen release and subsequent mixing with air in the garage in the two scenarios above.

Figure 31.5 displays the computed hydrogen distribution in the garage in the case of 34 g of hydrogen being released in 10 s. As can be seen, two separate plumes rise along two sides of the trunk within 7 s. Once the flow reaches the ceiling, a mushroom-shaped cloud forms. About 20 s after the beginning of hydrogen release, the cloud reaches its largest extension. The velocities in the rising plume are around 1 m s^{-1}. The momentum in the flow is redirected at the ceiling, causing an initially circular expansion of the burnable cloud and then reflections at the nearby walls. These reflections lead to a preferential flow along the ceiling towards the front of the car. Air is entrained by these convective gas motions, leading to a decrease in the combustible volume. Finally, the calculation predicts a stable stratification with most of the released hydrogen in a shallow layer underneath the ceiling.

For the slow release (0.34 g s^{-1} of hydrogen), the calculation predicts a completely different distribution process. Here, only one thin, tube-like H_2-air

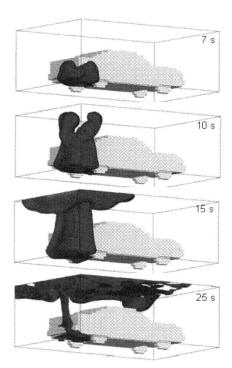

Fig. 31.5 Computed H_2 concentration field in a garage for case 1 in which 3.4 g s^{-1} of hydrogen was released for 10 s. The isosurface shown depicts the region with a burnable H_2-air mixture (>4% H_2).

plume extends from the source below the trunk to the ceiling. The plume is quasi-stationary with a slight pulsation in the flow rate and diameter. Similarly to the first scenario, a stably stratified hydrogen distribution develops in the garage, but with less gas motion along the ceiling. Shortly after the release of the total amount of gas (around 100 s), the combustible regions have nearly disappeared.

The simplistic calculation, treating the garage as a single volume with a perfectly mixed hydrogen and thus yielding a 0.6 vol.% H_2, will indicate no risk at all irrespective of the release rate. This example demonstrates the necessity for CFD modeling for reliable hydrogen safety investigations.

31.4.2
Investigation of Refueling Station Scenarios

The most recent and comprehensive European effort to develop a handbook for safe refueling station design and operation was undertaken within the EU-funded HyApproval project [26]. This project combined expertise from the petrochemical industry, suppliers of hydrogen filling station equipment, research organizations, and licensing authorities.

In the basic benchmark case (BBC) of the HyQRA internal project in HySafe, a virtual refueling station was used to compare different safety assessment approaches.

In the following, a summary is given of the main open questions and issues connected with the installation of hydrogen refueling stations in urban areas.

The cost scaling and the technical complexity of LH2 production favor a centralized large LH2 plant and the delivery of LH2 by trailers to the filling station. The main components for all design variations will include compressors, high-pressure buffer storage, and dispensers. Depending on the actual design, further components such as reformers, electrolyzers, and LH2 tanks have to be acccunted for. In most cases, conventional fuels will be stored and dispensed nearby.

For each subsystem of the filling station, potential accident scenarios have been identified. Possible leak sizes were classified into very small, small, medium, and large sized, which correspond to leakage from the valves, crack or pinhole, partial opening of a valve, and rupture of a storage vessel [27, 28].

Typically, the compressor will be installed inside a container to protect it against external influences such as the weather. It pressurized the hydrogen from about 10 to several hundred bar, for instance. Since the compressor is connected to high-pressure buffer tanks, leakage and back-flow from the buffer tanks are important scenarios. Further unwanted events are excessive pressure at the compressor outlet, compressed air ingress into the hydrogen line, and leakage of hydrogen due to vibrations. As possible safety barriers, pressure and vibration controls, hydrogen detection, and forced ventilation have been identified.

The buffer storage, usually placed in open air, contains the major hydrogen inventory handled in the refueling station. The main hazardous phenomena are hydrogen leaks with subsequent explosion or jet fire and bursting of hydrogen pressure vessels due to mechanical or thermal loads. Due to the high pressures and volumes of H_2 bottles, a burst would also create a large mechanical energy release. Insufficient data are currently available to predict whether bursting of one bottle would trigger the failure of other bottles in the bundle. The simulation of hydrogen release from a high-pressure storage system can be found elsewhere [29].

The main safety barriers against such scenarios are the ability to limit hydrogen leak rates, to isolate a leaky storage bottle, and to discharge the hydrogen inventory in a safe way. Moreover, the bottles must be safely vented in case of fire.

The dispenser consists of the refueling unit and the dispenser hose. The dispenser is generally covered by a weather protection shield. The use of a flexible hose and the connection/disconnection actions with the car significantly increase the probabilities of a hydrogen leak or a line rupture. One scenario of many others is the leakage at the filling nozzle due to the deterioration of the isolation valve [27]. Other unwanted events are hose rupture during/after refueling and back-flow from the vehicle tank. The dispenser scenarios are critical for the whole safety evaluation of the filling station, because customers are involved and located close to the release location. This scenario was also investigated experimentally by Tanaka et al. [30] and Takeno et al. [31]. Results from tests with premixed H_2-air clouds and H_2 jet releases have been published [32].

Possible precautious and safety barriers are hydrogen leak detection before connection and good nozzle and hose maintenance. Furthermore, an automatic filling procedure can be envisaged, where the customer is not involved and where an emergency shutdown procedure is implemented with well-defined actuation criteria.

31.4.3
Investigation of Tunnel Scenarios

Tunnel scenarios and the applicability of currently installed mitigation techniques have been analyzed in the HyTunnel project of HySafe (see the final report at www.hysafe.net/deliverable). A possible scenario is a vehicle collision in a tunnel involving a car fueled with LH2. If the vacuum of the thermal tank insulation is lost by mechanical damage of the outer tank structure, the LH2 tank inventory will be released in gaseous form into the environment within 10–15 min [33]. Figure 31.6 displays the numerical simulation of the accident. Here, it is assumed that the release point is below the trunk.

The first event once hydrogen has been released is the mixing of the cold gaseous hydrogen (20 K) with air (300 K). This will result in gas mixtures which are buoyant, rise upwards, and spread along the tunnel ceiling. The hydrogen distribution phase continues until an ignition occurs.

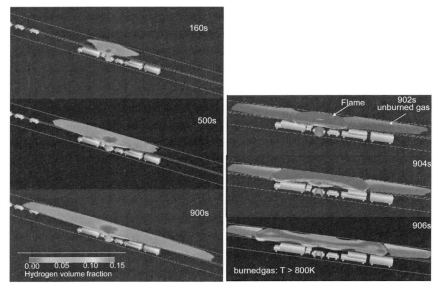

Fig. 31.6 Numerical simulation of a tunnel accident involving release of liquid hydrogen, mixing with air, and late ignition of the burnable H$_2$-air cloud.

Potential ignition sources include open fires, mechanical sparks, electrical sparks, hot surfaces, and electrostatic discharges. In the case of a successful ignition, a slow quasi-laminar deflagration will initially propagate from the ignition location to the surrounding mixture with flame speeds in the order of several meters per second. The expansion flow of the hot burned gases pushes unburned gas away from the ignition point. In the presence of obstacles, the flow may become turbulent. Depending on the hydrogen concentration, layer thickness, obstacle density, and geometric confinement, the slow deflagration might accelerate and even turn into a detonation. The conditions for these effects have been studied intensively in the HyTunnel experimental program.

From the broad study, the following research needs have been identified. Based on the EIHP scenario definitions, more realistic scenarios in tunnels should be studied including releases orientated downwards under the car, with delayed ignition of non-uniform mixtures. Scientifically grounded requirements for location and parameters of PRD have to be determined. The effects of impinging jet fires and conjugate heat transfer in conditions of blowdown are not known to a sufficient extent. Although there have been experimental programs for impinging jet fires on concrete walls, detailed measurement in the wall has not been addressed so far. Experimental programs accompanied by numerical simulations of releases into congested spaces with a deflagration-to-detonation transition have to be investigated in more detail. Anyway, one has to develop further the hydrogen safety engineering methodology and tune and apply it to tunnel scenarios, including special mitigation concepts such as sprays and mists, in the long term.

31.5
Outlook

Additional research activities will be needed in order to improve the knowledge and, in turn, mitigate the above potential hazards from the use of hydrogen in vehicle operation and in infrastructure installations. General issues which are largely independent of the specific vehicle design will be addressed. It would be very convincing if these topics could be resolved by the automobile community in a joint effort. Such joint efforts will further the cause of internationally accepted and harmonized hydrogen safety standards and procedures as initiated by EIHP and further supported by HySafe.

Issues related to the development of safe fuel cell cars have been presented [34]. With respect to vehicle safety, the following research topics will deserve closer attention in the future:
- safe venting of compressed hydrogen gas cylinders (35 and 70 MPa)
- optimum arrangement of H_2 storage vessels in the vehicle
- fire safety of hydrogen-powered vehicles with the primary goal to prevent bursting of the high-pressure hydrogen system
- guidelines for fire fighters in case of fire or accident
- optimum number and location of hydrogen detectors
- safety concept in case of a hydrogen leak detection in a running car
- tolerable H_2 leak rates in the vehicle for different operating conditions, including a parked car
- optimum position and activation criteria for pressure relief devices on the H_2 tank
- procedures to prevent penetration of hydrogen into the passenger compartment
- effectiveness of forced ventilation for reducing local H_2 concentrations in sensitive car areas
- maximum possible reduction of ignition sources
- development of standardized safety test procedures for new solid storage materials, such as nanocrystalline powders
- development of non-destructive testing methods for cryo-vessels and high-pressure tanks made from composite materials including highly accelerated lifetime testing.

Safety investigations for hydrogen infrastructure installations such as maintenance, service, repair, and filling stations should address the following research topics:
- determination of tolerable H_2 releases during vehicle repair, which pose no risk to the personnel

- design of effective and low-cost ventilation systems
- CFD analysis of leaking hydrogen scenarios, including complex surroundings near the vehicle, extension of the investigations described in [35] and including the special features of an LH2 leak including cold jets
- control of ignition sources and definition of a realistic conservative ignition model
- in the case of filling stations, the issues of protecting walls and safety distances.

More basic research is also needed to improve further the current knowledge of hydrogen properties:
- Measurement of ignitable space regions, given a certain leak size, shape, and mass flow rate, an extension of the work described in [36].
- Systematic investigation of active and passive safety systems, such as ventilators, catalytic recombiners, and flame arrestors.
- Modeling of ignition processes under realistic boundary conditions.
- Investigation of diffusion flame stability after ignition (limits for lift-off and extinction).
- Criteria for flame acceleration and detonation onset in H_2-air mixtures with concentration gradients and partial confinement (Note: the criteria described in Section 31.3.2 are valid for homogeneous and fully confined mixtures; they are, hence, very conservative with respect to practical accident conditions in mobile applications and should be extended to more prototypic conditions).
- Basic investigations of the gas behavior including its reactions at very low temperatures around 20 K and very high pressures.
- Effect of high-purity hydrogen on the relevant materials.

References

1 Molkov, V. (2007) Hydrogen safety research: state-of-the-art. In *Proceedings of the 5th International Seminar on Fire and Explosion Hazards*, University of Edinburgh, Edinburgh, UK, 23–27 April 2007.

2 Hattwig, M. and Steen, H. (eds) (2004) *Handbook of Explosion Prevention and Protection*, Wiley-VCH Verlag GmbH, Weinheim, ISBN 3-527-30718-4.

3 Astbury, G. R. and Hawksworth, S. J. (2005) Spontaneous ignition of hydrogen leaks. In *Proceedings of International Conference on Hydrogen Safety*, Pisa HySafe, Italy.

4 Wen, J. and Tam, V. (2009) Numerical study on spontaneous ignition of pressurised hydrogen release through a tube. In *Proceedings of 3rd International Conference on Hydrogen Safety*, Ajaccio HySafe, France.

5 Bragin, M. and Molkov, V. (2009) Physics of spontaneous ignition of high-pressure hydrogen release and transition to jet fire. In *Proceedings of 3rd International Conference on Hydrogen Safety*, Ajaccio HySafe, France.

6 Eckl, W., Eisenreich, N., Herrmann, M. M., and Weindel, M. (1995) Emission of radiation from liquefied hydrogen explosions. *Chem. Ing. Tech.*, **67**, 1015–1017

7 Veser, A., Grune, J., Stern, G., Breitung, W., and Dorofeev, S. (2003) Flame acceleration in a vented explosion tube. In *Proceedings of the 19th International*

Colloquium on the Dynamics of Explosions and Reactive Systems (ICDERS), Hakone, 27 July–1 August 2003, Paper 183 (CD-ROM).
8 Richmond, D. R., Fletcher, E. R., Yelverton, J. T., and Phillips, Y. Y. (1989) Physical correlates of ear drum rupture. *Ann. Otol. Rhinol. Laryngol.*, **98**, 35–41.
9 NASA Office of Safety and Mission Assurance (1997) *Safety Standard for Hydrogen and Hydrogen Systems*, Report NSS 1740.16, NASA Office of Safety and Mission Assurance, Washington, DC.
10 Keller, J., Moen, C., Houf, B., Schefer, R., Somerday, B., San Marchi, C., and Cox, R. (2005) *IX.3 Hydrogen Safety, Codes and Standards R&D, DOE Hydrogen Program FY 2005 Progress Report*, Sandia National Laboratories, Livermore, CA.
11 Suzuki, J., Tamura, Y., and Watanabe, S., *Fire Safety Evaluation of Vehicle Equipped with Compressed Hydrogen Gas Cylinders – Comparison with Gasoline and CNG Vehicles*, SAE 2006-01-0129, Japan Automobile Research Institute, Tsukuba.
12 Duijm, N. J. and Markert, F. (2007) *Safety-Barrier Diagrams for Documenting Safety of Hydrogen Applications*. Systems Analysis Department, Risø National Laboratory, Technical University of Denmark, Roskilde.
13 Casamirraa, M., Castiglia, F., Giardinaa M., and Lombardoa, C. (2009) *Safety Studies of a Hydrogen Refuelling Station: Determination of the Occurrence Frequency of the Accidental Scenarios*. Department of Nuclear Engineering, University of Palermo.
14 Kirchsteiger, C., Vetere Arellano, A. L., and Funnemark, E. (2007) Towards establishing an International Hydrogen Incidents and Accidents Database (HIAD), *J. Loss Prevention in the Process Industries*, **20** (1), 98–107.
15 LaChance, J., Tchouvelev, A., and Engebo, A. (2009) Development of uniform harm criteria for use in quantitative risk analysis of the hydrogen infrastructure. In *Proceedings of 3rd International Conference on Hydrogen Safety*, Ajaccio HySafe, France.
16 Kreiser, A. M., Fröhlich, G., Eichert, M., and Schatz, A. (1994) *Analyse von Störfällen mit Wasserstoff in bisherigen Anwendungsbereichen mit besonderer Berücksichtigung von LH2*, Report IKE 2-116, University of Stuttgart.
17 Zalosh, R. G. (1977) *Comparative Analysis of Hydrogen Fire and Explosion Incidents*, Report COO-4442-1, Factory Mutual Research Corporation, Norwood, MA.
18 Thomas, C. E. (1997) *Direct-Hydrogen-Fueled Proton-Exchange-Membrane Fuel Cell System for Transport Applications*, Report DOE/CE/50389-502, Report by Ford Motor Company, Dearborn, MI, for the US Department of Energy.
19 Ringland, J. T. (1994) *Safety Issues for Hydrogen-Powered Vehicles*, Report SAND94-8226, Sandia National Laboratories, Livermore, CA.
20 Cadwallader, L. C. and Herring, J. S. (1999) *Safety Issues with Hydrogen as a Vehicle Fuel*, Idaho National Engineering and Environmental Laboratory, Idaho, US, Report INEEL/EXT-99-00522.
21 Swain, M. R., Filoso, P., Grilliot E., and Swain, M. N. (2003) Hydrogen leakage into simple geometric enclosures. *Int. J. Hydrogen Energy*, **28** (2), 229–248.
22 Venetsanos, A., Papanikolaou, E., Delichatios, M., Garcia, J., Hansen, O. R., Heitsch, M., Huser, A., Jahn, W., Jordan, T., Lecome, J.-M., Ledin, H. S., Makarov, D., Middha, P., Studer, E., Tchouvelev, A. V., Teodorayk, A., Verbecke, F., van der Voort, M. M. (2009) An inter-comparison exercise on the capabilities of CFD models to predict the short and long term distribution and mixing of hydrogen in a garage. *International Journal of Hydrogen Energy*, **34** (14), 5912–5923.
23 Merlio, E., Groethe, M., Colton, J., and Chiba, S. (2009) Experimental study of hydrogen release accidents in a vehicle garage. In *Proceedings of 3rd International Conference on Hydrogen Safety*, Ajaccio HySafe, France.
24 Breitung, W., Necker, G., Kaup, B., and Veser, A. (2001) Numerical simulation of hydrogen release in a private garage. In *Proceeings of the 4th International Symposium on Hydrogen Power –*

HYPOTHESIS IV, 9–14 September 2001, Stralsund, Germany, Vol. 2, p. 368.
25 Travis, J. R., Royl, P., Redlinger, R., Necker, G., Spore, J. W., Lam, L. L., Wilson, T. L., Nichols, B. D., and Müller, C. (1998) *GASFLOW-II: A Three-Dimensional Finite-Volume Fluid Dynamics Code for Calculating the Transport, Mixing, and Combustion of Flammable Gases and Aerosols in Geometrically Complex Domains. Vol. 1, Theory Manual. Vol. 2. Users Manual*, Forschungszentrum Karlsruhe, Germany (1998), Reports FZKA-5994.
26 LBST GmbH (2008) HyApproval, www.hyapproval.org (accessed 2008).
27 Lim, S. and Perrette, L. (2006) *Risk Assessments and Accident Simulations as Per Matrix Table*, Report by INERIS, HyApproval Deliverable WP4, ST3, www.hyapproval.org (accessed 2006).
28 Perrette L. and Lim S. (2006) *Proposed List of Scenarios for the Modelling Task*, Report INERIS HyApproval Deliverable WP4, 4.X, www.hyapproval.org (accessed 2006).
29 Angers, B., Bènard, P., Hourri, A., Tessier, P., and Perrin, J. (2006) Simulations of hydrogen releases from high pressure storage systems. In *Proceedings of the 16th World Hydrogen Energy Conference* (ed. Veziroglu, T. N.), 13–16 June 2006, Lyon, IAHE, France.
30 Tanaka, T., Azuma, T., Evans, J. A., Cronin, P. M., Johnson, D. M., and Cleaver, P. (2005) Experimental study of hydrogen explosions in a full-scale hydrogen filling station. In *Proceedings of International Conference on Hydrogen Safety*, 8–10 September 2005, Pisa HySafe, Italy.
31 Takeno, K., Okabayashi, K., Kouchi, A., Nonaka, T., Hashiguchi, K., and Chitose, K. (2005) Phenomena of dispersion and explosion of highly pressurized hydrogen. In *Proceedings of International Conference on Hydrogen Safety*, 8–10 September 2005, Pisa HySafe, Italy.
32 Shirvill, L. C., Roberts, P., Butler, C. J., Roberts, T. A., and Royle, M. (2005) Characterization of the hazards from jet releases of hydrogen. In *Proceedings of International Conference on Hydrogen Safety*, 8–10 September 2005, Pisa HySafe, Italy.
33 Breitung, W., Bielert, U., Necker, G., Veser, A., Wetzel, F.-J., and Pehr, K. (2000) Numerical simulation and safety evaluation of tunnel accidents with a hydrogen powered vehicle. In *Proceedings of the 13th World Hydrogen Energy Conference* (eds. Mao, Z. Q., Veziroglu, T. N.), 12–15 June 2000, Beijing, China, Vol. 2, pp. 1175–1181.
34 Perrette, L., Paillère, H., and Joncquet, G. (2006) Presentation of the French National Project DRIVE, In *Proceedings of the 16th World Hydrogen Energy Conference* (ed. Veziroglu, T. N.), 13–16 June 2006, Lyon, IAHE, France.
35 California Fuel Cell Partnership (2004) *Support Facilities for Hydrogen-Fueled Vehicles. Conceptual Design and Cost Analysis Study*, Technical Report for California Fuel Cell Partnership (CaFCP), www.er.cafcp.org/pdf/H2-Vehicles-Facility-Study.pdf (accessed 2004)
36 Swain, M. (2004) *Codes and Standards Analysis DE-FC36-00GO 10606, A007*, Final Technical Report 04/15/03–04/14/04, University of Miami, Coral Gables, FL.
37 Léon, A. (ed.) (2008) *Hydrogen Technology, Mobile and Portable Applications*, Series Green Energy and Technology, Springer, Berlin, ISBN 978-3-540-79027-3.

Further Reading

Extensive literature on all aspects of hydrogen safety can be found in the public documentation of the HySafe project (www.hysafe.net); in particular, it is recommended to browse the Proceedings of the HySafe Conferences ICHS, and the Biennial Report on Hydrogen Safety, the so-called Hydrogen Safety Handbook. Further references may be found in the publications of the IEA HIA Task 19 (www.ieahydrogensafety.com/reports.htm) and the US DOE bibliography database (www.hydrogen.energy.gov/biblio_database.html). An overview regarding the state of the art with a focus on phenomena has been published [1] and a focus on consequence modeling is available [37].

32
Advancing Commercialization of Hydrogen and Fuel Cell Technologies Through International Cooperation of Regulations, Codes, and Standards (RCS)

Randy Dey

Abstract

This chapter discusses the importance of international cooperation in the development of regulations, codes, and standards on hydrogen and fuel cell technologies. To facilitate commercialization of these new technologies, it is essential to remove non-tariff barriers by harmonizing regulations, codes, and standards to facilitate global trade.

Keywords: regulations, codes, and standards, pre-normative research, harmonization, hydrogen, fuels cells, hydrogen refueling stations, International Organization for Standardization

32.1
Introduction

The world is steadily moving into the post-fossil fuel age. The new energy mix will comprise a number of energy solutions, including hydrogen. When combined with renewable energy such as wind, solar, and so on, hydrogen has the potential to "decarbonize" the energy system. It also offers a pathway for renewable energy to generate transportation fuel, reliable power and distributed energy.

This Chapter discusses the importance of international cooperation in the development of regulations, codes, and standards on hydrogen and fuel cell technologies. To facilitate commercialization of these new technologies, it is essential to remove non tariff barriers by harmonizing regulations, codes, and standards to facilitate global trade.

This subject matter should be of interest to decision-makers from energy, automotive, and infrastructure companies and also high-level government representatives. They will learn about updates on the development of regulations, codes, and standards, and the challenges and benefits of international cooperation. A special focus is placed on hydrogen refueling stations, which represent one of the key elements for accelerating the energy paradigm shift.

Hydrogen Energy. Edited by Detlef Stolten
Copyright © 2010 WILEY-VCH Verlag GmbH & Co. KGaA, Weinheim
ISBN: 978-3-527-32711-9

32.2
Hydrogen – a Part of the New Energy Mix

Due to considerations associated with climate change and security of energy supply, the world is steadily moving into the post-fossil fuel age. There is a general consensus that the emerging energy mix will comprise of a number of energy solutions, such as wind, solar, and other new energy technologies, including hydrogen. Hydrogen, being a universal energy carrier, provides a pathway for renewable energy to generate transportation fuel, reliable power, and distributed energy. It is one of the solutions that can be used to overcome the storage challenge faced by the use of intermittent renewable energy sources (e.g. electricity from wind or solar power).

Although hydrogen, fuel cell, and renewable energy technologies present environmental and societal benefits, business as usual will not ensure that the energy paradigm shift happens in the time frame that is needed to overcome the climate change challenges. *The leadership of government and industry will have to work together to enable the energy shift through RCS harmonization and other means!*

In the case of RCS, business as usual will not achieve the goal of having "one product, one standard" in the necessary time frame. Strong international cooperation will be required to avoid the market fragmentation that is often associated with the proliferation of national and regional standards.

32.3
Regulations, Codes, and Standards (RCS) – a Necessary Step to Commercialization

It is well recognized that a major barrier to commercialization of hydrogen and fuel cell technologies is the lack of globally harmonized RCS. Industry will be well served when a comprehensive set of international standards is available to guide the development and facilitate the commercialization of these new technologies on a worldwide basis.

The existence of non-harmonized standards for a given product in different countries or regions contributes to the so-called "technical barriers to trade" (TBT). Indeed, fair competition needs to be based on clearly defined common references that are recognized from one country to the next and from one region to another. International standards, developed by consensus among trading partners, serve as the language of trade and represent a key ingredient of the *World Trade Organization's* (WTO) *Agreement on Technical Barriers to Trade (TBT)*.

The *International Organization for Standardization* (ISO) (www.iso.org) together with the *International Electrotechnical Commission* (IEC) (www.iec.ch) have built a strategic partnership with the WTO. This partnership aims to avoid the proliferation of TBTs, which generally result from the preparation, adoption, and application of technical regulations and standards in a discriminatory manner.

The TBT agreement recognizes the important contribution that international standards make towards facilitating international trade. Where international standards exist or their completion is imminent, the TBT agreement requires that member countries use them as a basis for the standards that they develop.

The TBT agreement also aims at the harmonization of standards on as wide a basis as possible, encouraging all member countries to participate in the development of international standards.

Furthermore, the number of standardization bodies, which have accepted the *Code of Good Practice for the Preparation, Adoption and Application of Standards* presented in Annex 3 to the WTO's TBT Agreement underlines the global importance and reach of this accord. As of 1 October 2009, some 172 standardization bodies from 132 countries have accepted this Code of Good Practice.

32.4
International RCS Bodies – Responsible for the Standardization of Hydrogen and Fuel Cell Technologies

ISO and IEC play an important role in the development of international standards for hydrogen and fuel cell technologies. ISO/TC 197, which is one of the 208 technical committees of ISO, is responsible for the development of international standards for hydrogen technologies. IEC/TC 105, which is one of 94 technical committees of IEC, is responsible for the development of international standards for fuel cell technologies.

Both of these technical committees are very active. Each has a work program to meet the current stakeholders' needs in terms of international standards. The ISO/TC 197 work covers infrastructure, automotive and transportable hydrogen applications (see Appendix 1). The IEC/TC 105 work covers stationary, transportation, portable, and micro fuel cell applications (see Appendix 2).

The ISO and IEC work results in international agreements, which are published as International Standards. These international standards are market driven and globally relevant. They are based on the principle of consensus and are recognized as providing solutions that meet both the requirements of business and the broader needs of society.

These international standards could eventually form a significant portion of global regulations. For example, the ISO/TC 197 and IEC/TC 105 international standards have been used as the basis for the global regulations that apply to the transport of dangerous goods, which are being developed under the auspices of the UN Sub-Committee of Experts on the Transport of Dangerous Goods (SCETDG) and the International Civil Aviation Organization (ICAO). As an example, the UN/SCETDG made the decision to use ISO 16111 as the basis for the *UN Model Regulations* in December 2008. The provisions of the *UN Model Regulations* applicable to hydrogen stored in metal hydride assemblies are now written in such a way that they will allow shipment of hydrogen storage assemblies certified as conforming to ISO 16111 to become routine. Industry will no longer need special permits allocated on a case-by-case basis and the hydrogen stored in metal hydride assemblies will be able to travel across international borders. The removal of this TBT is a result of close cooperation between ISO and the UN/SCETDG. Similarly, the ISO/TC 197 and IEC/TC 105 standards could be referred to in the Global Technical Regulations (GTR) for hydrogen and fuel cell vehicles that are being developed by the *World Forum for Harmonization of Vehicle Regulations* (WP.29).

32.5
International Cooperation in RCS

Although all the instruments are already in place to facilitate global harmonization of RCS, further efforts are required. This is the acknowledgment that emanated from the *ISO Roundtable on Global Harmonization of RCS for Gaseous Fuels and Vehicles* that was held on 10 January 2007 in Geneva, Switzerland. The same conclusions were reached during the *Roundtable on Advancing Commercialization of Hydrogen and Fuel Cell Technologies, a Part of the New Energy Mix, Through International Cooperation of Regulations, Codes and Standards (RCS)* that was held as part of the 2009 European Future Energy Forum on 10 June 2009 in Bilbao, Spain.

The participants of both roundtables agreed that it was of the utmost importance that *the leadership of government and industry work together to enable the energy shift through RCS harmonization and other means!* It was recognized that it is important to create a willingness on the part of decision-makers to work towards the ultimate goal of *"one product, one standard"* that is so critical for the large-scale deployment of the hydrogen and fuel cell technologies.

To avoid the proliferation of national standards that differ from one country to the next, it was recommended that countries be encouraged to contribute to the development of ISO and IEC international standards and to use or adopt them as required by the WTO's TBT Agreement. Further, to avoid duplication of work, national Standards Development Organizations should be encouraged to bring their work to ISO/IEC and cooperate with ISO/IEC in the development of international standards. Lastly, the UN and the national regulatory bodies should be encouraged to recognize the value of the consensus-based ISO and IEC international standards by referring to them in global or national regulations.

The key to this effort resides in the commitment of high-level government and industry representatives to provide the necessary resources and drive in support of international cooperation to achieve global harmonization of RCS for hydrogen and fuel cell technologies. It is only through this kind of cooperation that a full set of globally harmonized RCS will be available when the technologies are ready to enter the market.

32.6
International Cooperation Between RCS and Pre-Normative Research (PNR)

International cooperation is also required between the global RCS bodies and the research organizations, specifically those involved in pre-normative research (PNR). On many hydrogen safety issues, there are gaps in the knowledge base that need to be addressed through research before they can be addressed in international standards. PNR work generally requires experimental activities that are resource based and time consuming. It is therefore essential to ensure that the PNR work is fully in line with the needs expressed by the RCS bodies to cover knowledge gaps. At the same time, it is also vital to have a process to ensure that the results of PNR are validated and then communicated to the RCS bodies.

In the current state-of-the-art, knowledge gaps still need to be addressed in many areas. For example, PNR work is required to understand better the degradation mechanisms that can affect composite cylinders used for the high pressure storage (70 MPa) of hydrogen. This knowledge is required to develop enhanced design requirements and testing procedures that can be incorporated into international standards with the view of ensuring the mechanical integrity of these cylinders throughout their service life. PNR work is also required to understand fully the impact of hydrogen fuel impurities on fuel cells and to develop appropriate methods for testing hydrogen fuel quality. The safe use of fuel cells indoors also presents some challenges. PNR work is required to close the knowledge gaps in hydrogen dispersion, fires, and explosions in confined spaces. The objective would be to develop prevention and mitigation measures that can be included as part of the indoor installation requirements for fuel cell systems. PNR work is also required to support the development of international standard on hydrogen refueling stations (HRS), which will be discussed in more detail in the next section.

32.7
Hydrogen Refueling Stations (HRS)

In terms of market entry, the roll out of hydrogen fuel cell vehicles is expected to gain momentum from 2015 onward in several regions of the world. To ensure the smooth operation of these initial vehicle populations, counting in the tens of thousands, an initial refueling infrastructure is required consisting of several hundred stations per region through 2020.

To support the development of the hydrogen refueling infrastructure, ISO/TC 197 has been working on the development of ISO 20100 *Gaseous Hydrogen – Fuelling Stations*. ISO/TC 197 WG 11, which is responsible for this work, has already developed a technical specification (TS) in 2008. It is continuing its work towards the development and eventual publication of an international standard which will support the planned build-up of refueling infrastructure.

A lot of effort is going into the work of WG 11. A task group is looking specifically at safety distances and hazardous locations. In terms of safety distances, the task group is looking at improving the ISO/TS 20100 safety distance table for each type/category of equipment using a risk-informed rationale. This approach, which is innovative and supported by pre-normative work, will allow more flexibility in the design of the fueling stations without compromising the safety of the installations.

Another task group is looking at defining the requirements applicable to the compressor and dispensing system. One of the interesting aspects that this task group has been working on is the preparation of a dispenser protection table. The dispenser protection table will identify safeguards necessary for each possible equipment malfunction. In this way, the standard will address safety of the users and protection of the downstream equipment on the vehicles.

ISO/TC 197 WG 11 is working on a 48 month schedule towards the publication of the international standard. It is expected that the international standard will be published in May 2012. The scope of their work is to cover the safety requirements applicable to the design, operation, and maintenance of standalone outdoor public and non-public and indoor warehouse fueling stations that dispense gaseous hydrogen used as fuel on-board land vehicles of all types.

To support the ISO/TC 197 work on the hydrogen refueling infrastructure, a number of PNR projects have been initiated. Some modeling and experiments have already been completed to support the new safety distance risk-based approach. Also, some work has been done on the methods of measurement of the response and recovery time of hydrogen sensors, which was initiated in support of the ISO/TC 197 WG 13 work on hydrogen detection apparatus.

In the environment of HRS, it is important that hydrogen sensors can reliably and accurately signal where hydrogen is present. The ISO 26142 standard is being developed to define the performance requirements and test methods of hydrogen detection apparatus which measure and monitor hydrogen concentrations in stationary applications. The PNR work is aimed at defining a reliable method for evaluating sensor response time and recovery time.

When the ISO 20100 standard is finally published as an international standard, it will be up to adopting countries to use it as a basis for the approval of the HRS that will be built on their territory. It is interesting to note that to facilitate the implementation of HRS successfully, Europeans have already recognized the need for harmonized minimum requirements for safe design, operation and maintenance of HRS and also the application of a harmonized permitting process.

In addition to the ISO work, a consortium of European companies completed in 2007 the *HyApproval Handbook*, which is intended to serve as a guideline to this approach. The key recommendation from *HyApproval* was to develop an EC regulatory framework for HRS based on the proven combination of essential requirements and harmonized standards (e.g., ISO 20100). Through this approach, the upcoming EC framework regulation will allow HRS built to the requirements to be accepted across Europe. This will facilitate the implementation of the hydrogen infrastructure, especially in Germany, where major companies signed the *Hydrogen Mobility* initiative on 10 September 2009.

32.8 Conclusion

The large-scale deployment of hydrogen and fuel cell technologies is anticipated around 2015. In the meantime, a collective effort needs to be made to prepare a comprehensive set of globally harmonized RCS that will facilitate the commercialization of fuel cell and hydrogen technologies when these products are ready. Since these international standards have to reflect the state-of-the-art, they should be well supported by PNR work. A special effort should be made on the part of the research organizations to make sure that their PNR work is in line with the expressed needs of RCS bodies (reflecting industry's needs).

It is only through a truly international spirit of cooperation that large-scale deployment of fuel cell and hydrogen products and systems can be achieved within the time frame needed by industry. Decision-makers have a key role to play as success involves moving away from current trends and business as usual. It is not, however, an impossible task; other industry sectors such as the telecommunications industry have successfully overcome this challenge. We should be inspired by their success.

Definitions

Authority	A body that has legal powers and rights (ISO Guide 2:2004, Clause 4.5). *Note:* an authority can be regional, national, or local.
Code	A document that recommends practices or procedures for the design, manufacture, installation, maintenance or utilization of equipment, structures or products (ISO Guide 2:2004, Clause 3.5).
Regulation	A document providing binding legislative rules, that is adopted by an authority (ISO Guide 2:2004, Clause 3.6).
Standard	A document, established by consensus and approved by a recognized body, that provides, for common and repeated use, rules, guidelines, or characteristics for activities or their results, aimed at the achievement of the optimum degree of order in a given context (ISO Guide 2:2004, Clause 3.2).

References

ISO Guide 2:2004 *Standardization and Related Activities – General Vocabulary.*

ISO 16111:2008 *Transportable Gas Storage Devices – Hydrogen Absorbed in Reversible Metal Hydrides.*

ISO/TS 20100:2008 *Gaseous Hydrogen – Fuelling Stations.*

ISO 26142 *Hydrogen Detection Apparatus – Stationary Applications* (to be published soon).

Appendix A	ISO/TC 197 Work Program and Publications
ISO 13984	*Liquid Hydrogen – Land Vehicle Fuelling System Interface*
ISO 13985	*Liquid Hydrogen – Land Vehicle Fuel Tanks*
ISO 14687-1	*Hydrogen Fuel – Product Specification – Part 1: All Applications Except Proton Exchange Membrane (PEM) Fuel Cells for Road Vehicles*
ISO 14687-2	*Hydrogen Fuel – Product Specification – Part 2: Proton Exchange Membrane (PEM) Fuel Cell Applications for Road Vehicles*
ISO 14687-3 [a]	*Hydrogen Fuel – Product Specification – Part 3: Proton Exchange Membrane (PEM) Fuel Cell Applications for Stationary Appliances*
ISO/PAS 15594	*Airport Hydrogen Fuelling Facility*
ISO 15869	*Gaseous Hydrogen and Hydrogen Blends –Land Vehicle Fuel Tanks*
ISO/TR 15916	*Basic Considerations for the Safety of Hydrogen Systems*
ISO 16110-1	*Hydrogen Generators Using Fuel Processing Technologies – Part 1: Safety*
ISO 16110-2 [b]	*Hydrogen Generators Using Fuel Processing Technologies – Part 2: Test Methods for Performance*
ISO 16111	*Transportable Gas Storage Devices – Hydrogen Absorbed in Reversible Metal Hydrides*
ISO 17268	*Compressed Hydrogen Surface Vehicle Refuelling Connection Devices*
ISO 20100	*Gaseous Hydrogen – Fuelling Stations*
ISO 22734-1	*Hydrogen Generators Using Water Electrolysis Process – Part 1: Industrial and Commercial Applications*
ISO 22734-2 [a]	*Hydrogen Generators Using Water Electrolysis Process – Part 2: Residential Applications*
ISO 26142 [b]	*Hydrogen Detection Apparatus – Stationary Applications*

[a] Under development.
[b] To be published soon.

Appendix B IEC/TC 105 Work Program and Publications

IEC/TS 62282-1	Fuel Cell Technologies – Part 1: Terminology
IEC 62282-2	Fuel Cell Technologies – Part 2: Fuel Cell Modules
IEC 62282-3-1	Fuel Cell Technologies – Part 3-1: Stationary Fuel Cell Power Systems – Safety
IEC 62282-3-2	Fuel Cell Technologies – Part 3-2: Stationary Fuel Cell Power Systems – Performance Test Methods
IEC 62282-3-201 [a]	Fuel Cell Technologies – Part 3-201: Stationary Fuel Cell Power Systems – Performance Test Methods for Small Polymer Electrolyte Fuel Cell Power Systems
IEC 62282-3-3	Fuel Cell Technologies – Part 3-3: Stationary Fuel Cell Power Systems – Installation
IEC 62282-4 [a]	Fuel Cell Technologies – Part 4: Fuel Cell System for Propulsion And Auxiliary Power Units (APU)
IEC 62282-5-1	Fuel Cell Technologies – Part 5-1: Portable Fuel Cell Power Systems – Safety
IEC 62282-6-100 [b]	Fuel Cell Technologies – Part 6-100: Micro Fuel Cell Power Systems- Safety
IEC 62282-6-200	Fuel Cell Technologies – Part 6-200: Micro Fuel Cell Power Systems- Performance
IEC 62282-6-300	Fuel Cell Technologies – Part 6-300: Micro Fuel Cell Power Systems – Fuel Cartridge Interchangeability
IEC/TS 62282-7 [b]	Fuel Cell Technologies – Part 7-1: Single Cell Test Method for Polymer Electrolyte Fuel Cell (PEFC)

a) Under development.
b) To be published soon.

Existing and Emerging Markets

33
Aerospace Applications of Hydrogen and Fuel Cells

Christian Roessler, Joachim Schoemann, and Horst Baier

Abstract

The expected climate changes force the aviation industry to reduce emissions. The fuel cell offers high efficiencies and hydrogen, as the fuel of choice, a much higher energetic value than fossil fuels. The requirements for fuel cell systems on flying platforms comprise low weight, high reliability, and insensitivity to temperature and density changes, in addition to high tilt angles. Fuel cell-powered air vehicles are only competitive where low power and high endurance are required. This is the case for small unmanned vehicles, which are already operated with suitable systems in a considerable number. As proven by prototypes based on motor gliders, manned air vehicles can also be powered only by fuel cells, but the flight performance is poor compared with their internal combustion engine equivalents. In the area of transport aircraft, the use of fuel cells as a primary power source is unrealistic. The technology is hence considered for auxiliary power sources or ground power supply. The type chosen for the recently flown and developed vehicles is almost exclusively the proton exchange membrane fuel cell. For applications in transport aircraft, the solid oxide fuel cell is also considered because of its ability for reforming kerosene, which will still be the fuel for the next decade.

Keywords: aerospace applications, aircraft, hydrogen, fuel cells

33.1
Introduction and Overview of Hydrogen and Fuel Cell Use

Hydrogen is often used *in space applications,* although in relatively low quantities in absolute terms because of the special field. Although in satellites liquid hydrogen is used, for example, for cooling of infrared sensors and instruments which due to unavoidable evaporation then also might limit the satellite's usage time, their power supply is mainly achieved, apart from very rare exceptions, via solar voltaic means provided by often fairly large solar cell arrays deployed in orbit. Necessary storage of energy is provided by rechargeable batteries. Because

Hydrogen Energy. Edited by Detlef Stolten
Copyright © 2010 WILEY-VCH Verlag GmbH & Co. KGaA, Weinheim
ISBN: 978-3-527-32711-9

of its favorable energy versus mass density and thrust efficiency, liquid hydrogen is used in the main propulsion system of all major new types of launcher systems, including Europe's Ariane 5 launcher, together with liquid oxygen as an oxidizer. This holds for both the first and second stages of the two-stage Ariane 5, where in its newest version the propulsion system of the second stage can be switched off and on in space in order to achieve in a single mission different orbits for different payloads. For the United States Space Shuttle, hydrogen fuel cells are also used to provide water with an on-board capacity of roughly 12 kg of water per hour, with a maximum mission time of the Shuttle close to 2 weeks.

Because of the much larger market volume and expected future growth, the focus in the following is on *applications in aircraft*. This relates to unmanned aerial vehicles (UAVs), to general aviation (such as business aircraft), and of course to large commercial aircraft.

Worldwide air traffic is expected to grow by about 5% p.a. [1], which means an increase of more than 50% by 2020. In contrast the, ACARE (Advisory Council for Aeronautics Research) goals to be obtained in 2020 are as follows [2]:

- 50% reduction of CO_2 emissions through drastic reduction of fuel consumption
- 80% reduction of NO_x emissions
- 50% reduction of external noise
- a green design, manufacturing, maintenance, and product disposal life cycle.

Together with more and more strict restrictions of local emissions at airports the need for new more efficient and fuel saving technologies is obvious.

A possibility of direct use of hydrogen as a substitute for kerosene in commercial aircraft has been investigated in the EU framework study Cryoplane. In brief, the findings from this study are as follows:

- A significant reduction (per definition) of climate gases can be achieved, while the amount of contrails might be increased.
- Due to the higher volume by a factor of roughly four compared with kerosene, a Cryoplane will have to have a larger cross-section, which increases aerodynamic drag. On the other hand, the fuel mass will be lower for an equivalent energy content.
- As in other applications, environmental benefit will be achieved only if the hydrogen is produced via renewable energy sources.
- Due to limited world-wide infrastructure and still very high cost, for commercial aircraft this will be an interesting option only in the very long term.

A very promising technology is the fuel cell and its potential attracted remarkable attention in industry and politics. Of its many characteristics, two benefits have been instrumental in driving the interest in using fuel cells as a future energy source: this energy conversion device operates with high efficiencies, and at the same time hardly intrudes on the environment, due to its low acid gas or solid emissions [3]. More general features of fuel cells, that are important for aviation applications, are listed in Table 33.1.

Table 33.1 Fuel cell characteristics.

Advantages	Disadvantages
High efficiency	High market entry cost
Low emissions	Unfamiliar technology to the power industry
Fuel flexibility	No infrastructure
Quiet	Low power to weight ratio
Siting ability	
Direct energy conversion (no combustion)	
No moving parts in the energy converter	
Demonstrated endurance/reliability	
Good performance at off-design load operation	
Modular installations to match load and increase reliability	
Remote/unattended operation	
Size flexibility	
Rapid load following capability	

Many research projects are being carried out on making fuel cells ready for use in aircraft. Their main application will be the supply of electricity. However, to compensate their weight disadvantages, system integration is critical and also other benefits and side products of the fuel cell must be used, such as water, low oxygen-containing exhaust air for inerting of the fuel tank, and maybe heat for functions such as de-icing. Unfortunately, the power-to-weight ratio is much lower than for today's jet engines so that fuel cells will not be feasible for propulsion in jet aircraft. However, they could be used to replace combustion engines in smaller aircraft such as motor gliders and UAVs in the future. Some demonstrators and research aircraft of this kind are already flying.

This chapter considers the possible aircraft applications of fuel cells in the future. First, eligible types of fuel cells are described, then the application possibilities for different sectors, namely unmanned vehicles, general aviation, sport utility aircraft, and commercial transport aircraft are discussed and already successful applications for each of these sectors are described.

33.2
Possible Fuel Cell Types for Aviation

The basic principle and different types of fuel cells were outlined by Bellis [4]. Of the variety of fuel cell types, two in particular show potential for application in aircraft: the polymer electrolyte fuel cell (PEFC), also called the proton exchange membrane fuel cell (PEM-FC), and the high-temperature solid oxide fuel cell SOFC [5].

The PEFC operates, at the present development state, in the temperature range below 100 °C. It shows a relatively high level of development and different systems will be available in a relatively short time. However, as the catalyst consists of platinum, which becomes passivated in its catalytic activity by CO, there is a high demand for cleaning in the reformation process of kerosene in order to achieve a tolerably low CO content of the fuel cell fuel in the ppm range. Therefore, one of the main tasks in the future should be the development of compact reformers and means for reducing fuel poisons such as CO and also sulfur. Before the reformation problems are solved, PEFCs are only suited for applications where pure hydrogen as fuel is feasible and acceptable. This is mainly the case in airplanes, where the fuel cell is the only power system including propulsion and therefore only hydrogen is used as fuel and for emergency systems.

The operating range of high-temperature SOFCs in already realized technical versions is currently around or above 800 °C. Recent developments are aimed at entering the range around 700 °C. CO in the fuel represents no problem for SOFCs and, due to the high temperature and the composition of the anode, the cell also has reforming qualities. Therefore, reforming and cleaning of kerosene for SOFCs is relatively simple. Despite the fact that the development state of SOFCs is inferior to that of PEFCs, the SOFCs have greater attraction for aircraft applications, where kerosene will still be the fuel for the next few decades, especially for large aircraft. Also, the combination of a SOFC with a gas turbine offers a very high efficiency potential of over 70%.

Before fuel cells become common in aircraft, further improvements in the following areas are necessary. In addition to reliability and availability, which are mandatory for aircraft applications, further key issues for the development of the fuel cell system for aircraft applications are high power density, low weight, suited for rapid heating up, which means also tolerance towards high thermal gradients and transients (SOFCs), and stability under mechanical stress due to vibrations and shocks of aircraft operation. The power-related design targets are 2 kW kg^{-1} and 2 kW l^{-1} for the stack and 1 kW kg^{-1} and 1 kW l^{-1} for the entire system [6]. Furthermore, the fuel cell must be operated under harsh environmental conditions, including low-pressure operation and rapid climatic changes, and under different inclination angles.

33.3
Application in Unmanned Aerial Vehicles (UAVs)

UAVs are available in a very wide variety. Already electrically powered small UAVs in particular are suited for first applications of fuel cells. By exchanging the batteries partially or totally with a fuel cell system, flight times can be significantly extended. Due to their low power requirement, weights and costs are in feasible ranges. Some research and demonstrator airplanes, which will be introduced later, have already been built.

However, the first application of fuel cells in the aeronautical sector was for a high-altitude ultra-long endurance airplane that would be capable of flying for several days. This is only possible with revolutionary new technologies.

33.3.1
High-Altitude (Ultra-) Long Endurance

The Helios program was carried out in the United States between 1999 and 2003 [7]. The Helios prototype (Figure 33.1) is a remotely piloted all-wing aircraft developed by NASA's Environmental Research Aircraft and Sensor Technology project in cooperation with AeroVironment, Inc. It is designed for high-altitude operation with the ultimate goal of reaching endurances from several days to months by applying a combination of solar panels and fuel cells.

The solar array produces electricity to supply the electric motor and an electrolyzer during daytime. The electrolyzer produces hydrogen and oxygen that are stored in the respective tanks on-board. The electrolyzer is supplied with water from a water tank. During night operation, the fuel cell is fed by hydrogen and oxygen of the respective tanks. It then delivers the electric power for the electric motor of the propulsion system. The water produced by the fuel cell is stored in the water tank during the night to be used by the electrolyzer during daytime. However, before this power system could be tested thoroughly, the aircraft crashed during a test flight in 2003 due to unexpected wind conditions [8].

On the 26 May 2005, AeroVironment successfully completed flight tests in Arizona of another unmanned platform with a similar aim to that of the Helios program, but without using solar energy. It was a scaled prototype of the Global Observer [9]. The fuel cell-powered UAV had a distributed electrical architecture in which liquid hydrogen fuel cells provided electricity to electric motors driving eight propellers. The Global Observer program continues, with the aim of developing a platform able to stay aloft at high altitude (20 000 m) for at least 1 week while carrying a 450 kg payload to perform surveillance, reconnaissance, and frontier monitoring missions.

Fig. 33.1 Helios UAV [7].

In 2008, the Defense Advanced Research Projects Agency (DARPA) launched the Vulture program to develop new UAV concepts able to stay aloft at high altitude for 5 years without interruption for intelligence, communications, surveillance, and reconnaissance missions over areas of interest [10]. Currently, the only systems capable of providing multiple years of coverage over a fixed area are geosynchronous satellites orbiting 22 233 miles above Earth. The Vulture vehicle's goal is to be capable of carrying a 1000 lb ($\sim$450 kg), 5 kW payload and have a 99% probability of maintaining its on-station position, which means maintaining sufficient speed to withstand the winds at 18 300–27 500 m. Amongst other technologies, SOFCs are being considered within this program.

33.3.2
Small Low-Altitude UAVs

In addition to the environmental and low noise aspects, fuel cell-driven electrical power trains offer energy density advantages compared with battery-powered propulsion systems. Lithium-based batteries have an energy density of around 200 Wh kg^{-1}. Today's fuel cell systems including the tank can reach 500 Wh kg^{-1}, which is 2.5 times higher. On the other hand, the power density of the fuel cells is much lower (333 W kg^{-1}) than that of today's lithium polymer batteries (up to 3500 W kg^{-1}). Therefore, the predestined application of fuel

Table 33.2 Summary of current fuel cell UAVs [11–26].

UAV	Gross mass (kg)	Wing span (m)	Endurance (expected) (h)	Fuel cell type	Power (W)	Fuel storage type	Fuel cell manufacturer
Spider Lion	2.54	2.2	3.3	PEM	100	Compressed hydrogen	Protonex Technology Corp.
Georgia Tech UAV	16.4	6.6	0.75	PEM	500	Compressed hydrogen	
Hyfish	6	1	0.25	PEM	1000	Compressed hydrogen	Horizon Fuel Cell Technologies
Puma	5.7	2.6	9	PEM		Chemical hydride	Protonex Technology Corporation
Pterosoar	5	5.5	(15.5)	PEM	650	Compressed hydrogen	Horizon Fuel Cell Technologies
Endurance	5.4	2.4	10.25	SOFC		Propane	Adaptive Materials, Inc.
Ion Tiger			26	PEM	500	Compressed hydrogen	Protonex Technology Corporation

cells is for small, slow-flying UAVs where the power demand is low and a significant endurance increase compared with battery-powered UAVs in this class can be achieved. Several UAVs of this type have already been built and are introduced here. A summary with the technical data and the fuel cell types used is given in Table 33.2.

Spider Lion
In a demonstration flight in November 2005, the Spider Lion UAV (Figure 33.2) flew for about 3.3 h, consuming 15 g of hydrogen [12]. The 100 W fuel cell system was designed and constructed at NRL largely using commercially available hardware and a fuel cell stack and components developed by Protonex. The Spider Lion UAS was developed by NRL as a high-impact research platform for testing fuel cell technology. Research and development continues, aimed at developing a fuel cell system capable of powering small military platforms currently in the field or in advanced development stages and requiring extended operation that is not achievable using current battery technology.

Georgia Tech UAV
A PEM-FC-powered UAV was designed during a research project at the Georgia Institute of Technology (Figure 33.3). The first flight was performed in August 2006. A detailed description of the fuel cell system, the airplane, and results from test flights is given in [14].

Hyfish
The Hyfish UAV (Figure 33.4) was developed with cooperation between DLR (German Air and Space Center), Horizon Fuel Cell Technologies, and Team Smartfish GmbH. It features an unconventional aerodynamic shape designed by Smartfish resembling the form of a fish. In a first test flight in April 2007, this air-fed fuel cell combined with an impeller propulsion system was able to propel the UAV to speeds of up to 200 km h–1 [15].

Fig. 33.2 Spider Lion UAV [11].

Fig. 33.3 Georgia Tech UAV [13].

Fig. 33.4 Hyfish UAV [15].

Fig. 33.5 Puma UAV [17].

PUMA

A derivative of the AeroVironment Puma (Figure 33.5) powered by a combination of fuel cell and batteries achieved a new long-duration flight record in March 2008. The aircraft managed to stay airborne for 9 h, breaking its previous year's record of 7 h [18].

Pterosoar

Oklahoma State University, California State University Los Angeles, and Horizon Fuel Cell Technologies, among others, participated in the Pterosoar project. In November 2007, the Pterosoar UAV (Figure 33.6) set a new micro UAV flight distance record of 128 km (78 miles) in over 3 h using only 25% of its hydrogen tank capacity. To qualify as a micro UAV, the gross weight of the Pterosoar

Fig. 33.6 Pterosoar UAV [19].

Fig. 33.7 Endurance UAV [21].

had to be slightly below 5 kg for the mentioned record run. With a full tank of hydrogen, the aircraft is supposed to have a range of roughly 500 km. The fuel cell employed in this aircraft is manufactured by Horizon Fuel Cell Technologies and is claimed to have an energy density of over 480 Wh kg^{-1} [20].

Endurance
The Solarbubbles UAV Team at the University of Michigan designed a remotely piloted UAV called Endurance (Figure 33.7). In October 2008, this UAV broke the endurance record of 9 h previously held by the AeroVironment Puma. The record flight of Endurance took 10 h 15 min while traveling roughly 159 km (99 miles) [22].

Ion Tiger
Researchers at the Naval Research Laboratory (Office of Naval Research) developed the Ion Tiger UAV (Figure 33.8). Previous battery-powered tests confirmed an endurance of roughly 3 h in March 2009 [25]. subsequently, using a polymer fuel cell, it set a new record with 26 h 1 min on its test flight on 16 and 17 November 2009 [26]. The applied fuel cell provides 500 W of electrical power (Protonex Technology Corporation) and is supplied by high-pressure hydrogen storage tanks. Ion Tiger is able to carry a payload mass of up to 2.27 kg (5 lb).

Fig. 33.8 Ion Tiger UAV [25].

33.4
Applications in General Aviation

Similarly to the situation in the unmanned sector, fuel cell-powered General Aviation (GA) aircraft are flying testbeds and far from batch production. In contrast to UAVs, there is a large number of commercially available aircraft that can be used to integrate fuel cells, thus saving the research teams from putting effort into designing a platform from scratch. Another advantage of conducting experiments on GA vehicles is the heavier payload that they can carry. Furthermore, having a pilot on-board makes autonomous control systems obsolete. Of course, this also adds a lot of the risk to the project, as a hazardous failure of the system might cause not only material damage but also physical injury.

For the choice of the platform, three points are decisive. It can be generally stated that actual fuel cell systems have a lower power density than internal combustion engines. Hence to avoid major structural changes to the platform, it needs to allow sufficient payload to account for the additional weight. The heavier an aircraft is, the more lift needs to be generated. Lift induces drag that requires more thrust, which implies more weight again. Consequently, the second requirement is that the platform is aerodynamically as optimized as possible, commonly expressed in a high lift-to-drag ratio. Furthermore, it has to be considered that all flying systems will be experimental. Therefore, a configuration that allows the unit to be glided down to the airfield in case of a propulsion power system failure enhances safety substantially. For these reasons, the two already flown fuel cell-powered systems are motor gliders.

The world saw the first two flights of manned fuel cell-powered aircraft in the last 2 years. The first flight of a manned fuel cell aircraft was performed by Boeing's Madrid branch in March 2008 [27]. However, the aircraft was only powered by the fuel cell in level flight; for takeoff and climb the hybrid system was supported by a lithium ion battery. The first aircraft to complete every flight phase powered by a fuel cell was the Antares DLR-H2, developed by the German Aerospace Center (DLR, Deutsches Zentrum für Luft- und Raumfahrt) and first flown in July 2009 [29].

The platforms used are both modified motor gliders. Boeing based its system (Figures 33.9 and 33.10) on an HK36 Superdimona from the Austrian manufac-

Fig. 33.9 Systems overview of the Boeing Fuel Cell Demonstrator airplane [31].

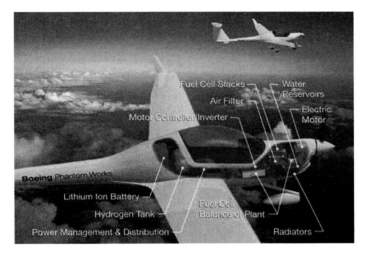

Fig. 33.10 Boeing Fuel Cell Demonstrator in flight [30].

turer Diamond Aircraft, whereas DLR used an Antares 20E from the German company Lange Aviation that was also involved in the project. The HK36 itself is bigger and consequently heavier than the Antares 20E, as can be seen in Table 33.3. Both aircraft were equipped with fuel cells with a continuous power of about 20 kW (DLR) and 15k W (Boeing). As a characteristic value for flight performance of propeller aircraft, the power loading, defined as the ratio between the power plant continuous power and the maximum weight, is commonly used. The standard version of the HK36 is equipped with a 100 hp (or 115 hp) internal combustion engine that generates of power load of 10.9 W N^{-1} (or 11.4 W N^{-1}). The standard version of the Antares 20E comes with a 42 kW brushless d.c. motor and lithium ion batteries [29]. This combination leads to a maximum power load of 5.7 W N^{-1}. Replacing those power plants with fuel cells drastically decreases the power load. The Boeing demonstrator uses a brushless d.c. motor with a maximum power output of 75 kW. With only the

Table 33.3 General aviation aircraft used as fuel cell research aircraft [27, 28, 32].

Property	German Aerospace Center		The Boeing Company	
	Antares 20E	Antares DLR-H2	HK 36 TC115	Fuel Cell Demonstrator
Maximum speed (km h^{-1})	Unknown	170	261	158
Empty weight (kg)	460	460	568	789
Maximum weight (kg)	660	750	770	860
Fuel cell continuous power (kW)	–	20	–	15
Fuel cell maximum power (kW)	–	25	–	20
Motor maximum power (kW)	42	42	85.8 (115 hp)	75
Power load (W N^{-1})	5.7	2.7	11.4	1.9
Power load hybrid system (W N^{-1})	–	–	–	11.7

fuel cell active, the power load is reduced to 1.9 W N^{-1}. In hybrid mode, the additional 50–75 kW of the auxiliary battery allows a power load of up to 11.7 W N^{-1} that is about equal to the design value [27]. The Antares DLR-H2, on the other hand, reaches a power load of 2.7 W N^{-1} that is clearly higher and closer to the value of the standard aircraft and thus explains the capability to take off and climb powered by the fuel cell. A comparison of the original and modified aircraft is given in Table 33.3.

The Diamond Superdimona in its commercial version is a two-seater, offering enough space inside the fuselage to accommodate all the necessary systems. Hence the exterior modifications were minor. This avoids investigating aeroelasticity for the wings and does not influence aerodynamics. The team around Nieves Lapea-Rey described in great detail the systems they used and modifications that were necessary to accommodate them in the airframe [27, 28]. The internal combustion engine, the fuel tank and the whole fuel system were removed. Instead, an electric d.c. brushless motor, the fuel cell system, a high-pressure hydrogen tank, and an auxiliary battery where added in addition to an electronic control system. These components were distributed throughout the fuselage, so that the balance of the aircraft was maintained and electrical components and wiring were separated as far as possible from the components and tubes containing hydrogen [27].

The fuel cell system consists of two PEM-FC stacks in series working at about 50% efficiency. Hydrogen is supplied from a 350 bar tank and ambient air is pressurized. The water that is created as a reaction by-product is re-used for the humidification of the membrane electrode assemblies (MEAs). Although the fuel cell system provides a maximum gross power of 24 kW, the net maximum power is about 20 kW as the rest is used to power the controllers, radiators, and the compressor [28].

Major problems that the team encountered in addition to the mentioned balance issue of the aircraft were the cooling of the fuel cell system and the motor as no heat is dissipated through exhaust gases and power management for the parallel operation of the fuel cell and the battery. Furthermore, safety issues were treated very seriously. Concerning the use of hydrogen on the one hand, the ATEX (EU Explosive Atmosphere Directive, from French: Appareils destinés á tre utilisés en ATmosphères EXplosibles) certificate is mostly not available for mobile devices. Therefore, an examination of whether a component can be considered spark free had to be done manually. On the other hand, the custom-made lithium ion battery used is highly sensitive to charge and discharge, rupture, fire, and temperature. Those issues were coped with by a battery management unit, a venting system, and a certified mounting into the aircraft [28].

Flight control systems were not changed at all. To avoid a critical fuel cell malfunction affecting them, a backup battery was installed for redundancy. Structural changes comprised new mounts for the components, a modification of the front cowling to allow for more room in the engine bay and include hydrogen outlets, and the complete removal of the co-pilot's seat, instrument panel, and control levers to free up room for the systems [27]. All changes need to be in accordance with EASA CS 22 regulations (certification specifications for sailplanes and powered sailplanes).

The Antares 20E is a one-seater aircraft (Figure 33.11). This makes it a lighter aircraft than the Superdimona, but it requires external installations to store the system components. The two external pods under the wing make the aircraft clearly recognizable. The additional drag caused by this installation is stated to be below 10% while 200 kg of additional mass can be accommodated. The fuel cell used is a PEM-FC very similar to the one that the team of Josef Kallo already used in the Airbus A320 ATRA research aircraft. With an overall weight of 60 kg, without hydrogen tank, the system delivers a maximum power of

Fig. 33.11 The Antares DLR-H2 [29].

about 25 kW and continuous power of above 20 kW. The maximum efficiency is about 52% [29].

Projects on the integration of fuel cells in general aviation aircraft include the Hydrogenius of the University of Stuttgart, which is expected for the 2011 NASA Green Flight Challenge. The project coordinator Rudolf Voit-Nitschmann was also responsible for the solar aircraft Icaré developed at the same university. In contrast to other projects, his team is developing a new motor glider with an estimated takeoff weight of 850 kg instead of using an existing platform [33].

The partners of the European Commission FP6 project ENFICA-FC (ENvironmentally Friendly Inter City Aircraft powered by Fuel Cells), led by the Polytechnic University of Turin, entered the final flight test phase in December 2009 [34]. For this project, a Rapid 200 by Jihlavan Aircraft is powered by a PEM-FC system [35].

33.5
Application to Commercial Transport Aircraft

Due to their low power density compared with jet engines, fuel cells are not suitable to replace the propulsion system. However, the trend in the aircraft industry is clearly to a more and more electric aircraft with a higher proportion of electrical systems replacing hydraulically or pneumatically driven components. Fuel cell systems are a promising solution for supplying the rising demand for electricity with enhanced energy efficiency in both flight and ground operations.

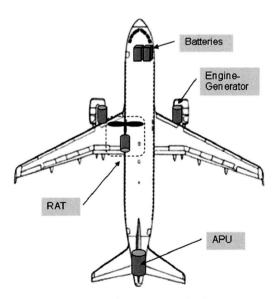

Fig. 33.12 Energy supply components for the A320.

33.5 Application to Commercial Transport Aircraft

Today there are four different systems onboard an aircraft to supply energy, shown in Figure 33.12. The batteries are mainly for starting the auxiliary power unit (APU). For the A320 aircraft there are two batteries of 28 V and 23 A h each. The APU is a small jet engine that is used for energy supply during ground operation and is normally switched off during flight. Additionally, the APU delivers air for the environmental control system (ECS) on the ground and pressurized air for starting the main engines. Although the main purpose of the engines is to generate propulsion, they also supply the electric power, the hydraulic system's energy demand, and bleed air for de-icing and ECS during flight. The extraction of bleed air from the core engine of modern high bypass ratio engines significantly reduces their efficiency. The operation of the main engine is therefore an optimized compromise of these different demands. The efficiency of the power generation of the main engine is in the order of 40% and that of the APU in the order of 20% [6]. For the rare event that the main engines and also the APU fail, the batteries can provide power for a short time until the ram air turbine (RAT) is deployed. The RAT is an emergency system that consists of a propeller that drives a hydraulic pump and a hydraulically driven generator. In the following, the possibility of substituting these systems with a fuel cell is discussed.

33.5.1
Substitution of the Batteries

The substitution of the batteries would of course be possible. However, substituting them with a system that also needs batteries, although smaller, for start-up seems inadequate. Furthermore, today's batteries are highly reliable and the complexity of the new system would increase.

33.5.2
Substitution of the Engine Generators

In this case, the engines can only be used for propulsion and the operation can be optimized for this purpose only. The fuel cell systems become the main power source during flight and also in park positions. The electric power is used as the main energy for all systems and for producing bleed air. The hydraulic system is changed to an electric system and local hydraulic components are electrically powered. Finally, the fuel cell can be used for the generation of usable water from its reaction water. The main energy saving can be expected by the better efficiency of the main engine and the much higher efficiency of the electric power generation in the fuel cell system (60–75%) with a SOFC system, coupled with a gas turbine.

On the other hand, the generator in an A320 aircraft is capable of generating 115 kW and weighs only 26 kg. A comparable fuel cell system would weigh as much as 400 kg [5]. For redundancy reasons, more than one system would be needed. Hence the disadvantages due to the much higher weight and volume demand of a fuel cell make this concept unfeasible. However, the generation of the necessary energy through a separate system is still a very promising approach which corresponds to substituting the APU described next. The engine generators could still be used as an emergency system.

33.5.3
Substitution of the APU

Instead of the APU, a fuel cell system as shown in Figure 33.13 would provide the necessary power not only on the ground but also during flight. All advantages described above would also be valid here. In addition, further advantages arise during ground operation by substituting the APU. Increasingly strict noise and emission regulations [CO, NO_x and hydrocarbons (HCs)] for airports lead to stringent limitations on travel schedules and flight movement numbers. The APU operation on the ground contributes to a great deal of these emissions. With a fuel cell, the noise and emissions would be reduced considerably. On the other hand, the weight and the necessary volume would still be much high-

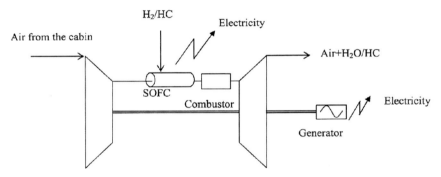

Fig. 33.13 SOFC with gas turbine for APU substitution [37].

Table 33.4 APU and fuel cell system comparison [5].

Property	APU	Fuel cell
System weight (kg)	180	1000
Total power (kW)	400 (132 electrical, 252 pneumatic, 16 other and losses)	400
Additional drive systems (kW)	16	16
Efficiency (full load) (%)	19	60–75
Dimensions (mm)	1350 × 850	1500 × 1500
Average operating time per flight (h)	1.5–2	Total flight

er, as shown in Table 33.4 for an A320 aircraft. However, taking into account the fuel savings of around 50 kg h^{-1} and production of water up to 400 kg for a mid-range flight [5], this disadvantage is reduced considerably depending on the flight time. In this scenario, only one fuel cell system is necessary and the main engine generators could still be kept for redundancy reasons. With rising integration and utilization of further synergy effects, more weight savings are possible. For example, the low oxygen exhaust air from the fuel cell could be used for inerting the fuel tank, making an additional currently used system, based on nitrogen tanks, obsolete [36]. Summarizing, the substitution of the APU with a fuel cell system is very promising and substantial research is being carried out in this area.

33.5.4
Substitution of the RAT

For an A 320 aircraft, the power of this system is 10–15 kW [5]. The RAT is an emergency system. By deploying a propeller, which extracts energy from the wind stream, hydraulic pressure for actuator movements and electricity for the avionics and emergency lights is generated. This causes a considerable amount of additional drag, reducing the glide ratio und thus the time and distance available for reaching the next airport. A fuel cell system would generate the power much more efficiently and would not cause this additional drag. As it is an emergency system, it must be operable independently from other systems. This would rectify the installation of a separate fuel system independent of kerosene, which could be pure hydrogen and oxygen tanks. This is the ideal precondition for a PEFC fuel cell.

The first fuel cell system in commercial aircraft is such an emergency system, which has been developed in collaboration between DLR, Airbus, and Michelin. The 20 kW H_2/O_2-operated system powered the control surfaces of the A320 DLR research aircraft during several successful flights [38]. The next step is to include more functions, such as continuous supplemental power supply, inerting the fuel tanks, and so on, up to a system capable of providing power for all loads including taxiing on ground by 2020 [36].

33.6
Conclusion

More efficient use of energy sources and less environmental impact of energy-using systems are general demands. In the same sense, aircraft of the future also have to show reduced exhaust and considerably improved efficiency. One exciting idea is the use of fuel cells for the supply of electricity for almost all functions on board, except propulsion of large aircraft. In the medium- and long-term UAV sector and general aviation, aircraft could be completely powered by a fuel cell system. For commercial transport aircraft, all mechanically, hydraulically, and pneumatically activated aggregates should be replaced by more efficient and better controllable electric systems, beginning perhaps with the currently used turbine-based APU.

It is a major challenge to develop suitable fuel cell systems, because high demands exist concerning reliability, power density, low weight, and long life time. There are different types of fuel cells under development or already available as prototypes. The low-temperature fuel cell PEFCs and the high-temperature SOFCs seem to show the highest potential from these developments. PEFCs are already used for applications where direct hydrogen supply is feasible and acceptable. However, because kerosene will remain the preferred fuel, at least for the next few decades, and SOFCs have less reforming and cleaning requirements to obtain a suitable fuel cell fuel, SOFC systems will attract more interest, despite their lower development state at present.

References

1 Airbus (2009) *Airbus Global Market Forecast 2009–2028*, http://www.airbus.com/en/corporate/gmf2009 (accessed 30 November 2009).
2 European Communities (2001) *ACARE Report European Aeronautics: a Vision for 2020*, Office for Official Publications of the European Communities, Luxembourg.
3 EG&G Technical Services (2003) *Fuel Cell Handbook*, 6th edn, DOE/NETL-2002/1179, US Department of Energy, Morgantown, WV.
4 Bellis, M. (2009) *Hydrogen Fuel Cells – Innovation for the 21st Century*, http://inventors.about.com/od/fstartinventions/a/Fuel_Cells_2.htm (accessed 30 November 2009).
5 Wiener, D. (2000) Anwendungsmöglichkeiten der Brennstoffzelle in der Luftfahrt, Diploma Thesis, Institute of Aviation, Technical University of Munich (in German).
6 Winkler, W. (2004) Fuel cells in aircraft and synergies, presented at H2 Expo 2004, Hamburg.
7 NASA (2009) *Helios Fact Sheet*, http://www.nasa.gov/centers/dryden/news/FactSheets/FS-068-DFRC.html (accessed 30 November 2009).
8 Noll, T. E., Brown, J. M., Perez-Davis, M. E., Ishmael, S. D., Tiffany, G. C., and Gaier, M. (2004) Investigation of the Helios Prototype Mishap, Vol. I: Mishap Report, NASA, Washington, DC.
9 AeroVironment (2009) *Global Observer of AeroVironment*, http://www.avinc.com/ADC_Project_Details.asp?Prodid=35 (accessed 26 October 2009).
10 Boeing (2008) *Vulture Program*, http://www.boeing.com/news/releases/2008/q2/080421d_pr.html (accessed 30 November 2009).
11 NRL (2006) *UAV Fact Sheets*, http://www.nrl.navy.mil/techtransfer/exhibits/pdfs/Info%20Sheet%20pdfs/UAV%20Info%20Sheets/Fuel%20Cell%202006.pdf (accessed 17 February 2009).
12 NRL (2005) *NRL Demonstrates Fuel Cell-Powered Unmanned Aerial System*, http://www.nrl.navy.mil/pao/pressRelease.php?Y=2005&R=59-05r (accessed 17 February 2009).
13 Georgia Institute of Technology (2006) *UAPT URETI Project Summary Presentation*, http://www.fcbt.gatech.edu/fuelcellairplane/Documents%20Folder/2_6_Education_PEM_FC_UAV_v2_2.pdf (accessed 12 March 2010).
14 Bradley, T. H., Moffitt, B. A., Thomas, R. W., Mavris, D., and Parekh, D. E. (2006) *Test Results for a Fuel Cell-Powered Demonstration Aircraft*, SAE Publication No. 2006-01-3092, SAE International, Warrendale, PA.

15 Horizon Fuel Cell Technologies (2007) *World's First Zero Emission, Hydrogen Fuel Cell Jet*, http://www.horizonfuelcell.com/hyfish.htm (accessed 16 February 2009).

16 Deutsches Zentrum für Luft- und Raumfahrt (2007) *Erfolgreicher Erstflug des "HyFish" – ein Brennstoffzellen-Flugmodell geht in die Luft*, http://www.dlr.de/desktopdefault.aspx/tabid-13/135_read-8329 (accessed 16 February 2009).

17 AeroVironment (2007) http://www.designation-systems.net/dusrm/app4/puma.html (accessed 16 February 2010)

18 AeroVironment (2009) *AeroVironment Puma Small UAS Achieves Record Flight of Over Nine Hours Using Fuel Cell Battery Hybrid System*, 30. 11. 2009, http://www.avinc.com/resources/press_release/aerovironment_puma_small_uas_achiev es_record_flight (accessed 30 November 2009).

19 Oklahoma State University (2007) *Micro-UAV distance record smashed*, http://www.gizmag.com/go/8287 (accessed 16 February 2009).

20 Horizon Fuel Cell Technologies (2007) *Horizon Fuel Cell Powers New World Record in UAV Flight*, http://www.horizonfuelcell.com/file/Pterosoardistancerecord.pdf (accessed 16 February 2009).

21 Oklahoma State University (2007) *Fuel Cell Powered UAV*, 17.2.2009, http://www.calstatela.edu/centers/mfdclab/research/fcuav.htm (accessed 17 February 2009).

22 Adaptive Materials Inc. (2008) *Adaptive Materials Inc. and Michigan Students Set Record Fuel-Cell-Powered, Radio-Controlled Airplane Flight*, http://www.adaptivematerials.com/internal.php?sid=5&nid=51 (accessed 18 February 2009).

23 University of Michigan (2008) *Fuel-Cell-Powered, Radio-Controlled Airplane, Endurance, Takes Flight*, http://www.ur.umich.edu/0809/Nov17_08/01.php (accessed 19 February 2009).

24 University of Michigan (2008) *Solar Bubbles Endurance UAV*, http://solarbubbles.engin.umich.edu/~solarbubbles/media/Pictures/FuelCell/pages/ IMG_1469.html (accessed 18 February 2009).

25 Naval Research Laboratory (2009) *Ion Tiger*, http://www.eurekalert.org/multimedia/pub/13174.php (accessed 7 April 2009).

26 Naval Research Laboratory (2009) Navy's Experimental UAV Flies 26 Hours, Sets New Record, 30. 11. 2009, http://www.dailytech.com/Navys+Experimental+UAV+Flies+26+Hours+Sets+New+Record/a rticle16976.htm.

27 Lapeña-Rey, N., Mosquera, J., Bataller, E., and Ortí, F. (2008) The first fuel cell manned aircraft, presented at COE 2008 Industry Workshop – Aerospace and Defense, Wichita, KS, 2008.

28 Lapea-Rey, N., Mosquera, J., Bataller, E., and Ortí, F. (2007) *The Boeing Fuel Cell Demonstrator Airplane*, SAE 2007 Aerotech Congress and Exhibition, Los Angeles, CA, Publication No. 2007-01-3906, SAE Internatiooal, Warrendale, PA.

29 Deutsches Zentrum für Luft- und Raumfahrt (2007) *Antares DLR-H2: Weltweit erstes pilotengesteuertes Flugzeug mit Brennstoffzellenantrieb*, http://www.dlr.de/tt/desktopdefault.aspx/tabid-4935/8219_read-13587/ (accessed 30 November 2009).

30 Boeing (2007) *Boeing Prepares Fuel Cell Demonstrator Airplane for Ground and Flight Testing*, http://www.boeing.com/news/releases/2007/q1/070327e_pr.html (accessed 6 December 2009).

31 Boeing (2008) *Boeing Successfully Flies Fuel Cell-Powered Airplane*, http://www.boeing.com/news/releases/2008/q2/080403a_nr.html (accessed 30 November 2009).

32 Lange Aviation (2009) *Lange Aviation Antares 20E – Technische Daten*, http://www.lange-aviation.com/htm/deutsch/produkte/antares_20e/technische_daten. html (accessed 6 December 2009).

33 Universität Stuttgart (2009) *Hydrogenius Project and Design*, http://www.ifb.uni-stuttgart.de/index.php/forschung/flugzeugentwurf/hydrogenius/246-projektdesign2 (accessed 1 December 2009).

34 University of Turin (2009) Environmentally Friendly Inter City Aircraft powered by Fuel Cells (ENFICA-FC), http://www.enfica-fc.polito.it/ (accessed 6 December 2009).

35 Romeo, G., Moraglio, I., and Novares, C. (2007) ENFICA-FC: Preliminary survey and design of 2-seat aircraft powered by fuel cells electric propulsion, presented at the 7th AIAA Aviation Technology, Integration and Operation Conference (ATIO), Belfast, Northern Ireland, 2007, AIAA2007-7754.

36 Friedrich, K.A., Kallo, J., Schirmer, J., and Schmitthals, G. (2009) Fuel cell systems for aircraft application, *ECS Transactions*, **25** (1) 193–202.

37 Eelman, S. (2002) *Identification of Potential Environmentally Friendly Concepts for Commercial Aviation, Technical Report 02/01, Institute of Aviation*, Technical University of Munich.

38 Airbus (2008) *Airbus International Press Release, Emission Free Power for Civil Aircraft: Airbus Successfully Demonstrates Fuel Cells in Flight*, 19 February 2008, http://www.airbus.com/en/presscentre/pressreleases/pressreleases_items/ (accessed 30 November 2009).

34
Auxiliary Power Units for Light-Duty Vehicles, Trucks, Ships, and Airplanes

Ralf Peters

Abstract

The demand on electricity in mobile applications increases in nearly all future prospects. Reasons for such a development are electric devices for more comfort and a guaranteed energy supply during idling mode. Today combustion engines and turbo jet engines were applied as auxiliary power units (APU) on-board of trucks and air planes. Fuel cells are envisaged as an environmental friendly and high efficient energy conversion system for future systems. Usually for logistical reasons, APUs must use the same fuel as the main engine. This will be kerosene or JET A-1 for air planes and diesel for trucks and ships. This contribution will show the requirements of different applications and will give an overview about today's developments.

Keywords: Auxiliary power unit, fuel processing, fuel cell, solid oxide fuel cell, polymer electrolyte fuel cell, vehicles

34.1
Operating Conditions for Auxiliary Power Units

A broad range of applications can be found in the literature for fuel cell systems as auxiliary power units (APUs). On-board electricity generation is an issue for nearly all transport systems by land, water, and air. Fuel cell-based APUs have to compete with batteries and electric generators coupled with internal combustion engines (ICEs). In addition to better efficiency, they offer additional features such as low emissions and low noise. The advantages and benefits of fuel cell-based APUs will be analyzed for each application.

Lutsey *et al.* [1] defined different applications for APUs in land transport, namely luxury passenger vehicles, law enforcement vehicles, contractor trucks and pick-ups, specialized utility trucks (garbage crushers, cement mixers, lifts, and moveable platforms), recreational vehicles (caravans, motor homes), refrigeration units, and line-haul heavy-duty trucks. The power class required for future applications of fuel cell-based APUs to 4–7 kW_e for luxury passenger vehi-

Hydrogen Energy. Edited by Detlef Stolten
Copyright © 2010 WILEY-VCH Verlag GmbH & Co. KGaA, Weinheim
ISBN: 978-3-527-32711-9

cles, recreational vehicles, and line-haul heavy-duty trucks. According to their analysis, law enforcement vehicles require less electricity, namely up to 2.5 kW$_e$, whereas refrigeration trucks need up to 30 kW$_e$ and pick-up trucks used by craftsmen, farmers, and contractors at remote work sites up to 20 kW$_e$. A very diversified area of application is that of specialized utility trucks, ranging from 5–35 kW$_e$ for platforms and manlifts to 60–75 kW$_e$ for cement mixers. The potential for annual vehicle sales in the United States equipped with APUs was estimated by Lutsey et al. in 2003 [1] as 1.2×10^6 luxury passenger cars, 300 000 light trucks, 70 000–80 000 law enforcement vehicles, 300 000 pick-ups, 70 000 specialized utility trucks, 190 000 recreational vehicles, 60 000 refrigeration units, and 100 000 line-haul heavy-duty trucks. Agnolucci [2] discussed the prospects for fuel cell APUs in the civil markets proposed by Lutsey et al. [1]. In his analysis, the total market is broken down to an addressable market, a potential market, and a capturable market. Conclusions from the market survey are that heavy-duty, recreational, and luxury vehicles have the best prospects. The largest capturable market is assigned to line-haul heavy-duty trucks.

34.1.1
Trucks

The requirements for APU operation in truck applications are well known. In addition to the better system efficiency, lower maintenance costs play an important role in economic calculations. The idling efficiency of an ICE amounts to 9–11% [3]. Maintenance costs associated with the APU are estimated to be between roughly $ 0.05 and $ 0.07 per hour of APU power [3, 4]. A typical line-haul sleeper truck is provided with electrically driven equipment resulting in a gross power of 12.6 kW$_e$. The main requirements arise from air conditioning (2.2 kW$_e$), several cooking devices such as microwave, toaster, and frying pan (5.8 kW$_e$), entertainment by computers and radio (0.6 kW$_e$), and the demand for the technical equipment of the truck (2.7 kW$_e$), for example a battery charger and a pump. Not all devices are in operation at the same time. The daily line-haul sleeper truck power requirements were estimated by Lutsey et al. [1] as an average of 2.95 kW$_e$. Following the data given by Sriramulu et al. [5], a demand of 5.2 kW$_e$ during the night and of 4.4 kW$_e$ for daily operation with air conditioning can be assumed. In addition to the data of Lutsey et al. [1], a cabin heating system may also be in operation.

Lutsey et al. [6] analyzed the potential fuel consumption and emission reductions from solid oxide fuel cell (SOFC) APUs in long-haul trucks by a Monte Carlo simulation method based on estimated boundary conditions and on survey data. According to their analysis, the annual idling time of a long-haul truck amounts on average to 1758 h per year with a fairly high standard deviation of 1155 h per year. The average accessory power amounts to 2.1 ± 1.3 kW$_e$. The idle fuel rate of the ICE amounts 3.5 ± 1.2 l h^{-1} compared to 0.57 ± 0.3 l h^{-1} for the SOFC APU. The annual fuel used for idling could be reduced from 11 000 l per year, that is, an average value (6100 l per year) plus standard deviation (5000 l per year), to 2000 l per year at high utilization by switching from ICE idling to

SOFC APUs. Results from an earlier United States truck survey [7] showed an average idling time of 6 h and 1700 h per year per truck. Idling times vary widely; for example, 10% of all class 7 and class 8 trucks consumed more than 13 000 l of diesel per year. Trucks are classified in the United States by their gross vehicle weight. Heavy-duty vehicles are assigned to classes 6–8, ranging from 8.8 t to more than 15 t for a class 8 truck.

Several technical solutions are available as add-ons to minimize the fuel consumption for idling, such as diesel-fired heaters, cooling systems, and thermal storage systems [6, 8]. Diesel auxiliary power units and generators are also commercially available [2, 6]. These current systems are heavy, expensive, and noisy and will not make a significant difference to emissions [8]. Another option to avoid truck idling discussed by several authors [2, 6, 8] is truck stop electrification. Jain et al. [8] calculated an infrastructure cost of US$ 1 million per 100 trucks, which seems to be uneconomic in comparison with the income from electricity sales. Perrot et al. [9] reported a payback time of about 3 years for the equipment on the truck which must be borne by the truck owner and the owner of the truck stop installation.

Jain et al. [8] performed a techno-economic analysis for fuel cell-based APUs. Their break-even estimates lead to payback times from roughly 3 months to 4.5 years depending on the idling time per hour or year and the fuel consumption per year. Important input data are the APU cost of $ 500 kW$_e^{-1}$ and the fuel cost of $ 0.37 l^{-1}. According to the data of Lutsey et al. [6], one can calculate an average payback period of 1.5 years. Finally, a payback period of 2 years should be gained. At an even higher investment cost of $ 1000 kW$_e^{-1}$, a higher fuel consumption of 5.7 l h^{-1} and 6 h usage per day, an APU reaches this goal for all fuel costs ranging between $ 0.26 and 0.66 kW$_e^{-1}$. Lutsey et al. [6] calculated a share of 10% of the truck population with a payback period of 2 years at $ 800 kW$_e^{-1}$ and 20% for the DOE goal of $ 400 kW$_e^{-1}$ [10] and a middle fuel price of $ 0.53 l^{-1}. The total capital cost for a 5 kW$_e$ APU was assumed with $ 7200 including an inverter, a heat pump, the fuel cell system ($ 400 kW$_e^{-1}$), and the installation cost. Sriramulu et al. [5] calculated the cost of an SOFC APU system of $ 840 kW$_e^{-1}$ and $ 580 kW$_e^{-1}$ on a near-term and long-term basis, respectively, leading to payback periods of roughly 1–3.5 years.

Contestabile [11] developed a dynamic simulation model which represents the mutual interactions between technology, markets, and policy. His calculations were based on 100 000 units per year and a cost target of $ 500 kW$_e^{-1}$. He analyzed the market entry of a diesel-based PEFC APU for different conditions in relation to a diesel-based SOFC APU and an ICE APU [11]. He assumed a cost for a diesel PEM fuel cell APU of >$ 10 000 at market entry and $ 1000 for mass production. Corresponding numbers for diesel SOFC APUs are >$ 50 000 and >$ 1500 for mass production in 2030. In relation to this, a diesel APU based on an ICE costs between $ 7000 and $ 9000 and a 40 kW h lithium ion battery $ 20 000. Those numbers are higher than those of Lutsey et al. [6] and of Baratto et al. [12] of $ 12 000–25 000 for SOFC systems. As the most important result, the cost of the fuel processor plays a significant role for the market share between SOFC and PEFC APUs. If both fuel processors are developed in parallel and reach their cost targets of $ 100 kW$_e^{-1}$ for SOFC and $ 150 kW$_e^{-1}$ for PEFC, the latter technology prevails. However, if the fuel processor costs remain high at $ 750 kW$_e^{-1}$ for PEFC and $ 500 kW$_e^{-1}$ for SOFC, diesel-based SOFC APUs can compete in a smaller market due to the higher cost. The potential market in Europe was assumed to be 100 000 units for the truck population [11]. A share of 35% adoption of fuel cell technology was taken into account at a cost of $ 4500 or € 3000. The potential market for the United States and Europe is estimated to be 450 000 diesel PEM APUs up to 2020. Lutsey et al. [6] stated that SOFC APUs could become economically feasible for 15% of the long-haul truck population, that is, about 60 000 trucks, up to 2015. A key parameter of these economic calculations is the fuel price, which varies in those studies between $ 0.37 l^{-1} [5] to $ 0.79 kW$_e^{-1}$ [6]. Therefore, Contestabile used a forecast model starting at $ 0.74 l^{-1} and an increase in oil price from $ 60 per barrel in 2007 to $ 105 per barrel in 2030, leading to $ 1.3 l^{-1} [11]. In mid-November 2009 the price of diesel was $ 0.74 l^{-1}, which is much lower than that in summer 2008 of $ 1.27 l^{-1} [13].

In addition to economic factors, fuel cell-based APUs offer the lowest emissions for electricity production. Lim [14] presented measured exhaust emissions from idling heavy-duty vehicles in a study of the United States Environmental Protection Agency (EPA). A class 8 truck during idling emits on average 144 g h^{-1} of NO_x and 8224 g h^{-1} of CO_2 and consumes 3.1 l h^{-1} of diesel. Emission test results [3] show a remarkable partial load behavior. During cruising at 55 mph (88.5 km h^{-1}) and with simultaneous electricity generation, 3.9 g h^{-1} of hydrocarbons (HCs), 57.4 g h^{-1} of CO, and 777 g h^{-1} of NO_x were emitted. These values can be reduced only in the transient period from driving to idling, and when idling at 600 rpm, for example, emissions amount to 15.3 g h^{-1} CO. Long idling at 1050 rpm in combination with electricity generation leads to 86.4 g h^{-1} of HCs, 189.7 g h^{-1} of CO and 225 g h^{-1} of NO_x, indicating strongly increased limited emissions. Nevertheless, CO_2 emissions are higher during driving, 60 000 g h^{-1}, compared with during idling, 4000–10 000 g h^{-1}. Lutsey et al. [6] calculated average emission savings of 0.29 t per year of NO_x and 0.08 t per year of HCs per truck, and Sriramulu et al. of 0.27 t per year of NO_x and 0.1 t per year of CO per truck [5]. In January 2007, a California regulation came in force which bans any truck (from class 3 up, i.e., 4.5 t truck weight) from idling or using an ICE-based APU for more than 5 min [15].

Baratto and co-workers analyzed fuel cell APUs by a series of complementary models [12, 16–18]. These studies were focused on SOFC technology. Baratto and Diwekar [16] performed a life cycle assessment which formed the basis for an impact assessment with regard to environmental and health impacts [18]. The latter included an exposure and a toxicity assessment leading to a risk characterization. A further input to this analysis came from a study on system performance and cost modeling [17]. The studies concluded with a multi-objective optimization [17] especially with a case study of the South California Air Basin [12]. The analysis of Baratto and Diwekar [16] showed that the main pollutant of

SOFC-based APUs is carbon dioxide and that the health risks for all effects, namely. cancer and chronic and acute hazard risks, are far below the safety limits. Large amounts of pollutants are released to the air, water, and soil during the system manufacturing and assembly. CO and NO_x are emitted as a 33% share for system manufacturing, 2–9% during diesel production, and about 63% during operation. Hydrocarbons, particles, and SO_x occur mainly (83–98%) during system manufacturing. Nevertheless, the emissions from fuel cell APUs, in comparison with diesel ICE APUs, lead for the life cycle to a reduction between 82 and 99% for CO, HCs, NO_x, and particles, and the CO_2 reduction amounts to 64%.

34.1.2
Passenger Cars

Originally, fuel cells were foreseen as the drivetrain option for the future. Research and development activities have shown that the targets defined for gasoline reforming in fuel cell propulsion systems cannot be reached. Consequently, the application of hydrocarbon reforming was modified after a negative decision made by the United States Department of Energy (DOE) in 2004 [19]. Therefore, the activities have been shifted to electricity production with APUs for cars, trucks, ships and aircraft. Table 34.1 shows the targets for APUs fixed by

Table 34.1 Targets for different APU applications [a].

Target	Passenger car [22]	Aircraft [37–39, 41, 42]	DOE [20], HDV [b] 2010/2015	EC [11] 2015
Electric power (kW)	10	100–400	3–30	
Power density (W l^{-1})	333	750	100	
Mass specific power (W kg^{-1})	250	0.5–1	100	
Durability (h)	5000		20 000/35 000	40 000 (HDV) [b] 5000 (LDV) [b]
Costs for 50 000/10 000 units (€ kW^{-1})	40		400	<500
Efficiency (%)	>35	40	35/40	35
Cold start	<1 s		15–30 min	<20 min [23] [c]
Dynamics (idle to 90%) (s)	<1			
Cycle capability	n.d.		150/250	>5000 [23] [c]
Emissions	<SULEV			
Load range	1:50			

a) The DOE targets [20] take care of the SECA goals for SOFC development [10].
b) HDV, heavy-duty vehicle; LDV, light-duty vehicle.
c) Stack targets which determine system performance.

the DOE [20]. APU applications are coupled to the Solid State Energy Conversion Alliance (SECA) program of the DOE [10], which focuses on SOFC technology. Nowadays, only a joint activity by Renault and Nuvera Fuel Cells is targeted on fuel cell drivetrains fed by a fuel processor [21].

Table 34.1 also includes targets fixed by Docter *et al.* [22] for APUs in passenger cars. As can be seen, the targets for cost and specific power are much more challenging than for trucks. Typically, APUs for cars might demand 5–10 kW$_e$ of electricity [22–25]. The mass-specific power target amounts to 250 W kg^{-1} and the specific power should reach 0.33 kW l^{-1} for the system [22] and 1 kW l^{-1} for the stack [23].

An important issue for the comparison between electricity generation by ICE-driven generators and by fuel cells is their different partial load characteristics. Grube *et al.* [25] showed for a passenger car that the use of an SOFC APU is more efficient than an innovative starter–generator concept only during the idling phase. During idling, an SOFC APU needs 203 g (kW h$_e$)$^{-1}$ compared with 380 g (kW h$_e$)$^{-1}$ for ICE idling, that is, 1.2 and 2.3 l h^{-1} for 5 kW$_e$. Simulating a coupled usage of mechanical energy from the ICE for propulsion and for electricity generation during different drive cycles offers an advantage for ICE concepts. The specific consumption in liters per 100 km decreases slightly at a constant engine speed (2000 rpm) and at higher torque caused by the air conditioning system. Due to an increased efficiency, this leads to only a slightly increased fuel demand. Without a precooling option for the passenger compartment, the fuel consumption during the European Driving Cycle (MVEG) leads to 7.9 l per 100 km for a conventional generator, 8.9 l per 100 km for starter–generator systems, and 10.3 l per 100 km for fuel cell APUs.

Diegelmann [26] analyzed several SOFC APU architectures: as replacement of the conventional generator set, as an add-on to a commercial system, with regenerative breaking, with start/stop automation, and with both optional systems. The APU is designed for 1.5 kW$_e$. Efficiency was estimated to be higher than 38% in a load range between 15 and 100% and offers a maximum of 41%. A decisive issue is the design of the start-up process. Diegelmann assumed a start-up time of 200 s and a start-up energy of 2000 kJ. If the energy consumption for the start-up process is considered, fuel consumption is increased relative to today's system during the New European Driving Cycle (NEDC). Without start-up energy, a slight benefit was calculated. In contrast, start/stop automation reduced fuel consumption by about 5% and in combination with regenerative breaking by about 6% during NEDC. Only for a special summer cycle [26] were advantages in fuel consumption of about 6.4% and 4.2% for a cold APU calculated. Diegelmann also discussed indirect benefits in fuel consumption by the combination of fuel cell-based APUs with start/stop automation and hybrids. Trade-offs between long electric distance capability and restricted storage capacity could be solved by an APU in a hybrid vehicle. Anyway, oversizing of the APU must be avoided due to the energy losses during start-up. Advantages are given for air conditioning during stand-still; about 36% fuel could be saved during a 30 min period. Finally, if efficiency is the only criterion, fuel cell-based

APUs for cars cannot prevail as a pure substitution product for an improved and commercially available technology.

34.1.3
Maritime Applications

Considering APUs for maritime applications, a wide range of demanded electric power can be found, from 300 W for yachts [27] to several megawatts for ocean-going ships [28]. Preferred fuel cell types are PEFCs [27, 29] and molten carbonate fuel cells (MCFCs) [30, 31]. The fuel for hydrogen production varies from hydrogen [32] to liquefied petroleum gas (LPG) [27], maritime diesel [28], and NATO F76 diesel [29, 33]. The broad variations of fuels, type of fuel cells, and power classes are due to the wide application range of sea transport. Bensaid *et al.* [30] discussed fuel cell APUs for tourist craft, leisure craft, offshore support vessels, research and survey vessels, fast ferries, ferries, passenger cruise vessels, coastal cargo vessels, and international cargo vessels. Driving forces for maritime applications are lower life cycle costs, reduced NO_x emissions, low vibrations and low noise levels. Generally, maritime transport accounts for about 3% of global petroleum consumption but contributes 14% of total NO_x and 16% of total SO_x emissions [30]. Efficiencies of MCFC systems with autothermal reforming (ATR) of diesel fuel were calculated to be between 32.7 and 39.1%. As an alternative process, a combination of cracking and subsequently steam reforming was proposed. In these cases, efficiencies amount to 44.6–50.6%. Specchia *et al.* [31] proposed, in addition to the closure of the water balance, the production of sanitary water and the use of sensible heat for heating purposes on-board. An addressable market can be estimated from the forecast study given by Karin *et al.* [34]. In the period 2011–2015, a total of 14 000 vessels for the United States market is predicted across different areas such as fishing vessels, oil tankers, passenger vessels, container ships, and others.

34.1.4
Aircraft

Fuel cells as APUs for airplanes were mentioned for the first time by Seidel *et al.* [35] in 2001 in a proceedings article which was later published as a full paper in 2004 [36]. Most of the available information in this area is in conference contributions and internal papers, downloadable via the Internet.

The A330-300 aircraft type with a take-off gross weight of 230 t demands 100 000 l of kerosene on average for a 10 000 km flight [37]. About 5% of the fuel consumed is required by on-board systems. Commercial APUs in ground operation work at an efficiency of 20%, and fuel cells should reach 40%. Specific weight targets for aerospace APUs are $1\,kg\,kW^{-1}$ and $1.5\,l\,kW^{-1}$. Daggett *et al.* [38] analyzed hybridized SOFC-turbine systems as APUs with 400 and 40 kW, respectively. For ground power generation they assumed a reduced fuel use of 75% by increasing the efficiency from 15 to 60%. During flight, the main engines produce electricity with an efficiency of 40–45%. It was planned to achieve an efficiency of 75% with hybrid systems based on SOFCs and turbines and to reduce fuel consumption to 40% of today's level. Srinivisan et al. [39] studied four different architectures for 450 kW SOFC-gas turbine hybrid systems. Their baseline airplane for 305 passengers demands 85 t of JET A-1 for a 14 260 km flight and 0.85 t for APU ground operation. During the flight profile, up to 550 kW$_e$ was required. If anti-icing is to be performed in future systems with electric heating, an additional demand of 400 kW$_e$ occurs during the climb and descent phases.

According to the data of Hiebel [40], about 500 kW are required for air conditioning, 250 kW for ice and rain protection, 100 kW for cabin systems, 300 kW for engine starting, 50 kW for the landing gear, and 150 kW for flight controls. These systems are powered by electrical, hydraulic, and bleed air power, mainly by the main engines (1 MW), and on ground by the APU (550 kW). Srinivisan et al. [39] also defined the baseline APU electrical loads as 450 kW for cabin pull down, 338 kW for sustaining a cool cabin, 450 kW during the passenger load, and 306 kW during the passenger unload phase. Subsequently, an electric power of about 306 kW is required during the service phase between two flights and 45 kW for the flight over-maintenance. Srinivisan et al. calculations were based on five and three flights per day for two short-range journeys of 926 and 1852 km, respectively, resulting in a total energy of 1550 and 2464 kW h per day, respectively. Gummalla et al. [41] performed calculations for a 300 kW SOFC system and a peak power of 260 kW for APU ground operation. Considering four flights per day, 3373 kW h must be delivered by the APU on the ground. SOFC system efficiencies are between 53 and 68% during cruising and between 39 and 48% on the ground [39]. The lowest efficiency corresponds to an architecture pressurized to only 1 atm; other system architectures work at higher pressures, delivering higher efficiencies. In the future, energy consumption might be reduced from 400 to 100 kW by an optimized power management system [37].

Airplane emissions account for 3% of total emissions worldwide and might increase to 4% [38]. Emission data for APU ground operation and for complete flights have been published [39, 41]. A commercial aircraft APU emits about 5.3 kg of NO_x, 6.2 kg of CO, and 0.4 kg of HCs per flight [39] and 1.9 t of NO_x, 2.4 t of CO, and 0.2 t of HCs per year [41]. SOFC APU emissions are much lower and can be neglected relative to the total emissions of a flight [39, 41]. For SOFC-gas turbine hybrid architectures, Srinivasan et al. and Gummalla et al. reported additional benefits for the emissions of the main engines during ground operation, namely 13.5–33% for NO_x, 3–16% for CO, and 3.5–15% for unburned hydrocarbons (UHCs). The projected benefits differ more between the two studies and less between the architectures. During a long-haul flight, such as 16 h and 14 200 km, the main engines emit during their 20 min operation on ground about 2% of the total CO and 10% of the total HC emissions [39]. NO_x emissions arise mainly during cruising. The yearly CO_2 emissions for APU ground operation amount to 1000 t of CO_2 per aircraft [39]. Emissions can be estimated with respect to an overall air carrier population in the United States of 6840 turbo-jets [44] with 6.8×10^6 t of CO_2 and 13 000 t of NO_x. This is in the same order for CO_2 and one order of magnitude lower for NO_x compared with long-haul heavy-duty trucks with 10.9×10^6 t of CO_2 and 190 000 t of NO_x [14]. Fuel cell-based APUs in aircraft would halve CO_2 emissions and would eliminate NO_x emissions. The specific CO_2 production amounts about 636 g CO_2 (kW h)$^{-1}$, which is only slightly higher than the rate from the United States electricity grid with 605 g CO_2 (kW h)$^{-1}$ in 2007 [45].

In addition to electricity production, fuel cell-based APUs can produce water and tail gases for tank inerting [37, 46]. Fuel cell-based APUs for aircraft offer multifunctional use with additional benefits compared with conventional solutions. Considering the market forecast by Agnolucci [2] and competing conventional technologies, the system design for a special application must deliver such additional advantages.

34.1.5
Special Markets

Other mobile applications of fuel cell-based APUs could be trains and buses. Fuel cell drive systems with hydrogen as energy carrier have been demonstrated for bus fleets in recent projects. Military applications were discussed for land transport by Patil *et al.* [47] and for submarines by Sattler [48].

34.1.6
Addressable Markets and Specific Targets

The prospects for market size for fuel cell-based APUs can vary widely. In order to draw a picture of addressable market sizes, population data for commercial applications [1, 32] and forecasts for future investigations have been taken into account [34].

Figures 34.1 and 34.2 show the relation between the APU power demanded for the propulsion power and the addressable market size. In most cases, APU power is approximately 10% of the installed power for propulsion. Forklifts can be an exception if the lifting power corresponds to that of propulsion [32]. If the APU power is in the region of 1%, an additional energy carrier such as methanol for a DMFC in a recreational vehicle can be an option. For large ships, additional fuel for the APU such as truck diesel instead of marine diesel oil could be necessary due to the difficulties caused by the reforming of heavy heating oils. With respect to aircraft, an additional fuel is not desired, and hydrogen could be a bridging solution for small systems.

Table 34.1 shows the targets for passenger cars, aircraft, and trucks. The primary Multi-Year Research, Development and Demonstration Plan was defined by the DOE for APUs in 2003 and was revised in 2006/2007 [20]. Due to the low cost targets of APUs for passenger cars in relation to the high cost of today's fuel cell systems and their limited use for idling purposes, passenger cars seem not to be suitable for a market entry. APUs in trucks and aircraft clearly

34 Auxiliary Power Units for Light-Duty Vehicles, Trucks, Ships, and Airplanes

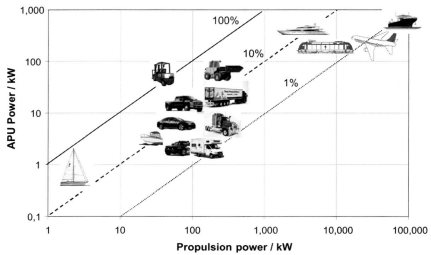

Fig. 34.1 APU power demanded for different applications in relation to the propulsion power. Basic data were taken mainly from [1, 32].

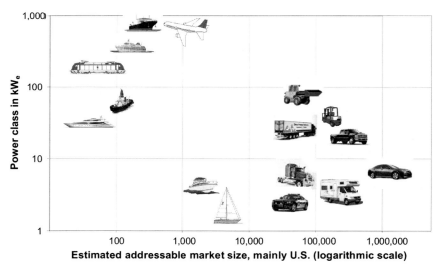

Fig. 34.2 Required power class for different applications in relation to the addressable market size. Data taken from [1, 32, 34]. Estimates for APUs in trains and leisure boats correspond to rates of 100 and 3% for today's train and leisure boat population in the United States (2006 [44])

show a multifunctional use and a plurality of benefits. The ambitious targets of aircraft APUs with regard to power density should be interpreted as a challenge which can be solved together with scale-up in the 100 kW power class. The target values for trucks must be reflected with the present development status (see Section 34.4).

34.2 System Design

The design of a fuel cell-based APU system requires information about the power class, the choice of fuel, and the variety of usages for the fuel cell system. An additional water use or a highly efficient heat recovery must be planned as part of the basic engineering. In order to operate a fuel cell, hydrogen or a hydrogen-rich gas is required, which can be converted from hydrocarbon fuels by chemical reactions. The choice of fuel for an APU is ideally the fuel which is already on-board for propulsion. In most cases, fossil fuels such as gasoline for cars, diesel for all kinds of trucks, buses and trains, kerosene for aircraft, and diesel fuel of different qualities for ships must be used. A few authors describe the use of LPG or dimethyl ether as fuel [27, 43]. Over the long term, part of the liquid energy carriers required today can be manufactured from biomass. If hydrogen is already on-board, for example in fuel cell-powered buses and cars or hybridized propulsion systems with fuel cells on-board, electricity generation will finally be realized by the system which uses hydrogen as fuel. At present, there is still no infrastructure for hydrogen as a future energy source. For the supply of a fuel cell it can be stored in pressurized tanks, in the liquid phase, or in metal hydrides. However, these storage variants suffer from low energy densities and narrow the range of operation for a defined storage capacity.

34.2.1 Challenges in Using Commercial Fuels

Commercial fuels which are in use for different applications lead to various challenges for fuel processing.
- Gasoline can contain up to 15% of olefins and up to 35–40% of aromatic compounds [22]. Higher concentrations of olefins and aromatics require more stringent reforming conditions such as higher operating temperatures [49].
- Jet fuel JET A-1 is specified with up to 3000 ppmw of sulfur, whereas a typical reforming catalyst can tolerate fuels with 10 ppmw of sulfur. Commercially available JET A-1 in Europe and the United States contains less than 1000 ppmw of sulfur, predominant species being benzothiophenes and their alkylated forms. Aromatic compounds are found primarily as monoaromatics at 15–25%.

- Diesel evaporates over a broad temperature range between 440 and 663 K and leaves a residue of up to 5% even at temperatures higher than 633 K. Diesel contains about 25–30% of aromatics and less than 100 ppmw of ash.
- In the marine market, different fuels are in use: marine gas oil (MGO), marine diesel oil (MDO), intermediate fuel oil, medium fuel oil, and heavy fuel oil. They more or less correspond to a series of blends of fuel oil No. 2, that is, diesel and with an increasing proportion of fuel oil No. 6, which is a residual fuel oil. MGO contains up to 1000 ppmw of sulfur and 100 ppmw of ash. Typically, marine gas oil evaporates at higher temperatures of 470–660 K. At 633 K, only 85% of the original fuel must be evaporated. Fuel oil No. 4 (MDO), a blend of No. 2 and No. 6, may contain up to 15% residual process streams and up to 5% polycyclic aromatic hydrocarbons [50].

With regard to the fuel quality, applications based on diesel seem to be the most attractive. Gasoline reforming may suffer from the high aromatic content with regard to long-term stability. Kerosene requires a complex desulfurization process or the introduction of a desulfurized fuel quality. Fuel cell-based APUs for ships should be envisaged only for marine gas oil as fuel. Other maritime fuel qualities show a high potential for plugging of nozzles, carbon deposition, and catalyst degradation by fuel residues.

34.2.1.1 Desulfurization of JET A-1

In general, a desulfurization process should be performed at the refinery. At present, kerosene as a precursor of the later product JET A-1 will not be desulfurized to 10 ppmw S as required by fuel cell systems. The dominant sulfur species in JET A-1, namely benzothiophenes and their derivatives, are easier to convert than those in diesel fuels, namely dibenzothiophenes and their alkylated forms. Desulfurization in small systems or on-board must use other technologies than conventional processes in refineries. Desulfurization should be carried out before the reformer in the liquid phase or after the reformer in the gas phase by H_2S adsorption [51]. The latter option requires a sulfur-tolerant reformer catalyst [52]. Possible processes for liquid-phase desulfurization have been described [53]. Two of them are the adsorption of sulfur-containing species [54] and a modified hydrofining process in the liquid phase [55]. Figure 34.3 shows these processes as basic flow sheets. Adsorption demands a regenerable adsorbent otherwise the material has to be removed after sulfur breakthrough. Regeneration requires additional process lines for the purge or regeneration gas.

Roychoudhury et al. [51] estimated for a reformate gas with 115 ppmv of sulfur, corresponding to 1150 ppmw of sulfur in the fuel, a breakthrough after 200 min. The modified hydrofining process requires a hydrogen source which could be hydrogen-rich reformate (about 30% H_2). JET A-1 containing 700 ppmw of sulfur will be pressurized up to 24 bar and heated up to 390 °C. The average sulfur content in the product during a 200 h experiment was 8.3 ppmw [55]. Hydrogen sulfide must subsequently be separated from the de-

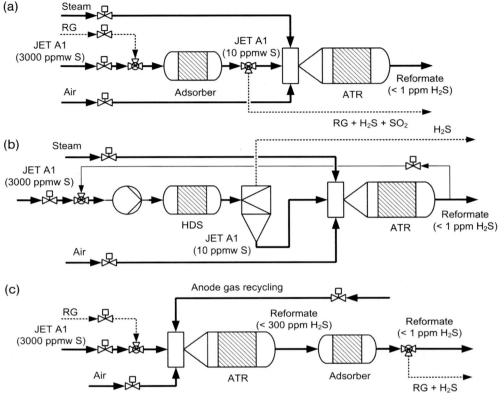

Fig. 34.3 Basic flow sheets for desulfurization of JET-A1. (a) Adsorption of sulfur-containing hydrocarbons, namely benzothiophenes [54]; (b) hydrofining with presaturated hydrogen in the liquid state [55]; (c) H_2S adsorption with a ZnO adsorber bed in the gas phase [51, 56].

sulfurized fuel using a hydrocyclone. It should be burned in the jet engine and must leave the system as SO_2.

34.2.1.2 Evaporation of Diesel

The most important issue in diesel reforming is the homogeneous mixing of diesel, steam, and air. Carbon deposition can be avoided in combination with a uniform flow field and a controlled temperature profile. Therefore, intensive computational fluid dynamics (CFD) modeling of mixing chambers and corresponding reforming experiments are necessary [57, 58]. Lindström et al. [58] showed that the cool-flame technology is not suitable for diesel evaporation due to insufficient conversion and unstable operation. By designing fourth and fifth generations of CFD-optimized reformers, Lindström et al. skipped prolonged recirculation zones and shortened the mean residence time of their ATRs from

>7 to 0.11–0.63 s. Porš et al. [57] measured a residence time of 70 ms for the mixing chamber of their eighth-generation ATR by visualization experiments using glass apparatus. Attention must be paid to the deposition of ash in small channels [59, 60]. More information about different mixing chamber designs and injector/nozzle concepts can be found in the literature [60].

34.2.2
Core Components

34.2.2.1 Fuel Processors

With regard to the basic chemical reactions, hydrogen production from hydrocarbons is based either on steam reforming (Equation 34.1) or on partial oxidation (Equation 34.2). Autothermal reforming constitutes a compromise between heated steam reforming and partial oxidation.

$$C_nH_mO_l + (n - l)H_2O \rightleftharpoons nCO + (m/2 + n - l)H_2 \quad (\Delta H_R > 0) \tag{34.1}$$

$$C_nH_mO_l + \frac{n-l}{2}O_2 \rightleftharpoons nCO + mH_2 \quad (\Delta H_R < 0) \tag{34.2}$$

In addition, two side reactions can be found, namely the water-gas shift reaction (WGS) (Equation 34.3) and methane synthesis (Equation 34.4), which is the opposite of methane steam reforming.

$$CO + H_2O \rightleftharpoons CO_2 + H_2 \quad (\Delta H_R < 0) \tag{34.3}$$

$$CO + 3H_2 \rightleftharpoons CH_4 + H_2O \quad (\Delta H_R < 0) \tag{34.4}$$

If the contact time of the gas molecules at the active sites of the catalyst is long enough, the gas composition in the product gas corresponds to chemical equilibrium. Typical residence times in the catalyst zone of an autothermal reformer are 100–200 ms, corresponding to a gas hourly space velocity of 20 000–35 000 (m^3 h^{-1}) m$^{-3}_{cat}$. The equilibrium gas composition of an ATR operated at 1073 K and 2 bar and applying an educt mixture with $O_2:C=0.47$ and $H_2O:C=1.9$ amounts to 9.14% CO, 8.49% CO_2, 0.02% CH_4, 31.2% N_2, 24% H_2O, and 27.15% H_2. Further information about these calculations can be found elsewhere [60].

The most important side reactions of the reforming process are those leading to undesired carbon formation, for example Boudouard Equation 34.5 and pyrolysis reactions Equation 34.6:

$$2CO \rightleftharpoons CO_2 + C \tag{34.5}$$

$$C_nH_m \rightarrow \text{"carbon deposit"} + m/2H_2 \tag{34.6}$$

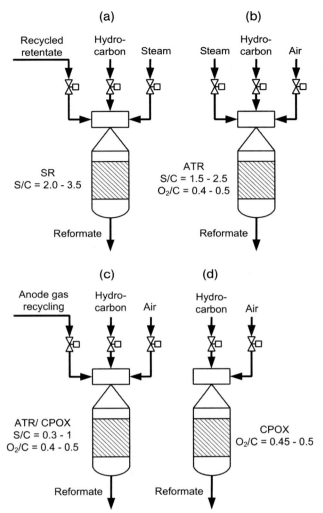

Fig. 34.4 Basic flow sheets for different reforming processes. (a) Modified steam reforming (SR) process with retentate recycling from a membrane [62]; (b) ATR process [59]; (c) modified catalytic partial oxidation (CPOX) process with anode gas recycling from SOFC [52]; (d) CPOX process [63].

A more detailed description of the mechanisms has been discussed by Rostrup-Nielsen [61].

Figure 34.4 shows four different options for hydrogen production with regard to APU application. Most developers choose the ATR or the catalytic partial oxidation (CPOX) process. The CPOX process can be modified in the case of SOFC systems by anode gas recycling to reach steam-to-carbon ratio between 0.3 and 1.0 [52]. ATR processes work at steam-to-carbon ratios between 1.5 and

2.5 and steam reformers between 2.0 and 3.5. It is advantageous to combine a steam reformer under a certain pressure with an H_2 separation membrane to achieve pure hydrogen for PEFCs [62]. Unfortunately, steam reformers require a larger amount of water than ATR, which can lead finally to a non-closed water balance.

34.2.2.2 Fuel Cells

Different types of fuel cells have been considered for APU applications. Figure 34.5 shows a schematic diagram of four different fuel cell types with some typical impacts from the fuel cell system. PEFC operates at temperatures around 353 K. Today, most stacks are optimized for H_2 operation in fuel cell propulsion systems. In the case of reformate operation, Pt/Ru must be used instead of pure Pt as anode material. Pt/Ru offers a higher CO tolerance than Pt but CO concentrations in reformate gas should still be lower than 50 ppm. Additionally, sulfur tolerance is extremely low. PEFC membranes require humidified air on the cathode side. The implications for system design will be explained by considering the basic flow sheet of the system. Due to the complexity of the system for the classic PEFC, new materials were envisaged for solid membranes at operating temperatures between 403 and 473 K. Recent papers [64–66] have reported new materials developed within the integrated project Autobrane funded

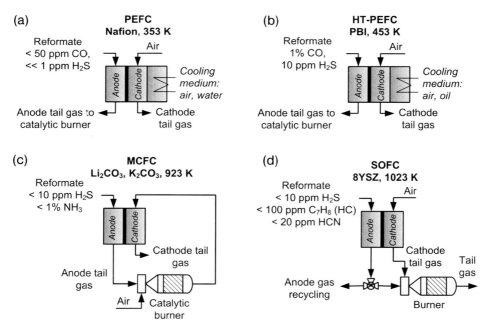

Fig. 34.5 Basic data for different fuel cell types: (a) PEFC; (b) HT-PEFC; (c) MCFC, after [31]; (d) SOFC.

by the European Commission. Polybenzimidazole (PBI) membranes with incorporated phosphoric acid are the most successful type of new material currently available. BASF Fuel Cell distributes them under the brand name CELTEC [67]. PBI membranes can tolerate higher amounts of carbon monoxide and hydrogen sulfide. In a reformate gas mixture, the H_2S concentration should be <10 ppmv at 1% CO and temperatures higher than 433 K [68].

High-temperature fuel cells such as SOFCs and MCFCs are under development for stationary applications. With regard to the power class of APUs for aircraft and ships, these fuel cell types have also been investigated. MCFCs operate at 923 K, use a molten carbonate as electrolyte, and can tolerate 10 ppm H_2S. Due to the electrochemical functionality – CO_3^{2-} ions diffuse from the cathode to the anode – the cathode air must contain carbon dioxide. This can be realized by the combustion of the anode tail gas with excess oxygen and the subsequent use of the tail gas as oxygen supply on the cathode side. Both high-temperature fuel cell types can convert CO electrochemically. Therefore, a gas cleaning step is not necessary. In the case of SOFC-based APU systems, part of the anode gas will be recycled to provide steam to the reformer. The other part can be burnt together with the cathode tail gas in a burner.

34.2.3
Basic Engineering and Modeling of Fuel Cell-Based APU Systems

34.2.3.1 Process Analysis
Depending on their application, a huge number of different fuel cell systems have been considered in the past. Figures 34.6–34.8 show basic configurations for systems for PEFC, HT-PEFC, SOFC, and MCFC systems. These basic flow sheets do not include any heat exchangers, evaporators, and pumps. A large set of parameters are used in the thermodynamic calculations and ultimately they determine the size of the apparatus and reactors. Therefore, it is sometimes difficult to evaluate results from different studies, for example in terms of efficiency.

Figures 34.6 and 34.7 compare the basic flow sheets for PEFC and HT-PEFC systems. Following a design by Roychoudhury et al. [69], a PEFC system includes two shift stages and three PROX stages to decrease CO concentrations from 9.9% to 50 ppmv. In addition, the PEFC requires a humidification system to avoid local drying-out effects of the membranes. The HT-PEFC can tolerate a feed gas with 1% CO and 10 ppm H_2S and should be operated with dry cathode air below the dew point [68]. Therefore, in addition to the lack of a humidification device, HT-PEFC systems can omit complex CO fine cleaning and an H_2S adsorption bed.

High-temperature fuel cell systems offer the advantage of a less complex subsystem for fuel processing. The simplest system can be designed for SOFCs including anode gas recycling and the use of cathode off-gas for the burner. Unfortunately, such a system is not suitable to gain any water due to the low dew point of the tail gas [46]. In order to achieve higher efficiency, several

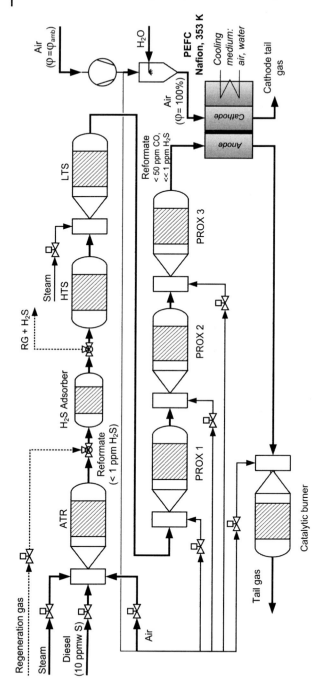

Fig. 34.6 Basic flow sheet of a PEFC system with three PROX stages for CO fine cleaning.

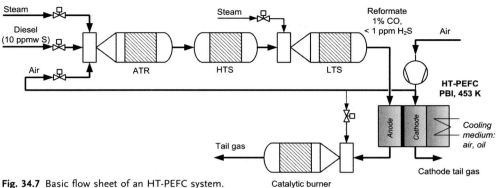

Fig. 34.7 Basic flow sheet of an HT-PEFC system.

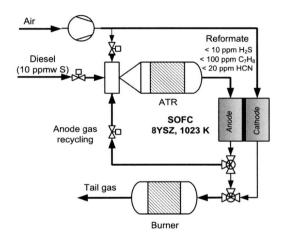

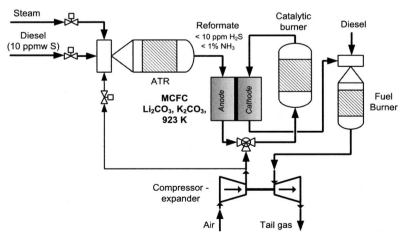

Fig. 34.8 Basic flow sheet of an SOFC system with anode gas recycling and MCFC-gas turbine hybrid system [31].

Table 34.2 Basic process design results for different fuel-gas turbine hybrid concepts [30, 70, 71].

Application	Fuel cell: power, P_{el} (kW$_e$)	Fuel	Fuel processing	Inlet conditions: turbine p/p_0; T_{in} (K)	η_{el} (%)
Aircraft [70]	PEFC: 60–80	H_2	None	1.2–4.4; 1073	47–53 (ground) 43.6–50.5 (cruise)
Aircraft [71]	SOFC: 185	JET-A1	ATR, S:C = 0.4, O_2:C = 0.5	1.3–4.7; 1023–1161	32–35 (ground) 32–42 (MES)[a] 46–53 (cruise)
Ship [30]	MCFC: 500	Diesel	ATR, S:C = 3, O_2:C = 0.35	3.3; 1233	50.6

a) MES: main engine start for aircraft application.

studies have evaluated hybrid systems consisting of micro gas turbines with a PEFC [70], SOFC [39, 41, 71], and MCFC [30, 31]. Figure 34.8 show such a pressurized system for an MCFC with an additional fuel burner to increase the inlet temperature to 1073 K of the gas turbine operated at 3.55/1.1 bar [31]. Table 34.2 shows calculated efficiencies for the PEFC, SOFC, and MCFC systems. Due to the different system architectures and load profiles which have been assumed for each of these studies [70, 71], the results cannot be compared directly. It is obvious that with ground operation and 50% partial load, efficiencies are much lower, 32% compared to 53% at full load and a system pressure of 3.6 bar. At ambient pressure, high efficiencies could be achieved in the case of 82% fuel utilization and a high cell voltage of 750 mV, namely 42.5% for a SOFC system [25].

Although high efficiencies are only one benefit from fuel cells, most studies compare the potential efficiencies for different fuel cell APU designs (see Table 34.3). A major issue is the selection of the most appropriate fuel processing. Ersoz et al. [72] showed advantages for steam reforming, showing 42% compared with 37% for ATR and 31% for partial oxidation (POX). Montel [73] calculated with statistical methods average values of 40% [steam reforming (SR)], 37% (ATR), and 33% (POX). In addition to the educt conditions such as steam-to-carbon and oxygen-to-carbon ratios, relevant parameters also include the cell voltage, cathode air ratio, and fuel utilization [46]. Cutillo et al. [74] preferred SR to ATR because the efficiency and H_2 concentration are higher; whereas Montel showed that the high steam-to-carbon ratios required for SR lead to large heat exchanger areas. The size of the required heat exchanger network can be evaluated by the Pinch-Point method [75]. Specchia et al. [76] presented efficiencies for ATR and SR of gasoline, diesel, and biodiesel. The efficiencies for ATR of gasoline and biodiesel amount to 36.5%, whereas the value for ATR of diesel is lower at 35.5%. The improvement with SR is best for gasoline at 2.5% and worst for diesel at 1.1%. Finally, the fuel choice plays a certain role in the design of an efficient system, but other parameters such as cell voltage and fuel

Table 34.3 Basic process design results for different types of fuel cells and fuel processing[a].

Fuel cell	P_e (kW)	Fuel	Process	Operating conditions	η_{el} (%)	Ref.
PEFC	50–100	Diesel	SR	3 bar; S/C = 3.5	42	[72]
PEFC	50–100	Diesel	ATR	3 bar; S/C = 2.5; $O_2/C = 0.5$	37	[72]
PEFC	50–100	Diesel	POX	1.5–3.5 bar; $O_2/C = 0.5$	31	[72]
PEFC	100	Diesel	ATR	S/C = 3; $O_2/C = 0.4$	33.8–36.6	[76]
PEFC	10	Diesel	ATR	2 bar; S/C = 2.25; $O_2/C = 0.38$	34.7–35.5	[74]
PEFC	10	Diesel	SR	2 bar; S/C = 3	37.1–38.7	[74]
PEFC	5	JET-A1	ATR	1.1–3 bar; S/C = 1.9; $O_2/C = 0.47$	22.4–35.9	[46]
HT-PEFC	5	JET-A1	SR	1.1–3 bar; S/C = 1.9	24.2–42.3	[46]
HT-PEFC	5	JET-A1	ATR	1.1–3 bar; S/C = 1.9; $O_2/C = 0.47$	19.8–38.5	[46]
SOFC	5	JET-A1	ATR	1.1–3 bar; S/C = 1.9; $O_2/C = 0.47$	22.2–40.3	[46]
SOFC	5	JET-A1	POX	1.1–3 bar; $O_2/C = 0.47$	21.7–39.1	[46]

a) The wide range in efficiencies of the studies by Pasel et al. [46] is caused by different cell voltages, namely 650–750 mV for a PEFC, and different fuel utilizations, for example 75–90% for an HT-PEFC.

utilization are decisive to achieve the targets in Table 34.1. Table 34.3 also shows the results from process studies performed by Pasel et al. [46], varying the decisive parameters by statistical methods. Their calculations were performed for various reforming processes in combination with PEFC, HT-PEFC and SOFC types of fuel cell. With the wrong set of parameters, efficiencies can be halved, and maximum efficiencies are gained only for a well-chosen set with a narrow range of values.

Additionally, Pasel et al. [46] calculated the potential water production and tail gas rates in the context of aircraft application. Due to the high air ratio for a SOFC system without internal reforming, the oxygen concentration in the cathode off-gas of 15% is too high for tank inerting, which demands a concentration less than 12%. Further, the anode and cathode off-gases should not be mixed before the burner because this will greatly restrict the thermodynamic potential of water condensing. In contradiction to the rating in [77], the results clearly show that water production is also possible for systems with kerosene reforming. Monthly average temperatures and humidity data for 245 locations worldwide were taken from the database of the German meteorological service.

System calculations for various PEFC and HT-PEFC systems with different reforming technologies and hydrogen as fuel were evaluated. The most promising ones were studied in more detail combining meteorological data and flight profiles. PEFC systems in combination with ATR of kerosene demand pressurization up to 2 bar to assure reliable water production. The most difficult locations exhibited a high temperature and an extremely low humidity in summer, namely $>40\,°C$ and $<20\%$ at Phoenix, AZ, USA and Riyadh, Saudi Arabia. The average potential water rates worldwide amount to $0.28–0.30\,l\,(kW_e\,h)^{-1}$ for an HT-PEFC with kerosene ATR at atmospheric conditions. Considering the yearly flight profile for a route in the Arabian region, only less than 2% of all data sets lead to a non-closed water balance for this system. It is obvious that water production rates of higher than $100\,l\,h^{-1}$ could be achieved only by large APU systems with approximately $350\,kW_e$. Hydrogen as fuel offers higher rates with an average value of $0.5\,l\,(kW_e\,h)^{-1}$ for an HT-PEFC system, but leaves the question of hydrogen storage.

34.2.3.2 Modeling

Static process analysis calculations result in a set of thermodynamic data for each system component. The size of these components can be derived together with heat and mass transfer models. During a dynamic simulation of the complete system, further improvements concerning the start-up strategy can be achieved. Dynamic simulations of fuel cells must include the performance of the complete system with all required reactors and heat exchangers. Such models allow the development of improved start-up strategies [78–86]. Most studies consider gasoline [78–86], others kerosene [82], diesel, JP-5 (fuel oil No. 5, >75% residual oil) [84], and natural gas [85, 86]. In most cases dynamic simulations were performed for PEFC systems. HT-PEFC and SOFC were considered by Samsun [82], the latter also by Sorrentino and Pianese [85]. Negative impacts of system components with unsuitable dynamics on the fuel processor performance were found especially for PEFC systems [79, 82]. Sommer et al. [79] analyzed the response behavior of a fuel processor system consisting of ATR and WGS during a load change from 10 to 90%. Because of the lack of water in one heat exchanger during the first 20 s, the reformate cannot be cooled as much as planned in the design stage. Consequently, a temperature peak of about 50 K in the high-temperature shift stage occurs, causing a CO peak of about 20%. The low-temperature shift stage is not able to decrease this concentration significantly. Such a high CO concentration cannot be treated in the PROX reactor.

Experiments with electrical devices for preheating the catalyst [82, 87] offer a heating-up time between 10 and 15 min, with the potential to reduce this to 5 min. Samsun [82] calculated potential start-up times of fuel processing systems including a start-up burner delivering suitable reformate qualities of about 3 min for SOFC, 4.3 min for HT-PEFC and 4.6 min for PEFC systems. Springmann et al. [78] proposed preheating by an electrically heated catalyst up to 673 K within 60 s. Reforming should be started in POX operation mode for 60 s. The reformer start-up was performed in 15 s while the WGS reactor required 6 min to reach the final state. Springmann et al. discussed several measures to shorten start-up times to 3 min while the shift section ignited after 60 s. Proposed measures are the oxidation of reformate in the high-temperature shift reactor section, electrical preheating of the shift reactor, variation of air ratios, and the availability of steam. Goebel *et al.* [88] reported on the development of a start-up burner that permitted a start time of 20 s to 30% power and 140 s to full power. Diegelmann [26] considered a start-up of the reformer in the POX mode and assumed a start-up energy of 2000 kJ for a start-up time of 200 s. Additionally, energy losses of 360 W are assumed for a blower, the control box, and actors. Only 64% of the heating value of the fuel is used for the warm-up process; 36% leaves the system with the tail gas.

Finally, dynamic simulation models can be used to develop control mechanisms for fuel cell systems [81, 85, 89] and fuel processors [83, 84, 86].

34.3
Present Status of Fuel Cell-Based APU Systems

In this section, APU systems for SOFC, MCFC and PEFC operation are considered in terms of their maturity in technological development. The development status of SOFC worldwide was described by Blum *et al.* [90] in 2005 and Singhal [91] in 2006. A list of SOFC developers worldwide was given by Blum *et al.* [90]. During recent years, industrial partners have ceased their activities or development teams have been merged. Currently, the companies Delphi Automotive, Fuel Cell Energy/Versapower, UTC Power, Rolls Royce and Siemens Energy are developing SOFC technology in the United States DOE SECA program [92]. Worldwide one can find development groups at Webasto/Staxera and ElringKlinger (Germany), Topsoe Fuel Cell (Denmark), Ceres Power (UK), Wärtsilä (Finland), Hexis and HTceramix (Switzerland), St. Gobain (France/USA), CFCL (Australia), and Kyocera, Toto, and Mitsubishi (Japan). These groups are supported by several universities and research centers such as nearly all United States national laboratories (LANL, ANL, PNNL, NETL, Oak Ridge, and Sandia) and in Europe by Forschungszentrum Jülich, DLR, and IKTS (Germany), Risoe (Denmark), VTT (Finland), CEA (France), and KIER (Korea).

Delphi is working in the SECA program together with PACCAR and Volvo on a truck APU. Blake [93] reported efficiency data for Delphi's APU system DPS300D as 35% at 2.5 kW and 31% at 4 kW (maximum). The system was first demonstrated with natural gas. The completed Gen 4 stack reaches a stack power density of 0.5 W cm^{-2} at 570 mA cm^{-2} and 870 mV. Higher power densities of 1 W cm^{-2} were reached only with H_2 in planar designs [90]; operation with reformate led to lower values [94]. The areal power density of tubular cells is much lower than that of planar cell designs [91], namely 0.2 W cm^{-2}. Cycle capability was shown by Delphi in 95 full thermal cycles at a minimum degradation of less than 0.5% [95]. The stack was operated with an H_2–N_2 mixture as part of a vibration test simulating 3.75 million miles of truck operation. Long-term measurements indicate a slight degradation in the last 1300 h of operation with synthetic reformate gas but additionally a negative sulfur effect was observed in the first 200 h. Finally, stack durability should reach degradation rates lower than 1% per 1000 h at operating times longer than 40 000 h for reformate operation, which has been shown only for single items and not for the triple target [90].

Delphi's reforming technology is based on an existing natural gas reformer which has been modified for gasoline and diesel. Two designs were developed, one as CPOX and the other as an endothermic reformer by recycling anode gas and accommodating its water content for steam reforming [96]. The latter shows a higher efficiency and a capability for recycling. The hydrogen concentrations amount to 21% H_2 for CPOX of gasoline and 30% for CPOX with anode gas recycling. The carbon monoxide concentration is fairly high, namely 22% for CPOX and 27% for the endothermic version, values that are excellent for SOFC operation and unfavorable for low-temperature fuel cells. Apart from H_2 and CO, only 0.5–0.6% of methane is formed and less than 0.1% of higher hydrocarbons. The CPOX version was tested for 100 h. A start-up could be performed in only 3 min. The endothermic version reached 500 h with diesel operation [97].

Webasto developed a 30-cell stack with a power density of 200 $W_e\,l^{-1}$ at 850 °C which was operated with diesel reformate for a short testing time of 14 h in 2006. The stack went through 15 full thermal and redox cycles [98]. The APU system delivers 1 kW_e after a heating-up period of 3 h. At system stop the stack went through a shut-down procedure with purging of all components and a switch-off of the reformer and the afterburner. A small purge flow of cathode air should hinder a cathode reduction by leaks. By this method, the stack cooled to 500 °C in 2.5 h and to 150 °C in 10 h [24]. The efficiency of the 1 kW system should be better than 20%.

Lindermeir *et al.* [63] described the development activities at Webasto for their CPOX reactor. For start-up a glow plug is heated with an electric power of 70 W_e. During initial testing, the heating-up period was extended to 10 min but a start-up within seconds should be possible. Tests were performed with a commercial diesel containing 8 ppmw sulfur and 24 wt% aromatics. Typical H_2 and CO concentrations in the dry state amount to 25% H_2 and 22% CO. During simulated anode gas recycling, CO_2 reforming and the reverse WGS reaction occur, resulting in increased formation of CO. In order to prevent catalyst degradation by carbon deposition, a precatalyst was used. Finally, due to the high entrance temperatures, carbon deposition leads to a reduced durability for the CPOX process. Two modified concepts were investigated: one with an intercooling zone between the precatalyst and the main reaction zone and a second concept splitting the diesel fuel for two reaction paths. Part of the diesel is burned completely at air ratios higher than 1 and mixed with the residual diesel to reach air ratios of O:C = 0.4–0.48. Using this concept, the reformer efficiency decreased to 78% but a complex regeneration process could be avoided. A prototype APU system based on the second concept was tested for 9 h.

A major disadvantage of SOFC operation with reformate is the loss of the cooling effect without internal methane reforming. Consequently, the released heat of reaction must be removed in SOFC mainly by warming the cathode inlet air. The enthalpy rate can be increased by either a higher flow rate or a higher temperature difference. Due to material stress, the latter option is limited to a temperature difference of about 100–150 K. Higher flow rates of cathode air lead to higher peripheral energy demands for compression. Therefore, Hansen [99] proposed a modified flow scheme with methanol or dimethyl ether as energy carrier and subsequent methanation after reforming and before the SOFC.

A list of MCFC developers worldwide was given by Moreno et al. [100]. MCFC systems for maritime APU applications are being developed by Fuel Cell Energy (United States), CFC Solutions (Germany) and Ansaldo Fuel Cells (Italy). The power density of MCFCs is significantly below that of SOFCs, namely up to 0.15 W cm^{-2} in combination with internal reforming. During the last 10 years, the lifetime of stacks has been extended from a few months to about 2 years or 10 000 operational hours [101]. Aicher et al. [33] reported a 20 kW$_e$ diesel fuel processor that should be scaled up to 350 kW$_e$ for a maritime MCFC APU. Part of the diesel is burned in an evaporator unit to generate steam because heat integration with the fuel cell was not planned. The educt mixture of the ATR is prepared using steam to carbon and oxygen to carbon ratios of 1.0–1.5 and 0.32–0.4, respectively. The dry product gas concentrations of hydrogen and carbon monoxide amount to 30% and 9–13%, respectively, with a decline of 7% and 3% during a 300 h test.

HT-PEFC stacks have been developed by Serenergy (Denmark), Volkswagen (Germany), and , research institutions such as Forschungszentrum Jülich, ZSW, and ZBT (Germany). Performance data for a 5 kW$_e$ stack reported by Steiger et al. [102] show a current density of 0.6 A cm^{-2} at 600 mV for hydrogen operation at 160 °C and 3 bar. Addition of 1 vol.% CO led to a performance loss of 6%. Wannek et al. [103] showed a U–I characteristic with 450 mV and 0.6 mA cm^{-2} at 180 °C for hydrogen and 450 mV and 0.4 mA cm^{-2} for reformate operation (1 bar). The decrease in cell voltage during operation with reformate, for example at 0.4 mA cm^{-2} and 160 °C, amounted to 10%. Experiments performed by Reiche [103] indicated that pressurizing to 3 bar leads to a doubled current density at 600 mV. Increased power densities between 0.47 and 0.87 W cm^{-2} were published [102, 104] relative to values between 0.19 and 0.27 W cm^{-2} at 1 bar [104, 105]. Durability experiments including start/stop cycling for H$_2$ and reformate operation were performed during 6000 h by Schmidt and Baurmeister [67]. Degradation is caused mainly by cathode catalyst corrosion and subsequent cathode flooding with electrolyte.

The PEFC is the most widely used fuel cell in transport applications. Cell power densities of 0.5 W cm^{-2} at a cell voltage of 0.7 V can be regarded as state-of-the art for a PEFC operated at 1.5 bar and 80 °C, and also for reformate operation. Stack power densities have been improved since the early 1990s from 0.2 to 1.5 kW$_e$ l^{-1} [94]. Kallo et al. [77] reported low-pressure performance investigations with an H$_2$ PEFC system. The stack voltage is reduced by about 30% at 0.62 bar relative to atmospheric conditions. Additionally, humidification at a low pressure of about 0.2 bar is difficult due to the water uptake from the gases and the low boiling temperatures. It is more efficient to use a supply with cabin air instead of an ambient air [71, 77]. Within the sixth framework program of the European Commission, the integrated project HYTRAN aims at the development of a system for traction power with an 80 kW direct hydrogen PEFC implemented in a passenger car and at the development of a PEFC-based APU including a microstructured diesel steam reformer [106]. Project partners are automotive manufacturers such as Volvo, Fiat, Renault, Volkswagen, and DAF, automotive suppliers such as Tenneco, and research organizations such as ECN (The Netherlands), PSI (Switzerland), IMM, RWTH Aachen (Germany), Imperial College London (UK) and Politechnico de Torino (Italy). So far, microreactors for propane steam reforming, gas cleaning, and diesel steam reforming have been developed on a 5 kW_e scale for PEFC systems [107, 108].

Due to the limited choice of fuel for APU applications, the design of the fuel processor plays a crucial role. Research and development activities are mostly carried out in smaller companies and research institutions. Important developers of fuel processing units worldwide are companies in the United States, namely Altex Technologies, Precision Combustion, Aspen Products Group, Idatech, Ceramatec, and Nuvera Fuel Cells (United States and Italy), and for example research organizations such as PNNL (United States), N-GHY (France), Hy-Gear (The Netherlands), OWI, ZBT, FHG-ISE, IMM, and Forschungszentrum Jülich (Germany). An overview of integrated fuel processors was given by Qi et al. [109]. Important development issues are costs and technical performance data such as complete conversion of a commercial fuel to a required durability and a compact design indicated by power density.

For the time being, not all of these targets can be aggregated in one unit. Long-term experiments with commercial premium diesel and with low-sulfur Jet A-1 (<10 ppmw S) were performed by Pasel et al. [59], resulting in a conversion of 99.7% after 1000 h of operation with diesel and 99% after 2000 h with Jet A-1. With regard to the measurement accuracy of concentrations, fuel conversion was determined by the carbon residues (C_2H_4, C_2H_6, C_3H_7, C_3H_8, $C_{4+}H_x$) in the dry gas phase and the organic carbon content of the condensate withdrawn from the cooled product gas. After 1100 h of operation with Jet A-1 the mass of total organic carbon (OC) in the condensate was determined as 70 $mg_{OC}\,l^{-1}$, from which a conversion of 99.986% was calculated. At the beginning, 0.5 $mg_{OC}\,l^{-1}$, corresponding to 99.99999% for fuel conversion, was measured, clearly indicating a creeping degradation with operation time. Roychoudhury [56] achieved for their sulfur-tolerant 5 kW_{th} reformer using a low-sulfur jet fuel (JP-8, <100 ppmw S) a conversion of better than 99.98% after 1100 h of operation, calculated from ethane and ethene gas phase concentrations lower than 5 ppmv and propane and propene concentrations lower than 10 ppmv. Using regular 400 ppmw S jet fuel (JP-8) leads to C_2 and C_3 concentrations of 24 ppmv and 8 ppmw, whereas 400 ppmw S jet fuels blended with dibenzothiophenes or benzothiophenes result in concentrations of about 960 ppmv C_2 and 100 ppmv C_3 [110]. Taking only these values into account, the conversion is 99.97 and 99%, respectively. Targets for conversion such as 99.99% after 5000 h must be checked according to the required APU lifetime and the tolerable hydrocarbon concentrations (C_{2+}) of subsequent apparatus and the fuel cell.

The evaluation criterion for compact design is power density, whereas the catalyst performance is often defined by the gas hourly space velocity in $m^3\,h^{-1}\,m_{cat}^{-3}$. Microreactors [107, 108] or microstructured reactors [69] are discussed as new innovative reactor types to reach the challenging APU targets. Kolb et al. [107] published microreactor dimensions for gas cleaning for a 5 kW_e PEFC APU system, leading to 0.83 $kW_e\,l^{-1}$ for the two WGS stages and to 1.7 $kW_e\,l^{-1}$ for a fine cleaning reactor based on preferential oxidation (PROX). Analogous design considerations by Roychoudhury et al. [69] for a three-stage PROX reactor in an ATR–isooctane system led to a higher power density of 60 kW_{th} (LHV H_2) l^{-1}, $\sim 25\,kW_e\,l^{-1}$, for their microstructured catalyst. The autothermal microreformer of IMM without both evaporator units – for isooctane and for steam – achieved a power density of 5 $kW_e\,l^{-1}$ [107]. Heat integration is an important issue of reformer development for APU systems. While the power density of a monolithic ATR catalyst at 30 000 h^{-1} led to 40 $kW_{th}\,l^{-1}$ or 18 $kW_e\,l^{-1}$, the whole ATR with integrated heat exchangers, evaporator, and mixing zones reached a power density of 2 $kW_e\,l^{-1}$ [111]. An integrated design for a 25 kW_{th} fuel processor including an autothermal kerosene reformer, a sulfur trap, and a WGS reactor resulted in power densities of the particular sections of 7, 8, and 15 kW_{th}/l^{-1} and in total 3 $kW_{th}/l^{-1} \sim 1.2\,kW_e/l^{-1}$ for the whole apparatus [56]. O'Connell et al. [108] reported a compact microreactor design for diesel steam reforming by combining reforming and catalytic combustion in one device with a power density of 3 $kW_e\,l^{-1}$. However, finally, microreactors with integrated heat exchangers were not the most compact design in each case: a conventional WGS design with two adiabatic stages and an intermediate cooling zone reached 10 $kW_{th}\,l^{-1}$ or 4 $kW_e\,l^{-1}$ for the whole apparatus [111] compared with 0.87 $kW_e\,l^{-1}$ as mentioned above. Future investigations on new reactor concepts should extend the testing time from a few hours (<50 h) to long-term experiments. Complete fuel conversion must be achieved for all operating conditions, but in addition analyzed byproducts such as C_{2+}–C_{7+} species in the range 200–3000 ppmv must be avoided [108].

Although certain compact reactor designs were achieved in the past, a complete 5 kW_e fuel processing system will require approximately 17 l, corresponding to a power density of 0.33 $kW_e\,l^{-1}$, following basic design considerations of Grube et al. [111]. Severin et al. [87] presented a conceptual design for a 3 kW_e PEFC-based APU system with 0.23 $kW_e\,l^{-1}$ for the fuel processing system without insulation (0.12 $kW_e\,l^{-1}$ with insulation) and 68 $W_e\,l^{-1}$ for the complete PEFC system. The diesel-fueled SOFC APU system of Delphi achieved a power density between 10 and 17 $W_e\,l^{-1}$ (2.5–4 kW_e) depending on the final design. Furthermore, older activities on complete APU systems using a PEFC with hydrogen as fuel are known from Freightliner for trucks and from BMW/International Fuel Cells for passenger cars [1, 112]. Idatech developed APUs for military application, applying a PEFC in combination with Pd/Cu membranes for CO fine cleaning [113].

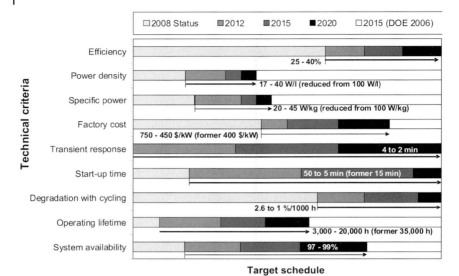

Fig. 34.9 Technical targets for 1–10 kW$_e$ fuel cell-based APUs operating in standard ultra-low sulfur diesel (Unites States, <10 ppmw S) published by the DOE in November 2009 [114] compared with those of the technical plan in 2006 [20]. Efficiencies are related to lower heating value; costs are defined on an annual production rate of 50 000 units per year; operating lifetime ends at 80% of original power.

34.4
System Evaluation

A comparison of the targets in Table 34.1 with the results obtained during the last 3 years leads to the conclusion that several goals could not be achieved with fuel cell-based APUs during the next 5 years. The DOE revised the set of targets with the aid of 18 entities from industry and national laboratories in November 2009. Figure 34.9 shows the time schedule of the targets for the coming years 2012, 2015 and 2020 in relation to the 2015 target from the Multi-Year Research, Development, and Demonstration Plan of the DOE in 2006–07 [20]. Each bar in Figure 34.9 begins at the status in 2008, which is determined by the actual SOFC system results from the SECA program.

The efficiency target was shifted from 2015 to 2020. In order to increase system efficiency from today's 25% via 30% in 2012 to the final goal of 40% in 2020, further improvements of the fuel cell performance such as higher fuel utilization and higher cell voltages at reasonable power densities (W cm^{-2}) are necessary. Estimates of the future size of APUs have been reduced during the period 2003–2009. In 2003, the target was defined with 150 W l^{-1} as power density and 170 W kg^{-1} as specific power for 2010. The first revision in 2006 was to 100 W l^{-1} and 100 W kg^{-1}. As can be seen from Figure 34.9, these targets were reduced in 2009 [114] to 40 W l^{-1} and 45 W kg^{-1} for 2020. The targeted factory costs were slightly increased for a 50 000 units per year production base from

$ 400 kW$_e^{-1}$ in 2015 to $ 500 kW$_e^{-1}$ in 2020. Considering power densities and factory costs, APUs for luxury passenger cars are out of range for a mid-term development period. Further changes were made for the operating lifetime, which is defined for 80% of the original power. In 2020, a durability of 20 000 h should now be achieved. New targets have been defined for system availability, degradation rate, and transients. Pressurized fuel cell systems in a hybrid architecture with gas turbines (see Table 34.2) offer high efficiencies, but today's stack technology is not mature enough to realize such large APU systems.

The targets for small APUs in Figure 34.9 are reasonable for APU application in trucks and leisure boats but not for large APUs in aircraft. APUs in aircraft must achieve higher power densities. SOFC-based APUs are currently not applicable for aircraft due to their low power density and their unsuitable tail gas composition with regard to tank inerting. Low-temperature fuel cell systems such as the PEFC or HT-PEFC promise a better performance in a multi-purpose aircraft system. Especially HT-PEFC systems should lead to a less complex system with acceptable production rates for water and tank inerting gases and a better tolerance towards impurities. PEFCs are mostly tested for stationary applications with synthetic fuels, for example with synthetic methane reformate, during 13 000 h [94]. Long-term experiments for HT-PEFC for synthetic reformate operation were investigated for 6000 h [67]. Considering aircraft application, the system performance can be checked during the major maintenance – known as the C-check – after 15–18 months and approximately 5000 operational hours.

34.5
Conclusion

Fuel cell-based APUs offer various advantages over conventional systems, especially if a multi-purpose use is sought. Successful fuel processing is a decisive aspect for the achievement of technical and cost targets. At present, truck and aircraft applications are mainly at the focus of research and development groups. With regard to the development status of SOFC systems in the 5 kW class, targets have been reduced. New technologies such as the HT-PEFC have the potential to catch up with SOFC systems. A successful market entry for fuel cell-based APUs must take place before 2020.

Acknowledgments

The author thanks the Department of Fuel Processing and Systems at the Forschungszentrum Jülich GmbH, Institute of Energy Research – Fuel Cells (IEF-3) for their excellent cooperation and the important results produced during the last 10 years of fuel cell research. R. C. Samsun, Th. Grube, and J. Pasel are thanked for valuable discussions and their technical assistance in preparing this chapter.

References

1 Lutsey, N., Brodrick, C.J., Sperling, D., and Dwyer, H.A. (2003) *J. Transportation Research Board*, **1842**, 118–126.
2 Agnolucci, P. (2007) *Int. J. Hydrogen Energy*, **32** (17), 4306–4318.
3 Brodrick, C.J., Lipman, T.E., Farshchi, M., Lutsey, N.P., Dwyer, H.A., Sperling, D., Gouse, S.W., Harris, D.B., and King, F.G. (2002) *J. Transportation Research Board Part D*, **7**, 303–315.
4 Stodolsky, F., Gaines, L., and Vyas, A. (2000) *Analysis of Technology Options to Reduce the Fuel Consumption of Idling Trucks*, Argonne National Laboratory, Argonne, IL, 2000, ANL/ESD-43.
5 Sriramulu, S., Isherwood, K., Lasher, S., Broderick, C.J., and Lutsey, N. (2004) *Proceedings of the Fuel Cell Seminar, San Antonio, TX, 1–5 November 2004* (CD), Courtesy Associates, Washington, DC.
6 Lutsey, N., Brodrick, C.J., and Lipman, T. (2007) *Energy*, **32** (12), 2428–2438.
7 Lutsey, N., Brodrick, C.J., Sperling, D., and Oglesby, C. (2004) *J. Transportation Research Board*, **1880**, 29–38.
8 Jain, S., Chen, H.Y., and Schwank, J. (2006) *J. Power Sources*, **160** (1), 474–484.
9 Perrot, T.L., Constantino, M.S., Tario, J.D., Kim, J.C., Hutton, D.B., and Hagan, C. (2005) Truck stop electrification as a long-haul tractor idling alternative. *Proceedings of the 84th Transport Research Board Annual Meeting of National Academies*, 9–12 January 2005, Washington, DC, available at http//www.epa.gov/otaq/smartway/documents/dewitt-study.pdf (accessed 16 November 2009).
10 Williams, M.C., Strakey, J., and Sudoval, W. (2006) *J. Power Sources*, **159** (2), 1241–1247.
11 Contestabile, M. (2009) Analysis of the market for diesel PEM fuel cell auxiliary power units onboard long-haul trucks and of its applications for the large-scale adoption of PEM FC. *Energy Policy*, doi: 10.1016/J.enpol.2009.03.044.
12 Baratto, F. and Diwekar, U.M. (2005) *J. Power Sources*, **139** (1), 197–204.
13 Energy Information Administration, Gasoline and Diesel Fuel Update (2009) http://tonto.eia.doe.gov/oog/info/gdu/gasdiesel.asp (accessed 20 November 2009).
14 Lim, H. (2002) *Study of Exhaust Emissions from Idling Heavy-Duty Diesel Trucks and Commercially Available Idle-Reducing Devices*, EPA420-R-02-025, US Enviromental Protection Agency, Office of Air and Radiation, Washington, DC.
15 Office of Administration Law, State of California, *California Code of Regulation § 2485, Airborne Toxic Control Measure to Limit Diesel-Fueled Commercial Motor Vehicle Idling* (2008) www.arb.ca.gov/msprog/truck-idling/2485.pdf (accessed 16 November 2009).
16 Baratto, F. and Diwekar, U.W. (2005) *J. Power Sources*, **139** (1), 188–196.
17 Baratto, F., Diwekar, U.M., and Manca, D. (2005) *J. Power Sources*, **139** (1), 205–213.
18 Baratto, F., Diwekar, U.M., and Manca, D. (2005) *J. Power Sources*, **139** (1), 214–222.
19 US Department of Energy (2004) *On-Board Fuel Processing Go/No-Go Decision*, www.eere.energy.gov/hydrogenandfuelcells/news_fuel_processor.html (accessed 16 November 2009).
20 US Department of Energy (2006) *Multi-Year Research, Development and Demonstration Plan, Hydrogen, Fuel Cells and Infrastructure Technologies Program, Revision 2006*, http://www.eere.energy.gov/hydrogenandfuelcells/mypp (accessed 16 November 2009).
21 Bowers, B.J., Zhao, J.L., Ruffo, M., Khan, R., Dattatraya, D., Dushman, N., Beziat, J.C., and Boudjemaa, F. (2007) *J. Hydrogen Energy*, **32** (10–11), 1437–1442.
22 Docter, A., Konrad, G., and Lamm, A. (2000) VDI-Ber., 2000 (1565), 399–411.
23 Lamp, P.; Tachtler, J., Finkenwirth, O., Mukerjee, S., and Shaffer, S. (2003) *Fuel Cells*, **3**, 146–152.
24 Lawrence, J. and Boltze, M. (2006) *J. Power Sources*, **154** (2), 479–488.
25 Grube, T., Höhlein, B., and Menzer, R. (2005) *Fuel Cells*, **7**, 128–134.
26 Diegelmann, C.B. (2008) PhD thesis, University of Munich.

27 Beckhaus, P., Dokupil, M., Heinzel, A., Souzani, S., and Spitta, C. (2005) *J. Power Sources*, **145** (2), 639–643.
28 Sattler, G. (2000) *J. Power Sources*, **86** (1), 61–67.
29 Krummrich, S., Tunistra, B., Kraaij, G., Roes, J., and Olgun, H. (2006) *J. Power Sources*, **160** (1), 500–504.
30 Bensaid, S., Specchia, S., Federici, F., Saracco, G., and Specchia, V. (2009) *Int. J. Hydrogen Energy*, **34** (4), 2026–2042.
31 Specchia, S., Saracco, G., and Specchia, V. (2008) *Int. J. Hydrogen Energy*, **33** (13), 3393–3401.
32 Lebutsch, P., Kraaij, G., and Weeda, M. (2009) *Analysis of Opportunities and Synergis in Fuel Cell and Hydrogen Technologies*, publishable report, R2H4007PU.3, http://195.166.119.215/roads2hycom/ (accessed 16 November 2009).
33 Aicher, T., Lenz, B., Gschnell, F., Groos, U., Federici, F., Caprile, L., and Parodi, L. (2006) *J. Power Sources*, **154** (2), 503–508.
34 Karin, Z., Leavitt, N., Costa, T., and Grijalva, R. (1999) *Marine Fuel Cell Market Analysis*, CG-D-01-00; US Coast Guard Research and Development Center, Washington, DC.
35 Seidel, J.A., Sehra, A.K., and Colantonio, R.O. (2001) NASA aeropropulsion research: looking forward, *Proceedings of the 15th ISABE*, Bangalore, 2–7 June 2001, NASA/TM-2001-211087, National Technical Information Service, Springfield, VA.
36 Sehra, A.K. and Whitlow, W. (2004) *Progress in Aerospace Sciences*, **40**, 199–235.
37 Heinrich, H.J. (2007) Presented at Symposium der Wasserstoffgesellschaft, Hamburg, 18 October 2007, http://www.h2hamburg.de/index.php?page=download (accessed 16 November 2009).
38 D. Daggett, J. Freeh, C. Balan, D. and Birmingham (2003) *Proceedings of the Fuel Cell Seminar*, Miami, FL, 3–7 September 2003 (CD), Courtesy Associates, Washington, DC.
39 Srinivisan, H., Yamanis, J., Welch, R., Tilyani, S., and Hardin, L. (2006) *Solid Oxide Fuel Cell APU Feasibility Study for a Long Range Commercial Aircraft Using UTC ITAPS Approach, Volume I: Aircraft Propulsion and Subsystem Integration Evaluation*, NASA/CR-2006-214458/VOL1, National Technical Information Service, Springfield, VA.
40 Hiebel, V. (2005) in *Proceedings of the European Strategic Research Agenda (SRA)*, Brussels, 17–18 March 2005, https://www.hfeurope.org/uploads/700/812/CELINA.pdf (accessed 16 November 2009).
41 Gummalla, M., Pandy, A., Braun, R., Carriere, T., Yamanis, J., Vanderspurt, T., Hardin, L., and W$_e$lch, R. (2006) *Fuel Cell Airframe Integration Study for Short-Range Aircraft, Volume 1: Aircraft Propulsion and Subsystems Integration Evaluation*, NASA/CR-2006-214457/VOL1, National Technical Information Service, Springfield, VA.
42 Glover, B. (2005) Presented at Aircraft Noise and Emissions Reduction Symposium, 24 Monterey, CA, May 2005.
43 Nilsson, M., Petterson, L.J., and Lindström, B. (2006) *Energy & Fuels*, **20** (6), 2164–2169.
44 Research Bureau of Transportation Statistics (2008) http://www.bts.gov/publications/national_transportation_statistics/2008/, US Department of Transport, Washington, DC (accessed 16 November 2009).
45 Energy Information Administration, Electric Power Annual 2007 (2008) http://www.eia.doe.gov/cneaf/electricity/epa/epaxlfilees1.pdf (accessed 20 November 2009).
46 Pasel, J., Samsun, R.C., Menzer, R., Peters, R., Stolten, D. (2009) in *Proceedings of the Lucerne PEFC Forum*, Lucerne, 29 June–2 July 2009 (ed. F. De Bruijn), European Fuel Cell Forum, Oberrohrdorf.
47 Patil, A.S., Dubois, T.G., Sifer, N., Bostic, E., Gardner, K., Quah, M., and Bolton, C. (2004) *J. Power Sources*, **136** (2), 220–225.
48 Sattler, G. (1998) *J. Power Sources*, **71** (1–2), 144–149.
49 Kopasz, J.P., Applegate, D., Ruscic, L. Ahmed, S., and Krumpelt, M. (2000) Effects of gasoline components on fuel

processing and implications for fuel cell fuels, *Procceedings of Fuel Cell Seminar*, Portland, OR, 30 October–2 November 2000, Courtesy Associates, Washington, DC.

50 Global Security, Diesel fuel by Pike, J. (2006) http://www.globalsecurity.org/military/systems/ship/systems/diesel-fuel.htm (accessed 16 November 2009).

51 Roychoudhury, S., Lyubovsky, M., Walsh, D., Chu, D., and Kallio, E. (2006) *J. Power Sources*, **160** (1), 510–513.

52 Krumpelt, M., Krause, T. R., Carter, J. D., Kopasz, J. P., and Ahmed, S. (2002) *Catalysis Today*, **77**, 3–16.

53 Peters, R., Latz, J., Pasel, J., and Stolten, D. (2008) *ECS Transactions*, **12** (1), 543–554.

54 van Rheinberg, O., Lucka., K., Köhne, H., Schade, T., and Andersson, J. T. (2008) *Fuel*, **87** (13–14), 2988–2996.

55 Latz, J., Peters, R., Pasel, J., Datsevich, L., and Jess, A. (2008) *Chemical Engineering Science*, **64**, 288–293.

56 Roychoudhury, S. (2009) *Proceedings of the 10th Annual SECA Workshop*, Pittsburgh, PA, 16 July 2009, http://www.netl.doe.gov/publications/proceedings/09/seca/ (accessed 16 November 2009).

57 Porš, Z., Pasel, J., Tschauder, A., Dahl, R., Peters, R., and Stolten, D. (2008) *Fuel Cells*, 2, 129–137.

58 Lindström, B., Karlsson, J. A. J., Ekdunge, P., De Verdier, L., Häggendal, B., Dawody, J., Nilsson, M., and Petterson, L. J. (2009) *J. Hydrogen Energy*, **34** (8), 3367–3381.

59 Pasel, J. Meissner, J., Porš, Z., Samsun, R. C., Tschauder, A., and Peters, R. (2007) *J. Hydrogen Energy*, **32** (18), 4847–4858.

60 Peters, R. (2008) Fuel processors. In *Handbook of Heterogeneous Catalysis* (eds G. Ertl, H. Knözinger, F. Schüth, and J. Weitkamp), Wiley-VCH Verlag GmbH, Weinheim, pp. 3045–3080.

61 Rostrup-Nielsen, J. R. (1984) Catalytic steam reforming. In *Catalysis, Science and Technology* (eds J. R. Anderson and M. Boudart), Springer, Berlin.

62 Schäfer, J., Sommer, M., Diezinger, S., Trimis, D., and Durst, F. (2006) *J. Power Sources*, **154** (2), 428–436.

63 Lindermeir, A., Kah, S. Kavurucu, S., and Mühlner, M. (2007) *Applied Catalysis B: Environmental*, **70**, 488–497.

64 Peron, J., Ruiz. E., Jones, D. J., and Rozière, J. (2008) *J. Membrane Science*, **314**, 247–256.

65 Parvole, J. and Jannasch, P. (2008) *J. Material Chemistry*, **18** (45), 5547–5556.

66 Perrin, R., Elomaa, M., and Jannasch, P. (2009) *Macromolecules*, **42** (14), 5146–5154.

67 Schmidt, T. J. and Baurmeister, J. (2008) *J. Power Sources*, **176** (2), 428–434.

68 Schmidt, T. J. and Baurmeister, J. (2006) *ECS Transactions*, **3** (1), 861–869.

69 Roychoudhury, S., Castaldi, M, Lyubovsky, M, LaPierre, R., and Ahmed, S. (2005) *J. Power Sources*, **152**, 75–86.

70 Campanari, S., Manzolini, G., Beretti, A., and Wollrab, U. (2008) *J. Engineering Gas Turbines Power*, **130**, 021701-1–021701-8.

71 Mak, A. and Meier, J. (2007) *Fuel Cell Auxiliary Power Study, Volume I: Raser Task Order 5*, NASA/CR-2007-214461/VOL1, National Technical Information Service, Springfield, VA.

72 Ersoz, A., Olgun, H., and Ozdogan, S. (2006) *J. Power Sources*, **154** (1), 67–73.

73 Montel, S. (2003) PhD thesis, University of Aachen.

74 Cutillo, A., Specchia, S., Antonini, M., Saracco, G., and Specchia, V. (2006) *J. Power Sources*, **154** (2), 379–385.

75 Pasel, J., Latz, J., Porš, Z., Samsun, R. C., Tschauder, A., and Peters, R. (2008) *ECS Transactions*, **12**, 589–600.

76 Specchia, S., Cutillo, A., Saracco, G., and Specchia, V. (2006) *Industrial and Engineering Chemistry Research*, **45** (15), 5298–5307.

77 Kallo, J., Schumann, P., Graf, C., and Friedrich, K. A. (2008) Presented at Fuel Cell Seminar, Phoenix, AZ, 2008, http//www.fuelcellseminar.com/assets/pdf/2008/presentations.aspx (accessed 20 November 2009).

78 Springmann, S., Bohnet, M.; Docter, A. Lamm, A., and Eigenberger, G. (2004) *J. Power Sources*, **128** (1), 13–24.

79 Sommer, M., Lamm, A. Docter, A., and Agar, D. (2004) *J. Power Sources*, **127** (1–2), 313–318.
80 Ahmed, S., Ahluwalia, R., Lee, S.H.D., and Lottes, S. (2006) *J. Power Sources*, **154** (1), 214–222.
81 Pischinger, S, Schönfelder, C., and Ogrzewalla, J. (2006) *J. Power Sources*, **154** (2), 420–427.
82 Samsun, R.C. (2008) *Kerosinreformierung für Luftfahrtanwendungen*, Forschungszentrum Jülich, Zentralbibliothek, Jülich.
83 Chrenko, D., Coulié, J., Lecoq, S., Péra, M.C., and Hissel, D. (2009) *Int. J. Hydrogen Energy*, **34** (3), 1324–1335.
84 Tsourapas, V., Sun, J., and Nickens, A. (2008) *Energy*, **33**, 300–310.
85 Sorrentino, M. and Pianese, C. (2009) *J. Fuel Cell Science Technology*, **6**, 1–12.
86 Lin, S.T., Chen, Y.H., Yu, C.C., Liu, Y.C., and Lee, Y.C. (2006) *Int. J. Hydrogen Energy*, **31** (3), 413–426.
87 Severin, C., Pischinger, S., and Ogrzewalla, J. (2005) *J. Power Sources*, **145** (2), 675–682.
88 Goebel, S.G., Miller, D.P., Pettit, W.H., and Cartwright, M.D. (2005) *Int. J. Hydrogen Energy*, **30** (9), 953–962.
89 Korsgaard, A.R., Nielsen, M.P., and Kær, S.K. (2008) *Int. J. Hydrogen Energy*, **33** (7), 1921–1931.
90 Blum, L., Steinberger-Wilckens, R., Meulenberg, W.A., and Nabielek, H. (2005) In *Fuel Cell Technologies: State and Perspectives* (eds N.M. Sammes, A. Smirnova, and O. Vasylyev), Springer, Berlin, pp. 107–122.
91 Singhal, S.C. (2006) *Advances in Science and Technology*, **45**, 1837–1846.
92 U.S. Department of Energy (2009) *10th Annual Solid State Energy Conversion Alliance (SECA) Workshop*, Pittsburgh, PA, 14–16 July 2009, http://www.netl.doe.gov/publications/proceedings/09/seca/index.html (accessed 16 November 2009).
93 Blake, G.D. (2008) Presented at DOE Peer Review 2008, http://www.hydrogen.energy.gov/pdfs/review08/fc_44_blake.pdf (accessed 16 November 2009).
94 de Bruijn, F. (2005) *Green Chemistry*, **7**, 132–150.
95 Kerr, R. (2009) Presented at the 2009 SECA Annual Meeting, 14–16 June 2009, Pittsburgh, PA.
96 Shaffer, S. (2005) Presented at the 2005 SECA Review Meeting, Pacific Grove, CA, 20 April 2005.
97 Mulot, J., Niethammer, M., Mukerjee, S., Haltiner, K., and Shaffer, S. (2008) *Proceedings of Fundamentals & Development of Fuel Cell Conference*, 10–12 December 2008, Nancy, France.
98 Stelter, M., Reinert, A., Mai, B.A., and Kuzncov, M. (2006) *J. Power Sources*, **154** (2), 448–455.
99 Hansen, J.B. (2005) Oxygenates as fuels for SOFC auxiliary power units, presented at the 15th International Symposium on Alcohol Fuels, 9 May 2005, San Diego, CA.
100 Moreno, A., McPhail, S., and Bove, R. (2008) *International Status of Molten Carbonate Fuel Cell (MCFC) Technology*, http://143.130.16.49/(en)/publ/pdf/fuelcell_mcfc.pdf (accessed 16 November 2009).
101 Selman, J.R. (2006) *J. Power Sources*, **160** (2), 852–857.
102 Wannek, C., Dohle, H., Mergel, J., and Stolten, D. (2008) *ECS Transactions*, **12** (1), 29–39.
103 Reiche, A. (2006) Presented at Fuel Cell Seminar, San Antonio, TX, 2006, http//www.fuelcellseminar.com/assets/pdf/2006/W_ednesday/3B/Reiche_Annette_1010_5 02&507.pdf (accessed 16 November 2009).
104 Steiger, W., Seyfried, F, and Huslage, J. (2007) *Motortechnische Zeitschrift*, **12**, www.atzonline.de (accessed 16 November 2009).
105 Wannek, C., Lehnert, W., and Mergel, J. (2009) *J. Power Sources*, **192** (2), 258–266.
106 Hytran, Hydrogen and Fuel Cell Technologies for Road Transport (2007) http://www.hytran.org/ (accessed 16 November 2009).
107 Kolb, G., Schürer, J., Tiemann, D., Wichert, M., Zapf, R., Hessel, V., and Löwe, H. (2007) *J. Power Sources*, **171** (1), 198–204.

108 O'Connell, M., Kolb, G., Schelhaas, K.P., Tiemann, D., Ziogas, A., and Hessel, V. (2009) *Int. J. Hydrogen Energy*, **34** (15), 6290–6303.

109 Qi, A., Peppley, B., and Karan, K. (2007) *Fuel Processing Technology*, 88, 3–22.

110 Roychoudhury, S., Junaedi, C., Walsh, D., Mastanduno, R., Spence, D., DesJardins, J., and Morgan, C. (2008) *Proceedings of the Fuel Cell Seminar*, Palm Springs, CA, 28–30 Octyober 2008 (CD), Courtesy Associates, Washington, DC.

111 Grube, T., Menzer, R., Samsun, R.C., Pasel, J., and Peters, R. (2006) *VDI-Ber.*, (1975), 379.

112 Grupp, D., Forrest, M., Mader, P., Broderick, C.J., Miller, M., and Dwyer, H. (2004) Internal Report UCD-IST-RR-0416, http://hydrogen.its.ucdavis.edu/publications/pubpres/2004pubs/ (accessed 20 November 2009).

113 Simpkens, E. and Ferry, E. (2008) Presented at Fuel Cell Seminar, Phoenix, AZ, 2008, http//www.fuelcellseminar.com/assets/pdf/2008/presentations.aspx (accessed 20 November 2009).

114 U.S. Department of Energy, Distributed Stationary Fuel Cell Systems (2009) http://www1.eere.energy.gov/hydrogenandfuelcells/fuelcells/systems.html (accessed 22 November 2009).

35
Portable Applications and Light Traction
Jürgen Garche

Abstract

It is reported about low-power FC systems (<250 W) and high-power systems (<5 kW) for portable applications and light tractions. Based on the application it is derived the demand on fuel cells. For portable applications and light traction PEMFCS; direct liquid fuel cells (mainly DMFC) and SOFCs are discussed.

Furthermore about the fuel supply is reported. Mainly liquid fuels are applied, which are used directly (alcohols, on ethanol, formic acid, sodium borohydride, dimethylether) or via reforming to H2-reach gases (alcohols, LPG). But also hydrogen is used stored in high-pressure cylinders or hydrides.

The whole FC system (stack, gas processing, management system, and power conditioning) is described. Examples for portable power generators (500 W–5 kW) as backup power, grid-independent generators and auxiliary power units (APUs) are given. Also low-power portable applications are described in the ~25 W region (e.g. mobile and cordless phone, pagers, walkmans) and in the 25–250 W range (e.g., notebook, professional camcorder, toys).

In the same way military applications are discussed, which have special requirements on the thermal and noise signature and the robustness of the systems.

Because the power requirement of some special vehicles (e.g. scooters, motorbikes, forklifts, boats recreation vehicles) is in the range of the power of portable fuel cell, in this chapter are discussed also light traction applications.

Keywords: fuel cell, portable systems, light traction, power generator

35.1
Introduction

Fuel cells are attractive for portable applications mainly in two areas:
- low-power systems (<250 W) for portable mostly integrated electronic/electric devices as replacement for consumer batteries with the advantage of a longer operating time;

- high-power systems (< 5 kW) for portable grid-independent power generators as auxiliary power units (APUs) or backup generators as replacement for ICE emergency generators. These high-power systems are also the basis for light traction applications, such as motorbikes, scooters, wheelchairs, boats, and fork lifts, where they replace mainly lead-acid batteries.

General portable applications with fuel cells, compared with fuel cells in mobile or stationary applications, have the advantages of high production volumes and that cost limitations are easier to meet. Fuel cells in portable applications, however, will not contribute significantly to energy savings and to a reduction in global environmental problems through reduced emissions. However, this innovative fuel cell technology is of high economic importance, for example as a replacement for batteries in a multi-billion euros market.

Portable applications are discussed as the first commercial fuel cell field with a door-opener function for the whole technology. Therefore, the portable applications are sometimes also called early market applications. An early market introduction will help build up manufacturing volume and operating experience and provide a near-term incentive to fuel cell developers and component suppliers to invest in technical development and production facilities to provide the base from which low-cost production of mass-market stationary and automotive applications can be established.

35.2
Demand on Fuel Cells for Portable Applications

The demand on fuel cells for portable applications is as follows:
- high power and energy density
- low costs
- long lifetime
- comply with safety standards
- environmental acceptability
- comply with operating conditions
  - ambient temperature
  - ambient humidity
  - fast start-up time
- fuel availability.

Typical target specifications for low-power electronic systems (< 50 W) and high-power (500 W–5 kW) systems are given in Table 35.1.

These targets are long-term targets. Short- and medium-term targets for low-power systems are 30–100 W kg^{-1} and 30–100 W l^{-1} and for high-power systems 100–200 W kg^{-1} and 100–200 W l^{-1}.

Table 35.1 Target specifications for different portable application [1].

Specification	Units	Low-power systems	High-power systems
Specific power	$W\,kg^{-1}$	300	500
Power density	$W\,l^{-1}$	200	400
Energy density	$W\,h\,l^{-1}$	1000	–
Electrical efficiency (ratio of electrical power to HHV)	%	–	>35
Cost (military/industry use)	$€\,W^{-1}$	–	3
Cost (commercial use)	$€\,W^{-1}$	1–2	0.5
Lifetime (military/industrial use)	h	–	5000
Lifetime (commercial use)	h	5000	2000
Start-up time (hybridization with battery/supercap)	–	Instantaneous	Instantaneous

35.3
Fuel Cell Technology

Caused by the demand for short start-up times and ambient operating temperatures, mostly low-temperature fuel cells are used for portable applications, namely the proton exchange membrane fuel cell (PEMFC) and the direct liquid fuel cell (DLFC).

Whereas the PEMFC in mobile and stationary applications is used almost without exception in a traditional bipolar design, for portable applications other constructions are also applied (Figure 35.1).

The planar design allows the construction of lightweight fuel cells with a flat design, air breathing, and general lower auxiliaries. The only disadvantage is the restriction to low current due to the higher ohmic losses due to the side-by-side connections of cells.

For electronic devices, micro PEMFCs were developed based on microelectromechanical system (MEMS) technology with Si wafers. Alternatively, low-temperature co-fired ceramic (LTCC) is also used. However, both materials are almost electrical insulators, so for current collection a conductive layer must be coated on to these substrates. Printed circuit board (PCB) technology offers an interesting platform to get the current collector on to the fuel cell body itself.

The DLFCs have the advantage of the use of liquid fuels, which is especially an advantage for low-power portable application. There are different types of DLFCs, where the highest development state is the direct methanol fuel cell (DMFC), which is already partially commercialized. The methanol is used either as vapor, liquid, or in solution. Liquid solutions are generally preferred due to more convenient thermal management and effective CO_2-methanol separation at the anode exhaust. In relation to the H_2-PEMFC, the DMFC has slower electrokinetics of methanol electrooxidation and shows a high methanol crossover, which lead to an efficiency loss and a reduced cathodic potential caused by for-

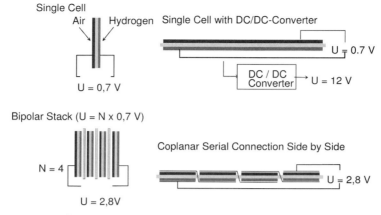

Fig. 35.1 Different PEMFC designs.

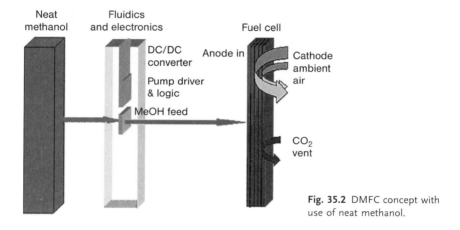

Fig. 35.2 DMFC concept with use of neat methanol.

mation of a mixed potential. The crossover effects could be reduced by a low methanol fuel concentration. With methanol concentrations of <1 mol l^{-1} single cell power densities exceed 100 mW cm^{-2} in the typical operating temperature region of 30–60 °C. However, in miniature devices, higher fuel concentrations should be utilized to minimize size. Therefore, much effort is made to use neat methanol. One concept is shown in Figure 35.2, where an *in situ* methanol dilution system, using water produced at the cathode, is deployed. The water required for the process on the fuel side is transferred internally within the fuel cell from the site of water generation on the air side of the cell. This internal flow of water takes place without the need for any pumps, complicated recirculation loops, or other micro-plumbing tools.

Relatively high power densities are reached only with a Pt catalyst content of 3–6 mg cm^{-2}, which, however, leads to high costs. Therefore, alkaline DMFCs with a low Pt concentration and an anion-exchange membrane are also under development. The problems with carbonization, however, are severe.

Other DLFCs are based on ethanol, formic acid, sodium borohydride, and dimethyl ether, but they have not reached the development state of DMFCs.

If, however, start-up times of up to 1 h are acceptable, then the solid oxide fuel cell (SOFC) can also be used. The SOFC has the advantage of a high specific power, high efficiency, and the relatively easy use of hydrocarbon fuels, such as liquefied petroleum gas (LPG). Portable SOFCs with an operating temperature of about 500 °C are under development in both the $\leq$2 W and also the 50 W power regions (see Figure 35.3). So far, portable SOFC systems have been successfully developed only on the laboratory scale; commercialization still has a long way to go.

The fuel cell system is composed not only of the fuel cell stack and the gas processing unit but also of control and management systems (water management, operating temperature, mass flow) and power conditioning. The latter has to respond fast to changes in the power demanded of the device. The transient time for the system to reach its operating temperature or to react to load changes has to be minimized. A hybrid system with a small rechargeable battery or an ultra-capacitor is an option to deliver the power while the fuel cell reaches its preferred operating conditions. Furthermore, the battery or ultra-capacitor can recover power peaks. For portable application, all these components must be miniaturized and inexpensive. The same is valid for standardization of the physical interfaces of portable systems and the fittings.

Research topics on the material side for portable fuel cell systems do not differ very much from those for systems in the other application areas. However, the relatively low operating temperature (PEMFCs, DMFCs) requires very active catalysts and catalysts with a low CO-sensitivity in the case of reformer systems.

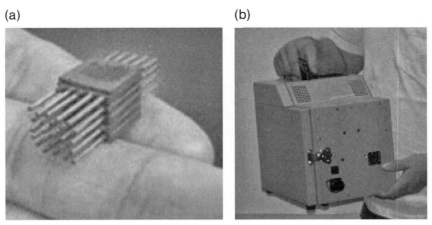

Fig. 35.3 Portable SOFC applications: (a) 2 W cube-type sub-millimeter tubular SOFC driven by wet hydrogen [2]; (b) 50 W driven by desulfurized JP8–kerosene fuel [3].

Further, advanced manufacturing methods and improved engineering and component multi-functionality are important topics for the improvement of the system packaging. Other important improvements are related to balance of plant components, namely improved pump efficiencies and reduced noise emissions of ancillary subsystems.

35.4
Fuel

Since the energy is stored in the fuel cell as a fuel reservoir rather than integrated into the power source, fuel cells are expected to provide higher energy density (W h l^{-1}) and specific energy (W h kg^{-1}) due to the high energy that can be stored in hydrogen and liquid fuels.

For the fuel technologies considered in the previous section, the fuels mostly used are hydrogen and the liquid fuels methanol, ethanol, and sodium borohydride. Hydrogen from high-pressure cylinders or hydrides is feasible for portable generators. However, these cylinders could not be scaled down efficiently and therefore they are mostly impractical for small electronic devices. Therefore, liquid fuels are often used for these systems directly, with advantages of use of conventional storage systems, no high-pressure tanks, no gas monitoring systems, and quick refilling.

These fuels are used also indirectly via reforming in a reformed H_2-PEMFC, because the power of the H_2-PEMFCs is much higher than that of DLFCs. The development of light and compact reformer systems in the lower power class is considered of special importance. In the very low-power class, the MEMS technology is also used for reformer construction. The gas processing has a significant balance of plant (BOP), consisting of a heating system, oxidant storage, pump, sensor and distribution components, control system, and partially a d.c.–d.c. converter for power conditioning. The BOP often constitutes 30% or more of system cost and mass. Furthermore, it requires a significant amount of energy from the device, thus lowering the total system efficiency. There is also some time delay in the response at the start of hydrogen production. Therefore, it is necessary to consider the pros and cons of the direct and the indirect use of liquid fuels for special applications.

Another approach is the thermal (with physical storage) or hydrolytic (with chemical storage) release of H_2 from hydrides (metallic, chemical) or metals (e.g., zinc, aluminum, sodium) [3].

On the other hand, the consumption of hydrogen in larger high-power applications is relatively high, which leads to logistic problems. Therefore, for high-power applications fuels are used that are logistically easier to handle, such as methanol, LPG, gasoline, diesel, or kerosene, which have to be reformed to an H_2-rich gas. Very compact fuel reformers with an enhanced heat transfer mechanism have been developed which show a good cold start-up behavior of about 10 min without noticeable CO fluctuation.

In addition to these fuels, anhydrous liquid ammonia is also an interesting carrier of hydrogen. Ammonia is converted via low-temperature (350 °C) crackers into hydrogen-rich gases [4].

For the wide use of portable applications we need a widespread distribution network of hydrogen and liquid fuels worldwide. Especially for military applications, a single fuel (e.g., jet fuel) that required that is readily available worldwide [5]. Furthermore, there handling and transport regulations, legal issues, and standardizations of fuel specification and fuel containers need to be established. Constraints regarding the use and transport of fuel containers especially in airplanes need to be eliminated. A recycling and reuse concept for fuel containers has to be developed.

35.5
Applications

35.5.1
Portable Power Generators (500 W–5 kW)

Unfortunately, the term "portable" is not really well defined. The German Fire Department uses the concept that portable equipment must be able to be carried by four fire fighters even in mountainous areas.

35.5.1.1 Backup Power Generators

These are used for assurance for critical systems and ensuring a reliable power supply for industries such as telecommunications. It is a very demanding application of fuel cells where high energy and power densities are essential. Additionally, they have to operate with very short start-up times under ambient conditions and sustain a wide dynamic operational range, especially when used in uninterruptible power systems (UPSs) (start-up time < 20 s for UPSs). Therefore, mainly PEMFC systems are suitable; for low-power applications, DMFCs could also be used. To provide instant start capability, fuel cell systems are paired with batteries or ultracapacitors and typically include power electronics and hydrogen fuel storage. Compared with batteries, fuel cells offer longer continuous run times and greater durability in harsh outdoor environments under a wide range of temperature conditions. PEMFCs can also offer significant cost advantages over both battery–generator systems and battery-only systems when shorter run time capability of up to 3 days is sufficient [6].

There are already (pre)commercial products with H_2-PEMFCs, for example:
- Premion T 4000 from P21 with a rated power of 4 kW
- PureCell® Model 5 fuel cell systems from UTC Power with a rated power of 5 kW
- ElectraGen fuel cell system of Idatech with a rated power of 3 and 5 kW
- GenCore5 fuel cell systems of Plug Power with a rated power of 5 kW (see Figure 35.4).

Fig. 35.4 GenCore® with optional hydrogen storage module.

Whereas the Premion and PureCell systems have a weight of < 100 kg, the ElectraGen and GenCore systems weigh in the region of 220 kg without fuel supply. In the latter case, the expression "portable" seems to be borderline.

The present poor power densities of these systems (20–100 W l^{-1}) are partly caused by the lack of dedicated and integrated system components. The costs amounted to $\leq €\,10\,W^{-1}$ and the lifetime is about 1000 h. The systems have an efficiency of $\geq 35\%$. The operating temperature area is in the region of –40 to 60 °C.

35.5.1.2 Grid-Independent Generators

These are used for on-site services in areas that are off-grid (distributed generation), for example for energy supply for signaling or weather stations, and for systems for warning lights and signs. In contrast to backup power generators, in these applications the fuel cell has to work for a longer time and not only in emergency cases. Therefore, the lifetime of the fuel cell is an important parameter. The advantage of these systems in relation to ICE grid-independent generators is the possible inside use and the low noise. The parallel with electricity-developed heat is not used in grid-independent applications.

In the military sector, the applications for fuel cell systems are numerous (see Section 35.5.4).

In the lower power region, the grid-independent generators are driven by hydrogen or even by methanol. In the 5 kW class, the consumption of hydrogen will be very high. Therefore, liquid fuels or natural gas if available should be used via a reformer.

Figure 35.5 shows an example of a hydrogen grid-independent PEMFC generator produced by ZSW-UBZM. This unit, which delivers at 24 V a power of

Fig. 35.5 Ulmer Stromschachtel grid-independent 1.1 kW PEMFC generator (ZSW-UBZM).

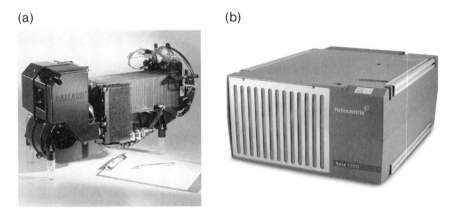

Fig. 35.6 Nexa grid-independent 1.2 kW PEMFC generator: (a) Ballard; (b) Heliocentris

1.2 kW$_{cont}$ and 1.8 kW$_{peak}$, has dimensions of 750×250×530 mm and a weight of 31 kg.

A further example is the 1.2 kW NEXA unit of Ballard (Figure 35.6). This unit was further developed by Heliocentris. The Heliocentris Nexa® 1200 is based on the FCgen 1020 ACS stack from Ballard. The parasitic power consumption of the old system could be reduced by up to 50% by atmospheric air supply in combination with the air-cooled stack. The system has dimensions of 400×550×220 mm and a weight of 22 kg.

Whereas the above systems are driven by hydrogen, the 650 W Emerald system of Voller runs on LPG.

More about this topic can be found in the chapter Off-Grid Power Supply and Premium Power Generation in this book.

35.5.1.3 Auxiliary Power Units (APUs)

These provides energy for vehicle on-board functions other than propulsion, for example for boats, heavy-duty trucks, or airplanes. The technical designs of APUs and grid-independent generators are similar, hence sometimes both terms are used synonymously. Further information can be found in the chapter APUs for LDV, Trucks, Ships, and Airplanes in this book.

35.5.2
Low-Power Portable Applications (≤25 W)

The low-power fuel cell packs in the region of ≤25 W are related to power cellular phones, cordless phones, pagers, walkman, radio, discman, tape recorders, and so on. The development target is to use the integrated fuel cell power pack device. Caused by the still too high volume of the power packs, mostly the pack is used as a charger for the battery of the electronic devices or it is externally docked with the electronic equipment (e.g., with notebooks). The power required for electronic devices is shown in Table 35.2.

Cellular phones, however, also have short and heavy current peaks. To meet this demand, often a hybrid configuration of a fuel cell with a small battery or ultracapacitor is used.

Volume constraints are the most important development challenges. The miniaturization of systems and subsystems is an important target, and the passive operation of the fuel cell stack via smart designs is the main way to reach this target. Water management is perhaps the most complicated problem for miniature PEFCs and DMFCs as it includes fuel concentration control, membrane humidification, and water recovery and recirculation.

In this low-power class, mostly the DMFC is used, because the supply of liquid fuels is easier than fuel supply via hydrogen cartridges (pressure, metal hydrides) for PEMFCs.

Table 35.2 The power demand of different portable applications [7].

Item	Power required (W)
Cellular phone	1
Personal digital assistant (PDA)	1
Notebook personal computer	20–30
Flashlights and toys	1–10
Tablet personal computer	10
Playstation portable (PSP)	2
Digital multimedia broadcast-receiving (DMB) phone	3
iPhone	2
Robot	10–15
Digital camera	1

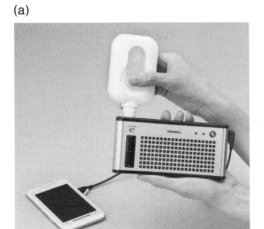 

Fig. 35.7 Low-power DMFC systems: (a) Dynario (Toshiba); (b) Xtreme 24-7 (Medis).

Many components, especially for the subsystems, need to be developed further with improved H_2- and liquid fuel (methanol)-compatible materials. The simplicity of design, meaning the elimination of components, would lead to benefits in reliability, manufacturing, and costs of fuel cell systems. Important developments in the subsystems include low-power miniature electronics (e.g., converters), miniature fluid handling systems (valves, etc.), sensors and controls, liquid-gas separation and handling, and thermal management. Figure 35.7 shows two low-power DMFC packs.

Toshiba has developed the Dynario, which is a d.c. 5 V, 400 mA DMFC charger of weight 280 g (without fuel) and dimensions W150×H21×L74.5 mm. The Dynario can replenish the batteries in devices like cell phones and digital cameras via a USB. According to Toshiba's Web store, 3000 units (US$328) will be sold up to October 2009. A refill bottle with 50 ml of methanol costs US$7.

The Medis Xtreme 24-7 is a d.c. 3.8–5.5 V, 1 A DMFC charger, which operates between 0 and 40 °C, of weight 185 g and dimensions W68×H57×L74.5 mm. A price of US$49.99 has been announced.

MTI MicroFuelCells, Samsung, Ultracell, and others are also developing such DMFC power packs. MTI has demonstrated a power density of 84 mW cm^{-2} and energies of 1800 W h kg^{-1} and 1.4 W h cm^{-3}. Ultracell has developed a 25 W XX25^{TM} DMFC system of weight 1.24 kg and dimensions 23×15×4.3 cm.

Voller is developing a PEMFC system for use in remote and mobile environments, running on LPG with electrical output up to 650 W.

35.5.3
Low-Power Portable Applications (25–250 W)

Fuel cell systems in the low-power range 25–250 W are required for somewhat larger electronic devices (e.g., notebook, professional camcorder), and also for some toys and power tools. Furthermore, they are used for independent electrical energy supply for leisure (APUs), e.g., for sailing boats and caravans. A special application area is the replacement of batteries where long, durable operation is needed and charging or exchange is difficult (e.g., sea buoys, lighthouses, weather stations, telecom stations). Here, fuel cell technology is applicable even at today's cost as soon as robust systems become technically available.

However, the fuel cell power packs in this power area also demand stringent miniaturization efforts. Nevertheless, in some cases it may not be necessary to integrate the energy converter into the device itself. This then corresponds to a charger design.

In general, the replacement of batteries by fuel cells leads to a more pronounced thermal load (electric efficiency 50% versus 80% in batteries). This could be a problem especially with higher power electronic equipment such as notebooks. Smart thermal integration designs are required, for example a combination of H_2-PEMFCs cells with heat-consuming metal hydrides (heat is required for hydrogen release). Furthermore, the development of compact and highly efficient thermal barriers is an important topic.

Not only in the low-power region (≤ 25 W) but also at higher powers (25–250 W) the methanol fuel cell is the dominant fuel cell technology. Examples of methanol FC systems are shown in Figure 35.8.

A few companies have already commercialized methanol FC systems in the 25–250 W range. Especially Smart Fuel Cell (SFC) offers a wide range of DMFC systems in its EFOY class of 25, 38, 50, and 90 W, with weight 6.8–7.9 kg and dimensions L43.5×W20×H27.6 cm. A further example is the 250 W Protonex M250-B unit (military version M250-CX) of weight 18 kg and dimensions

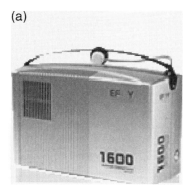

Fig. 35.8 Examples of methanol FC systems in the 25–250 W class: (a) DMFC system: 65 W EFOY 1600 (SFC); (b) reformed methanol FC system: 250 W M250-B (Protonex).

L56.4×H35.5×D30.5 cm. This methanol PEMFC system works with an advanced methanol reforming technology.

35.5.4
Military Applications [8]

In general, military fuel cell systems do not differ very much from the above-described civil systems, but there are special requirements on the thermal and noise signature and the robustness of the systems. If these demands can be satisfied, the military will accept a much higher cost.

The proliferation of mission-critical electronic equipment in the modern battlefield has created increasing requirements for advanced, high-performance fuel cells. The applications range from a soldier-portable power source to the power supply of devices such as sonar, radio sets, noctoviser, and radar, to unmanned equipment such as aerial drones. At the focus of applications are the soldier-portable power source (20–50 W) and the field-based charging station for batteries.

The requirement of very low thermal radiation excludes SOFCs and favors DMFCs/PEFCs. Therefore, methanol (direct and indirect) and LPG are used as fuel for these low-temperature fuel cells. However, for larger units the short-term requirement will be to utilize the present fuel infrastructure of conventional battlefield fuel, which is based on diesel, gasoline, and so on, but these fuels require an efficient reformer. Most reformer technologies for hydrocarbon fuel, however, are too bulky and heavy for portable fuel cell systems. There is a lot of effort being made to use conventional military fuels for the high-temperature SOFC, which does not need reformate gases with a low CO concentration as do PEMFCs.

35.5.4.1 Soldier-Portable System
These systems normally have a built-in battery to start up the fuel cell system. Most fuel cell systems operate directly on methanol; the French and American armies are also considering H_2-PEMFCs. Normally neat methanol is used. At operating temperatures $>40\,°C$, however, methanol–water mixtures are to prevent drying of the membrane at high air flows and also to avoid high vapor pressures in the fuel cartridge.

A critical issue is the nature of the infrared signature of the fuel cell system. Further improvements to reduce the exhaust temperatures are required. The noise level should be $<40\,dB(A)$ at a distance of 1 m. Furthermore, operating altitudes $>1500\,m$ could be a problem, caused by reduced oxygen partial pressure.

Examples of methanol fuel cell systems are shown in Figure 35.9. Both systems shown are driven by methanol. The Jenny system is a DMFC system with a continuous power of 25 W, of weight 1.7 kg and dimensions 252×171×74 mm. The DMFC start-up time is very short. The UltraCellXX25 system is a reformed methanol PEMFC system with a continuous power of 25 W, of weight 1.24 kg and dimensions 230×150×43 mm. Due to the reformer, the start-up time is 20 min.

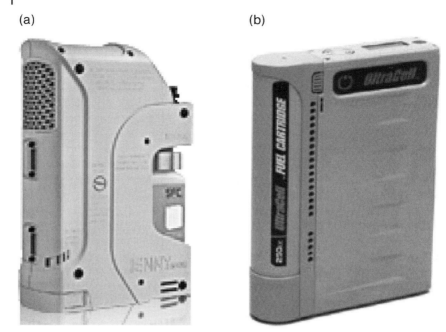

Fig. 35.9 Soldier-portable methanol fuel cell systems:
(a) Jenny 25 W DMFC (SFC); (b) UltraCellXX25 25 W RMFC (UltraCell).

35.5.4.2 Portable Fuel Cell-Based Battery Chargers

Aside from larger units that would be fixed to a vehicle, smaller recharger units (~ 250 W) are also available (Figure 35.10). The fuel storage, which is integrated into most soldier-worn systems, will be separated for most of the larger systems. The separation of storage and fuel cell systems for mounting in vehicles allows better on-board space economy.

With respect to fuel, again systems using methanol have achieved the highest technology readiness level (TRLs). Systems of this kind are supplied, for instance, by Ida-Tech and Prontonex. Alternatively, DMFC systems in the 250 W range are under development.

35.5.5
Light Traction [9]

The power requirement of some special vehicles is, at 100 W–5 kW, in the range of the power of portable fuel cell systems. Therefore, these special vehicle applications – also called light traction – are often discussed together with the portable applications. To this group of light traction vehicles belong forklifts, airport tugs, recreation vehicles, tour carts, wheelchairs, boats, scooters, motorbikes,

Fig. 35.10 250 W DMFC battery charger (SFC).

maintenance robots in industry institutions, and so on. For these applications there are early market opportunities since fuel cells may provide clear system benefits, such as. no emission for indoor systems and fast recharge, especially for material handling vehicles.

General conditions for light traction applications are, in addition to high energy and power density, in particular a short start-up time and highly dynamic operation. The latter requirements call for low-temperature fuel cells (H_2-PEMFC, DMFC) for light traction applications. For higher power applications, such as scooters, the PEMFC is used either alone or in combination with batteries (hybrids) to avoid peak power problems. DMFC–battery hybrids could also be used for power applications. The DMFC, especially as a hybrid, is a good alternative to the H_2-PEMFC as long the hydrogen infrastructure remains undeveloped.

The largest market, however, is the scooter and motorbike sector, especially in Asia. Table 35.3 shows the performance requirements for a scooter. To meet these requirements, pure fuel cells or fuel cell-battery hybrids are used as power sources for scooters and motorcycles (see Table 35.4).

Table 35.3 Performance requirement of a mid-range power scooter.

Specification	Requirement
Maximum motor power	4–6 kW
Range per refueling at 30 km h^{-1}	200 km
Fuel consumption	2.4 l per 100 km
Curb weight	< 130 kg

Table 35.4 Fuel cell/fuel cell–hybrid specifications for motorcycle driving (Taipei cycle).

Specification	Fuel cell (FC)	Hybrid 1	Hybrid 2
Maximum FC electrical power (kW)	5.9	3.2	1.1
Maximum FC heat generation (kW)	7.0	3.5	1.2
Efficiency at maximum power (%)	41.2	43.2	43.6
Assisting battery power (kW)	0	2.6	4.6
Fuel cell stack weight (kg)/volume (l)	7.6/7.8	5.4/5.3	4.1/3.2
Battery weight (kg)/volume (l)	0/0	3.1/3.0	5.6/5.4
Total drive system weight (kg)/volume (l)	61/43	63/43	60/30
Average fuel cell power (W)	674	698	577
Average total power output (W)	674	709	726
Fuel cell conversion efficiency (%)	56	53	47

Fig. 35.11 Aprilia Atlantic Zero Emission with 2×3 kW H2-PEMFC

Fig. 35.12 1.9 kW DMFC scooter of Forschungszentrum Jülich

Examples of fuel cell scooters are shown in Figures 35.11 and 35.12. The 6 kW H_2-PEMFC powers the scooter to a top speed of around 85 km h^{-1} and the riding range is up to 140 km. The four-wheel scooter is driven by a DMFC. The stack power amounts to 3.1 kW and the drive module 1.9 kW.

In addition to scooter development, fuel cell motorcycles are also under development.

Fig. 35.13 200 W H$_2$-PEMFC bikes (Masterflex).

The fuel systems already discussed above (DMFC, H$_2$-PEMFC) in the low-power region (25–250 W) are the basis for system integrators such as Stalleicher, City Com, GUF, ElBike, AkSale, Veloform, van Raam, Meyra, and Masterflex, who are developing electric-assisted motorcycles, scooters, tricycles, rickshaws, commuter vehicles, and so on.

In addition to motorcycles, cargobikes are also under development. Both are developed by, for example, Masterflex (Figure 35.13). The basis for the powertrain is a 200 W PEMFC of weight 8 kg and dimensions 14×26×48 cm, and an H$_2$ catridge, which allowed a range of up to 250 km. A cargobike with a payload of 150 km could be used, for example, in postal services.

In the European Community project The HYCHAIN MINI-TRANS, more than 150 small urban vehicles have been demonstrated, including small utility cars and minibuses, wheelchairs, scooters, and cargobikes, all powered by hy-

Table 35.5 Lifecycle cost comparison of PEMFC- and battery-powered forklifts [10].

Item	3 kW PEM fuel cell paired with integral NiMH battery, for pallet trucks			8 kW fuel cell paired with integral ultracapacitor, for sit-down rider trucks		
	Battery-powered (2 batteries per truck)	Powered with no tax incentive	Powered with $1000 kW^{-1} tax incentive	Battery (2 batteries per truck)	Powered with no tax incentive	Powered with $1000 kW^{-1} tax incentive
Net present value of capital costs (US$)	17 654	23 835	21 004	43 271	63 988	56 440
Net present value of O&M costs (including the cost of fuel) (US$)	127 539	52 241	52 241	76 135	65 344	65 344
Net present value of total costs of system (US$)	145 193	76 075	73 245	119 405	129 332	121 784

drogen fuel cells. More than 2000 reusable pressurized hydrogen cylinders, varying in size from 2 to 20 l, are used in the project. Drivers will exchange empty cylinders for full ones at designated distribution sites. The project is taking place in the Rhne-Alpes area in France (Grenoble Alpes Métropole Agglomeration Community), Castilla y León in Spain (cities of Soria and León), North-Rhine Westphalia in Germany (region of Emscher-Lippe Agglomeration Community) and the city of Modena in Italy.

Battery-powered material handling equipment, such as forklifts, are used in warehouses worldwide; about 2.5 million forklifts are operating in North America and Europe. The main problem with battery-driven forklifts is the battery charge, which takes about 8 h and need a charging station. The replacement of batteries by fuel cells would solve these problems and would be also cheaper.

For this application, H_2-PEMFCs are used. To avoid the H_2 logistics problem, on-site electrolyzers could be used or DMFC systems. Possible peak power problems could be mitigated, for example, by batteries and ultracapacitors.

References

1 European Commission (2005) *Strategic Research Agenda*, European Hydrogen and Fuel Cell Technology Platform, Brussels, July 2005; http://ec.europa.eu/research/fch/pdf/hfp-sra004_v9–2004_sra-report-final_22jul2005.pdf#view=fit&pagemode=none (accessed 20 February 2010).

2 Suzuki, T., Yamaguchi, T., Fujishiro, Y., and Awano, M. (2008) Cube-type micro SOFC stacks using sub-millimeter tubular SOFCs. *J. Power Sources*, **183**, 544–550.

3 Browning, D., Jones, P., and Packer, K. (1997) An investigation of hydrogen storage methods for fuel cell operation with man-portable equipment. *J Power Sources*, **65**, 187.

4 Kordesch, K., Gsellmann, J., and Aronson, R. (1999) Intermittent use of a low-cost alkaline fuel cell-hybrid system for electric vehicles. *J. Power Sources*, **80**, 190–197.

5 Cowey, K., Green, K.J., Mepsted, G.O., and Reeve, R. (2004) Portable and military fuel cells. *Curr. Opin. Solid State Mater. Sci.*, **8**, 367–371.

6 Mahadevan, K., Judd, K., Stone, H., Zewatsky, J., Thomas, A., Mahy, H., and Paul, D. (2006) *Identification and Characterization of Near-Term Direct Hydrogen proton Exchange Membrane Fuel Cell Markets*, Contract No. DE-FC36GO13110, US Department of Energy, http://www1.eere.energy.gov/hydrogenandfuelcells/pdfs/pemfc_econ_2006_report_final_0407.pdf (accessed 20 February 2010)

7 Kundu, A. and Jang, J.H. (2009) Application-portable, portable devices: fuel cells. In *Encyclopedia of Electrochemical Power Sources* (ed J. Garche), Elsevier, Amsterdam, Vol. 1, p. 39.

8 Cremers, C., Tübke, J., and Krausa, M. (2009) Application – portable, military: batteries and fuel cell devices: fuel cells. In *Encyclopedia of Electrochemical Power Sources* (ed J. Garche), Elsevier, Amsterdam, Vol. 1, p. 13.

9 Z. Qi (2009) Application– transportation; light traction: fuel cells. In *Encyclopedia of Electrochemical Power Sources* (ed J. Garche), Elsevier, Amsterdam, Vol. 1, p. 302.

10 US Department of Energy *Early Markets: Fuel Cells for Material Handling Equipment* (2008) http://www1.eere.energy.gov/hydrogenandfuelcells/education/pdfs/early_markets_forklifts.pdf (accessed 20 February 2010)

Stationary Applications

36
High-Temperature Fuel Cells in Decentralized Power Generation

Robert Steinberger-Wilckens and Niels Christiansen

Abstract

Decentralised power generation (DG) can contribute to increases in efficiency of the entire power generation and distribution network. It reduces grid losses by moving the generation closer to the customer, thus also allowing the use of the waste heat generated in electricity production. At the same time it offers competitive advantages to industrial customers in supplying cost effective peak production, grid stabilisation and uninterruptible power supply.

Total power generation efficiency, and subsequently also CO_2 balances, though, is only increased if the DG electrical efficiency meets minimum standards. These are defined by the grid characteristics within which the DG is operated. High temperature fuel cells offer a high value due to their high electrical conversion efficiencies of above 50%, reaching up to over 60% with the Solid Oxide Fuel Cell (SOFC).

High efficiency, though, is only achieved with adequate system architectures. Worldwide, a number of manufacturers and developing groups are working on fuel cells for distributed electricity generation, with increasing success.

Keywords: high-temperature fuel cells, decentralized power generation, electricity production, combined heat and power

36.1
Introduction

Combined heat and power (CHP), or even polygeneration, are means of dramatically increasing the energy conversion efficiency in the field of electricity generation, at the same time generally reducing greenhouse gas emissions and improving the efficiency of electric transmission networks. CHP may be employed in single family housing (discussed in another chapter) or in larger units for industrial, utility, and district heating use. "Distributed" (or "decentralized") generation (DG) is often seen as a synonym for combined heat and power,

Hydrogen Energy. Edited by Detlef Stolten
Copyright © 2010 WILEY-VCH Verlag GmbH & Co. KGaA, Weinheim
ISBN: 978-3-527-32711-9

although in some cases it may prove economically sensible to generate electricity in distributed units even when the heat is completely rejected.

Whereas the motivation to employ micro-CHP units in the range 1–5 kW$_{el}$ in residential buildings is mostly driven by the desire to reduce CO_2 emission levels, the installation of industrial-scale units has more of an economic background. "Industrial scale" will indicate here units for commercial (business units, workshops, etc., from approximately 10 kW$_{el}$ upwards), industrial (100–1000 kW$_{el}$), and power generation (1 MW$_{el}$ upwards) applications. For many industrial companies, the cost of electricity is divided into two components: the cost per unit of electricity and the cost of power provision. The latter takes account of the cost incurred with the delivering utility to provide reliable power at any time, whereas the former is directly related to the cost of electricity production and distribution. Companies with high peak loads will suffer from high payments for grid connection, which become especially excessive if the load capacity factor is low. Employment of DG technology is therefore driven by three factors: reducing the grid connection cost, reducing the cost of electricity, and increasing the reliability of electricity supply. Only the second aspect takes the utilization of the heat in a CHP scheme into full account.

The typical gas or diesel engine CHP units suffer from problems of high maintenance effort and (relatively) low electricity conversion efficiency, whereas the noise encapsulation necessary for residential applications is less important.

Fuel cells, and especially high-temperature fuel cells, offer benefits through their high efficiency (even at low load levels), low noise, high modularity, and multi-fuel capabilities. The main motivation for the employment of fuel cells in distributed generation contexts, however, will be high efficiency and provision of high-temperature off-heat, both of which constitute added values compared with conventional technology. Moreover, high-temperature fuel cells have the ability to convert methane to heat and electricity directly, without fuel processing, which makes them an ideal combination with bio-fermenter gas (biogas). Natural gas from the grid, however, will require conversion of the higher hydrocarbons and removal of sulfur-containing odorants upstream of the fuel cell stack.

This chapter summarizes the current status of worldwide activities in high-temperature fuel cells (solid oxide and molten carbonate) for distributed generation and combined heat and power. It concentrates on major industrial players and research developers building, manufacturing, or delivering high-temperature fuel cells in the power range from 10 kW$_{el}$ upwards.

36.2
Distributed Generation as a Tool to Improve the Efficiency of Electricity Provision

Worldwide, central electricity generation based on coal, nuclear, and other fossil sources has an average net efficiency of 32% [1]. The remaining 68% of the fuel energy is rejected to the atmosphere. Average European and German efficiencies differ only marginally from this figure. Another 5–10% of the electricity generated is lost during transport in the supply grids [2]. Central electricity generation thus has an overall efficiency below 29% and is one major cause of CO_2 emissions and waste of depletable, fossil fuels.

The industrial interest in on-site DG introduced above therefore coincides with the potential to supply more efficiently produced electricity. Moreover, the location of the generating equipment near to the point-of-use allows for various uses of the waste heat, which can now be fed into heating, hot water, and process heat distribution systems. The total efficiency of the fuel use will now greatly surpass the electrical efficiency and can – given that the heat can be put to full use – reach up to 100% (LHV) total efficiency. Due to European Directives, most EU countries today demand a minimum total efficiency of 70% in order for CHP units to benefit from some favorable regulations on taxes and price guarantees [3, 4].

The main equipment applied in CHP and DG is gas or diesel internal combustion engines and gas turbines. In the region between 1 kW_{el} and 1 MW_{el}, these have a very limited electrical efficiency ranging from 25 to 35%. Only large gas turbines of multi-MW_{el} size reach high efficiencies up to 40%. Diesel engine generators of 1 MW and more can display efficiencies above 40% [5]. In addition, all these units have high requirements on maintenance: gas turbine blades have a lifetime of 25 000 h [6], whereas ignition engine units are incrementally completely renewed within an operating cycle of 10 years [7]. The reject heat from an engine is sufficient to supply heating. Gas turbines can supply slightly higher exhaust temperatures that might be used, for instance, in steam generation or for feeding the regeneration unit in adsorption chillers.

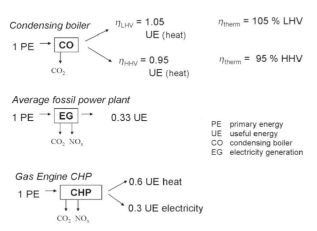

Fig. 36.1 Schematic diagrams of a conventional heating boiler (condensing boiler, upper diagram) and the average German electricity grid conversion efficiency (middle diagram) as systems of reference. The lower diagram shows the combined production of electricity and heat in a CHP unit.

Fig. 36.2 Comparison of the processes in Figure 36.1 with respect to their delivery of useful energy (energy services). Calibrating to the energy delivery, the total primary energy effort for both scenarios can be compared as 1:1.56, i.e., bringing a 38% decrease in primary energy effort for the gas engine CHP case.

Table 36.1 CO_2 emissions of the systems according to Figures 36.1 and 2, based on German supply grid data.

Process	Fuel	CO_2 emission (g per kWh useful energy) (LHV)
Condensing boiler	Natural gas	235 (heat)
German electricity supply grid	Various	560 (electricity)
Gas engine CHP	Natural gas	392 (heat) 783 (electricity) 260 (combined)

Nevertheless, especially gas engine CHP units have found widespread application in DG. Although the electrical efficiency will in many cases even be below the average efficiency of electric power generation, a clear case can be made for gas engine CHP with respect to reductions in CO_2 emissions and fossil fuel savings. Figure 36.1 shows an energy flow analysis of a state-of-the-art heating boiler (condensing boiler) and the average electricity generation as a reference for further consideration. At the bottom is shown the energy balance of a CHP unit as an alternative to the two above it. The processes are compared with respect to their end-use (useful) energy delivery in Figure 36.2. The summation of primary energy input clearly reveals the superiority of the CHP application. The CO_2 emission balance is given in Table 36.1. Although the use of the CHP device to produce *only* heat or electricity is out of discussion, the combined use of both energy outputs leads to significant reductions in fossil energy use and thus CO_2 emissions.

This macro economic model, however, does not acknowledge the economic considerations of the CHP operator, who may be driven by completely different motivations. Limiting ourselves to industrial-scale applications of DG in the range 10 kW_{el}–1 MW_{el} for the purposes of the discussion presented here, it is found that the economy of DG applications is strictly ruled by the relation between the cost of electricity bought from the grid and the cost of natural gas (or other fuel). Generally, the cost of electricity will be higher than that of gas by a factor of 3–4 [8]. The capital depreciation and all other operating costs of the CHP unit have to be made up from this cost difference, including also any savings from a reduced power rating of the grid connection. In recent years, the conversion of natural gas to heat and electricity in industry has met limitations due to a relative increase in gas prices. Therefore, many units, especially in Germany, were closed down after the year 2000 for lack of competitiveness.

36.3
Fuel Cells in Distributed Generation

High-temperature fuel cells offer interesting options to improve the effectiveness of industrial-sized CHP units. The reasons are many: the prospects for high efficiency, low emissions, low noise, scalability, reliability, and potentially low costs. However, the main attraction of high-temperature fuel cells is their high system efficiency and co-generation networking capabilities, even for small capacities and at part load operation. The technology is currently moving from a research and test phase into one of industrial development and demonstration. Among the most interesting applications are stationary distributed power generation (DG, including CHP), small residential combined heat and power units (micro CHP), and various transportation applications [auxiliary power units (APUs)]. With a properly designed fuel processing system, high-temperature fuel cells can use all available fuel types. However, the complexity, cost, and efficiency of the fuel processing are strongly dependent on the choice of fuel, fuel processing route, and fuel cell type.

The molten carbonate (MCFC) and solid oxide (SOFC) types of high-temperature fuel cell operating at temperatures above 600 °C have significant fuel flexibility and can reform methane (the main component of natural gas) directly within their cells or operate directly on CO-containing gas or even on ammonia. Their basic principles have been widely explained in the literature [9]. Both MCFCs and SOFCs offer a higher fuel-to-power efficiency than the low-temperature variants. In stand-alone systems, MCFCs and SOFCs have electrical system efficiencies above 50–55% based on natural gas. SOFCs even have the potential to reach a net electrical system efficiency of 60% [10] across the full range of power discussed here. In a combined cycle system, where the exhaust gas from the high-temperature fuel cell is used to drive a gas turbine, the overall system electrical efficiency may reach over 70%. The efficiencies of gas turbines and diesel engines decrease at part load, whereas the efficiency of fuel cell-based systems will be almost independent of part load up to very high turn-down ratios.

These high efficiencies are possible due to a marked increase in cell performance in recent years, resulting in significantly reduced overpotentials (increased cell efficiencies). The fossil fuel and carbon balances according to Figure 36.2 and Table 36.1 are therefore even more favorable for the CHP case. An SOFC CHP unit can reach fossil fuel savings of over 53% (Figure 36.3) while delivering the same energy services. Hand-in-hand with the increased electrical

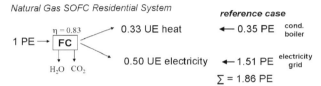

Fig. 36.3 Modification of Figure 36.2 with parameters for a high-efficiency SOFC system. Primary energy use is decreased by 53% for the identical energy service. Making more specific use of the high-temperature quality of the fuel cell exhaust heat can further increase the energy efficiency.

energy output, the influence of the spread between gas and electricity prices becomes less pronounced and operation of a CHP unit more economic.

The typical operating temperature between 600 and 850 °C and the use of nickel catalysts in the anode result in the ability of high-temperature fuel cells to convert methane directly into hydrogen. This "internal reforming" capability is often termed "direct fuel cell" technology. It offers the advantage of minimization of gas processing for natural gas operation, since the only conversion step necessary is the reforming of the higher hydrocarbons in the natural gas (in comparison with low-temperature fuel cells, which require reforming, shift stage, and fine cleaning). The reforming on the anode is endothermic and by extracting heat reduces the amount of cathode air for cooling purposes. Whereas MCFCs operate at around 650 °C, the wider temperature window from 500 to 900 °C of SOFC technology makes it possible to design a simpler and more optimal balance of plant (BOP) for the fuel cell system with more flexibility in operating possibilities, although below 600 °C it has to be ascertained that the kinetics of internal reforming are still sufficient, if need be by adding further catalysts.

With high-temperature fuel cells, the off-heat further offers the possibility to exploit polygeneration options more freely. This is the parallel production of electricity, heat, and other products such as steam and hydrogen or heat supplied to air conditioning and cooling equipment. Whereas the additional use of heat to produce steam or running an adsorption chiller is obvious, the production of hydrogen is less so. It can be reasoned that running the fuel cell at lower power than would correspond to the fuel flow will result in an excess of fuel leaving the fuel cell. On the other hand, the methane reforming capabilities of the anode material are generally so good that most of the excess fuel will be converted to hydrogen anyway, although this is not used in the electrochemical conversion in the fuel cell itself. It will therefore leave the anode compartment with the exhaust gas. By separating the hydrogen out of the gas stream, a multi-purpose source of hydrogen is established, thus for instance coupling CHP implementation with a hydrogen filling station [11]. The exhaust can also be adjusted to a syngas composition with carbon monoxide concentrations that can be used by downstream chemical processes, for instance as discussed in [12].

Similar considerations apply if it were desired to separate CO_2 from the exhaust gas stream as input to chemical or biological processes [13]. Due to the feed of pure fuel (methane) to the fuel cell, the exhaust gas from the system afterburner will be composed of water and carbon dioxide alone, which again lends itself to a fairly simple process of carbon capture by drying and compressing the exhaust gas.

In the context of DG for the sake of grid stabilization, peak load production, and decentral power generation, high electrical efficiency is the main specification of interest. For a CHP unit in industrial, residential, or district heating applications, the overall efficiency (including heat use efficiency) and the potential for polygeneration will play an equal role to the efficiency of power generation. In any case, DG lays a much stronger emphasis on the electrical efficiency than is currently being discussed for residential units.

36.4
Designing for High Efficiency

From the point of view of heat management, high-temperature fuel cell systems can not only make use of many internal heat flows for preheating air and fuel streams, but also utilize the high temperature of operation for internal reforming of methane, as discussed above. A very simple base case system design for natural gas fuel consists of the primary unit components as depicted in Figure 36.4 [14]. The heat-consuming internal reforming of methane in combination

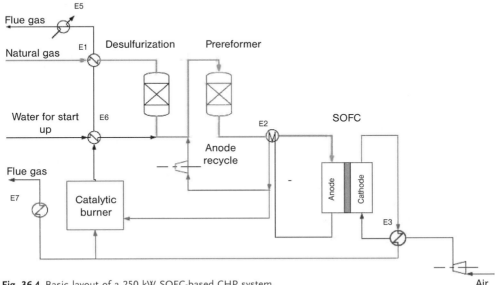

Fig. 36.4 Basic layout of a 250 kW SOFC-based CHP system.

Table 36.2 Composition of simulated fuel (Danish natural gas).

Component	Content (vol.%)
CH_4	88.1
C_2H_6	6.4
C_3H_8	2.8
C_4H_{10}	1.0
C_5H_{12}	0.18
C_6H_{14}	0.06
N_2	0.3
CO_2	1.3
H_2S	10 ppm

with the waste-heat producing electrochemical cell reaction at high temperatures lower the demands on the fuel processing system and lead to higher efficiency of SOFCs compared with low-temperature fuel cells.

The process can be described as follows. Natural gas is preheated to 400 °C and desulfurized in a zinc oxide bed. A very important criterion for the design of fuel processing systems is the need to avoid carbon formation. The adiabatic pre-reformer quantitatively converts higher hydrocarbons to form a mixture of methane, hydrogen, carbon monoxide, and carbon dioxide and thus eliminates the risk of carbon formation. A typical composition of Danish natural gas is given in Table 36.2. A final heat exchanger will increase the inlet fuel gas temperature to around 650 °C. On the cathode side, air is compressed in a blower and preheated in a heat exchanger to about the same temperature. The cathode off-gases, consisting of oxygen-depleted air, are sent partly to a catalytic burner together with the anode off-gases. The rest of the cathode off-gases together with flue gases will then deliver heat to a heating, ventilation, and/or air conditioning system within a building, for example. The burner exhaust gases are also used for steam generation during the start-up phase and for fuel preheating. System simulations of a 250 kW SOFC system based on natural gas fuel as described above yield a net electrical efficiency (including all internal system losses) of approximately 56% and a total efficiency of 88%.

In a previous study at Haldor Topsøe [14], it was calculated that approximately half of the waste heat from the oxidation reaction, however, is used to drive the internal reforming reaction, where methane in the anode feed is reformed with water to generate hydrogen. This reaction not only reduces the size of the heat exchanger (E3 in Figure 36.4) and reduces the parasitic loss for air compression, but also "upgrades" waste heat to chemical energy. This is a major reason for the higher electrical efficiency of a stationary SOFC CHP system.

The energy flows of the base case are represented in the Sankey diagram in Figure 36.5, showing the energy content of the streams in kW and kJ s^{-1}. Only part of the enthalpy from the reaction of oxygen and hydrogen is available as external work (exergy or Gibbs free energy). The rest is dissipated in the fuel cell

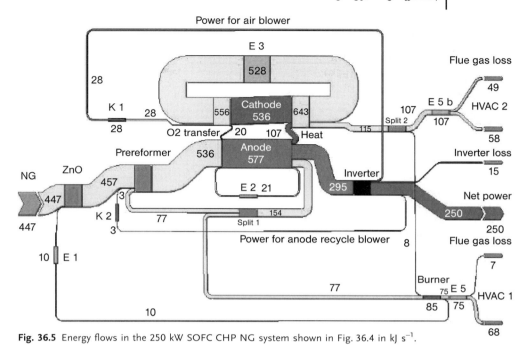

Fig. 36.5 Energy flows in the 250 kW SOFC CHP NG system shown in Fig. 36.4 in kJ s^{-1}.

Table 36.3 SOFC system efficiencies for various fuels.

Fuel	Electrical efficiency (%) (LHV, net)	Total efficiency (%) (90 °C)
Natural gas	55	84
Bio-gas with 50% CO_2	54	80
Methanol	53	85
DME	53	83
Ammonia	55	84
Diesel CPO (5 kW)	41	85

as heat and mainly removed with the cathode air through heat exchanger (E3). This heat transfer is the main source of exergy losses in the plant.

The complexity and efficiency of the fuel processing system depend strongly on the choice of fuel, fuel processing route, and fuel cell type. In general, the higher the C/H and C/O ratios of the fuel, the more difficult is the fuel processing, especially taking into account the sulfur most often found in commercial hydrocarbons. Normally the fuel processing options are limited to those based on steam reforming, cracking, catalytic partial oxidation (CPO), or auto-thermal reforming (ATR).

Table 36.3 shows a comparison of the system efficiencies calculated for a 250 kW SOFC CHP system based on different fuels [15]. Methanol and dimethyl ether (DME) be produced can relatively easily from biomass. The numbers for MeOH and DME involve a new process layout, developed by Haldor Topsøe. The range of fuels has now been extended to include ethanol and coal syngas [16] by development of a coke-resistant ethanol reforming/methanation catalyst and leveraging catalyst know-how from coal gas treatment.

36.5
Developments in the United States

One of the first companies to develop SOFCs for distributed power generation was Westinghouse. They followed a tubular concept with a more or less seal-less design [17]. The tubes have a diameter of 25 mm (2 in) and a length of 180 cm (150 cm active length). They build on a tube of cathode material on which the electrolyte and the anode are applied. The tubes are interconnected by a nickel felt and grouped in "bundles" with 24 cells; 48 bundles are combined into a "stack" of up to 1152 tubes and 100 kW_{el} overall rating [18]. The Westinghouse development was acquired by Siemens together with the acquisition of the parent company in 1998.

Although the tubes are operated at high temperatures around 1000 °C due to the use of a moderately thick electrolyte layer (40 µm) and the required high activity of the electrodes, long lifetimes of up to 70 000 h with very low degradation could be proven for single tubes in a ceramic test housing [19]. Only two prototypes of a 100 kW_{el} system were built, one of which was installed in Arnhem, The Netherlands, in 1999. This unit was operated successfully for 10 000 h and then moved to a site of RWE in Essen, Germany, in 2001. After termination of tests scheduled at these premises during some months, the CHP system was finally removed and relocated to Turin, Italy, at the Siemens Turbocare plant [20]. There it was refurbished, including replacement of about one-third of the tubes, and completed a total of 30 000 h of systems operation. Due to the termination of Siemens' development of the original tubes, the system finally had to be shut down in 2007 [21]. A second field test unit, SFC200, rated at 125 kW_{el}, was intended to be installed in Hannover, Germany, on the Industry Fair ground. After a number of delays, the unit was installed in January 2007 [22], but suffered damage to the air preheater, after which the unit was shipped back to the company premises in Pittsburgh for component checks and was never reinstalled. Two units were built for pressurized operation. Increasing the pressure in a fuel cell, according to the well-known Nernst equation, results in a logarithmic increase in the cell voltage. A pressure around 4 bar will result in approximately 15% better performance, after which further increases in pressure have only a marginal effect. The Siemens systems underwent up to 3500 h of testing before being abandoned [23].

Siemens adopted several concepts of improved power density tubes in which the current path and compactness of the tube bundles were increased. These "high power density" (HPD) tubes used a "flattened" design with current-transporting ribs across the air transport channel [24]. Finally, the Delta 9 design further modified the layout to encompass triangular gas transport channels [25]. Neither HPD nor Delta were tested at stack power ratings suitable for DG and test systems remained in the 5 kW$_{el}$ class.

The tubular concepts require an involved manufacturing on three-dimensional structures, which leads to elevated production cost levels. Siemens, however, passed the SECA cost benchmark [26]. Although the base technology appears to provide for long lifetimes, the high overall volume of the packaging of a tubular structure and the manufacturing involved prevented Siemens from suitably reducing the manufacturing costs. Thermomechanical problems were caused by thermal gradients between air inlets and outlets and by the insufficient support of the tube ends, which effectively hang free in the fuel compartment. After having completed the SECA Phase 1 with a 5 kW system built in cooperation with Fuel Cell Technology (FCT), a now defunct Canadian company, Siemens entered the SECA Phase 2 (see below), but put the Pittsburgh operations up for sale in 2008 [27].

General Electric (GE) was active in SOFCs as early as 1969 [28]. In 2001, GE acquired the Honeywell (Allied Signal) SOFC technology [29]. GE Energy and GE Global Research teamed up in further developing the base technology, including their own interconnect steel, ATR reformer, and high-performance inverter development [30]. Disk-shaped cells of diameter up to 30 cm made by tape calendaring were demonstrated in a stack in 2003 [31]. Stacks achieved lifetimes of up to 3000 h. Work concentrated on 3–10 kW CHP systems and GE [also under GE Hybrid Power Generation Systems (GE HPGS)] successfully delivered a SECA 6 kW-rated system to NETL in 2006 [32]. Although SECA Phase 2 was initially entered in order to develop large, coal gasification-based systems [33], it was decided to revert to fundamental research in order first to prove the lifetime potential of SOFC technology before venturing into engineering of stacks and systems. The technology development group was subsequently closed in 2007.

Global Thermoelectric (Calgary, Canada), acquired the base technology of anode supported SOFCs from Forschungszentrum Jülich (Germany) in 1997 [34]. Rapidly progressing through several development improvements [35], Global achieved a leading technology position based on their own developments. In 2003, the SOFC operations were bought by the American company Fuel Cell Energy and transferred into a joint venture between FCE, the Gas Institute, EPRI, and the University of Utah [36] named Versa Power. Versa successfully accomplished the SECA Phase 1, also supplying stacks to the Cummins SECA group. Today, Versa is the main supplier of large stack units in the SECA Phase 2, working on a 10 kW$_{el}$-plus stack unit [37], which is one of the largest known worldwide and has already accomplished over 5000 h of continuous testing. Given the success of this development, Versa is poised to be one of the first companies to build megawatt-class SOFC systems in the near future. SECA Phase 2 aims at combining SOFC technology with coal gasification plants in an attempt to increase efficiency, reduce water requirements in power generation, and allow for relatively simple CO_2 capture from the flue gases [38].

Ion America began work on SOFCs, polygeneration of electricity, heat, and hydrogen, and reversible fuel cells well before 2005 [39, 40]. The company became Bloom Energy and is now active in the development of 100 kW$_{el}$ class SOFC systems, including demonstration units delivered to eBay, Google, and San Francisco Airport [41]. Little is known of their activities, except that the 100 kW$_{el}$ systems are built of a multitude of stacks of 1–2 kW$_{el}$ rating. As the development progresses, it is expected that more information will be available in the course of 2010.

Concerning MCFCs, M-C Power in the United States were working on 250 kW sized units in the 1990s and up to 2001 [42], but without reaching the demonstration phase.

Beginning in 1970, Energy Research Corporation (ERC) had been developing battery and MCFC technology [43]. From 1990, they entered a license and technology transfer agreement with the Daimler (EADS) subsidiary MTU in Germany. The 1990s saw the first 250 kW units being installed and tested [44]. In 1999, the battery division was split off and ERC was renamed FuelCell Energy. FCE has established a base technology that can today be considered as the world leader in MCFC technology. FCE "direct fuel cell" (DFC) stacks are built at a rating of 300 kW$_{el}$ and combined in systems of 300 kW$_{el}$, 1.4 MW$_{el}$ and 2.8 MW$_{el}$ with 1, 5, and 10 stacks, respectively, that run on internally reformed natural gas [45]. In addition to grid support applications (75% of installations), the main market currently consists of units operating on bio-gas from fermentation (11%), thus using a renewable source of methane, often in combination with sewage treatment. FCE currently has 95 MW$_{el}$ of operational or contracted units in the field in 55 installations worldwide. A cooperation with POSCO in South Korea has since 2008 resulted in orders totaling 51 MW$_{el}$ to be delivered from 2009 [46]. The net efficiency of the units is around 50%. Degradation rates are 1 mV per 1000 h at 800 mV and 170 mA cm^{-2} [47]. Combination with a Capstone micro gas turbine has resulted in efficiencies (LHV) of 58% in grid connection [48]. This so-called DFC/T unit was operated for 8000 h in 2006–2007.

36.6
Asian and Pacific Developments

In Japan, the main emphasis of SOFC development has been on residential units (cf. the chapter on residential applications). Nevertheless, within the development program(s) run by NEDO, several groups have been and are working on larger units for business and industrial applications [49].

Mitsubishi Materials (together with Kansai Electric Power) have been progressing their 1 kW-class stack development with disk-shaped cells with a lanthanum gallate electrolyte and 750 °C operating temperature. The 5 and 10 kW-class units are made up of 8 and 16 stack modules, respectively, rated at 800 W each. At 770 mV cell voltage, these stacks deliver around 50% d.c. efficiency (HHV) and 208 mW cm^{-2} (270 mA cm^{-2}). A total of 3000 h was achieved with the first (2007) system test. The second 300 h test in 2008 brought 41% a.c. net efficiency at a voltage degradation of approximately 1% per 1000 h [50], which, however, was higher than targeted.

Mitsubishi Heavy Industries (MHI), in contrast, have been working with JPOWER on 150 kW_{el}-class pressurized SOFC systems for industrial CHP based on segmented tubular cells since 1991 [51]; 5 kW "cartridges" are built up from 104 cells and integrated into 25 kW "sub-modules." Between 2007 and 2009, a total of 10 000 h of operation were achieved with three sub-modules rated together at 102 kW and operating at ambient pressure. The stack efficiency was 45.4% (LHV) and voltage degradation was 2.3% per 1000 h at 900 °C [51].

MHI alone have been developing a 200 kW-class pressurized hybrid system under NEDO contracts since 2004, when a 75 kW class unit was successfully tested. In 2009, the first 3000 h of test runs with a 200 kW-class system were reported as accomplished [52]. Maximum output was 229 kW at a net efficiency of 52% (LHV).

The MOLB ("Mono-Block Layer Built") SOFC design was developed from 1996 between MHI and Chubu Electric Power. It consists of a "corrugated" design of 20×20 cm square, which integrates functions of interconnect and electrode. They presented two units generating around 30 kW_{el} at the World Exposition in Aichi in 2005 [53]. The operating temperature is high at 1000 °C and recently there have been no further reports about the success of this development [49]. Activities have reverted back to basic laboratory work in order to increase power output of the cells [53].

The TOTO/Hitachi tubular SOFC development aims at a 20 kW class system. So far, 1000 h of operation have been achieved on stacks with around 48% d.c. efficiency (HHV). System development seems to be still ongoing and an 8 kW (10 kW class) demonstration unit for an office building is planned [49]. A thermally self-sustaining module consisting of 20 stacks achieved 6.5 kW at 0.2 A cm^2 and 50% LHV operating at 900 °C. A further development by Acumentrics Japan and NSC of a 10 kW-class unit [54] has not been reported on more recently [49].

Ceramic Fuel Cell (CFCL) in Australia pursued a 10 kW-class design in 2000 [55]. This was abandoned, however, when it became apparent that the all-ceramic concept would not be successful in operating the planned 40 kW demonstration unit. CFCL is now successfully active in building 2 kW_{el} units for residential CHP displaying high net system efficiencies of around 60% (LHV) [56].

In South Korea, major interest in DG has built up with a focus on supplying heat and electricity to multi-family housings. A number of companies are looking especially into SOFC technology. POSCO Power has established a close relation with FuelCell Energy (see above) and acquired major options on deliveries to South Korea [46].

36.7
European Developments

Following Siemens' move to acquire Westinghouse and the tubular SOFC concept and drop their proprietary planar electrolyte-supported concept end of the 1990s and Dornier's closure of SOFC activities, a number of smaller European developers adopted the planar SOFC concept. This has now successfully evolved into a high-performance technology, especially with the anode-supported variant.

Development groups for planar technology at a scale above 10 kW$_{el}$ include the companies Topsøe Fuel Cells (TOFC) (Denmark), ProtoTech (Norway), and Wärtsilä (Finland).

TOFC was founded in 2004 as a daughter company of Haldor Topsøe (HTAS). In 2009, TOFC started a pilot-scale facility for the production of planar anode-supported cells and stacks. The TOFC stack concept aims at cost efficiency by using thin sheet metallic interconnects and minimizing stack component materials. Their stacks have been operated with a stable performance for more than 12 000 h. Currently, TOFC is supplying stack modules in the 10 kW size range to the US SECA phase II program and also to other technology partners. Multi-stack arrays are being developed for larger systems. For smaller capacities, it is of paramount importance to have a very tight integration, both mechanically and thermally, of all the hot components of the SOFC system. TOFC has therefore initiated a program to develop a so-called PowerCore unit comprising fuel processing, the stack, feed-effluent heat exchangers, and catalytic burner(s). A complete 5 kW PowerCore unit based on methanol has been constructed and tested successfully under dynamic load-following operating conditions. In similar systems based on diesel fuel, strategies to counteract the impact of sulfur are under investigation.

Wärtsilä is a company with a steel and shipbuilding background which acquired the stationary diesel engine business from Sulzer in 1997 [57]. Their main interest is the application of SOFC systems as auxiliary power units (onboard electricity generation) on ships and as multi-hundred kW$_{el}$ to MW$_{el}$-sized DG units [58]. With TOFC as a stack supplier [59], a complete 20 kW demonstration unit based on pre-reforming of natural gas with anode recycling was operated for more than 1000 h with 24 stacks of the 75 cell (12×12 cm^2) type in 2007 [60]. Another 24-stack prototype, now based on methanol, was built by Wärtsilä in the project METHAPU, a further unit was commissioned for bio-gas use [61]. The methanol is methanated upstream of the anode, using a proprietary HTAS catalyst. More recently, within the EU project LargeSOFC, Wärtsilä has built a 50 kW$_{el}$ system due for operation in 2009–2010 [62]. This will be the first SOFC system worldwide in this power class operating at ambient pressure and using planar cells. Follow-up test units of up to 150 kW$_{el}$ are expected within the next few years.

Prototech has been developing stacks and systems intended for the power range 200 kW$_{el}$/200 kW$_{th}$ and up to MW$_{el}$ units. So far, systems with 3 kW$_{el}$ have been tested for up to 1800 h [63].

Rolls Royce Fuel Cell Systems (UK) (RRFCS) is working on medium-sized industrial power generation units up to and above 250 kW, and eventually up to several megawatts of electrical power [64]. The basic component is a flattened tube with a sequence of cells printed on it. In contrast to the Siemens concept, fuel flows within the tube and air outside. The design is termed "integrated planar" since it uses characteristics of both tubular (gas supply) and planar (printing on flat surface) designs. The tube units are combined to sub-stacks that are again arranged around a "hub". This module is pressurized and combined with a turbo-charger [65]. In this way, RRFCS makes use of the higher voltage achievable with pressurized operation (see above). RRFCS is targeting the United States power generation market with the help of SOFCo, which it acquired in 2007 [66]. Recently, considerable funding was awarded by the DOE within the SECA Phase 2 program [38].

On the research side, Forschungszentrum Jülich has been working on a 20 kW_{el} technology demonstrator which is due for completion in 2010. The main design feature is the combination of all "hot" balance of plant components into a so-called "integrated module" with a common thermal insulation [67].

Together with Fuel Cell Energy in the United States (see above), the German MTU Onsite Energy are the world leaders in molten carbonate technology. Both use identical stack technology, while an MTU-led group of companies established the "HotModule" design for application in Europe, in which the system design was simplified in order to lower costs and the stack was arranged in a horizontal position [68]. The latter allows for larger stacks with up to 60 layers, not limited by the pressure of their own weight. MTU systems have achieved up to 30 000 h of operation on a single stack and 47% net efficiency (LHV) at a power level of 237 kW. The standard unit size with MTU was 250 kW with 300 and 400 kW_{el} units currently under development (500 and 600 cells, respectively), that can be grouped into 2 MW_{el} systems [69]. More than 16 field test units have proven the reliability and durability of stacks [70]. More recently, MTU has been moving towards their own stack design, the EUROCELL [71], which is to be specifically designed for horizontal installation instead of the upright position favored by FCE. Within the HotModule, the horizontal position leads to long-term problems with the molten electrolyte and/or the sealing, which may prevent lifetimes far beyond 30 000 h from being achieved.

Although some interest in MCFC technology existed in the 1980s in Europe (for instance at ECN, The Netherlands [72] and the Dutch Fuel Cell Corporation, BCN, jointly looking at 400 kW_{el} units for commercial and hospital applications [73]), this was abandoned later due to a lack of industrial partners for commercialization.

Ansaldo (AFCo; before 2001: Ansaldo Ricerche) in Italy is today the third remaining worldwide major developer of MCFC technology [74]. AFCo have developed a similar technology to that of FCE/MTU [75], but built on a European supplier base and using several coupled stacks (TwinStacks) integrated into a single container. Even at temperatures between 650 and 700 °C the power density is relatively low (0.07–0.09 $W\,cm^{-2}$) [76]. So far, six stack units of different sizes have been commissioned with a maximum stack life of 12 000 h and system operation of 30 000 h. In January 2005, a full system was established and testing began with and without micro turbine integration. The system is designed for 500 kW-class units that can be combined into multi-MW$_{el}$ plants [77].

36.8
Economic Prospects in DG Fuel Cell Development

Interestingly, the high potential of SOFC technology has also attracted considerable financial investment. While TOFC was able to acquire EU and Danish state funding for their new production facilities (partly through the EU LIFE program), RollsRoyce Fuel Cell Systems brought in funds from a Singapore investor which allowed the takeover of the SOFCo activities in the United States.

TOFC is pursuing three market segments; micro combined heat and power (microCHP, 1–5 kW), auxiliary power units for the transportation segment (3–10 kW), and distributed generation (20–500 kW). These markets are all transitioning from the development phase into the test and demonstration phase. This calls for a markedly increased number of stacks for development, testing and demonstration. To meet this growing demand, TOFC is constructing a new cell and stack production facility with an annual capacity of 5 MW. The facility comprises all unit operations, from ceramic powder processing to assembly of final stacks. The processes are based on up-scaling of existing, well-known methods, which are already widely used industrially. The plant is located together with TOFC's offices and research laboratories in Lyngby, Denmark, thus ensuring short communication paths and feedback from production to development. Start-up occurred in late 2008. Two Danish projects in the PSO program supported by Danish Utilities have been initiated. One is aimed at demonstrating a 1 kW$_{el}$ micro-CHP unit and the other is a 10 kW$_{el}$ DG stack development.

It may be added as a comment that it is interesting to note the high activities in fuel cell companies, but at the same time to observe a lack of marketable products and commercial activities with the potential for positive cash flow. Obviously, in establishing MCFC and SOFC technology in the market, the technology readiness level has to be continuously improved. This brings increasing attention to product design and customer performance aspects. On the other hand, costs have to be reduced, and this will only be realized by introducing mass manufacturing processes – at least for the cells and interconnects. The companies MTU, TOFC, H.C. Starck and CeramTec have built up, or are in the process of building up, adequate production capacity. This a is a risky business, given that market entry still remains to be accomplished, cost reduction being one prerequisite in doing so – which brings up the usual "chicken-and-egg" problem.

36.9
Outlook

Fuel cell technology for distributed generation has reached a status that will allow market-compatible equipment to be produced within the next few years. Whether or not a realistic market entry will be possible at this point remains to be shown. The costs of stacks and units available today are adequate for prototype technology but not for marketable products. It depends on the further development of manufacturing technology, the ability of industry to convince financial institutions to invest, and last but not least the installation of feed-in tariffs for high-efficiency-generating technology, amongst others fuel cells – similar to the German Photovoltaics tariffs – whether fuel cells will finally make it to the distributed generation markets and establish the cutting edge technology status required to attain a leading position in installed generation capacity within the next decade.

References

1 Grausa, W. and Worrella, E. (2009) Trend in efficiency and capacity of fossil power generation in the EU. *Energy Policy*, **37** (6), 2147–2160.
2 Brown, M.H. and Sedano, R.P. (2004) *Electricity Transmission – A Primer*, National Council on Electric Policy, Washington, DC; *U.S. Climate Change Technology Program – Technology Options for the Near and Long Term*, a Compendium of Technology Profiles and On-going Research and Development at Participating Federal Agencies, published by the DoE, November 2003, Washington, DC.
3 European Commission (2004) *Directive on the Promotion of Cogeneration*, Directive 2004/8/EC, European Commission, Brussels.
4 German Federal Government (2004) *Mineralölsteuergesetz*, German Federal Government, Bonn, 21 January 1992 version of 1 January 2004.
5 Herdin, G.R. (2000) Increasing gas engine efficiency. In *Proceedings of the World Energy Engineering Congress*, Atlanta, GA.
6 Jordal, K., Assadi, M., and Gentrup, M. (2002) Variations in gas-turbine blade life and cost due to compressor fouling – a thermoeconomic approach. *Int. J. Appl. Thermodyn.*, **5**, 37–47.
7 Herdin, G.R (2002) *Wasserstoff als Antriebsenergie für konventionelle Ottomotoren [Hydrogen as a Fuel for Conventional 4-Stroke Engines]*. Report for P&T Technologies.
8 Harrison, J.D (2001) Micro combined heat and power potential impact on the electricity supply industry. In *Proceedings of "Electricity Distribution 2001", CIRED 16th International Conference*, IEE Conference Publication No. 482.
9 Larmine, J. and Dicks, A. (2001) *Fuel Cell Systems Explained*, John Wiley & Sons Ltd, Chichester.
10 Payne, R., Love, J., and Kah, M. (2009) Generating electricity at 60% electrical efficiency from 1 to 2 kWe SOFC products. *ECS Trans.*, **25** (2), 231–240.
11 Colella, W., Rankin, A., Sun, A., Margalef, P., and Brouwer, J. (2009) Thermodynamic, environmental and economic modeling of hydrogen co-production integrated with stationary fuel cell systems. Presented at the DOE Hydrogen Merit Review Meeting, 18 May 2009.
12 Kim, I., Pillai, M., McDonald, N., Shastry, T., and Barnett, S. (2009) Co-generation of electricity and syngas under electrochemical partial oxidation using novel SOFCs. *ECS Trans.*, **25** (2), 271–280.

13 NAPE (2008) *Polygeneration in Europe*, Milestone Report 6.8 of the POLYSMART Project, EU Contract No. 019988.

14 Pålsson, J., Hansen, J.B., Christiansen, N., Nielsen, J.U., and Kristensen, S. (2003) Solid oxide fuel cells – assessment of the technology from an industrial perspective. In *Proceedings of the Risø Energy Conference*, Risø.

15 Hansen J.B. and Christiansen N. (2007) Solid oxide fuel cell development at Topsøe Fuel Cell A/S. In *Proceedings of the World Hydrogen Technology Convention 2007*, Montecatini.

16 Hansen, J.B., Rostrup-Nielsen, J., Højlund Nielsen, P., and Rostrup-Nielsen, T. (2006) Optimum processing of fuel gas from coal gasification for SOFC/gas turbine power plants. In *Proceedings of the Fuel Cell Seminar*, Honolulu.

17 Singhal, S.C. (1993) in *Transactions of the ECS: SOFC IV* (eds M. Dokiya, O. Yamamoto, H. Takagawa, and S.C. Singhal), PV95-1, Electrochemical Society, Pennington, NJ, pp. 195–207.

18 Kendall, K., Minh, N.Q., and Singhal, S.C. (2003) Cell and stack designs. In *High Temperature Solid Oxide Fuel Cells – Fundamentals, Design and Applications* (eds S.C. Singhal and K. Kendall), Elsevier, Oxford, 2003..

19 Hassmann, K. (2000) Produktentwicklung Festelektrolyt-Brennstoffzellen (SOFC) [Product development SOFC]. In *Themen 1999/2000: Zukunftstechnologie Brennstoffzelle*, Forschungsverbund Sonnenenergie, Berlin.

20 Orsello, G., Casanova, A., and Hoffmann, J. (2008) Latest info about operation of the Siemens SOFC generators CHP100 and SFC5 in a factory. In *Proceedings of the 8th European Fuel Cell Forum*, Lucerne, July 2008, Paper B0204.

21 Gariglio, M., De Benedictis, F., Santarelli, M., Cah, M., and Orsello, G. (2009) Experimental activity on two tubular solid oxide fuel cell cogeneration plants in a real industrial environment. *Int. J. Hydrogen Energy*, **34**, 4661–4668.

22 News Item (2007) *Siemens Power Generation*, HZwei, 7 (April), 8.

23 Hassmann, K. (2001) SOFC power plants, the Siemens–Westinghouse approach. *Fuel Cells*, **1**, 78–84.

24 George, R.A. (2002) In *Proceedings of the 3rd Annual Solid State Energy Conversion Alliance (SECA) Workshop*, Washington, DC, 2002.

25 Huang, K. (2007) Development of delta-type SOFCs at Siemens stationary fuel cells. In *Proceedings of the 2007 Fuel Cell Seminar*, San Antonio, TX, 15–19 October 2007.

26 Surdoval W. (2007) U.S. DOE fuel cell programs. Presented at the Advanced Fuel Cell Workshop, University of Florida, 27 May 2007.

27 Siemens (2008) Siemens Power to put SOFC unit on block. *Fuel Cell Today*, retrieved from www.fuelcelltoday.com/online/news/articles/2008-07/Siemens-power-to-put-SOFC-unit-o (accessed 28 December 2009).

28 Tadmon, C.S., Spacil, H.S., and Mitoff, S.P. (1969) *J. Electrochem. Soc.*, **116**, 1170.

29 Minh, N. (2000) Honeywell solid oxide fuel cells – markets and technology status. In *Proceedings of the SECA Workshop*, Baltimore, MD, 1–2 June 2000.

30 General Electric (2004) GE presentation at the SECA Annual Workshop and Core Technology Program Peer Review, Boston, MA, 11–13 May 2004.

31 Minh, N. (2003) GE SECA solid oxide fuel cell program. Presented at the Fourth Annual Solid State Energy Conversion Alliance Meeting, Seattle, WA, 15–16 April 2003.

32 Zhou, Q., Goodman, J., Powers, J., and Campbell, T. (2007) GE SECA prototype system development and testing. *ECS Trans.*, **7** (1), 167–172.

33 Alinger, M. and Taylor, S. (2007) SECA SOFC program at GE Global Research. Presented at the 8th Annual SECA Workshop and Peer Review Meeting, San Antonio, TX, 7–9 August 2007.

34 Weaver, G. (2002) *World Fuel Cells – An Industry Profile with Market Prospects to 2010*, Elsevier, Oxford.

35 Borglum, B., Fan, J.-J., and Neary, E. (2003) Following the critical path to commercialization: an update on Global

Thermoelectric's SOFC technology and product development. In *Proceedings of the 8th ECS SOFC Symposium, SOFC VIII*, Paris, May 2003, pp. 60–69.
36. McConnell, V. (2007) Versa Power's SOFC could scale to MW for SECA, and work in transport hybrids. *Fuel Cells Bull.*, (9), 12–16.
37. Borglum, B., Tang, E., and Pastula, M. (2009) The status of SOFC development at Versa Power Systems. *ECS Trans.*, **25** (2), 65–70.
38. Surdoval, W. (2009) The status of SOFC programs in USA 2009. *ECS Trans.*, **25** (2), 21–27.
39. Ferguson, J. (2005) Chattanooga fuel cell demonstration project. Presented at the 5th Hydrogen Review Meeting, DOE.
40. Sridhar, K.R. (2005) Ion America – a distributed generation company. Podcast, 20 September 2005; available at http://www.ostp.gov/pdf/sridhar_ionamerica_pcast_20sep05.pdf (accessed 28 December 2009).
41. Seeking Alpha (2009) eBay Installing Bloom Energy Fuel Cells. Seeking Alpha, 21 October 2009; available at http://seekingalpha.com/article/167936-ebay-installing-bloom-energy-fuel-cells (accessed 28 December 2009).
42. Tarman, P.B. (1996) Fuel cells for distributed power generation. *J. Power Sources*, **6**, 87–89.
43. Willis, H.L. and Scott, W.G. (2000) *Distributed Power Generation: Planning and Evaluation*, Marcel Dekker, New York.
44. Williams, M.C., Parsons, E.L., and Mayfield, M.J. (1994) The U.S. molten carbonate fuel-cell development and commercialization effort. In *Proceedings of the Intersociety Energy Conversion Engineering Conference (IECEC)*, Monterey, CA, 7–12 August 1994.
45. Ghezel-Ayagh, H., Walzak, J., Patel, D., Jolly, St., Lukas, M., Michelson, F., and Adrani, A. (2008) Ultra high efficiency direct fuelcell systems for premium power generation. In *Proceedings of the 2008 Fuel Cell Seminar*, Phoenix, AZ, paper RDP33-2.
46. FCE (2009) *FuelCell Energy Fact Sheet*, company information, FCE, Danbury, CT.
47. Farooque, M., Berntsen, G., Carlson, G., Leo, T., Rauseo, A.F., and Venkataraman, R (2007) Direct fuel cell improvements based on field experience. In *Proceedings of the 2007 Fuel Cell Seminar*, San Antonio, TX.
48. Ghezel Ayagh, H., Walzak, J., Junker, S.T., Patel, D., Michelson, F., and Adriani, A. (2007) DFC/T power plant: from submegawatt demonstration to multimegawatt design. In *Proceedings of the 2007 Fuel Cell Seminar*, San Antonio, TX.
49. Hosoi, K. and Nakabura, M. (2009) Status of national project for SOFC development in Japan. *ECS Trans.*, **25** (2), 11–20.
50. Nishiwaki, F., Kinugasa, A., Kato, M., Inagaki, T., Hirata, K., Sato, M., and Eto, H. (2009) Development of disk-type intermediate temperature SOFC by KEPCO and MMC. *ECS Trans.*, **25** (2), 257–266.
51. Haga, T., Komiyama, N., Nakatomi, H., Konishi, K., Sutou, T., and Kikuchi, T. (2009) Prototype SOFC CHP system (SOFIT) development and testing. *ECS Trans.*, **25** (2),71–76.
52. MHI (2009) MHI achieves 3,000-hour operation, unprecedented in Japan, for SOFC-MGT combined-cycle power generation system. *JCN Newswires*, 2 October 2009.
53. Mitsubishi Heavy Industries (2009) http://www.mhi.co.jp/en/power/technology/sofc_system/contents/development_sit uation.html (accessed 15 December 2009).
54. Kawakami, A., Matsuoka, S., Watanbe, N., Saito, T., Ueno, A., Ishihara, T., Sakai, N., and Yokokawa, H. (2007) Development of two types of tubular SOFCS at TOTO. In: *Advances in Solid Oxide Fuel Cells II: Ceramic Engineering and Science Proceedings* (eds N.P. Bansal, A. Wereszczak, and E. Lara-Curzio), Vol. 27, Issue 4, American Ceramics Society, Westerville, OH, pp. 3–13.
55. Foong, E. (2003) CFCL's first commercial SOFC technology. *Fuel Cell Catalyst*, **3** (3), 2.
56. Love, J., Amarasinghe, S., Selvey, D., Zheng, X., and Christiansen, L. (2009) Development of SOFC stacks at Ceramic Fuel Cells Limited. *ECS Trans.*, **25** (2), 115–124.

57 Wärtsilä (2009) company information, available at http://www.wartsila.com/ch,en,aboutus,0,generalcontent,268ED391-F702–4B7D-86 8B-D53F823F4A39,FF1887CC-BB47-4717-AC81-B0464BD908B9„.htm (accessed 15 December 2009).

58 Wärtsilä (2008) *Wärtsilä Solid Oxide Fuel Cell Units for Stationary and Marine Applications*, company brochure, Wärtsilä, Helsinki.

59 Christiansen, N., Holm-Larsen, H., Juel, M., Hendriksen, P. V., Hagen, A., Ramousse, S., and Wandel, M. (2009) Status of development and manufacture of solid oxide fuel cells at Topsøe Fuel Cell A/S and Risoe/DTU. *ECS Trans.*, **25** (2), 133–142.

60 Laine, J. and Fontell, E. (2008) Status of the SOFC system development at Wärtsilä. In *Proceedings of the 2008 Fuel Cell Seminar*, Phoenix, AZ.

61 Sandström, C.-E., Phan, T., Mahlanen, T., and Fontell, E. (2007) Specific targeted research project METHAPU "validation of renewable methanol based auxiliary power system for commercial vessels". In *Proceedings of the 2007 Fuel Cell Seminar*, San Antonio, TX, November 2007.

62 Rosenberg, R., Kiviaho, J., Göös, J., Jansson, P., Tarnowski, O.C., Jacobsen, J., Blum, L., and Steinberger-Wilckens, R. (2009) Large-SOFC, Towards a large SOFC power plant. In *Proceedings of the 2009 Fuel Cell Seminar*, Palm Springs, CA, November 2009.

63 Waernhus, I. and Vik, A. (2009) Field testing of SOFC at CMR Prototech. *ECS Trans.*, **25** (2), 267–270.

64 Nichols, D. K., Agnew, G., and Strickland, D. (2008) Outlook and application status of the Rolls-Royce Fuel Cell Systems SOFC. In *Proceedings of the Power and Energy Society General Meeting – Conversion and Delivery of Electrical Energy in the 21st Century*, IEEE, Vols. 20–24, pp. 1–3.

65 Agnew, G. D., Collins, R. D., Jörger, M. B., Pyke, S. H., and Travis, R. (2007) The components of a Rolls-Royce 1 MW SOFC system. *ECS Trans.*, **7** (1), 105–112.

66 Ohio Department of Development (2008) *Third Frontier Annual Report*, Ohio Department of Development, Technology Division, 19 February 2008.

67 Blum, L., Peters, R., David, P., Au, S. F., Deja, R., and Tiedemann W. (2006) Integrated stack module development for a 20 kW system. In *Proceedings of the 6th European SOFC Forum*, Lucerne.

68 Berger, P. (2004) Brennstoffzellen zur Biogasverwertung – Stand der Technik. In *Proceedings of the Symposium "Innovationen in der Biogastechnologie"*, Deggendorf, 2 December 2004.

69 Rolf, St. (2007) The HotModule system, advancement in stationary fuel cells. In *Proceedings of the 2007 Fuel Cell Seminar*, San Antonio, TX.

70 Steinberger-Wilckens, R. (2006) European fuel cell technology. In *Proceedings of the 6th European SOFC Forum*, Lucerne, 2006.

71 Bischoff, M. (2006) Molten carbonate fuel cells: a high-temperature fuel cell on the edge to commercialization. *J. Power Sources*, **160**, 842–845.

72 Kraaij, G. J., Rietveld, G., Makkus, R. C., and Huijsmans, J. P. P. (1998) Development of second generation direct internal reforming molten carbonate fuel cell stack technology for cogeneration application. *J. Power Sources*, **71**, 215–217.

73 Kortbeek, P. J. and Ottervanger, R. (1998) The "advanced DIR-MCFC development" project, an overview. *J. Power Sources*, **71**, 223–225.

74 Ansaldo (2006) Ansaldo to expand MCFC roll-out with new design. *Fuel Cells Bull.*, (7), 5–6.

75 Arato, E., Bosio, B., Massa, R., and Parodi, F. (2000) Optimisation of the cell shape for industrial MCFC stacks. *J. Power Sources*, **86**, 302–308.

76 Amorelli, A., Wilkinson, M. B., Bedont, P., Capobianco, P., Marcenaro, B., Parodi, F., and Torazza, A. (2003) An experimental investigation into the use of molten carbonate fuel cells to capture CO_2 from gas turbine exhaust gases. In *Proceedings of the 2nd Conference on Carbon Sequestration*, Alexandria, VA.

77 Ansaldo (2005) *MCFC "Series 500" Twinstack® powered first-of-a-kind "MCTWINS"*, company information, Ansaldo, Genova.

37
Fuel Cells for Buildings

John F. Elter

Abstract

Buildings account for about one third of the primary global energy demand, and as such, a major source of energy related greenhouse gas emissions. In this chapter, fuel cells are considered as an alternative and more efficient source of electricity and heat for building applications. Starting with a discussion of the importance of the customer's perspective, a brief overview of the different types of fuel cells applicable to buildings is given. These include those based upon polymer and solid oxide electrolytes. Newer alternatives including alkaline polymer and solid acid systems are mentioned. A in-depth discussion of the recent advancements made in low temperature and high temperature PEM electrodes and membranes is provided as a context from which fuel cell systems for buildings is discussed. A brief discussion of some aspects of fuel reforming and system control is provided, with relevant references, along with some examples of nearly commercial fuel cell systems.

Keywords: fuel cells, buildings, residential cogeneration, reforming, system control, combined heat and power

37.1
Introduction

The world is finally coming to grips with the issue of climate change and its anthropogenic causes. Each nation is examining its sources of greenhouse gases (GHGs) and developing strategies for reducing them to levels that enable sustainable development. Figure 37.1 illustrates the flow of CO_2 emissions in the United States in 2007, from their sources to their distribution across the end-use sectors. The left-hand side shows CO_2 by fuel sources and quantities and other gases by quantities; the right-hand side shows their distribution by sector.

The use of carbon-free fuels, such as hydrogen produced from renewable resources, or bio-fuels that close the loop on the carbon cycle, are two such strategies for dealing with the issue of GHG emissions. Others, as suggested by the

Hydrogen Energy. Edited by Detlef Stolten
Copyright © 2010 WILEY-VCH Verlag GmbH & Co. KGaA, Weinheim
ISBN: 978-3-527-32711-9

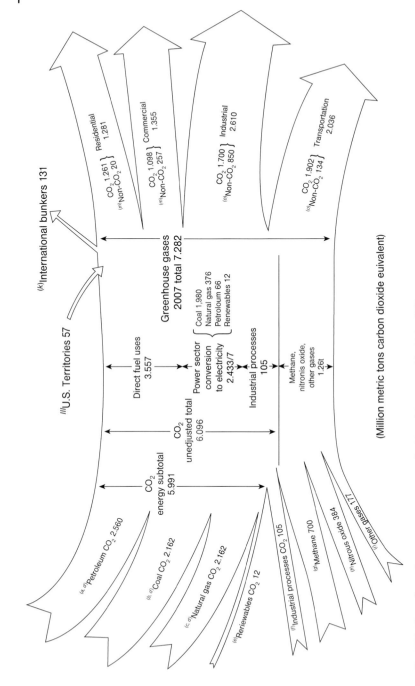

Fig. 37.1 Flow of greenhouse gas emissions in the United States in 2007 (EIA, 2008).

ASME (American Society of Mechanical Engineers) in its 2009 general position paper on reducing carbon dioxide emissions in the energy sector, include revolutionizing the carbon footprint of fossil power, realizing the potential of renewable electric power, expanding the utilization of nuclear power, reinventing transportation, greening the manufacturing sector from cradle to grave, and designing and operating buildings with much greater efficiency, approaching a "zero energy" goal. Each of these strategies has within it formidable technical, economic, and policy-related problems that need to be addressed. Here we will discuss some of those aspects associated with the built environment.

Residential, commercial, and institutional buildings account for about one-third of the primary global energy demand. Buildings represent a major source of energy-related GHG emissions. In the United States, for instance, emissions from buildings, including those emissions from both fuel combustion and use of electricity derived from CO_2-emitting sources, account for nearly 37% of total CO_2 emissions. Electricity accounts for about 45% of energy use in the buildings in the United States, and about 70% of all electricity generated is consumed in buildings. Because buildings are one of the longest-lived assets, their initial design and construction practices can impact long-term energy consumption and efficiency options.

In 2000, there were 83 million buildings in the United States having a floor space of almost 15 billion square meters, with annual new construction growing at a rate of about 1–3%. Economic growth in the developing nations will drive dramatic increases in building construction. It is estimated that 50% of all new buildings will be constructed in China and India, with China alone adding over 2 billion square meters per year. At this rate, in the next decade, China will construct new buildings having a floor space equivalent to all of the current United States building stock.

Efforts to reduce future GHG emissions due to buildings are focused primarily on advances in energy efficiency and employment of sustainable technologies. Typically, over 80% of the life cycle energy use is associated with operation of the building. Therefore, efforts to stimulate more sustainable practices in building design, construction, and operation have gained considerable momentum.

Intrinsically, buildings represent an energy requirement that is decentralized, and logically one could ask if it makes technological and economic sense to meet this demand with some form of decentralized generation (DG). Indeed, in some cases, decentralized generation (and storage) of energy may allow one not only to meet local electricity demands, but also to capture the benefits of the waste heat normally lost to the environment in centralized generation. With DG, one is potentially able to capture locally the waste heat and utilize it for heat load requirements. These so-called combined heat and power (CHP) systems offer the opportunity for exceptional total energy efficiency. In addition, they offer the associated benefits of reduced total fuel consumption and therefore reduced GHG emissions. In fact, renewed interest is beginning in the ability of small-scale DG systems as a means of dealing with the increased energy demand for the built environment, and considerable of research is ongoing into addressing the opportunity with these so-called micro-CHP or µCHP systems.

Peacock and Newborough (2006), for example, studied the impact of µCHP systems on *energy flows* in the UK electricity supply industry. Recently, they examined the effect of heat saving measures on the CO_2 savings attributable to µCHP systems in UK dwellings (Peacock and Newborough, 2008). Penht (2008) studied the *environmental impacts* of distributed energy µCHP systems by carrying out a detailed life cycle analysis (LCA) and an assessment of local air quality impacts for a variety of µCHP technologies, including Sterling engines, reciprocating engines, and fuel cells. For the same three µCHP systems, Hawkes and Leach (2007) examined the economic and environmental *attraction of alternative operating strategies*, finding that the least cost operating strategy varies with season. Matics and Krost (2008) examined the *control* of µCHP systems for residential applications using computational intelligence techniques, and Ren et al. (2008) examined the *optimal sizing* of residential CHP systems for use in Japan. Hawkes and Leach (2008) examined some of the *government policy* instruments needed for the support of micro combined heat and power (µCHP) in the UK, again for individual dwellings, and Karger and Bongartz (2008) investigated the *relevance of external determinants* for the adoption of stationary fuel cells, mainly those involving infrastructure and general political conditions, among various user groups within the European Union. Staffell et al. (2008) and others have examined the cost targets for domestic CHP units over a range of fuel cell types. Taken together, these recent studies illustrate the complexities involved in the design, development, and deployment of µCHP systems. This chapter deals with some of these complexities in greater detail, focusing in particular on µCHP systems involving fuel cells, particularly for domestic applications.

37.2
Voice of the Customer

Ultimately, the attraction of fuel cells for application in buildings will depend on the ability of the system to satisfy the needs of the customer. This is true in the development of any product, and obtaining the "voice of the customer" (VOC) is one of the critical initial activities that need to be undertaken. Examples abound of how product development efforts have failed either because the VOC was not adequately captured, or it was captured too late in the development cycle. Over-specification of performance can unnecessarily result in additional costs, making the fuel cell system unattractive, or the lack of complete specification can result in a failure to meet critical requirements. Techniques such as the use of focus groups, Kano analysis and quality function deployment are available not only to capture the VOC, but also to deploy it down to the factory floor.

In the case of fuel cell systems for buildings, there are at least four primary types of potential customers who need to be satisfied: the installer, the private

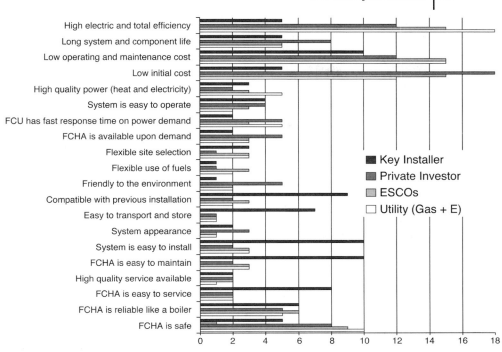

Fig. 37.2 Capturing the voice of the customer for μCHP systems.

investor, the energy supplier, and the utility. Each has its own perspective on the requirements of the system. The requirements themselves can be extensive, covering fuel, power, power quality, environment, industrial design, standards, emissions, safety, communications, operation, duty cycle, reliability, ease of installation and service, diagnostics, and remanufacturing/recycling requirements. Financial goals take the form of capital, maintenance, and operational costs.

An example of capturing the voice of the customer for μCHP systems for multi-family dwellings in the EU is shown in Figure 37.2. In this case the various customers were asked, in focus groups, to list their requirements in order of priority. It is seen that the priorities of the various criteria differ by customer. The key findings from this study support the ideal value proposition requirements for a μCHP system solution in this particular application, namely that the system needs high overall efficiency to save energy costs, reduces GHGs, is water independent, runs on natural gas, has a reasonable payback period, provides for annual reduction in operating costs, is a noiseless and vibration-free source of power, can easily be retrofitted into existing installations, and provides a standardized technical solution for service, in addition to remote control of its operation and service.

Capital, operational, and maintenance costs associated with the CHP application need to be able to compete effectively with the available incumbent technol-

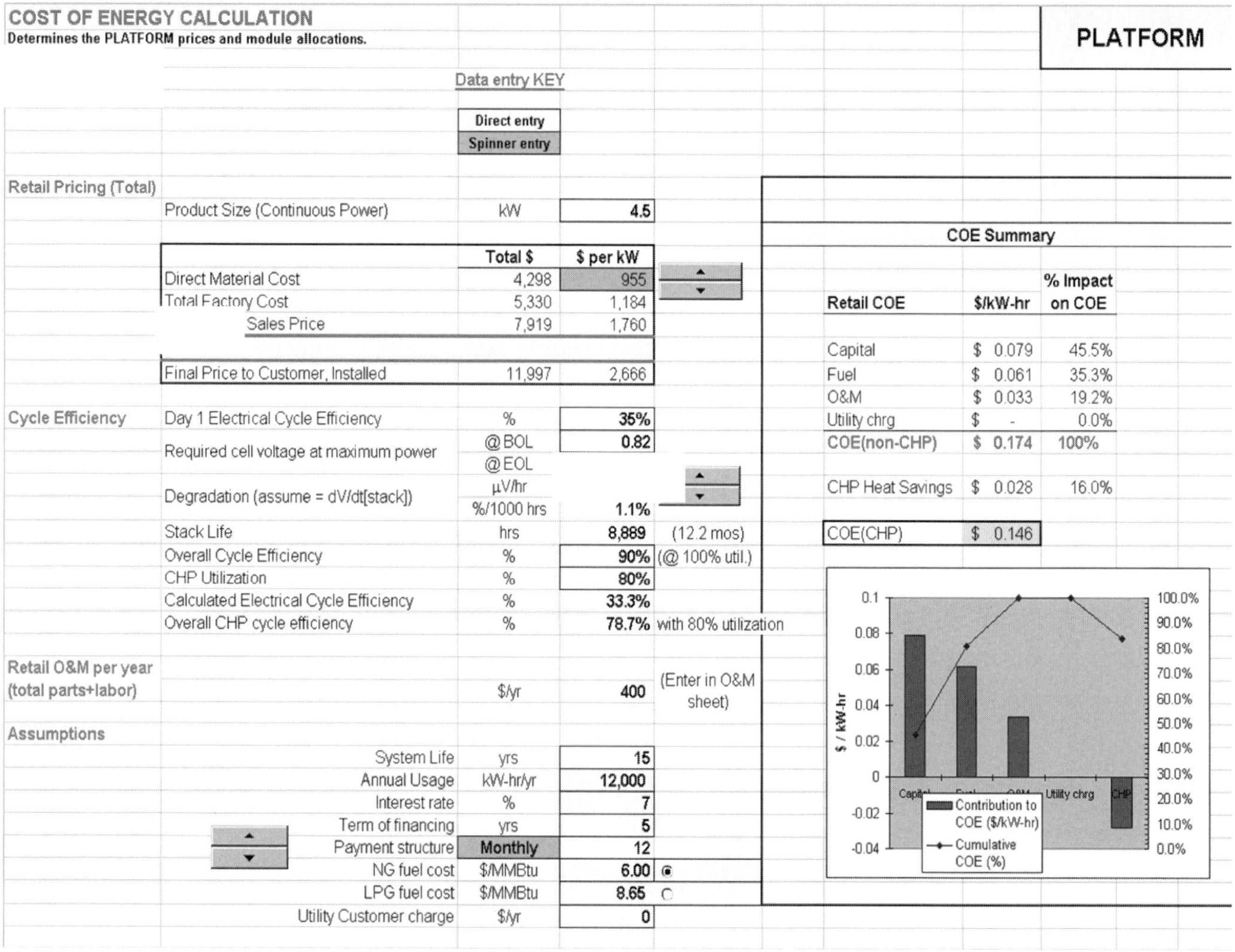

COST OF ENERGY CALCULATION
Determines the PLATFORM prices and module allocations.
PLATFORM
Data entry KEY
Direct entry
Spinner entry
Retail Pricing (Total)
Product Size (Continuous Power) kW 4.5
Total $ $ per kW
Direct Material Cost 4,298 955
Total Factory Cost 5,330 1,184
Sales Price 7,919 1,760
Final Price to Customer, Installed 11,997 2,666
Cycle Efficiency
Day 1 Electrical Cycle Efficiency % 35%
Required cell voltage at maximum power @BOL 0.82 @EOL
Degradation (assume = dV/dt[stack]) µV/hr %/1000 hrs 1.1%
Stack Life hrs 8,889 (12.2 mos)
Overall Cycle Efficiency % 90% (@ 100% util.)
CHP Utilization % 80%
Calculated Electrical Cycle Efficiency % 33.3%
Overall CHP cycle efficiency % 78.7% with 80% utilization
Retail O&M per year (total parts+labor) $/yr 400 (Enter in O&M sheet)
Assumptions
System Life yrs 15
Annual Usage kW-hr/yr 12,000
Interest rate % 7
Term of financing yrs 5
Payment structure Monthly 12
NG fuel cost $/MMBtu 6.00
LPG fuel cost $/MMBtu 8.65
Utility Customer charge $/yr 0
COE Summary
Retail COE $/kW-hr % Impact on COE
Capital $ 0.079 45.5%
Fuel $ 0.061 35.3%
O&M $ 0.033 19.2%
Utility chrg $ - 0.0%
COE(non-CHP) $ 0.174 100%
CHP Heat Savings $ 0.028 16.0%
COE(CHP) $ 0.146
$ / kW-hr
0.1
0.08
0.06
0.04
0.02
0
-0.02
-0.04
100.0%
90.0%
80.0%
70.0%
60.0%
50.0%
40.0%
30.0%
20.0%
10.0%
0.0%
Utility chrg
CHP
Contribution to COE ($/kW-hr)
Cumulative COE (%)

Fig. 37.3 Sample cost of energy calculation.

ogies. In developing the system, use is generally made of a "cost of energy" (COE) model, which allows one to express the economic operating window as a function of capital and maintenance costs as a function of fuel cell efficiency, fuel costs, and utilization. An example of a detailed cost of energy calculation is shown in Figure 37.3 for a particular set of operating conditions. Varying the input parameters in the COE model allows one to define the allowed tradeoff between capital and annual maintenance costs as a function of the system efficiency, payback time, and so on. Of course, one of the key factors in determining the economic viability is to match properly the size of the fuel cell system and its mode of operation to the heat and electric load profiles. This requires a simulation tool that can provide hourly demand data for the specific site under consideration. References to the use of these will be given below.

The United States Department of Energy (DOE) has published suggested performance targets for stationary fuel cell systems. Overall, the target values indicate required fuel cell stack lives of 40 000 h, degradation rates less than 1% per 1000 h, and total systems cost in the neighborhood of 500 kW^{-1} installed. These are tough economics to achieve to with today's technology, especially for the relatively small-sized systems needed for domestic applications. Part of the reason for the difficulty in meeting these requirements is that fuel cell systems, although conceptually simple, can in fact be complex. In order to understand this, we need a brief introduction to how fuel cells operate, how they fail, and how they drive system complexity.

37.3
Fuel Cell Basics and Types

Fuel cells are energy conversion devices that transform chemical energy into electrical energy, in much the same way that a battery electrochemically converts chemical potential energy into electrical kinetic energy. Unlike a battery, in the case of a fuel cell the chemicals consumed in the reaction are supplied from external sources. The term "electrochemical" refers to the fact that the chemical reaction does not proceed uncontrolled, but is manipulated by directing the flow of the electrons involved in the chemical transformation. A familiar example of an electrochemical reaction process is the case of an electrolyzer using electricity to split water into hydrogen and oxygen. In this case, hydrogen and oxygen gases are generated through the application of a potential applied to two electrodes, with an ion-conducting material separating the gases. A hydrogen fuel cell operates in reverse. It permits the generation of electricity and water by controlling the combination of hydrogen and oxygen:

$$H_2 + O_2 \rightarrow H_2O \tag{37.1}$$

Ideally, this reaction would take place at a potential of 1.229 V under standard conditions (25 °C and 1 atm) if were it not for losses due to irreversible internal processes. In fact, the reversible potential E_r of this reaction can be expressed in terms of the change in the Gibbs free energy, ΔG, as

$$G = -nFE_r \quad (37.2)$$

where n is the number of electrons involved and F is the Faraday constant, 96 485 C of charge per mole of electrons. The derivation of Equation 37.2 is a straightforward application of thermodynamics, recognizing that in this case the work done is the product of the amount of charge transferred and the potential. For the case of standard conditions, we know from thermodynamics that $\Delta G = 237$ kJ mol^{-1} when the water is produced as a liquid [corresponding to the higher heating value (HHV)] and $n=2$ for the H_2–O_2 fuel cell reaction. Of course, not all of the energy in the reaction is turned into electricity. In every chemical reaction entropy is produced, and because of that some portion of hydrogen's HHV is converted into heat. The enthalpy change in the reaction is given by

$$\Delta H = \Delta G + T\Delta S \quad (37.3)$$

From thermodynamic tables, we find that at standard conditions $\Delta H = 286$ kJ mol^{-1}. It follows that roughly 49 kJ mol^{-1} are converted into heat and the theoretical efficiency η_{th} of a fuel cell operating at standard conditions is

$$\eta_{th} = \Delta G / \Delta H = 237/286 = 83\% \quad (37.4)$$

From this theoretical efficiency, there are various losses which are due to irreversible processes in the cell itself. These losses, which are intrinsic to the cell and are materials dependent, reduce the fuel cell efficiencies to around 40–60%, depending on the type of fuel cell and its application.

There are numerous references that discuss fuel cells in detail, including studies by Srinivasan (2006) and Vielstich *et al.* (2004). Here only a very cursory

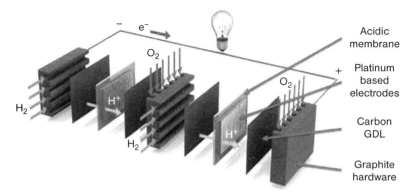

Fig. 37.4 Stack cell materials and components in PEM fuel cells. Courtesy of Cellera, Inc.

overview is given. A schematic diagram of a fuel cell (FC) based on an acidic polymer electrolyte membrane (PEM) is shown in Figure 37.4.

In an FC, the two reactants, in this case a hydrogen-rich stream and oxygen, which is brought in with air, are each directed macroscopically by flow channels and microscopically by a gas diffusion medium to come into contact with porous electrodes. A membrane is provided that serves as the electrolyte, providing ionic conductivity. The membrane not only maintains the separation of these reactant gases, but also prevents the flow of electrons. The porous electrodes provide a support for the catalysts needed to facilitate the reactions at the anode and the cathode, and provide conductive pathways for the ions and electrons. At the anode electrode, hydrogen gas molecules adsorb on the supported catalyst material, which is used to help strip the hydrogen of its electron, oxidizing it to become a hydrogen cation, or proton. This proton is transported through the polymer electrolyte membrane (PEM) to the cathode electrode. Here, oxygen molecules are fed into the active layer of the cathode electrode, usually by supplying air. Like the active layer at the anode, appropriate catalysts are supported on electrically conducting porous materials, such as carbon. Once on the catalyst surface, the oxygen's electronic bonds are broken, forming oxygen anions. The oxygen anions then combine with protons arriving from the membrane and electrons that have been driven through an external circuit by the difference in electrochemical potential of the anode and cathode. The point on the catalyst surface at which protons, electrons, and oxygen atoms combine to form water is generally called the triple phase boundary (TPB). All of the necessary components, the gas diffusion medium, the electrodes, and the membrane, are manufactured as a "membrane electrode assembly" (MEA) by various suppliers. Since a typical FC operates at a voltage of less than 1 V, multiple cells are electrically connected in series to form a "stack" of cells that can be sized for a particular application. The integrated stack design must accommodate a means for managing the products of the FC reaction: electricity, heat, and water. More will be said of this later.

The electrochemical equation governing the hydrogen oxidation reaction at the anode of a single cell is

$$H_2 \rightleftharpoons \leftrightarrow 2H^+ + 2e^- \qquad E^0 = 0 \text{ V} \tag{37.5}$$

This is a thermodynamically reversible reaction and, with platinum as the catalyst, is a standard reference known as the reversible hydrogen electrode (RHE), for which the potential is universally chosen as 0 V. The oxygen reduction reaction (ORR) at the cathode, on the other hand, is thermodynamically irreversible, and is generally expressed in terms of the dominant four-electron reaction

$$O_2 + 4H^+ + 4e^- \rightarrow 2H_2O \qquad E^0 = 1.229 \text{ V (vs RHE)} \tag{37.6}$$

In fact, the ORR is an extremely complicated reaction that is dependent upon the electronic structure and number of the active sites on the catalyst surface, which in turn are influenced by many factors, including the size and shape of the catalyst, the presence of any species competing for reaction sites, the presence of oxides formed during potential cycling, and the effect of the catalyst support. In addition to the four-electron reaction, there is the possibility of a two-electron peroxide pathway:

$$O_2 + 2H^+ + 2e^- \rightarrow H_2O_2 \tag{37.7}$$

In low-temperature FCs, the peroxide pathway is problematic because if the peroxide is not quickly decomposed into water, it can form radicals that chemically attack the membrane, causing premature failure. More will be said of this later.

The ORR is a major source of efficiency loss in a, FC, but it is not the only source of so-called "overpotential". In general, there are also resistive and mass transport losses. The resistive losses are associated with the finite conductivities of the electrolyte and the electrodes and contact resistive losses at interfaces, whereas mass transport losses are generally associated with the lack of adequate fuel or air reaching the reaction sites. For low-temperature systems, mass transport losses are primarily due to the build-up of liquid water in the electrode or gas diffusion layer (referred to as "flooding"). In both low- and high-temperature FCs, mass transport losses are also associated with the tortuous pathways that the reactants must take in the porous electrode structures used to create and maintain the triple phase boundaries needed for the electrocatalytic reactions. A typical "polarization curve" for an FC is shown in Figure 37.5, in which are shown the various regions in which the so-called activation, ohmic resistive, and concentration polarization or mass transport loses affect the performance. Most of the activation loss is associated with the sluggish kinetics of the ORR at the cathode.

In general, then, a single FC consists of two catalyzed electrodes, an electrolyte, a means of supplying reactants to the catalyzed porous electrodes, methods of directing the flow of electrons to the external load, and finally, a means of

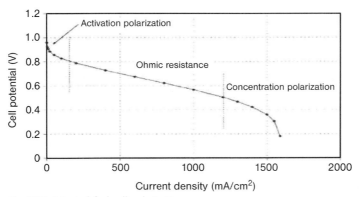

Fig. 37.5 A typical fuel cell polarization curve.

dealing with the byproducts of the reaction, water and heat. There are many types of FCs whose properties and characterization depend mainly on the chemistry of the electrolyte. Electrolytes can be acidic or basic, and can be in liquid or solid form. For example, phosphoric acid fuel cells (PAFCs) utilize phosphoric acid as their proton conductor. The phosphoric acid is usually retained in an SiC matrix, and these systems operate at temperatures around 200 °C. PAFC systems are commercially available from United Technologies Corporation in the 400 kW size range.

Others, such as alkaline fuel cells (AFCs), use potassium hydroxide as the electrolyte, with the temperature of operation being dependent on the concentration. The alkaline cell uses OH^- species to connect anode and cathode electrodes electrochemically. Here again, the electrolyte is retained in an appropriate matrix, and temperatures range from < 120 °C for low electrolyte concentrations up to 250 °C for higher concentrations. One very attractive aspect of ACFs is that they can use a wide range of non-precious metal catalysts, such as Ni and metal oxides. They are, however, generally intolerant to CO_2 which is present in air, and any liquid alkaline system must make an accommodation for dealing with this issue, such as by providing CO_2 scrubbers to reduce the concentration to acceptable levels.

Molten carbonate fuel cells (MCFCs) have an electrolyte composed of a combination of alkali metal (Li, Na, and K) carbonates, which is held in a ceramic matrix. The electrolyte conducts CO_3^{2-} anions between the anode and cathode, and the cell runs at 650 °C, a temperature at which the carbonate salts become conducting and noble metal catalysts are not required. Fuel Cell Energy designs and manufactures MCFCs in the 1 MW size range.

Solid oxide fuel cells (SOFCs) use a solid non-porous metal oxide, usually yttria-stabilized zirconium (YSZ), as the electrolyte. These cells, which can have a variety of designs based on the nature of the electrolyte support, operate at high temperatures (750 °C and above) where conduction by oxygen ions takes place. These systems are still in pre-commercial development by a dozen or so companies. They have a wide range of potential applications that can capitalize on their high temperature, including the ability for internal reforming of natural gas to achieve a hydrogen-rich fuel, in addition to being able to use high-temperature waste heat in useful thermodynamic cycles. More will be said of these systems below.

Polymer electrolytes, such as the PEMs, are actually hybrid in the sense that although they are solid polymers, they depend upon a high level of hydration to conduct protons effectively. In effect, the polymer becomes a host for hydrated moieties within the polymer structure that permit proton conduction through mechanisms that involve proton exchange between them. Sophisticated molecular dynamics models have been developed to elucidate the proton transport mechanisms and their dependence on the nanoscale structure of these moieties (Spohr, 2004). In low-temperature (60–80 °C) PEM systems based on perfluorosulfonic acidic (PFSA) ionomers, Nafion is used as the baseline standard upon which to measure improvements in membrane performance and durability. In this case, proton conductivity is a strong function of the level of hydration of the membrane, and the cell and stack design must accommodate the ability to manage this hydration level over a wide range of operating and environmental conditions. Usually this is accomplished by ensuring that the reactant streams are humidified. Low-temperature PEM FC MEAs are manufactured by a variety of suppliers, including 3M, Gore, Johnson-Matthey, and Asahi Chemical.

It should be noted that other, newer membrane electrolyte systems have nonetheless been under development for some time. In particular, higher temperature PEM systems are based on polybenzimidazole (PBI)-type materials. These systems rely on doping of the material with a proton-conducting acid such as phosphoric acid (H_3PO_4) to achieve reasonable levels of proton conductivity. The H_3PO_4 acts as the proton conductor and has a very low (but finite!) vapor pressure at elevated temperatures. The ability of these systems to operate at higher temperatures, in the range 160–200 °C, enables the FC to tolerate higher levels of CO in the fuel stream, as would be the case when running the FC on reformed natural gas. High-temperature proton exchange membranes (HTPEMs) based on PBI-like materials are available as Celtec® P MEAs manufactured by BASF Fuel Cells. Their membrane is based on a sol–gel process to prepare PBI membranes with high molecular weight and high acid doping levels. Schmidt and Baurmeister (2008) discussed the fabrication of these MEAs, their system enabling characteristics, and their performance under steady-state and start–top modes of operation. In spite of the fact that the systems using PBI are less sensitive to the level of hydration in the membrane than in the lower temperature PFSA systems, care must be taken to maintain the proper acid concentration levels in the membrane to ensure long-term operation. One major drawback to the H_3PO_4-based systems is the lower power density of these systems. This is believed to be due to the adsorption of H_3PO_4 and its phosphate anion on the Pt catalyst. This adsorption results in the need for higher platinum loadings and larger stacks to achieve power levels equivalent to those based on perfluorosulfonic acid membranes used in low-temperature PEM FCs. This lower power density results in larger stacks for PBI-based PEM systems.

Cellera Technologies (Caesaria, Israel) is engaged in the development of a new membrane technology which allows the use of non-platinum catalysts and thereby removes a major component of current PEM FC costs. The ability to replace precious metal catalysts by catalysts based on inexpensive metals, such as Co, Fe, and Ni, is made possible by transitioning from the proton-conducting membrane and ionomer used in the PEM fuel cell to an OH^- ion-conducting membrane and ionomer which provides an alkaline counterpart to the acidic materials that have been used to date in polymer electrolyte fuel cell stacks. The alkaline ion-conducting polymer also provides a much more benign chemical environment, which allows, in turn, the use of electropositive metal catalysts at much lower risk of instability caused by corrosion processes prevalent in the highly acidic environment of the PEM fuel cell. Furthermore, this benign chemical environment also opens the door to other potential important benefits, including the ability to use light and highly manufacturable metal bipolar plates/

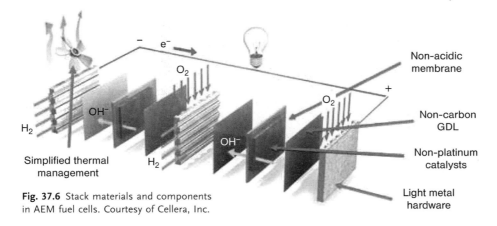

Fig. 37.6 Stack materials and components in AEM fuel cells. Courtesy of Cellera, Inc.

flow. Figure 37.6 provides a schematic depiction of cell components and materials for anion-exchange membrane (AEM) fuel cells.

The reason for the relatively late onset of significant interest in AEM FC technology is the list of difficulties, including some real and some perceived, on the way to reducing the technology to practice. It has been widely believed that the quaternary ammonium hydroxide ($R_4N^+OH^-$) functional group, which is the one used in most AEMs, is "self-destructive", because the OH^- ion is likely to attack the R_4N^+ cation and/or the tether carrying the cation. This suspected instability of the ionomeric material itself was probably the most important deterrent. In addition, the specific conductivity of an OH^--conducting ionomer was suspected of being a factor of at least 3–4 lower than that of the H^+-conducting ionomer, thereby setting a limit on power output. It was also suggested that, as the AEMs developed to date have been based on hydrocarbon, rather than fluorocarbon, backbones, the preparation of effective and stable membrane-electrode assemblies should present a significantly tougher challenge than the case of poly-PFSA ionomers which facilitate preparation of ionomer dispersions in alcohol-water mixtures. Finally, a central question concerns the effect of CO_2 from the air feed on the OH^--conducting ionomer. Since conversion of the OH^- ion in the alkaline ionomer to bicarbonate (and/or carbonate) ion is a very likely process, the cell performance could suffer from negative effects on both ionomer conductivity and, particularly, the kinetics of electrode processes.

First indications of some answers for the strong challenges on the way to AEM-FC reduction to practice came several years ago when it became known that Tokuyama, a Japanese company specializing for a long time in membrane technology for electrodialysis and desalination, started to move ahead with development of AEMs in OH^- form, targeting fuel cell applications. In addition, work at Los Alamos National Laboratory showed that proper selection of the functional group for the AEM could secure stability of the ionomer, maintaining the functional group population over 1000 h under demanding chemical conditions. Work at Celera so far has proved that factors limiting AEM-FC longevity do not include the chemical stability of the alkaline membrane, that is, this key component of the AEM-FC has reached a quality that provides a good basis for further technology development. Furthermore, it has been found that membrane resistivity is not more than two times higher than that of Nafion. Finally, it was shown that the effects of CO_2 from the air on the performance of the AEM-FC can be virtually eliminated by using a combination of tools upstream of the stack and in the stack. Unlike ordinary filtration, the approach developed for CO_2 handling upstream of the stack totally avoids any need for routine replacement(s) of system component(s), thereby fully preserving the low maintenance feature expected of a fuel cell-based power source. Cellera has so far demonstrated power densities of 200 mW cm^{-2} at a cell voltage of 0.4 V with no platinum catalysts for a hydrogen–air stack, no addition of liquid electrolyte, and no humidification from an external source (S. Gottesfeld, personal communication, 2009). Given the level of effort to date, these results give a good indication of the potential of this new technology in future fuel cell systems.

One other type of fuel cell is worth mentioning because of its unique properties, namely is the solid acid fuel cell (SAFC). This type of fuel cell is based on the fact that certain types of solid acids undergo a polymorphic structural transition from an ordered state to a disordered state. Accompanying this order-disorder transition is a dramatic increase in proton conductivity. The best material turns out to be cesium diphosphate (CsH_2PO_4), which undergoes its superprotonic transition at 228 °C. Thus, this FC chemistry involves a solid-state system that is intermediate between the PEM systems and the high-temperature SOFC systems. It offers the advantages of other higher temperature systems, namely an increased tolerance to CO, a higher quality of waste heat, and reduced system complexity. However, it has challenges in that it tends to have lower power densities and, since it is soluble in water, some unique failure modes. Chisholm *et al.* (2009) provided a review that captured both the opportunities and the major challenges that lie ahead with this technology.

37.4
Recent Advances

Because of the fact that FCs have been around for some time, yet are not widely commercialized, critics have asked the question, "What has changed?" The answer is that in the past few years various factors have changed. For one, there have been tremendous advances in computational science driven by the capability brought on through the convergence of "nano" and "info" technology to follow Moore's law (researchers are now working on advanced computer chips that have nodes below 22 nm, enabling some 4 billion transistors per chip). These advances in computation have allowed theoreticians to explore electrochemical and electromechanical processes at the micro-, meso-, and nanoscales, thereby improving our theoretical understanding of the basic processes involved. Furthermore, the nanoscience and nanomaterials knowhow that has been developed along the way is now being applied in a variety of fields, not the least of which is the development of nanoscale materials and structures that have direct bearing on energy generation, conversion, and storage. In addition, scientists and engineers are no longer regarding FCs as laboratory curiosities, but instead are solving the real-world problems associated with specific commercial applications. In the past few years, the amount of understanding of how FCs operate, and fail, in real-world applications has grown tremendously, driven in large part by government support from basic research to full-scale demonstrations (see, for example, Feitelberg et al., 2005).

The ability to predict behavior analytically has provided substantial insights into FC operation and performance. Researchers and engineers have developed models of single fuel cells in an attempt to describe the current-voltage response of the FC to the critical parameters through the specifications of the material properties of the cell components. These models, which are remarkably successful in describing the overall performance of a single cell, are macro-homogeneous in the sense that they use local average properties, such as electrode porosity, conductivity, and so on, to describe the material. There are excellent reviews of the models that have been developed to elucidate the various aspects of FC behavior under a variety of conditions, such as those by Weber and Newman (2004) and Wang (2004). In addition, there are specialized models that deal with the root cause of specific failures, such as platinum dissolution (Darling and Meyers, 2003) and carbon corrosion (Reiser et al., 2005; Meyers and Darling, 2006) due to high potentials generated during startup and shutdown (Tang et al., 2006), and so on. In fact, knowledge of the cell and stack failure modes, and their dependence on the process critical parameters and material specifications, is fundamental to designing and developing commercially viable fuel cell systems. It is one thing to describe how a fuel cell system works, but it is another to prevent it from failing prematurely.

It has been mentioned that one of the sources of low-temperature PFSA membrane degradation is the formation of peroxide. A model (see, for example, Shah et al., 2009) of the degradation process suggests that hydrogen peroxide forms by distinct mechanisms in the cathode and anode. This peroxide forms radicals through Fenton reactions involving metal ion impurities. The radicals then participate in the decomposition of reactive end groups in the membrane, to form, among other species, hydrogen fluoride, which can be detected in the product water. Higher fluoride release rates correlate with higher rates of membrane degradation. The degradation occurs through the "unzipping" of the polymer backbone and the cleavage of the polymer side chains. The two critical parameters which accelerate this degradation mechanism are dry conditions and high temperatures. In addition, there is a built-in "death spiral" in PEM fuel cells in the sense that if "drying" conditions exist, the conductivity of the membrane falls, internal resistances increase, and temperatures rise even more, further drying out the membrane until "pinholes" form in the thinning membrane, allowing gas crossover and cell failure. Therefore, PFSA-like materials must be kept under good temperature control and be maintained fully hydrated.

Indeed, the key to long-term low-temperature PEM FC performance is "water management" schemes that provide for proper humidification for both high conductivity and long membrane life. Such schemes must include means for efficient removal of product water to avoid "flooding" and localized fuel starvation on the one hand, and drying conditions on the other. The importance of water and thermal management is now well recognized, and there has been a considerable effort not only to understand the process and material property requirements to meet this objective, but also to devise means for achieving it.

Much effort has been expended in determining the mechanisms by which the kinetics of the reactions at both the anode and the cathode can be improved. For the case of the hydrogen oxidation reaction, pure platinum is used, and it is usually supported on a porous carbon material. In those cases where the fuel is contaminated by CO, as in the case of reformed natural gas, which will be discussed below, it has been found that the a CO-tolerant catalyst consists of a platinum–ruthenium alloy. In this case the Pt–Ru catalyst helps oxidize the CO to CO_2, which is facilitated by the bleeding of small amounts of air into the reformate. The effect of the amount and type of air bleed (continuous versus pulsed) has been studied and reported (Du et al., 2008). Again, simplified models have been developed to describe the effect of air bleed on CO poisoning and its effect on cell potential.

Because of the economic benefits of improving the kinetics of the ORR reaction, considerable effort has gone into the research and development of membrane electrode assemblies that incorporate the advantages of new nanomaterials and nanoscale architectures intended to reduce Pt loadings significantly while maintaining high power densities. Chief among these are electrodes that incorporate Pt alloys with transition metals of various types as a means of increasing catalytic activity, thereby allowing a reduction in platinum loading. Currently, Pt_3Co alloys supported on carbon are achieving a roughly twofold improvement over the state-of-the-art Pt/C electrodes. Intermetallic architectures of Pt and other materials have been fabricated by a variety of techniques, some of which yield so-called "core–shell" and "skeleton–skin" structures. Some of these, which involve monolayers of Pt particles on Pd cores supported on carbon, are reported to have achieved very significant ($\sim$20-fold) increases in mass activities. These research activities certainly hold great promise for overcoming the obstacles to widespread commercial viability of fuel cells.

Recent advances have been made in the understanding of heterogeneous ORR catalysis using the combination of nanoscale fabrication techniques, molecular and atomic level simulation tools, and nanoscale characterization techniques. For example, the important theoretical work of Kitchen (Kitchen et al. 2004), Mavrikakis (Mavrikakis et al. 1998), Neurock (Taylor et al. 2006) and others have now given us a deeper understanding not only of the thermodynamics involved in the ORR, and the likely cause of the overpotential-associated ORR, but also how variations in the electronic structure determine trends in the catalytic activity of the ORR across the Periodic Table. This work, coupled with the insights provided by the seminal work of Adzic (Adzic et al. 2001) by Markovic (Markovic et al. 1997) and Stamenkovic (Stamenkovic et al. 2006), and by others in the synthesis and performance characterization of Pt alloys and core–shell architectures, has significantly advanced the state-of-the art of catalysts for use in both HOR and ORR fuel cell reactions. However, there is much more to be done in order to close the gap between where we stand today and where we need to be in the future.

Although the exact mechanisms underlying the oxygen reduction reaction are still debatable, Nørskov et al. (2004) have used density functional theory (DFT) to advance significantly our understanding of the thermodynamics of the various possible reaction pathways and their dependence on surface properties. In combination with detailed DFT calculations, they were able to provide a detailed description of the free-energy landscape of the electrochemical ORR over Pt(111) as a function of applied bias. In doing so, they found that adsorbed oxygen and hydroxyl are very stable intermediates at potentials close to equilibrium, and the calculated rate constant for the activated proton/electron transfer to these adsorbed intermediates accounts for the observed kinetics. On this basis, they were able to account for the trends in the ORR rate for a large number of different transition metals. In particular, they were able to construct "volcano" plots of maximum catalytic activity as a function of the oxygen and hydroxyl adsorption energies, describing known trends and the observed effects of alloying. It is clear that significant progress has been made in the past several years in our basic understanding of the electrocatalytic processes at the nanoscale. An excellent summary of this and other work has been given in a recent book on fuel cell catalysis by Koper (2009).

In parallel with this significant gain in theoretical understanding, progress has been made experimentally in reducing the platinum loadings, particularly in the hydrogen/air fuel cells used in automotive applications. The current state of the art is such that platinum loadings are around 0.6 mg of platinum per square centimeter of MEA active area (0.6 mg cm^{-2} MEA). At the same time, power densities have increased up to 0.7 W cm^{-2} MEA corresponding to a Pt specific power density of roughly 1 g$_{Pt}$ kW^{-1}. Gasteiger et al. (2005) discussed the strategies needed to achieve the nearly fivefold reduction in loading from these values in order to achieve automotive cost targets. These strategies, which involve increasing the power density at high voltages by reducing the Pt loading by increasing the Pt activity, and by reducing mass transport losses in the cell, should also be relevant to stationary applications.

Much of the early progress made in the reduction of catalyst loadings came through the use of high surface area carbon as support, which allowed better dispersion of the platinum nanoparticles, thereby increasing the density and effectiveness of the TPBs. One major drawback to the use of carbon supports is that, under certain operating conditions, the carbon can corrode (oxidize). Certain transient conditions such as start-up and shut-down of the fuel cell can exacerbate the situation, as can situations involving fuel starvation. The resulting loss of the support and high potentials cause the catalyst particles to agglomerate, oxidize, dissolve, and be transported out of the electrode. These phenomena have

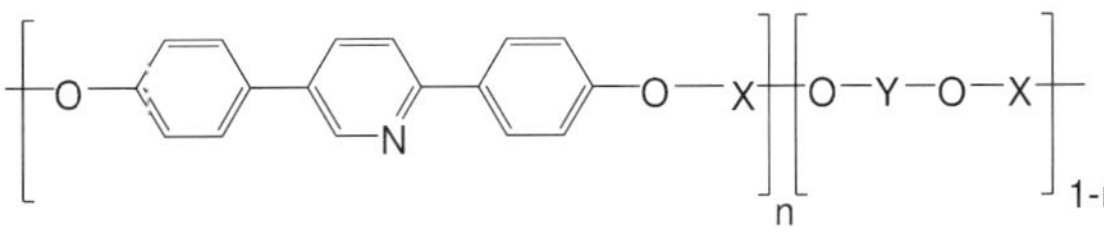

Fig. 37.7 Chemical structure of aromatic polyethers bearing pyridine units in the main chain. X = 4-biphenylsulfone, 4-biphenylphenylphosphine oxide, benzophenone, octafluorobiphenyl; Y = phenyl, 4,4′-biphenyl, 3,3′;,5,5′-tetramethyl-1,1′-biphenyl, 4,4′-isopropylidenebiphenyl, 4,4′-hexafluoroisopropylidenebiphenyl.

been observed experimentally and their mechanistic understanding validated with models, as proposed for example by Meyers and Darling (2006).

The problem of carbon corrosion is now being addressed through several avenues. The first is the development of carbonless nanostructured thin-film electrodes, such as those being developed by 3M. In this approach, continuous coatings of catalyst (Pt and its alloys) are vacuum coated on to organic "whiskers". These carbonless electrodes have shown excellent high voltage stability, as discussed by Debe *et al.* (2006). Because these electrode structures are 20–30 times thinner than conventional Pt/C electrodes, they are more susceptible to mass transport losses due to flooding at high current densities, and care must be taken to match the electrode with the proper gas diffusion media. Other approaches to achieve stable electrode supports involve the use of corrosion-resistant carbon nanotubes (Wang *et al.*, 2006) and more stable metal oxides, such as TiO_2 (Ioroi *et al.*, 2005). Both of these approaches are still in the research phase. It is expected, however, that commercially viable approaches to deal with the issues of electrode durability will continue to make progress as the rate of learning accelerates.

In addition to new electrode materials and structures, new membranes are also being developed. For example, new high-temperature polymer electrolyte membranes based on aromatic polyethers containing polar pyridine moieties in the main chain have been developed by Advent Technologies for use in HTPEM fuel cells (Figure 37.7) (Kallitsis *et al.* 2009). The newly developed polymer products are called Advent TPS®. Such materials combine the excellent film-forming properties with high mechanical, thermal, and oxidative stability and the ability to be doped with phosphoric acid. Highly proton-conducting membranes were produced after treatment with amounts of phosphoric acid that could be controlled by varying the pyridine-based monomer content (Gourdoupi *et al.* 2003). The pyridine groups strongly retain the phosphoric acid molecules, due to their protonation, thus inhibiting leaching out.

The fuel cell performance based on TPS MEA operating at 180 °C with pure hydrogen or reformate with different CO contents and air feed gases is shown in Figure 37.8. As expected, because of the high operating temperature, this system tolerates up to 2 vol.% CO poisoning without a significant decrease in performance.

However, in comparison with the current–voltage performance with PFSA systems, it is seen that this performance, which is characteristic of phosphoric

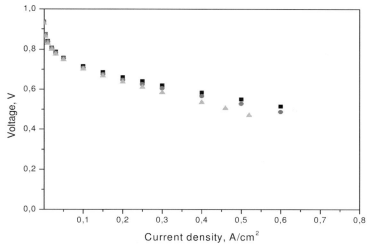

Fig. 37.8 Polarization curves for Advent TPS MEA operating at 180 °C using pure hydrogen (■) or reformate containing 1% CO (●) and 2% CO (▲) and air feed gases at ambient pressure (λ_{H_2}=1.2 and λ_{air}=2).

acid systems, is inferior. As discussed, this effect is thought to be due to the presence of phosphoric acid and/or its anions that have adsorbed on the surface of the catalyst, thereby requiring the oxygen to diffuse through this layer to reach active sites on the platinum surface.

37.5
Fuel Cell Systems

Babir (2005) discussed the overall design of fuel cell systems for stationary applications. The fuel cell stack by itself is virtually useless until it is supported by the "balance of plant": the additional components that provide the means of reforming and supplying fuel, managing heat and water (including humidification), conditioning the power output from the stack, and providing for system control. To these critical subsystems one needs to add those required for providing for the overall structure of the system and its installation, maintenance, and safety. A schematic representation of the basic architecture of a general fuel cell system is shown in Figure 37.9, which shows the reactant processing module (RPM) that delivers both air and fuel to the power generation module (PGM), which includes the stack. If the stack requires a humidified reformate with <5 ppm of CO and <5 ppb of H_2S, as in the case of a low-temperature PEM system, then the RPM must include these functions, and also those associated with the exhaust gases leaving the stack. Likewise, the PGM includes a means for managing the product water and ensuring that the membrane is maintained in a

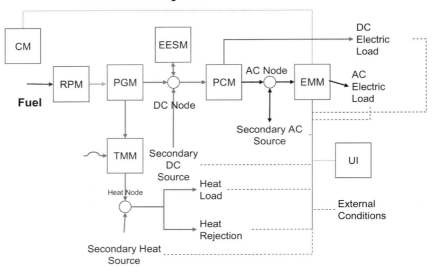

Fig. 37.9 Fuel cell system architecture.

healthy state relative to the level of hydration. The thermal management module (TMM) manages and directs the flow of heat generated in the stack and maintains the stack temperature within specified ranges. The power conditioning module (PCM) delivers both regulated DC and AC power, and the electrical energy storage module (EESM) provides power at start-up and through rapid transients. Also shown explicitly are the other supporting modules such as controls module (CM) for providing the hardware and software for system control, an energy management module (EMM) for providing the input needed to coordinate the fuel cell output loads, and a user interface module (UI) that interfaces into the system.

Given this architecture, one can begin to appreciate the potential complexity of a fuel cell system, given the need for high levels of thermal and material integration between modules, and how that complexity can be reduced by the simplification and or even elimination of major components. Of course, the most obvious solution is to eliminate the RPM by delivering pure hydrogen to the stack.

37.5.1
Reforming Considerations

Whereas the fuel cell itself utilizes hydrogen, in most cases hydrogen in its pure state is not normally available and it must be obtained by "reforming" a hydrocarbon fuel such as methane, propane, liquid petroleum gas (LPG), or kerosene to extract the hydrogen. There are basically three alternative processes for extracting hydrogen from these logistical fuels: steam reforming, partial oxidation, and autothermal reforming. Steam reforming is an endothermic process: heat must be supplied to the reactor. This is normally accomplished by combusting an additional amount of fuel in a separate, but thermally integrated, reactor. Usually this reactor burns the unused hydrogen which flows through the anode chamber of the fuel cell stack, and is referred to as the "anode tail gas oxidizer" (ATO). The steam reformation reaction is reversible, and the product gas is a mixture of hydrogen, carbon monoxide, carbon dioxide, and water vapor. In the case of methane (assuming that sulfur compounds have been removed), the steam reforming step is

$$CH_4 + H_2O \rightarrow CO + H_2O \tag{37.8}$$

To remove the high levels of CO, this stream is then fed to a "shift" reactor, which reduces the CO concentration and makes additional hydrogen:

$$CO + H_2O \rightarrow CO_2 + H_2 \tag{37.9}$$

Unlike steam reforming, partial oxidation is an exothermic process. It is essentially a combustion, but with a less than stoichiometric amount of oxygen. As in steam reforming, the reformate gas must go through a shift reaction to produce more hydrogen, and through a preferential oxidation (PROX) step to reduce the CO content to an acceptable level (<5 ppm). Partial oxidation produces less hydrogen than steam reforming, and unlike steam reforming, the product gas contains relatively large amounts of nitrogen from the air used in the process.

Because partial oxidation is an exothermic process, and steam reforming is an endothermic process, these two can be combined in the so-called "autothermal" reforming process. The overall reaction of an autothermal process for methane, with the shift reaction included, is

$$CH_4 + \chi O_2 + (2-2\chi) H_2O \rightarrow CO_2 + (4-2\chi) H_2 \tag{37.10}$$

where χ is the number of moles of oxygen per mole of fuel and its value determines if the reaction is exothermic, endothermic, or thermoneutral. In practice, the value is chosen to avoid the risk of carbon formation in the reactors, and the amount of steam added is usually in excess of that theoretically required. In addition to these processes, a desulfurization process is needed to remove sulfur compounds present in the fuel. All of these process steps use specially designed catalysts to facilitate the reaction steps. Catalyst designs are in many cases proprietary, although in most cases references to catalyst materials used in similar reactions can be found in the literature.

In practice, the key parameters characterizing reformer operation are the steam to carbon ratio and the oxygen to carbon ratio. In fact, Feitelberg and Rohr (2005) have shown that these two variables, when combined with specifications on the fractional conversion of fuel in the reformer and on the dew point of the reformate, define an "operating line" for the fuel processor. This operating line defines the required relationship between steam to carbon and oxygen to carbon ratios for a reformer that meets fuel conversion and reformate dew point specifications.

Most Japanese residential μCHP fuel cell systems use steam reformers. As discussed by Feitelberg *et al.* (2005), Plug Power's 5 kW reformer-based stationary fuel cell systems have generally employed autothermal reformers. In this regard, Feitelberg and Rohr (2005) indicated that the maximum theoretical efficiency of a steam reformer-based system is only about 1% higher than the maximum theoretical efficiency of a typical autothermal reformer-based fuel processor. Adachi *et al.* (2009) recently reported the results of their effort to design an autothermal reforming fuel processor for a 1 kW CHP system for residential applications. It was bench tested and found to yield a reformate containing 48% of hydrogen and less than 5 ppm of CO.

A fully operational fuel cell system is a complex device that has numerous subsystems that interact in nonlinear ways, each with its own particular set of requirements, operating characteristics, and failure modes. It is because of this overall system complexity, driven mainly by the choice of the technology and the requirements placed on it by the application, that fuel cell systems can end up being more costly than intended. In order to minimize the system complexity and cost, it is necessary to consider the balance of plant (BoP) up front in the selection process and to devise ways of eliminating components through careful value analysis/value engineering (VA/VE). The biggest driver, of course, will be the selection of the basic technology as defined by the electrolyte.

How does one technology differ from another in its ability to meet the customer requirements while at the same time minimizing the cost of complexity? The answer to this question can only be obtained by an in-depth evaluation of the technical alternatives when viewed from a total system perspective. Once this evaluation has been made, against quality, cost, and delivery criteria, a down-selection process to the most appropriate technology will enable one to begin the design and development process. There are various ways to conduct the down-selection process, ranging from the fairly simple (Pugh) to the complex (Analytical Hierarchy Process). Usually, there is uncertainty in the process, since in most cases the technology is not "ready" in the sense that all failure modes are known or can be anticipated. Hence one technology platform, once selected, will determine the inherent cost and complexity of going forward. To minimize this uncertainty, one performs a critical analysis of the key attributes and requirements that the fuel cell system must deliver.

In the case of fuel cell systems for buildings, three fuel cell technologies are worth evaluating: low-temperature PEM, high-temperature PEM, and solid oxide. These systems will be discussed briefly here, not with the intent to select one over the other, but rather to discuss some of the considerations that should be made in the selection process. Common to all of these, of course, is the stochastic nature of both the electrical and the heating loads, which must be carefully taken into account in sizing system components. In some cases it is not immediately obvious how to deploy the fuel cell system to meet the desired objectives. In particular, one of the main considerations is whether the system should be "heat load following" or "electric load following", or a combination of both. This depends, of course, on a number of factors, including the building infrastructure, geographical location, and feed-in tariffs, if any. Most CHP systems are designed as heat-following devices to maximize system utilization, although systems that are designed for least-cost operation are capable of following algorithms that address both needs.

37.5.2
Low-Temperature PEM (LTPEM) Systems

A typical LTPEM CHP system has an overall system diagram similar to that shown in Figure 37.10. In fact, however, the actual process and instrumentation diagram (P&ID) for an LTPEM CHP system can become complex, depending on the performance specification that the design is intended to satisfy. Key drivers of complexity are associated with the need to maintain the quality reformate, the health of the stack, the efficiency of the system, and the avoidance of failures by controlling the key critical parameters (pressure, temperature, flow rate, and composition) so that they all lie within an operating window that allows for degradation over time. That is, each of these critical parameters must be controlled within specified limits at each operating point and over the intended range of operation. In addition, the required turn-down ratio (maximum/minimum power) and the range of environment extremes (ambient temperature, relative humidity, and altitude) can make this a difficult if not formidable task.

A number of papers have been published on the use of LTPEM systems for application in single and multiple family dwellings. Radulescu *et al.* (2006) reported some early experimental results on five identical LTPEM CHP systems running in France on steam-reformed natural gas. The units were far from optimized and exhibited low efficiencies. Similar results were reported by Gigliucci *et al.* (2004) in Italy. Although there have been serious follow-on demonstration programs in Europe and elsewhere using PEM fuel cells designed for single and multiple family dwellings (Feitelberg *et al.*, 2005), these projects have for the most part been terminated with the conclusion that improvements are needed in component reliability, cost reduction, and thermal integration.

Perhaps the most extensive deployment of LTPEM systems for CHP applications has been in Japan, where the conditions for fuel cell µCHP development are well suited: high electricity prices, large and growing natural gas infrastructure, world-class industrial base, and an export-oriented industrial policy. The Ministry of Economy, Trade, and Industry (METI) has proposed a comprehensive three-stage commercialization program. The first stage, consisting of technology demonstrations, resulted in hundreds of small, 1 kW CHP systems being deployed in residential homes. The second stage, planned for the period

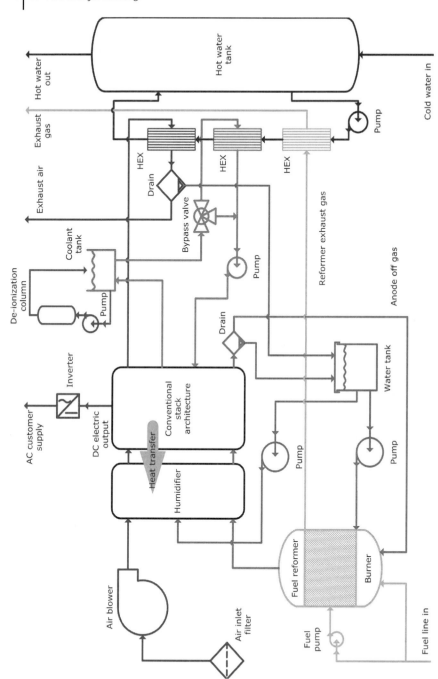

Fig. 37.10 Conventional low-temperature (LTPEM) system. Courtesy of Intelligent Energy.

2005–2010, was to accelerate introduction of practical fuel cell systems through performance improvements and cost reductions, resulting in the installation of over 2 GW of stationary systems. The third and final stage was envisioned as the diffusion stage, in which market growth was to happen naturally without government subsidy. Early on in this program, Ebara Ballard, a joint venture between Ebara Corporation and Ballard Fuel Cells, announced a 1 kW CHP system that achieved a total system efficiency of 92% [on a lower heating value (LHV) basis] and a 34% a.c. electrical efficiency.

What has actually happened since then is considerably less than what was originally anticipated. As of 2008, after spending more than 100 million on the demonstration programs, supporting the work of five companies, including Toyota and Toshiba, Japan's regional energy suppliers have installed just a few thousand units for field testing. Recently, Ballard Power Systems has dissolved the joint venture with Ebara for producing electricity and hot water in Japanese homes, indicating that it would take longer to develop the technology, requiring much greater investment.

This turn of events raises serious questions about the viability of low-temperature fuel cells for CHP applications. In particular, it raises questions about both cost and durability. It is the present author's opinion that although durability is certainly a major concern, the critical issue lies in the ability achieve the cost targets required for commercial success. This is true in part not only for the stack, but also for the balance of plant. In fact, the stack for the most part scales with system size. The BoP, however, scales less well as the system size is reduced, and this intrinsic hardware cost is difficult to overcome, even with projected economies of scale.

One of the key enablers to reduce the cost of the BoP is to reduce the system complexity by eliminating parts. This can be done by relaxing performance specifications, focusing on smaller or niche markets. For example, requiring operation at $-40\,°C$ drives the need for internal heaters in case of an unplanned system shut-down. Avoiding freezing conditions altogether allows even more system simplification. Requiring a 10:1 turn-down drives the need for additional sensors and actuators to avoid conditions of cell flooding and fuel starvation, whereas limiting the turn-down ratio to <5:1 can result in reduced control requirements and the associated hardware.

Another way to reduce cost is through advanced technology and value engineering, in which parts are combined and serve multiple purposes. An example is the system under development by Intelligent Energy. In this case, Intelligent Energy's fuel cell systems have been simplified compared with the conventional format by the use evaporative cooling (EC) stack technology with a low operating pressure. This reduces the number of stack components (and therefore the cost), simplifies the balance of plant, and reduces the complexity of the air delivery system. So far, Intelligent Energy's stationary fuel cell systems have been focused at a power level of 10–15 kW. Stack cooling is managed using direct water injection. This water partly evaporates, providing the necessary cooling. It also eliminates the need for humidified reactants and thereby associated high thermal energy transfers do not take place. No water cooling channels are required between cells and consequently each cell has a single sheet stainless-steel bipolar plate with no welded parts and no cooling channels. The number of components can be reduced and manufacturing complexity minimized, resulting in a cell packing density twice as high as in conventional designs, thus maximizing the power density. However, even though VA/VE can be expected to yield cost reductions in the neighborhood of 20% on a given design, if a more significant cost reduction is required it can only be achieved through technological innovation.

37.5.3
High-Temperature PEM (HTPEM) Systems

As discussed earlier, HTPEM systems based on PBI offer several advantages over LTPEM systems. Since they run at significantly higher temperatures, they can tolerate higher levels of CO in the reformate stream. For example, LTPEM systems certainly require <10 ppm CO, even with small amounts of air bleed, whereas HTPEM systems that run at 180–200 °C can tolerate concentrations up to 10 000 ppm. This allows the system designer to eliminate the air bleed, low-temperature shift, and preferential oxidation steps in the reforming process, reducing the number of parts and system complexity. The system can be further simplified by eliminating or simplifying active humidification. It is claimed that PBI-based systems do not require active humidification. Assuming that this is the case, the system can be simplified even further and a system with greatly reduced complexity can be achieved.

HTPEM systems are under development by both Plug Power and ClearEdge Power. A compact HTPEM system for CHP applications under development by Plug Power is shown in Figure 37.11. In this system, an auxiliary peak heater is included in the BoP. These systems are being designed as co-generation systems for residential and small commercial applications. These higher temperature systems enjoy the benefit of better heat integration with the building and, due to their tolerance for higher levels of CO, simplification in the overall system design. Both use phosphoric acid-doped PBI-type high-temperature MEAs, and both systems are considerably less complex than their low-temperature PFSA counterparts. Much of this has to do with simplification of not only the reforming step, but also the balance of plant associated with external humidification. It should be noted, however, that just as maintaining hydrated membranes through water management is required for the lower temperature PFSA systems, in an analogous way the higher temperature PBI-based systems require "electrolyte management," and in this sense they cannot be run absolutely dry. The reactants need to be humidified to maintain phosphoric acid concentrations within reasonable limits in order to maintain protonic conductivity and mechanical integrity. There are limits on the amount of acid loss and maximum operating temperature. Hence, although these systems have the potential for a simpler balance of plant, avoidance of failure modes due to acid loss and chemi-

Fig. 37.11 Plug Power HTPEM system.

cal and mechanical degradation of the membrane over time require careful system design. Furthermore, stress relaxation of the membrane must be accommodated in the stack design (Surorov et al., 2008).

Models of high-temperature systems have been developed and exercised against heating and electrical demands. For instance, Korsgaard et al. (2008a,b) proposed a novel HTPEM system consisting of a natural gas steam reformer, including a water gas shift reactor for improved system performance, integrated into a system with an anode tail gas burner, which is used to capture heat for the reforming process. In this case, the fuel cell stack is placed inside a liquid reservoir, where it is cooled by natural convection. The control of this system was simulated using a Matlab® Simulink model. The model was subjected to the consumption pattern of 25 Danish single family houses with measurements of heat, power, and hot water needs. Three scenarios were analyzed ranging from heat following only (with grid compensation for electricity) to heat and power following with net export of electricity during high and peak load hours. Average electrical efficiencies were above 40% and total efficiencies above 90% were obtained, both on an LHV basis.

In examining the simplification enabled by the HTPEM system, it is clear that the cost shifts more towards the stack. In an LTPEM system, costs usually split roughly equally between stack, reformer, and the remaining balance of plant, including controls and electronics. In HTPEM systems, a greater percentage of the system cost is born by the stack. In this case, the sluggish ORR kinetics add to this cost base in that the stack needs to have a larger active area in order to compensate for the increased activation losses with the ORR (Neyerlin et al., 2008). In addition, platinum loadings tend to be somewhat higher than in LTPEM systems, placing an additional burden on the need for cost-effective stack and system designs. Although there is considerable experience with phosphoric acid systems, there is far less field experience with the phosphoric acid-doped PBI type of systems, as these are currently in a pre-commercial state of development. Consequently, there is considerable learning that needs to occur with regard to longer term failure modes, especially around the MEA, stack, and its components. It is clear, however, that the HTPEM systems offer a distinct advantage over the LTPEM systems for residential and light commercial applications.

37.5.4
Solid Oxide Fuel Cell Systems

SOFCs have a well-known set of attributes that make them potential winners in the pursuit of low-cost fuel cell systems aimed at stationary applications. As already mentioned, SOFC materials become ionic conductors at high temperatures, enabling high electrical efficiency. By virtue of their operation at high temperatures, they are tolerant to normal impurities that plague normal low-temperature fuel cells. This tolerance to impurities provides the basis for flexibility in fuels that can be used. Impurities such as CO in normal PEM fuel cells become fuels in SOFCs. One of the unique attributes of SOFCs is their ability to permit reforming internal to the stack. Hence they allow for the design of relatively simple systems, providing the potential for improved reliability and efficiency, and reduced system cost. Finally, they do not depend on the use of expensive precious metal catalysts, again providing the opportunity for low-cost stacks. These, combined with high-temperature waste heat, offer the possibility of extremely efficient, low-cost, multifunctional systems for a variety of applications, ranging from megawatt-scale systems for centralized power down to 1 kW systems for domestic μCHP applications.

Siemens has developed a 100 kW tubular SOFC system for distributed generation and Sulzer Hexis has deployed several hundred 1 kW units in Europe for residential μCHP applications. In these cases, cost and durability have been obstacles to commercial success. SOFCs for residential co-generation applications of are now being developed by Kyocera (Japan), Ceramic Fuel Cells (Australia), and Ceres Power (UK). Another interesting application is being pursued by Delphi (USA). Here the SOFC is being applied toward auxiliary power units (APUs) that provide power for transport trucks and recreational vehicles.

SOFCs are not without their challenges, however. Formidable materials challenges are brought on by the need to operate at high temperatures, including component sealing, failures due to excessive thermal stresses associated with start-up and shut-down, and long-term materials stability. Accordingly, significant research is under way to optimize the materials sets for the porous anode and cathode electrodes, metallic interconnects, and nonporous electrolytes. Attempts are also being made by relieving some of the constraints through novel cell architectures. Whereas most SOFCs are now planar and "anode supported,"

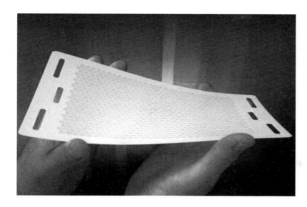

Fig. 37.12 NexTech SOFC design.

other possibilities are being examined in an effort to alleviate the materials-related problems.

An interesting and novel SOFC architecture is being pursued by NexTech. The NexTech design is an electrolyte-supported system, which allows for very thin anode materials. Most planar cells are anode supported, which results in very thick anode catalyst layers in comparison to NexTech's design. These thicker anode catalyst layers result in brittle cells that are fragile. In contrast, NexTech's cells are somewhat flexible (Figure 37.12). It is felt that this level of flexibility will allow higher yields, more robust manufacturing methods, and more durable stacks. The thinner anode catalyst layer allows for reduced resistance to the flow of fuel into the catalyst layer. Since water (steam) is formed on the anode in SOFCs that conduct oxygen ions, thick electrodes hamper the diffusion of fuels flowing against the escaping steam. With thin anode electrodes, this mass transfer limitation is reduced, allowing for higher fuel utilization and increased power density. As a side benefit, the thinner anode has more tolerance towards redox cycling. In addition, the fact that the anode material is not used as the support material provides an additional degree of freedom in anode material design flexibility. This allows, for example, the selection of materials that have improved sulfur tolerance. Another unique attribute of the "FlexCell" technology is that the active electrolyte layer in the repeat units is actually fairly thin, again resulting in high performance due to higher conductivity. Finally, the dense perimeter allows for increased latitude in design of seals.

Worldwide Energy, on the other, hand is pursuing a cell architecture in which the anode, cathode, and electrolyte materials are protected by porous metal tubular interconnects. This system is seen to be very rugged with respect to both mechanical and thermal stresses, easy to seal, and inexpensive to manufacture. Mass transport losses occur at high current densities, but the system, which is being developed with several of the DOE's national laboratories, is in the early stages of development. Current efforts are aimed at materials optimization.

It is important to note that the materials issues around SOFCs are nontrivial. Hence, whereas ideally the attributes of SOFC systems make them prime candidates for µCHP applications involving buildings, especially where there is a constant base load so that there few starts and stops, the materials issues at the higher temperatures represent an increased level of risk relating to reliability. The SOFC devices are fundamentally different in that the mechanism involves conduction of oxygen ions. As such, the performance of these systems depends critically on the microstructure of the electrode/electrolyte interfaces. The current state of the art involves lanthanum strontium manganite (LSM) cathode materials on YSZ electrolyte. However, the electrochemical performance of such devices is not well understood. Consequently, much effort is now being focused on understanding the influence of the detailed microstructure on the electrochemical behavior of SOFCs (Smith *et al.*, 2009) Of course, the well-known effects of chrome poisoning from stack and reformer components need to be accommodated in the selection of materials (Singhal and Kendall, 2003). The incorporation of the reformer into the stack, a well-known potential advantage of the SOFCs, although significantly simplifying the system, adds to the complexity of the fuel cell materials, as now additional constraints must be imposed on their selection.

Braun *et al.* (2006) reported on the evaluation, through simulation, of several different types of configurations for SOFC-based µCHP systems in residential applications. In this case, they considered an anode-supported planar fuel cell geometry, and found that maximum efficiency is achieved when cathode and anode gas recirculation is used along with internal reforming of methane. System electric efficiencies of 40% (HHV) and CHP efficiencies of nearly 80% were achieved. Their analysis indicates that the heat loss from small-scale SOFC systems can have an adverse effect on CHP efficiencies. Interestingly, they found that methane-based systems were more efficient than those running directly on hydrogen, mainly due to the contribution from the thermal energy in the recirculated exhaust gas. Similar conclusions were drawn by Kazempoor *et al.* (2009) in their development of a full system model of an SOFC system.

37.6
System Control

Methods of controlling low-temperature PEMFC systems have been reported by Pukrushpan *et al.* (2005). The authors described in detail, using model-based control schemes, the control of air and fuel to meet dynamic load profiles found in automotive applications. They also discussed the case when the fuel processing system is coupled to the fuel cell system and the additional complexities that arise. Tanrioven and Alam (2006) reported the results of an effort to model control (voltage and active power control) and simulate a PEMFC-based power supply system for residential applications. Using a Matlab software package, they were able to show that the voltage at the load point and active power demand versus load variations could be achieved with a proportional–integral-type controller.

The control of a reformer-based fuel cell system for µCHP applications is very complex and nontrivial. For reformer-based systems, the control system needs to deal with tightly coupled interacting subsystems with widely varying time constants, yet integrate these into a system that can respond to varying load profiles, both heat and electric, over a wide range of environments. In addition, there is minimum direct sensing of critical parameters, and in many cases the state of the system has to be inferred. To overcome these challenges, very often advanced controls need to be designed and implemented. These include enhanced classical proportional integral and derivative (PID) controls, adaptive model predictive control (MPC), observer-based control state estimation, and other optimum control schemes. Frequently used control algorithms, such as PID and MPC, are generally implemented with generic functions (coding structure) that can be easily configured and called into service.

Very often system operations, such as start-ups, stops, and transients, are controlled with "state machines" that govern and control transitions from one state to another. The state machine is a systematic way to gather information, process the information, make decisions, and execute. The state machine is systematic and transparent, which makes it easier to optimize machine operation, start-up, and shut-down sequences. The controls and software are ideally structured in modules with a multi-layer hierarchy. Control functions of each layer can then be well defined and coordinated through system-level control algorithms. This provides a basic framework for global optimization. The module or lower level functions can be coordinated with system/supervisory level control functions such as *efficiency* and *system health* (especially stack) controls. This permits the global optimization of the overall system performance and proactively prevents potential issues that the system may run into otherwise. The system-level controls also dynamically adapt themselves to achieve the best performance with the current system (aging effect) and operating (ambient and load) conditions.

The control systems should be supplemented with on-board diagnostics to assist troubleshooting, to analyze system degradation, and to calculate system/component life statistics. This significantly shortens the troubleshooting process, reduces service cost, and prompts timely preventive maintenance. The diagnostics performed at the start-up of the system prevent catastrophic failures of modules, thereby reducing replacement costs and increasing the machine availability. The automatic start-up features also improve the machine availability and utilization, especially if web-based remote diagnostics are enabled. Plug Power reformer-based systems have, for example, been designed with a remote diagnostics capability and have been able to feed back vital performance data and statistics for reliability engineers to analyze fuel cell performance in real time. This capability is invaluable not only during the development stages of a product, but also as a means of avoiding unnecessary service calls in the mature product offering.

37.7
Conclusion

From discussing each of the candidate fuel cell system types for building applications, it is clear that each has its strengths and weaknesses. LTPEM systems have the positive attributes of rapid load cycling and shorter warm-up times as compared with HTPEM and SOFC systems. SOFC systems, on the other hand, have relative advantages in the areas of electrical efficiency, heat quality, system complexity, and fuel flexibility in comparison with both LTPEM and HTPEM systems. HTPEM systems offer some of the attributes that address the shortcomings of both the SOFC systems (start/stop cycling, start-up time, rapid load cycling) and the LTPEM systems (system costs and complexity, tolerance to impurities). It would appear that if the failure modes intrinsic to the HTPEM system can be avoided, and if the cost savings due to system simplification can compensate for the increase in stack size required due to reduced ORR kinetics, then the HTPEM system would be the preferred system for μCHP applications. However, if the SOFC system can be made reliable and its shortcomings with regard to start/stops and load cycling can be avoided through appropriate base load sizing, then the SOFC technology would be superior. It is therefore important, when comparing fuel cell system technologies, to do so in the context of the specific application.

No fuel cell technology is currently commercially viable for use in building applications, other than those used in large-scale applications. This is because cost, complexity, and reliability do not scale linearly with system size. These obstacles in cost, complexity, and reliability must be removed before there can be widespread use of fuel cells in building applications – and they will. The economic and environmental benefits of applying fuel cells to the decentralized task of providing heat and electricity for buildings are environmentally significant and economically compelling. It is just a matter of time. Our rapidly growing understanding of the underlying electrochemical and electromechanical processes at the nanoscale will lead to the necessary improvements needed to serve the built environment.

References

Adachi, H., Ahmed, S., Lee, S. H. D., Papadias, D., Ahluwaia, R. K., Bendert, J. C., Kanner, S. A., and Yamazaki, Y. (2009) A natural gas fuel processor for a residential fuel cell system, *J. Power Sources*, **168**, 244–255.

Adzic, R. R., Zhang, J., Sasaki, K., Vukmirovic, M. B., Shao, M., Wang, J. X., Nilekar, A. U., Mavrikakis, M., Valerio, J. A., and Uribe, F. (2007) Platinum monolayer fuel cell electrocatalysis. *Top. Catal.*, **46**, 249.

Barber, F. (2005) *PEM Fuel Cells*, Elsevier, Amsterdam.

Braun, R. J., Klein, S. A., and Reindl, D. T. (2006) Evaluation of system configurations for solid oxide fuel cell-based micro-combined heat and power generation in residential applications. *J. Power Sources*, **158**, 1290–1305.

Chisholm, C., Boysen, D., Papendrew, A., Zecevic, S., Cha, S. Sasaki, K. A., Varga, A., Giapis, K. P., and Haile, S. M. (2009) From laboratory breakthrough to technological realization: the development path for solid acid fuel cells. *Electrochemical Society Interface*, **18** (3), 53–59.

Darling, R. M. and Meyers, J. P. (2003) Kinetic model of platinum dissolution in PEMFCs. *J. Electrochem. Soc.*, **150**, A1523–A1527.

Debe, M. K., Schmoedkel, A. K., Ventsrom, G. D., and Atanasoski, R. (2006) High voltage stability of nanostructured thin film catalysts for PEM fuel cells. *J. Power Sources*, **161**, 1002–1011.

Du, B., Guo, Q., Qi, Z., Mao, L., Pollard, R., and Elter, J. F. (2008) In *Materials for the Hydrogen Economy* (eds R. H. Jones and G. J. Thomas), CRC Press, Boca Raton, FL, pp. 251–308.

EIA (2008) *Emissions of Greenhouse Gases*. DOE/EIA Report 0573, Energy Information Administration, Washington, DC, 2007; released 3 December 2008; available at http://www.eia.doe.gov/oiaf/1605/ggrpt/index.html (accessed 1 November 2009).

Feitelberg, A. S. and Rohr, D. F. Jr (2005) Operating line analysis of fuel processors for PEM fuel cell systems. *Int. J. Hydrogen Energy*, **30**, 1251–1257.

Feitelberg, A. S., Stathopoulos, J., Qi, Z., Smith, C., and Elter, J. F. (2005) Reliability of Plug Power GenSys™ Fuel Cell Systems. *J. Power Sources*, **147**, 203–207.

Gasteiger, H., Kocha, S. S., Sompalli, B., and Wagner, F. T. (2005) Activity benchmarks and requirements for Pt, Pt-alloy and non-Pt oxygen reduction catalysts for PEMFCs. *Appl. Catal. B*, **56**, 9–35.

Gigliucci, G., Petruzzi, L., Cerelli, E., Garzisi, A., and LaMendola, A. (2004) Demonstration of residential CHP system based on PEM fuel cells. *J. Power Sources*, **131**, 62–82.

Gourdoupi, N., Andreopoulou, A.K, Deimede, V., and Kallitsis, J. K. (2003) Novel proton-conducting polyelectrolyte composed of an aromatic polyether containing main-chain pyridine units for fuel cell applications. *Chem. Mater.*, **15**, 5044–5050.

Hawkes, A. D. and Leach, M. A. (2007) Cost effective operating strategy for residential micro-combined heat and power. *Energy*, **32**, 711–723.

Hawkes, A. D. and Leach, M. A. (2008) On policy instruments for support of micro combined heat and power. *Energy Policy*, **36**, 297–282.

Ioroi, T., Siroma, Z., Fujiwara, N., Yamazaki, S., and Yasuda, K. (2005) New synthesis and electrochemical oxygen reduction of platinum nanoparticles supported on mesoporous TiO_2. *Electrochem. Commun.*, **7**, 183–188.

Kallitsis, J. K., Geormezi, M., and Neophytides, S. (2009) Polymer electrolyte membranes for high-temperature fuel cells based on aromatic polyethers bearing pyridine units. *Polym. Int.*, **58**, 1226–1233.

Karger, C. R. and Bongartz, R. (2008) External determinant for the adoption of stationary fuel cells – infrastructure and policy issues. *Energy Policy*, **26**, 798–810.

Kazempoor, P., Dorer, V., and Ommi, F. (2009) Evaluation of hydrogen and methane-fueled solid oxide fuel cell systems for residential applications: system design alternative and parameter study. *Int. J. Hydrogen Energy*, **34**, 8630–8644.

Kitchen, J. R., Nørskov, J. K., Barteau, M. A., and Chen, J. G. (2004) Modification of the surface electronic and chemical properties of Pt(111) by subsurface 3d transition metals. *J. Chem. Phys.*, **120**, 10240.

Koper, M. T. M. (ed.) (2009) *Fuel Cell Catalysis*, John Wiley & Sons, Inc., New York.

Korsgaard, A. R., Nielson, M. P., and Kr, S. K. (2008a) Part one: a novel model of HTPEM-based micro-combined heat and power fuel cell system. *Int. J. Hydrogen Energy*, **33**, 1909–1920.

Korsgaard, A. R., Nielson, M. P., and Kr, S. K. (2008b) Part two: control of a novel HTPEM-based micro combined heat and power fuel cell system. *Int. J. Hydrogen Energy*, **33**, 1921–1931.

Markovic, N., Gasteiger, H., and Ross, P. N. (1997) Kinetics of oxygen reduction on Pt (hkl) electrodes: implications of the crystalline size effect with supported Pt electrocatalysts. *J. Electrochem. Soc.*, **144**, 1591–1597.

Matics, J. and Krost, G. (2008) Micro combined heat and power home supply: prospective and adaptive management achieved by computational intelligence techniques. *Appl. Thermal Eng.*, **28**, 2055–2061.

Mavrikakis, M., Hammer B., and Norskov J. K. (1998) Effect of strain on the reactivity of metal surfaces. *Phys. Rev. Lett.*, **81**, 2819–2820.

Meyers, J. P. and Darling, R. M. (2006) Model of carbon corrosion in PEM fuel cells. *J. Electrochem. Soc.*, **153** (8), A1432–A1442.

Neyerlin, K. C., Singh, A., and Chu, D. (2008) Kinetic characterization of a Pt–Ni/C catalyst with a phosphoric acid doped PBI membrane in a proton exchange membrane fuel cell. *J. Power Sources*, **176**, 112–117.

Nørskov, J. K., Rossmeisl, J., Logadottir, A., Lindqvist, L., Kitchin, J. R., Bligaard, T., and Jónsson, H. (2004) Origin of the overpotential for oxygen reduction at a fuel cell cathode. *J. Phys. Chem. B*, **108**, 17886–17892.

Peacock, A. D. and Newborough, M. (2006) Effect of micro-combined heat and power systems on energy flows in the UK electricity supply industry. *Energy*, **31**, 1804–1818.

Peacock, A. D. and Newborough, M. (2008) Effect of heat saving measures on the CO_2 savings attributable to micro-combined heat and power (CHP) systems in UK dwellings. *Energy*, **33**, 601–612.

Pehnt, M. (2008) Environmental impacts of distributed energy systems-the case of micro generation. *Environ. Sci. Policy*, **11**, 25–37.

Pukrushpan, J. T., Stefanopolou, A. G., and Peng, H. (2005) *Control of Fuel Cell Power Systems*, Springer, London.

Radulescu, M., Lottin, O., Feidt, M., Lombard, C., LeNoc, D., and LeDoze, S. (2006) Experimental and theoretical analysis of the operation of natural gas cogeneration system using a polymer exchange membrane fuel cell. *Chem. Eng. Sci.*, **61**, 743–752.

Reiser, C. A., Bregoli, L., Patterson, T. W., Yi, J. S., Yang, J. D., Perry, M. L., and Jorvi, T. D. (2005) A reverse-current decay mechanism for fuel cells. *Electrochem. Solid-State Lett.*, **8**, A273.

Ren, H., Gao, W., and Ruan, Y. (2008) Optimum sizing for residential CHP system. *Appl. Thermal Eng.*, **28**, 514–523.

Schmidt, T. and Baurmeister, J. (2008) Properties of high temperature PEFC Celtec® P 1000 MEAs in start/stop operation mode. *J. Power Sources*, **176**, 428–434.

Shah, A. A., Ralph, T. R., and Walsh, F. C. (2009) Modeling and simulation of the degradation of perfluorinated ion-exchange membranes in PEM fuel cells, *J. Electrochem. Soc.*, **156**, B465–B484.

Singhal, S. and Kendall, K. (eds) (2003) *Solid Oxide Fuel Cells*, Elsevier, Amsterdam.

Smith, J. R., Chen, A., Glostovic, D., Hickey, D., Kundinger, D., Duncan, K. L., deHoff, J., Jones, E. K. S., and Wachman, E. D. (2009) Evaluation of the relationship between cathode microstructure and electrochemical behavior of SOFCs. *Solid State Ionics*, **180**, 90–98.

Spohr, E. (2004) Molecular dynamics simulation of proton transfer in a model Nafion pore. *Mol. Simul.*, **30**, 107–115.

Srinivasan, S. (2006) *Fuel Cells: from Fundamentals to Applications*, Springer, Berlin.

Staffell, I., Green, R., and Kendall, K. (2008) Cost targets for domestic fuel cell CHP. *J. Power Sources*, **181**, 339–349.

Stamenkovic, V., Mun, B. S., Mayrhofer, K. J. J., Ross, P. N., Markovic, N. M., Rossmeisl, J., Greely, J., and Nørskov, J. (2006) Changing the activity of electrocatalyst for oxygen reduction by tuning the surface electronic structure. *Angew Chem. Int. Ed.*, **45**, 2897.

Suvorov, A. P., Elter, J., Staudt, R., Hamm, R., Tudryn, G., Schadler, L., and Eisman, G. (2008) Stress relaxation of PBI based membrane electrode assemblies. *Int. J. Solids Struct.*, **45**, 5987.

Tang, H., Qi, Z., Ramani, R., and Elter, J. F. (2006) PEM fuel cell cathode carbon corrosion due to the formation of air/fuel boundary at the anode. *J. Power Sources*, **158**, 1306–1312.

Tanrioven, M. and Alam, M. S. (2006) Modeling, control and power quality evaluation of a PEM fuel cell-based power supply system for residential use. *IEEE Trans. Ind. Appl.*, **42**, 1582–1589.

Taylor, C. D., Wasileski, S. A., Filhol, J. S., and Neurock, M. (2006) First principles modeling of the electrochemical interface: consideration and calculation of a tunable surface potential from atomic and electronic structure. *Phys. Rev. B*, **73**, 65402.

Vielstich, W., Lamm, A., and Gasteiger, H. A. (eds) (2004) *Handbook of Fuel Cells*, John Wiley & Sons, Ltd, Chichester.

Wang, C. Y. (2004) Fundamental models for fuel cell engineering. *Chem. Rev.*, **104**, 4727–4766.

Wang, X., Li, W., Chen, Z., Waje, M., and Yan, Y. (2006) Durability investigation of carbon nanotube as catalyst support for proton exchange membrane fuel cell. *J. Power Sources*, **158**, 154–159.

Weber, A. Z. and Newman, J. (2004) Modeling transport in polymer-electrolyte fuel cells. *Chem. Rev.*, **104**, 4679–4726.

Transportation Applications

38
Fuel Cell Power Trains

Peter Froeschle and Jörg Wind

Abstract

The world is facing severe challenges due to the limitation of fossil energy resources and rising green-house-gas concentration in the atmosphere. Hence the automotive industry is searching for new concepts to fulfil the requirements of the transport sector for the future. One of the most promising approaches for future road transport is the electric vehicle featuring the fuel cell system technology in combination with an electric engine. With this technology most disadvantages of vehicles with an internal combustion engine can be solved without bigger disadvantages concerning driving performance, range and price. Therefore this technology has medium- and long-term the potential to achieve a significant market share. After several decades of developing prototypes and test vehicles of fuel cell vehicles, the automotive manufacturers are now on the verge of commercialize this seminal technology. In the last years a couple of OEMs introduced new generations of fuel cell vehicles, produced in small series, with the aim to validate the technology and to improve the customer acceptance. The aim of the article "Fuel Cell Power Trains" is to describe the layout and functionality of the fuel cell hybrid power train as well as to overview the involved OEMs with their current fuel cell vehicles. Thereby is shown the application of the fuel cell technology not only in passenger cars but also in buses and delivery vehicles.

Keywords: fuel cells, power train, hybrid power train, hydrogen fuel cell, electric-powered vehicles

38.1
Introduction

The world is facing severe challenges due to the limitation of fossil energy resources and rising greenhouse-gas concentrations in the atmosphere. All energy-consuming sectors have to contribute to a reduction of energy consumption and greenhouse gas emissions. Therefore, the automotive industry is working on new drive trains for a sustainable mobility of the future. Due to the changing basic conditions and regulations concerning environmental pollution, limited resources, and higher prices of fossil fuels, the automotive industry is searching for new concepts to fulfill the requirements of the transport sector for the future. One of the most promising approaches for future road transport is the electric vehicle featuring fuel cell system technology in combination with an electric engine. With this technology, most disadvantages of vehicles with a conventional combustion engine can be solved: electric drive trains, especially when the electricity is supplied by a fuel cell system, offer a number of benefits [1, 2]:

- Electric drive trains are the most efficient power trains; in combination with a fuel cell system, they offer twice the efficiency of internal combustion engines (see Figure 38.1).
- Electric drive trains offer a very high torque even at the lowest speed.
- Electric drive trains have no local emissions.
- Electric drive trains are very quiet.
- Use of a fuel cell system offers a longer driving range and very short refueling times compared with electric vehicles with batteries as electricity supply.
- Of all possible fuels for fuel cell vehicles, hydrogen is considered the most suitable.
- Hydrogen can be produced from a large variety of primary energy sources, including renewable energies.
- Thus the dependence of transport on limited resources such as fossil fuels can be overcome.
- Well-to-wheel analyses show that fuel cell vehicles with most pathways to produce hydrogen result in an overall reduction of energy consumption and greenhouse gas emissions compared with vehicles with internal combustion engines (see Figure 38.2).
- Hydrogen offers the possibility of storing renewable electricity by conversion of the electricity to hydrogen.

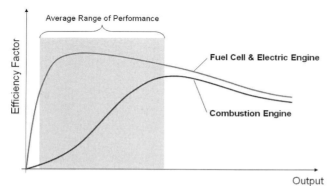

Fig. 38.1 Efficiency of fuel cell power trains compared with combustion engines. Courtesy of Daimler AG.

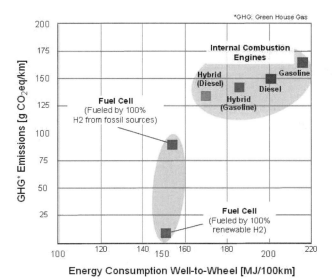

Fig. 38.2 Total energy balance – well-to-wheel classification. Courtesy of Daimler AG.

In order to use these advantages over conventional combustion engines, various automotive manufacturers started to develop fuel cell vehicles at the end of the 20th century. In the meantime, multiple concept cars, prototypes, and test vehicles have been developed and tested. This resulted in major development steps and a high maturity and market readiness of fuel cell vehicles. In addition to the application of the fuel cell drive train in passenger cars, the technology has also been introduced in buses and small trucks.

After nearly 15 years of research, development, and testing, a couple of automotive manufacturers are now on the brink of producing and selling fuel cell vehicles commercially. However, in spite of all the technological advances, there are still some challenges left, especially the necessary cost reduction of fuel cell technology and the creation of a sufficient hydrogen infrastructure.

38.2
Layout and Functionality of the Fuel Cell Hybrid Power Train

The fuel cell hybrid power train differs significantly from a power train with an internal combustion engine. The components are described in detail in Sections 38.2.1–38.2.4. The layout of a fuel cell hybrid power train is shown in Figure 38.3.

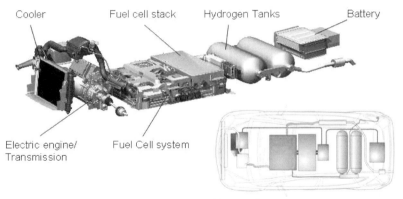

Fig. 38.3 The most important components and layout of a fuel cell hybrid power train. Courtesy of Daimler AG.

38.2.1
Fuel Cell System

The heart of the drive train is the fuel cell system, the most important components of which are the fuel cell stack, the air supply system, the humidifier, and the anode supply (hydrogen). A power management system establishes an electric connection between the fuel cell system and the vehicle. Via a corresponding converter, it provides high-voltage components such as the electric engine and the traction battery, and also electric ancillary units.

The main component of the fuel cell system is the stack, which provides the electricity for the electric drive of the car through the reaction of hydrogen and oxygen. The hydrogen required is stored in the hydrogen tanks, while the oxygen is taken from the ambient air. The stack is composed of up to several hundred electrolyte–electrode units [called a membrane electrode assembly (MEA)] which are electrically connected in series by gas-tight separator plates (also called bipolar plates). In the case of the proton exchange membrane (PEM) fuel cell, a very fine polymer membrane (20–50 m thickness) is used as the electrolyte. The membrane separates hydrogen and oxygen gases. It is covered on both sides by a very thin platinum coating, acting as a catalyst. It dissociates the hydrogen into positively charged protons and negatively charged electrons. The protons pass through the membrane and react with the oxygen, forming water. Electrons, however, cannot travel through the membrane and are held back. The surplus of electrons on the hydrogen side and the corresponding deficit on the oxygen side result in the formation of a cathode and an anode. Connecting the two creates an electric current, which is utilized to power the car [3]. The design and functionality of a PEM fuel cell is shown in Figure 38.4.

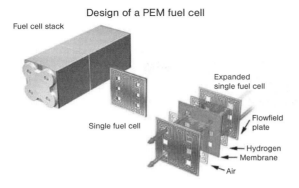
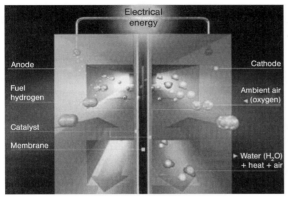

Fig. 38.4 Design and functionality of a PEM fuel cell. Courtesy of Ballard Power Systems.

38.2.2
Hydrogen Storage System

Hydrogen storage systems must satisfy many requirements, such as safety aspects, weight, dimensions, durability, costs, refill handling, recycling of the used materials, and robustness against mechanical strength. All these requirements must be considered during development.

Several technologies for storing hydrogen have been developed. Storage in pressure tanks or as liquid hydrogen is the most advanced. For road transport application, storage as compressed hydrogen is the most promising method. Modern fiber compound tanks realize pressures of up to 700 bar (see Figure 38.5). The mid- and long-term targets are improvement of the storage characteristics, installation of international standards and cost reduction of the tank systems [4].

Fig. 38.5 Diagram of hydrogen storage and a tank-filling system. Courtesy of Daimler AG and Daimler Global Media Site, http://media.daimler.com/.

38.2.3
Battery

Combination of a fuel cell system with a small battery offers several advantages. This vehicle concept is called "fuel cell hybrid drive train." Mainly used for this application are nickel-metal hydride or lithium-ion batteries. These types of accumulators currently have the best specific power and energy density.

The additional traction battery adds complexity to the drive train; however, the advantages prevail: only with a battery is it possible to store recuperated braking energy and the electric engine can be supported with additional power during acceleration, and furthermore it is possible to assemble a smaller fuel cell unit [5].

38.2.4
Electric Engine

Today, electric engines exist as different types and are optimized for application in road vehicles. They are ideal for road vehicles due to their very good rotation speed/torque characteristics. Further advantages are the high efficiency, the low noise emission, and the high torque which is available across nearly the entire rotational speed range (also by poor rotation speed). In addition, a multilevel transmission is normally not necessary.

Requirements on electric engines are as follows [6]:
- compact construction
- high efficiency
- low weight or rather high power density
- low noise emission
- low price
- controllability over a wide range of torque and rotation speed
- low maintenance requirements.

38.3
Technological Leaders of Fuel Cell Drive Train Development

38.3.1
Passenger Cars

The application of fuel cell systems for the propulsion of passenger cars is the most advanced concerning series production. Several automotive manufacturers are developing fuel cell vehicles together with their system suppliers.

In September 2009, the leading vehicle manufacturers in fuel cell technology (Ford Motor Company, General Motors Corporation/Opel, Honda Motor Co., Ltd, Hyundai Motor Company, Kia Motors Corporation, the alliance of Renault SA and Nissan Motor Co., Ltd, Toyota Motor Corporation, and Daimler AG) made a joint statement regarding the development and market introduction of electric vehicles with fuel cells through a Letter of Understanding. The signing automobile manufacturers strongly anticipate that from 2015 onwards a significant number of electric vehicles with fuel cells could be commercialized. This number is estimated to be a few hundred thousand units worldwide. In the following sections the automotive manufacturers with the highest development level are described with their current fuel cell models [7].

38.3.1.1 Daimler AG
With the Necar1, Daimler presented its first fuel cell vehicle in 1994 and was one of the first automotive manufacturers to develop an electric vehicle with the complex and challenging fuel cell technology. Over the next few years, numerous concept cars and prototypes were built to prove the feasibility of the fuel

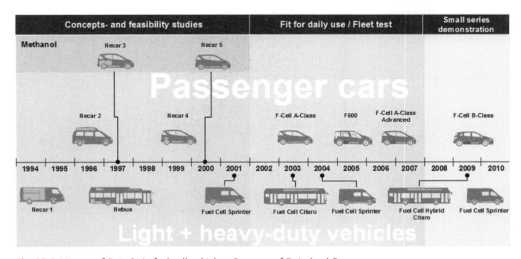

Fig. 38.6 History of Daimler's fuel cell vehicles. Courtesy of Daimler AG.

cell technology. The history of Daimler's different generations of fuel cell buses, vans and passenger cars is illustrated in Figure 38.6.

At the beginning of 2003, the first fleet of fuel cell cars was built and, for the first time, field tested in everyday use by customers. So far, the different prototypes and test vehicles (passenger cars, buses, and vans) have accumulated more than 4 000 000 km, which makes Daimler one of the automotive manufacturers with the greatest experience of fuel cell technology.

In September 2009, Daimler presented their latest generation of fuel cell passenger cars at the International Motor Show (IAA) in Frankfurt. The B-Class F-CELL (shown in Figure 38.7) is based on the B-class series passenger car and will be produced in a small series. Selected customers in Europe and the USA will have the possibility of driving the vehicles from 2010.

The special components of fuel cell technology have been placed in the sandwich under floor of the B-Class car so that customers will have no limitations concerning the interior space (Figure 38.8). Further improvements of the B-Class F-CELL are as follows:

Fig. 38.7 The B-Class F-CELL, the latest generation of Daimler's fuel cell passenger cars. Courtesy of Daimler Global Media Site, http://media.daimler.com/.

 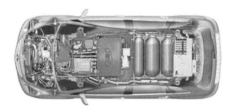

Fig. 38.8 The packaging of the B-Class F-CELL. Courtesy of Daimler Global Media Site, http://media.daimler.com/.

- Cold-start feasibility down to −15 °C.
- A higher range due to the improved tank system (higher tank volume and 700 bar pressure instead of 350 bar) and higher efficiency of the fuel cell system.
- Improved battery output with the new lithium-ion technology.
- The system module concept of the fuel cell system, which allows damaged parts to be changed very easily and quickly and simplifies the packaging when adding additional components.
- The fuel cell stack, which has a longer lifetime and better power performance and has become lighter than its predecessor.
- The optimization of the software by incorporating start/stop strategy designs to disconnect the fuel cell stack during the idle phase.

38.3.1.2 General Motors Company

General Motors (GM) began research and development of fuel cell technology in the 1960s. The GM Electrovan was the world's first fuel cell vehicle, presented by GM in 1966 in Detroit (see Figure 38.9). However, only the development of the PEM fuel cell in the 1990s allowed for more powerful fuel cell systems.

After several different concept cars, prototypes, and test vehicles, GM presented the Chevrolet Equinox Fuel Cell as their latest generation fuel cell car in 2007. The most important components of the power train are the new fuel cell system, the 700 bar hydrogen storage system, the 35 kW nickel-metal hydride battery pack, and the electric engine with a maximum power output of 94 kW and a maximum torque of 320 Nm. The new and improved drive train allows for a range of more than 320 km, a top speed of 160 km h^{-1}, and acceleration from 0 to 100 km h^{-1} in less than 12 s.

More than 100 Equinox Fuel Cell vehicles are currently in operation in the United States (Los Angeles, New York, and Washington, DC), Japan, Korea, China, and Germany (Berlin) (Figure 38.10). In Europe, the vehicles are represented by Opel as the HydroGen4. GM's global fuel cell vehicle fleet takes part in different real market test activities such as the Department of Energy program in the United States and the Clean Energy Partnership (CEP) in Germany.

Fig. 38.9 First and current generation of GM's fuel cell vehicles. Courtesy of GM Media Site, http://media.gm.com/.

Fig. 38.10 A test fleet of over 100 Chevrolet Equinox Fuel Cell vehicles is currently in operation. Courtesy of GM Media Online Opel Germany, http://archives.media.gm.com/de/opel/de/.

Fig. 38.11 The engineers of the Honda FCX Clarity had the possibility to design a totally new car around the fuel cell power train. Courtesy of Honda Media Information FCX Clarity, http://www.hondazeroemissions.eu/, and Honda Media Newsroom, http://www.hondanews.com/.

In addition to corporate customers, private individuals have the possibility to apply for an Equinox Fuel Cell as part of GM's project "Driveway." So far, more than 80 000 people have applied for this opportunity to test one of the zero emission vehicles. After 2 years of operation, the entire fleet has accumulated more than 1 000 000 miles [8].

38.3.1.3 Honda Motor Co. Ltd

After developing different prototypes and test vehicles, Honda presented the Honda FCX in October 2002. Over the next few years, 35 units of this fuel cell electric vehicle were leased to special customers who operated them under different conditions in Japan and the United States.

After this successful test, Honda presented the Honda FCX Clarity, which was produced in a small series in June 2008. This car is also the first vehicle with a fuel cell power train which has its own purpose design (Figure 38.11).

About 20 vehicles are operated for a monthly lease fee in the United States and Japan by selected partners. Due to the delay in the establishment of a hydrogen infrastructure (such as hydrogen fueling stations), Honda is also been in developing decentralized hydrogen generation technologies. One is the Home Energy Station, which generates hydrogen from natural gas to supply hydrogen to fuel cell vehicles at home. The other is the solar powered hydrogen station, which uses Honda's own solar cells to generate electricity to split water into hydrogen and oxygen via electrolysis.

A technical highlight is the V Flow fuel cell stack, which is mounted vertically in the centre tunnel using gravity to facilitate more efficient drainage of the water by-product. The result is better fuel cell performance and a smaller and lighter stack. With the introduction of the new fuel cell stack, which has improved output/volume density by 50% and output/weight density by 67%, the FCX Clarity has a 20% higher fuel efficiency and a 30% longer operating range than the previous FCX model [9].

38.3.1.4 Hyundai Kia Automotive Group

The Hyundai Kia Automotive Group has been developing fuel cell vehicles since the 1990s. After the first generation of fuel cell vehicles based on the Santa Fee presented in 2001, Hyundai introduced the improved Tucson FCEV in 2004. This sport utility vehicle (SUV) is equipped with a UTC fuel cell and a high-voltage battery which supply the 80 kW electric engine with the necessary energy. The hydrogen storage tanks with a volume of 152 l allow a cruising range of about 300 km. From 2004 to 2009, over 30 Tucson FCEVs participated within a demonstration programme of the United States Department of Energy and currently oversees pilot operations of 66 fuel cell electric vehicles in Korea and abroad.

In 2008, the next generation of fuel cell vehicles was presented with the Hyundai i-Blue. This futuristic vehicle has its own purpose design and is based on a crossover utility platform. Advances of the concept car are the new 100 kW fuel cell stack and a improved lithium-ion battery. The 115 l capacity of compressed hydrogen is sufficient for a range of 600 km [10, 11].

The automotive manufacturer Kia Motors, which belongs to the Hyundai Kia Automotive Group, has also developed fuel cell vehicles in cooperation with the Hyundai Motor Company. In 2008, Kia presented the Borrego FCEV at the Los Angeles International Auto Show. This SUV is equipped with a 115 kW fuel cell system, a new battery lithium-ion polymer, a 110 kW electric engine and a 700 bar tank system. The 202 l capacity of compressed hydrogen allows a range of over 650 km. With the use of lightweight aluminum body shell components, the Borrego FCEV has a power-to-weight ratio that is similar to that of the new production model SUV. Due to technical improvements, it was able to solve the problem of cold start feasibility, so that the new vehicle can start in temperatures down to −30 °C. In 2010, Kia will introduce a fleet of Borrego FCEVs [12].

38.3.1.5 Nissan-Renault Alliance

Nissan started the development of fuel cell technology in 1996 and since that time several prototypes and test vehicles have been introduced. In 2003, the X-Trail FCV was presented, equipped with a 63 kW stack, produced by UTC Fuel

Cells, a lithium-ion battery, 85 kW electric engine, and a 250 bar hydrogen storage system which allows a range of over 350 km. Several of the vehicles were commercialized for limited leasing from 2003.

Two years ago, Nissan and Renault formed an alliance to advance fuel cell technology. On the basis of this coalition, the partners developed the concept car Scenic ZEV H2 and presented it in 2008. Some of the specific components, such as the fuel cell stack, the hydrogen tanks, and the traction battery, were developed by Nissan and implemented in the Renault Scenic [13, 14].

In 2008, Nissan presented the next generation of fuel cell stack. The in-house development shows several advances compared with the previous generation. Higher durability electrode material results in a 50% reduction in the amount of platinum required. Mainly due to this improvement, the new stack achieves a 35% cost reduction. The combined enhancements in the cells resulted in double the power density, which permitted downsizing of the fuel cell stack size by one-third and significant cost reductions, without sacrificing performance. Compared with the previous generation, the power output of the new generation of stack has been increased 1.4-fold from 90 to 130 kW, which can power larger vehicles. The stack size has been reduced by 25% from 90 to 68 l, which allows for improved packaging flexibility. At the end of 2008, Nissan started to test the improved stack and also other advancements such as a 700 bar hydrogen storage system [15].

38.3.1.6 Toyota Motor Corporation

Like the other automotive manufacturers, Toyota started to develop vehicles with fuel cell technology in the early 1990s. Since then, they have developed several concept cars, prototypes, and test vehicles. Especially the FCHV series (Fuel Cell Hybrid Vehicle generation 1–5) demonstrates the improvements and advancements that could be achieved over the years.

The FCHV is based on the Toyota Highlander. The utilization of an SUV as the base has the advantage that there is more space available to accommodate the fuel cell-specific components such as the fuel cell system, traction battery, and hydrogen storage tanks.

The FCHV-adv (advanced) is the latest generation of Toyota's fuel cell vehicles and was presented in 2008 (Figure 38.12). With its new 700 bar hydrogen storage system, a more efficient fuel cell system, and improved energy management, the range of the FCHV-adv could be boosted to about 800 km. Therefore, it is the first fuel cell vehicle which is fully competitive in range compared with vehicles with conventional combustion engines.

Through optimization of the MEA design, the problems with cold-start feasibility could be solved, so that the Toyota FCHV-adv can start and operate in cold regions at temperatures as low as $-30\,°C$. Since 2002, selected customers in Japan and the United States have had the possibility to lease the FCHV [16].

38.3 Technological Leaders of Fuel Cell Drive Train Development

Fig. 38.12 The Toyota FCHV-adv, the latest generation of Toyota's fuel cell passenger cars. Courtesy of Toyota Presse Deutschland, http://www.toyota-media.de/.

Fig. 38.13 Ten of the new Citaro FuelCELL-Hybrid buses will be operated within the public transport system in Hamburg. Courtesy of Daimler AG and Daimler Global Media Site, http://media.daimler.com/.

38.3.2 Buses

The application of fuel cell technology in city buses offers good possibilities to reduce energy consumption and greenhouse gas emissions. As city buses run in urban centers and have many acceleration and brake cycles during their operation, the technology can demonstrate its full range of advantages within this vehicle type.

Daimler AG presented and operated a fleet of fuel cell buses in the European Union's successful CUTE and HyFLEET:CUTE projects, which were carried out from 2003 to 2009. In the projects, 36 Mercedes-Benz Citaro city buses equipped with a second-generation fuel cell system performed outstandingly for 12 public transport agencies on three continents. In more than 140 000 h of operation, during which they covered a total of more than 2.2×10^6 km, the fuel cell buses demonstrated their ability to function reliably under operating conditions. Their availability – between 90 and 95% – provided an impressive demonstration of the environmentally friendly fuel-cell drive's suitability for everyday use in regular-service urban buses.

The Citaro FuelCELL-Hybrid is the latest generation of fuel cell buses and was presented for the first time in Hamburg in November 2009 (Figure 38.13).

About 30 of the buses will be operated in fleet trials in Hamburg and other European metropolitan areas.

The first fuel cell hybrid bus from Mercedes-Benz is an innovative vehicle concept which combines elements of already proven Mercedes-Benz fuel cell buses, the diesel–electric Citaro G BlueTec Hybrid bus, and technical enhancements to form a new, trend-setting drive concept. In addition to the considerably increased service life of the fuel cells, which has now risen to at least 6 years or 12 000 h of operation, the service requirements of the fuel cell and also of the batteries and the electric motors have been reduced drastically, so that they are practically maintenance free for life.

The two fuel cell systems (60 kW each) of the third generation, which supply the two electric wheel hub engines (60 kW continuous/80 kW peak) with energy, are of the same type used in the B-Class F-CELL. Thanks to improved fuel cell components and hybridization with lithium-ion batteries, the Citaro FuelCELL-Hybrid consumes almost 50% less hydrogen than the preceding generation. The capacity of the batteries (27 kW h) allows the electric motors to be supplied with a constant 120 kW and is sufficient for the Citaro FuelCELL-Hybrid bus to run for several kilometers on battery power alone. The 35 kg of hydrogen, which are stored in 350 bar pressure cylinders on the roof of the bus, allows an operating range of about 250 km.

Three years after introducing its first fuel cell bus to the world, Hyundai Motor Company presented the second-generation hydrogen fuel cell bus at the Seoul Motor Show in 2009. Based on the most recent Low Floor Aero City bus platform, the vehicle accommodates 26 seats. The bus features numerous technical advances, including two parallel connected fuel cell systems with an output of 100 kW each, three 100 kW electric engines, and six 350 bar hydrogen tanks located on the roof of the bus, providing the bus with an operating range of 360 km (in city mode). The next step will come at the end of 2010 when a fleet of the second-generation buses will be operated in daily service in cities around Korea [17].

The Canadian bus manufacturer New Flyer Industries has been developing fuel cell buses since the early 1990s. In 1994, New Flyer Industries, along with Ballard Power Systems, developed one of the world's first hydrogen fuel cell buses. The latest generation of fuel cell hybrid bus, a 12 m low-floor bus, is equipped with a 130 kW heavy-duty fuel cell module from Ballard Power Systems. The 60 kg of compressed hydrogen, which is stored in a 350 bar tank system, and the energy out of a nickel-metal hydride battery are sufficient for a range of approximately 500 km. Twenty of the buses will be in operation during the Olympic Winter Games in 2010 in Vancouver and Whistler [18, 19].

In 2001, Hino Motors and its holding company, Toyota Motor Corporation, presented the first prototype of a fuel cell hybrid bus. Three buses of the latest generation were introduced in 2006 and are in operation on public routes around the Central Japan International Airport (Centrair) south of Nagoya. The bus, which has a capacity of over 60 passengers, was developed jointly by Toyota and Hino, using technology and know-how cultivated by both companies. Toyo-

ta developed the fuel cell hybrid system and Hino the vehicle body and other components of the bus. The seven 350 bar high-pressure tanks with a capacity of about 20 kg of hydrogen supply the two fuel cell systems, which have an output of 90 kW each. The fuel cell system together with the nickel-metal hydride battery supply the two 80 kW electric engines, which have a torque of 260 Nm each, with energy [20].

The bus manufacturer Van Hool NV produces, in cooperation with the fuel cell manufacturer UTC Power, fuel cell hybrid buses for the American and European markets. Since 2005–2006, the company has sold several fuel cell hybrid buses to the American market (California and Connecticut). In Europe, the first fuel cell hybrid bus was introduced in June 2007 by a Belgian transportation company.

The latest generation of fuel cell hybrid buses, a two-axle bus for the American market, is based on the Van Hool A300L and has a length of over 12 m and a passenger capacity of 28 seats. The bus is lighter than previous bus models, has a higher operating efficiency, and uses advanced lithium-ion battery systems. Hydrogen tanks on the roof of the bus provide a range of up to 300 miles. The bus contains a UTC Power Pure Motion Model 120 fuel cell power system with an output of 120 kW. Van Hool has orders for 16 of the new buses for use in the California and Connecticut urban transit fleets, which will be delivered from late 2009 through 2010 [21–23].

38.3.3 Delivery Vehicles

The application of fuel cell drive trains in trucks is, at the current state of the technology, reasonable only in light- and medium-duty commercial vehicles. Apart from a few studies that investigated the possibilities of integrating the fuel cell drive concept in heavy- and medium-duty transport applications, there is only one serious development of fuel cell hybrid trucks that has already been realized.

So far, Daimler AG has developed two generations of light-duty vehicles. The Mercedes-Benz and Dodge Sprinter were introduced in 2001 (first generation) and 2004 (second generation) and in everyday operations by international parcel services in Germany and the United States (see Figure 38.14). The second gen-

Fig. 38.14 Fuel cell hybrid Sprinter in operation by international parcel services. Courtesy of Daimler Global Media Site, http://media.daimler.com/.

eration, which has been in operation for 2 years, contained a 68 kW fuel cell system and a 20 kW nickel-metal hydride battery, which delivered the energy for the 80 kW electric engine. The 350 bar hydrogen storage system with a capacity of 5.6 kg of hydrogen allowed a range of over 150 km. During their operation, the three Sprinters that were introduced accumulated over 2400 h of operation and 64 000 km without any major problems.

At the end of 2010, one fuel cell hybrid Sprinter of the third generation will be introduced. In the test vehicle, the current fuel cell system, which is also contained in the B-Class F-CELL, will deliver the necessary energy for the electric engine.

38.4
Next Milestones on the Way to Commercialization

In addition to technical breakthroughs, which are needed for the commercialization of fuel cell vehicles, advances concerning the development of a hydrogen infrastructure are especially necessary. Without sufficient hydrogen filling stations, the commercialization of fuel cell vehicles is pointless. As home refueling units for local production and fueling of hydrogen are not competitive so far concerning price and sophistication, future customers will depend on a sufficient network of hydrogen filling stations.

For the development of the infrastructure, joint efforts of governments, automotive manufacturers, and especially the major oil and gas companies are necessary. Recently announced Letters of Understanding and first infrastructure projects partly funded by the German Government are a first step in the right direction, but the efforts must be intensified to achieve an area-wide hydrogen infrastructure in time for the planned commercialization of fuel cell vehicles in 2015.

More technical improvements must also be realized before the intended market entry. These improvements range from durability, efficiency, and cost/weight reduction to related codes and standards. The new technology will only be successful in the markets if they are fully competitive towards all customer needs. The main challenge in reaching this target is the price, especially for fuel cell systems, high-voltage batteries, power electronics, and hydrogen storage systems. Future cost reductions will be achieved through several measures:

- As fuel cell drive train and battery production increases, economies of scale will lower the average cost per system since fixed costs will be shared between an increased number of components.
- Technology advances can also contribute to reduce costs. Factors leading to these cost reductions can be the reduction or replacement of expensive materials or enhanced and simplified system architecture of the new technologies whereby a reduction in material use can be achieved.
- As the market for the special components of the electric mobility grows, the business model for entering the market will become more attractive and the

entire industry will grow in size. As the number of available suppliers increases, industry conditions will tempt vendors to use price cuts or other competitive measures to boost unit volumes.

Further improvements in fuel cell technology must be made with regard to hydrogen storage. Adequate storage of hydrogen is essential for the range that can be reached by the fuel cell vehicle. Potential for improvement exists within the construction of the tanks and also the pressure at which the hydrogen can be stored within the tank. So far, the construction of hydrogen storage systems is very complex and expensive when trying to fulfill all necessary safety aspects. In the future, the construction must be simplified without losing sight of the safety aspects.

In the development of high-voltage batteries, new approaches to improve the usable energy content of the cells are required.

38.5
Future Outlook

Due to limited fossil resources and the major challenges concerning environmental pollution, the transport sector has to strike new paths. Electric mobility has the potential to solve the existing problems and to contribute greatly to the mobility of tomorrow. The different drive train concepts within electric mobility, such as battery electric vehicles, plug-in hybrid vehicles, and fuel cell vehicles, do not compete against each other but rather complement one another. Whereas battery electric vehicles have the most efficient drive train and are optimal for local zero-emission mobility in urban areas, fuel cell vehicles are also suitable for inter-urban traffic and long distances. Therefore, hydrogen-powered vehicles will play an important role in the future of mobility. There is certainly a lot left to be done until the commercialization of fuel cell vehicles becomes a reality.

References

1 Eichlseder, H. and Klell M. (2008) *Wasserstoff in der Fahrzeugtechnik*, Vieweg + Teubner Verlag/GWV Verlag, Wiesbaden, p. 139.
2 Grasman, R. (2008) Sustainable mobility with fuel cell and battery electric vehicles. Presented at the Dow Jones Clean Tech Summit, 8–9 October 2008, Frankfurt.
3 Braess, H.-H. (2006) *Vieweg Handbuch Kraftfahrzeugtechnik*, Vieweg Friedr. und Sohn Verlag, Wiesbaden, p. 120.
4 Braess, H.-H. (2006) *Vieweg Handbuch Kraftfahrzeugtechnik*, Vieweg Friedr. und Sohn Verlag, Wiesbaden, p. 125.
5 Braess, H.-H. (2006) *Vieweg Handbuch Kraftfahrzeugtechnik*, Vieweg Friedr. und Sohn Verlag, Wiesbaden, p. 126.
6 Gerl, B. (2002) *Innovative Automobilantriebe – Konzepte auf der Basis von Brennstoffzellen, Traktionsbatterien und alternativen Kraftstoffen*, Verlag Moderne Industrie, Landsberg/Lech, pp. 42–43.

7. Daimler (2009) *Automobile Manufacturers Stick up for Electric Vehicles with Fuel Cell*, Press Release, 9 September 2009, Daimler AG, Stuttgart.
8. General Motors Company, http://www.gm.com/experience/technology/fuel_cells/ (9 February 2010).
9. Press Information, *FCX Clarity*, November 2009, Honda Motor Co., Ltd, Torrance, CA.
10. Hyundai Motor Deutschland (2005), *Umfangreiche Forschungen an den Antrieben der Zukunft*, Press Release, 9 August 2005, Hyundai Motor Deutschland GmbH, Neckarsulm.
11. Hyundai Motor Company (2009) http://worldwide.hyundai.com/web/innovation/fuel_cell.html (9 February 2010).
12. Kia Motors (2008) *Kia Reveals Next Generation Borrego Fuel Cell Vehicle and Production at Los Angeles Auto Show*, 21 November 2008, Press Release, Kia Motors, Seoul.
13. Nissan Motor Company, Ltd., http://www.nissan-global.com/EN/TECHNOLOGY/INTRODUCTION/DETAILS/XTRAIL-FCV/index.html (9 February 2010).
14. Nissan (2008) *Nissan Doubles the Power Density of Next Generation Fuel Cell Stack*, Press Release, 6 August 2008, Nissan Motor Company, Ltd, Yokohama.
15. Heise Zeitschriften Verlag (2008) http://www.heise.de/autos/artikel/Renault-Nissan-zeigt-Konzeptauto-Scenic-ZEV-H2-mit-Brennstoffzelle-455139.html (9 February 2010).
16. Toyota (2008) *Toyota Fuel Cell Technology*, June 2008, Toyota Motor Corporation, Aichi.
17. PressPortal.com.au (2009) http://www.pressportal.com.au/news/120/ARTICLE/4782/2009-04-07.html (9 February 2010).
18. New Flyer (2007) *New Flyer Awarded BC Transit Fuel Cell Bus Order*, Press Release, 3 August 2007, New Flyer Industries, Winnipeg.
19. BC Transit (2007), *Contracts for Hydrogen Bus Fleet Finalized*, Press Release, 3 August 2007, BC Transit, Victoria, BC.
20. Toyota and Hino Motors (2006) *Toyota–Hino Fuel Cell Bus to Serve Centrair and Vicinity*, Press Release, 18 July 2006, Toyota Motor Corporation, Aichi and Hino Motors, Ltd, Tokyo.
21. van Hool (2009) *Van Hool@Busworld 2009 – Public Transportation Invests in Environment-Friendly Alternative Drives*, Press Release, 1 October 2009, van Hool, Koningshooikt.
22. van Hool (2007) *Van Hool and Its Partners are Presenting the Hybrid Fuel Cell Bus for Europe at the UITP Exhibition in Helsinki*, Press Release, 21 May 2007, van Hool, Koningshooikt.
23. UTC Power, *AC Transit Orders 4 More UTC Power Fuel Cell Systems for Its Next-Generation Fuel Cell Buses in California*, Press Release, 11 March 2009, UTC Power, South Windsor, CT.

39
Hydrogen Internal Combustion Engines

H. Eichlseder, P. Grabner, and R. Heindl

Abstract

Over 100 years of development maturity, dual fuel capability and insensitivity to fuel purity distinguish the internal combustion engine as an interesting concept for direct conversion of hydrogen into useful mechanical work – particularly as a bridging technology while an H_2 infrastructure could be constructed.

Within this chapter the state of the art in hydrogen engines is presented by example of vehicles from different manufacturers. The physical properties of hydrogen make it an excellently suitable fuel for use in internal combustion engines and lead to a broad range of possible mixture preparation concepts. A brief introduction of these concepts is given together with a theoretical comparison of their respective potentials, advantages and challenges.

Consequently, aside from existing vehicle solutions extensive theoretical work is going on at universities and research facilities, yielding to increase power output and engine efficiency. A few results and examples intend to give a brief overview of the recent hydrogen combustion system development for the ICE.

Keywords: hydrogen internal combustion engine, performance potential, emission, efficiency, port fuel injection, direct injection

39.1
A History

Considerably before the invention of the spark-ignition (SI) four-stroke engine by Nicolaus August Otto in 1876, and indeed somewhat before the demonstration of an atmospheric gas engine by Barsanti and Matteucci in 1854, which may well have inspired Otto, more than 200 years ago François Isaac de Rivaz presented the first hydrogen engine in 1807 in Paris. On 30th January he was awarded patent No. 731 for the propulsion of a vehicle powered by an atmospheric engine fueled by hydrogen gas. In his patent sketch, Rivaz, who was Swiss, indicated the use of electrical ignition, yet he also spoke of ignition through "compression" of oxygen, an early allusion to the possibility of ignition

Hydrogen Energy. Edited by Detlef Stolten
Copyright © 2010 WILEY-VCH Verlag GmbH & Co. KGaA, Weinheim
ISBN: 978-3-527-32711-9

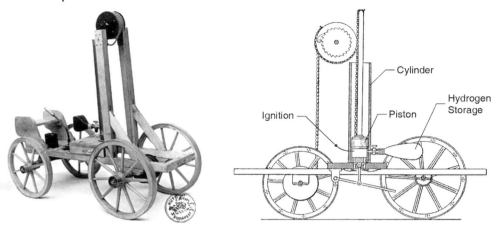

Fig. 39.1 First hydrogen-powered vehicle (François Isaac de Rivaz, 1807) [2].

(a) Hippomobile (b) Norsk Hydro

Fig. 39.2 (a) Hippomobile (Jean Joseph Étienne Lenoir, 1860); (b) Hydrogen commercial vehicle (Norsk Hydro, 1933) [2].

by means of the heat of compression [1]. Although he was only able to cover a few hundred meters in his vehicle in 1813, this was nevertheless an astonishing accomplishment for the time (Figure 39.1).

In addition to Rivaz, others working on hydrogen propulsion included Jean Joseph Étienne Lenoir and what is today Norsk Hydro Power Company. Lenoir's Hippomobile of 1860 was propelled by a horizontally installed single-cylinder engine powered by hydrogen generated by electrolysis. In 1933, Norsk Hydro built a vehicle with an on-board ammonia reformer for generating hydrogen (Fig. 39.2) [3].

As early as the 1920s and 1930s, the German engineer Rudolf Arnold Erren started to push forward the use of hydrogen in internal combustion engines. Following excess deliveries, Erren and his colleagues converted several thousand vehicles in Europe and the United States to direct hydrogen injection. On 19th December 1939 he was granted a US Patent (Internal Combustion Engine using Hydrogen as Fuel, No. 2 183 674) for the hydrogen-powered internal combustion engine. Figure 39.3 shows the drawings included with the patent application [3].

This is just a small sample of the early history of the development of hydrogen propulsion for vehicles. If one now turns to more recent activities, one can

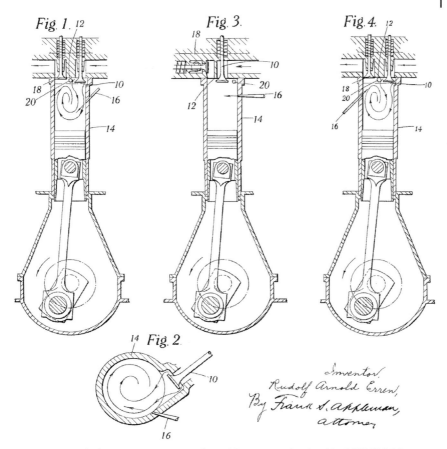

Fig. 39.3 Erren hydrogen engine (excerpt from US patent application No. 2 183 674) [4].

see a stronger research emphasis on hydrogen in vehicle applications from the end of the 1960s. This coincides, somewhat unsurprisingly, with the period during which the United States had invested a great deal of energy in the production, storage, and handling of liquid hydrogen for space exploration. In the 1970s and 1980s, various groups in the United States, Japan, and Germany developed and test-drove prototypes. In additon to laboratories in the United States, those working on hydrogen propulsion included the Musashi Institute of Technology (today Tokyo City University, Japan) and DFVLR (today DLR, Germany). Among vehicle manufacturers, BMW, Daimler Benz and MAN in Germany conducted intensive research in this area [3].

Whereas the original motivation arose principally from fuel properties such as ease of mixture preparation, broad ignition limits, and low ignition energy requirement, the subsequent drivers were a war-related limitation of supply of raw materials [5] and awareness of finite petroleum reserves. Over recent years, the particular drivers for hydrogen use have arisen from the desire for independence from fossil fuel resources and the absence of CO_2 emissions when energy is released. Here the majority of work has concentrated on electricity generation by means of fuel cell systems. Given the internal combustion engine's comparatively high power density coupled with favorable manufacturing costs and over 100 years of development maturity, in addition to the dual fuel capability of reciprocating piston engines, the direct conversion of hydrogen into useful mechanical work using an internal combustion engine represents an altogether interesting alternative.

The advantages listed above distinguish the hydrogen internal combustion engine as an outstanding bridging technology until a functional hydrogen infrastructure has been built. Moreover, in comparison with the fuel cell, it also enjoys the advantage of insensitivity to the composition and purity of the fuel used. The use of hydrogen-rich gases is an additional possibility, which is discussed in more detail in the following text.

39.2
State of the Art

Ever since the start of the 1990s, hydrogen has been booming as a drive system for vehicles. Above all in the area of fuel cells, a large number of demonstrator vehicles have been presented and potential production launch dates discussed, yet so far none has come to fruition. Meanwhile, with regard to internal combustion engines, companies engaging in more intensive research in this sector were BMW, Ford, MAN, Mazda, and Quantum [3], and they presented convincing demonstrations of their level of achievement through prototype, demonstrator, and even (very) limited production vehicles. Figure 39.4 shows by way of ex-

Fig. 39.4 Small-series and research vehicles with H_2 Internal Combustion Engines.

Table 39.1 Physical properties of hydrogen and gasoline (at 1.013 bar, 273 K). Modified from [3] and [6].

Property	Units	Hydrogen	Gasoline
Density, ρ	kg m^{-3}	0.09	730–780
Ignition limits in air[a]	vol.%	4–76	1–7.6
Minimal ignition energy	mJ	0.02	0.24
Self-ignition temperature	°C	585	230–450
Laminar flame velocity at $\lambda=1$	cm s^{-1}	230	40
Density of stoichiometric mixture, ρ_{mix}	kg m^{-3}	0.94	1.42
Stoichiometric air demand, L_{st}	–	34.3	14.1
Lower heating value, LHV	MJ kg^{-1}	120	42.1
Mixture calorific value, H_{mix}	MJ m^{-3}	3.2[b]	3.8[b]
		4.5[c]	3.9[c]

[a] at 293 K.
[b] Port injection.
[c] Direct injection.

ample the vehicles of BMW (Hydrogen 7), Ford (Shuttlebus E-450), MAN (hydrogen buses) and Mazda (RX-8 Hydrogen).

Through its realization of a limited series of 100 Hydrogen 7 model vehicles, developed for the first time under series conditions and produced in series production facilities, BMW reached a significant milestone.

When designing or modifying an internal combustion engine for hydrogen fuel, the aspects of particular relevance are those exceptional fuel properties having an influence on the system of mixture preparation, engine control, ignition system, mechanics of the powertrain, crankcase ventilation, supercharging system, and so on. The significant differences in the properties of gaseous hydrogen compared with gasoline are shown in Table 39.1. Even though the comparison of the ratio between the lower heating value LHV and the stoichiometric air demand L_{st} shows an advantage for hydrogen, the data indicates an 18% lower mixture calorific value in the external mixing hydrogen mode when compared with the gasoline situation. This effect has to be primarily attributed to the particularly low density of gaseous hydrogen, which implies a lower density for the air-fuel mixture ρ_{mix}, and thus a lower mixture calorific value H_{mix}.

A typical property of hydrogen is related to its ignition capabilities within a wide range of air-hydrogen mixture ratios. This allows the engine to be operated in a quality controlled mode even with completely homogeneous mixtures. The unthrottled mode of operation is beneficial to the engine's overall efficiency at part load conditions, because leaner combustion is more efficient (thermodynamic properties of a mixture) and throttling losses can be avoided. Moreover, the high flame propagation velocity in air–hydrogen mixtures offers outstanding combustion properties [6].

Of particular interest now are the engines of the vehicles listed above which, with regard to mixture formation and gas exchange, make use of a variety of

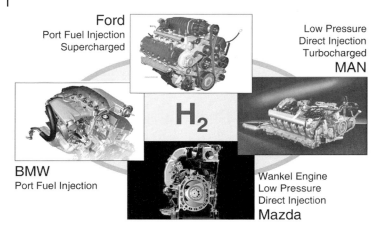

Fig. 39.5 Engines of small-series and research vehicles with H₂ internal combustion engines.

concepts covering practically the entire range of potential solutions (Figure 39.5):
- *External mixture preparation* in naturally aspirated operation (BMW Hydrogen 7), with mechanical supercharging (Ford Shuttle E-450), and turbocharging (Toyota Prius modified by Quantum).
- *Internal mixture preparation* with low-pressure direct injection combined with turbocharging (MAN Omnibus) along with a naturally aspirated Wankel engine (Mazda RX-8 Hydrogen).

Given that it is the commonest layout to date, intake manifold injection at ambient temperature will now be explained in greater detail using the BMW Hydrogen 7 powertrain by way of example. The engine, derived from a series 12 cylinder gasoline unit, is capable of bivalent operation. In the Hydrogen 7, switching between gasoline or hydrogen fuel is achieved by operating a switch on the steering wheel while driving, there being no sudden change in torque, while the rated performance data are equal (maximum power 191 kW, maximum torque 390 Nm; by way of comparison, the monovalent gasoline engine delivers 327 kW and 600 Nm). The remarkable feature of this vehicle is liquid hydrogen storage, with a tank capacity of around 7.8 kg.

On this engine, hydrogen is delivered at ambient temperature and a slight excess pressure of around 1 bar. Pressure regulation is by means of an electromagnetic control valve. The gaseous fuel is fed via a gas manifold (rail) to the injector valves and injected selectively into each cylinder via the inlet ports at a supercritical pressure ratio. The combustion chamber then receives a relatively homogeneous fuel–air mixture [7].

But what about the emissions? What about the formation of NO_x? It is evident that, due to the lack of carbon in the hydrogen fuel, there are no HC, CO, and CO_2 emissions except for those traces produced by the combustion of a

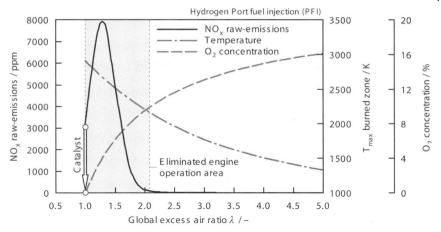

Fig. 39.6 Possible strategy to avoid NO$_x$ emissions [8].

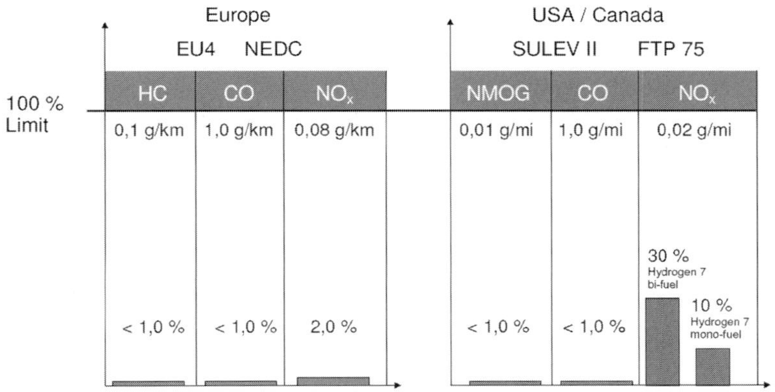

	Europe				USA / Canada		
	EU4 NEDC				SULEV II FTP 75		
100 % Limit	HC	CO	NO$_x$		NMOG	CO	NO$_x$
	0,1 g/km	1,0 g/km	0,08 g/km		0,01 g/mi	1,0 g/mi	0,02 g/mi
							30 % Hydrogen 7 bi-fuel
	< 1,0 %	< 1,0 %	2,0 %		< 1,0 %	< 1,0 %	10 % Hydrogen 7 mono-fuel

Fig. 39.7 Emissions in driving cycle for the BMW Hydrogen 7 (2007) [9].

small amount of the engine lubricating oil. But due to the nitrogen in the atmospheric air there might be NO$_x$ formation. Intensive testing has indicated that operating a hydrogen combustion engine in a lean homogeneous mode, at λ values larger than 2, does not produce any NO$_x$ (Figure 39.6). Burning the air-rich mixture means that the temperature of NO$_x$ formation (~2200 K) is not reached in the combustion zone. If this temperature is exceeded at higher loads with $\lambda=1$–2, NO$_x$ is being formed [6].

In order to reduce nitrogen oxides arising during combustion, at high load the Hydrogen 7 engine runs at an excess air ratio $\lambda \leq 1$ in combination with a three-way catalytic converter. Here the hydrogen is itself an outstanding reducing agent. In the low load range, the engine runs lean ($\lambda > 2$) in order to avoid exceeding the combustion temperature that is critical for NO$_x$ [7]. In combination with this operating strategy, it was possible for the BMW Hydrogen 7 to achieve the emission values shown in Figure 39.7 during the relevant driving cycles.

Fuel economy and emissions testing of two BMW Hydrogen 7 Mono-Fuel demonstration vehicles at Argonne National Laboratory's Advanced Powertrain Research Facility (APRF) revealed even lower drive-cycle emissions. The BMW Hydrogen 7 Mono-Fuel demonstration vehicles were tested for fuel economy and emissions in the Federal Test Procedure FTP-75 cold-start test and also the highway test. The results showed that these vehicles achieve emissions levels that are only a fraction of the Super Ultra Low Emissions Vehicle (SULEV) standard for nitrogen oxides (NO_x) and carbon monoxide (CO) emissions. For non-methane hydrocarbon (NMHC) emissions the cycle-averaged emissions are actually 0 g per mile, which require the car to actively reduce emissions compared with the ambient concentration. In addition to cycle-averaged emissions and fuel economy numbers, time-resolved (modal) emissions as well as air-fuel ratio data were analyzed to investigate further the root causes of the remaining emissions traces. Switching between the two main operating modes was found to be a major source of the remaining NO_x emissions. The fuel economy figures on the FTP-75 test were 3.7 kg of hydrogen per 100 km, which, on an energy basis, is equivalent to a gasoline fuel consumption of 17 miles per gallon (mpg). Fuel economy figures for the highway cycle were determined as 2.1 kg of hydrogen per 100 km or 30 miles per gallon of gasoline equivalent (GGE) [10].

39.3
New Concepts

Concerning the question of what challenges could remain for future concepts, given such outstanding achievements in terms of emissions, it is necessary to acknowledge both the low specific power and the question of further efficiency improvement so pertinent to every technical solution. For further development of specific power, it is necessary to consider mixture formation concepts in greater detail.

39.3.1
External Mixture Preparation – Port Fuel Injection (PFI)

39.3.1.1 External Mixture Preparation at Ambient Temperature
The compelling advantage of external mixture preparation with hydrogen at ambient temperature is system simplicity alongside low hydrogen supply pressure requirements. For injection into the inlet port, the excess pressure of a liquid storage tank is sufficient, for example, typically ranging between 0.5 and 5 bar. Injection always takes place close to the inlet valves. Systems using central mixture formation with hydrogen have been unsuccessful, given the increased tendency for backfiring into the inlet manifold. It is possible to classify the concepts on the basis of their injection strategies. On the basis of the duration of

injection, one can differentiate between continuous mixing, in which hydrogen is injected throughout the entire working cycle, and sequential injection, which takes place on the basis of individual cylinders, ideally synchronized with aspiration. Given that it is possible to take advantage of the entire working cycle for the delivery of fuel, at least at high rotation speeds, the demands placed on the injection valves with regard to switching times are on the low side, although the cross-sectional area to be controlled is around 500 times greater than that of a comparable gasoline valve as a consequence of the low fuel density [3].

39.3.1.2 External Mixture Preparation at Cryogenic Temperature

Another possibility is offered by the delivery of cryogenic hydrogen to the inlet system (cryogenic external mixture formation). Here, thanks to cooling of the entire intake charge, it is theoretically possible to achieve approximately equally high mixture calorific values as would be the case for direct injection into the combustion chamber (cf., Figure 39.8). The reduced temperature leads to an increase in mixture density and thus to the mixture calorific value. Given the assumptions made, potential power is around 15% higher than that for gasoline operation.

Additionally, cooling of the fresh charge can have a positive influence on the occurrence of combustion anomalies, above all backfiring and pre-ignition. Further details and trial results relating to cryogenic port injection (CPI) are available in [11].

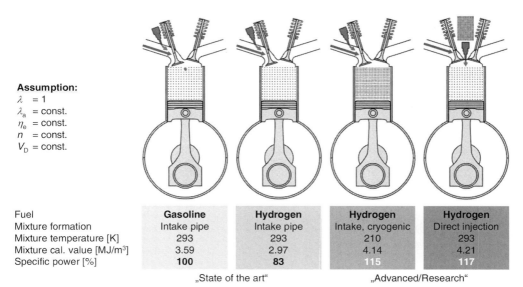

Assumption:
$\lambda = 1$
$\lambda_a = $ const.
$\eta_e = $ const.
$n = $ const.
$V_D = $ const.

Fuel	Gasoline	Hydrogen	Hydrogen	Hydrogen
Mixture formation	Intake pipe	Intake pipe	Intake, cryogenic	Direct injection
Mixture temperature [K]	293	293	210	293
Mixture cal. value [MJ/m³]	3.59	2.97	4.14	4.21
Specific power [%]	100	83	115	117

„State of the art" „Advanced/Research"

Fig. 39.8 Potential of various mixture formation concepts for hydrogen in comparison with gasoline manifold injection [12].

39.3.2
Internal Mixture Preparation – Direct Injection (DI)

In contrast with external mixture preparation, for internal mixing the fuel is delivered directly into each respective cylinder. Classification of direct hydrogen injection mixture formation concepts is based on the time of injection. Here a distinction is drawn between concepts involving early injection shortly after inlet valve closure and later injection close to the time of spark ignition. Assuming that injection occurs after inlet valve closure, backfiring can be ruled out. Pre-ignition, another form of combustion anomaly in which the fuel–air mixture ignites during the compression phase with inlet valves already closed, is not generally ruled out in the case of early internal mixing, although it is less likely given the more inhomogeneous mixture present. Later injection will allow this unwanted phenomenon to be ruled out reliably also, given the absence of an ignitable H_2–air mixture in the combustion chamber for a large proportion of the compression phase. No clear borderline exists between these two processes, it is more appropriate to talk of a sliding transition coupled with retardation of injection. It is possible to differentiate between variants on the basis of the injection pressure required. Systems with early internal mixture formation work with hydrogen supply pressures in the range from around 10 to 40 bar. With late injection, in order to ensure a supercritical pressure ratio and thus an injection period and amount that are independent of counter-pressure, supply pressures of at least 50 bar are required, depending on the compression ratio. If injection is also required during combustion, the injection pressure required rises to values in the range 100–300 bar.

A further advantage of internal mixture preparation relates to the achievable power density. Because an air-displacing effect in the inlet tract is avoided, the mixture calorific value for stoichiometric operation is around 42% higher than that for external hydrogen delivery at ambient temperature (cf., Figure 39.8). This implies a potential power level under otherwise identical conditions that is theoretically 17% greater than that of a conventional gasoline engine. Of this potential attributable to direct hydrogen injection in comparison with external mixing and gasoline operation, the realizable proportion depends on, amongst other things, efficiency and volumetric efficiency. The performance disadvantage of external mixture formation, resulting from air displacement and combustion anomalies, can actually be recovered thanks to direct injection and indeed converted into an advantage of around 10–15% in comparison with conventional gasoline operation [3].

39.3.3
Comparison of Potential

Using a comparison between mixture calorific values and power levels calculated from them, and subject to the boundary conditions prevailing, it is possible to conduct a comparison of potential covering a variety of hydrogen mixture preparation systems in relation to gasoline manifold injection. Using equation 39.1, the brake mean effective pressure BMEP can be derived from the volumetric efficiency λ_a, mixture calorific value H_{mix} and effective thermal efficiency η_e:

$$BMEP = \lambda_a H_{mix} \eta_e \qquad (39.1)$$

The mixture calorific value can be calculated for an air aspirated engine using equation 39.2 and for a mixture aspirated engine using equation 39.3:

$$\overline{H}_{mix} = \rho_{air} \frac{LHV}{\lambda L_{st}} \qquad (39.2)$$

$$H_{mix} = \rho_{mix} \frac{LHV}{1 + \lambda L_{st}} \qquad (39.3)$$

where LHV is the lower heating value, ρ_{air} the density of aspirated air, ρ_{mix} the mixture density, λ the excess air ratio, and L_{st} the stoichiometric air ratio.

Now the effective power P_e of a four-stroke internal combustion engine having a constant displaced volume V_D and constant rotation speed n is directly proportional to the mean effective pressure BMEP according to equation 39.4:

$$P_e = BMEP \, V_D \frac{n}{2} \qquad (39.4)$$

Hence if, as a first approximation, one assumes the volumetric efficiency and effective thermal efficiency to be constant, then the potential power is directly proportional to the mixture calorific value. However, in reality, efficiency will be strongly influenced by combustion process, gas exchange, and friction, yielding

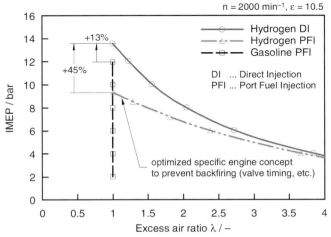

Fig. 39.9 Full load behavior with direct injection [13].

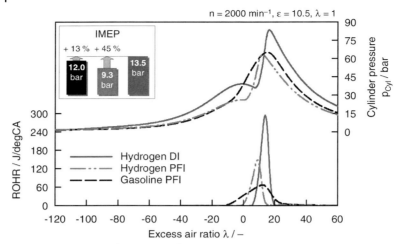

Fig. 39.10 Comparison between combustion processes for different concepts at $n = 2000$ min^{-1}.

essentially large differences. However, for a theoretical comparison of potential, this assumption is sensible.

These results are shown in Figure 39.8, with gasoline manifold injection used as a basis for comparison. Alongside this technology, one can also identify external hydrogen mixture formation as state of the art. The effects mentioned above lead to a reduction in potential power of around 17%. Among processes currently under development, the use of both external cryogenic mixture formation and direct injection allows this loss to be avoided and indeed give access to around 15–17% greater potential power.

Figure 39.9 shows a comparison between mean effective pressures achieved in practice at a rotation speed of 2000 min^{-1} on a passenger car engine configuration under investigation (500 cm^3 displacement per cylinder). The theoretically calculated differences are on essentially the same scale in practice. With external hydrogen mixing it is only possible to realize excess air ratios approaching $\lambda = 1$ without accompanying backfiring into the inlet manifold through timing and geometry optimization. Hence it is necessary to create optimum conditions for gas exchange as a function of rotation speed (no reverse flow of hot residual gas into the inlet tract and generally small amounts of residual gas in the cylinder).

Figure 39.10 shows a comparison between different curves of rate of heat release (ROHR) and cylinder pressure for operation using gasoline and hydrogen with external and internal mixture preparation measured on a research engine, where the rapid combustion with hydrogen and resulting rate of pressure increase are particularly striking. The combustion process that causes this can be influenced considerably in the case of internal mixture formation by using the timing of injection to change the mixture composition.

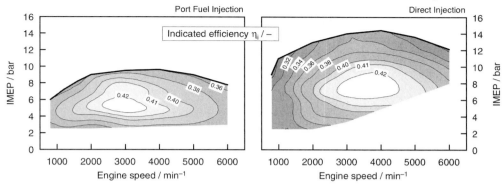

Fig. 39.11 Comparison of efficiency with external and internal mixture formation [13].

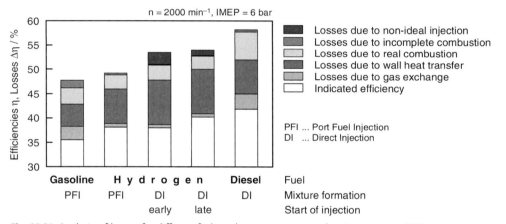

Fig. 39.12 Analysis of losses for different fuels and passenger car combustion concepts [14].

Using a combination of direct injection and supercharging it was indeed possible to achieve a specific power level exceeding $100\,\text{kW}\,\text{dm}^{-3}$ [12]. Alongside the higher mean effective pressure achievable with direct injection, Figure 39.11 also clearly shows the large extent of higher indicated efficiency.

In order to make the best possible use of favorable material properties, comprehensive investigations were conducted at Graz University of Technology aimed at optimizing the direct injection combustion process. Initial trials with hydrogen direct injection were followed by a process of incremental optimization of efficiency, emissions and power. For the purposes of evaluation and further development, it proved useful to perform, amongst other things, thermodynamic analysis of the working process on the basis of loss distribution.

Figure 39.12 shows a comparison of the loss distributions of particular gasoline, hydrogen, and diesel engines with $500\,\text{cm}^3$ cylinder capacity at $2000\,\text{min}^{-1}$ and an indicated mean effective pressure ($IMEP$) of 6 bar. For the hydrogen engine, illus-

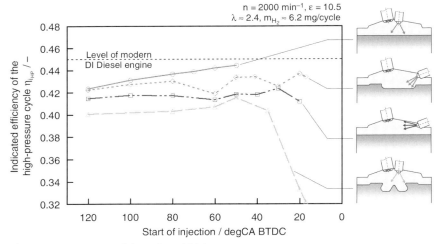

Fig. 39.13 Comparison of the indicated high-pressure efficiency for jet-guided and wall-guided operation at $n = 2000$ min^{-1} and $IMEP = 6$ bar [14].

Fig. 39.14 Hydrogen fork lift truck with direct injection engine [15].

tration is further subdivided into external and internal, early and late mixing. The comparison shows that the indicated efficiency of the hydrogen engine lies between those of the gasoline and diesel engines. It is evident from the diagram that different loss levels arise, assuming the efficiency of an idealized engine with real charge induction. For these losses to be minimized in a targeted way, precise knowledge of the individual losses and their causes is enormously important. Most significant are the losses through wall heat transfer. A reduction in the amount of wall heat flux to coolant would be possible if, for example, the fuel–air mixture burns while being confined as tightly as possible within the middle of the combustion chamber and high combustion temperatures and charge motion near the walls can be avoided. In order to approximate this "perfect stratification" approach, various jet and wall-guided processes were investigated and compared, particularly with regard to efficiency potential (Figure 39.13).

Given the flexibility in the choice of injection timing and consequent influence on charge stratification, it is possible to influence not only efficiency but also raw NO_x emissions within wide limits. An interesting application of a hydrogen engine with direct injection was described by VW [15]. This covers a supercharged engine with direct injection, which was specially developed for fork lift truck propulsion (Figure 39.14). In order to meet very high requirements for dynamic load response and stay within a relatively narrow engine speed range, a mechanical supercharger was chosen. For the combustion process, a $\lambda=1$ operating mode was implemented with torque control by throttling. The NO_x emissions that arise in stoichiometric operation can be reduced by means of conventional catalytic exhaust gas after-treatment. The working pressure of the H_2 injectors is 20 bar, ensuring largely complete consumption of the hydrogen stored as gas in a pressure vessel. With a compression ratio of 9, the four-cylinder inline engine with 2 litre displacement achieves maximum power of 75 kW (at 3000 min^{-1}) and maximum torque of 320 Nm (at 2000 min^{-1}).

39.4
Future Perspectives

39.4.1
Alternative Combustion Concepts

As presented above on the basis of selected examples, independently of the mixing strategy, only SI engines have been used for the prototypes and limited series vehicles demonstrated to date, and indeed in documented research activities relating to H_2 internal combustion engines for use in passenger cars. The twin requirements to position a sufficiently combustible mixture near the spark plug at the time of ignition and to enable high combustion efficiency during flame propagation imply a large proportion of pre-mixing even with strong mixture stratification, thereby limiting elevation of the compression ratio because of the tendency for knock to occur.

In a way analogous to the conventional diesel engine, in the event that very late direct injection allows a non-premixed combustion to be realized with hydrogen, the risk of knock can be removed and thus the compression ratio can be raised. However, in order to benefit from a resultant increase in efficiency, wall heat losses must be minimized at the same time, as otherwise they would more than compensate for the benefit of raised compression and could even lead to a loss of efficiency [16]. It is possible to counter this risk by means of appropriate combustion chamber and injector geometry design.

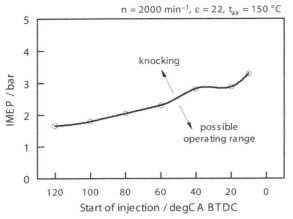

Fig. 39.15 Operating range for H_2 CI with inlet air pre-heating [18].

The wide ignition limits of hydrogen in air, in the range $0.13 \leq \lambda \leq 10$, would apparently qualify this fuel as being suitable for a diesel combustion process. Here the great challenge for realizing compression ignition (CI) arises from the high spontaneous ignition temperature of hydrogen, $T_{ig} = 853$ K, at stoichiometric conditions. Assuming direct injection at ambient temperature, the temperature at the end of compression would need to be around 1100 K in order to guarantee sufficient heat input [17]. Simply raising the compression ratio does not allow such temperatures to be achieved, with the result that additional measures must be taken, such as inlet air pre-heating [18].

In practice, CI operation proves to be scarcely applicable to passenger car use, even with high levels of inlet air pre-heating, because it is insufficiently controllable and limited to low engine loads. In this context, Figure 39.15 shows the maximum load achievable with natural aspiration for $\varepsilon = 22$ and inlet air pre-heating to $t_{air} = 150\,°C$. Here, increasing the amount of inlet charge to raise load leads immediately to the appearance of knock, while running at low load is aggravated by misfiring. In contrast with highly transient passenger car application, H_2 CI could, however, constitute a strategy for steady load operation that is worthy of pursuit.

In contrast, passenger car application requires an operating strategy that retains the concept of non-premixed combustion and yet, in terms of ignition, moves away from the conventional diesel cycle combustion process. One such possibility is the use of surface ignition by means of a glow plug. Here, following late direct injection, the outer boundary of the cloud of hydrogen, which is already mixed with air, is ignited by contact with the surface of the hot glow plug. The resulting exothermic reaction then generates the conditions for subsequent, rapidly progressing non-premixed combustion (Figure 39.16).

An H_2 combustion process such as this is very robust. It is possible to dispense with measures such as inlet air pre-heating and combustion anomalies are practically excluded. In addition, the combustion process is compatible with

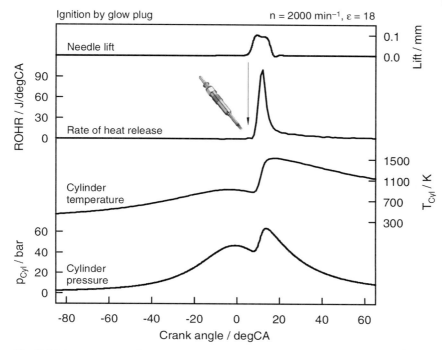

Fig. 39.16 Operating point for H_2 non-premixed combustion and glow plug ignition [19].

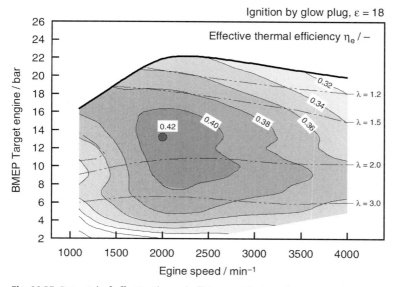

Fig. 39.17 Potential of effective thermal efficiency with glow plug ignition [19].

Fig. 39.18 Mercedes 200 NGT for hydrogen–methane blends in arbitrary ratios (HYCAR 1) [20].

almost unlimited supercharging and can cover the entire range of performance map relevant to passenger car application. Figure 39.17 shows the relevant efficiency potential within the engine map, based on the frictional losses of a modern passenger car diesel engine and a turbocharger optimized for the operating point. Here the best efficiency point with hydrogen is at the same level as the most efficient passenger car turbo-diesel engines.

One disadvantage of this concept arises from the coupling of the time of injection to the start of combustion, a link that could be broken by changing from a glow plug to a spark plug as the source of ignition.

This would then render plausible the idea of a combination of homogeneous, lean pilot combustion with a main combustion phase of the non-premixed type, possibly CI. However, the feasibility of such a concept for a passenger car powered by hydrogen must first be proven experimentally using an appropriate design of powertrain.

39.4.2
Bridging Technology

Alongside operation with pure hydrogen, there is particular interest in using gases containing hydrogen in vehicles with internal combustion engines because that would ensure an outstanding bridging technology for the period during which an H_2 infrastructure could be constructed.

A relevant example here is the Mercedes NGT 200, built as a result of a joint research project initiated by Graz University of Technology and HyCentA (Figure 39.18). Installed in this vehicle is a 350 bar pressurized tank for hydrogen or hydrogen-methane blends at arbitrary ratios, complete with the necessary safety features. The engine was modified to run on hydrogen–methane mixtures, which required that the electronic engine management be revised also to cover quality-controlled (throttle-free) operation [20]. The vehicle also has a single approval to drive on public roads in Austria.

39.5
Conclusion

For more than 100 years, the reciprocating internal combustion engine has been contributing significantly to the (individual) mobility of mankind. Its automotive application has never been seriously challenged by alternative drive systems. Also, in the scope of a future hydrogen economy, the internal combustion engine has a real chance to maintain its unique position. This uniqueness will be characterized by its high power density with regard to volume and weight, its high efficiency, nearly emission-free operation and an economic production. There is a great chance for the hydrogen-fueled direct injection internal combustion engine to be the basis of our individual mobility for the future.

References

1 Sass, F. (1962) *Geschichte des deutschen Verbrennungsmotorenbaues von 1860 bis 1918.* Springer, Berlin.
2 TÜV SÜD Industrie Service GmbH (2009) *H2Mobility: Hydrogen Vehicles Worldwide*; available at: www.h2mobility.org (accessed 1 November 2009]
3 Grabner, P. (2009) *Potentiale eines Wasserstoffmotors mit innerer Gemischbildung hinsichtlich Wirkungsgrad, Emissionen und Leistung.* PhD thesis, Graz University of Technology.
4 Erren, R.A. (1939) *Internal Combustion Engine using Hydrogen as Fuel.* US Patent 2 183 674.1939.
5 Eichlseder, H., Klüting, M., and Piock, W.F. (2008) *Grundlagen und Technologien des Ottomotors. Der Fahrzeugantrieb*, Springer, Vienna.
6 Freymann, R. and Eichlseder, H. (2003) *The State of the Art and Future Perspectives of the Application of Hydrogen IC Engines.* Presented at the Celebratory Conference 150 years of the Internal Combustion Engine – Engines of Sustainable Development, Naples.
7 Enke, W., Gruber, M., Hecht, L., and Staar, B. (2007) *Der bivalente V12-Motor des BMW Hydrogen 7.* MTZ 68(06), pp. 446–453.
8 Eichlseder, H., Wallner, T., Freymann, R., and Ringler, J. (2003) *The Potential of Hydrogen Internal Combustion Engines in a Future Mobility Scenario,* SAE Paper 2003-01-2267.
9 Braess, H. (2007) *Wasserstoff F&E Aktivitäten bei der BMW Group.* Presented at A3PS Conference Vienna.
10 Wallner, T. (2009) *Opportunities and Risks for Hydrogen Internal Combustion Engines in the United States.* In *VKM-THD Reports*, Vol. 92-II, The Working Process of the Internal Combustion Engine, pp. 679–695.
11 Heller, K. and Ellgas, S. (2006) *Optimisation of a Hydrogen Internal Combustion Engine with Cryogenic Mixture Formation.* In *VKM-THD Reports*, Vol. 88, 1st International Symposium on Hydrogen Internal Combustion Engines, pp. 49–58.
12 Gerbig, F., Heller, K., Ringler, J., Eichlseder, H., and Grabner, P. (2007) *Innovative Brennverfahrenskonzepte für Wasserstoffmotoren.* In *VKM-THD Reports*, Vol. 89, The Working Process of the Internal Combustion Engine, pp. 382–405.
13 Grabner, P., Eichlseder, H., Gerbig, F., and Gerke, U. (2006) *Opimisation of a Hydrogen Internal Combustion Engine with Inner Mixture Formation.* In *VKM-THD Reports*, Vol. 88, 1st International Symposium on Hydrogen Internal Combustion Engines, pp. 59–70.
14 Eichlseder, H., Grabner, P., Gerbig, F., and Heller, K. (2008) *Advanced Combus-*

tion Concepts and Development Methods for Hydrogen IC Engines. FISITA Paper F2008-06-103.

15 Willand, J., Grote, A., and Dingel, O. (2008) *Der Volkswagen-Wasserstoff-Verbrennungsmotor für Flurförderzeuge*. ATZ offhighway, pp. 24–35.

16 Wimmer, A., Wallner, T., Ringler, J., and Gerbig, F. (2005) *H2-Direct Injection – A Highly Promising Combustion Concept*, SAE Paper 2005-01-0108.

17 Prechtl, P. and Dorer, F. (1999) *Wasserstoff-Dieselmotor mit Direkteinspritzung, hoher Leistungsdichte and geringer Abgasemission, Teil 2: Untersuchung der Gemischbildung, des Zünd- und des Verbrennungsverhaltens*. MTZ 60(12), pp. 830–837.

18 Heindl, R., Eichlseder, H., Spuller, C., Gerbig, F., and Heller, K. (2009) *New and Innovative Combustion Systems for the H2-ICE: Compression Ignition and Combined Processes*, SAE Paper 2009-01-1421.

19 Eichlseder, H., Spuller, C., Heindl, R., Gerbig, F., and Heller, K. (2010) *Brennverfahrenskonzepte für dieselähnliche Wasserstoffverbrennung*, MTZ 71(01), pp. 60–66.

20 Eichlseder, H. and Klell, M. (2010) *Wasserstoff in der Fahrzeugtechnik*, Vieweg+Teubner, 2nd. ed.

40
Systems Analysis and Well-to-Wheel Studies

Thomas Grube, Bernd Höhlein, Christoph Stiller, and Werner Weindorf

Abstract

Energy systems analyses provide powerful assessment frameworks for research and development projects involving new energy technologies. These analyses rely on the description of state-of-the-art technologies and incorporate scenarios on future developments in the energy sector in order to identify the benefits and weaknesses of the technologies under consideration. This chapter presents selected assessment highlights relating to hydrogen and fuel cell technologies for transportation. With respect to fuel cell system cost, noble metal requirements are of utmost significance. The corresponding balances for fuel cell vehicles will be introduced. Moreover, the relevance of dynamic powertrain simulation for the evaluation of energy process chains in transportation will be discussed. In a further part, the methodology and results of well-to-wheel analyses of energy use and greenhouse gas emissions will be discussed, followed by an assessment of relevant hydrogen-focused well-to-wheel studies of different world regions as well as a comparison and interpretation of key findings.

Keywords: systems analysis, well-to-wheel, fuel cells, platinum group metals, dynamic powertrain simulation, vehicle simulation

40.1
Introduction

The motivation for establishing hydrogen as a new energy hub and the introduction of fuel cells as clean and highly efficient energy converters for mass markets has been substantiated with numerous scientific publications and with international R&D and deployment programs [1–3]. Drastically reduced emissions of greenhouse gases (GHGs) and of locally active air pollutants and less dependence on energy feedstock imports highlight the expectations connected with these new technologies. In parallel, the utilization rate of energy services, such as transportation, heat, light, communications, and mechanical work, is predicted to increase

Hydrogen Energy. Edited by Detlef Stolten
Copyright © 2010 WILEY-VCH Verlag GmbH & Co. KGaA, Weinheim
ISBN: 978-3-527-32711-9

in all sectors over the next few decades. The following factors confirm the complexity of the description of future energy system evolution:
- constraints derived from the severe environmental impacts of energy use with respect to climate change
- energy feedstock and materials availability
- the expected increase in energy use
- emerging new technologies.

Energy systems analysis reflects such developments and provides an independent description on state-of-the-art technologies and takes a look into the foreseeable future. Scenarios for future primary energy mixes and sector-related energy utilization are used to provide more information on the benefits and weaknesses of the new technologies under consideration, while always integrating the entire process chain from the primary energy source to the final energy use. With respect to hydrogen and fuel cell technologies, the following systems analysis-related tasks to be studied can be described:
- systems and subsystems currently under development
- competitive approaches to providing energy services and how they are likely to change over time
- examination of future energy scenarios and forcing functions
- development of understanding of how proposed technologies could fit into national systems.

A broad range of methods and assessment frameworks exist that can be employed for assessments and evaluations in this context. Some prominent examples are life cycle assessment, well-to-wheel analysis, life-cycle and external cost assessments, process analysis of energy conversion and material balances. The level of detail of the assessment method applied is defined by motivation and the analysis target.

This chapter presents some selected assessment highlights related to the transportation sector. Section 40.2 concerns the platinum group metal requirements for fuel cell systems as one important example of material balances. Section 40.3 deals with the relevance of dynamic powertrain simulation, which is used to determine vehicle fuel economy, on the one hand, and for specific development tasks, such as the identification of operational strategies of hybrid powertrains, on the other. The method and results of well-to-wheel balances providing the basis for comparing assessments of vehicle and fuel systems are presented in Section 40.4.

40.2
Platinum Group Metal Requirements for Fuel Cell Systems

Alternative powertrain systems under consideration include gasoline and diesel hybrid electric vehicles (HEVs), plug-in hybrid vehicles (PHEVs), hybrid fuel cell vehicles (FCVs), and battery electric vehicles (BEVs). In this chapter, we concen-

trate on FCVs and their material and cost balances. Worldwide efforts aimed at optimizing the conversion of hydrogen into electricity using polymer electrolyte fuel cells (PEFCs) in FCVs include the following:
- improving durability, reliability and feasibility
- increasing specific power (W kg^{-1}) and power density (W l^{-1})
- ensuring heat removal, air supply, and water management
- optimizing fuel cell technology (membrane/electrolyte)
- optimizing the catalyst and minimizing platinum group metal (PGM) requirements
- reducing material and manufacturing costs of certain PEFC components
- reducing the overall costs of fuel cell vehicles (FCVs).

The solutions developed so far have allowed a first mini-series to be tested. More recently, in a joint statement, several car manufacturers and other industrial companies, and also the German government, have announced plans to introduce FCVs on the market in 2015. With regard to the last two optimization criteria for FCV developments listed above, the following information based on the literature will help to describe and evaluate the current state of the art.

40.2.1
FZJ and PSI (2001)

In 2001, a group of workers from Forschungszentrum Jülich (FZJ, Germany) and the Paul Scherrer Institute (PSI, Switzerland) discussed the efficiencies and platinum group metal requirements of PEFC powertrains [4]. Specific data on PGM requirements were derived for various fuel cell characteristics taken from the literature and using data from the relevant publications. The essential criteria for further analyses were the determination and definition of a maximum specific power (kW$_{el}$ m^{-2}) in order to determine the stack area needed, which also depends on operating parameters such as media pressure, temperature, and gas composition. If the stack area is known, it is possible to determine the absolute value of the PGM requirement for the fuel cell stack of a vehicle based on the necessary PGM loading (mg$_{Pt}$ m^{-2}).

On the basis of data in the literature and some assumptions, it was found that the specific PGM requirement of PEFCs should be approximately 0.13 g kW$_{el}^{-1}$ or 9 g of PGM for a PEFC with a maximum power of 70 kW$_{el}$ based on a 0.1 mg cm^{-2} anode and cathode in an 8 kW$_{el}$ m^{-2} membrane electrode assembly (MEA) instead of 1–1.5 g kW$_{el}^{-1}$ of PGM (2001) or 70–105 g of PGM for a 70 kW$_{el}$ PEFC.

Based on these calculations, 1 million fabricated fuel cell passenger cars would require >10 t of PGM in comparison with <2–4 (6) t for current internal combustion engines with three-way catalytic converters fulfilling the EURO IV and the American ULEV standards. A large proportion of the global demand for PGM originates in the automobile industry: platinum 32% of 177 t, palladium 61% of 261 t, and rhodium 24% of 64 t. The large-scale introduction of fuel cell passenger cars would therefore require significant expansion in the production of primary PGM in the introduction phase. Increased amounts of secondary metals would only ease the situation again thereafter. The price of platinum has nearly doubled since 2002 [5] to about US$ 1000 per ounce in 2008 [6]. The PGM supply for the automobile industry increased during the same period by about 20% [6].

40.2.2
Massachusetts Institute of Technology (MIT) (2008)

Kromer and Heywood [7] described a comparative assessment of electric propulsion systems in the 2030 US light-duty vehicle fleet. In this context, they refer to the fact that fuel cell technology requires improvements across a number of different dimensions. They assume that durability and reliability issues have been largely addressed with respect to performance and costs. They discuss two primary possibilities for fuel cell cost reduction: stack power density to decrease the amount of active material and PGM loading in addition to reducing the size and complexity of the balance of plant. Table 40.1 shows the results of the two scenarios, "Baseline" and "Conservative", compared with the "Present Day" in terms of stack costs (US$ kW_{el}^{-1}), balance of plant (BOP), and the entire system. The calculations assumed platinum costs of US$ 900 per ounce and a mass production of FCV with high durability (miles) in 2030.

For an 80 kW electric peak power system with a PEFC in the conservative scenario, this means a platinum demand of 32 g [assumption: DOE long-term (2015) performance target of 1500 mA cm^{-2} at 0.65 V or 10 kW_{el} m^{-2}] or about US$ 1000 for the platinum, about US$ 3300 for the BOP, and US$ 6100 for the total system.

Table 40.1 Specific costs of PEFC systems and assumptions for specific stack power and PGM loading according to Kromer and Heywood [7].

Scenario	Assumptions	Cost (US$ kW_{el}^{-1})		
		Stack	BOP	Total
Present day	PEFC=420 W kg^{-1} Pt=0.75 g cm^{-2}	67	41	108
Baseline	PEFC=650 W kg^{-1} Pt=0.2 mg cm^{-2}	22	30	52
Conservative	PEFC=520 W kg^{-1} Pt=0.4 mg cm^{-2}	35	41	76

40.2.3
FZJ (2009)

The currently high cost of fuel cells is determined by expensive materials and low production volumes. A detailed understanding of cost structures for FCV powertrains reveals unexploited potential that can reduce costs in the future. This requires a method of predicting costs that can be applied with little effort, which not only offers a sufficient degree of detail but is also very accurate. Similar results were presented by Werhahn [6] using a new calculation method. Instead of determining costs on the level of the aggregated system, the cost forecast was applied directly to the individual components. Due to the great influence of the production rate on the manufacturing costs, an additional dependence on the number of units was also integrated. Based on an assumed production rate of 10 000 FCVs per year, Figure 40.1 presents the cost distribution of an 80 kW PEFC system including the stack and balance of plant according to Werhahn [6]. More than 60% of the costs for the full fuel cell system come from the MEA, which includes 110 g of PGM at a specific price of € 55 g_{Pt}^{-1}. The specific costs for the system are about € 170 kW_{el}^{-1}. No information was given on the durability of the fuel cell system.

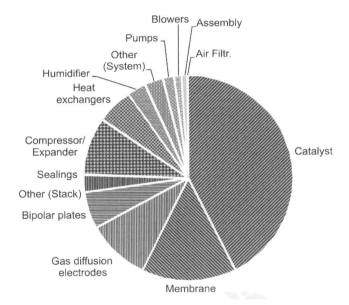

Fig. 40.1 Cost breakdown of an 80 kW$_{el}$ PEFC system at an assumed production rate of 10 000 units per year according to Werhahn [6].

Table 40.2 Durability, PGM loading, and relative cost of PEFC systems for different time horizons.

Time	Durability (miles)	Platinum (g)	Relative cost (%)
Present day HydroGen4-I	30 000–80 000	80	100 BOP + stack ~50 stack
2015: next generation 10 000 FCVs per year	120 000	30	~25 BOP + stack ~15 stack
Subsequent generations 1×10^5–1×10^6 FCVs per year		<10	<10 BOP + stack <5 stack

40.2.4
General Motors Europe (2009)

At the *f-cell* Conference in Stuttgart 2009, von Helmolt [8] described the next steps on the cost and durability path: a comparison of fuel cell system durability in miles and fuel cell costs for the scenarios "Today", "2015" and "2020–2022" (Table 40.2).

40.2.5
Conclusion

The evaluation of data from the literature shows that the production costs of fuel cell systems for FCV powertrains will be determined mainly by the catalyst costs for the electrodes in addition to the costs for the balance-of-plant components. The costs for the total fuel cell system could be reduced by a factor of five by 2015 compared with the situation today, assuming an improved PGM material situation (~30 g of platinum for an FCV) at a production rate of 10 000 FCVs per year. The total PGM cost for a PEFC with a maximum power of 80 kW_{el} assuming 10 kW m^{-2} and 4 g m^{-2} in accordance with [7] would mean – at a platinum price of US$ 1000 per ounce – about US$ 1000 per stack or about US$ 12 kW_{el}^{-1}.

40.3
Dynamic Powertrain Simulation

A detailed understanding of the vehicle fuel economy of alternative powertrains is essential because it is the final step in process chains of transportation fuels, which means that any losses here have a great impact on the overall well-to-wheel balance. Life cycle assessments and life cycle cost evaluations rely to a large extent on the results of fuel economy assessments. Methods assessing efficiencies using steady-state approaches have a limited value here. The evaluation of the energy requirement of vehicles is more complex than for other steps in process chains with respect to highly dynamic operation. Points of operation in the lowest partial area load up to full load at high acceleration or at high velocity during motorway driving, and also transient operation, are relevant for the assessment. This makes dynamic vehicle simulation on the basis of driving profiles mandatory. Moreover, powertrain simulation supports the identification of suitable powertrain configurations, and operational strategies when hybrid configurations are considered. In addition, powertrain component scaling is required, which is also relevant for materials balances (cf. Section 40.2) and helps to identify R&D tasks on the component level. In the literature, vehicle fuel economy is estimated using

- constant powertrain efficiencies neglecting the influence of drive cycles
- system efficiency evaluated per time step of a drive cycle on the basis of constant component efficiencies [9, 10]
- dynamic models with fundamental or map-based component models [7, 11].

The following sections consider dynamic powertrain simulations in relation to alternative powertrain development. Various activities in this field have been introduced in the past, ranging from commercial products to simulation tools developed in research institutions and universities that have partly been made available for public use.

40.3.1
Simulation Approaches

The physical basis of dynamic powertrain simulation is first provided by the forces balance of the road wheel system. Sets of equations, such as those given by Wallentowitz [12], use vehicle design data in order to evaluate the total force as the sum of the resistance forces at the wheel of a vehicle as given below.

Rolling resistance force $F_R(t)$:

$$F_R(t) = f_R(t) m_V g \qquad (40.1)$$

with

$$f_R(v) = c_{R0} + c_{R1} v(t) + c_{R2} v(t)^4 \qquad (40.2)$$

Air drag resistance force $F_D(t)$:

$$F_D(t) = c_W A_V \frac{\rho_A}{2} v(t)^2 \qquad (40.3)$$

Acceleration force $F_a(t)$:

$$F_a(t) = (e m_V + m_D) a(t) \qquad (40.4)$$

Total force $F_t(t)$:

$$F_t(t) = F_R(t) + F_D(t) + F_a(t) \qquad (40.5)$$

In this context, the two general approaches to simulation models need to be discussed: forward-facing and backward-facing simulation. Backward-facing simulation infers the system operation from the actual effect, in other words, the velocity change defined by the time steps of a velocity profile (e.g., a drive cycle). From the velocity change, $F_t(t)$ is calculated using the design data of the vehicle, followed by the conversion of $F_t(t)$ into the torque and of the vehicle's translatory speed $v(t)$ into the rotational speed $\omega(t)$ of the drive axle. All subsequent calculations are carried out backwards through the powertrain until the energy demand or fuel consumption has been determined for each time step. This approach implies that the vehicle design in conjunction with the powertrain configuration and scaling allows the vehicle actually to follow the desired velocity change. Backward-facing simulations have been used in the past in order to reduce simulation run times. However, the suitability of backward-facing simulations is limited if controller circuits and hybrid strategies are to be developed, and if performance limits, such as acceleration and climbing capabilities, are to be evaluated. Moreover, if transient behavior of powertrain components is to be considered, forward-facing simulation is preferable.

Forward-facing simulation considers realistic cause-and-effect chains. Starting with the driver behavior, which is usually represented by a proportional-integral (PI) controller, the difference between the actual and the desired velocity is evaluated, resulting in acceleration or deceleration commands. The powertrain "reacts" within its defined physical limits, for example, the maximum torque characteristic of internal combustion engines or electric motors. The power flow through the powertrain is consistent with real-world systems, allowing for the implementation of transient effects and controller algorithms and for the evaluation of performance limits.

Simulation run-time is also discussed in conjunction with the level of detail of the component models. Apart from simulation models based on the physical and/or chemical processes, operational maps could also provide the required data for system components. On account of simulation accuracy, the simulation run-time is then reduced. Some workers (e.g., see [13]) have proposed that fundamental models be used for the steady-state-based derivation of operational maps with task-dependent consideration of influence parameters, such as pressure, temperature, or humidification for polymer electrolyte fuel cells. Other components, such as electric motors and internal combustion engines, are usually based on performance maps (e.g., [14]).

If hybrid powertrains are considered, the operational strategy chosen has a great impact on the fuel economy of vehicles. It decides the share of power provided by the main aggregate (fuel cell or internal combustion engine) that can be delivered using the direct and, thus, more efficient path to the transmission and how much power needs to be stored intermediately in the electric storage. Apart from constraints that lie in the dynamic properties of the power generator, the optimization of the operational strategy is very specific for the vehicle design and for the driving cycles and load situations that are used for the evaluation.

The computational basis is in many cases the widely accepted Matlab/Simulink® environment [14–16]. Simulink already applies a graphical standard that is easier to use as executable program code. Some models, such as ADVISOR and AVL CRUISE, feature a specially designed graphical user interface (GUI).

40.3.2
Examples of Vehicle Simulators

One of the most popular examples of vehicle simulators is the Advanced Vehicle Simulator ADVISOR [17], first developed by NREL in 1994 as a tool to support the transportation-related activities of the US Department of Energy (DOE). It basically uses a backward-facing approach with some forward-facing functionality. In 1999, it claimed that simulation run times were 2.6–8 times faster than strictly forward-facing simulators [14]. ADVISOR has been commercially distributed by AVL List GmbH since 2004. The component description relies mainly on operational maps. Results of dynamic powertrain simulations with ADVISOR were given by Kromer and Heywood [7], where a comparative assessment of various powertrain options for 2030 was presented (see Figure 40.2) for a combined drive cycle integrating the US Federal Test Procedure (FTP) and the Highway Fuel Economy Test (HWFET). It can be seen that the PHEV-30 (30 miles all-electric range) and the battery electric vehicle (BEV) show the lowest tank-to-wheel energy use.

The vehicle simulator FCVsim specifically focuses on the dynamic simulation of direct hydrogen fuel cell systems (DH-FCVsim) in load-following or hybrid configurations. Special emphasis is given to the fuel cell stack and system component modeling. A detailed description of the fuel cell subsystem was presented by Moore *et al.* [13].

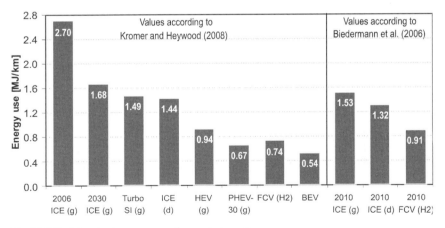

Fig. 40.2 Tank-to wheel energy use of powertrain options, after Kromer and Heywood [7] and Biedermann *et al.* [11]; g, gasoline; d, diesel; H2, hydrogen.

LFM (light, fast, and modifiable) is a forward-facing simulation tool based on Matlab/Simulink that was designed for rapid performance analysis of hybrid powertrains [16]. More specifically, the model should allow for easy model modifications, making parametric studies with automated simulation runs possible. Fuel cell, electric motor, and transmission are implemented as operational maps. The realization of the operational strategy incorporates a hybrid controller and a load combiner. Brown et al. [16] quantitatively compared the simulation results with vehicle measurements of a bus with 22 seats with respect to (i) energy use, (ii) energy recovered, (iii) battery state of charge, and (iv) fuel cell energy output. Higher error margins for the six drive cycles examined were found for the energy recovered with up to 13% and state of charge with up to 23%. The latter is also responsible for increased error margins of the fuel cell output, as this is partly determined by the battery state of charge (SOC) via the operational strategy.

Argonne National Laboratory's Powertrain Systems Analysis Toolkit (PSAT) is a forward-facing simulation tool based on Matlab/Simulink/Stateflow. It has been used in the context of hybrid vehicle development, for example, for the determination of the battery requirements of PHEVs [18]. A further topic was the assessment of the evolution of hydrogen-fueled vehicles in comparison with conventional vehicles up to 2045 [19]. Cao et al. [15] validated the component models and controls with a Toyota Prius in the Hymotion plug-in configuration which includes a 5 kWh_{el} lithium ion battery pack. It was found that the relative difference in fuel consumption was less than 8.8% in charge-depleting mode and 1.7% in charge-sustaining mode. PSAT has been supporting DOE's FreedomCar Initiative and the Vehicle Technologies Program.

Activities at Forschungszentrum Jülich involve comparing assessments of alternative powertrain configurations. The results of dynamic powertrain simulations in the context of a well-to-wheel balance were presented as estimates for 2010 [11]. Selected values are given in Figure 40.2. A further topic involving dynamic powertrain simulation was the assessment of options for onboard power generation, comparing fuel cell auxiliary powertrains with conventional and advanced engine-coupled power generation options ([20], cf. section *APUs for Road Vehicles, Ships, and Airplanes*).

40.3.3
Conclusion

It was shown that a broad range of vehicle simulation activities exists; only a selection was presented here. In addition to comparing assessments of different powertrain configurations with respect to vehicle fuel economy and emissions, such tools support the actual powertrain development because they help to identify suitable configurations, particularly for hybrid powertrains, and find optimized component sizes in terms of power and energy storage capacity. Finally, the development of operational strategies greatly benefits from dynamic powertrain simulations.

In general, comparing fuel economy figures on the basis of dynamic powertrain simulations is difficult, as time horizons, drive cycles used, and battery SOC correction methods are different when applied to hybrid vehicle configurations. Moreover, vehicle design specifications, such as performance targets and vehicle mass and drag coefficients, are not consistent. With respect to well-to-wheel analysis, more consistent and internationally agreed assumptions for vehicle specifications and drive cycles could improve the quality and reproducibility of the results.

40.4 Well-to-Wheel Studies

The driving forces for the development of alternative fuels are, on the one hand, the anxiety about the security of the oil supply, on which the transport sector still depends heavily, and, on the other, a reduction in transport-related emissions of GHGs and air pollutants. In this respect, hydrogen and fuel cells are competing with a number of other fuels and conversion technologies, such as synthetic fuels, biofuels, and natural gas, and improved gasoline and diesel engines. With regard to powertrains, internal combustion engines still dominate. In addition to the improved efficiency of these engines, vehicle concepts based on electric drives and relying to varying extents on batteries as a source of motion energy are also being developed.

In recent years, full energy supply chain analysis has developed into a well-accepted method supporting consensus processes on the advantages and disadvantages of transportation fuel alternatives, specifically when the focus is on environmental constraints. Earlier calculations revealed that a rigorous and transparent analysis of full energy supply chains "from the cradle to the grave" must replace the comparison of individual processes along energy supply chains, such as fuel cells and internal combustion engines.

Although in principle stationary and transport-specific energy chains can be analyzed, this chapter focuses on the latter, which are also referred to as well-to-wheel (WTW) analyses. The primary focus of WTW analysis is on the global environmental impact, in other words, on GHG emissions expressed as CO_2 equivalents. Other issues of interest are (i) primary energy demand (resource utilization), (ii) local pollutant emissions, and (iii) full energy or fuel supply costs. WTW analysis covers the entire fuel supply chain: feedstock extraction, feedstock transportation, fuel production, fuel distribution, and fuel use in a vehicle.

This section first reviews the basic methodologies and assumptions behind well-to-wheel analysis. This methodological review is then followed by a screening of the most relevant activities with regard to WTW analysis, focusing on hydrogen and fuel cells. Finally, some key results of these activities are briefly compared and differences interpreted.

40.4.1
Methodology of WTW Analysis

WTW analysis is a specific form of life cycle assessment (LCA), namely an LCA of transportation fuels. An LCA typically also takes other factors of a product or an energy carrier into consideration in addition to global GHG emissions. It also accounts for the provision of all construction materials required for the necessary processing plants, plant decommissioning, and the manufacture of the vehicles. In most cases, a detailed general LCA analysis is not needed at the level of policy discussion in order to reach a broad consensus on alternative fuels or drive systems. As a subset of WTW analysis, a well-to-tank (WTT) analysis is often used to separate the environmental and economic effects of fuel supply and vehicle use.

The overall WTW energy, Energy$_{WTW}$, use is calculated by

$$\text{Energy}_{WTW} = \text{Energy}_{TTW}[\text{MJ}_{\text{final fuel}}\,\text{km}^{-1}] \times \sum \text{Energy}_i[\text{MJ}_i/\text{MJ}_{\text{final fuel}}] \quad (40.6)$$

where Energy$_{TTW}$ = fuel consumption of the vehicle and Energy$_i$ = input of primary energy source i to generate the final transportation fuel.

The energy inputs can originate in a multitude of different processes along the energy supply chain. To get to the overall energy use, the energy intensities of all processes related to the final product or service (e.g., per MJ of final fuel, or per kilometer driven) must be calculated based on the efficiencies along the chain, and the specific energy uses must then be aggregated.

The input of primary energy includes the energy content retained in the final fuel, feedstock losses through conversion processes, and side energies (see Figure 40.3).

The overall WTW GHG emissions, GHG$_{WTW}$, expressed in grams of CO_2 equivalent are calculated by

$$GHG_{WTW} = GHG_{WTT}[\text{g MJ}^{-1}] \times \text{Energy}_{TTW}[\text{MJ km}^{-1}] + GHG_{TTW}[\text{g km}^{-1}] \quad (40.7)$$

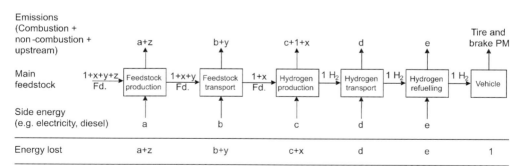

Fig. 40.3 Example of accounting for energy losses along an energy supply chain [21]; PM, particle matter.

where GHG_{WTT}=the aggregate GHG emissions of the well-to-tank chain, related to the final fuel energy content (the aggregation approach is similar to that described for energy use) and GHG_{TTW}=the GHG emissions arising from using the final fuel in the vehicle (e.g., burning fuel while releasing CO_2).

Pollutant emissions are calculated in the same way as GHG emissions.

WTW and WTT discussions are often accompanied by an assessment of further closely related issues, such as
- analysis of technology and cost learning, typically by applying standard development curves or other specific background information from the literature
- analysis of primary fossil or nuclear energy availability, (regional) renewable energy potentials, and identification of competing resource requirements
- availability of regional potentials to apply CO_2 capture and storage (CCS).

In order to handle high uncertainties regarding process input data such as efficiencies, emissions, or costs, probabilistic analyses, such as a Monte Carlo analysis, are usually applied.

WTW analyses must be applied for all relevant time steps in order to understand the evolution of environmental effects and possibly costs in the short to long term. This is particularly important when innovative processes are being considered that are characterized by technology development and cost curves with high gradients.

In addition to energy use, GHG emissions, and air pollutant emissions, further environmentally relevant factors could also be taken into account, such as water use, water pollution, biodiversity, and societal aspects. However, accounting for this quantitatively is much more complex and arbitrary. Societal impacts, for example, are difficult to quantify, and the impact of water use is worse in regions or seasons of water scarcity. Therefore, a qualitative evaluation of these factors is often preferred.

40.4.1.1 Impact of Greenhouse Gases

GHGs are generally considered to include carbon dioxide (CO_2), methane (CH_4), and nitrous oxide (N_2O) (other greenhouse gases are CFCs, HFCs and SF6, but these are not relevant in this context.). Their global warming potential is expressed in CO_2 equivalents. Table 40.3 shows the global warming potential for a time period of 100 years according to the Intergovernmental Panel on Cli-

Table 40.3 Global warming potential of different GHGs [22].

GHG	CO_2 equivalents
CO_2	1
CH_4	23
N_2O	296

mate Change (IPCC). These potentials are used in the majority of WTW analysis projects.

A special case is the treatment of CO_2 emissions from biofuels, which accrue upon combustion but represent a closed loop because crops absorb the same amount of CO_2 from the atmosphere when growing. In some studies, this is handled as a CO_2 credit on plant harvesting, which is compensated with emissions released during burning (e.g., in the GREET model for the United States, and also the Japanese approach [23]); others treat the crop cultivation and combustion as CO_2 neutral. Although this clearly influences the balance of specific steps in the energy supply chain, it does not influence the aggregate results [21].

40.4.1.2 Accounting of Primary Energy

In the CONCAWE/EUCAR/JRC study [10], the "efficiency method" was used to calculate the energy requirements. This method is similar to the procedure adopted by international organizations (IEA, EUROSTAT, ECE). The efficiency of electricity generation from nuclear power can be calculated from the heat released by nuclear fission, and is about 33%. In the case of electricity generation from hydropower or other renewable energy sources that cannot be measured in terms of a calorific value (wind, solar energy), the energy input is assumed to be equivalent to the electricity generated, which leads to an efficiency of 100%. An exception to this is, for example, the GREET model, which assumes 100% primary energy efficiency for nuclear fission (disregarding the waste heat generated), which leads to a strong reduction in primary energy use for nuclear pathways [21]. Ishitani et al. [23] also disregarded the waste heat from nuclear fission.

40.4.1.3 Accounting of Coproduction Processes

Many processes produce not only the desired product but also "by-products", such as combined heat and power generation. In such cases, there are two principal methods to calculate the energy use and emissions of the desired product while also accounting for the by-product:

1. *Displacement (or substitution) method:* The energy use and emissions of the main product are calculated by subtracting the energy and emissions of the product that are realistically displaced by the by-product from the total energy and emissions. Here, an assumption has to be made on how the displaced energy would otherwise be produced [e.g., in a combined heat and power (CHP) process, the by-product heat could displace heat from a low-emission natural gas boiler or from a high-emission coal oven]. The substitution method has also been used in a WTW analysis [10] and is proposed in the EU Renewable Energy Directive [24] for excess electricity from CHP plants. According to the Directive, excess electricity shall replace electricity using the same fuel as the CHP plant itself.

2. *Allocation by energy, mass, or market value:* Another method of accounting for by-products is to allocate the inputs and emissions to the products based on the energy content, mass, or market value of the products. Allocation by energy is proposed in the EU Renewable Energy Directive [24] for the assessment of GHG emissions regarding the supply of biofuels when dealing with by-products that are not electricity.

The displacement method should only be applied if the by-product volume is smaller than that of the main product, otherwise negative emissions can occur (assuming high emissions of the displaced process).

40.4.2
Most Relevant Hydrogen-Focused Well-to-Wheel Studies

WTW and WTT analyses have been applied recently to a number of important high-level consensus-finding processes among industry and its relevant automobile and energy branches and public representatives throughout the world. Although applying different software-based tools in Europe, Japan and the United States, international information exchange has proven that the WTW and WTT methodologies produce converging results [21].

40.4.2.1 Europe
In Europe, WTW and WTT analyses for the road transport sector emerged from the German Transport Energy Strategy (TES), an initial group of seven automobile and energy companies that successfully developed a consensus on a phased introduction of alternative fuels and drive systems for Germany, with hydrogen and fuel cells being the most relevant medium- and long-term options [25]. Using the United States and European General Motors WTW studies as a stepping stone [26–28], the automotive and energy industries then joined forces in a European partnership to develop successfully a broad European consensus (CONCAWE/EUCAR/JRC study [10]). This study is still ongoing, with updates including revisions and new pathways being released almost every other year. The main assumptions of this study were also used for HyWays, the European hydrogen roadmap project [29]. The calculation tool that was used is E3database [30], a relational database-based calculation tool for the supply and use of energy carriers, products, and services. It calculates cumulative energy demands, material balances (e.g., for the construction of an industrial plant), emissions of air pollutants and GHGs, and economic aspects (investment costs, operation and maintenance costs, depreciation conditions, etc.).

40.4.2.2 United States
After various early WTW studies with relevance to hydrogen pathways in the 1990s, General Motors' well-to-wheel study in 2001 was the first major study

that involved a consortium of major industrial stakeholders (such as BP, Exxon-Mobil, and Shell). It contained a total of 13 energy chains, five of them using hydrogen as a vehicle fuel. The study was updated in 2005, mainly to include pollutant emissions and the powertrain assumptions.

All recent United States WTW studies were based on the GREET model, developed by Argonne National Laboratory (ANL) with Michael Wang as the main contact person [31]. GREET allows researchers and analysts to evaluate various vehicle and fuel combinations on a full fuel-cycle/vehicle-cycle basis and is free to use. The first version of GREET was released in 1996, and ANL has continued to update and expand the model since then. ANL has used GREET to evaluate various engine and fuel systems for the DOE, other government agencies, and industry. The DOE's Hydrogen Program includes hydrogen pathway analysis activities with models devoted to the economic analysis of pathways (models H2A and HDSAM), while factors related to energy efficiency and emissions are continuously updated in GREET. The latest prominent publication of WTW results was included in the Hydrogen Posture Plan [32], comparing several hydrogen pathways in terms of energy efficiency, emissions, and economics.

40.4.2.3 Japan

In Japan, WTW investigations began with a study conducted by the Japanese Electric Vehicle Association from 1999 to 2002 focusing on tank-to-wheel [33]. Following this, the Japan Hydrogen and Fuel Cell Demonstration Project (JHFC) conducted an extensive WTW study between 2003 and 2005, where the published data on efficiencies were evaluated together with data acquired from the demonstration projects within JHFC. The study included a wide variety of pathways for hydrogen and alternative fuels (biofuels). Hydrogen sources were mostly hydrocarbons (kerosene, LPG, city gas, natural gas, coke oven gas, heavy oil), but also included renewable energies (wind, solar, biomass). Both liquid and gaseous distribution and onsite generation were assessed. The results showed that the measured efficiencies were slightly above the calculated efficiencies in most cases [33].

Furthermore, Toyota commissioned an extensive study in 2004, which was conducted by Mizuho Research Institute [23]. The study was supervised by an external advisory committee. Many well-to-tank pathways for hydrogen and alternative fuels were assessed; the vehicle technologies studied were ICE, hybridized ICE and fuel cells. The study focused mainly on fossil energy sources with the only renewable options being some second-generation biofuels.

40.4.2.4 Others

Many other WTW analyses from different parts of the world, conducted by universities and private research organizations, are widely available (e.g., (34, 35]).

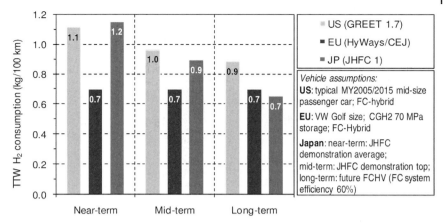

Fig. 40.4 Vehicle hydrogen consumption assumptions from different studies [21, 33].

40.4.3
Comparison and Interpretation of Results

In the following, some results of the major WTW activities mentioned above are compared and discussed briefly. The data have been taken from the HyWays-IPHE project, where assumptions, data and results from the European Union (EU) (HyWays/CEJ) and the United States (GREET) were extensively compared [21]. Other data have been taken from a publication on the Japanese JHFC WTW results [33].

Figure 40.4 shows the assumptions made on the hydrogen consumption of fuel cell vehicles of different generations and time horizons. Whereas the EU study [10] assumed constant consumption for fuel cell hybrids from 2010 onwards, the American and Japanese studies forecast a considerable reduction in fuel use in the near to long term. Reasons for these differences are, on the one hand, varying sizes of the reference vehicles and dissimilar driving cycles, and on the other, the fact that the EU study did not take into account that demonstration vehicles could have a higher consumption than mass-produced vehicles.

Figure 40.5 shows the results for WTW CO_2 emissions (other GHGs such as CH_4 and N_2O emissions could not be obtained from the Japanese study). The overlap between pathways published was small. Most JHFC pathways, where information was available, related to onsite reforming of liquid hydrocarbons, and also hydrogen from coke oven gas or an NaOH by-product. The EU and American studies, on the other hand, used natural gas, biomass, coal, and electricity as inputs for different production and distribution schemes. The comparison shows large differences for conventional vehicles, which again are due to different driving cycles and vehicle sizes. This is also reflected in the results for on-site steam methane reforming (SMR), where the correlation of the WTT part is generally good, but the different assumptions on the hydrogen consumptions of

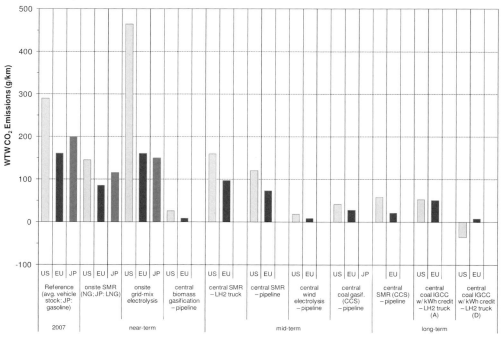

Fig. 40.5 WTW CO$_2$ emissions from different studies [21, 33].
(A) By-product allocation by energy; (D) by-product allocation by displacement (cf. Section 40.4.1.3)

cars and on the feedstock result in a considerable variation of the results. The deviation in emissions of hydrogen produced electrolytically from grid mix electricity is very high for the United States, whereas values similar to the reference vehicles were obtained for the EU and Japan. The main influence factors here are the regional electricity production mix, which is coal dominated in the United States, and also the higher vehicle consumption assumed there.

The difference in vehicle consumption is also the main reason for the GHG differences between hydrogen from biomass gasification, central SMR, and wind power electrolysis. These figures are higher in the United States than in the EU. For the coal gasification and SMR pathways, where carbon capture and storage (CCS) was applied, further differences between the EU and the United States were caused by varying assumptions on the carbon capture ratio and process efficiencies. The results for the integrated coal gasification combined cycle (IGCC) process also vary considerably depending on whether the coproduct electricity is accounted for by the allocation by energy (A) or by the displacement (D) method. Once again, this is mainly due to the difference in the GHG intensity of the displaced grid mix electricity.

40.4.4
Conclusion

The fundamental principles of WTW studies are generally straightforward, but their application for the analysis of different fuel/vehicle pathways is complex due to the large amount of detailed information needed. The differences between studies are therefore generally not due to major differences in approach, but to the detailed assumptions made.

For instance, as results depend on system boundaries in time and space, it is essential to fix the boundaries with respect to both dimensions. Most processes for the production of fuels generate coproducts (such as gases, gas oil, and heat), and the treatment of these coproducts is critical to the results of WTW studies (depending on how the allocation and substitution methods are used). Even higher uncertainties can be expected for biofuel analyses than for hydrogen pathway analyses due to the large variability and relatively poor scientific understanding of emissions associated with crop production, and in particular with the emission impact of land use and land use change. Therefore, biofuel pathways are a contentious area, where results are highly sensitive to the input assumptions.

Abbreviations

APU	Auxiliary power unit
BEV	Battery electric vehicle
BOP	Balance-of-plant components
CHP	Combined heat and power
CCS	Carbon capture and sequestration
CFC	Chlorofluorocarbon
DOE	US Department of Energy
FCV	Fuel cell vehicle
FTP	Federal Test Procedure
GHG	Greenhouse gas
HEV	Hybrid electric vehicle
HFC	Hydrofluorocarbon
HFCV	Hydrogen fuel cell vehicle
HWFET	Highway Fuel Economy Test
ICE	Internal combustion engine
IGCC	Integrated gasification combined cycle
LCA	Life cycle assessment
LH2	Liquefied hydrogen
LNG	Liquefied natural gas
MEA	Membrane electrode assembly
NG	Natural gas
PEFC	Polymer electrolyte fuel cell

PGM	Platinum group metals
PHEV	Plug-in hybrid electric vehicle
PM	Particle matter
SMR	Steam methane reforming
SOC	Battery state of charge
ULEV	Ultra-low emissions vehicle
WTT	Well-to-tank
WTW	Well-to-wheel

Symbols

ρ_A	Air density
ω	Rotational speed
A_V	Cross-sectional area of the vehicle
c_{R0}, c_{R1}, c_{R2}	Rolling resistance coefficients
c_W	Air drag coefficient
e	Reduced moment of inertia
F_a	Acceleration force
F_D	Air drag resistance
F_R	Rolling resistance
f_R	Velocity-dependent rolling resistance coefficient
F_t	Total force at the wheel of the vehicle
g	Gravitational constant
m_D	Mass of the driver
m_V	Mass of the vehicle
v	Velocity of the vehicle

References

1 BMVBS, BMBF, and BMWi (2006) *National Innovation Programme for Hydrogen and Fuel Cell Technology (NIP)*. Federal Ministry of Transport, Building and Urban Development, Berlin.

2 European Commission, Directorate-General for Research (2004) *Hydrogen Energy and Fuel Cells – A Vision of Our Future*. Final Report of the High Level Group, EUR 20719 EN, Brussels (2003) and Hydrogen and Fuel Cells Technology Platform (2005) Steering Panels Strategic Research Agenda and Deployment Strategy, Final Reports, Brussels (2005).

3 US Department Energy (2007) *Multi-Year Research, Development and Demonstration Plan: Planned Program Activities for 2005–2015*, updated edition, DOE, Washington, DC.

4 Höhlein, B., Grube, Th., Stolten, D., Rodatz, P., Röder, A., and Wokaun, A. (2001) Efficiencies and platinum group metal requirements of fuel-cell power trains. *Proceedings of the First European PEFC Forum*, Lucerne, 26 July 2001.

5 Carle, G. (2002) *Brennstoffzellen für den Automobilbau im Wettbewerb*, Report, ETH Zürich, October 2002.

6 Werhahn, J. (2008) Kosten von Brennstoffzellensystemen auf Massenbasis in Abhängigkeit von der Absatzmenge. *Energy and Environment*, Vol. 35,

Forschungszentrum Jülich, Jülich, ISBN 978-3-89336-569-2.

7 Kromer, M. A. and Heywood, J. B. (2008) *A Comparative Assessment of Electric Propulsion Systems in the 2030 US Light-Duty Vehicle Fleet*. SAE Technical Paper Series, 2008-01-0459, SAE International, Warrendale, PA.

8 von Helmolt, R. (2009) Fuel cell or battery electric vehicles? Similar technology, different infrastructure. *Proceedings of the f-cell*, Stuttgart, 2829 September 2009.

9 Campanari, S., Manzolini, G., and de la Iglesia, F. G. (2009) Energy analysis of electric vehicles using batteries or fuel cells through well-to-wheel driving cycle simulations. *J. Power Sources*, **186**, 464–477.

10 CONCAWE, EUCAR, and JRC (2007) *Well-to-Wheels Analysis of Future Automotive Fuels and Powertrains in the European Context*, http://ies.jrc.ec.europa.eu/WTW (accessed 31 October 2009).

11 Biedermann, P., Grube, Th., and Höhlein, B. (eds) (2006) Methanol as an energy carrier. *Energy Technology*, Vol. 55, Forschungszentrum Jülich, Jülich, ISBN 3-89336-446-3.

12 Wallentowitz, H. (1996) *Längsdynamik von Kraftfahrzeugen*. Forschungsgesellschaft Kraftfahrwesen Aachen mbH, Aachen, ISBN 3-925-194-32-0.

13 Moore, R. M., Hauer, K. H., Friedmann, D., Cunningham, J., Badrinarayanan, P., Ramaswamy, S., and Eggert, A. (2005) A dynamic simulation tool for hydrogen vehicles. *Journal of Power Sources*, **141**, 272–285.

14 Wipke, K. B., Cuddy, M. R., and Burch, S. D. (1999) ADVISOR 2.1: a userfriendly advanced powertrain simulator using a combined backward/forward approach. *IEEE Transactions on Vehicular Technology*, **48** (6), 17511761.

15 Cao, Q., Pagerit, S., Carlson, R., and Rousseau, A. (2007) PHEV hymotion Prius model validation and control improvements. *Proceedings of the 23rd International Electric Vehicle Symposium (EVS23)*, Anaheim, CA, December 2007.

16 Brown, D., Alexander, M., Brunner, D., Advani, S. G., and Prasad, A. K. (2008) Drive-train simulator for a fuel cell hybrid vehicle. *J. Power Sources*, **183**, 275–281.

17 Markel, T., Brooker, A., Hendricks, T., Johnson, V., Kelly, K., Kramer, B., O'Keefe, M., Sprik, S., and Wipke, K. (2002) ADVISOR: a systems analysis tool for advanced vehicle modeling. *J. Power Sources*, **110**, 255–266.

18 Rousseau, A., Shidore, N., Carlson, R., and Freyermuth, V. (2007) Research on PHEV battery requirements and evaluation of early prototypes. *Proceedings of AABC 2007*, Long Beach, CA, 16–18 May 2007.

19 Delorme, A., Rousseau, A., Sharer, P., Pagerit, S., and Wallner, T. (2009) Evolution of hydrogen fueled vehicles compared to conventional vehicles from 2010 to 2045, *Proceedings of the SAE World Congress*, Detroit, April 2009, SAE Paper 2009-01-1008, SAE International, Warrendale, PA.

20 Grube, Th., Höhlein, B., and Menzer, R. (2007) Assessment of the application of fuel cell APUs and starter-generators to reduce automobile fuel consumption. *Fuel Cells*, **7** (2), 128–134.

21 European Commission (2007) *HyWays-IPHE Benchmarking of Hydrogen Roadmap Models and Tools, Specific Support Action, 6th Framework Programme*; see also http://www.hyways-iphe.org (accessed 31 October 2009).

22 Solomon, S., Qin, D., Manning, M., Chen, Z., Marquis, M., Averyt, K. B., Tignor, M., and Miller, H. L. (eds.) (2007) *Climate Change 2007: The Physical Science Basis*, IPCC Forth Assessment Report (AR4), Cambridge University Press, Cambridge, United Kingdom and New York, NY, USA.

23 Ishitani, H., Ikematsu, M., Nishumura, F., Kitahara, T., Matsumoto, K., Nakanishi, K., Kobayashi, S., Hoshi, H., and Kaji, Y. (2004) *Well-to-Wheel Analysis of Greenhouse Gas Emissions of Automotive Fuels in the Japanese Context*. Well-to-Tank Report, November 2004, Mizuho Information and Research Institute, Tokyo.

24 European Commission (2009) *Directive 2009/28/EC of the European Parliament and of the Council of 23 April 2009 on the Promotion of the Use of Energy from Renewable Sources and Amending and Subsequently Repealing Directives 2001/77/EC and 2003/30/EC*, European Commission, Brussels.

25 TES (2001) *Transport Energy Strategy (TES) – a Joint Initiative from Politics and Industry (Germany), Second Task Force Status Report to the Steering Committee*, 13 June 2001.

26 Wang, M., He, D., Weber, T.R., Skellenger, G.D., Masten, D.A., Finizza, A., and Wallace, J.P. (2001) *Well-to-Wheel Analysis of Energy use and Greenhouse Gas Emissions of Advanced Fuel/Vehicle Systems North American Analysis*, General Motors Corporation, Detroit, MI.

27 Choudhury, R., Wurster, R., and Weber, T.R. (2002) *Well-to-Wheel Analysis of Energy Use and Greenhouse Gas Emissions of Advanced Fuel/Vehicle Systems a European Study*. General Motors Corporation, Detroit, MI.

28 Brinkman, N., Wang, M., Weber, T., and Darlington, T. (2005) *Well-to-Wheels Analysis of Advanced Fuel/Vehicle Systems a North American Study of Energy Use, Greenhouse Gas Emissions, and Criteria Pollutant Emissions*. General Motors Corporation, Detroit, MI.

29 European Commission (2007) *HyWays – An Integrated Project to Develop the European Hydrogen Energy Roadmap, 6th Framework Programme*; see also http://www.hyways.de (accessed 31 October 2009).

30 Ludwig-Bölkow-Systemtechnik GmbH (2009) *E3database, a Tool for the Evaluation of Energy Chains*, Ludwig-Bölkow-Systemtechnik GmbH, München-Ottobrunn 2009; see also www.e3database.com.

31 Argonne National Laboratory. *The Greenhouse Gases, Regulated Emissions, and Energy Use in Transportation (GREET) Model*, Argonne National Laboratory, Argonne, IL 2009; see also http://www.transportation.anl.gov/modeling_simulation/GREET/.

32 US Department of Energy (2006) *Hydrogen Posture Plan An Integrated Research, Demonstration and Development Plan*. DOE, Washington, DC.

33 Ishitani, H. (2006) Well-to-wheel efficiency analysis. *Proceedings of the 4th JHFC Seminar for FY 2005*, Tokyo, 6 March 2006.

34 Huang, Z. and Zhang, X. (2006) Well-to-wheels analysis of hydrogen based fuel-cell vehicle pathways in Shanghai. *Energy*, **31**, 471–489.

35 Svensson, A.M., Møller-Holst, S., Glöckner, R., and Maurstad, O. (2007) Well-to-wheel study of passenger vehicles in the Norwegian energy system. *Energy*, **32**, 437–445.

41
Electrification in Transportation Systems

Arndt Freialdenhoven and Henning Wallentowitz

Abstract

Environmental questions and shortcomings in hydro carbon resources are the driving forces for the electrification in transportation systems. The report describes these starting points of the discussion and studies the design of battery-electric and fuel-cell powered cars. They are both using electricity for driving.

The questions regarding the energy sources (batteries, fuel-cells) are discussed, the new requirements for these electric operated cars are described. The requirements are touching the whole car and especially the aggregates like steering, braking, heating, etc. As a consequence changes in industrial structures will happen.

This finally opens the discussion about large scale changes in conventional car companies and the chances for "new-comers". Co-operations seem to be an adequate solution for the future business.

Keywords: electrification, battery electric vehicles, electric mobility

41.1
Driving Forces for Electric Mobility

The development of alternative vehicle engines and especially the introduction of electric mobility are basically driven by two factors: first, the limited availability of fossil fuels is a stimulating driving force for intensified efforts regarding the development of alternative power train concepts, and second, by a variety of laws, the legislator directly influences research and development activities and ultimately the choice of the cars by the customers.

41.1.1
Availability of Crude Oil

The limited availability of fossil fuels has a lasting effect on the development of alternative concepts. Because this factor is influenced by many parameters, diverging opinions exist about the availability of fossil fuels in the future [1–3].

Hydrogen Energy. Edited by Detlef Stolten
Copyright © 2010 WILEY-VCH Verlag GmbH & Co. KGaA, Weinheim
ISBN: 978-3-527-32711-9

The decisive factor for the development of future power trains, however, is the oil price, which, in an economic feasibility study, suggests the advantage of conventional power trains so far. On the supposition that the oil price conforms to basic laws of the market, supply and demand of oil set the price of the oil. However, we might face a shortage of oil, hence the development of alternatives is very reasonable.

As a result of global industrialization, the need for oil and accordingly oil production have increased steadily in the past, whereas the number of successful explorations of new oil sources has decreased since the 1970s. Resulting from this, since the 1980s the need for oil has been higher than the amount of new oil sources discovered. Especially the kind of oil in place and the extraction technology determine the economic efficiency of oil production. Because of technical limitations, such as borehole stability, only one-third of the available oil can be unearthed in present oil fields. Tar sands, and oil shale to an even larger extent, can only be mined if greater expense is accepted, so that profitability is only guaranteed against the background of higher oil prices.

The need for oil has increased as a result of industrialization and especially because of the fast-growing emerging markets, such as China and India, and is characterized by a high dynamic. This development will lead to a situation in which the worldwide oil production reaches its maximum, the so called Peak Oil. Taking into consideration the development of future power trains, the Peak Oil is very important, since reaching the maximum amount of discharge will lead to a jump in the price of oil, in an economic view. The oil supply will no longer be able to accommodate the demand. Not only the law of supply and demand, but also increasing costs along with an increase in exhaustion of oil fields, will lead to higher oil prices.

Against the backdrop of limited availability of fossil fuels, a further influencing factor on vehicles with an alternative drive train can be seen in the efforts of great industrial nations to become independent in terms of environmental and energy issues. Securing the national energy supply, especially the supply of oil, is given top priority by industrial nations.

41.1.2
Legislative Regulations

Legislative conditions try to achieve a better environmental compatibility of vehicles. However, it has to be stressed that passenger cars produce only 12% of the total human-produced or so-called anthropogenic carbon dioxide emissions. Hence the total share of passenger cars in CO_2 emissions throughout the world is less than 0.5%

41.1.2.1 Low-Emission Vehicles Program
With the aim of reducing local emissions, the state of California has taken a pioneering role in the United States. In 1990, the Air Resources Board of the California Environmental Protection Agency (CARB) adopted the so-called low-

emission vehicles (LEV) program, which has already been revised several times. Currently the LEV 2 standard is in effect, which roughly distinguishes between three exhaust categories: low-emission vehicle (LEV), ultra-low-emission vehicle (ULEV) and super ultra-low-emission vehicle (SULEV). These groups are characterized by different thresholds for the emission of nitrogen oxides, carbon monoxide, hydrocarbons and sooty particles. According to the latest modifications, each vehicle manufacturer must offer a dedicated rate of zero emission vehicles (ZEVs) for sale in California. However, further vehicle categories were defined, which could be taken account of in the required proportion of ZEVs. This requirement represents a significant factor influencing alternative drive trains, which encourages vehicle manufacturers to intensify research and development in the field of low-emission vehicles.

41.1.2.2 CO_2 Regulations in Europe

Also in the European market, CO_2 emissions have a strong influence on the development of automotive power trains [4]. After the European vehicle manufacturers broke their negotiated agreement of reducing CO_2 emissions to 140 g km^{-1} by 2008, the European Union suggested a law for CO_2 emissions of the vehicle fleet of each vehicle manufacturer. At the end of 2008, the European Parliament, the European Member States and the European Commission agreed upon the introduction of a binding plan by stages.

The aims of the plan will be pursued by an annual increase in the proportion of new cars whose average CO_2 emission may not exceed a threshold that results from the average weight of the fleet itself. Until 2015 it will be gradually required that the average CO_2 emission of all vehicles does not exceed the specific vehicle fleet threshold. From 2012, vehicle manufacturers will be fined if the average CO_2 emission of the fleet is higher than the specific value that results from the average weight. Depending on the difference between these two values, the penalty can reach different amounts. If this regulation had been in effect in 2008, the BMW Group would have had to pay 1.77 billion euros. A far-reaching aim is to limit the average CO_2 emission to 95 g km^{-1}, which equates a consumption of about 4 l of fuel per 100 km for a gasoline engine.

41.1.3
Local Regulations

In addition to the legislative regulations at the international level in North America and Europe described so far, there also exist a few local limitations regarding road traffic aimed at an improvement of environmental friendliness by restricting the amounts of CO_2 emissions allowed.

41.1.3.1 London Congestion Charge

The efforts to improve local air quality in Europe were originally inspired by the implementation of the London Congestion Charge in 2003. This is a charge which has to be paid in order to pass through London's inner city area on working days. The intention behind this project was, on the one hand, environmental awareness and, on the other, a reduction of the volume of traffic to shorten the traveling time of commuters. As electric vehicles are exempt from the charge, this project led to a higher demand for suitable vehicles in London.

41.1.3.2 Environment Zones in Germany

In Germany especially the large portion of diesel-powered vehicles with their high fine dust emissions had a negative effect on air quality. Different legislative regulations are being implemented to reduce fine dust pollution. In addition to prohibition of access for commercial road vehicles, numerous so-called environment zones were established. These are inner city areas which vehicles with a bad classification of air pollutants are not allowed to enter.

In almost every German state, environment zones have now been established. They are concentrated especially in large cities such as Berlin, Düsseldorf, Cologne, Munich, and Stuttgart. Only vehicles with a good classification of air pollutants are allowed to cross the environment zones and they must have a windscreen sticker providing information on their classification. In total there are four different classes, and only for the three with the lowest emissions is a windscreen sticker available.

All measures related to German environment zones aim at better air quality and a reduction in fine particulate air pollution. The environment zones try to handle the toxic load of fine particulates, but they do not regulate the toxic load of carbon dioxide.

41.2
Design of Battery Electric Vehicles (BEVs)

Today, BEVs are regarded as the future drive concept. It differs from other concepts particularly in the advantage of local emission-free driving. This drive concept will lead to a complete substitution of the combustion engine, so that a basically rethink has to occur, especially in branches of industry in which combustion engines have been produced.

41.2.1
Vehicle Concept

Whereas the structure of the conventional power train has basically influenced the body structure of current vehicles, the structure of the electric power train provides new opportunities for structuring. Generally, two approaches of design exist, the Conversion and the Purpose Design. The Conversion Design is based on the modification of series-production vehicles with an internal combustion engine and is a lower cost concept. New product development of vehicles con-

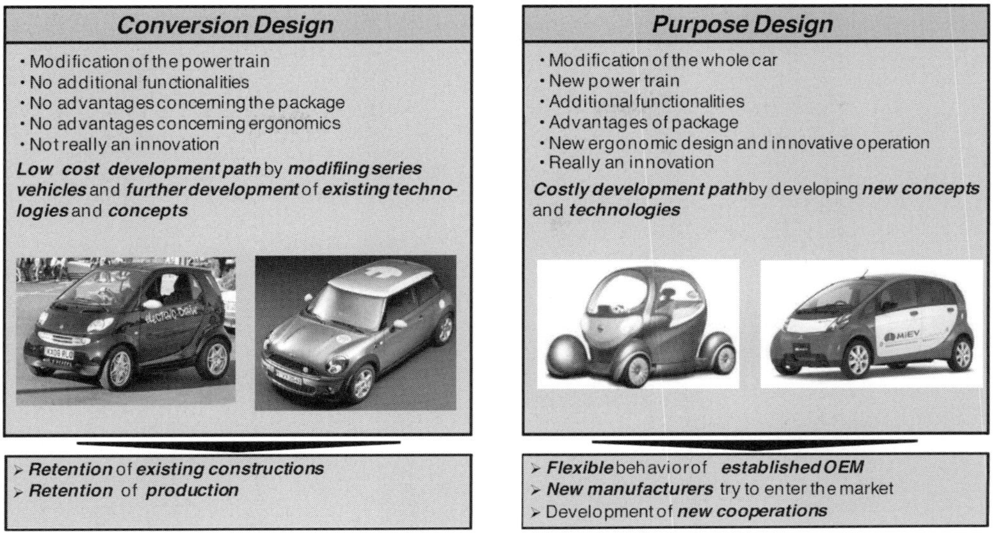

Conversion Design
- Modification of the power train
- No additional functionalities
- No advantages concerning the package
- No advantages concerning ergonomics
- Not really an innovation

Low cost development path by *modifying series vehicles* and *further development* of *existing technologies* and *concepts*

➤ *Retention* of *existing constructions*
➤ *Retention* of *production*

Purpose Design
- Modification of the whole car
- New power train
- Additional functionalities
- Advantages of package
- New ergonomic design and innovative operation
- Really an innovation

Costly development path by developing *new concepts* and *technologies*

➤ *Flexible* behavior of *established OEM*
➤ *New manufacturers* try to enter the market
➤ Development of *new cooperations*

Fig. 41.1 Summary of Conversion and Purpose Design.
Source: Institut für Kraftfahrzeuge, RWTH Aachen University

sidered the Purpose Design, by which additional functionalities and also ergonomic design and innovative operation and display systems are integrated. Furthermore, the vehicle components are rearranged, so that the advantages of package volume can be optimally used. Figure 41.1 gives an overview of important aspects concerning these two design approaches.

However, one of the severe problems for Purpose Design is the low volume of new cars. Perhaps ultimately the purpose electric car may become a converted combustion engine-powered car?

41.2.2
Power Train

41.2.2.1 Electric Motor
Electric motors are suitable for operating in vehicles because of their generation of power output and torque (Figure 41.2). Right from the start the electric motor consistently provides the maximum torque until a specific number of revolutions leads to attenuation of the magnetic field.

The use of electric motors in the power train increases the degrees of freedom, because the clutch and the conventional transmission are no longer required. Even the use of differentials is not necessary, if wheel-hub motors are integrated, so that the adjustment between the wheels can be controlled by an electric system. Further additional functions resulting form the use of electric motors include torque vectoring and an electronic stability program.

For generating the required driving power, various types of electric motors can be used. Among the common direct current (d.c.) motors there exist further

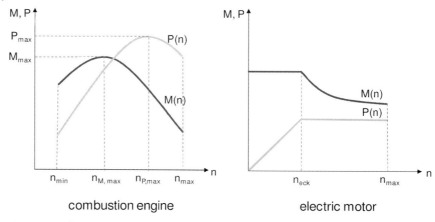

Fig. 41.2 Performance of the combustion engine and the electric motor [5].

constructions of electric motors such as the synchronous motor, the asynchronous motor, and the reluctance motor. The main differences in the constructions are the current feed, the position of the coils which produce the magnetic field, and the development of the magnetic fields.

The functionality of all electric motors is based on the physical law that two magnetic fields align themselves if they are oriented opposite each other. Compared with other variants, the construction and the control of d.c. motors are less complex. For the operation of more modern electric motors, multiphase current is needed. Hence the operation of synchronous and asynchronous motors in combination with a d.c. supply needs an inverter and a corresponding control unit.

In contrast to d.c. motors, asynchronous motors do not have collector rings, carbon brushes, or comparable elements, hence they are very tough and need minimal maintenance. Whereas the rotational magnetic field of asynchronous motors is generated by electromagnetic induction, the synchronous motor generates the magnetic field by the use of wiper contact or by permanent magnets, fixed to the rotor, which lead to additional costs. Additionally, the design of synchronous motors compared with asynchronous motors is more complex. On the basis of these facts, asynchronous motors are preferred in electric drive trains. The control electronics of both engine types attain similar complexity.

The properties of different kinds of electric motors are illustrated in Table 41.1. The d.c. motor has comparatively poor technical properties, but has a well-engineered state of development. The asynchronous motor also has a high state of development and has advantages over the d.c. current motor in terms of technical properties. The different designs of the synchronous motor and the reluctance motor still have development potential and their properties in a technical sense are very good. Consideration of the engine operating points shows that for daily city traffic a power of about 15 kW is sufficient. This power lies in the present power range of electric motors.

Table 41.1 Characteristics of electric motors [6] [a].

Characteristic	Direct current		Synchronous		Asynchronous	Transverse flux	Reluctance
	Non-permanent magnet	Permanent magnet	Non-permanent magnet	Permanent magnet			
Power density	0	+	+	++	+	++	++
Reliability	0	+	+	+	++	+	++
Efficiency	– –	–	+		0	++	+
Controllability	++	++	+	+	0	+	++
Overload capability	+	+	+	+	+	+	+
Noise level	–	–	+	+	+	+	+
Overload protection	–	–	+	++	+	+	+
Stage of development	++	++	0	0	+	–	0

a) ++, Very good; +, good, 0, medium, –, bad; – –, very bad.

41.2.2
Control Unit

The control unit is one of the essential parts of the electric power train. It transforms the direct current coming from the battery into the required current. For different kinds of electric motors, specific transformations are necessary. By the use of d.c. motors, the control unit only has the function of adapting current and voltage to the operating point of the electric motor. For the control of asynchronous and synchronous motors, a three-phase alternating current has to be generated, the frequency of which has to be adjusted to the rotation speed of the motor. The control unit generating an alternating current is much more complex than the control unit providing direct current, hence the control of alternative current leads to higher costs.

In contrast to combustion engines, electric motors generate the maximum torque over a broad speed range. In addition, electric motors can generate a higher power than their nominal power for a short time without becoming damaged. This same capability of overload allows electric motors to cover excessive loads on demand if strong acceleration is required.

41.2.3
Battery

The electric power train in a passenger car has to be powered by energy storage, which supplies electric energy. There are different technologies available, such as capacitors, flywheels, and batteries, which can be used to store energy that

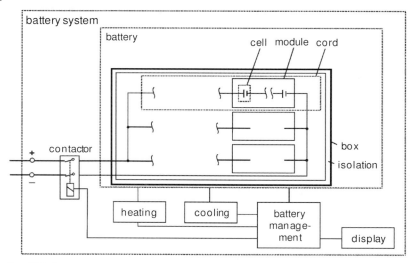

Fig. 41.3 Battery system.

can be converted on-board into electric energy. For the electric drive, the battery in the form of series-connected secondary cells is of special importance. To achieve the necessary on-board power supply voltage, modules consisting of series-connected solitary cells are bunched to increase capacity (see Figure 41.3).

Requirements on Batteries
Available types of batteries have contrasting properties, in which the applicability of energy storage depends on different requirements, especially the power density and the energy density of batteries (Figure 41.4).

On the one hand, the energy storage has to supply a certain amount of energy to satisfy the customer's requirements concerning range, and on the other, the energy storage has to supply energy in a certain time interval, especially in acceleration phases. Further requirements that the battery has to meet include a low weight, a short charging period and a long lifetime.

Battery Cells
For application in a BEV, battery cells such as sodium-nickel chloride batteries, lead acid batteries, nickel-cadmium batteries, nickel-metal hydride batteries, and lithium-ion batteries are generally suitable. The sodium-nickel chloride battery, also referred to as a Zebra battery, is a battery working at a high temperature (300 °C). The energy density of Zebra batteries is relatively high and the batteries are maintenance free and recyclable. Deficiencies of this battery type are a low power density, a limited life cycle, and the permanent generation of a high operating temperature of about 300 °C, which needs additional energy. This energy is thus lost for driving purposes.

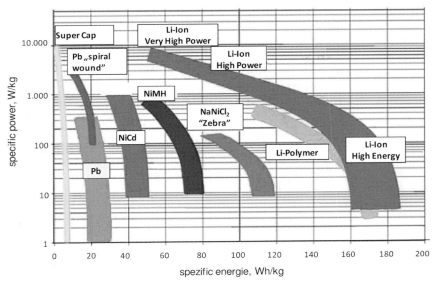

Fig. 41.4 Energy and power density of energy storage [6].

Lead acid batteries are commonly used in small electric vehicles because of the cost efficiency. However, this battery type is not promising because of its low power density.

The nickel-cadmium battery is also well engineered, but the danger of poisoning and the complexity of disposal are two characteristics which is a barrier to the spread of this technology in the future. In contrast to the nickel-cadmium battery, the nickel-metal hydride battery consists of non-toxic materials and in addition its energy density and the number of possible load cycles are higher. In spite of its comparatively short life cycle and its sensitivity to certain ranges of temperature, nickel-metal hydride batteries are used in several vehicles, for example the Toyota Prius. However, nickel-metal hydride batteries do not have a promising future because of their low potential concerning the battery capacity.

The lithium-ion battery is currently being intensively researched. The reason for this is its high potential regarding energy density and therefore high battery capacity. Furthermore, lithium-ion batteries are about 30% smaller and 50% lighter than common nickel-metal hydride batteries and they can be charged faster, are more durable, and have a higher power density. The market for this battery type varies widely and was developed in the past (mobile phones, laptops, etc.). Advantages of this technology are lower costs, a high power density, good cycle stability, a short charging period, and acceptable safety. All this factors provide the reasons for the promising use of this technology.

User acceptance concerning the cruising range of future vehicles is difficult to assess, so that alternative battery technologies also have to be considered. Lithium-sulfuric acid batteries are non-polluting and more capable than conventional lithium-ion batteries, but suffer from poor safety and a short durability. These battery types are used, for example, in laptops. Zinc-air batteries have a high energy density and can be constructed as coin batteries. However, these batteries are not yet rechargeable and are used primarily in hearing aids. The redox-flow battery is a type of rechargeable flow battery. The main advantages of the redox-flow battery are that it can offer almost unlimited capacity simply by using larger and larger storage tanks, it can be left completely discharged for long periods with no ill effects, and it can be recharged simply by replacing the electrolyte if no power source is available to charge it. This battery type is currently used in mobile radio based stations and in wind power stations. High-performance capacitors have a favorable effect on high performance requirements, and they are the perfect complement to batteries with a high energy density.

Safety
Among the previously mentioned factors concerning the evaluation of battery types, safety aspects must be considered. Several safety-critical states can appear when using lithium-ion batteries. Mechanical damage could lead to an underrun of the battery. Charging the battery for too long, discharging the battery deeply, or a heat load on the battery are safety hazards. The so-called thermal runaway is considered to be exceedingly dangerous. This means that an increase in temperature in one cell could cause an explosion of that cell. For operating safely, a temperature range between 0 and 65 °C is recommended.

Charging
Charging the battery represents a new challenge. In European households, a power connection with 16 A and 230 V is available, which means a maximum power of about 3 kW. If a battery with a capacity of 15 kW h needs to be completely charged by this power connection, it would require 5 h. This is why several automobile manufacturers and electric energy service providers agreed on a uniform charging socket by which a power of 25 kW can be supplied. With this system, the charging time can be reduced to about 30 min.

Dimensioning
The energy density and the cruising range are directly linked. This linkage can be estimated for different vehicle categories by using a 150 kg battery pack. In doing so, a range of about 120 km results for a small-sized BEV equipped with a lithium-ion battery and driving the New European Driving Cycle (NEDC). An increase in the battery size can increase the range, but of course it requires a larger installation space and has a higher weight, both of which influence the dynamics of the vehicle.

Thermal Management
The substitution of the internal combustion engine has an effect on several auxiliaries, whose functionalities are based on the combustion engine. In this respect, especially the heat management has to be revised. In addition to the pres-

ent tasks, the heat management on the one hand must be able to warm up the battery until the operating temperature is reached and on the other it must be able to cool the battery in times of ordinary operation. The thermal management should ideally consider each battery cell and keep her in a comfortable temperature range of 20 to 40 °C.

41.2.4
Fuel Cells

Another electro-chemical based energy source for the future is the fuel cell. While the battery needs for the production of electrical energy chemical products, which are already in the battery-cells, the fuel cell needs a continuous feeding with fuel, similar to the combustion engine. For the near future the hydrogen fed fuel cell will be the most important one, as the other types (methanol fuelled ones) are even more complex in operation. Therefore the car industry has decided to focus on PEM fuel cells first.

The efficiency of the fuel cell is not limited to the Carnot-efficiency, but it is determined by design features. To use fuel cells in practical applications, several auxiliaries are necessary. Figure 41.5 shows a simple example of a fuel cell system. This is a system which uses pure hydrogen, the necessary oxygen is taken from the ambient air.

This general systems architecture is usable for mobile systems, especially as an energy converter in automobiles. The hydrogen is taken out of a tank and after humidification it is piped to the anode. Compared to the energy content of normal petrol with 8.8 kWh/l the hydrogen has only 0.003 kWh/l. This means, the energy content per litre is nearly three thousand times less. This disadvan-

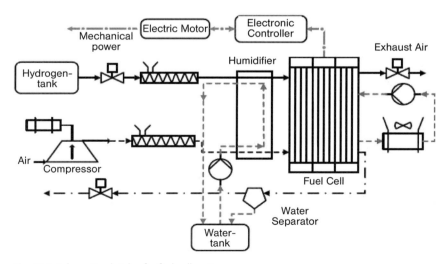

Fig. 41.5 Schematic sketch of a fuel cell system.

tage of hydrogen can partly be compensated by pressurising the hydrogen up to 700 bar in the pressure tank. With the limited space in the car the driving distance is also limited, but better then those of the battery powered cars. An alternative is the usage of liquid hydrogen in a deep temperature tank in the car. But then the energy losses by warming up the liquid hydrogen are not negligible. In addition, the needed energy to produce the deep cooled hydrogen is nearly half of the total energy which is available to generate the hydrogen.

For oxidising the hydrogen, the ambient air is used, which is directed to the cathode by pressure. The air is also humidified. After the electro-chemical reaction in the fuel cell the exhaust gas of the cathode is piped to a condenser, which is liquidating the water from the exhaust gas. This water is used again for the humidification of the gases.

A part of the electrical energy, which is produced by the fuel cell, is used to power the compressor. The other power output of the fuel cell is converted in a way that the necessary power for operating the car will be available.

Figure 41.6 shows a fuel cell system, installed into a car. This is the design of the Daimler B-Class F-Cell, the next generation of fuel cell vehicles [7].

It becomes obvious, that the former engine compartment and the high floor of the vehicle is an ideal space for installing the equipment.

In addition Figure 41.6 clarifies, that the fuel cell vehicle should have a high voltage battery to start the system and to use recuperation of the braking energy. Thus, the fuel cell car might for a certain amount be a battery electric vehicle as well.

Most of the fuel cell prototypes of today follow the hybrid idea and they are using an additional energy source, mainly to cover peak power demands. The additional energy source is mostly a battery. This in addition enables the system to recuperate braking energy. But it needs additional weight and space. Without such an additional storage device there will be problems with respect to the dynamical behaviour of the car, that means the reactions to throttle inputs will be slow.

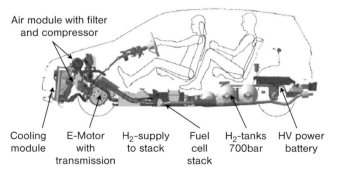

Fig. 41.6 Daimler B-Class F-Cell [7].

For the electric power of the fuel cell stack and for the water-balance the pressure of the gases, used in the fuel cell, plays an important role. The fuel cell efficiency and the power density increases, when the working pressure is above the ambient pressure. The increase of the working pressure has a direct influence to the needed power for the compressor. The research work with respect to the cathodes is focussing to find the most efficient working pressure and to improve the water-balance.

The used compressors are those which are working according the displacement principle. These are piston compressors or rotational compressors. The fuel cell needs the flexibility with respect to the pressure ration with low loads. These days, fuel cell systems are mostly using piston compressors. The disadvantage is the noise of the compressor which normally follows the required power.

The water-management of the fuel cell is another critical point, as the performance of the fuel cell and the membrane durability depends on this management. Therefore the humidifiers are used, which must deliver steam. Pure water would reduce the fuel cell performance or even destroy the fuel cell.

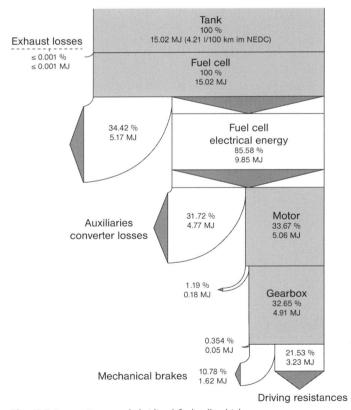

Fig. 41.7 Losses in a non hybridised fuel cell vehicle.

Figure 41.7 shows the losses in a vehicle which is only equipped with a fuel cell (empty mass 1590 kg, 70 kW fuel cell stack) and which drives according to the NEDC.

It is obvious, that only 32.6% of the energy which has been stored in the hydrogen is delivered to the wheels. As there is no recuperation, nearly 11% of the energy is wasted by braking (in the NEDC test). Therefore the above mentioned hybrid design is necessary.

The other main element in Figure 41.6 are the hydrogen storage units. To store pressurised hydrogen, the containers are normally manufactures out of carbon fibre, for deep cooled hydrogen they are made out of a material mix.

41.2.4.1 Hydrogen storage

The hydrogen reservoirs are shown in Figure 41.8. They are up to now produced by carbon fibre and the hydrogen is stored with up to 700 bar (Figure 41.8a). This high pressure makes it evident, that a reasonable amount of energy is needed to bring the hydrogen into the reservoir.

The other used possibility to store hydrogen in cars is the cryogenic solution, when the hydrogen is deep cooled to −250 °C to reach the liquid phase (Figure 41.8b). In case the hydrogen would have been produced from oil, more of half of the energy content would have already been used to store the hydrogen in the car.

Other hydrogen storage devices, which has been worked on earlier, are again under consideration. These methods are using metal-hydride storage devices. They have the disadvantage to be very heavy. May be, carbon nano-tubes are better options, but this is not realised in commercial volume up to now.

41.2.4.2 Safety

The above given description shows a reasonable complexity of the fuel cell systems in cars. Therefore another "open point" for further consideration will be the operational safety of hydrogen powered cars. Starting with filling the tanks

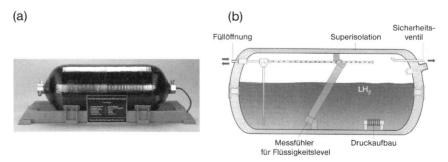

Fig. 41.8 Hydrogen storage devices; (a) pressured hydrogen; (b) deep cooled hydrogen.

(pressurized or deep cooled), new safety requirements have to be developed. In addition the behaviour during crashes must be designed safety-related. There may no fire coming up!

41.2.4.3 Application of Electric Vehicles

The different equipment in the battery-electric vehicle and the fuel-cell electric vehicle shows the problems, both types of electric vehicles have to overcome in the future. The complexity of this kind of transportation is evident. Therefore the usage of these modern cars will be different. In [7] there was a picture published which summarizes the future application in an interesting way. In addition to the electric vehicles, cars with combustion engines and hybrid cars are considered.

Here, the energy content and the weight of the battery is the limiting factor for transportation usage. Therefore it is usually distinguished for the application of electric powered vehicles:
1. battery drives for urban applications only,
2. fuel cell drives for urban and interurban drives,

and, if hydrogen will be available at filling stations,
3. fuel cell drives even for longer distance traffic.

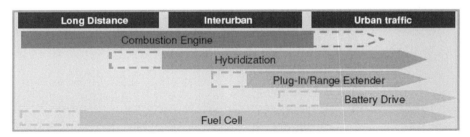

Fig. 41.9 Car usage with respect to distance and electric vehicle application with respect to vehicle class [7].

The filling stations for hydrogen are also under development [8]. Several filling stations have been installed already. But even here we will have a long way to go, to have a all over the country supply with hydrogen. The same must happen with the electric supply stations [9]. But it seems to be, that this equipment will be available much earlier.

41.2.5
Chassis

The electrification of the power train technically means more than just substitution of the power train. Further components and systems, such as the steering system and the braking system, are also influenced by this development.

41.2.5.1 Steering System
The ordinary hydraulic power steering is powered by a pump that is driven via a belt from the internal combustion engine. The power consumption of the pump is almost independent of the power which is actually needed by the hydraulic power steering. That means that the hydraulic pump was replaced by an electric motor with electric control. This electric power steering can generally be used in BEVs. However, in the meantime many vehicles with a combustion engine have an electromechanical power steering system. This development was one of the results of experiences from the first development phase of electric vehicles in the 1990s in Germany (Rügen-Versuch). This electromechanical power steering supports the steering movement with an electric motor operating on the steering column or the steering rack. Furthermore, this system is able to work depending on the driving situation if it is software controlled.

41.2.5.2 Braking System
The braking system also has to meet new requirements. In contrast to the conventional vacuum-based braking system, especially future systems will have to recuperate energy by using braking energy. In order to recuperate energy, the electric motor is used as a generator. Because for safety reasons the maximum necessary braking power has to be higher than the driving power, the additional availability of conventional braking systems is necessary. In the future, these brake systems will support the electric motor in order to generate sufficient braking forces. This leads to characteristics of the curves shown in Figure 41.10. For generating this braking force distribution, an electronic control unit is required.

With replacement of the internal combustion engine, the vacuum from the inlet manifold, which is used by the power brake unit, is no longer available. In electric vehicles, this low pressure can be generated by a vacuum pump powered by an electric motor. This solution is an opportunity for using conventional components in conversion-designed vehicles. The use of these conventional components certainly has adverse effects because it faces the problem of blend-

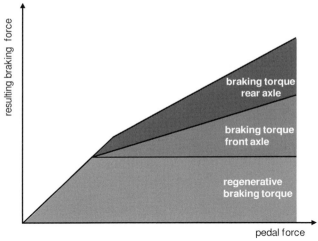

Fig. 41.10 Distribution of the braking forces [10].

ing. This problem can be fixed by the use of electric-hydraulic braking systems, which are already in use in purpose-designed vehicles, such as the hybrid vehicle of Toyota/Lexus.

41.2.5.3 Chassis Systems

The electric mobility also leads to changes in the design of chassis systems. During the last few years, complex constructions of chassis systems were developed. The aim was to ensure high safety and comfort even at high speed. This lead to the development of axles with several arms. Such multi-link suspensions can be found in compact cars these days.

Related to the introduction of electric vehicles, different utilization profiles must be considered. Electric vehicles especially have advantages in urban traffic. In this respect other requirements especially concerning the maximum speed have to be met, so that it is possible to return to simpler designs. There is a wide range of improvements for lightweight and simple designs with new materials and also alternative production processes.

41.3
Requirements on Players

41.3.1
Future Core Competence of the Traditional OEM

The development of BEVs is associated with the elimination of the conventional combustion engine, which will lead to a shift of power and competence within the automotive industry. So far, expert knowledge about the thermal processes inside the combustion engine has been very important. Prospective core competence would be a knowledge of energy storage and power electronics. There is little chance that the production of energy storage will be conducted by vehicle manufacturers, because this is an economy of scale. However, it seems probable that the integration of solitary battery cells will be conducted by vehicle manufacturers, so that they will have to gain knowledge about the management of batteries. In BEVs, gearboxes and clutches are no longer required, so the power train is far less complex. In the current competition, vehicle manufacturers distinguish themselves by offering different gearboxes and engines, so especially due to the omission of the gearbox the vehicle manufacturer will lose an important distinguishing feature. New possibilities of product differentiation regarding BEVs arise from an intelligent distribution of energy within the vehicle and from the power electronics.

41.3.2
Models of Cooperation

Whereas the electrification of the power train is a major challenge for the whole automotive industry, it offers a great chance of achieving this task in close collaboration with other actors such as cell producers and utilities.

41.3.2.1 Chances and Risks of Cooperating

Especially the high costs of R&D and investments relating to electric mobility are fraught with uncertainty. The costs and even the risk for each market player can be reduced by cooperation. However, their own efforts in doing R&D concerning new technologies, such as electric motors or lithium-ion batteries, are very important to gain core competences, which are necessary to rise to future challenges. Without these efforts, the manufacturer is facing a loss of prestige and more value added is generated by the external suppliers, thus weakening the position of the Original Equipment Manufacturers (OEM) with respect to the suppliers. Despite everything, certain risks such as increased complexity, costs of coordination, and a loss of know-how are not negligible.

One example of cooperating companies is the strategy alliance in the field of batteries of Daimler AG. Being concerned with engineering batteries, the German company Evonik Industrie AG wants to research, develop, and produce dedicated battery systems for electric vehicles in close collaboration with Daimler AG. Part of this collaboration is a joint venture. The company Deutsche Accumotive has been commissioned to produce lithium-ion batteries. The aim is to sell the batteries to third parties and to find a third partner, which proves the ability to accomplish the system integration.

As electric mobility cannot be sold without an infrastructure that gives the opportunity for charging the batteries, more and more general partnerships between OEM and electricity utilities are being agreed. Within such cooperation, a

41.3 Requirements on Players

Table 41.2 Cooperations between OEM, battery cell manufacturer and energy supplier [11].

Battery cell manufacturer	Vehicle manufacturer	Energy supplier
Panasonic / Panasonic EV Energy Co., Ltd.	Toyota	eDF
Johnson Controls / SAFT	Ford	Edison
	Renault	Project Better Place
NEC	Nissan	
SANYO	Volkswagen	e·on
LG Chem	GM	AEP / Edison
Johnson Controls / SAFT	smart (Mercedes)	RWE, eDF
SANYO	Honda	
ETC (Battery and FuelCells Sweden AB)	Volvo	Vattenfall
	BMW	Vattenfall

few fleet test and pilot projects are now taking place. By doing this, the utilities can gain experience concerning the power supply and the accounting system, while the OEM can test the electric vehicles. In addition, important information about the endurance of components and about user acceptance can be gained. An overview of current cooperations between producers of battery cells, energy utilities, and vehicle manufacturer is given in Table 41.2.

It became apparent that for companies pursuing a strategy which is concerned with electric mobility, part of this strategy has to involve cooperation. Otherwise, the high costs and the risks of the revolutionary technology leap cannot be covered.

41.3.3
Electric Mobility as a Chance for New Players

For new vehicle manufacturers, it is very difficult to break into the automotive market. The traditional vehicle manufacturers possess a lot of knowledge about the production of vehicles, which makes it very difficult for new manufacturers to offer products identical in quality and to develop a comprehensive network

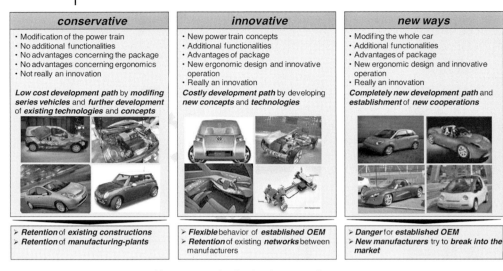

Fig. 41.11 Possible scenarios for the development of BEVs [4].

concerning sales and services of their vehicles. The technology leap opens up chances for new manufacturers who are developing new technologies of electric motors or energy management, because the traditional manufacturer's knowledge is at nearly the same (low) level.

The possibility that new players could penetrate the market represents a hazard for the traditional vehicle manufacturer. Maybe this will lead to the faster development of electric mobility. It is also possible that new vehicle manufacturers will find their way into the market by cooperating with traditional OEM. By doing so, the new manufacturer could gain knowledge about standard components and the traditional manufacturer would gain insights, for example into the integration or the management of battery systems. The cooperation between Daimler AG and Tesla Motors is an example of such a cooperation. Figure 41.11 shows the chances for new vehicle manufacturers and possible positions concerning the traditional vehicle manufacturer in the two different scenarios.

41.4
Conclusion

The future development of legislative regulations and the future trend of the oil price are the major driving forces that will bring alternative engines more and more into focus. Especially the tightening legislative measures will force the vehicle manufacturers to produce eco-friendly vehicles. An accepted position is that the battery electric mobility has the highest potential for meeting future requirements.

The rollout of the electric mobility requires diverse approaches to the development of electric vehicles concerning the power train, the chassis, and the body. Regarding energy storage, the lithium-ion battery has the highest potential, but in this case much applied research and development has to be done. Concerning the general development of the vehicle, the Conversion Design is a suitable way of gaining experience of electric mobility. However, in the long run the Purpose Design has to be applied, but then a reasonable vehicle volume must be demanded by the customers.

In the future, all players have to gain knowledge in terms of electricity and mechatronics. In order to rise to the future challenges, all players have to go into appropriate partnerships. Then an effective collaboration will brings BEVs into a situation where the daily use is competitive with current vehicles with combustion engines.

References

1 BGR Studie (2009) *Energierohstoffe 2009*, BGR, Hannover.
2 International Energy Agency (IEA) (2008) *Medium-Term Oil Market Report 2008*, IEA, Paris.
3 McKinsey & Company (2006) *Drive Studies – The Future of Automotive Power*, McKinsey & Company, Munich.
4 Wallentowitz, H., Freialdenhoven, A., and Olschewski, I. (2009) *Strategien zur Elektrifizierung des Antriebstranges Technologien, Märkte und Implikationen*, Vieweg+Teubner, Wiesbaden.
5 Henneberger, G. (1999) *Umdruck Elektrische Antriebe und Steuerungen*, Institut für elektrische Maschinen, RWTH Aachen University, Aachen.
6 Wallentowitz, H. (2008) *Unkonventionelle Fahrzeugantriebe, Schriftenreihe Automobiltechnik, Vorlesungsumdruck 2008*, Institut für Kraftfahrwesen, RWTH Aachen University, Aachen.
7 Niestroj, A. (2009) *F-Cell B-Class, Implementation of Concept Blue Zero in series production*, INEA Conference Automobile and Future: "Hydrogen and engine technology", Wolfsburg, 22.–24. Oct. 2009
8 Schnell, P. (2009) *Die Wasserstoffaktivitäten der Total*, INEA Conference Automobile and Future: "Hydrogen and engine technology", Wolfsburg, 22.–24. Oct. 2009
9 Groll, M. (2009) *RWE E-Mobility, Markterwartungen, Geschäftsmodelle, Umsetzungsstatus*, INEA Conference Automobile and Future: "Electric Drives and Regenerative Energies", Wolfsburg, 22.–24. Oct. 2009
10 Wallentowitz, H., Freialdenhoven, A., and Olschewski, I. (2008) *Strategien in der Automobilindustrie – Technologietrends und Marktentwicklung*, Vieweg+Teubner, Wiesbaden.
11 BCG Consulting Group (2008) *Das Comeback des Elektroautos? Nachhaltige Strategien in der Automobilindustrie, 10/2008*, BCG Consulting Group, Boston, MA.

Index

a
adsorption 431
aerospace applications 661
aircraft 661
airplane 681
alkaline electrolysis 243
auxiliary power unit (APU) 681

b
700 bar refuelling 609
battery 661, 793
– electric vehicles 853
biohydrogen 169
biomass 169, 307, 465
biosynthesis 169
borohydrides 415
Bruggeman relationship 89
buildings 755
buses 793
by-product hydrogen 609

c
carbon
– dioxide footprint 291
– materials 431
carbon-support corrosion 3
catalyst
– degradation 3
– development 3
– supports 41
cell
– degradation 227
– voltage 41
chassis 853
coal 449
– conversion 291
combined heat and power 735, 755
combustion engine 793
commercialization 597

complex hydrides 415
compressed gas tank 121
computer modeling 89
crude oil 853
cryogenic liquid truck 121
cumulative energy demand 551

d
decentralized power generation 735
delivery vehicle 793
dense ceramic membranes 321
desulphurization 291, 661, 681
diaphragm materials 243
direct injection 811
DMFC concept 715
dynamic power train simulation 831

e
economic analysis 567
economics 489
electric
– engine 793
– mobility 853
– motor 853
– power train 551
– powered vehicles 793
– vehicle 551
electricity
– generation 465
– production 735
electrification 853
electrocatalysis 3
electrolysis 207, 243, 465
electrolyte 17
electrolyzers 243
emission efficiency 811
energy 489
engine generator 661
environment 489
exergy 489

Hydrogen Energy. Edited by Detlef Stolten
Copyright © 2010 WILEY-VCH Verlag GmbH & Co. KGaA, Weinheim
ISBN: 978-3-527-32711-9

f

finance 567
fluidized-bed gasifier 307
fossil fuels 291, 449
fuel cell 89, 149, 189, 207, 227, 291, 351, 377, 465, 489, 551, 577, 597, 649, 661, 681, 715, 755, 793, 831, 853
– direct ethanol fuel cell 41
– direct liquid fuel cell 41
– direct methanol fuel cell 41
– efficiency 61
– high-temperature fuel cell 17, 61, 735, 755
– hydrogen fuel cell 793
– hydrogen polymer electrolyte fuel cell 89
– low temperature fuel cell 89, 755
– molten carbonate fuel cell 61
– planar solid oxide fuel cell 227
– polarization curve 755
– polymer electrolyte fuel cell 681
– polymer electrolyte membrane (PEM) fuel cell 17, 793
– powered vehicle 3
– proton exchange membrane fuel cell 3
– solid oxide fuel cell 61, 681, 755
– system efficiency 207
– vehicles 449, 609
fuel processing 681

g

gaseous
– hydrogen compressor 121
gasification 307
greenhouse gases 755, 831

h

harmonization 649
heat 207
heating boiler 735
high-pressure electrolysis 243
high-temperature electrolysis 227, 243
hydrogen 489, 597, 649, 661
– adsorption 431
– assessment 351
– distribution 121, 149
– energy 533, 567, 577
– extraction 377
– fuel 61
– fueled vehicles 449
– fueling station 121
– industrial hydrogen 149
– infrastructure 121, 609
– internal combustion engine 811
– liquefaction 377
– liquid hydrogen 377
– networks 121
– oxidation 3
– plant 291
– power train 551
– production 149, 169, 189, 207, 227, 271, 291, 351, 465, 533, 609
– refueling stations (HRS) 649
– renewable hydrogen 465
– separation 321
– storage 121, 377, 395, 415, 431, 533, 853
– storage density 377
– storage system 793
– supply 149
– transportation 121, 533
– vehicles 351, 629
hydrothermal gasification 307
HYVOLUTION 169

i

infrared communication 609
infrastructure 351, 629
interdisciplinary triangle 489
internal mixture preparation 811
International Organization
  for Standardization 649
ionic liquids 17

k

kinetics 243, 395

l

life cycle
– analysis 551
– assessment 489
light traction 715

m

market introduction 577
membrane
– electrode assembly 17
– materials 17, 351
metal hydrides 395, 415
metal/metal oxide cycles 189
metal-organic frameworks 431
microporous ceramic membranes 321
mixed proton-electron conductivity 321
modeling 89
municipal development 597

n

national fuel strategies   449
natural gas   449
non-thermal processing   169
nonionic electrolyte conduction   227

o

organic hydrides   415

p

partial oxidation   291
passenger cars   681
performance potential   811
photofermentation   169
physisorption   431
pipelines   121
platinum
– catalysts   3
– group metals   831
pollution factors   489
polymer electrolyte membrane   271
port fuel injection   811
portable systems   715
– backup power generator   715
– grid-independent generator   715
power generator   715
power train   793, 853
– hybrid power train   793
pre-normative research   649
pressure-composition isotherm   395
proton conductivity   17

r

reforming   291, 307, 681, 755
refueling   377
regional development   597
regulations, codes, and standards (RCS)   649
research and development targets   533
residential cogeneration   755
resources   351

s

safety   629
saturation jump   89
secondary energy carrier   449
ships   681
social behavior   567

socioeconomics   567, 609
solid
– oxide electrolyzer cell   227
– storage   415
stack material   755
stacks   17
steam
– electrolysis   207
– reforming   291, 551
storage capacity   431
sulfur cycles   189
sustainable development   489
system control   755
systems analysis   351, 831

t

tank filling system   793
templated SiO_2 membranes   351
thermochemical cycles   189
thermodynamics   61, 243, 395
thermophilic fermentation   169
transport parameters   89
transportation   465
trucks   681

u

unmanned aerial vehicle (UAV)   661

v

vehicle   681
– simulation   831
voltage
– efficiency   207
– losses   3
volumetric energy density   377

w

water
– electrolysis   207, 227, 271
– management   89
well-to-wheel   831
wind   207

z

zeolite   431
– membranes   351
Zero Regio   609